HUMAN GEOGRAPHY

| A Spatial Perspective |

NATIONAL GEOGRAPHIC

Bednarz | Bockenhauer | Hiebert

Human Geography: A Spatial Perspective, 1st Edition

Sarah Witham Bednarz, Mark Henry Bockenhauer, Fredrik T. Hiebert

SVP, Higher Education Product Management: Erin Joyner

VP, Product Management, Learning Experiences: Thais Alencar

Product Director: Maureen McLaughlin

Product Manager: Vicky True-Baker

Product Assistant: Emily Loreaux

Learning Designer: Leslie Taggart

Content Manager: Sean Campbell

Digital Project Manager: Mark Hopkinson

Director, Product Marketing: Jennifer Fink

Portfolio Marketing Manager: Taylor Shenberger

Content Acquisition Analyst: Ann Hoffman

IP Project Manager: George Brown

Production Service: Straive, Inc.

Designer: Chris Doughman

Cover Image Source: Extreme-Photographer/Getty Images

© 2024 Cengage Learning, Inc. ALL RIGHTS RESERVED.

No part of this work covered by the copyright herein may be reproduced or distributed in any form or by any means, except as permitted by U.S. copyright law, without the prior written permission of the copyright owner.

Unless otherwise noted, all content is Copyright © Cengage Learning, Inc. "National Geographic," "National Geographic Society," and the Yellow Border Design are registered trademarks of the National Geographic Society® Marcas Registradas.

> For product information and technology assistance, contact us at
> **Cengage Customer & Sales Support, 1-800-354-9706
> or support.cengage.com.**
>
> For permission to use material from this text or product, submit all requests online at **www.copyright.com**.

Library of Congress Control Number: 2022922090

Student Edition: ISBN:978-0-357-85200-2

Loose-leaf Edition: ISBN:978-0-357-85201-9

Cengage
200 Pier 4 Boulevard
Boston, MA 02210
USA

Cengage is a leading provider of customized learning solutions with employees residing in nearly 40 different countries and sales in more than 125 countries around the world. Find your local representative at **www.cengage.com**.

To learn more about Cengage platforms and services, register or access your online learning solution, or purchase materials for your course, visit **www.cengage.com**.

Printed in the United States of America
Print Number: 01   Print Year: 2023

About the Authors

Sarah Bednarz
Professor Emerita of Geography
Texas A&M University

Sarah Witham Bednarz is professor emerita of geography at Texas A&M University. She served as president of the American Association of Geographers (2015–2016). Previously, she was associate dean for academic affairs in the College of Geosciences (2008–2014). Dr. Bednarz's research interests are in the intersection of teaching and learning geospatial technologies and spatial and geographic thinking. She was PI on two major curriculum and educational research projects; co-authored the national geography standards, *Geography for Life* (1994 and 2012); participated in the National Research Council Learning to Think Spatially project, and helped to develop the National Assessment of Educational Performance (NAEP) framework in geography. In 2013 she co-chaired the Geography Education Research Committee (GERC) of the Road Map for 21st Century Geography Education Project. A second interest has been in higher education policies. As a member of the Healthy Departments Committee, she organized professional development workshops for department leaders and presented sessions on planning for promotion and tenure and mid-career success strategies. She served on the Leadership Team for the Texas A&M ADVANCE grant and was an active participant in the Strategies and Tactics for Recruiting to Improve Diversity and Excellence (STRIDE) component of the grant designed to reduce bias in search and award processes. Dr. Bednarz holds a University Professorship for Teaching Excellence, received the Gilbert H. Grosvenor Honors for Geographic Education (2007) and was an inaugural AAG Fellow. She was educated at Mount Holyoke College (A.B. 1973); the University of Chicago (M.A.T. 1974); and Texas A&M University (PhD 1992).

Mark Bockenhauer
Professor of Geography
St. Norbert College

Mark Henry Bockenhauer is Professor of Geography at St. Norbert College in De Pere, Wisconsin, where he teaches courses in geography and environmental studies. He has 35 years of experience in K-12 professional development in geography, and is Coordinator of the Wisconsin Geographic Alliance. He served as president of the National Council for Geographic Education (2007). Dr. Bockenhauer's scholarship and outreach have focused on geographic education and curriculum development. He was co-PI on a $2.1 million-dollar federal grant from the National Oceanographic Partnership Program to enhance the scientific literacy of K-12 teachers with integrated oceanography, marine science, and geography (1996–1998). Dr. Bockenhauer served as geographer-in-residence at the National Geographic Society (2002), co-authored several books and atlases for NGS—including *Our Fifty States* (2004) and the *World Atlas for Young Explorers* (2007), and directed NGS field institutes for national groups of K-12 educators in Hawaii, Alaska, and California. He also directed two NGS geography mentor institutes, an NGS natural hazards workshop for K-12 and museum educators, and piloted the NGS Grosvenor Teacher Fellow Program aboard the National Geographic Explorer cruise to Svalbard. He served on the Professional Development and Instructional Materials Committee of the Road Map for 21st Century Geography Education Project (2011–2013). He has co-authored two textbooks for National Geographic Learning/Cengage Learning: the middle-level *World Cultures and Geography* (2013) and *Human Geography: A Spatial Perspective for high school Advanced Placement* (2021). He has received the highest St. Norbert College awards for teaching (1999), community service (2009), and scholarship (2022). He was educated at the University of Wisconsin-La Crosse (B.S. Geography 1979), Southwest Texas State University (M.A. Applied Geography 1987), and the University of Wisconsin-Milwaukee (PhD Geography 1996).

Fred Hiebert
Archaeologist-in-Residence National Geographic Society

Fredrik T. Hiebert has been an archaeologist and explorer with the National Geographic Society since 2003. He has traced ancient trade routes overland and across the seas for more than 40 years. Hiebert has led excavations at ancient sites across Asia, from Egypt to Mongolia. His excavations at a 4,000-year-old Silk Road city in Turkmenistan made headlines around the world. He also conducts underwater archaeology projects in the Black Sea and in South America's highest lake, Lake Titicaca, in search of submerged settlements. He rediscovered the lost Bactrian gold in Afghanistan in 2004 and was the curator of National Geographic's exhibition *Afghanistan: Hidden Treasures from the National Museum, Kabul*, which toured major museums in the United States and internationally. As National Geographic's Archaeologist-in-Residence, he extends the enthusiasm for archaeology and the geography of place in lectures, presentations, films, and museum exhibits. In his role as a consultant on National Geographic Learning's Secondary Social Studies programs, he has provided a unique lens into the role archaeology plays in our understanding of geography and history. Hiebert was educated at University of Michigan (BA, 1984), Harvard University (MA 1989, PhD 1992), and received an Honorary Doctorate of Humane Letters from SUNY, Geneseo (2016). Prior to joining National Geographic, Dr. Hiebert held the Robert H. Dyson Jr. Chair of Archaeology at the University of Pennsylvania. In addition to numerous media awards for exhibitions and documentaries, Hiebert received the Chairman's Award from the National Geographic Committee for Research and Exploration in 1998.

Content Reviewers

Bob Amey
Bridgewater State University

Brenda Barr
Senior Consultant on the Geo-inquiry Process Former Educator

Sylvia Brady
Metropolitan State University of Denver

Jeffrey L Brewer
University of Cincinnati

Jonathan Burkham
University of Wisconsin-Whitewater

Annalie Campos
Oakland University

Perry Carter
Associate Professor of Geography
Texas Tech University

Kevin Curtin
University of Alabama

Elizabeth Dudley-Murphy
University of Utah

Brian Frain SJ
Rockhurst University

Ava Fujimoto-Strait
Sam Houston State University

Bob Gumbrecht
Colorado Mountain College

Sarah Hein
Edison State Community College

Linda Lea Jones
Texas Tech University

Elizabeth M Larson
Arizona State University

Benjamin McDaniel
North Hennepin Community College

Parisa Meymand
Chief Academic Officer
Siena Catholic Schools of Racine
Senior Content Consultant

Amal Ali Moustafa
Salisbury University

Christopher Murphy
Community College of Philadelphia

Katherine Nashleanas
University of Nebraska-Lincoln

Alex Oberle
Professor of Geography
University of Northern Iowa

Anita Palmer
National Geographic
Education Fellow
CEO/Education Consultant GISetc

Timothy S Pruett
Towson University

Debra Sharkey
Cosumnes River College

The National Geographic Society has been our partner on *Human Geography: A Spatial Perspective* from Concept through creation.

Our collaboration with each of the following has been a pleasure and a privilege: National Geographic Maps, National Geographic Education and Children's Media, and National Geographic Programs—with special thanks to Mike Ulica, President and COO, National Geographic Society, Lina Gomez, VP, Strategy and Operations, National Geographic Education, and Jack Dangermond, Founder and President of Esri and National Geographic Society Board Member.

National Geographic Explorers, Photographers, and Affiliates

Aaron Vincent Elkaim
Photographer

Adjany Costa
Explorer

Amy Toensing
Photographer

Andrés Ruzo
Explorer

Anna Antoniou
Explorer

Aziz Abu Sarah
Explorer

Bill Parkinson
Explorer

Caroline Gerdes
Explorer

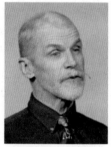
Charles Fitzpatrick
Esri

Clinton Johnson, Esri
Collaborator

Danielle Sanders
Collaborator

Daniel Raven-Ellison
Explorer

David Guttenfelder
Photographer

Eduardo Neves
Explorer

Enric Sala
Explorer-in-Residence

Farhod Maksudov
Explorer

Fred Hiebert
Archaeologist-in-Residence

Guillermo de Anda
Explorer

George Steinmetz
Photographer

Hindou Oumarou Ibrahim
Explorer

Jason De León
Explorer

Jennifer Burney
Explorer

Jerry Glover
Explorer

Jim Enote
Explorer

John Stanmeyer
Photographer

Justin Dunnavant
Explorer

Lehua Kamalu
Explorer

Lillygol Sedaghat
Explorer

Lynsey Addario
Photographer

Maria Fadiman
Explorer

Maria Silvina Fenoglio
Explorer

Meghan Dhaliwal
Explorer

Michael Frachetti
Explorer

Michael Wesch
Explorer

Nora Shawki
Explorer

Pardis Sabeti
Explorer

Paul Salopek
Explorer

Sandhya Narayanan
Explorer

Stephanie Pau
Explorer

T. H. Culhane
Explorer

Thiago Sanna Freire Silva
Explorer

Tristram Stuart
Explorer

William Allard
Photographer

Zachary Damato
Explorer

Brief Contents

Unit 1
Thinking Geographically 2

Chapter 1 The Power of Geography: Geographic Thinking 6

Chapter 2 Geographic Inquiry: Data, Tools, and Technology 30

Unit 2
Population and Migration Patterns and Processes 62

Chapter 3 Patterns of Population 66

Chapter 4 Population Growth and Decline 90

Chapter 5 Migration 116

Unit 3
Cultural Patterns and Processes 156

Chapter 6 Concepts of Culture 160

Chapter 7 Cultural Change 188

Chapter 8 Spatial Patterns of Language and Religion 212

Unit 4
Political Patterns and Processes 250

Chapter 9 The Contemporary Political Map 254

Chapter 10 Spatial Patterns of Political Power 276

Chapter 11 Political Challenges and Changes 292

| Unit 5 |

Agriculture and Rural Land-Use Patterns and Processes 322

Chapter 12 Agriculture: Human-Environment Interaction 326

Chapter 13 Patterns and Practices of Agricultural Production 360

Chapter 14 Agricultural Sustainability in a Global Market 382

| Unit 6 |

Cities and Urban Land-Use Patterns and Processes 422

Chapter 15 Urban Settlements 426

Chapter 16 The Urban Landscape 456

Chapter 17 Urban Living 478

| Unit 7 |

Industrial and Economic Development Patterns and Processes 518

Chapter 18 The Growth and Diffusion of Industrialization 522

Chapter 19 Measuring Human Development 550

Chapter 20 Globalization, Interdependence, and Sustainability 578

Cape Town, South Africa

World Political Map xviii
World Physical Map xx
Preface xxii
Foreword 1

Unit 1
Thinking Geographically 2

National Geographic Explorer Enric Sala: Safeguarding Pristine Seas 4

Chapter 1
The Power of Geography: Geographic Thinking 6

1.1 What Is Human Geography? 7
 Case Study: New Orleans—Site versus Situation 14
 National Geographic Explorer Adjany Costa: Conserving the Delta 16
1.2 Spatial Patterns: Scale and Region 17
 Case Study: India—Regional Differences in Scale 22
1.3 Globalization and Sustainability 23
Chapter 1 Summary & Review 28

Chapter 2
Geographic Inquiry: Data, Tools, and Technology 30

2.1 Thinking Like a Geographer: Processes of Inquiry 31
2.2 Geographic Data and Tools 32
 Case Study: Detroit—GIS Helps Find Safer Routes 36
2.3 Understanding Maps 38
 National Geographic Explorer Lehua Kamalu: Navigating with Nature 46
2.4 The Power of Data 47
 National Geographic Collaborator Clinton Johnson, Esri: Mapping Social Justice 51
Chapter 2 Summary & Review 52

Unit 1
Writing Across Units, Regions & Scales 54
Maps & Models Archive 56

Shenzhen, China

Unit 2
Population and Migration
Patterns And Processes 62

National Geographic Explorer Paul Salopek: Out of Eden A Walk Through Time 64

Chapter 3
Patterns of Population   66

- **3.1** Where People Live 67
  - National Geographic Explorer Lillygol Sedaghat: Transforming Trash 73
- **3.2** Consequences of Population Distribution 74
  - Case Study: Population Distribution at the Country Scale 76
- **3.3** Population Composition 78
- **3.4** Measuring Growth and Decline 81
  - Examining Population Pyramids at Different Scales 87
- Chapter 3 Summary & Review 88

Chapter 4
Population Growth and Decline   90

- **4.1** Why Populations Grow and Decline 91
- **4.2** Theories of Population Change 96
  - National Geographic Explorer Pardis Sabeti: Cracking the Genetic Code 102
  - Case Study: Zika Virus in South and North America ... 103
- **4.3** Population Policies 104
  - Case Study: China's Population Policies 105
- **4.4** Consequences of Demographic Change 108
- Chapter 4 Summary & Review 114

Chapter 5
Migration   116

- **5.1** Why Do People Migrate? 117
- **5.2** Types of Migration 121
  - National Geographic Explorer Jason De León: Documenting the Stories of Migrants 126
  - Case Study: Migration from Central America 127
- **5.3** Refugees and Internally Displaced Persons 128
  - Case Study: Surviving War in Syria 132
- **5.4** Migration and Policy 133
  - Feature: Walls That Divide Us 137
  - National Geographic Collaborator Danielle Sanders: Journalism and the Great Migration 138
- **5.5** Effects of Migration 140
- Chapter 5 Summary & Review 146

Unit 2

Writing Across Units, Regions & Scales 148

Maps & Models Archive 150

La Paz, Bolivia

Unit 3
Cultural Patterns and Processes 156

National Geographic Explorer Fred Hiebert: Discovery and Preservation 158

Chapter 6
Concepts of Culture 160

- 6.1 An Introduction to Culture 161
  - Case Study: Wisconsin's American Indian Nations 164
- 6.2 Cultural Landscapes 165
  - Case Study: Tehrangeles 169
  - National Geographic Explorer Jim Enote: Mapping the Zuni Landscape 171
- 6.3 Identity and Space 172
- 6.4 Cultural Patterns 178
  - National Geographic Explorer Sandhya Narayanan: Studying Indigenous Languages 182
- Chapter 6 Summary & Review 186

Chapter 7
Cultural Change 188

- 7.1 Cultural Diffusion 189
  - Feature: A Portrait of Relocation Diffusion 190
  - Case Study: African Culture in Brazil 194
  - National Geographic Explorer Justin Dunnavant: Retracing Afro-Caribbean Maritime Routes 195
- 7.2 Processes of Cultural Change 196
  - Case Study: Fútbol—A Globalizing Force 203
- National Geographic Photographer William Allard: Introducing People Across Cultures 204
- 7.3 Consequences of Cultural Change 206
- Chapter 7 Summary & Review 210

Chapter 8
Spatial Patterns of Language and Religion 212

- 8.1 Patterns of Language 213
- 8.2 The Diffusion of Language 218
  - Case Study: French or English in Quebec? 227
  - National Geographic Explorer Maria Fadiman: Preserving Indigenous Knowledge 228
- 8.3 Patterns of Religion 229
  - Case Study: Shared Sacred Sites 232
- 8.4 Universalizing and Ethnic Religions 233
- Chapter 8 Summary & Review 240

Unit 3

Writing Across Units, Regions & Scales 242

Maps & Models Archive 244

Panmunjom, South Korea

Unit 4
Political Patterns and Processes ... 250

National Geographic Explorer Aziz Abu Sarah: Reconciliation Through Narrative ... 252

Chapter 9
The Contemporary Political Map ... 254

- **9.1** The Complex World Political Map ... 255
- **9.2** Political Power and Geography ... 259
- **9.3** Political Processes Over Time ... 262
  - Case Study: The Kurds ... 264
- **9.4** The Nature and Function of Boundaries ... 265
  - Case Study: The DMZ in Korea ... 267
  - National Geographic Photographer David Guttenfelder: Revealing Mysteries ... 268
- Chapter 9 Summary & Review ... 274

Chapter 10
Spatial Patterns of Political Power ... 276

- **10.1** Organization of States ... 277
  - Case Study: Political Control and Nunavut ... 283
- **10.2** Electoral Geography ... 284
  - Case Study: Gerrymandering and Race ... 287
- Chapter 10 Summary & Review ... 290

Chapter 11
Political Challenges and Changes ... 292

- **11.1** Devolution: Challenges To State Sovereignty ... 293
  - Case Study: Irredentism in Ukraine ... 296
  - National Geographic Explorer Michael Wesch: Technology's Impact on Society ... 300
- **11.2** Supranationalism: Transcending State Boundaries ... 301
  - Case Study: Brexit ... 305
- **11.3** Forces That Unify and Forces That Divide ... 306
  - National Geographic Explorer Anna Antoniou: Understanding a Divided Cyprus ... 311
- Chapter 11 Summary & Review ... 312

Unit 4

Writing Across Units, Regions & Scales ... 314

Maps & Models Archive ... 316

Utah, USA

Unit 5
Agriculture and Rural Land-Use
Patterns and Processes ... 322

National Geographic Explorer Jerry Glover: Agriculture for a Hungry Future ... 324

Chapter 12
Agriculture: Human-Environment Interaction ... 326

12.1 Agriculture and the Environment ... 327
  National Geographic Explorer Stephanie Pau: Understanding Tropical Forests ... 332
12.2 Agricultural Practices ... 333
  Feature: Rural Survey Methods ... 334
  Feature: Rural Settlement Patterns ... 336
  National Geographic Photographer George Steinmetz: Viewing the World from Above ... 344
12.3 Agricultural Origins and Diffusions ... 346
12.4 Advances in Agriculture ... 351
  Case Study: Women and Africa's Green Revolution ... 357
Chapter 12 Summary & Review ... 358

Chapter 13
Patterns and Practices of Agricultural Production ... 360

13.1 Agriculture Production Regions ... 361
  Feature: The Changing Dairying and Ranching Industries ... 364
13.2 The Spatial Organization of Agriculture ... 366
  National Geographic Explorer Tristram Stuart: Eating Ugly ... 370

13.3 The Von Thünen Model ... 371
13.4 Agriculture as a Global System ... 373
  Case Study: Coffee Production and Consumption ... 379
Chapter 13 Summary & Review ... 380

Chapter 14
Agricultural Sustainability in a Global Market ... 382

14.1 Consequences of Agricultural Practices ... 383
  National Geographic Explorer Hindou Oumarou Ibrahim: Mapping Indigenous Climate Knowledge ... 391
  Case Study: Building Africa's Great Green Wall ... 393
14.2 Challenges of Contemporary Agriculture ... 394
  Feature: Precision Agriculture ... 397
14.3 Feeding the World ... 401
  Case Study: Food Deserts ... 404
  National Geographic Explorer Jennifer Burney: Local Changes, Global Consequences ... 407
14.4 Women in Agriculture ... 408
Chapter 14 Summary & Review ... 412

Unit 5

Writing Across Units, Regions & Scales ... 414

Maps & Models Archive ... 416

xiv

Delhi, India

Unit 6
Cities and Urban Land-Use
Patterns and Processes 422

National Geographic Explorer Daniel Raven-Ellison: Geography for the People 424

Chapter 15
Urban Settlements — 426

15.1 The Origin and Influences of Urbanization 427
- National Geographic Explorer Michael Frachetti; National Geographic Collaborator Farhod Maksudov: Unearthing Secrets of a Silk Road City 430
- National Geographic Explorer Eduardo Neves: Rediscovering Amazonian Plaza Villages 431

15.2 Factors That Influence Urban Growth 432
- Case Study: Re-urbanizing Liverpool 437

15.3 The Size and Distribution of Cities 438
- National Geographic Feature: City of the Future 442

15.4 Cities and Globalization 446
- Case Study: How Shanghai Grew 449
- National Geographic Feature: The Shape of Cities 450

Chapter 15 Summary & Review 454

Chapter 16
The Urban Landscape — 456

16.1 The Internal Structure of Cities 457
- Case Study: Informal Housing in Cape Town 464

16.2 Urban Housing 465
- Case Study: Land-Use Change in Beijing 467

16.3 Urban Infrastructure 469
- National Geographic Explorer T. H. Culhane: Empowering People with Clean Energy 471

Chapter 16 Summary & Review 476

Chapter 17
Urban Living — 478

17.1 Designing for Urban Life 479

17.2 Causes and Impacts of Urban Changes 486
- Case Study: The Effects of Redlining in Cleveland 489

17.3 Creating Sustainable Urban Places 498
- Feature: New York City's High Line 499
- National Geographic Explorer Zachary Damato: A Wild Mile in the City 501
- National Geographic Explorer Maria Silvina Fenoglio: Transforming Urban Rooftops 503
- Case Study: Milan and Urban Sustainability 507

Chapter 17 Summary & Review 508

Unit 6
Writing Across Units, Regions & Scales 510

Maps & Models Archive 512

Cowley, Oxford, England

Unit 7
Industrial and Economic Development
Patterns and Processes 518

National Geographic Explorer Lynsey Addario: Educating Afghan Girls 520

Chapter 18
The Growth and Diffusion of Industrialization 522

18.1 Processes of Industrialization 523
  Case Study: The Fourth Industrial Revolution 532
18.2 How Economies are Structured 533
  Case Study: Damming The Xingu River 538
  National Geographic Photographer Aaron Vincent Elkaim: Documenting Impacts of the Belo Monte Dam 539
18.3 Patterns of Industrial Location 541
Chapter 18 Summary & Review 548

Chapter 19
Measuring Human Development 550

19.1 How is Development Measured? 551
19.2 Measuring Gender Inequality 556
  Case Study: "Women-Only" Cities 562
  National Geographic Photographer Amy Toensing: Widow Warriors 563
19.3 Changing Roles of Women 565
  Case Study: The Development of the Grameen Bank 571

19.4 Theories of Development 572
Chapter 19 Summary & Review 576

Chapter 20
Globalization, Interdependence, and Sustainability 578

20.1 Trade Relations and Global Corporations 579
20.2 Connected Economies 584
  Case Study: The Financial Crisis of 2007–2008 585
20.3 Developing a Sustainable World 592
  Case Study: Reducing Waste in Fisheries 594
  National Geographic Explorer Thiago Sanna Freire Silva: Sustainable Land Management 595
  National Geographic Explorer Andrés Ruzo: Sustainable Ecotourism 599
Chapter 20 Summary & Review 600

Unit 7
Writing Across Units, Regions & Scales 602
Maps & Models Archive 604

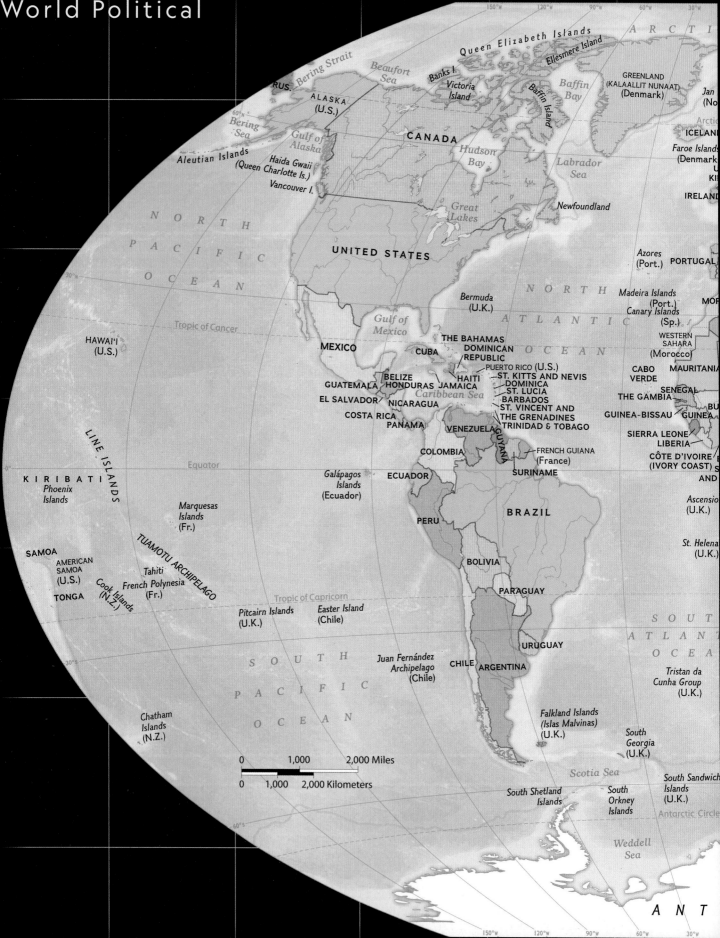

Preface

Human geography focuses on the interactions between humans and the environments in which they live. Geographers interpret the world through a lens that allows them to make connections at a range of scales to better understand our human stories. We have written this book to help instructors use a geographic lens to investigate the movement of people, cultures, and ideas; the political organization of countries; and the development of agriculture, settlements, and linked economies. Our goal is to support you in developing in your students an informed global awareness and to help them discover the factors that influence the world's livability and sustainability.

Organization of the Text

We are confident that *Human Geography: A Spatial Perspective* will be an accessible and thought-provoking exploration of geographic thinking for you and your students, placing human geography within the broader discipline of geography. We hope the book reflects the deep love we have for our discipline and its practitioners. We have included insights from a wide variety of prominent geographers, including quoted references. After introductory chapters on the power of geographic thinking and methods of inquiry, six units and 18 chapters span the core patterns and processes of human geography: population and migration, concepts of culture, spatial aspects of political behavior, agriculture and rural land-use, cities and urban landscapes, and industrial, economic, and human development.

- Chapter 1 introduces Human Geography and the unique perspectives of geographic thinking. We discuss the importance of spatial patterns and the processes that produce them, and establish key themes used across the text of globalization and sustainability.

- Chapter 2 explores thinking like a geographer in greater depth and how geographers collect and use quantitative and qualitative data, maps, and geospatial technologies as tools of analysis at a range of scales.

- Chapter 3 examines where and why people live on Earth and the consequences of uneven patterns of distribution. We then introduce ways to consider the composition of population again at a range of scales and the visual tools geographers use to examine population growth and decline.

- Chapter 4 considers the dynamics of population, both growth and decline, and the roles culture, economics, and public policies play in shaping the consequences of demographic changes. Two models are used to capture changes in population over time, the Demographic Transition Model and the Epidemiological Transition Model.

- Chapter 5 continues to consider a key force in population dynamics, migration. Why people migrate, both voluntarily and involuntary, the policies that affect migration, and the intended and unintended consequences of migration are central concepts in the chapter.

- Chapter 6 begins an examination of cultural patterns and processes by exploring the essential role cultural landscapes play in human geography. Related to landscapes, we focus on the ways identities shape spaces and human attachment to places.

- Chapter 7 considers the dynamics of cultural change, examining relocation and expansion diffusion processes and the consequences of changes within places. We consider this chapter especially valuable for students struggling to understand rapid cultural changes occurring in modern times.

- Chapter 8 concludes the unit on culture by using spatial patterns of language and religion to illustrate the roles migration, diffusion, and processes of cultural change play in the interactions of people over space and time.

- Chapter 9 begins the unit on political patterns and processes with an overview of the complexity of contemporary political boundaries and the forces of power relationships that shape political processes and their spatial manifestations.

- Chapter 10 expands the discussion of space and power by looking at trends in the organization of states internationally and at local-level electoral geographies, particularly the ways political power may be used and abused to represent different constituencies.

- Chapter 11 concludes the unit by examining the forces at work in unifying and dividing political entities. Nationalism, supranationalism, and devolution are all challenging state sovereignty. The chapter provides students with key concepts needed to understand contemporary global politics.

- Chapter 12 begins the unit on agriculture and rural land use with an examination of the effects of human-environment interaction. Agricultural practices are explained and the origins and diffusion of such practices explored through innovations in agriculture over time and space. Sustainability is a theme throughout the unit.

- Chapter 13 presents an overview of agricultural production regions and how agriculture, like any economic activity, has a distinct spatial organization encapsulated by models and theories such as bid-rent theory and the von Thünen model. Contemporary agriculture is presented as an integrated global system.

- Chapter 14 concludes the unit exploring the sustainability of agriculture in a global market. A special focus on women in agriculture and the challenges of contemporary agriculture raise issues related to concerns about food security at a range of scales.

- Chapter 15 introduces the patterns and processes of urbanization and city land-use. The factors that influence urban growth and decline, including site and situation, and central place theory and are used to explain the development of systems of settlements in the world. The influence of contemporary urban processes on urban landscapes is examined.

- Chapter 16 examines the internal structure of cities and the roles infrastructure and housing play in making urban areas attractive and sustainable. The evolving shapes of cities illustrate processes of urbanization and the roles of transportation in ongoing patterns of land-use.

- Chapter 17 explores urban living and ways innovative design ideas are shaping the livability of cities. The causes and impacts of urban change focus attention on the challenges cities face throughout the world, including gentrification and urban renewal. The continued legacy of systemic racism and practices such as redlining on cities in the United States is discussed.

- Chapter 18 begins the unit on industrial and economic development with an examination of changes in industrial processes over time and the consequences of these changes. The structures of economies are explained as a way to compare levels of economic development in different world regions. The principles and models that influence industrial location provide insight into global patterns of production and consumption.

- Chapter 19 introduces ways that human and economic development can be measured as a way to examine theories of development and the world economy. A special focus is on measuring gender inequality and the changing roles of women in the contemporary world.

- Chapter 20 concludes the unit with an examination of globalization, interdependence, and sustainability through close looks at trade relations, evolving systems of connected economies, and their effects on efforts to achieve sustainability.

Although we have organized the book in this sequence, we encourage faculty to re-organize and establish their own narrative and order. Where possible, links across chapters of important concepts have been made to support synthesis and understanding. It is a challenge to understand processes of urbanization without recognizing the forces that drive migration; equally it is a challenge to understand population without some core idea of the role culture plays in demographic change. Learning human geography simply has to be an iterative activity.

Features of this Book

Each chapter includes relevant images, maps, models, photos, and other graphics, rich case studies, and features that profile the amazing work of National Geographic explorers, fellows, writers, and photographers—all to support the geographic content presented. We have focused consistent attention throughout on impacts of globalization, as well as on human-environment interaction—including climate change and sustainability. The maps and graphics we have chosen are many and engaging, clearly showing the power of the spatial perspective. We have included hundreds of photos that strikingly capture ideas, people, and places discussed in the text. In writing this book over the past year (2021–2022) we have consciously included evolving results of the COVID-19 pandemic as appropriate in each unit. It has been a struggle to feature up-to-date data skewed as it is by the pandemic. In some cases, we felt that older data on maps and graphs better represented contemporary conditions.

As college instructors, we know getting students to read and understand texts is a challenge. We believe that this book will appeal to students for three reasons. First, it is written in a clear and easily understandable style designed to engage the reader. Core concepts and ideas are emphasized and illustrated by a limited selection of meaningful and appropriate examples, making the material more memorable. Clear and precise learning objectives guide students as they read. Second, by featuring a diverse range of National Geographic explorers, fellows, collaborators, photographers, and writers who share their work connected to the content of each unit, human geography comes alive and becomes relevant to the contemporary lives of students. We hope the professionalism of these change agents can inspire you and your students. The case studies are wide-ranging, topical, and timely, again designed to link the daily lives of students to the course content. Third, the abundance of photos, maps, graphs, and other visuals make the material approachable to today's students who are visually literate. The maps have been specifically designed to comply with accessibility standards for students with red-green color blindness.

Using the Text

That said, we strongly recommend that instructors help students to understand the degree to which geographers communicate through maps and other spatial representations. Explicit instruction on map interpretation and how to use the rich array of visuals as a support to learning the material will benefit most students. We also recommend taking the time to establish for students how it is that geographers see the world. The first unit of the book is Thinking Geographically, and we feel it is important to make the point that gaining this new worldview may be especially valuable for young members of society in these challenging times. For that reason, we have focused consistent attention on a central geographic theme of human-environment interaction, including climate change and sustainability.

Course Solutions

MindTap for Human Geography

MindTap for Bednarz, Bockenhauer, and Hiebert, *Human Geography* helps students learn geographic concepts and processes on their terms. They can read or listen to textbooks and study with the aid of instructor notifications, flashcards, and practice quizzes. Students can also create custom flashcards, highlight key sections in their textbook they want to remember, complete homework assigned by their instructor, watch videos and animations to help strengthen their understanding of lecture and reading material, and apply their learning through a variety of critical thinking activities.

Students can track their own scores and stay motivated toward their goals. Whether they have more work to do or are ahead of the curve, they'll know where they need to focus their efforts. They can learn at their own pace and at a time convenient to them using the Cengage Mobile App.

Features of MindTap for Human Geography

MindTap for Human Geography includes activities for students who may start with different levels of preparedness for college and this course.

- **Brushing Up on the Basics** tutorials in the Course Orientation folder are a particular benefit to students who are underprepared in skills related to graphing, data interpretation, and scientific process. The Course Orientation folder also includes country identification map tests that you can use to determine students' prior knowledge at the beginning of the course and the progress they have made by the end of the course.

- **Talk About It** activities before each chapter help students understand how chapter topics relate to their everyday lives and create an engaging classroom environment by sharing students' responses to polling questions.

- **Concept Visualizations** using maps, videos, graphs, and photos help students visualize complex foundational processes and concepts. Paired with auto-graded assessment questions that provide immediate feedback, these media activities deepen student understanding of core course concepts.

- **Key Terms Reviews** prepare students to use the terminology of the discipline and ask students to connect terms and definitions with authentic examples.

- **Check Your Understanding** quizzes give students an opportunity to test their knowledge of chapter topics.

- **GIS Map Analysis** activities ask students to think critically about map data in ArcGIS and develop spatial reasoning skills through application and observation. Access to ArcGIS is provided via MindTap; separate access is not required.

- **Case Study Supports**, two in each chapter, help students connect the specific information from the text's case studies with geographic principles to extend their understanding.

- **Data Interpretation Labs** ask students to apply critical thinking to quantitative information from two or more maps, graphs, charts, and tables, promoting 21st century literacy.

Instructor Resources

Instructor Companion Site

Everything you need for your course in one place! This collection of book-specific lecture and class tools is available online via www.cengage.com/login. Access and download PowerPoint presentations, images, instructor's manual, videos, and more.

Cengage Learning Testing Powered by Cognero

Cengage Learning Testing Powered by Cognero is a flexible, online system that allows you to:

- author, edit, and manage test-bank content from multiple Cengage Learning solutions
- create multiple test versions in an instant
- deliver tests from your Learning Management System, your classroom, or wherever you want

Acknowledgments

We wish to thank the team of people from Cengage Academic without whom this volume would not have been possible, particularly Erin Joyner, Senior Vice President, Vicky True-Baker, Senior Portfolio Product Manager, and Leslie Taggart, Learning Designer. We also wish to thank Sean Campbell, Senior Content Manager, for his good humor and most helpful comments and suggestions. Several of the exciting maps and graphics throughout the book are due to the excellent work of Glen Pawelski and his staff at Mapping Specialists, Ltd. We are also deeply indebted to Marcie Goodale, Senior Director, National Geographic Learning/Cengage for her vision, guidance, and perseverance. We also wish to acknowledge Gilbert M. Grosvenor, National Geographic Society, for his longstanding commitment to geography and geography education. Finally, we must thank our families for their patience as we worked on this project.

Foreword | Gilbert M. Grosvenor

> Physical geography tells you where you are. Human geography explores the interaction of humans with the world around them. It helps you envision the future and better understand the past—human geography gives humans dimension.
>
> Gilbert M. Grosvenor
> July 2019

In 1979, I joined an expedition to the North Pole and jumped at the once-in-a-lifetime chance to scuba dive under massive ice blocks that thrust nearly 40 feet beneath the ocean surface. Afterward, I surveyed that harsh icescape and understood for the first time the importance of the Arctic and Antarctic as bellwethers for the future of Planet Earth. Loud cracking sounds accompanied the inevitable collisions of ice floes as nature constantly powered a million square miles of thick ice across the Arctic Ocean. I never thought of the Arctic in the same way after that.

At the time I was editor of *National Geographic* magazine and would eventually become president of the National Geographic Society, but my history with National Geographic goes back even further. My great-grandfather, inventor Alexander Graham Bell, was National Geographic's second president. He defined geography as "the world and all that's in it," and that idea has driven the organization for more than 132 years.

In fact, geography is a great tool for examining the complicated interaction between humans and the natural environment. In an era in which climate change and issues of sustainability are commanding greater attention, geography has become a more scientific, technologically oriented discipline—it gives us a good way to understand a complex, rapidly changing world.

This Human Geography textbook couldn't be more closely connected to National Geographic's mission to understand and preserve our world. The seven key topics that you'll read about in this text have formed the backbone of the work we do in exploration, storytelling, and geography education for decades.

It's been my good fortune to work with many brilliant and dedicated women and men over the years, including marine biologists, archaeologists, mountain climbers, astronauts, photographers, and cartographers. As inspirational as their contributions to science and culture have been, I am convinced that your future contributions will rival theirs for lasting impact—you are the new Explorers. Through this Human Geography program, National Geographic hopes to be your partner in that exciting endeavor.

Those moments under the Arctic ice changed my life. You have something equally amazing in store for you, too—you have only to seek it. Start by asking those geographic questions: Where in the world? And why?

I and my colleagues at National Geographic wish you an astonishing journey.

This photo reveals two of Gil Grosvenor's passions: photography and sailing on Bras d'Or Lake in his beloved Baddeck, Cape Breton, Nova Scotia, where he grew up amidst several generations of Grosvenors.

Continuing National Geographic's Legacy

Whether you're more familiar with the yellow-bordered magazine or the online interactive version, for more than a century National Geographic has told stories of intrepid explorers, groundbreaking photographers, and global geographic phenomena.

Like the issue on migration shown here, many of those stories connect to the topics you'll study in this Human Geography course. In roles from photographer to editor-in-chief, Gil Grosvenor's vision shaped the direction of those stories.

Unit 1
Thinking Geographically

The Human Impact

Greenmarket Square in Cape Town

Humans have left an indelible imprint on every part of the globe, as evidenced here by the sprawling communities and built-up shoreline of Cape Town, South Africa. At the same time, Earth's features and processes have molded profound and enduring aspects of human societies—our cultures, economies, and politics. Human geography focuses on the interactions between humans and the environments in which they live.

Geographers interpret the world through a lens that allows them to make connections among local, national, regional, and global issues to better understand our human stories.

In this book, you use this geographic lens to investigate the movement of people, cultures, and ideas; the political organization of countries; and the development of agriculture, settlements, and linked economies. You build an informed global awareness and discover the factors that influence the world's livability and sustainability.

Chapter 1
The Power of Geography: Geographic Thinking

Chapter 2
Geographic Inquiry: Data, Tools, and Technology

Unit 1 Writing across Units, Regions, & Scales

Unit 1 Maps & Models Archive

National Geographic Explorer **Enric Sala**

Safeguarding Pristine Seas

Biodiversity makes our life possible. The rich diversity of species on Earth helps to keep our water clean and stabilize our climate and contributes to food security. Enric Sala, National Geographic Explorer-in-Residence and founder of the Pristine Seas project, has made it his mission to preserve the biodiversity of Earth's oceans.

Learning Objective
LO 1.6 Explain why sustainability is an important theme in human geography.

A Mission to Protect A team of scientists gathered by Enric Sala left for an expedition in 2005 to the Line Islands, a cluster of coral outcrops in the Pacific Ocean roughly 1,000 miles south of Hawaii. Some of the islands that dot this vast ocean area are U.S. territories. In this remote patch of ocean, Sala's team conducted groundbreaking research on one of the few coral reef ecosystems that remain largely untouched by human activity. These studies provided scientific support for what would become an ecological triumph.

In January 2009, President George W. Bush signed into existence the Pacific Remote Islands Marine National Monument, which placed several U.S. territories, including the American Line Islands, off limits to commercial fishing and other for-profit activities. This protected zone was expanded in 2014 by President Barack Obama to more than 745,000 square miles—covering a greater area than all U.S. National Parks on land combined.

In 2008, Sala founded the Pristine Seas project to expand his work in exploration and conservation. Sala's methods, models, and data are intended to establish the threshold of ecosystem health that can be used to inform conservation priorities and efforts of governments and organizations. Pristine Seas expeditions have investigated some of the most isolated places on the planet in locations ranging from the South Pacific to the high Arctic. "These remote, untouched places are the only baseline we have left for what the oceans used to be like. They are like the instruction manual for the ocean," says Sala. The Pristine Seas team includes not only researchers but also filmmakers and policy and communications specialists who support the mission to discover, inform, and advocate for changes in the ways humans interact with the oceans. As of early 2022, Pristine Seas has inspired the protection of more than 2.5 million square miles in 24 of the largest marine reserves on the planet.

Challenges to Conservation Enric Sala admits, "It is difficult to be optimistic about the ocean in my lifetime." Indeed, while Pristine Seas has cause to celebrate notable successes in protecting a number of fragile habitats, 2.5 million square miles is a tiny fraction of the oceans' total surface area. Notwithstanding all of the zones protected by the efforts of Pristine Seas and other groups, 97 percent of Earth's oceans are still open to fishing. Much of this vast area is vulnerable to overfishing and pollution, which Sala calls "ecological sabotage."

At the same time, fishing is a means of survival for millions of people. Recognizing this fact, Sala calls for governments to better manage fisheries, improve fish farming known as aquaculture, and enforce laws against marine pollution. He argues that marine conservation actually enhances the sustainability of the fishing industry, citing the example of a fishing community in Kenya where incomes doubled because marine reserves had helped restore the health of sea life in the region's waters. "We know what works," Sala says, "we just need the political will and the vision to protect much more of our waters."

Geographic Thinking

How might the creation of relatively small marine reserves affect natural resources and sustainability in the rest of the world's oceans?

In November 2021, the Oceanographic Institute awarded Enric Sala the Prince Albert I Grand Medal, the highest international recognition for leaders whose scientific expertise is matched by their determination to promote protection of the ocean.

Sharks hunt off the Galapagos Islands, where Pristine Seas helped to create the Darwin and Wolf Marine Sanctuary, protecting the highest abundance of sharks known in the world.

Pristine Seas Expeditions & Protected Areas

● Protected area   ● Completed expedition

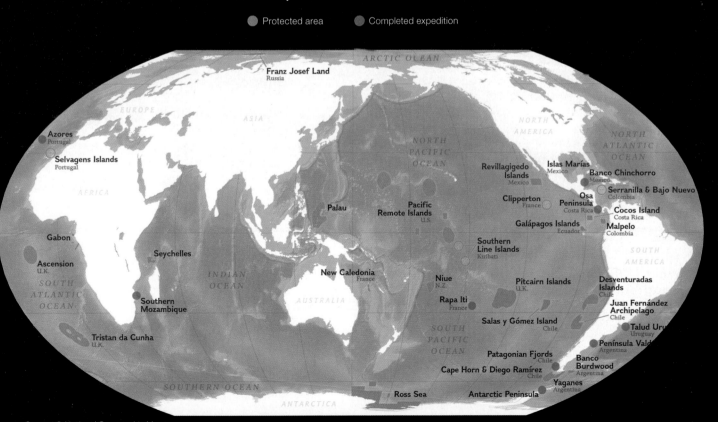

Source: © National Geographic Maps

Chapter 1

The Power of Geography: Geographic Thinking

Critical Viewing Rows of buildings line the Charles River in Boston, Massachusetts, a city that was founded in 1630 along the banks of the river and Boston Harbor. ▍Explain what geographers today might learn by studying the location of the city of Boston.

Geographic Thinking Why does geography matter?

1.1
What Is Human Geography?

Case Study: New Orleans—Site versus Situation

National Geographic Explorer
Adjany Costa

1.2
Spatial Patterns: Scale and Region

Case Study: India—Regional Differences in Scale

1.3
Globalization and Sustainability

1.1 What Is Human Geography?

Exploring why things are located where they are can offer insights into how human activity shapes the world. Human geographers study the ways in which people use, adapt to, and change Earth, as well as how they are influenced by it. Establishing the concepts and perspectives that inform a human geographer's work will give you a context for the rest of this book.

Studying Human Geography

Learning Objectives
LO 1.1 Explain the fundamental differences between physical and human geography.
LO 1.2 Describe the features that distinguish geography from other disciplines.

Whether you're aware of it or not, you are regularly engaged in geographic thinking. The simple act of traveling around your community requires you to know where your destination is, to plan a route, to consider distance and traffic, and to estimate how long your trip will take. When you perform these calculations, you are thinking geographically.

Geography is an integrative discipline that brings together the physical and human dimensions of the world in the study of people, places, and environments. Its subject matter is Earth's surface and the processes—continuous actions taking place over time—that shape it, as well as the relationships between people and environments and the connections between people and places.

The discipline of geography is divided into two major areas. **Physical geography** is the study of natural processes and the distribution of features in the environment, such as landforms, plants, animals, and climate. For example, a physical geographer might study the movement of glaciers in different eras, investigate how a process like erosion changes a riverbed, or research change in a forest's growth rates related to climate change. **Human geography** is the study of the events and processes that have shaped how humans understand, use, and alter Earth. A human geographer researches how people organize themselves socially, politically, and economically and how they interact with and impact the natural environment.

Because geographers work on many of the same questions and problems as experts in other fields in the physical and social sciences, they face the challenge of differentiating geography from those fields. One distinguishing feature is geographers' focus on the relationship between humans and environments. Other disciplines tend to focus on either one or the other. Geography also recognizes the importance of where events and phenomena occur, focusing on how processes vary depending on location. For instance, a society will develop differently in a rural environment than in an urban setting. A third distinguishing feature is geographers' focus on geographic scales. You learn more about the importance of scale later in the chapter.

In 2002, geographer Reginald Golledge summarized the unique nature of geographic thinking: "The way geographers reason[ed] about space, and particularly their penchant for representing complex spatial meanings in a clearly understandable form (spatially based maps, graphics, and images) emphasized that geographic thinking and reasoning gave a perspective that was not matched by any other single academic discipline." Thinking geographically offers advantages to geographers—and to us all. In 2004, Susan Hanson coined the term "geographic advantage," meaning "to understand relationships between people and the environment; to understand the importance of spatial variability and the place dependence of processes; to understand that processes operate at multiple and interlocking scales; and, to integrate space and time through spatial and temporal analysis."

Geographic Perspectives

Learning Objectives
LO 1.3 Compare the spatial and ecological perspectives of human geography.
LO 1.4 Describe the key spatial concepts of human geography.

Both branches of geography analyze complex issues and relationships from two key perspectives, or points of view: the spatial perspective and the environmental or ecological perspective. These perspectives help geographers interpret and explain spatial patterns and processes on Earth, and understand the interactions between environments and human societies.

The **spatial perspective** refers to where something occurs. In the same way that history is concerned with time and the chronological aspects of human life, geography is concerned with the spatial aspects—where things are located and why they are located there. When human geographers take a spatial perspective, they are studying how people live on Earth, how they organize themselves, and why the events of human societies occur where they do.

The second key perspective is the **ecological perspective**, which refers to the relationships between living things and their environments. Looking at an issue from an ecological perspective involves studying the interactive and interdependent relationships among living things, ecosystems, and human societies. This perspective helps explain human societies' dependence on diverse ecosystems for essential resources such as food and water. Taken together, the spatial and ecological perspectives help human geographers understand the many relationships between humans and environments. The awareness that these and other perspectives exist is fundamental to a geographer's understanding of the world's people and places.

The essential elements of geography can be summed up neatly in the following three questions: Where? Why there? Why care? When geographers apply these questions, they are thinking geographically. Asking where something is located is the starting point for any geographic inquiry. Asking why it is located there pushes geographers to analyze the reasons behind events, processes, and interactions. And asking why someone should care helps them establish the importance and relevance of their inquiries. (Geographic questions are discussed further in Chapter 2.) Certain spatial concepts help geographers answer these questions. These concepts include location, place, space, flows, pattern, distance decay, and time–space compression.

Location and Place It should be no surprise by now that where things are found is an important geographic concept. **Location** is the position that a point or object occupies on Earth. Location can be expressed in absolute or relative terms. **Absolute location** is the exact location of an object. It is usually expressed in coordinates of longitude and latitude. The city center of Budapest, Hungary, for instance, is located at 47.50° N, 19.04° E. With the proper means of transportation and a global positioning system (GPS) receiver, you could use these coordinates to get to Budapest from any other location on Earth. **Relative location** is a description of where a place is in relation to other places or features. A geographer might describe Budapest's relative location as 134 miles straight-line distance southeast of Vienna, Austria, or might replace distance with time and say it's just a 2.5-hour bullet train ride from the Austrian capital to the Hungarian capital. A geographer might also describe that Budapest straddles the Danube River in the middle of the Carpathian Basin in north central Hungary.

The term **place** is related to but different from location. A place is a location on Earth that is distinguished by its physical and human characteristics. The physical characteristics of a place include its weather and climate, landforms, soils, water sources, vegetation, and animal life; the human characteristics include its languages, religions, and other aspects of culture, political systems, economic systems, population distribution, architecture, and quality of life.

When people say they feel a strong "sense of place," they are referring to the emotions attached to an area based on their personal experiences. The sense of place that people have for their hometown or certain buildings—a school, town square, riverside park, or baseball stadium, for example—is stronger than it is for places they don't know, and it is tied to their sense of identity. Because humans create the concept of place in their encounters with the world around them, someone who grew up in Boston may strongly identify with the city or a particular neighborhood within it, just as a person who grew up in Berlin or Beijing may strongly identify with that city. Their surroundings, including the natural landscape and built environment of the city, the people they know there, and the familiar cultures of the place, influence their perceptions. The attachment that Bostonians, Berliners, or Beijingers form with the history, architecture, landforms, and people of their cities contribute to their identities.

Places change over time. As a society's values, knowledge, resources, and technologies evolve, the people within that society make decisions that alter the place they occupy. Cities grow in population, construction covers the land, wetlands are filled in, and farmlands are turned into suburban neighborhoods and shopping centers. Other

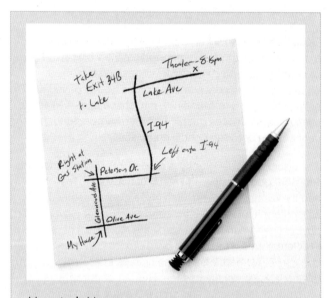

Mental Maps

If you think about a place you go to regularly, you can probably imagine and describe the route you take to get there. And you can likely draw a reasonably accurate map of your school, neighborhood, or town from memory. These internalized representations of portions of Earth's surface are called **mental maps**. Your mental map of different areas of the world depends on many factors. Your experiences, your age, where you live, and other factors contribute to the accuracy of your mental maps. For instance, what do you picture when you think of New England? How about the South, or the Pacific Northwest? Now compare your mental map to actual maps of these areas. You most likely have a clearer mental map of the area you live in than one that's far away.

places might shrink in population as people move away and, over time might disappear altogether. Decisions about how to organize society and how to interact with other places cause changes as well. The relationships between places affect all involved, politically, economically, and culturally. Over a long enough period of time, empires rise and fall, climates change, and society evolves enough to change life significantly for the people who live in a place. If you've visited or seen pictures of cities that have been around for a long time, however, you've probably noticed that elements of the place's history and geography—its original sense of place—usually remain.

Human geographers focus on two factors that influence how humans use a particular place. The first factor is **site**, which refers to a place's absolute location, as well as its physical characteristics, such as its landforms, climate, and resources. The second factor is **situation**, which refers to a place's location in relation to other places or its surrounding features. Situation describes a place's connections to other places, such as transportation routes (like roads, rail lines, and waterways), political associations, and economic and cultural ties.

When describing the site of the Spanish city of Barcelona, a geographer would say – perhaps in addition to its absolute location – that it is located on a coastal plain with the Besós River to the north and the Llobregat River to the south. It lies between a rocky outcrop and a semicircle of mountains. It has a mild Mediterranean climate. Barcelona's situation, on the other hand, is that it is a port city on the Mediterranean Sea, which historically controlled the western portion of that sea along with Mallorca and Valencia. And because there are few navigable rivers in the region, Barcelona was well situated as a major hub on the trade route from France to southeastern Spain.

Space, Pattern, and Flow As you've read, when geographers think geographically, they are considering the arrangement of things in **space**. Space in this instance refers to the area between two or more things on Earth's surface. Studying the ways in which things are **distributed**, or arranged within a given space, can help human geographers describe and analyze the organization of people, places, and environments on Earth. Density and pattern are key concepts in the examination of distribution.

Density is the number of things—people, animals, or objects—in a specific area. For example, a geographer might compare the population density of a large city to that of a rural area. Manila, the capital city of the Philippines, has over 171,000 people per square mile. A rural area like the province of Davao del Sur, on the other hand, has about 850 people per square mile. Based on this statistic, what conclusions might a geographer be able to draw about the lives of a person in Manila and a person in Davao del Sur?

Pattern—how things are arranged in a particular space—is another factor of distribution. Depending on how humans settled and developed a place, and what their needs are for it, its features might be arranged in a neat, geometric pattern, or they might be arranged in a pattern that seems more random. Studying the patterns of phenomena in space can help geographers understand different processes, such as patterns of agricultural production, urban settlement, or the distribution of fast-food restaurants in a town. In Unit 5 you learn about different types of rural settlement patterns and the reasons why each pattern developed. Patterns can be observed in urban areas as well. Many old cities in Europe, for example, are made up of narrow, winding roads laid out long before automobiles that are, of course, inconvenient for car and truck traffic. But Denver, Colorado, was built on a grid system in more recent times, with many streets that intersect at right angles, which makes the city easier to navigate. How might getting around and dealing with traffic patterns differ between London and Denver?

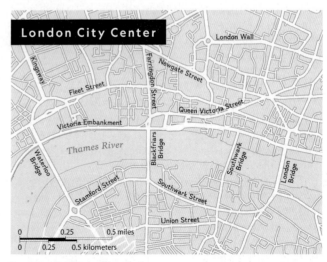

The street maps of Denver and London illustrate contrasting urban patterns—a city like London that has evolved over a long period of time looks very different from a relatively newer city like Denver. The winding roads of London bear little resemblance to Denver's planned angular, gridlike layout.

Obviously, geographers are not studying a world at rest. Any given space changes over time as things move from one place to another. The study of the **flow** of people, goods, and information—including the economic, social, political, and cultural effects of these movements on societies—is an important aspect of human geography. You learn about the flow of people in Chapter 5 on migration, the flow of culture in Chapter 7 on cultural change, and the flow of goods in Chapter 20 on trade.

Human-Environment Interaction

Learning Objectives
LO 1.5 Explain how theories and principles of human-environment interaction have developed and changed over time.
LO 1.6 Explain why sustainability is an important theme in human geography.

Regardless of where people live, they depend on, adapt to, and modify the natural environment. They make decisions about how to live based on environmental features, and they make changes to the environment as a result of those decisions. Humans have always changed the landscapes they settled—in myriad ways, such as using land for agriculture, tapping into natural resources, and building structures in which to live or work. But technologies and building techniques have given modern humans the ability to alter their environment in almost unlimited ways.

Human geographers work to understand spatial patterns and processes on Earth, realizing that the planet is composed of living and nonliving elements that interact in complex webs of relationships within nature and between nature and human societies. Humans are part of the interacting and interdependent relationships in Earth's ecosystems and are one among many species that constitute the living part of Earth. Human geographers study how these human–environment interactions and changes affect both humans and the environment itself, and they might consider both short-term and long-term consequences of these interactions. Their views on the causes and effects of human societies' interactions with the natural environment have evolved over time.

Compare Models: Distance Decay and Time-Space Compression

A model is a representation of reality that presents significant features or relationships in a generalized form. Models help geographers analyze spatial features, processes, and relationships. One example is the distance decay model. **Distance decay** is a key geographic principle that describes the effect of distance on interactions. The principle states that the farther away one thing is from another, the less interaction the two things will have. Cartographer and geographer Waldo Tobler's first law of geography states that while all things on Earth are related to all other things, the closer things are to one another, the more they are related. Think about an earthquake, flood, or political revolution. The closer you are to any of these phenomena, the more likely you will be affected by it. Distance decay is connected to friction of distance, a concept that states that distance requires time, effort, and cost to overcome. Friction of distance applies to economic, political, religious, and cultural movements as well, but because of modern advancements in technology and transportation, it has less impact today than in the past.

Time-space compression is a key geographic principle that is related to friction of distance. It describes the processes causing the relative distance between places to shrink. Modern transportation has greatly reduced travel times, and the internet and other forms of communication have made it easier to communicate with people anywhere on the planet and to send money around the world through online banking transfers.

Through these technologies, humans have effectively caused the distances between places to seem shorter, as they are able to cross those distances more quickly and exchange goods and information more easily. The map below illustrates time–space compression. Why might Europe and North America seem closer together to a person today than they seemed to Columbus?

Transatlantic Travel Times

Theories of Interaction In the 18th, 19th, and some of the 20th centuries, many geographers and others subscribed to a theory of human–environment interaction that has since been discredited, largely because it inaccurately favors the accomplishments of certain societies over others. This theory—**environmental determinism**—argues that human behavior is largely controlled by the physical environment. According to the theory, natural attributes like a region's topography, climate, and soil fertility dictate how a society develops as it adapts to the environment. Environmental determinism has been widely criticized, however, because it argued that the environment most suited to human development is that of western Europe and North America. Environmental determinism was used as a tool to legitimize racism, colonialism, and imperialism. This theory fails to take into account the fact that civilizations in other regions, such as North Africa, Central America, and Asia, arose as early as or earlier than those in Europe and North America and were more advanced technologically and highly influential culturally for long periods of human history. Impacts of racism, colonialism, and imperialism are discussed further in later chapters.

Modern geographers favor **possibilism**, a theory that argues that humans have more agency, or ability to produce a result, than environmental determinism would suggest. According to possibilism, individuals are active, not passive, agents. The environments in which they live offer individuals and societies opportunities and challenges. People and societies react to those opportunities and challenges in different ways depending on the decisions they make, their ingenuity, and the technologies available to them. The environment certainly places some limitations on human activity, but individuals and societies have a range of options in deciding how to live within a physical environment. Consider societies and settlements that have developed in desert environments. People have diverted rivers and built dams to irrigate land for agriculture and have constructed aqueducts to transport drinking water. They have built farms and cities in places that were once too barren and dry to support human life.

B. L. Turner II notes that while geographers might examine physical processes or human processes in isolation, he contends that the "geographic" follows from their integration. He cites two reasons why what he termed

Climate Change in the Anthropocene

The United Nations (UN) defines climate change as "long-term shifts in temperatures and weather patterns," and states that these shifts may be natural, such as through variations in solar radiation reaching Earth. The UN explanation continues: "But since the 1800s, human activities have been the main driver of climate change, primarily due to burning fossil fuels like coal, oil and gas. Burning fossil fuels generates greenhouse gas emissions that act like a blanket wrapped around the Earth, trapping the sun's heat and raising temperatures." Emissions of greenhouse gases, so named because they trap heat like a glass-windowed greenhouse does, include carbon dioxide and methane. These gases cause climate change and come from many sources, using gasoline or other oil-based fuels for cars, trucks, trains, planes, and ships and using coal or natural gas for heating buildings or generating electricity. Cutting forests and clearing land also release carbon dioxide, while landfills for garbage and certain types of agriculture are major sources of methane emissions.

As a result of these human-caused emissions, Earth is now about 1.1°C (2°F) warmer than it was in the late 1800s. The decade 2011–2020 was the warmest on record. While some people think of climate change as mainly meaning warmer temperatures, temperature rise is only part of the story. Because Earth is a system in which everything is connected, changes in one area can cause or influence changes in others. Consequences of climate change include water scarcity and droughts, severe wildfires, melting polar ice, flooding, rising sea levels, catastrophic storms, and declining biodiversity. Climate-change emissions come from every region of the world and affect everyone, but some countries produce much more than others. The UN finds that the 100 least-emitting countries generate 3 percent of total emissions, while the 10 countries with the largest emissions contribute 68 percent.

Scientists have referred to the recent time period in which human activity has had a significant impact on the planet's climate and ecosystems as the "Anthropocene Epoch," an unofficial unit of geologic time (from the Greek words *anthropo*, for "human," and *cene*, for "new"). Archaeologists have documented that human effects on the environment have been incrementally growing over the last 10,000 years, with major increases coming about 6,000 years ago (the origins of cities) and then 3,000 years ago (the advent of iron smelting). These impacts, of course, have greatly accelerated since the Industrial Revolution beginning in about 1800, and have been growing exponentially since 1950. Anthropocene climate change is affecting people across the planet, in terms of human health, the ability to grow food, housing, safety, and work. (UN-led efforts to reduce greenhouse gas emissions, to adapt to impacts of climate change, and to finance needed adjustments are discussed in other sections of the book.)

"human–environment geography" grew "substantially" in the United States during the latter part of the 20th century: the first was "increased public awareness of environmental issues," and the second was "the 'formal' diversification of geography at large." This diversification of geography has included many important topics in human–environment interaction: land and natural resource use, sources and impacts of pollution and environmental degradation, natural hazards, and climate change (see box Climate Change in the Anthropocene).

Research and teaching in human geography include how climate change affects regions and groups of people in different and clearly uneven ways. Glen MacDonald states that climate change will "disproportionately affect equatorial regions and closely adjacent areas" and challenged his fellow geographers in a 2018 speech to point out and address unequal geographies, including those of climate change: "This is the time for geographers to be at the forefront and to make good on our long-held claims that the multifaceted geographical perspective is of fundamental importance. In doing so we can also create a nexus that draws together the diverse elements of geography into a more cogent whole." Some of the groups that are already more vulnerable to climate impacts include people living on small islands and those in poorer countries. Conditions like sea-level rise have progressed to the point where entire communities have had to relocate. Extreme droughts are putting farmers and others at risk of famine, and the number of what some authorities term "climate refugees" is on the rise. For example, geographer John Rennie Short points out a "cruel turn of events" for the continent of Africa, with changing rain patterns, flooding, extreme heat, and drought making the region one of the most impacted now and in the future by climate change, while having among the lowest per capita carbon footprints. Human geographers also research and help develop solutions for how countries and groups of people can adapt to and become resilient in the face of changing climates.

Sustainability An important concept in thinking about human–environment interaction is **sustainability**, the use of Earth's land and natural resources in ways that ensure they will continue to be available in the future. Sustainable use requires consideration of whether a particular use of a natural resource or of land is renewable, meaning nature produces or replenishes it faster than people use it, or nonrenewable, meaning people consume or diminish it faster than nature produces or replenishes it. Nonrenewable resources are limited in supply, while renewable resources are essentially limitless. Solar and wind energy, for instance, are renewable and thus sustainable energy resources—available perpetually, while coal and other fossil fuels are nonrenewable—because deposits of these minerals were formed millions of years ago and ultimately only finite amounts are accessible. Sustainable land use is exhibited by the centuries-old rice terraces of Japan and the Philippines, with careful preservation of soils and application of natural nutrients by community farmers, while typical industrial agriculture, with its reliance on fossil fuel–derived synthetic fertilizers and fossil fuel-powered machinery, compacts, erodes, and depletes soils, and may not be sustainable. The effects of a society's use of natural resources are important to consider as well. For example, what advice do you think a geographer concerned with sustainability might give to a government deciding what laws to pass to fight climate change? Sustainability and climate change are discussed further in several units of this book.

Geographic Thinking

1. Explain whether the address of a restaurant is an absolute location or a relative location.

2. Describe how geographic concepts help to explain the distribution of phenomena on Earth.

3. Describe how technology "shrinks the world" using the time–space compression model.

4. Compare the theories of environmental determinism and possibilism.

Critical Viewing Lettuce and other produce grows in a garden behind center field at AT&T Park (which became Oracle Park in 2019) in San Francisco. Restaurants inside the ballpark use produce grown here to serve directly to baseball fans. ▌Explain how this feature of the ballpark is an example of sustainability.

Case Study

New Orleans—Site versus Situation

The Issue New Orleans' proximity to resources, advantageous natural features, and transportation routes is ideal, but the land upon which it sits offers many challenges.

Learning Objective
LO 1.7 Contrast site and situation in geography.

By the Numbers

50% of the city of New Orleans is at or below sea level

454,845
Population pre-Katrina

383,997
2020 population

Source: *The Atlantic*, United States Census Bureau.

In the Early 18th Century, the French colonists who settled Louisiana needed a location for the colony's capital. They considered a variety of sites—some inland and some on the coast. After much consideration, they decided on an option that was slightly inland, on the Mississippi River. This would become the site of the city of New Orleans. Centuries later, residents of the city and the country are still grappling with the choice the original settlers made.

They knew the site wasn't ideal. Located on a sharp bend on the east bank of the Mississippi River, at the head of the delta leading to the Gulf of Mexico, and just south of Lake Pontchartrain, New Orleans floods easily and regularly and is subject to severe storms from the Gulf of Mexico. However, the city's situation was perfect. Its location at the southern end of the Mississippi meant that New Orleans would be connected to a huge area of the lands to the north. The Mississippi is the largest river system on the continent and has river links to two-thirds of the continental United States. Much of the country's commerce still travels down the river right past the city. The city's founders decided that the advantages of the situation outweighed the disadvantages of the site. The city's situation is still valuable today—because of its location near the mouth of the Mississippi and its access to the Gulf of Mexico, New Orleans remains one of the busiest ports in the United States. (Site and situation are discussed further in Chapter 15.)

In order to deal with the site disadvantages, the city's early residents built artificial levees, or embankments, to keep the river from flooding the streets, and in the mid-1800s engineers figured out how to drain the wetlands between the river and Lake Pontchartrain, allowing the city's borders to spread out into low-lying terrain. Today roughly 50 percent of the city lies below sea level, a shallow bowl surrounded by more modern—but certainly not perfect—levees to keep the water out.

The imperfections of those levees were revealed in late August 2005, when Katrina, a destructive and deadly hurricane, slammed into New Orleans. The 10 inches of rain that Katrina dumped, combined with a devastating storm surge, overwhelmed the levees that held back the waters of Lake Pontchartrain and nearby Lake Borgne. After the levees failed, water poured into the city, eventually flooding 80 percent of its land.

More than a million people evacuated the region before the storm, but tens of thousands remained because they could not or would not leave. Thousands of stranded residents took shelter at the Louisiana Superdome and the New Orleans Convention Center, but a lack of food and drinkable water and absence of sanitation created a public health emergency.

By September 6, the city had been almost completely evacuated, and fewer than 10,000 people remained. Much of the city was eventually rebuilt, and some of the displaced returned, but the population today remains lower than it was when Katrina struck. Sixteen years later in August 2021, a Category 4 storm named Hurricane Ida hit New Orleans, but little damage was sustained because the improved levee system that was built post-Katrina held. With climate change causing more frequent and more severe tropical storms, these levees will likely be tested again, as New Orleans residents continue to live with the site disadvantages of their river city. ∎

Geographic Thinking

Explain why New Orleans' founders decided that the advantages of the location's situation outweighed the disadvantages of its site.

Floodwaters flow over a failed levee along the Inner Harbor Navigation Canal near downtown New Orleans following Hurricane Katrina, while a trapped resident waves a white flag for help. A major shortcoming of New Orleans' site is found in its low-lying land along Lake Pontchartrain and the Mississippi River. Massive flooding devastated the region after Hurricane Katrina struck in 2005.

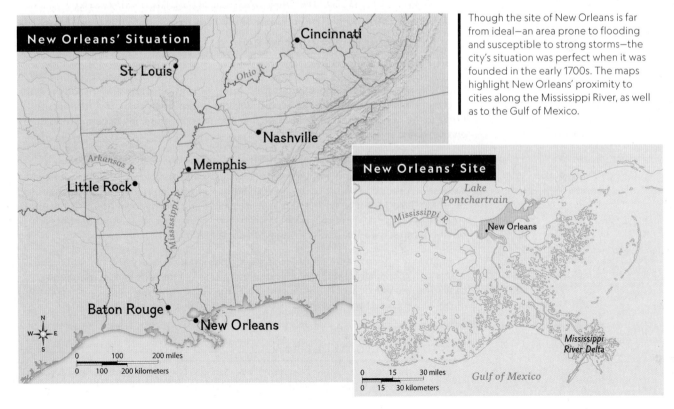

Though the site of New Orleans is far from ideal—an area prone to flooding and susceptible to strong storms—the city's situation was perfect when it was founded in the early 1700s. The maps highlight New Orleans' proximity to cities along the Mississippi River, as well as to the Gulf of Mexico.

The Power of Geography: Geographic Thinking **15**

National Geographic Explorer **Adjany Costa**
Conserving the Delta

Costa and the Okavango team are featured in *Into the Okavango*, a 2018 National Geographic documentary film covering their 1,500-mile-long expedition on the Okavango River filmed in Angola, Botswana, and Namibia.

Learning Objective
LO 1.6 Explain why sustainability is an important theme in human geography.

As climate change worsens and world population grows, sustainability becomes an ever more important topic in human geography. It is also a major concern of biologist Adjany Costa, the assistant director for National Geographic's Okavango Wilderness Project. Costa feels that the most effective human-environment interaction is to have as little impact on the environment as possible.

The Okavango River Basin is the largest freshwater wetland in southern Africa and provides water for a million people. Its delta in northern Botswana is rich with biodiversity and is home to the world's largest remaining elephant population, plus lions, cheetahs, wild dogs, and hundreds of species of birds.

Costa is a member of a team of scientists who have embarked on a series of canoe and mountain bike expeditions into the least known, most inaccessible areas of the watershed. As Costa explains, the basin's situation—its relationship with the surrounding areas—informs her work: "It's adjacent to a protected area in Namibia and two national parks, so it would create this whole square of conservation of land and ocean that are independent of each other but can still work together in regard to conservation."

In addition to research, another major part of Costa's job is advocacy and education. She meets with community leaders to educate them about the benefits of conserving the basin. The ultimate goal of the team's work is to help establish a sustainable management plan that will protect the Okavango watershed's source rivers forever.

Geographic Thinking

Identify and explain the reasons why it is important to sustain the Okavango River Basin.

1.2 Spatial Patterns: Scale and Region

Human geographers examine issues from different angles. They might get a broad overview of the effects of a process on a large area and then move on to study how the same process affects a small space. They group areas together into cohesive units to identify and organize the space they study. These varying views help them interpret Earth's complexity.

Zooming In and Out

Learning Objectives
LO 1.8 Identify the different scales used for analysis in human geography.
LO 1.9 Explain how using different scales of analysis allows geographers to understand how events and processes influence one another.

Think about an issue being covered in the news today. Is it a local, regional, national, or global issue? When answering a question like this, you are taking **scale** into account. This concept is different from scale on a map, which tells you how distance on the map compares to distance on the ground. Scale here refers to the area of the world being studied. Geographers use different scales of analysis as a framework for understanding how events and processes influence one another. For instance, a geographer might study the effects of air pollution in a city's industrial zone, in the entire city, or in the country as a whole. Examining the effects of pollution at these different scales of analysis can help geographers gain a better understanding of the impacts of atmospheric processes on pollution or of the human health and environmental effects of industrial air pollution.

The U.S. and Mississippi food insecurity maps reveal that on a national scale, about 12.5 percent of the U.S. population struggles to put food on the table. Examining the issue at a regional or local scale, however, it becomes clear that the problem is more serious in certain areas of the country. At the state level, Alabama, Arkansas, Louisiana, and Mississippi have food-insecurity rates between 17 and 19.9 percent. Focusing on the county level within a state reveals that in several counties in Mississippi, more than 30 percent of the population suffers from food insecurity. At an even more local level, Issaquena County in western Mississippi does not have a single grocery store, which means that many residents buy much of their food day-to-day at local convenience stores. With limited access to fresh foods,

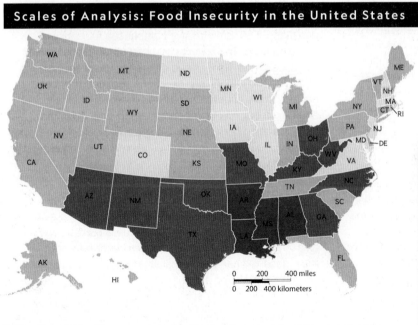

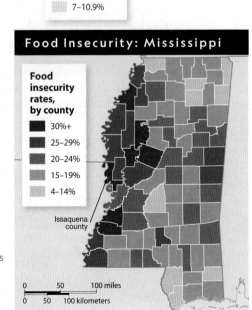

Reading Maps The maps show food-insecurity rates on a national scale and a local scale. The shading on the U.S. map reveals that the percentage of people who struggle with hunger is higher in certain areas of the country, such as the South. Focusing on the county level in the Mississippi map reveals some of the highest rates of food insecurity in the nation. ▍ Explain what using different scales of analysis reveals about food insecurity in the United States.

some end up eating less healthy, processed food rather than food with higher nutritional value. What happens at one scale affects processes at other scales, and data at different scales is necessary to fully understand issues such as food insecurity.

Observations at a local scale can also reveal details that might not be apparent at a regional scale. For instance, at the regional scale, an analysis of the population of the six-state region of New England reveals that 76.7 percent of the population is White, 9.6 percent is Hispanic, 6.6 percent is African American, and about 6 percent is Asian. Looking at the issue at a more local scale, however, reveals that Suffolk County, where Boston is located, is much more diverse, with about 45 percent White, 22 percent African American, 23 percent Hispanic, and 9 percent Asian. In short, looking at diversity at a regional scale, New England is one of the least diverse regions in the United States. But when they look at diversity at a local scale focused on Suffolk County, geographers find a very different—and much more diverse—picture. It is important to note that even this more local scale can obscure actual spatial patterns. Suffolk may be a diverse county, but as geographers drill down even further, they find that the countywide diversity does not apply to all Boston neighborhoods. The city is actually quite segregated. Some neighborhoods are more than 80 percent White, others are more than 80 percent African American, and still others are more than 65 percent Hispanic.

Geographers' understanding of scale drives their research questions and data collection, and their findings then inform policymakers, hopefully to make better decisions. An issue that has a major effect on the planet, such as climate change, can be more fully understood by analyzing it at a variety of scales. For instance, scientists have determined that, globally, the planet is warming. As they change focus to a regional scale, however, the impacts differ depending on a variety of factors. Canada's Arctic region is warming at twice the global rate, potentially causing heat waves across other sections of the country and increasing risk of wildfires and drought. These impacts, of course, will be seen and felt to varying extents at more local scales across the huge country.

In regions where climate change is occurring more slowly, its effects are more subtle. Sea-level rise caused by global warming, for instance, is not occurring uniformly across the planet. In certain regions, such as the Eastern Seaboard of the United States and in the Gulf of Mexico, sea levels have risen at higher-than average-rates. In other regions, like the U.S. West Coast and the oceans around Antarctica, sea levels have risen at lower-than-average rates. Understanding that the impacts of climate change will not be uniform helps governments prepare for problems that specifically affect their areas.

Geographic Thinking

1. Explain how using different scales of analysis helps geographers and other scientists understand the ways climate change is affecting the planet.

2. Describe how the analysis of the population of New England differs at a regional and local scale.

Unifying Features

Learning Objective
LO 1.10 Compare the three types of regions (formal, functional, and perceptual) as defined by human geographers.

A **region** is an area of Earth's surface with certain characteristics that make it distinct from other areas. Regions are human constructs, meaning people decide how they appear. The boundaries between regions are typically not clearly defined and are often transitional, overlapping, and contested. In other words, regional boundaries can be fuzzy. For instance, the United States' Southwest is thought of as one distinct region, and El Norte (The North) in Mexico is also considered to be its own region, but because of the consistency with which the people and cultures cross the international border, in some ways it is a single region. This is evident in the cities of El Paso, Texas, and Ciudad Juárez, Mexico, which share a regional economy where cultures and traditions blend and people cross the border, back and forth, on a regular basis.

Regions are a valuable tool for human geographers because they serve as an organizing technique for framing detailed knowledge of the world and for asking geographic questions. They are effective comparison tools as well. Knowing about the features of different regions is useful in discussing similarities and differences between parts of the world. Regions can be of any size, and they can act as a scale of analysis between the local, the national, and the global, helping geographers to synthesize their understanding of the world. Geographers define three types of regions based on the features that an area shares: formal, functional, and perceptual.

Formal Region A **formal region** is an area that has one or more shared traits. It is also referred to as a uniform region. The shared trait can be physical, such as a landform like a mountain range or a climate area like a desert. It can be cultural, such as a language or religion. Or it can be a combination of natural and cultural traits, defined by data such as measures of population, income, ethnicity, elevation, topography, or precipitation. For instance, a country is a formal political region whose shared characteristics include its territory, government,

Formal Region: The Pampas

The Pampas of South America are grasslands that cover an area of 300,000 square miles. The region is defined by its moderate climate and is one of the richest grazing areas in the world.

laws, services, and taxes. A smaller example of a formal political region is a state or a province within a country. The continent of Africa, with its distinct boundaries, is a formal region as well. The Southwest/El Norte region described previously is made up of physical characteristics marked by a generally dry climate, desert environments and wildlife, and rugged mountains. The region is also defined by its human population—a deep Native American history, shared Latino and "Old West" culture, and distinctive art, music, and cuisine.

The Rocky Mountains make up a formal physical region in the United States, as do the Great Plains and the Atlantic Coastal Plain. Formal regions might be defined by their agricultural possibilities, such as temperate regions with substantial growing seasons. The Pampas region of South America, shown on the "The Pampas" map, is defined by its moderate climate, generally flat and grassy landscape, and an agriculture known for cattle ranching. The Corn Belt, an area of the Midwest United States with rich soils and mostly level terrain where corn and soybeans are the dominant crops, is a formal economic and agricultural region. The Pyrenees Mountains create a formal region along the French–Spanish border in Western Europe. People in this rugged region, the Basques, have developed their own culture over thousands of years.

At a more local scale, a city, like Albuquerque or Cincinnati, qualifies as a formal region as well. And within cities, shared traits can define small formal regions—ethnicities, economic activities, architectural styles, and more. The ethnic neighborhoods found in many large cities, for instance, may be considered formal regions. In the borough (administrative division) of Brooklyn within New York City, a large number of Hasidic Orthodox Jews live in the neighborhoods of Crown Heights, Williamsburg, and Borough Park. These neighborhoods form a region shaped by a shared religion and culture. Every major city and even smaller ones in the United States and across the globe have distinctive business districts or ethnic neighborhoods that residents and often visitors recognize by name: Boston's Back Bay, London's Docklands, Cape Town's Bo-Kaap Area, and Shanghai's Pudong District are examples. Can you think of others? Is there a district or neighborhood—a formal region in your community—that compares?

Functional Region A **functional region** is defined as an area organized by its function around a focal point, or the center of an interest or activity. The focal point of a functional region is called a **node**. The node is the focus of the region, such as the downtown of a city. Nodes serve a particular function—often a political, social, or economic purpose—and have internal connections that tie the region together. For instance, the central business districts of some cities form the focal point for the cities' economic activity. Workers commute to this district, usually downtown, along the internal connections of roads and rail lines from other areas of the city, or from the **suburbs**, the residential areas surrounding a city. The central business district acts as the node for the functional region that consists of the metropolitan area. (Urban structure, functions, and definitions are discussed further in Unit 6.)

Functional regions exist at a range of scales and can apply to a variety of geographic activities. The "Airline Flight Routes" map shows the functional region created by a major airline's flights from Hartsfield-Jackson Atlanta International Airport to locations throughout the Americas. The airport acts as the region's node—a major hub for national and international flights. All flights from some smaller airports in the U.S. South go through Hartsfield-Jackson, making it the only connecting point for air travel between these locations.

The hub-and-spoke design of many public transportation systems form functional regions as well. The hub at the center of the system is a node, at which or near a great deal of government, economic, and cultural activity occurs. From there, rail lines branch off toward outer areas of the city and its suburbs. An example is the rapid transit system of Washington, D.C., depicted in the "D.C. Metro System" map, which has six rail lines that connect the outskirts of

The Power of Geography: Geographic Thinking

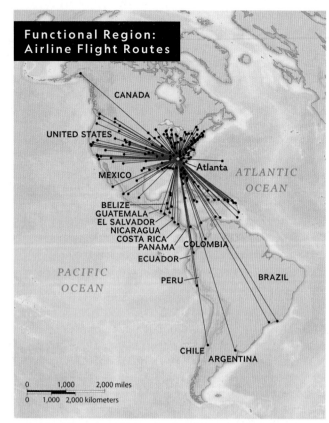

Delta Airlines flies to cities throughout the Americas from the Hartsfield-Jackson Atlanta International Airport. These cities form a functional region with the massive hub airport as the node.

The Metrorail system connects Washington, D.C., to its surrounding suburbs. L'Enfant Plaza and Metro Center serve as nodes from which all six color-coded lines branch off.

The Midwest is a perceptual region, which in this version comprises 12 states. It is defined in part by people's perceptions of the region—for example, as a largely rural area with a hard-working, friendly population.

the city to the city center and to other surrounding areas as well. For example, a commuter might take the Red Line daily from an outer station like Bethesda in the northwest or Takoma in the northeast to Metro Center, a system hub near the National Mall. Another commuter might take the Green Line from Anacostia or another station in the southeast and change to a Yellow Line (or Blue Line) train at another system hub, L'Enfant Plaza, heading to work at the Pentagon or Washington National Airport.

Cities with ports, or large commercial shipping facilities, form functional regions with their surrounding areas, called hinterlands. The ports act as nodes of these regions. Goods come in on ships and are distributed to the appropriate processing plants and shipping centers in the hinterlands. At these facilities, goods are received, produced, processed, and shipped out, either through the port or into the interior of the country via trains or trucks. A more local example of a functional region is the service area of a pizza shop. At the node is the shop, which might limit its delivery range to a two-mile or some other radius. The edge of the service area is the limit of the pizza shop's functional region.

Perceptual Region A **perceptual region**, also called a **vernacular region**, is a type of region that reflects people's impressions, feelings, and attitudes about a place. A perceptual region, therefore, is defined by people's perceptions of the area—that is, their subjective understanding of the world as influenced by their culture and experience. What are the characteristics that make up the Midwest region of the United States, for example? Many people might think first about agriculture and farms—especially corn and dairy products—and polite, hard-working, and down-to-earth people. The Midwest is generally not thought to be very ethnically or racially diverse, and politicians and pundits sometimes belittle it as "flyover country," meaning that it's nothing but a place you fly over to get from one coast to the other. Each of these characteristics and more influence people's perceptions of the Midwest. Key among factors that may influence how people perceive a region like the Midwest include where the person lives or has lived, whether or not the person has traveled to or learned about the region in question (or has traveled much or little anywhere), and even the person's age and other life experiences.

People often disagree on the boundaries of perceptual regions. Someone from the East Coast might perceive the Midwest to include Ohio, Indiana, Michigan, and parts of Pennsylvania, but people living in Minnesota, Wisconsin, Nebraska, or Kansas might feel strongly that they're Midwesterners. And while these regions may help to impose a personal sense of order and structure on the world, they often do so on the basis of stereotypes that may be inappropriate or incorrect. Of course, the Midwest does have enormous swaths of rural areas and huge agricultural production, but it also contains large, ethnically and racially diverse cities like Chicago, Detroit, St. Louis, and many more.

Outside of the United States, Eastern Europe is an example of a perceptual region. According to the United Nations, Eastern Europe consists of 23 countries including Czechia, Poland, Romania, Ukraine, and the countries of both the Baltic and the Balkan peninsulas. But the region exists in most people's minds based on its political, historical, and cultural characteristics. The area makes up most of what was known as the Eastern Bloc during the Cold War—the countries that were aligned with the Soviet Union and ruled by Communist governments in the years after World War II. Culturally and historically, the region was influenced by several empires over the centuries. Together these characteristics define a region that exists separately in people's perceptions from Western Europe.

Geographic Thinking

3. Quebec is a province in Canada in which 83 percent of the population speaks French as a first language. Identify Quebec's region type.

4. Compare the functional region of a pharmacy in a dense city with few cars and drivers to the functional region of a pharmacy in a sparsely populated suburb with more cars and drivers.

5. Describe the role that cuisine, or style of food, might play in the understanding of a vernacular region.

Case Study

India—Regional Differences in Scale

The Issue India has experienced impressive economic growth this century, but large portions of its population aren't experiencing a fair share of the benefits.

Learning Objective
LO 1.9 Explain how using different scales of analysis allows geographers to understand how events and processes influence one another.

By the Numbers

$468 billion
GDP in 2000

$2.66 trillion
GDP in 2020

25%
Poverty rate in rural areas

Source: World Bank

The contrast between poverty and wealth is evident in Mumbai, India, where skyscrapers rise behind an informal housing settlement on the city's outskirts.

Since the Turn of the 21st Century, India has experienced astonishing economic growth, seeing its gross domestic product (GDP) increase from $468 billion in 2000 to $2.66 trillion in 2020, a nearly 470 percent increase in the size of its economy (which actually represented a pandemic-caused decline from 2019 GDP). To put that into perspective, the U.S. economy has grown by 104 percent during the same time frame.

However, as geographers focus on more local scales of analysis, they find that this new wealth is not distributed evenly throughout the country. Indians in some places and regions have become very rich, while many parts of the country remain very poor. Much of the wealth is concentrated in just a handful of states, such as Maharashtra, Kerala, and Tamil Nadu. Since economic growth began to accelerate in India two decades ago, the wealth divide between certain regions has continued to expand.

Before the acceleration of economic growth, incomes between different states were converging. Since the acceleration, incomes have diverged, with the average person in the three richest states having three times more wealth than the average person in the three poorest states. Looking at patterns within states reveals a rural-urban divide—a large proportion of the new wealth is being generated in cities like Mumbai and Delhi. Mumbai, a port city located on the Arabian Sea in southwestern India, is considered to be the financial and commercial center of India. The country's central bank is located in Mumbai, as is a government-owned life insurance corporation, investment institutions, and the Bombay Stock Exchange. In addition to its robust service sector, the economy of Delhi, which is where the country's capital is located, has created many jobs in trade, finance, public administration, and professional services. Both are among the world's largest cities in population.

In the less wealthy regions of India, fewer people are living in abject poverty, but the growing wealth divide concerns geographers. Most of the country's wealth—about 77 percent of the total national wealth—is held by just 10 percent of the population. Many accusations of corrupt dealings between India's politicians and the rich have been made, giving rise to anger and protests among the less wealthy. Without political reforms to fight corruption and expand social services to people who are being left behind, geographers worry the problems will continue to get worse.

Geographic Thinking

Explain why geographers—and policymakers—might have concerns about growing inequality in India.

1.3 Globalization and Sustainability

Time-space compression describes how technology is causing people around the globe to become more connected—able to interact with and travel to far-off locations faster than ever before. Human geographers study how this process is changing the world, in ways both positive and negative, and consider how the needs of human societies today can be met without using up resources that will be needed in the future.

Global Versus Local

Learning Objectives
LO 1.11 Describe factors that drive contemporary globalization.
LO 1.12 Explain Wallerstein's world-system theory and its three-tiered structure.

Over the past half century, the countries of the world have become increasingly connected and integrated through **globalization**, the expansion of economic, cultural, and political processes on a worldwide scale. Globalization is an important, overarching theme in human geography. A number of factors have contributed to globalization. Lower production costs and advances in transportation technology have expanded many companies' reach outside of the borders of their home country. The internet gives people in countries across the world the ability to easily communicate with one another, allowing for the spread of cultural ideas faster than ever before. Social media apps allow people to share their views instantaneously, becoming instrumental in the spread of political movements such as the Arab Spring, a series of pro-democracy protests that took place in North Africa and Southwest Asia (also referred to as the Middle East) in 2011. Similarly, social media enabled the Black Lives Matter protests sparked by the 2020 killing of George Floyd in police custody to become not just a social movement in the United States, but indeed a global conversation about race, poverty, law enforcement, and more.

Government policies have played an essential role in globalization as well. Trade deals throughout the world have lifted restrictions and made the movement of goods and jobs across borders happen more easily. Beginning in 1994, for

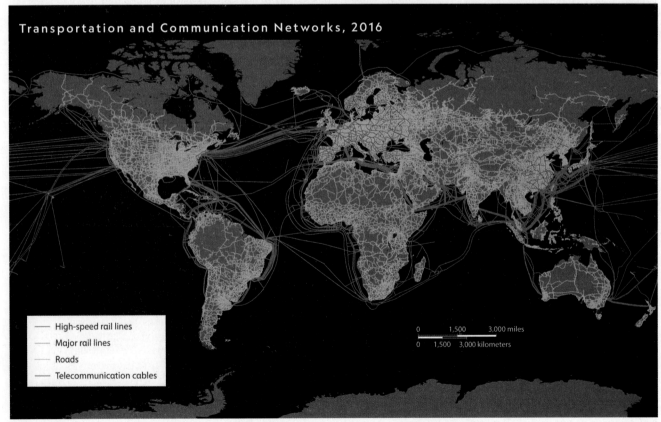

Reading Maps The map highlights ways in which locations in the world have become increasingly interconnected by transportation and communication networks. ▌ Choose two locations on the map and describe the connections between them.

instance, the North American Free Trade Agreement (NAFTA) allowed companies in the United States, Canada, and Mexico to sell goods and hire workers in any of the three countries. In 2018, the countries renegotiated NAFTA and drafted a similar pact called the U.S.-Mexico-Canada Agreement. In Europe, the European Union formed in 1993, in part to allow people and goods to easily pass from country to country. International trade has had detractors in recent years—the Trump administration in the United States imposed tariffs on China and other countries starting in 2018, for example—but the overall trend since World War II has been toward more international trade, and this trend has completely reshaped the global economy.

Globalization is a process that affects all aspects of human life today. It is related to the geographic concepts of location, space, place, and flows, and human geographers study the influence that it has on the patterns they observe. As you study each unit in this book, think about the advantages, challenges, and issues surrounding globalization.

Wallerstein's World System Theory Throughout your study of human geography, you'll come across tools that can help you understand geographic concepts. A **theory** is a system of ideas intended to explain certain phenomena. In the 1970s, sociologist Immanuel Wallerstein developed the **world system theory** to describe the spatial and functional relationships between countries in the world economy. The theory helps to explain the history of uneven economic development among countries and the reasons why certain regions have held onto political and economic power over long periods of time. It is based on the idea that interdependence between countries has created a world system with an economy that is a single entity with a single market and division of labor. In other words, companies are not limited to buying, selling, or hiring within the borders of the country in which they're located. They can buy commodities or products anywhere, and open factories and sell products around the world.

World system theory categorizes countries into a three-tiered structure: core, periphery, and semi-periphery. Wealthier countries with higher education levels and more advanced technology are considered part of the **core**. Core countries are highly interconnected, with reliable transportation and communication networks, sophisticated infrastructure, and highly skilled labor forces that all support economic activity. They have stable governments and strong political alliances. Core countries are economically (and thus politically) dominant, and they control the global market. Countries that have less wealth, lower education levels, and less sophisticated technology are considered part of the **periphery**. Peripheral countries tend to have less-stable governments and poorer services such as health care. They are less connected than core countries, with inferior transportation and communication networks, inadequate infrastructure, and less-skilled labor forces for supporting economic activity. Countries where both core and periphery processes occur are labeled **semi-periphery**. Semi-peripheral countries are in the process of industrializing and modernizing. They are often active in manufacturing and the exporting of goods, with increasingly skilled workforces. They have better connections than peripheral countries, with improving transportation and communication networks. Semi-peripheral countries have the potential to grow into core countries.

World system theory states that the three types of countries form a power hierarchy, with core at the top, periphery at the bottom, and semi-periphery in between. The strong central governments, trade partnerships, and skilled labor of core countries allow them to control and benefit from the world economy. They exploit peripheral countries for cheap labor and natural resources. The weaker, less stable governments and poor infrastructure of peripheral countries mean they have little power outside their borders. Semi-peripheral countries act as an economic and political link between the core and the periphery. They can be exploited by core countries but might exploit peripheral countries.

Because of this exploitation, it is difficult for peripheral countries to improve their situation. Historically, the colonial powers of Europe in the 17th, 18th, and 19th centuries were core countries. They exploited their colonies, which were part of the periphery, for crops, labor, and raw materials. Many of the colonizers of this period have remained core countries to the present day.

The core-periphery model doesn't only apply at the global scale. It can also be observed at the national, regional, state, and city level. For instance, while the United States is a core country, core characteristics are not distributed evenly throughout. A state like California is known for its wealthy cities and innovative tech-driven economy but also has much poorer rural areas. New York City has core areas like Manhattan, where the real estate, media, and financial sectors generate significant amounts of wealth, as well as peripheral areas – not far away – like the Bronx, which is far less wealthy, with an economy driven by retail, hospitality, and service industries.

Core-Periphery Terminology

For decades, geographers have used terms to describe the differences among global economies. You may be familiar with the term *developed countries*, which is sometimes used to describe industrialized countries with strong economies, and *developing countries*, sometimes used to describe less industrialized countries with weaker economies. Those terms became a common way to divide the world beginning in the 1950s. However, critics argue that this terminology is problematic for several reasons, including its focus on economic growth and its assumption that developing or traditional societies need to "catch up" to Western ideals. Instead, the terminology of the core-periphery model is used throughout this course to describe global differences. Using core-periphery terminology helps to distinguish the status and hierarchy of the world's diverse countries. It also helps to better illustrate the relationships that exist between countries.

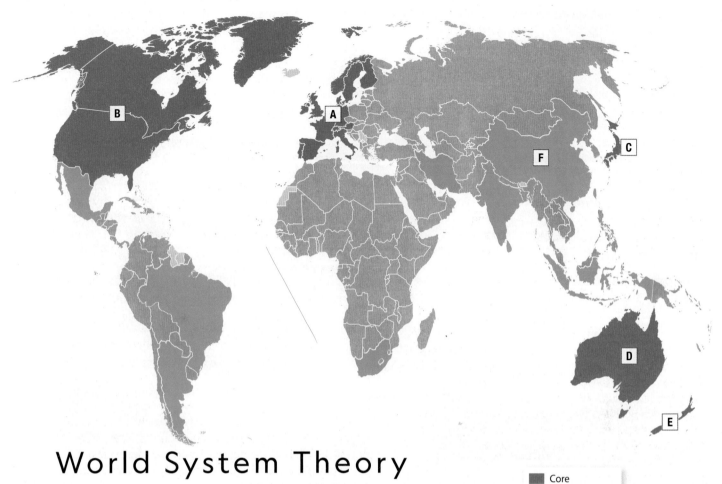

World System Theory

At the top of the hierarchy, core countries control and benefit from the world economy. Peripheral countries are exploited for cheap labor and natural resources. Semi-peripheral countries act as a political and economic link between the two.

▌ Explain how core countries are connected economically to periphery countries.

Core
Semi-periphery
Periphery
Other

Core Countries include Western Europe A and most of North America B, along with Japan C, Australia D, and New Zealand E.

China F has the world's second-largest economy, but its per capita income ranks 71st and the country is considered part of the **Semi-Periphery**.

Many countries that were colonized in the 17th, 18th, and 19th centuries are today part of the **Periphery**.

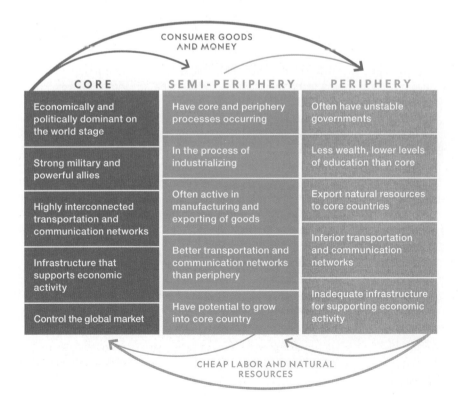

The Power of Geography: Geographic Thinking

Water is distributed to residents in the city of Bouaké in the Ivory Coast in West Africa after the lake that provided water to 70 percent of the city's population dried up. Organizations work to come up with sustainable solutions to prevent the depletion of essential resources.

Sustainability

Learning Objective
LO 1.6 Explain why sustainability is an important theme in human geography.

As you have read, human–environment interaction includes the study of the impacts that human societies have on nature. Whether these impacts are sustainable is essential to the survival of humanity. Sustainability, as you have learned, is the use of land and natural resources in such a way that they will not be irreversibly depleted. Like globalization, sustainability is an essential theme of human geography, related to the geographic perspectives and key concepts to which you've been introduced. Geographers study sustainability and sustainable practices and promote the idea that sustainability should drive decisions about how humans react to and influence other geographic processes. Geographer Michael Meadows contends that his discipline, because of its integrative "advantage" over most, "could be considered as *the* science for sustainability." Meadows emphasizes that the "geographical approach" is appropriate and necessary to forge a sustainable future: "Geography aims to integrate the study of both natural and human realms and their interactions, focusing on space, places, and regions, addressing and questioning both short-term and longer-term processes and their resultant patterns."

Globalization and economic expansion have made the world more connected, but the advantages have disproportionately gone to core countries. Climate change, depletion of the world's resources, and wealth inequality are worldwide problems that continue to grow. To help people face these challenges and to save the planet from the ravages of climate change, global leaders have been taking steps to encourage governments and industries to operate and grow more sustainably. In 2015, the United Nations launched its 2030 Agenda for Sustainable Development, a plan that lays out 17 goals to increase peace, freedom, and prosperity around the world. **Sustainable development** is development that meets the needs of the present without compromising the ability of future generations to meet their own needs. The UN recognizes that sustainability requires consideration of the availability of natural resources, innovations to make better use of renewable resources, and efforts to reduce pollution and waste, but that it is also necessary to ensure that the attainment of these goals must be a global effort—shared across core, peripheral, and semi-peripheral countries. The UN's goals also include the elimination of poverty and hunger, increased access to quality education, gender equality, and more.

You learn more about these goals in Unit 7 but, like globalization, sustainability is an idea that is important to all aspects of human geography. It is a theme that drives much of the work that human geographers do. As you read through this textbook and learn the ways in which humans are interacting with and impacting their environments—whether through agriculture, economic development, political or cultural activity, in rural or urban settings—keep sustainability in mind.

Geographic Thinking

1. Explain why it might be difficult for a peripheral country to become a part of the core.

2. Describe how world system theory is related to globalization.

3. Explain why sustainability is an important human geography theme.

Chapter 1 Summary & Review

Chapter Summary

Geography can be divided into two major areas: physical geography and human geography. Physical geography is the study of natural processes and the resulting distribution of features in the environment. Human geography is the study of the processes that have shaped how humans understand, use, and alter Earth's surface.

- Spatial patterns refer to the arrangement and placement of objects and events on Earth's surface.
- Geographers analyze complex issues and relationships from two key perspectives: the spatial perspective and the ecological perspective.
- Human geographers use spatial concepts to study how humans interact with the environment: absolute and relative location, place, space, flows, patterns, distribution, distance, time–space compression, and distance decay.
- Human geographers study how human–environment interactions and changes affect both humans and the environment.
- Human geographers' views of human–environment interaction have evolved over time, including environmental determinism and possibilism.
- Human geographers study how Anthropocene climate change affects regions and groups of people in different and clearly uneven ways.
- Sustainability is an important concept of human–environment interaction and includes the replacement of nonrenewable resources with renewable resources.

Geographers study events at different scales as a framework for understanding how processes influence one another. A region is an area of Earth's surface with shared characteristics that make it distinct from other areas.

- Scales of analysis include global, regional, national, subnational, and local.
- Geographers use different scales to reveal details that might not be apparent at one scale.
- Regions are human constructs with subjective boundaries.
- A formal region is an area that has a shared physical or cultural trait, such as a type of landform or a language.
- A functional region is an area organized around a connection or focal point, such as a central business district in a city or a port and its surrounding hinterlands.
- A perceptual region, or vernacular region, is a type of region based on people's understandings of places and mental maps, such as the boundaries of the Midwest.

Over the past half century, globalization has created an increasingly connected and integrated world. Sustainable development is development that meets the needs of the present without compromising the ability of future generations to meet their own needs.

- World system theory explains the way globalization functions in practice. It categorizes countries into three tiers: core, periphery, and semi-periphery. Core countries have wealth, high education levels, and advanced technology. Peripheral countries have less wealth, low education levels, and less developed technology. Semi-peripheral countries have a mix of core and periphery characteristics.
- Sustainable growth has become a challenge as economic expansion and globalization have increased. The geographical approach is key to addressing issues of sustainability of many types and at all scales.

Review Questions

Use complete sentences to answer the questions.

1. **Apply Conceptual Vocabulary** Consider the terms *place* and *region*. Write a standard dictionary definition of each term. Then provide a conceptual definition—an explanation of how each term is used in the context of this chapter.

2. Describe the absolute location of your town or city. Then describe the relative location of your town or city.

3. Explain how the terms *site* and *situation* are related.

4. Describe the spatial perspective using the terms *space* and *distribution*.

5. How are the concepts of *time–space compression* and *distance decay* related?

6. Use the city of Las Vegas, Nevada, which was built in a desert, as an example to explain the theories of environmental determinism and possibilism.

7. How can conducting an analysis at different scales, as with impacts of climate change, be informative?

8. Describe a technological advancement that has contributed to globalization and explain how it has had an impact. Use a specific event or product if you can.

9. Explain how ethnic or racial diversity can differ at different scales of analysis.

10. Define the concept of *sustainability* and explain why it has become a growing challenge.

11. Compare the terms *node* and *functional region*. How are the terms related?

12. Define the term *perceptual region* and provide an example.

Interpret Maps

Study the map and caption and then answer the following questions.

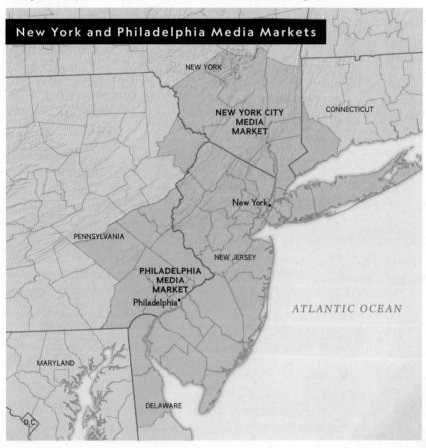

A media market is a region covered by local television and radio stations. The people living within a particular media market receive the same programs from their local stations.

13. **Identify Data & Information** Identify the type of region represented in this map. How do you know?

14. **Analyze Visuals** Describe how the two media markets might have an impact on daily life in South Jersey and North Jersey.

15. **Analyze Geographic Concepts** Explain why the boundaries of the regions shown in the map are more distinct than they would be if the map depicted the extent of a sports team's fans.

16. **Synthesize** Explain why the reach of New York and Philadelphia's media markets might have less impact today than it would have had 30 years ago.

Chapter 2

Geographic Inquiry: Data, Tools, and Technology

Critical Viewing As part of a mapping project initiated by UNICEF, Fatima Wariou, a university student from Niamey, Niger, captures GPS coordinates in the Sahara Desert. Other participants of the MAP4DEV group study the movement of the sand toward an inhabited area nearby. ▮ Explain how collecting data helps these geographers to analyze the progression of the sand.

Geographic Thinking What tools do geographers use to depict spatial relationships?

2.1
Thinking Like a Geographer: Processes of Inquiry

2.2
Geographic Data and Tools

Case Study: Detroit—GIS Helps Find Safer Routes

2.3
Understanding Maps

National Geographic Explorer
Lehua Kamalu

2.4
The Power of Data

National Geographic Collaborator
Clinton Johnson, Esri

30 Chapter 2

2.1 Thinking Like a Geographer: Processes of Inquiry

Geographers use spatial analysis to explain patterns of human behavior and understand how places and societies are organized. They seek to understand where things are, why they are there, and why this matters.

Thinking Like a Geographer

Learning Objectives

LO 2.1 Describe the concepts of geographic thinking.

LO 2.2 List examples of questions geographers may ask during the inquiry process.

As you read in Chapter 1, geographers think spatially in terms of location, space, place, arrangement, flow, regions, and human–environment interaction. Understanding the complexities of the world, or even of a small community or neighborhood, involves careful observation from a geographic perspective of the environment, aspects of culture, politics, economics, and more from a geographic perspective. Therefore, thinking like a geographer often includes the integration of ideas shared with many disciplines; it has been said that "geography works well with others." Geographic thinking calls for asking questions, collecting and organizing data from a myriad of sources, identifying patterns, analyzing outcomes, making connections, drawing conclusions, and, importantly, finding and sharing answers in useful ways. Only then can a person, a group, a community, an organization, or a country make informed decisions and take appropriate action. Maps, globes, graphs, photographs, and satellite imagery provide geographers with enormous amounts of data about the world that is analyzed to understand the processes driving human geography. Mapping also provides a way to organize and display these vast amounts of spatial information.

Following an inquiry process provides geographers with a systematic way to examine complex issues at various scales—local, regional, or global. Geographer Reginald Golledge, who offered his view of the unique nature of geographic thinking in Chapter 1, makes the case for asking and answering the "why of where" questions involved in geographic inquiry: "In short, geographic thinking and reasoning has provided a basis for understanding—or reasoning out—*why* there are spatial effects, not just finding *what* they are. Furthermore, it enables us to reveal patterns in spatial distributions and spatial behaviors that may not be obvious to a casual observer in the real world (e.g., the pattern of shopping centers in a city) and consequently helps us understand the reasons for occurrences of episodic behaviors (e.g., obligatory and discretionary activities) in terms of spatial processes."

Geographic questions are at the heart of this inquiry process. Geographic questions include: Where is it located? Is there a pattern? Why is it there? And importantly, what does it matter? For a human geographer, the "it" can be whatever is observed—an event, an incident, people, places, things, the episodic activities Golledge already mentioned, and more. Human geographers may also ask additional questions about spatial distributions, including: What is significant about its location? Why is "it" *not* found everywhere, but only in certain locations? How is its location related to the locations of other events, incidents, people, places, things, activities, or environments? If it is found in certain locations, can it be predicted also to be found in others? There might be additional questions that a geographer might ask—and attempt to answer using this spatial inquiry process.

Geographic Thinking

1. Describe geographic thinking.

2. Explain how thinking like a geographer benefits decision making.

3. Identify three geographic questions about any environmental, social, or economic issues that interest you, and then explain how your questions might change when considering different scales of analysis.

Critical Viewing An aerial photograph is a tool that gives geographers information about the site and situation of a place. The photo shows a school nestled in the Santa Susana Mountains north of Los Angeles, California. ▮ Identify some questions geographers may have asked about the location when considering whether or not to build a school there.

2.2 Geographic Data and Tools

Geographic data is information about the characteristics of locations on Earth. Geographers use data to explore and better understand places and the processes that influence these locations.

Collecting Data

Learning Objectives
LO 2.3 Compare qualitative and quantitative data used in human geography.
LO 2.4 Identify various groups and individuals that collect, analyze, and use geographical data.

Geographers use a variety of methods for collecting data. Geographic information is any data with a location tied to it, such as a street address or its elevation. It is information at a given spot whether it is about the human world, natural world, or anything else. Approximately 80 percent of the world's data is spatial.

The methods by which data are collected include observing and systematically recording information, reading and interpreting maps and other graphic representations of spaces and places, reading reports and policy documents, and interviewing people who can provide both information and perspectives about places and issues. The data collected through these methods are **quantitative** or **qualitative**. Information measured by numbers is called quantitative data. The population of a city or the temperature recorded there is quantitative data. Qualitative data are interpretations of data sources such as field observations, media reports, travel narratives, policy documents, personal interviews, landscape analysis, and visuals such as art or photographs. Skills involved in analyzing quantitative and qualitative data involve seeking patterns, relationships, and connections.

The data geographers collect has to be at the appropriate scale and align to the nature of the research questions. So, for example, answering a question about the migration patterns out of a country to other countries might require population data at the country level but likely not at the neighborhood level. But answering a question about ethnic change or patterns of poverty in a city would indeed require neighborhood-level data.

Thinking geographically, suppose you want to determine the effects of the use of pesticides in your community (pesticides include a range of chemicals to rid a field, lawn, or garden of pests, including insecticides to kill insects and herbicides to kill weeds). A number of data sources could be considered. You could first collect information about the insecticides and herbicides used locally—the benefits and risks of their use, and data about where and how often they are used. Initial data might be gathered from media reports and other online sources to determine the effects of the chemicals on the environment and humans. Field observations and personal interviews with local farmers and gardeners could provide invaluable insights about the location and frequency of use of the chemicals.

Land-use maps and aerial photography could also be analyzed to see if areas of high use are adjacent to waterways or neighborhoods. In addition, you might seek policy documents from government agencies such as the Environmental Protection Agency, Food and Drug Administration, or the county agricultural agent about the regulations and guidance regarding the use of insecticides and herbicides. All of these sources provide evidence about the impact of humans and their use of chemicals on the land that could be used to frame an argument and possibly take action on this issue. You might decide that the data you've collected from these or other sources might best be displayed and shared by constructing a map to show spatial patterns of pesticide use and possible impacts on the environment or human health in your community.

Who Collects Data? Countless organizations, both public and private, collect and analyze data. The U.S. Census Bureau, for example, conducts a **census** of the U.S. population every 10 years. A census is an official count of the number of people in a defined area. The U.S. Census Bureau also conducts dozens of other surveys, including the American Community Survey, which gathers information about educational attainment, employment, income, language proficiency, migration, and housing. The Census Bureau also gathers information from American businesses as part of its Economic Census. All information collected by the U.S. Census Bureau is available to the public through written reports as well as online where a search feature enables users to gather information at a range of scales, about a particular state, county, city, or zip code.

Many agencies of the federal government collect and make available varied types of data. For example, the U.S. Department of Agriculture (USDA) carries out a census of agriculture every five years, a "complete count" of farms and ranches and the people who operate them, land use and ownership, operator information, agricultural practices, income, and more. The USDA also provides and analyzes data on meat production, dairy products, and crops to ensure the quality and availability of food and to help American farmers and businesspeople make informed decisions. In addition, the USDA Economic Research Service gathers and shares a wealth of data related to economics and agriculture, including food security and agricultural commodity chains.

Both U.S. government agencies and private organizations collect useful information about elections at a range of spatial scales to observe patterns of voting behavior. Results of past elections are kept by the Federal Election Commission and the National Archives, and state and local governments also keep election records. Organizations like Ballotpedia collect and make available federal, state, and local political, election, and policy data in a nonpartisan "digital encyclopedia." Another nonpartisan group, American National Election Studies, shares data from its surveys on voting, political participation, and public opinion.

International-scale data is collected by a wide range of organizations, institutions, companies, and also governments. The United Nations System alone has dozens of programs, funds, agencies, and other entities that collect an enormous variety of data. For example, the Food and Agriculture Organization collects country and region-level data related to hunger and food insecurity, the World Health Organization collects data related to global health, and the World Bank collects data of many types related to living standards.

In addition to making use of the data gathered by organizations, geographers may conduct their own data-gathering efforts—quantitative, qualitative, or both—based on the specific research question or questions they want to answer. Data gathering takes multiple forms, such as written surveys or in-person or phone interviews that gather information about people and their experiences or opinions. Individuals also conduct field observations using photography, sensors, and scientific **probeware**, which are probes and sensors integrated with computer software to collect real-time data and to record information about characteristics of specific locations like temperature, light, elevation, and distance. Travel narratives, government or company reports, and social media blog posts that describe the physical and cultural characteristics of a place may be useful as well.

Geographic Information Systems

Learning Objectives
LO 2.5 Identify the characteristics of geographic information systems (GIS).
LO 2.6 Describe the real-world applications of geographic information systems.

Geospatial technologies encompass the modern tools used to analyze data about specific locations across the globe. Governments, organizations, businesses, and individuals use the technologies to find precise locations, collect and share data, create maps, and analyze and track changes in characteristics of places on Earth's surface. The development of sophisticated mapping software systems called **geographic information systems (GIS)** has immensely helped geographers and others with their work. GIS captures, stores, organizes, and displays precisely located geographic data that can then be used to configure both simple and complex maps. Such maps are created by organizing layers of information to form a combined image. Each type of information is stored in a separate layer that represents a specific theme and dataset, such as roads, population, land use, voting district boundaries, and much more. The layers that are selected to display will vary depending on the goals of the project and the research question that geographers seek to answer.

A wide range of spatial data is easily compared and analyzed using GIS. A GIS map can display information about the physical geography of the land, such as elevation or **topography**, which is the shape and features of land surfaces. It can also display demographic information about the people who live in a certain place, such as age, ethnicity, income, or family size. Combining the data from these layers makes it easy for geographers to make connections, for instance, understanding how natural resources impact the economic activities that take place in a region.

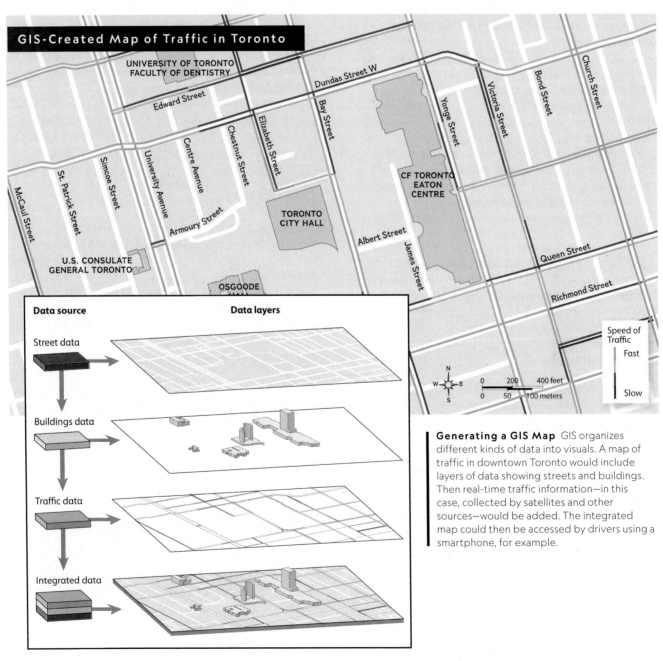

Generating a GIS Map GIS organizes different kinds of data into visuals. A map of traffic in downtown Toronto would include layers of data showing streets and buildings. Then real-time traffic information—in this case, collected by satellites and other sources—would be added. The integrated map could then be accessed by drivers using a smartphone, for example.

34 Chapter 2

GIS maps support **geovisualization**, which is the process of creating visuals for geographic analysis using maps, graphs, and multimedia. This process allows users to analyze geospatial data interactively, aiding visual thinking and providing insights into the issues geographers are researching. One common use of GIS involves comparing natural features with human activity. GIS could be used to evaluate environmental risks to a community such as flood potential or human-made risks such as industrial pollution levels. Such information can help communities plan for future sustainable development. Present and future impacts of climate change, as well as efforts to adapt to these impacts, are tailor-made for GIS analysis—at scales from local to global.

Geospatial technologies collect and analyze immense amounts of data—data accessible to anyone with internet access—leading to a revolution in spatial decision making. The geospatial revolution encompasses nearly every aspect of human life, from the relatively mundane, or common, everyday activities, to the most critical of decisions. Today, in an instant, individuals and organizations can send, receive, and broadcast information about where they are, where they have been, and where they are going. Maps created out of this geospatial data have a wide variety of uses. Locally, these maps provide information on everything from recommending restaurants to finding the nearest hospital to tracking crime. Many U.S. cities are using geospatial data to address problems with public transportation or food access in underserved communities. On an international scale, multilayer geospatial maps can aid relief efforts after an earthquake, track global trade and shipping, document the potential impacts of climate change, support the deployment of troops during conflict, or assist in the drawing of territorial boundaries as a part of peacemaking.

Geospatial technologies and the wide availability of online interactive maps and other geovisualizations were critical to understanding and fighting COVID-19. Public health officials, governments, organizations, businesses, and even families and individuals were able to track the spread of COVID-19. Geospatial technologies were vital for all to make informed decisions about closures, school, work, and travel. GIS was critical in efforts at all levels from global to local to help staff support and maintain medical and other critical services, businesses, and infrastructure.

Geographic Thinking

1. Describe the difference between quantitative and qualitative data and provide an example of each.
2. Explain what GIS is and how it is used to understand spatial patterns and relationships.

Other Remote Sensing Tools

Learning Objectives
LO 2.7 Describe the uses of remotely sensed data.
LO 2.8 Explain the uses of global positioning systems (GPS).

A variety of geospatial technologies gather data; some do so remotely, or without making physical contact. This method of collecting data is called **remote sensing**. Most remote sensing used by geographers relies on satellites, aircraft-based sensors, or drones to collect data.

Satellites take images of sections of Earth at regular intervals to determine changes that occur on the surface. Then the

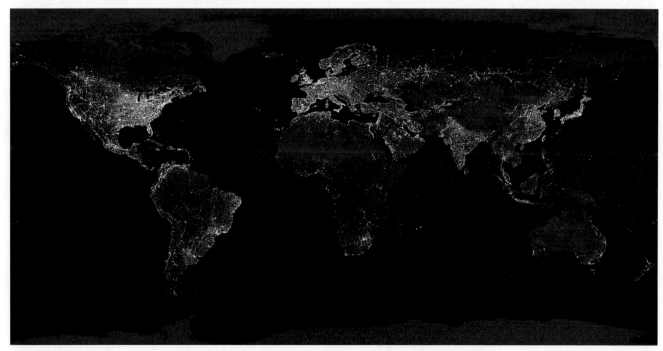

Geographers use data acquired from satellite images to study environmental and developmental changes. This image of Earth at night, compiled from more than 400 satellite images, provides insights into the location of urban populations and economic development using the nighttime light cities emit.

Geographic Inquiry: Data, Tools, and Technology

Case Study

Detroit—GIS Helps Find Safer Routes

The Issue Getting to and from school safely was a problem for some students in Detroit, as an economic decline contributed to abandoned houses and urban blight.

Learning Objective
LO 2.6 Describe the real-world applications of geographic information systems (GIS).

At Wayne State University's Center for Urban Studies in Detroit, middle and high school students learned how to use GIS to map their neighborhoods and discover safer routes to and from school.

By the Numbers

639,111
population of Detroit in 2020

53,000
students enrolled in Detroit Public Schools Community District in 2019

30,000+
estimated number of abandoned houses in Detroit in 2018

Sources: World Population Review; Chalkbeat; Detroit Metro Times

Geographic Information Systems have a wide range of real-world applications. They assist in displaying and analyzing data to make evidence-based decisions in a range of situations. In Detroit, Michigan, for example, GIS has been used to map safer school routes. Since 1999, the Detroit Public Schools Community District has used GIS to divide the district into patrol sectors and blocks. The city collects and analyzes data from each of the sectors to identify safe routes to school for students.

Nonprofit organizations and the school district also enlisted students to address urban blight, or areas in disrepair. Studies show that vacant structures are a strong predictor of assault risk. The number of vacant houses in Detroit had grown following an economic decline. One assessment put the number around 22,000, but an investigation by a Detroit newspaper suggested the number was at least 30,000. Because abandoned buildings often lack easily identifiable addresses, no one had been able to pinpoint the locations of the vacant structures. Even as the city struggled to board up or sell the vacant homes, new abandoned buildings appeared. In an innovative program called Mapping Out a Safer Community, middle and high school students used GIS software and handheld GPS to map the vacancies. The technology allowed them to identify the exact locations of abandoned buildings and vacant properties. The students then compiled their data, presented it to local government officials, and suggested areas to target for code enforcement. Using the students' information, the city boarded up abandoned buildings.

The student initiative was just one part of a broader data-gathering initiative, however. Members of the AmeriCorps Urban Safety Project walked the school routes and conducted surveys of parents, students, and school personnel to identify hazards. In addition to abandoned buildings, they mapped issues related to lighting, sidewalks, and dangerous intersections. This information has been used as part of a broader, federally funded program called Safe Routes to School that continues to work to ensure safe travel for Michigan students to and from school.

Geographic Thinking

Think about the mental mapping you did in Chapter 1. What would you expect to find if you followed Detroit's example in mapping routes to your campus?

U.S. Army Corps of Engineers analyze existing and new data about the island's electrical power grid so they could make repairs and provide generators to those without power even in remote areas. Airplane-mounted sensors are also used to measure the gradual sinking of land along the Gulf Coast to assess risk to local communities. As the technology advances and becomes less expensive, drones are making remotely sensed data more accessible than ever. They enable cartographers and other geographers to take detailed measurements between features or places on Earth's surface. Drones collect data that are then brought into GIS to determine changes in land use or environmental conditions. For example, farmers can use the data to get a bird's-eye view of the condition of their crops. The use of drones to identify a cluster of diseased plants in a large field or areas in need of water or fertilizer helps farmers treat targeted areas of their land and save resources.

Another source of geographic data is the **global positioning system (GPS)**, an integrated network of at least 31 satellites in the U.S. system that orbit Earth and transmit location data to handheld receivers. Essentially, a GPS receiver uses the time it takes to receive a transmitted signal to measure the distance to each satellite. The receiver uses this data to pinpoint the exact location of the receiver. The accuracy of the information allows people to determine the precise distance between two points, making GPS especially useful for navigation purposes. Pilots of airplanes and ships use GPS to stay on course. Smartphones and automobiles are also equipped with GPS receivers, enabling motorists to receive instructions for the fastest or most direct route to a desired destination. GPS-based mapping systems provide users with both maps and verbal directions to follow while traveling. GPS also uses information collected from other receivers to determine the speed of travelers and where traffic is stopped or slowed. GPS is used for geospatial applications beyond GIS.

A forestry conservation analyst from the World Wildlife Fund (WWF) uses an unmanned drone to map an area of the western Amazon rain forest in Brazil. Unmanned drones can be used in more locations than traditional aircraft and they provide the same data-collection capacity at a fraction of the cost. This makes them an increasingly valuable data-collection tool.

remotely sensed images are brought into GIS along with other data for comparison and analysis. Comparing satellite images can help identify phenomena such as trends in urban development or the shrinking of the polar ice caps. Satellites are also used for real-time decision making. For example, satellites can track the path, size, and intensity of a hurricane and the speed it is traveling. This data can be used to help predict where a hurricane will land. After an event, it also can be used to show the extent of damage, enabling aid to be directed appropriately. Days after Hurricane Maria devastated much of Puerto Rico in 2017, satellite images showed dramatic changes to the landscape and helped identify the hardest-hit areas.

Remote sensors mounted on aircraft or drones are another source of data. In addition to satellite images, aerial photographs were taken of the land months after Maria's damage. The photographs, along with GIS, helped the

One of the challenges that geographers face today is the enormous amount of available data. In addition to the GPS satellites that provide positioning data and satellites that collect images of Earth, there are hundreds of satellites collecting information about population, migration, soils, ocean currents, and more. Online mapping services collect and share even more data through aerial photographs and street-level cameras. Internet-based supercomputer systems are being developed to help geographers manage, analyze, and share this data.

Geographer Michael Goodchild sees this new era of "Big Data" as both a challenge and an opportunity for geographers and for GIS research. He observes that the "velocity" of data collection (often in real or near real time, available immediately or soon after being collected) is now much faster, and that the "variety" of data sources (networks of sensors, video surveillance cameras, and crowdsourced data from citizens) is now much broader than ever before. This increase in both velocity and variety have resulted in an unprecedented "volume" of data for geographers and geospatial technologies to handle, as well as unmatched opportunities: "Big Data in GIS will also be about prediction, not about when but about where (and sometimes when, too)." Goodchild anticipates Big Data requiring "an entirely new set of tools for integration and synthesis, in effect making useful data out of a morass of disparate observations." He concludes that the result will be enhanced "spatial prediction" to track hurricanes, forecast disease impacts, predict home values, find optimal store locations, and more.

Geographic Thinking

3. Identify three ways geographers collect data.
4. Describe how drones have impacted the acquisition of geospatial data.
5. Explain why it is important to collect data at the appropriate scale.
6. Describe one way that geographers could use GPS in their work.

2.3 Understanding Maps

Maps are the way geographers depict relationships of time, space, and scale. Maps are indeed among the geographer's most important tools because they display data in a spatial way. Geographers use many different types of maps to help them answer the three geographic questions you've already learned about: *Where? Why there? Why care?*

Mapmaking

Learning Objectives
LO 2.9 Describe the uses and purposes of maps.
LO 2.10 Identify the role scale and projection play in map construction.

Maps are the fundamental tool most uniquely identified with geography. People have used maps to depict information for thousands of years, and they continue to use increasingly sophisticated maps today. As you know, maps come in all shapes, sizes, and formats and have a wide variety of

Critical Viewing This map from Ortelius's atlas shows the world in two dimensions. ▮ Explain what this map reveals about geographers' knowledge of the world in 1570. Then compare it to today's maps.

uses and purposes. One of the most common uses of maps is to locate something, such as a country, a river, or a town. The purpose of many maps, such as road maps or subway maps, is to communicate how to get from one place to another. Centuries ago, **cartographers** created maps to help explorers follow the routes of those who came before them and to estimate how long it might take to travel to uncharted lands. These explorers then collected critical data for the creation of new maps.

Patterns in our world are seldom random; spatial features tend to be clustered, dispersed, or linear. For example, geographers use maps to illustrate the clustering or dispersal of patterns of the distribution of populations. To describe the spacing of places or people, the terms **absolute distance** and **relative distance** are used. Absolute distance is distance that can be measured using a standard unit of length, like miles or kilometers. Relative distance is measured in terms of other criteria such as time or money. For instance, it takes approximately 7 hours to fly from New York City to Paris, France, or a ride share in New York City from Manhattan to LaGuardia Airport costs $50. The terms **absolute direction** (the cardinal directions north, south, east, and west) and **relative direction** (left, right, up, down, front, or behind based on people's perceptions) are used to describe direction and location when interpreting maps.

Maps are important problem-solving tools. For example, depicting the spread of a disease epidemic on a map can often be the first step in finding its cause and stopping further outbreaks. Mapping cholera cases in the mid-1800s helped doctors recognize that outbreaks of the disease tended to happen near water supplies. That mapped data helped identify the use of contaminated water as the cause. About 150 years later, geographer Korine N. Kolivras used GIS to analyze dengue fever outbreaks in Hawaii. Dengue fever is a disease carried by a particular

Map Scales

On a walk through a city, such as Charlotte, North Carolina, you might use a highly detailed map that shows only the downtown area. To drive up the Atlantic coast, however, you would use a map that covers a large area, including several states. These maps have different scales.

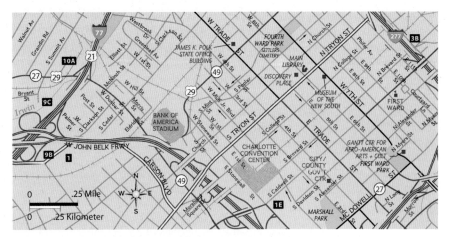

Large-Scale Map This detailed map shows only the city of Charlotte, North Carolina. The map scale shows that a half inch on the map represents a quarter mile on Earth's surface.

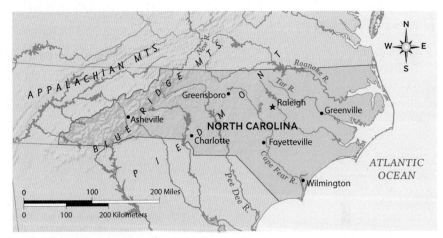

Medium-Scale Map This map shows the entire state of North Carolina. It includes fewer details than the large-scale map and shows a larger area. The map scale shows that three-quarters of an inch on the map represents 100 miles on Earth's surface.

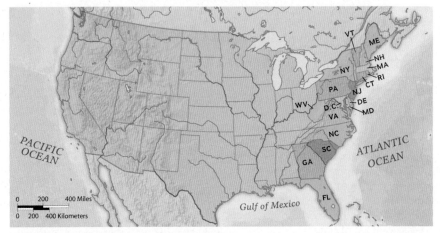

Small-Scale Map This map identifies the Atlantic coast states from Florida to Maine. It covers a large area and shows even fewer details than the medium-scale map. The map scale shows that a half inch on the map represents 400 miles on Earth's surface. ▮ About how far is it from the southern tip of Florida to northern Maine?

Geographic Inquiry: Data, Tools, and Technology

type of mosquito, so Kolivras's work included analyzing the breeding conditions needed for this mosquito to thrive. By mapping the precipitation, vegetation, and other related variables, Kolivras was able to predict the places at greatest risk of dengue fever and other mosquito-borne illness.

As representations of the entire world or part of the world, maps are selective in the information they represent. It is impossible to fit every feature or piece of data onto a single map, so mapmakers must decide how much of Earth to show and how to show it. These decisions are driven by the purpose of the map. Maps have many different purposes—to show location, distance, or some other spatial relationship. Maps can also be used to measure change over time. Comparing a map of Boston from 1775 with one of the city today reveals how humans have altered the landscape over time.

It is important to understand that maps are what historical geographer J. Brian Harley terms "social constructions." Just as essays, books, and documents are written by individuals with varying points of view and can be read in different ways and interpreted for different purposes, maps should also be critically analyzed and not accepted as objective representations of reality. Harley contends that "Both in the selectivity of their content and in their signs and styles of representation, maps are a way of conceiving, articulating, and structuring the human world which is biased towards, promoted by, and exerts influence upon particular sets of social relations."

Cultural geographers Denis Cosgrove and Stephen Daniels agree that one should develop a "critical awareness and skepticism about maps as well as other graphics and images" and that the study of mapmaking and map reading

Map Projection Types

Within the broad categories of projections, including conformal, cylindrical, and equal-area, are four common projection types: Robinson, Mercator, Gall-Peters, and azimuthal. Each projection has advantages and limitations and distorts the sizes and shapes of Earth's landmasses in different ways. The different projections also handle direction differently.

The Robinson projection has curved lines of longitude and straight lines of latitude, which means directions are true only along the parallels (including the equator) and the central meridian. Its unique, globelike appearance makes the Robinson projection useful for many different types of maps.

The Mercator and Gall-Peters projections show true direction, which is direction measured with reference to the north geographic pole. These two projections are often used for navigation. The azimuthal projection is well suited for maps of the Arctic and Antarctic.

Robinson Projection The shapes of the continents become more distorted farther away from the equator or the map's central meridian.

Mercator Projection The continents' shapes are maintained and direction is displayed accurately, but the sizes of the continents are very distorted.

"must include explicit instruction about how to interrogate a map—to consider the conditions under which it was produced, whether it may portray a particular point of view, and what, if any, messages it conveys about power and perspective."

Geographer and cartographer Mark Monmonier describes a frequently cited example of a map as a social construction, one that "lies": "During the Cold War, those who wished to emphasize the danger faced by the United States from the threat of communism often portrayed global spheres of influence using a Mercator projection. The size of countries on this type of map is exaggerated as one moves toward the poles. Thus, using it to depict Eastern Europe, China, and the USSR (often in a bright symbolic red color) made this northerly region seem larger and more menacing than it would have on a map that represented areas more realistically." Cosgrove and Daniels point out that "Choosing where to center a map also conveys the point of view of the mapmaker" and that centering "this same Mercator projection on Europe strongly supported Europeans' view" of their global supremacy. This Eurocentric view makes it appear that two-thirds of the world is in high latitudes, just where Europe lies, while simultaneously diminishing the mapped size of their more southerly colonial possessions. Evaluating maps critically by considering aspects such as the source of the data and the intent of the cartographer, as well as the time period and the culture in which it was created, will enhance your understanding of human geography—and may help you if you create your own maps.

Map Scale Maps can show information at almost any scale, from the entire world to a neighborhood, to a school, or even a classroom. A **map scale** is the mathematical

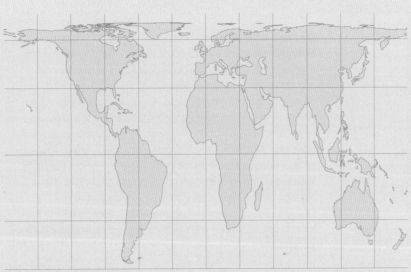

Gall-Peters Projection The relative size of the continents is more easily displayed than with other projections, but the shape of the continents is distorted.

Azimuthal Projection A flattened disk-shaped portion of Earth is shown from a specific point.

Map Projection	Advantages	Limitations
Mercator	• Shows true direction • Good for navigation purposes	• Distorts area • Size is distorted increasingly near the poles
Gall-Peters	• Shows true direction • Area is relatively precise	• Distorts shape • Continents appear elongated
Robinson	• A globelike appearance that "looks real" • Distorts size and shape, but not too much	• Imprecise measurements • Extreme distortion at the poles; flat on the poles and compressed near the equator
Azimuthal	• Preserves direction • When used from the point of the North Pole, no country is seen as center	• Distorts shape and area • Only shows one half of Earth

Geographic Inquiry: Data, Tools, and Technology

relationship between the size of a map and the part of the real world it shows. It allows you to measure absolute distance. The scale can be expressed in three ways: as a representative fraction, a written scale, or graphically. A representative fraction is often expressed as a ratio, for example a scale of 1:1,000,000 means 1 unit on the map represents 1 million of the same units on Earth's surface. An example of a written scale is 1 inch representing 200 miles. A graphic scale is expressed with a scale bar showing the relationship between the distance on the map and the distance on Earth's surface.

The scale of a map is an important clue to the level of detail portrayed on the map as well as the purpose of the map. As the scale of analysis varies, so does the kind and amount of information shown on the map. For example, a city map shows streets, buildings, and landmarks. A map of a state or province shows less detail—cities, rivers, and highways—and covers a larger area. A map of an entire country or continent shows even less detail; perhaps just the major natural features and national borders. The scale of a map impacts the analysis of the map and therefore its purpose. A certain pattern may be obvious at one scale, but as you zoom out, other patterns may become clear. When making decisions or conducting spatial analyses, it is important to use a map at the appropriate scale that shows pertinent details. The scale of a map must fit the purpose.

Map Projections Cartographers are tasked with using just two dimensions to represent a three-dimensional object—Earth. A sphere cannot be flattened onto a piece of paper or screen without altering its original shape. Over time, cartographers have developed various mathematical equations to handle the distortion, or misleading impressions, of Earth's surface that occur during the mapmaking process.

A map projection is any method used to represent the world or part of the world in two dimensions. Different projections distort spatial relations in shape, area, direction, or distance. All map projections create distortion, but the types and degrees of distortion vary considerably. The purpose of the map should guide the type of projection used. A conformal projection distorts area but keeps the shapes intact, giving the impression that some continents are larger than they actually are. Cylindrical projections also distort shapes, but they preserve direction. Equal-area projections, on the other hand, attempt to distribute the distortion of area equally throughout the map; however, in so doing, they distort the shapes of landmasses.

Geographic Thinking

1. Describe one example of absolute distance and one example of relative distance.
2. Compare the three ways scale is expressed on maps by explaining how they are alike and different.
3. Explain why the Robinson projection is one of the most commonly used map projections. Explain the pros and cons of using the Mercator projection.

Types of Maps

Learning Objectives
LO 2.11 Compare the features and functions of reference maps and thematic maps.
LO 2.12 Explain how different types of thematic maps organize and show human geography data.

There are two major categories of maps: reference and thematic. **Reference maps** are generalized sources of geographic data and focus on location. Highway or road maps are a common example, showing a state's or region's roads, cities, rest areas, parks, and other places of interest to drivers and tourists (once typically stored in a car's glovebox). **Thematic maps** have a theme, topic, or specific purpose and focus on the relationship among geographic data. While a reference map might show streets or highways and other general city or state features, a thematic map, depending on its scale, might show the spread of disease across a city, distribution of agricultural production in a state or province, or trade patterns among countries across the globe.

Reference Maps Reference maps illustrate the boundaries, names, and other unique identifiers of places and regions. They focus on the location of geospatial elements such as countries, cities, lakes, rivers, and other features of a landscape. Physical maps, which primarily show landforms and other natural features, and political maps, which primarily show boundaries between governmental units like countries or states, are examples of reference maps. Reference maps often show absolute location in terms of latitude and longitude. For example, the absolute location of Portland, Oregon, is 45.52° N latitude and 122.68° W longitude. You can locate Portland on any map marked with latitude and longitude using these coordinates.

Beginning in 2005, online mapping services began to use satellite imagery, aerial photography, street maps, and panoramic views of streets to enable users to see reference maps of almost any place on Earth at almost any scale. Mapping services take advantage of GPS-enabled software systems to provide real-time traffic conditions and route planning that includes estimated travel time. In addition to travel by car, mapping services offer reference maps for travel on foot, by bicycle, and by public transportation.

Thematic Maps Thematic maps are maps focused on a particular topic or theme. For example, early human migrations out of Africa and their rate of dispersion to other parts of the world can be shown on a thematic map. Thematic maps can show the distribution, flow, connection of, or relationship among one or more attributes. A thematic map might focus only on population density, or it might also include multiple attributes, such as ethnicity and election results. Showing too many attributes on the same map can confuse its message.

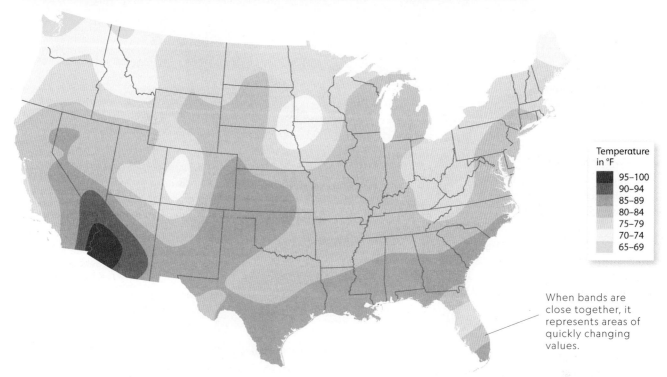

Isoline Map Lines connect data points of the same value. Isoline maps are used to show particular characteristics of an area. On this temperature map, isolines represent bands of similar surface temperatures.

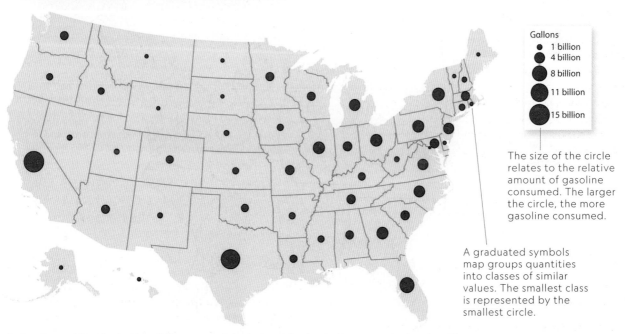

Graduated Symbols Map Different-sized symbols are used to indicate quantitative data. Bigger circles or icons represent a larger numerical value of a particular attribute. A graduated symbols map is useful for showing population, earthquake magnitude, or, as in this map, gasoline consumption.

Geographic Inquiry: Data, Tools, and Technology

Contiguous United States Population, 2018

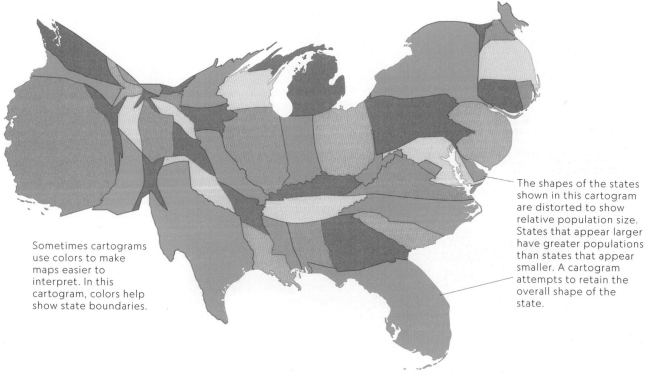

Sometimes cartograms use colors to make maps easier to interpret. In this cartogram, colors help show state boundaries.

The shapes of the states shown in this cartogram are distorted to show relative population size. States that appear larger have greater populations than states that appear smaller. A cartogram attempts to retain the overall shape of the state.

Cartogram Statistical data and geographic location are combined to communicate information at a glance. Cartograms show the relative size of an area based on a particular attribute, like population or energy consumption. Sometimes geographic regions are distorted to convey quantity or extent.

Milk Cow Inventory of the United States, 2017

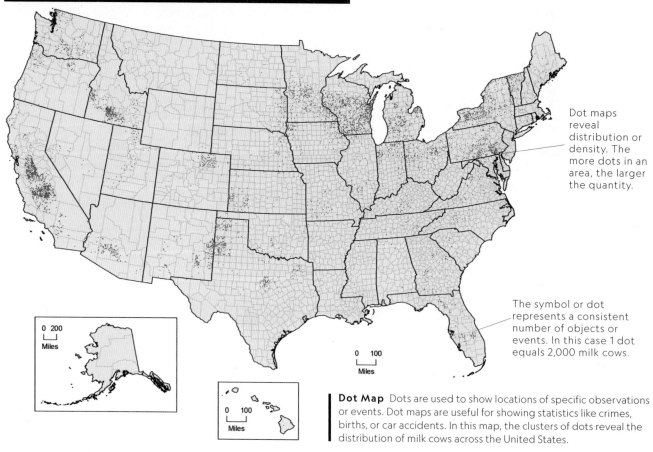

Dot maps reveal distribution or density. The more dots in an area, the larger the quantity.

The symbol or dot represents a consistent number of objects or events. In this case 1 dot equals 2,000 milk cows.

Dot Map Dots are used to show locations of specific observations or events. Dot maps are useful for showing statistics like crimes, births, or car accidents. In this map, the clusters of dots reveal the distribution of milk cows across the United States.

2020 U.S. Presidential Election

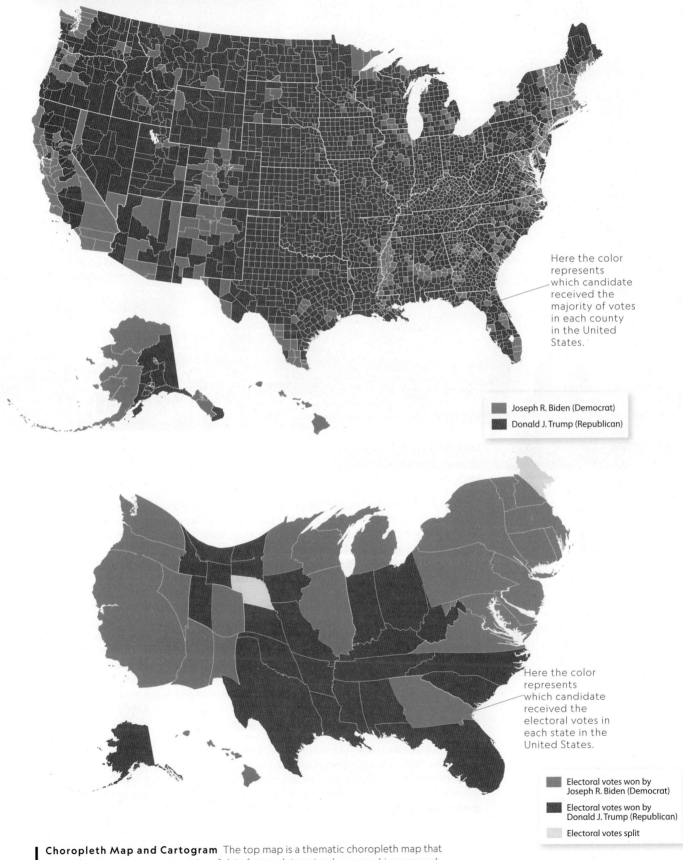

Here the color represents which candidate received the majority of votes in each county in the United States.

- Joseph R. Biden (Democrat)
- Donald J. Trump (Republican)

Here the color represents which candidate received the electoral votes in each state in the United States.

- Electoral votes won by Joseph R. Biden (Democrat)
- Electoral votes won by Donald J. Trump (Republican)
- Electoral votes split

Choropleth Map and Cartogram The top map is a thematic choropleth map that uses colors to represent categories of data for predetermined geographic areas such as census tracts, counties, states, provinces or countries. Choropleth maps are useful for communicating quantitative data such as demographics or election results. The bottom map is a cartogram showing the relative size of election results at the state level.

Geographic Inquiry: Data, Tools, and Technology

Lehua Kamalu is the Voyaging Director of the Polynesian Voyaging Society, a nonprofit educational organization that preserves the art and science of traditional Hawaiian navigation.

Learning Objective
LO 2.11 Explain how geographic data are used by individuals, businesses, organizations, and governments to make informed decisions.

National Geographic Explorer **Lehua Kamalu**

Navigating with Nature

Maps have guided ocean voyagers for centuries, but ancient Polynesian navigators needed no maps to cross thousands of miles of open ocean to land on tiny islands. Lehua Kamalu leads voyages around the world using these ancestors' subtle and sophisticated methods.

The knowledge used by generations of voyagers to navigate around Polynesia, a vast section of the Pacific Ocean dotted with small islands and archipelagos, was nearly lost to history. In the 1970s, the founders of the Polynesian Voyaging Society built a traditional canoe and invited Mau Piailug, a navigator from Micronesia, to visit Hawaii and share his knowledge. Mau navigated the double-hulled canoe, named *Hōkūle'a*, on a 2,300-mile voyage from Hawaii to Tahiti and also trained a new generation of navigators.

Many voyages later, in 2018, Lehua Kamalu became the first woman to serve as captain and lead navigator on *Hōkūle'a*'s sister canoe, *Hikianalia*. With a highly trained crew of a dozen people, she sailed 2,300 miles from Hawaii to California. She explains that her "map" for any voyage is based on what is happening in the environment: "What I do as a navigator is spend my day and night looking out over the horizon to see what signs nature can show me about which direction I'm going in." These signs include the exact location of sunrise and sunset, the positions of the moon and stars in the night sky, predictable patterns of wind and ocean waves, the shapes and colors of clouds, and the migration routes of marine animals and birds. Anything that she and her crew observe in nature can be used as a navigational tool.

Besides *Hōkūle'a* and *Hikianalia*, more than 20 traditional canoes have been built and now sail around the Pacific, educating people about the possibilities for living and working in harmony with nature. Even with her years of experience, Kamalu recognizes that there is always more to learn. She begins each voyage with "a humility that reflects a deep respect for the ocean and the navigators who have traversed its seas for hundreds of years before me."

Geographic Thinking

Could the navigational tools used in traditional Polynesian navigation be represented on a map? Explain.

Basemaps form the foundations of both reference and thematic maps. Many thematic maps use a basemap showing coastlines, city locations, and political boundaries. The map's theme is then layered onto this basemap. Political divisions, cities, or natural features provide reference points to help users understand the data that is presented on a thematic map, which can focus on a variety of topics.

Most geographic data relate to specific points, lines, and areas. The way maps display these types of data affects map analysis. Clusters are best illustrated in maps that use dots or graduated symbols, for instance. Isoline maps connect data points of equal value, like elevation, temperature, or precipitation. Choropleth maps use color or shading to display quantitative data in preset regions. Graduated symbols represent differences in number, size, or extent of something in an area, like populations of a state or traffic volume by county. Greater numbers are represented by larger symbols.

A cartogram is a unique type of map that conveys information by making the areas on a map proportional to the variable being mapped. As one example, a cartogram might redraw the spatial features of the U.S. states according to population distribution, so that New Jersey or Massachusetts appears much larger than Wyoming or Montana.

Geographic Thinking

4. Choose one of the thematic maps from this section. Based on specific details, describe one conclusion you can draw from the map.

5. Explain similarities and differences between dot maps and graduated symbols maps. Why might one or the other be preferable for different types of data?

2.4 The Power of Data

Geographic data help people make informed decisions at all scales—from individuals making personal decisions to businesses determining their marketing strategy, and from communities engaged in development planning to countries and international organizations looking to solve the world's most pressing problems.

How Data Are Used

Learning Objective
LO 2.13 Explain how geographic data are used to gain a deeper understanding of a situation or problem.

Geographic data are used to help people understand problems, consider options that lead to making decisions, and measure the effects of those decisions. For example, understanding common behaviors—where people work and shop and their commuting habits—helps city planners determine the future location of roads or make decisions about land use. Data can convince people to take action. Examining data about people's shopping habits can help determine where a shopping center should be located, and how many parking spaces will be needed there.

U.S. health officials typically use data to help public health departments make decisions when planning for the annual flu season. Data from the Centers for Disease Control and Prevention (CDC), the leading national public health institute in the United States, are used to provide feedback, inform policy, and make recommendations for new and better flu vaccines. The CDC receives and characterizes the genetic makeup of thousands of influenza viruses each year from across the United States and around the world. Compiling and analyzing these data allows the CDC to track when and where flu activity is occurring, determine what flu viruses are circulating, detect changes in the viruses, and measure the impact of the virus on hospitalizations and deaths. In addition, using data from past seasons helps to determine the severity of the virus each season. The CDC shares this information with health officials and agencies to help them make better decisions about what goes in each year's flu vaccine. It also helps evaluate viruses for their pandemic potential, allowing health agencies to better prepare for and prevent the spread of illness.

Throughout the COVID-19 pandemic, the CDC, WHO, other public health agencies, governments, and private industry

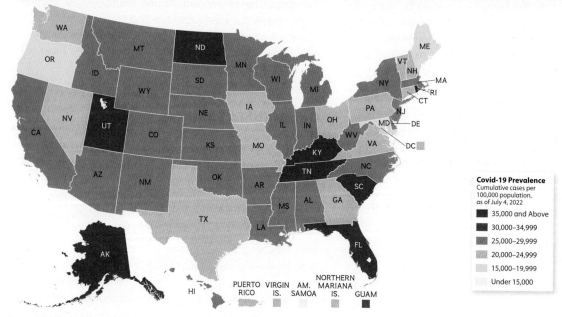

Cumulative COVID Rates in the U.S., July 2022

Sources: Johns Hopkins Center for Systems Science and Engineering (CSSE), U.S. Centers for Disease Control (CDC), and Utah Department of Health.

This choropleth map shows the cumulative level of COVID-19 infections, by state, between the start of the pandemic in early 2020 through July 4, 2022. This is quantitative data converted to a ratio showing cases per 100,000, to account for the huge variation in population among states.

used data of many types and at many scales from across the world to combat the virus. These data were used to document, map, and predict the spread of the virus, to determine disease rates, and to institute mandates and other barriers to disease transmission, as well as to create vaccines and to carry out vaccination campaigns. These same public health entities also use, and will continue to use in the future, spatial data from the COVID-19 pandemic to plan for and react to future disease outbreaks.

Making Decisions with Geographic Data

Learning Objective
LO 2.14 Explain how geographic data are used by individuals, businesses, organizations, and governments to make informed decisions.

Improving health and fighting disease are but two uses of data. Individuals and businesses also make use of a variety of geographic data, formally and informally.

Personal and Organizational Decision Making Geographic data influence where people decide to live. People in the market for a home consider its walkability and proximity to work, shopping, schools, parks, and more. They also may look at property taxes and school districts, as well as crime statistics, floodplain or earthquake data, and other risks. People may study sources that provide information on the neighborhood, public transportation, and commute times.

Individuals looking to sell their home will also employ geographic data to determine their property's value. For example, consulting data on the prices other homes in the neighborhood have been sold for can help predict a home's value and asking price.

Businesses make location decisions in ways similar to those of individuals. When deciding where to locate, businesses typically review demographic data on potential customers, the workforce, tax rates, and more. Businesses, depending on their type, might check out the ease of customer access via streets or public transit using maps and other spatial data. They might want to see what other types of businesses are in the area. Some organizations and businesses need data and maps that are specialized to carry out their operation. For instance, a home insurance company may look at floodplain maps, while a delivery company or a limousine service will employ GPS and GIS to optimize their routing and timing.

Other groups seek to expand data related to quality of life, mapping public health, education, and public safety services, for instance. OpenStreetMap has become a catalyst for such purposes. OpenStreetMap was founded in 2004 by software developers who wanted to create an open source network for mapmaking. Today, it maintains data about locations—roads, parks, railway stations, and more—all over the world. OpenStreetMap has become important in mapping lands of Native Americans and other Indigenous peoples. It has also played an important role in saving lives. Following a devastating earthquake that struck Haiti in 2010, and again for the 2021 earthquake, volunteers used satellite imagery to create detailed digital maps of roads, buildings, and other features, facilitating recovery and relief services. Similar efforts have resulted following other disasters. After Hurricane Harvey hit Houston, Texas, in 2017, people stranded in their homes or neighborhoods used online mapping services as a tool for assistance. An app enabled them to mark their location; some apps enabled people to add notes indicating the type of help required, such as "WOMAN IN LABOR!"

No matter what the need for or purpose of data—for one's own use, one's business, or one's organization—a critical interpretation of this data is called for. Critical reading of data involves a questioning process to understand it more completely and to make better decisions. Just as one should critically read documents and maps, one should also carefully evaluate data: Who or what was the source of the data? How was the data collected? Is it recent or old—is the data new or "dated"? Is the source credible? As a consumer of data, and certainly as a student of geography, one must attend to these questions. For any data considered for research, for a report, or for a presentation, it is necessary to understand the data used and to trust its source.

Governmental Decision Making Governments at all levels—local, regional, state, federal—use GIS data for myriad purposes. Researchers indicate that as much as 80 percent of data stored by the government has a spatial component.

Local governments use GIS data for addressing local problems. U.S. municipalities like cities, towns, and even some villages maintain spatial data and GIS mapping for a multitude of administrative and management purposes: roads, water, sewer, and other municipal infrastructure; zoning (the permitting of certain land uses in certain areas) and taxing of property; emergency services; recreation; and more. For instance, a city's fire department will use GPS on its fire trucks and emergency vehicles, plus GIS map layers for streets and another with lots and buildings, to make dispatching and routing more efficient and to save lives. A police department may use similar GIS data displaying streets, lots, and buildings, along with other specific spatial information, such as a GIS layer with neighborhood socioeconomic attributes, another showing where various types of crimes have occurred (what the department might map out as "crime hotspots"),

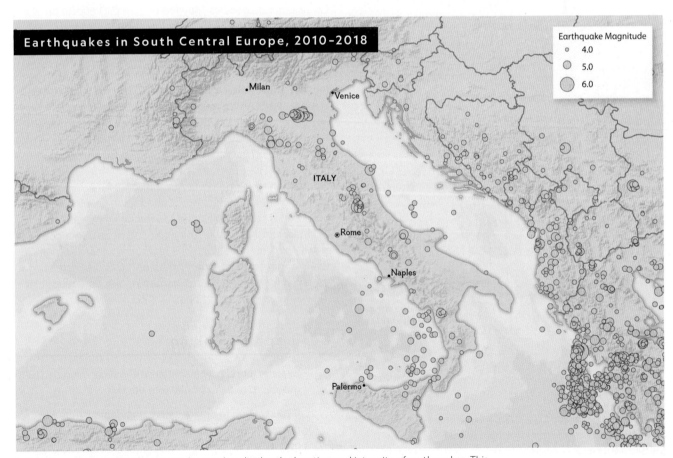

Reading Maps GIS software can be used to display the location and intensity of earthquakes. This map shows earthquakes that occurred in Italy and surrounding countries between 2010 and 2018. Each circle corresponds to an earthquake of varying intensity. ▌ Describe how the information in this map might be used by individuals, businesses, and governments.

Geographic Inquiry: Data, Tools, and Technology

and another indicating the presence of streetlights or surveillance cameras. In addition to helping the police department understand its community better and assign officers and neighborhood outreach more effectively, this use of spatial data and GIS can also help the municipality analyze the effectiveness of social services, street lighting, and other crime-reduction strategies.

In Los Angeles County as in other local jurisdictions, GIS produces maps used to address populations experiencing homelessness, or those who are "unhoused," the term that housing advocates and some agencies now use. The housing authority has mapped the distribution and characteristics of the unhoused, as well as various risk factors. In addition to considering where unhoused populations are today, decision-makers have mapped where people were when they became unhoused and where people are likely to become unhoused. GIS maps enable housing decision-makers to recognize and analyze spatial patterns, using the data to weigh options for prevention efforts, for locating shelters, and for organizing other resources for unhoused populations.

National governments across the world use spatial data and GIS, as do many of their regional, state, and provincial governments. The U.S. government uses GIS to help manage a vast range of programs and to evaluate outcomes of policy decisions. Agencies that administer federal lands, including the USDA (national forests, national grasslands), the Department of the Interior (national parks, wildlife refuges, Bureau of Land Management holdings, subsurface mineral holdings, and more), and the Department of Defense (DOD; military reservations) all gather and maintain spatial data and employ GIS to manage and protect federal lands entrusted to their care. Of course, branches of the DOD have long employed spatial data and GIS to defend the country. GPS was originally created and designed with military use in mind, and the branches of the U.S. military depend on GPS, essential spatial data, and sophisticated GIS systems to analyze, model, and visualize all aspects of their operational environments. U.S. intelligence services like the Central Intelligence Agency and the National Security Agency also depend on critical spatial data and cutting-edge GIS platforms to gather, scrutinize, and deliver information to the country's decision-makers.

Other federal agencies may focus their GIS efforts on disaster prevention and mitigation, like the Federal Emergency Management Agency (FEMA, part of the Department of Homeland Security). FEMA gathers, uses, and shares location-based data with other relevant federal, state, and local agencies "to prepare for, protect against, respond to, recover from, and mitigate all hazards." Spatial data published by the U.S. Geological Survey helps officials assess the risk of earthquakes in specific regions and develop strategies for mitigating such risks. The Environmental Protection Agency uses GIS to monitor air quality by overlaying spatial information with environmental data like ground-level ozone (an air pollutant). The data are used to identify areas with poor air quality and inform jurisdictions and people in those areas. In all of these agencies and many others at multiple levels of government, spatial data and GIS are increasingly essential for staff to make informed decisions and to fulfill their organizational missions.

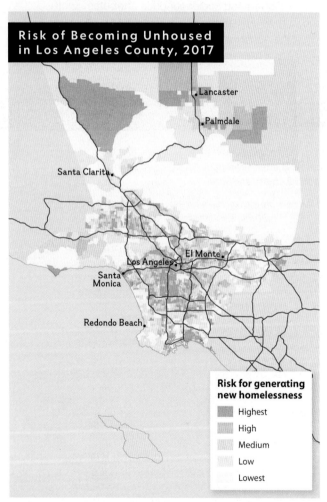

The risk for persons becoming unhoused (homeless) varies in different parts of Los Angeles. The darker colors represent higher levels of risk.

Geographic Thinking

1. Explain how showing spatial patterns can help decision making. Use an example from the text.

2. If you wanted to create a map that demonstrated to the public the seriousness of a certain city's unhoused population problem, what type of thematic map would you use? Explain your thinking.

National Geographic Collaborator **Clinton Johnson, Esri**

Mapping Social Justice

Clinton Johnson leads the Racial Equity and Social Justice team at Esri, the global leader in GIS technology. Prior to joining Esri, he worked for the City of Philadelphia, where he used GIS programming to guide municipal projects from the Office of Innovation and Technology.

Learning Objective
LO 2.14 Explain how geographic data are used by individuals, businesses, organizations, and governments to make informed decisions.

GIS maps translate large amounts of data into easy-to-understand pictures, revealing information that is otherwise hard to see. At the tech company Esri, Clinton Johnson harnesses the power of GIS to promote racial equity and social justice.

When Johnson first saw redlining maps for his hometown, he suddenly understood his own experience in a new light. He had assumed he was imagining things when he moved from one Philadelphia neighborhood to another and found the summer heat less intense in his new home. However, the maps revealed the long-term effects of redlining, the practice of refusing loans to residents in predominantly Black neighborhoods. As a result of redlining, those neighborhoods received less care than wealthier areas, including less investment in trees and other green spaces. And urban areas with fewer trees are hotter in the summer. "Those maps," Johnson says, "were pictures that told the story of my life."

The original redlining maps were printed on paper in the 1930s. By moving that information into GIS, it can be layered with another map showing current vegetation. Add two more layers showing summer high temperatures and the percentage of people in each neighborhood who suffer from heat-related health issues, and the GIS map tells a compelling story about the impact of historical racial injustice on people's lives today.

The visual stories told by GIS maps help people understand a situation and take effective action. For example, Johnson's blog describes how, in 2020, the GIS team in King County, Washington, used the technology to identify areas where people lacked adequate Wi-Fi coverage at a time when many schools were teaching online due to the COVID-19 pandemic. Their maps helped service providers expand access in those areas. In 2019, Johnson founded a grassroots organization that evolved into NorthStar of GIS, a nonprofit working to create a more racially just world through a more racially just GIS, geography, and STEM (science, technology, engineering, and mathematics). NorthStar strives to increase representation, inclusion, and belonging for people of African descent by using GIS technology to tell their own stories and advance racial justice.

Geographic Thinking

Explain how GIS mapping can help promote racial justice.

Chapter 2   Summary & Review

Chapter Summary

Geographers use a spatial perspective to interpret the world.

Geographic inquiry provides a systematic way to investigate and understand the world through the patterns, processes, and interactions between human and natural systems. Geographic questions are at the heart of this inquiry process, starting with: Where? Why there? and Why care?

Maps are among a geographer's most important tools. Maps depict data spatially, tying locations to aspects of the Earth at various scales. Mapping provides a way to organize, analyze, and display vast amounts of spatial information.

- There are two main categories of maps:
  - Reference maps focus on location and phenomena.
  - Thematic maps focus on the spatial variation of one or more characteristics.
- Map scale shows the relationship of the size of the map to the size of the area it represents on Earth's surface. It determines the level of detail of a map.
- All maps are distorted because of the problems of representing a three-dimensional spherical object (Earth) in two dimensions. Maps distort shape, area, direction, or distance.
- Each type of map projection has both advantages and disadvantages. The projection used depends on the purpose of the map.
- Types of thematic maps include dot maps, isoline maps, graduated symbols maps, choropleth maps, and cartograms.

Today's technologies enable complex data to be gathered in real time by individuals and organizations. Geographers collect a range of data that can help individuals, businesses, organizations, and governments make informed decisions.

- Geographers gather data through a variety of remote sensing methods, including satellite images and aerial photographs, as well as through field observation and interviews and written accounts including media reports and policy documents.
- Geographic information systems (GIS) capture, store, analyze, and display geographic data. The data can be used to create map layers that are combined on a single map. Location is the key aspect that links these data layers.
- The global positioning system (GPS) enables geographers (and others) to determine precise locations for and distance between points on Earth's surface.

Review Questions

Use complete sentences to answer the questions.

1. **Apply Conceptual Vocabulary** Consider the term *distortion*. Write a standard dictionary definition of the term. Then provide a conceptual definition—an explanation of how the term is used in the context of the chapter.

2. How might a cartographer show the topography of a region or country?

3. Describe how remote sensing tools have improved the work geographers do. Provide an example.

4. What is the primary difference between GPS and GIS?

5. Differentiate between reference maps and thematic maps. Give an example of each.

6. How does map scale affect cartography?

7. How are the terms *map projection* and *distortion* related?

8. What is the Mercator projection and how does it distort?

9. Explain similarities and differences between choropleth maps and isoline maps. Give a common use for each.

10. Do you think the U.S. census contains mostly quantitative or qualitative data? Explain.

11. Describe the absolute distance and the relative distance between your school and where you live.

■ **Interpret Maps**

Study the cartogram and then answer the questions.

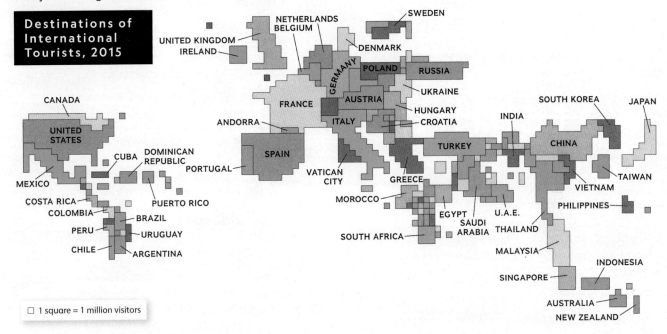

Destinations of International Tourists, 2015

□ 1 square = 1 million visitors

12. **Describe Geographic Concepts** What aspects of the map tell you it is a cartogram?

13. **Explain Spatial Relationships** Why is the United States represented as so much larger than Canada?

14. **Evaluate Models & Theories** What is the advantage of using a cartogram to show this data?

15. **Synthesize** How might different countries use this map when considering how to promote tourism in the future?

Unit 1 | Writing Across Units, Regions, & Scales

They Are Watching You

by Robert Draper

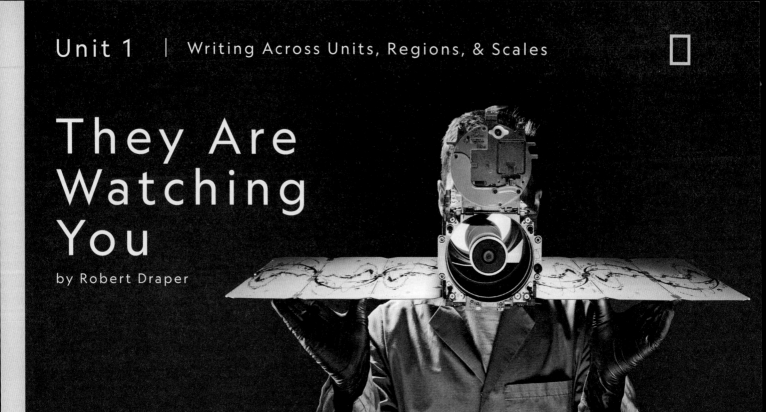

The Dove satellite being held by a senior spacecraft technician at the San Francisco–based tech company Planet is camera equipped and able to snap two images per second. Satellites are important geographic tools that are used to investigate even small changes on Earth.

Expanding networks of satellites are providing unprecedented views of humans' influence on the land, the climate, and ourselves—documented in real time. The technology in question can monitor Earth's entire landmass every single day. It's the brainchild of a San Francisco–based company called Planet, founded by two idealistic former NASA scientists named Will Marshall and Robbie Schingler. At NASA they had been captivated by the idea of taking pictures from space, especially of Earth—and for reasons that were humanitarian rather than science based.

They experimented by launching ordinary smartphones into orbit, confirming that a relatively inexpensive camera could function in outer space. "We thought, What could we do with those images?" Schingler said. "List the world's problems: poverty, housing, malnutrition, deforestation. All of these problems are more easily addressed if you have more up-to-date information about our planet."

A Changing View of Earth In storybook fashion, Marshall and Schingler developed their first model in a garage in Silicon Valley. The idea was to design a relatively low-cost, shoebox-sized satellite to minimize military-scale budgets often required for designing such technology—and then, as Marshall told me, "to launch the largest constellation of satellites in human history." By deploying many such devices, the company would be able to see daily changes on Earth's surface in totality.

In 2013, they launched their first satellites and received their first photographs, which provided a far more dynamic look at life around the world than previous global mapping imagery. "The thing that surprised us most," said Marshall, "is that almost every picture that came down showed how Earth was changing. Fields were reshaped. Rivers moved. Trees were taken down. Buildings went up. Seeing all of this completely changes our concept of the planet as being static."

What Satellites Can Do Today, Planet has more than 200 satellites in orbit, with about 150 so-called Doves that can image every bit of land every day when conditions are right. The company works with the Amazon Conservation Association to track deforestation in Peru. It has provided images to Amnesty International that document attacks on Rohingya villages by security forces in Myanmar. At the Middlebury Institute's Center for Nonproliferation Studies, recurring global imaging helps the think tank watch for the sudden appearance of a missile test site in Iran or North Korea.

Those are pro bono clients. Its paying customers include Orbital Insight, a Silicon Valley–based geospatial analytics firm that interprets data from satellite imagery. With such visuals, Orbital Insight can track the development of road or building construction in South America, the expansion of illegal palm oil plantations in Africa, and crop yields in Asia. In the company's conference room, James Crawford, the chief executive, opened his laptop and showed me aerial views of Chinese oil tanks, with their floating lids indicating they were about three-quarters full. "Hedge funds, banks, and oil companies themselves know what's in their tanks," he said with a sly grin, "but not in others', so temporal resolution [the amount of time needed to collect data] is extremely important."

Meanwhile, Planet's marketing team spends its days gazing at photographs, imagining an interested party somewhere out there who might benefit from the images. An insurance company wanting to track flood damage to homes in the Midwest. A researcher in Norway seeking evidence of glaciers eroding. But what about . . . a dictator wishing to hunt down a dissident army?

Here is where Planet's own ethical guidelines would come into play. Not only could it refuse to work with a client having malevolent motives, but it also doesn't allow customers to stake a sole proprietary claim over the images they buy. The other significant constraint is technological. Planet's surveillance of the world at a resolution of 10 feet is sufficient to discern the grainy outline of a single truck but not the contours of a human.

The Future of Planet On a bracing autumn evening in San Francisco, I returned to Planet to see the world through its all-encompassing lens. More than a dozen clients would be there to show off how they're using satellite imagery—what it meant, in essence, to see the world as it's changing.

I zigzagged among semicircles of techies gathered raptly around monitors. Everywhere I looked, the world came into view. I saw, in the Brazilian state of Pará, the dark green stretches of the Amazon jungle flash red, prompting automatic emails to the landowners: *Warning, someone is deforesting your land!* I saw the Port of Singapore teem with shipping activity. I saw the croplands of southern Alberta, Canada, in a state of flagging health. I saw oil well pads in Siberia—17 percent more than in the previous year, a surprising sign of stepped-up production that seemed likely to prompt frantic reassessments in the world's oil and gas markets.

Planet's hosts halted the show-and-tell to say a few words. Andy Wild, the chief revenue officer, spoke of the new frontier. It was one thing to achieve, as Wild put it, "a daily cadence of the entire landmass of the Earth." Now the custodians of this technology had to "turn it into outcomes." Tom Barton, the chief operating officer, said, "I hope one year from now, we're here saying, 'We really did change the world.'"

> Source: Adapted from "They Are Watching You" by Robert Draper, *National Geographic*, February 2018.

Write Across Units

Unit 1 explored the tools geographers use in their work. This article focuses on one of those tools—satellite imagery—and the astonishing array of spatial and temporal data it can yield for geographical study. Use information from the article and this unit to write a response to the following questions.

Looking Back

1. What sets Planet's satellite network apart from other data collection tools you learned about? Unit 1

Looking Forward

2. What types of data could Planet's satellites provide to a government agency seeking to understand the effects of migration within a region? Unit 2

3. How might the existence of mass data gathering and communication technology such as satellite networks influence cultural changes? Unit 3

4. What political advantages could a government gain by possessing satellite-gathered data about its own territories? Unit 4

5. How might satellites help researchers trace the effects of technology on agriculture globally? Unit 5

6. How could planners use satellite data to design a public transportation system for a city? Unit 6

7. What types of satellite data could help a government create and implement policies for sustainable development? Unit 7

Write Across Regions & Scales

Identify two locations in the world experiencing issues related to a topic mentioned in Unit 1, such as income inequality or crime.

Determine your question. Then write a paragraph for each of the steps in the process, explaining how you would collect information, visualize the data, create a presentation, and act on what you learned. Draw on evidence from the unit and article to plan your project.

Think About

- How comparing data from more than one location might inform the way you present your findings and choose to act on them

- The advantages and disadvantages of satellite imagery and other forms of data collection

Unit 1

Chapter 1

Distance Decay Model

A model is a representation of one aspect of reality, such as a geographic relationship, in a generalized form. It is important to examine models and compare them to real-world data to determine the degree to which they explain geographic effects in different contexts or at different scales.

The distance decay model describes a fundamental relationship in human geography: the impact of distance on interactions between locations. Specifically, the model states that the farther away two things or places are from each other, the less interaction they will have. Friction of distance—the idea that distance requires time, effort, and cost to overcome—explains some key factors that contribute to distance decay. Here, the first graph illustrates the distance decay model in its generalized form. The second graph illustrates the effects of distance decay on consumers deciding how far they will travel to shop. ▌ Explain the degree to which the distance decay model explains the geographic effects of distance on individuals' shopping behavior.

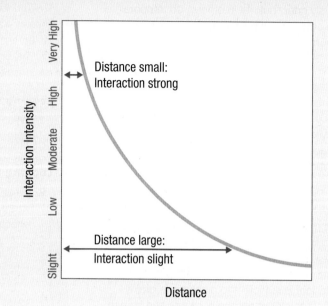

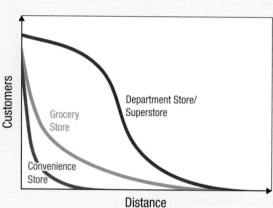

Chapter 1

Time-Space Compression

Some concepts are visually depicted in different ways. Here, time-space compression is represented by a shrinking globe, while the visual you studied in Chapter 1 highlights travel times across the Atlantic. The concept of time-space compression reflects the forces, such as improvements in transportation and communication, that can overcome the friction of distance. By compressing the amount of time it takes to travel or transmit information, these forces give the impression of lessening the space between distant locations. ▌ Compare this visual with the one in Chapter 1 and identify which you think most clearly illustrates the effects of time-space compression. Explain.

Advances in Transportation and Communication

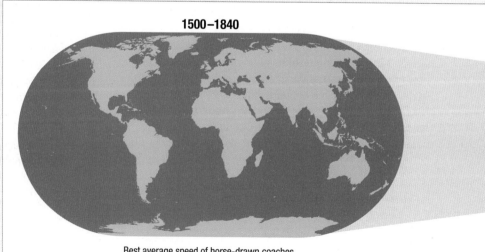

Best average speed of horse-drawn coaches and sailing ships was 10 mph

56

Maps & Models Archive

Chapter 1

Wallerstein's World System Theory

A theory is a system of ideas that attempt to explain observed phenomena. Like models, theories explain geographic effects to varying degrees, depending on where and how they are applied.

Immanuel Wallerstein developed the world system theory to explain the global economic phenomena that he observed. Wallerstein's theory views the entire globe as a single economic system bound together through a complex network of trade and communications. It describes not only the economic ties between countries but also patterns of power across the globe.

Core countries derive the greatest benefits from the world economy and tend to dominate politically as well. Countries in the semi-periphery and periphery have less wealth and often less-stable governments, and thus find themselves in positions of lesser power. ▌Explain the degree to which the world system theory explains the economic influence of a smaller core country such as France.

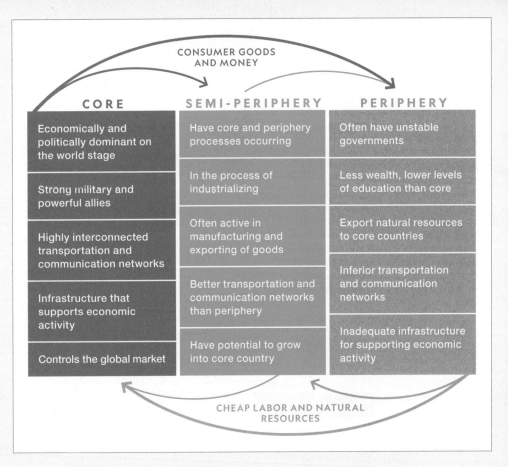

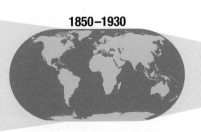

1850–1930
Steam locomotives averaged 65 mph, fastest steamships could travel at 30+ mph

1950s
Propeller aircraft 300–400 mph

1960s
Jet passenger aircraft 500–700 mph

1990s
Spread of the internet makes worldwide communications instantaneous

Chapter 1

Geographic Scale

The term *scale* refers to the size of an area being studied. Examining a phenomenon at different scales reveals new perspectives that can lead to a deeper understanding. For example, geographers studying population density—the number of people relative to the amount of land—look at maps on a variety of scales.

These maps show the population densities in Asia on a national scale and the population of Pakistan on a district-wide scale. ▌ Compare the maps and describe what each reveals about the population density of Pakistan. What additional information could you expect to learn from a density map of Islamabad?

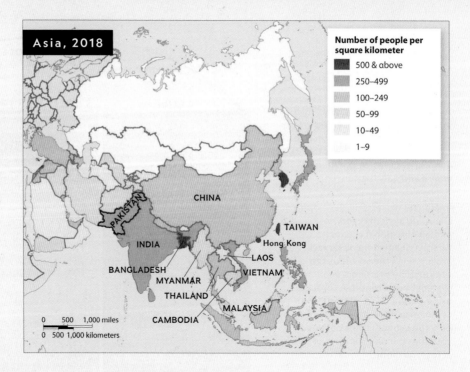

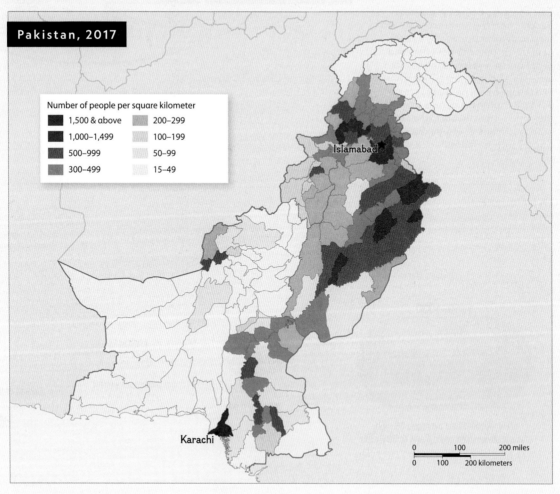

58

Maps & Models Archive

Chapter 2

Smartphone Maps

The maps many people interact with most often are found in their smartphone's app, that is based on GPS navigation. One advantage of smartphone maps is their ability to display an area at a seemingly infinite variety of scales. The two smartphone maps provided show the same location at different scales. ▌Compare the maps and describe what each scale reveals to the map user.

225 Baker St NW, Atlanta, GA

Chapter 2

Map Projections

Our perceptions about countries and continents are strongly influenced by map projections. Equal-area projections accurately portray the area of a landmass, but they distort the shapes of various continents. Mercator projections preserve the continents' shapes and show accurate direction, making this projection ideal for plotting straight-line courses on navigation charts. However, the Mercator projection distorts the true area of landmasses, an effect that increases with distance from the equator. ▌Compare the way the equal-area projection and the Mercator projection each depict the United States and Canada.

Mercator Projection

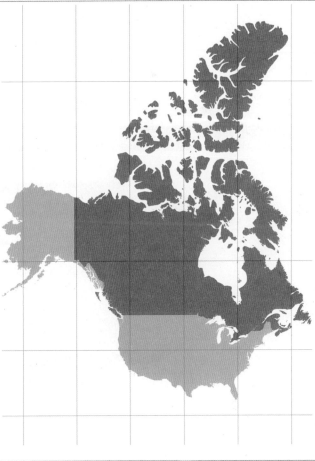

Equal-Area Projection

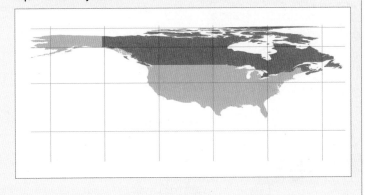

Unit 1

Chapter 2

Comparing Types of Maps

At times, it is useful to portray the same set of data on more than one type of map. Both maps show the percentage of the long-term average precipitation that has fallen over the period of one month in November. The long-term average precipitation is calculated over a 50-year base period. The dark brown color indicates areas that received 50 percent or less of the long-term average precipitation. Dark green indicates areas that received 150 percent or more than the average. ▌ Compare the dot and isoline maps. Describe what each map helps viewers understand about precipitation during November 2019.

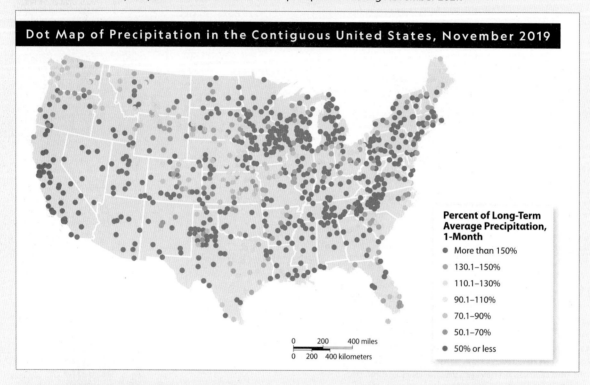

Dot Map of Precipitation in the Contiguous United States, November 2019

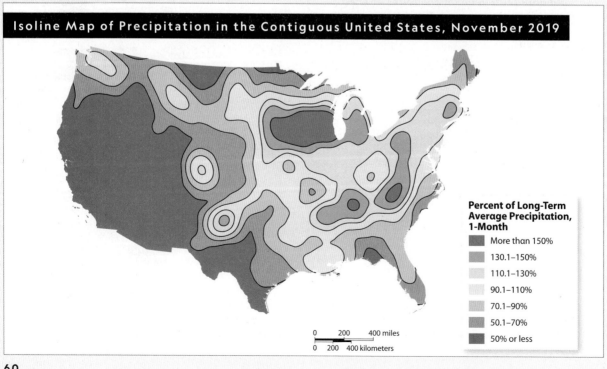

Isoline Map of Precipitation in the Contiguous United States, November 2019

Chapter 2

Map Layers of a Thematic Map

This thematic map depicts where wildfires burned in the Amazon Basin in 2019. Its base map is the physical features of South America. One GIS map layer shows the political borders of each country. ▌Identify other map layers used to create this map. Explain how the map layers could be used by decision-makers in South America.

Unit 2
Population and Migration
Patterns and Processes

Where and How We Live

Sai Yeung Choi Street in Mongkok, Hong Kong

Issues involving world populations and their movement across the globe pop up in news sources nearly every day. Whether the concern is addressing the needs of changing populations, caring for expanding numbers of migrants, or coping with the cultural impacts of different ethnicities, languages, or religions jostling for space, it's clear to geographers, other scientists, and policy makers across the spectrum: understanding population change and managing finite resources are foundational challenges.

The city of Shenzhen, in southeastern China, links Hong Kong to the Chinese mainland. Here, the juxtaposition of the modern city gleaming on the horizon with rice paddies in the foreground illustrates the urban expanding into the rural, and the tensions between accommodating a growing population and feeding it.

It's the human stories rising from a cultural landscape like this one that capture our imaginations—and should drive our policies.

Chapter 3
Patterns of Population

Chapter 4
Population Growth and Decline

Chapter 5
Migration

Unit 2 Writing across Units, Regions, & Scales

Unit 2 Maps & Models Archive

National Geographic Explorer **Paul Salopek**

Out of Eden
A Walk Through Time

Geographers, historians, and archaeologists study the political, economic, and cultural effects of human migration and human interaction. To tell this story of interaction and impact, National Geographic Explorer and Pulitzer Prize-winning journalist Paul Salopek is walking 24,000 miles across the world on a continuous, multiyear journey of discovery. He began in Ethiopia in 2013 and reached China in late 2021. Salopek's online dispatches and writings for print reflect what he calls the practice of "slow journalism," a reliance on forming personal bonds with people who share compelling stories with him about their everyday lives.

Learning Objective
LO 3.1 Identify the factors that influence population distribution at different scales.

To Walk the World I am on a journey. I am in pursuit of an idea, a story, a chimera, perhaps a folly. I am chasing ghosts. Starting in humanity's birthplace in the Great Rift Valley of East Africa, I am retracing, on foot, the pathways of the ancestors who first discovered the Earth at least 60,000 years ago. This remains by far our greatest voyage. Not because it delivered us the planet. No. But because the early *Homo sapiens* who first roamed beyond the mother continent—these pioneer nomads numbered, in total, as few as a couple of hundred people—also bequeathed us the subtlest qualities we now associate with being fully human: complex language, abstract thinking, a compulsion to make art, a genius for technological innovation, and the continuum of today's many races.

If you ask, I will tell you that I have embarked on this project, which I'm calling the Out of Eden Walk, for many reasons: to relearn the contours of our planet at the human pace of three miles an hour. To slow down. To think. To write. To render current events as a form of pilgrimage. I hope to repair certain important connections burned through by artificial speed, by inattentiveness. I walk, as everyone does, to see what lies ahead. I walk to remember.

The Natural History of Compassion The oldest hominins ever found outside of Africa were unearthed atop a rocky promontory in the Republic of Georgia, in the lush southern Caucasus. Their bones lay—gnawed on, in some cases, by giant prehistoric hyenas—beneath a medieval town. Under mossy ruins that include a church and a fortress. Under a cross. Under the sword. I think about this primordial contrast in human aspiration while walking the archaeological site.

At its most basic level, "survival of the fittest" holds that any individual whose DNA gets replicated most within a population wins the evolutionary lottery. It doesn't matter one whit how: Morality isn't biology. Indeed, selfishness is rewarded. Cheating or violence is fine. Thus, a rogue who only looks out for Number One . . . would, in theory, survive longest and produce the most babies. He or she would become a gold medalist in the genetic Olympics. And yet, we are not venal thugs. At least, not all the time. We grapple with our self-centered natures. Occasionally, we even lay down our lives for the weak, vulnerable, and downtrodden. For scientists this behavior is mystifying.

The most famous example is Shanidar 1, a Neanderthal man found in a cave in Iraq. His right arm was atrophied. He was partially deaf and blind. He was so crippled by arthritis and injuries, he could barely move. Still, he reached the ripe old Neanderthal age of about 40—only, it would seem, with the laborious support of other people in his clan. A useless old man was cared for.

I am walking the world. I knock on unfamiliar doors. I call out to the tents of unknown people. I slog onwards . . . into a vast and rumpled topography of human want and compassion. Do not be afraid.

Geographic Thinking

Explain the historical and contemporary geographic effects of migration, as highlighted in Salopek's "slow journalism."

Top: National Geographic Fellow Paul Salopek (left) walks with his Ethiopian guide, Ahmed Elema, as they leave Herto village in the Afar region of northwestern Ethiopia. *Left:* The earliest clues to altruism have been found at Dmanisi, a site crowned by a medieval church and fortress. *Right:* A pair of Salopek's sturdy hiking boots typically last for about 1,000 miles.

Chapter 3
Patterns of Population

Critical Viewing Stone houses are built into a steep gorge in a farming village in western Iran. ▌ How might population growth or decline change this human-built landscape?

Geographic Thinking Where do people live, and why?

3.1
Where People Live

National Geographic Explorer
Lillygol Sedaghat

3.2
Consequences of Population Distribution

Case Study: Population Distribution at the Country Scale

3.3
Population Composition

3.4
Measuring Growth and Decline

Feature: Examining Population Pyramids at Different Scales

3.1 Where People Live

People live in distinct patterns across the globe. Some of these places are densely settled, while others remain less populated. Geographers identify several physical, environmental, and human factors that help explain global population distributions. By studying these factors, geographers can make predictions about how populations will interact with Earth's different environments and change over time.

Why Study Population?

Learning Objective
LO 3.1 Identify the factors that influence population distribution at different scales.

According to the United Nations, the global population reached 8 billion people by the end of 2022. Two-thirds of that population can be found in four regions: East Asia, South Asia, Southeast Asia, and Western Europe. During the second half of the 20th century, the world's population increased at a faster rate than ever before. And today, more people are alive than at any other time in Earth's history.

These facts compel geographers to analyze the complex reasons that populations grow and spread, and the challenges that result. They analyze where people live, differences among populations, changes within populations, and the reasons for each. **Population distribution**—where people live within a geographic area—affects the cultural, political, environmental, and economic aspects and conditions in any given area. To better understand fundamental patterns of population and the processes that affect populations, geographers study population distributions at local, regional, national, and global scales.

Global distributions of people are influenced by physical, environmental, and human factors. These factors may cause people to move to newly inhabited areas or to continue living in the same place for decades, centuries, or longer. At every scale—from local to global—there are identifiable factors that influence population distribution. These interconnected factors are liable to change, which means that populations will change with them. With a record number of people on the move, especially refugees, movement and mobility are major factors in contemporary population trends.

Population Distribution Patterns

Geographers analyze the patterns of population distribution to better understand the processes of population growth and movement over time. Patterns of population distribution can be described using a variety of terms, such as *uniform*, *clustered*, *linear*, *dispersed*, or *random*. For example, a population is said to follow a uniform pattern of distribution if it is spread out evenly over an area. Clustered populations are grouped or clumped together around a central point. A linear pattern describes a population that appears to form long and narrow lines, and a population that is spread out is dispersed. Random patterns describe populations that are distributed without apparent order or logic.

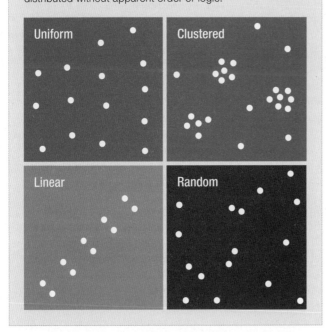

Physical and Environmental Factors

Learning Objective
LO 3.2 Explain how environmental factors influence population distribution.

Humans depend on the environment for their survival, so they are more likely to settle where there are moderate climates, rich soils, and adequate water supplies. Most people tend to live in areas that are not too hot, too dry, too wet, or too cold. Coasts and waterways are often densely populated because they offer economic advantages and convenient transportation routes. Populations are most heavily concentrated at low elevations and diminish rapidly as the elevation increases.

Today, the majority of the world lives at elevations of 500 feet or less, and 80 percent of the world's population lives below elevations of 1,500 feet. Except for a few large cities such as Mexico City at 7,300 feet and Quito, Ecuador, at 9,300 feet, statistics demonstrate that most people do not live at higher elevations. It's not coincidental that the four regions—East

World Population Distribution, 2016

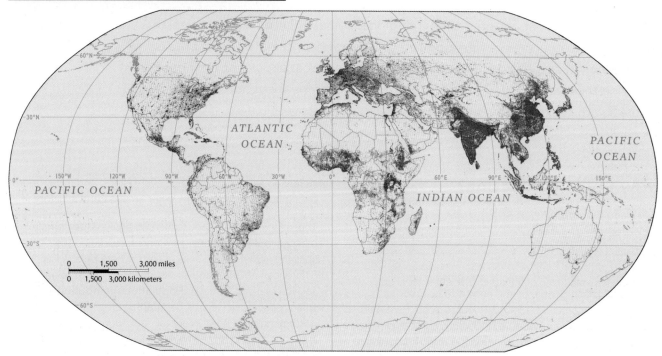

The red areas on this map represent where people live on Earth. The majority of people live in areas with low elevations, temperate climates, and accessible water.

Asia, South Asia, Southeast Asia, and Europe—where two-thirds of the world's inhabitants live are generally low-lying areas with fertile soil and a moderate climate.

Climate **Climate** refers to the long-term patterns of weather in an area that greatly affect population distribution in direct and indirect ways. Unlike weather, which changes from day to day and even from hour to hour, climate represents the overall averages of temperature, precipitation, wind, and other atmospheric conditions year after year. The weather conditions that define a climate are normally established by averaging the weather of a specific location over decades. Climate helps shape the soil, vegetation, and agricultural opportunities of an environment. But it is important to remember that climates may change. Climate change is a change in the typical or average weather of a region or place, like a city. This could be a change in a region's average annual rainfall, for example. Or it could be a change in a city's average temperature for a given month or season. Climate change is also a change in Earth's overall climate. This could be a change in Earth's global average temperature, or it could be a change in Earth's previously typical, or average, precipitation patterns. These changes will influence—possibly in great measure—the distribution of population.

A region's growing season, or period of the year when temperature and rainfall allow plant growth, is determined by the region's climate and is particularly important for the support of human life. Extreme climatic conditions can limit the concentration of population in an area, while areas with **temperate climates**—those with moderate temperatures and adequate precipitation amounts—are usually more densely populated.

As technology continues to develop, humans engineer new ways to live in previously uninhabitable areas. For example, rising sea levels may mobilize communities to find ways to relocate buildings to higher ground or to build seawalls. However, some places with extreme climates don't easily allow for such adjustments, like the arid Sahara in North Africa or the subarctic and polar climates in much of Canada. These areas tend to be less populated because of the extreme impact of the climate on everyday life as well as the climate's effect on the viability of agriculture. Climate change, however, is compelling geographers, scientists, engineers, policymakers, and people everywhere to think about and design ways to cope with, adjust, adapt, and plan for increasingly unusual conditions that may become normal in many places and regions.

Landforms The natural features of Earth's surface, or **landforms**, also influence population distribution. People prefer lowlands due to the ease of building, planting crops, and transporting goods. Flat, low-lying areas like the Ganges Valley in India have deeper soil more suitable for growing crops. Habitable land has adequate water sources, relatively flat terrain, and the potential to produce food.

Areas adjacent to rivers—river valleys, deltas, and plains—typically have rich soils that agricultural communities rely on. These alluvial soils are created when the flow of rivers and streams slows, allowing particles of soil and other matter suspended in the water to settle, particularly in the lower reaches and deltas of rivers. River valleys across the world have long supported dense populations; early civilizations formed along the Nile, Tigris-Euphrates, Indus, Ganges, and

Huang He rivers. Today, alluvial soil still supports agriculture as an essential, profitable sector in Egypt, despite the fact that the desert covers 95 percent of Egypt's total area.

Rocky, steep regions, such as the highlands of Tibet, are typically less populated due to the difficulty of building structures and transportation links such as roads and railways. Mountainous regions are remote and difficult to access and to live in; the air contains less oxygen than at sea level. These same areas present challenges agriculturally because higher elevations result in much colder temperatures. With every 1,000-foot rise in elevation, the average temperature drops by approximately 3°F.

Water Accessibility Ready access to fresh, ample water is essential for human survival and development and thus is a key factor in population distribution. People use surface water supplies like rivers, lakes, and oases and can also access underground water sources called aquifers. Today, people continue to inhabit areas with adequate water supplies. Water is needed not only for agriculture but also for industrial use, sanitation, hygiene, cooking, and, of course, drinking. Areas that are dry or that suffer from regular drought are difficult for people to inhabit; however, technology and innovation have helped some people live in drier areas.

In China, a number of physical and environmental factors, including water accessibility, influence population. Population distribution in China has remained relatively steady over time. The country has an enormous—though unevenly distributed—population within its borders, ranking it just ahead of India in total population with about 1.45 billion in 2022. Within China, much of the population is concentrated in the eastern half of the country along the Huang He, Yangtze, and Zhujiang River valleys.

Historically, the Yangtze River valley has been a strong economic center because of the river and its branches, which allowed for transport and trade. Today, the valley and its waterways continue to support exceptionally high numbers of people. Major manufacturing centers are located along the Yangtze River due to the ease of transport, and this region is also important to China's agricultural production. Large numbers of people make their livings by moving goods and people along the Yangtze's many lakes, canals, and tributaries.

Many other highly populated areas in China benefit from physical factors such as the rich soil deposited from running water that makes up alluvial plains, which support intensive agriculture. This agriculturally rich economy—like the population it sustains—is rooted in the past and has carried its influence forward to the present.

Human Factors

Learning Objective
LO 3.3 Explain how human factors such as economics, politics, culture, and history influence population distribution.

In addition to physical and environmental factors, changing human factors—such as economics, politics, culture, and history—influence the distribution of human populations on many scales. Geographers identify patterns that reflect how each of these factors influences population distribution.

Economic Factors People tend to live where they can earn a living through agriculture, natural resource extraction, manufacturing, sales, and a variety of other activities. The degree of opportunity for economic activity within an area influences population distribution. An increase in technologically advanced and global economic activities has resulted in a significant redistribution of population.

A driving force in population distribution that is often influenced by economics is **human migration**, which occurs when people make a permanent move from one place to another. Though the reasons behind migration are complex and rarely exclusive, people are often influenced to migrate for economic reasons, such as job opportunities, better working conditions, and higher pay.

One historically significant example of migration influenced by economic factors includes the movement of Europeans to North and South America, Africa, Australia, and New Zealand as part of imperialism and colonialism in the late 1800s and early 1900s. Additionally, the forced migration of millions of enslaved people to the Americas to support developing colonial economies had far-reaching effects on population distribution as well.

The presence of natural resources in an area can also attract people to live there to extract, process, or transport the resources. Areas that provide economic opportunity may become densely populated even though the environment presents conditions that are otherwise challenging for human habitation. One example of how economic conditions prevail over environmental limitations is Norilsk, Siberia. Nearly 200,000 people in this city, which is located far north of the Arctic Circle, brave extremely harsh weather conditions every winter. Why do they live there? The region has the world's largest known deposits of nickel, an industrially important metal.

Political Factors People who are dissatisfied with their government or political system may voluntarily migrate to another region within the same country. Unstable political circumstances or war may also compel citizens to leave their home countries or regions. For example, from the 1930s to the 1950s, the Russian government forced thousands of people to move from the western regions of the country east to Siberia to develop the natural resources and economy of that region. Here, political and economic factors combined to influence population.

Similarly, in Cambodia in 1975, leaders of the Khmer Rouge forced more than 2 million residents in the capital city of Phnom Penh to migrate to the rural countryside. Under the four-year rule of the Khmer Rouge, an estimated 1.5 million to 2 million people were executed, starved, or worked to death.

Though the climate of Saudi Arabia is severe—extreme heat in summer and low rainfall—humans have found ways to make agriculture viable. Farms use center-pivot irrigation, a type of irrigation in which rotating sprinklers turn desert into farmland. Farmers must drill deep beneath the sand to unearth reserves of water used to grow vegetables, fruits, and grains to feed the population.

Governments sometimes create policy to redistribute population and encourage migration. The Brazilian government encouraged the development of central Brazil by establishing Brasilia in 1960 as the new capital city and central economic point. In the 1980s, Nigeria built its first planned city. The government chose to move its capital from coastal Lagos to inland Abuja, based on Abuja's central location within Nigeria, its accessibility, and its climate.

Cultural Factors Housing availability, safety, access to transportation, and a feeling of belonging and community are cultural factors that may serve as initial reasons to live in a certain place. Once an area is settled, it tends to stay settled and increase its population as long as the initial reasons for habitation persist.

Cultural factors that influence changes in population include religion, the roles and status of women—including the socialization of men and women within their cultural norms—and familial attitudes regarding marriage and children. For example, some societies have a strong preference for sons, which affects population composition. You will read about several of these cultural factors later in this chapter.

Historical Factors History, the significant events that took place in the past, is often intertwined with other physical, environmental, and human factors. The population distribution of the past can influence the population distribution of today and tomorrow. For example, humans began migrating out of Africa at least 60,000 years ago. By the end of the Pleistocene around 11,700 years ago, climate changes promoted milder conditions in more areas, resulting in major population movements and shaping the settlement patterns of humankind.

Another important migration took place between 300 and 700 C.E. during an era known as the Migration Period. During this migration, people from Central Asia moved into Europe and drove groups from the region now known as Germany into what was formerly the Roman Empire. Factors such as climate change, extreme weather, economic interests, and political conflict all may have influenced this migration. The population distribution that resulted from the Migration Period influenced population concentrations in Europe, northern Africa, and western Asia from that point forward.

The historical duration of settlements in a region and that region's present population concentration are often linked. Many of the densely populated areas of the world have an exceptionally long history of human habitation, while sparse populations in certain areas tend to have a less established history of human habitation. However, the human, physical, and environmental factors that influenced population distribution in an area may change, even suddenly, which will cause a trend in population distribution to also change.

Measuring Population Density

Learning Objectives
LO 3.4 Define *arithmetic density*, *physiological density*, and *agricultural density* as ways to measure population density.
LO 3.5 Compare measures of population density.

By 1800, Earth's population had reached 1 billion. People concentrated in Asia, then Europe, and then Africa. Asia still has the most concentrated population today: 60 percent of the world's 8 billion people live in Asia. Africa is now the second most populous continent, and Europe is the third. The world's four largest population concentrations—East Asia, South Asia, Southeast Asia, and Western Europe—lie on the Eurasian landmass.

Population density, which is the number of people occupying a unit of land, is an important aspect of population. Geographers use three methods to calculate population density: arithmetic density, physiological density, and agricultural density. Each method uses a different land unit to provide key information about the pressure population exerts on the land.

Arithmetic Density
Arithmetic density measures the total number of people per unit area of land. It is calculated by dividing the total population by the total land area. Though this method is used most frequently to calculate population density, arithmetic density is also called crude density because it does not account for land that is difficult to live on or uninhabitable. It is the most general of the three density measures because it provides an average density with no information about distribution patterns, such as how dispersed or clustered a population is on the land.

The island of Taiwan is a good example of how population density can vary at different scales and why geographers calculate population density in different ways. Though Taiwan has one of the highest population densities in the world, about 75 percent of its people live on just one-third of its land area. Roughly 30 percent of Taiwan's entire population lives in the greater urban area of Taipei, its largest city. The remaining two-thirds of the island is covered with high mountains, where just one-fourth of Taiwan's people live in remote rural areas and towns. To understand the interactions between Taiwan's environment and inhabitants, geographers must use several methods to measure the population density.

Particularly interesting to geographers is the direct link between population density and the available resources in an environment. This connection helps explain the degree to which interactions between people and their environment occur. In an area with a high population density, such as the urban region of Taipei, the population may put great strain on the resources of the environment and its capacity to support life there.

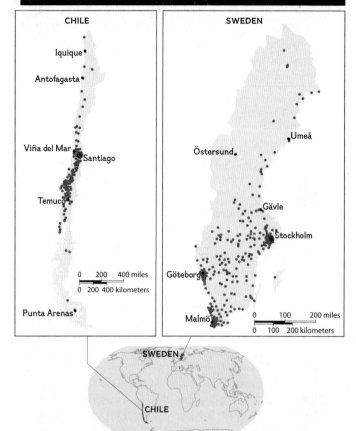

The maps show how the population is dispersed in Chile and Sweden. Both countries have low arithmetic densities, but this method provides only an average density. The maps more accurately reveal the dense settlement patterns that exist in certain areas of each country.

Physiological Density
Physiological density is the total number of people per unit of **arable land**, which is land that can be used to grow crops. Physiological density provides insight into whether people can sustain themselves agriculturally. The higher the physiological density, the greater the pressure on the land and its resources.

One country with greatly varying arithmetic and physiological densities is the United Arab Emirates (UAE). These densities differ due to the country's lack of arable land. The United Arab Emirates is located within an arid region, and nearly the entire country is desert. As of 2020, the total land area of the UAE was about 32,278 square miles—but only 0.5 percent of that land was arable. The UAE's population in 2020 was about 9.7 million. Calculating the country's arithmetic density results in a measurement of only 300 people per square mile. The country's population divided by arable land area, however, measures a much higher density value—nearly 60,103 people per square mile. This example illustrates how the physiological density reveals the pressure a population exerts on the land and, for example, whether a country has to import food.

Such high population density within portions of the UAE is a challenge. For example, 85 percent of the UAE's population lives on the west coast of the Musandam Peninsula, home to

COMPARING MEASURES OF DENSITY (people per sq km in 2018)

	Brazil	India	Japan	Kuwait	Sweden	Uganda	United States
Total Population	208,847,000	1,296,834,000	126,168,000	4,622,000	10,041,000	40,854,000	329,256,000
Arithmetic Density	64	1,022	865	672	58	439	87
Physiological Density	655	1,943	7,585	167,951	889	1,276	522
Percent Arable Land	9.7	52.6	11.5	0.4	6.5	34.4	16.6

The table compares the arithmetic density and physiological density of seven countries that span the globe. ■ How do the two measures of density help you understand the relationship between each country's population and the environment in which the population lives?

Sources: National Geographic Atlas of the World (Eleventh Edition), World Bank

one of the region's critical commercial and financial centers. Though farming in the UAE is extremely limited and can only be supported with irrigation, the country has been able to develop sustainability efforts within heavily crowded urban areas.

Sweden is another example of a country where the physiological density is quite different from its arithmetic density. Approximately 15 percent of Sweden's total land area lies north of the Arctic Circle and is sparsely populated. Only 6.5 percent of Sweden's land is arable, and most of the roughly 10 million people in Sweden farm and live in the southern portion of the country. In this area, the climate is milder and supports a growing season lasting about 240 days. The tundra climate of northern Sweden—located within the Arctic Circle—is a physical factor that influences population distribution in the country. Because Sweden's population distribution is far from uniform and is concentrated in the south where the climate is milder, it is important to calculate its population density in different ways to understand the interactions between its people and its environment. Sweden's arithmetic density is a low 58 people per square mile, while its physiological density is about 15 times higher, at 889 people per square mile.

In both the United Arab Emirates and Sweden, physiological density provides a basic indicator of how much food-producing land is available in each country. However, neither density explains how fertile and productive the farmland really is. Central Illinois and northern Wisconsin have similar physiological densities, but land in central Illinois is extremely fertile and productive, while farmland in northern Wisconsin is of poorer quality. Comparing arithmetic and physiological densities helps geographers understand the capacity of the land and its resources to support the needs of the population living on it, yet as the central Illinois and northern Wisconsin examples show, there are sometimes details that neither measure shows.

Agricultural Density Agricultural density is the measure of the total numbers of farmers per unit of arable land and can be used to further understand the varying types and characteristics of farming from country to country and region to region. The highest agricultural densities are found in parts of Asia. Agricultural density can reveal more about a country's wealth than its population distribution. But even this measurement doesn't supply data and detail about how much technology and money can be put into cultivation. Countries with similar physiological densities but different economic conditions tend to produce vastly different amounts of food or other crops. Not all arable land is equal.

A higher agricultural density, in which there are high numbers of farmers per land unit, suggests that most of the farming taking place is providing crops and livestock for only the farmers' families and close community. This practice, called **subsistence agriculture**, is common in peripheral and semi-peripheral countries where there are fewer mechanical resources available to work the land and more people are needed to care for crops and livestock. This intensive type of farming, which can include modifications such as irrigation that greatly impact the environment, requires specialized agricultural practices to get the greatest yield from small fields. It may even involve the growth and harvest of two or more crops from just one plot of land.

The different economic structures of countries may be reflected in their agricultural density measurements. A lower agricultural density, like that of the United States, is typically a result of high levels of mechanization. A country's access to technology and other resources allows for fewer people to farm extensive land areas, while still feeding many people. Though resources such as farm machinery, high-yield seeds, agricultural chemicals, and geographic information systems can be seen as beneficial in terms of yielding more crops, they can also have negative long-term effects on the environment and its ability to sustainably support populations. Subsistence agriculture and the consequences of agricultural technologies and innovations for the future are discussed in detail in Unit 5.

Agricultural density data help geographers understand how much food is being produced. Comparing physiological density and agricultural density helps geographers analyze the overall impact on the environment and the pressures that may occur in densely populated places where settlements are encroaching on arable land. In some countries, such as Rwanda where there is a large population and limited arable land, the pressures on arable land and food supplies are high.

Geographic Thinking

1. Explain how physical and environmental factors influence population distribution.

2. Identify an example of how human factors influence population distribution.

3. Describe how population distribution differs from population density.

National Geographic Explorer **Lillygol Sedaghat**
Transforming Trash

Sedaghat has documented the ways in which Taiwanese people are working to conquer the output of waste from Taiwan's dense population, including women who collect recyclables via scooter.

Learning Objective
LO 3.1 Identify the factors that influence population distribution at different scales.

Taiwan, a self-governing island off China's coast, was once known as "Garbage Island" but is now a world leader in recycling. Fulbright-National Geographic Storyteller Lillygol Sedaghat's work shows how Taiwan created a sustainable environment.

Sedaghat believes that everyone can learn something from Taiwan's story. Taiwan had a population of approximately 23.6 million in 2020. The terrain is largely mountainous and uninhabitable, so much of the population is concentrated along its western edge, including in Taipei City, the capital. Over the past 60 years, Taiwan transitioned from an agricultural society to an industrial one. Rapid industrialization resulted in increased consumerism and, therefore, more waste. With no formal waste management system in place until the 1980s, the fallout from trash accumulation was staggering. Yet a group of mothers pushed Taiwan to change its ways, designing an integrated waste management system that requires people in major cities today to separate their waste into burnables, recyclables, and compost so that once-used material can be recaptured and repurposed.

In fact, Taiwan is now a world leader in the circular economy (CE). Sedaghat explains the CE model as "a new theory that takes into consideration environmental, health, and social benefits with an economic model that maximizes resources throughout their life cycle." Resources are not wasted, and materials are continuously reused: the goal is zero waste. (You'll read more about sustainable development initiatives like the CE model in Unit 7.)

Sedaghat used visual art and digital media to document recycling methods and processes used in Taiwan. Examples of her work include graphics to explain the plastics recycling process, music videos featuring sounds from Taipei City's waste management system, and blogs describing the sights and smells of Taipei's first incinerator. In May 2019, Sedaghat joined a women-led expedition team to study plastic pollution in the Ganges River as part of National Geographic's *Planet or Plastic*? initiative. The expedition studied the amounts of and ways in which plastics enter the ocean in river systems, from the original source to the sea, through the lenses of people, land, and water.

Geographic Thinking

Explain how physical and human factors may have contributed to Taiwan's waste management success.

3.2 Consequences of Population Distribution

Population distribution and density affect society and the environment. A population depends on, modifies, and adapts to its environment. This relationship shifts as populations grow and concentrate in some places. The impact that populations have on society and the environment may intensify to a point that is no longer sustainable.

Impact of Population Distribution on Society

Learning Objective
LO 3.6 Describe the social impacts of patterns of population distribution and density.

The distribution and density of Earth's population reflects its landforms, soils, vegetation, climate types, available resources, and levels of economic development. People are the driving forces behind factors that impact the environment, such as pollution, resource depletion, and land degradation. Geographers want to know where people live and the consequences of population distribution—such as crowding, isolation, unequal access to services and resources, and environmental impacts—which ultimately affect the quality of human life and human society as a whole.

The more people who are born in or move to an area, the greater the need for access to adequate housing, jobs, and fresh water and services like sanitation and health care. Providing services to clustered populations is easier than providing them to dispersed populations. The distances people are required to travel for work or social services— as well as the available modes of transportation—impact people's ability to take advantage of social services, like health care, and to support themselves.

Social services are more efficient when population distribution is clustered because the operating costs of services such as police, fire, medical, and waste collection are lower when serving a smaller, densely populated geographic area. For example, fire response times are greater for rural areas than urban areas due to a dispersed population as well as fewer personnel in areas where people are spread out. Populations that are dispersed throughout especially hard-to-travel terrain will likely receive fewer social services than populations dispersed on easily navigable terrain that has plenty of travel route options.

However, clustered population distributions with especially high densities necessitate greater levels of public services and resources. In areas where populations are clustered, for example, fires can pose a bigger threat to more inhabitants. This is especially true in crowded areas of decay where neglected buildings are more susceptible to catching fire. The higher the hazard level, the greater the risk, which requires more and specialized resources to mitigate.

Increases in population densities can also lead to disparities in economic growth between the areas where populations are clustered and well connected versus the areas where populations are scattered and poorly linked. As you have learned, these areas, across world regions or even within a country, can be categorized as core and peripheral, and land relationships between these areas are called core–periphery relationships. There may be stark contrasts in wages, opportunities, and access to health care between the core and periphery. Unlike scattered, sparsely populated areas, evenly and densely populated areas usually have more economic development and therefore more power.

Impact of Population Distribution on the Environment

Learning Objectives
LO 3.7 Describe the environmental impacts of population.
LO 3.8 Define carrying capacity.

As a population grows, greater pressure is placed on arable land, water, energy, and natural resources to provide an adequate supply of food. It becomes harder to maintain a balance with the environment. Greater population densities can strain resources, such as clean water, and the possibility of exhausting an environment's resources is a real concern.

The maximum population size an environment can sustain is called its **carrying capacity**. A key concept originating from biology, carrying capacity represents the threshold at which the population of a species levels off at an environmentally determined maximum because of a shortage of resources. Geographers use this term when discussing the limits that human populations face in various regions and at a range of scales. The carrying capacity of Earth, however, has yet to be determined. An area with high population density may not be considered overpopulated if the area has a high carrying capacity, which could be the result of fertile soil or modern farming methods. As a contrasting example, the huge island of Greenland has an exceptionally low carrying capacity because of its harsh, cold climate.

Greater population densities can lead to environmental degradation, which is the deterioration of the environment. When more and more humans inhabit a specific area, they may deplete and pollute the resources there. This has profound ecological consequences.

Urban areas with dense populations increasingly impact the environment through pollution and resource use. Burgeoning cities consume more resources and create pollution from additional vehicles, heating and cooling sources, industrialization, and the construction of buildings and infrastructure. High levels of consumption in densely populated areas also have environmental impacts through the use of large amounts of energy and the generation of excess amounts of waste. Some city officials and researchers are developing ways to use resources more efficiently and lessen the environmental impact of densely populated urban areas.

By analyzing the population density and carrying capacity of an area, geographers can better understand how a changing population interacts with its environment. Resources such as air quality, food supply, biodiversity, and safe water can be monitored to find ways to extend an environment's carrying capacity or prevent a population from exceeding the environmental limit.

Geographic Thinking

1. Describe how population density can negatively and positively influence society.

2. Identify which of the three methods for calculating population density would be most helpful in assessing a country's carrying capacity.

High population density and its impact on the environment is evident in the bustling, smog-filled Port of Shanghai, located on the east coast of China at the mouth of the Yangtze River. It is the world's busiest container port. In 2020, Shanghai handled a record 739.5 million tons of cargo.

Case Study

Population Distribution at the Country Scale

The Issue Population density data can be misleading in providing an accurate description of the population distribution of an area.

Learning Objective
LO 3.5 Compare measures of population density.

By the Numbers

Egypt

99,413,000
Population as of 2020

257 PEOPLE PER SQ MI
Arithmetic density

9,182 PEOPLE PER SQ MI
Physiological density

Canada

35,882,000
Population as of 2020

9 PEOPLE PER SQ MI
Arithmetic density

194 PEOPLE PER SQ MI
Physiological density

Sources: National Geographic Atlas of the World (Eleventh Edition), World Bank

When Comparing Population Density Data for different geographic areas, keep in mind that arithmetic density data are most useful for small areas, such as neighborhoods. However, for larger areas, such as countries, arithmetic density data are less likely to provide meaningful measurements because they don't take into account patterns of population distribution across geographies of similar scale. Take the example of comparing the countries of Egypt and Canada.

Egypt has a low arithmetic density—only 257 people per square mile in 2020—but changing the method to calculating physiological density reveals a much higher number: 9,182 people per square mile. This value reflects the high concentration of people along the fertile Nile River Valley, which is approximately 5 percent of Egypt's land area. Almost all of the country is desert, and 95 percent of the population lives along the Nile River and its delta on Egypt's only arable land. People are further concentrated in the capital city of Cairo, which is located close to the Nile Delta. It is an expansive urban environment disproportionately larger than any other Egyptian city. By comparison, Alexandria, the second largest city in Egypt, is only 30 percent the size of Cairo. Egypt's high agricultural density, which exceeds 695 farmers per square mile, shows that Egypt requires many rural laborers to work what little arable land it has. This data also suggests that mechanical farm equipment and the type of farming practiced is less developed. Egypt's high physiological density indicates that the country is under pressure to provide people with a sufficient food supply and that importing food is a necessity.

Although Canada doesn't have people concentrated along a single river, or even in one comparatively large city, its population is concentrated along its southern border just as the Egyptian population is concentrated along the Nile River. One of the most sparsely populated countries in the world, Canada had an arithmetic density of 9 people per square mile in 2020. Much of its enormous territory is largely inhospitable and uninhabited. The arithmetic density, in this case, masks the vast, sparsely inhabited area. Canada's physiological density is 194 people per square mile, which indicates the small percentage of arable land. Ninety percent of the population lives within 100 miles of Canada's border with the United States. Environmentally, the conditions there are milder, which results in a temperate climate and longer growing seasons. The soils in these southern regions are generally fertile. People concentrated in southeastern Canada have access to the Great Lakes and the Saint Lawrence Seaway and benefit from many connections with the U.S. economy.

Both Egypt and Canada contain large, basically uninhabited areas, with most of their populations clustered in a small percentage of land area. However, sheer size also matters: Canada has nearly 10 times as much land area as Egypt does, but its total population is just over one-third of Egypt's. Canada's arithmetic density is among the lowest in the world, and its physiological density is low as well—in fact, lower than Egypt's arithmetic density. These measurements show both Canada's vast size and relatively small population. Indeed, Canada is about four times as wealthy as Egypt, putting it in the global core while Egypt is in the periphery. Population density numbers tell only part of a country's story. ∎

Geographic Thinking

Describe how the population density maps provide a truer picture of each country's population density than arithmetic density data may provide.

Both Egypt and Canada have population densities concentrated in a small portion of each country. In Egypt, millions of people live in greater Cairo, surrounded by a desert that includes the Giza pyramids.

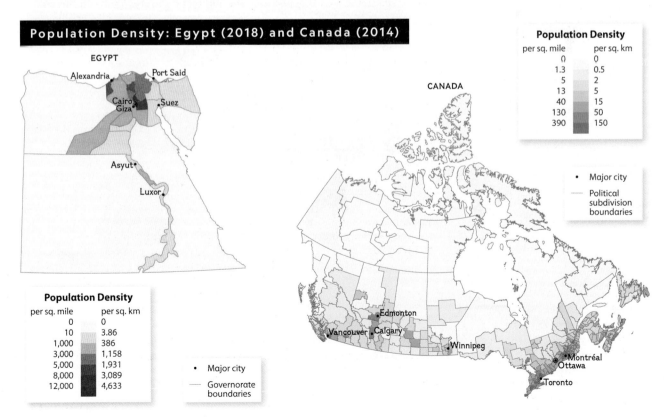

Patterns of Population 77

3.3 Population Composition

The structure of a population is its makeup, or groups that compose the whole, such as a population's number of men, women, and different age groups. Geographers study population structures, statistics, and trends and represent this data using diagrams, graphs, and maps. By analyzing the data, they can see a past, present, and future demographic story about a specific place. They use their analyses to evaluate similarities and differences among and within countries—and at many scales from neighborhoods to world regions—to identify cultural, political, and economic patterns across the globe.

Dependency Ratio

Learning Objective
LO 3.9 Explain how dependency ratios impact government policy and economic development at different scales.

The age structure of a population has significant government policy and economic implications. For example, a population with many young people needs a sufficient number of schools and eventually jobs to accommodate them. Conversely, countries with a large proportion of older people must develop retirement systems and medical facilities to serve them. Geographers analyze these and other age-related demographic studies using ratios for specific age groups. The **dependency ratio** is the number of people in a dependent age group (under age 15 or age 65 and older) divided by the number of people in the working-age group (age 15 to 64), multiplied by 100.

Consider a population of 100 people that has 60 people in the working-age range. This population also has 30 children and 10 people age 65 and older. In this example, the dependency ratio would be 66.7. This ratio indicates that the population in the dependent age groups is two-thirds of the population in the working-age range. The dependency ratio is used as an indicator to measure the demand placed on the working-age population to provide for the dependent population.

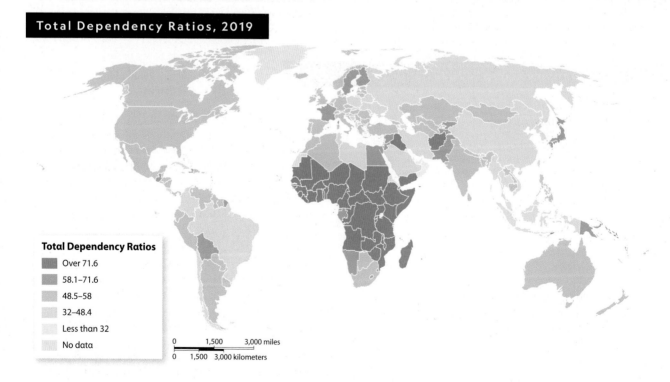

Total Dependency Ratios, 2019

Total Dependency Ratios
- Over 71.6
- 58.1–71.6
- 48.5–58
- 32–48.4
- Less than 32
- No data

Reading Maps This map shows each country's dependency ratio, which is the combined youth and old-age populations relative to the working-age populations. Countries with higher dependency ratios have relatively fewer people who are of working age. ▎Which regions have the highest dependency ratios? Identify some of the economic challenges these regions might face.

Related measurements include youth dependency ratios, which compare the number of children under age 15 with the working-age population. Old-age dependency ratios are comparisons between the number of people 65 and older with those in the working-age group. However, it is important to note the assumptions when calculating dependency ratios. In reality, the workforce includes people older than 64 and younger than 15. It is also incorrect to assume that everyone age 15 to 64 is working.

Dependency Ratios and Economic Development The size of the dependent population relative to the size of the working-age population is a key factor influencing economic development. Dependency ratios are useful when comparing societies over time and when comparing different areas. They are also used at all levels (national, regional, local) to indicate current and future productivity at a large-scale level. Monitoring these ratios provides insight into the policies that countries may need as their population ages and changes.

The United Nations releases a dependency ratio for every country in the world. Each ratio covers a five-year period beginning in 1950. In 2020, the United States had a total dependency ratio of 53, which was the result of a youth-dependency ratio of 28 and an old-age dependency ratio of 25.5. That old-age dependency ratio was almost double the 1950 ratio of 12.6 because so many Baby Boomers—people born between 1946 and 1964—had reached retirement age.

A higher dependency ratio indicates fewer people of working age and fewer people in the workforce earning income and paying taxes. This means that in a country with a high dependency ratio, people of working age—and the overall economy—encounter greater pressures in supporting the aging and youthful population groups. As the percentage of nonworking people rises, people who are working may face increased taxes to compensate for a larger dependent population. Ideally, an economy should have a smaller dependency ratio, with a constant flow of people entering the workforce and a smaller number exiting the working-age population each year.

Data beneath a Dependency Ratio The global population is projected to increase to 9.7 billion by 2050, but it is also projected to stop growing and peak at about 10.9 billion by 2100. In most of the world's countries, fewer babies are being born and the number of older adults continues to rise. In fact, the fastest-growing segment of the world population is people age 65 and older. Dependency ratios are on the rise, and countries with aging populations will continue to struggle with how to support their elderly. Although this scenario applies to most countries, others—especially countries in Sub-Saharan Africa—continue to see rapid population growth, with one of the largest segments being the youngest. In this case, countries with growing younger segments could at some point have the same dependency ratio as countries with growing older segments.

Japan and Namibia are examples of countries with similar dependency ratios. In 2020, Japan's ratio was approximately 69, and Namibia's was about 68. Though the difference between these ratios isn't significant, the data beneath the ratios tells a different story. Japan has many elderly dependents, whereas Namibia has many youth dependents. These differences present distinct challenges. Japan's youth dependency ratio was 21, and its old-age dependency ratio was 48. Namibia's youth dependency ratio was nearly 62, and its old-age dependency ratio was just over 6. This example shows that geographers must study dependency ratios in various ways and at varying scales to achieve an accurate demographic picture. Using only these overall numbers to compare the two countries would prevent an accurate analysis of the issues masked by the apparently similar dependency ratios.

Examining age dependency at different scales within the United States is important as well. As you have read, in 2020, the country had a total dependency ratio of 53, with an old-age dependency ratio of 25.5. Looking beyond the national rate to the dependency ratio of specific states shows patterns of age structure in certain areas. For instance, at the time of the 2020 census, Florida had an old-age dependency ratio of 35.8 for 2020. Florida has the highest percentage of older adults in the United States. As major retiree destinations, both Arizona and Florida are home to large populations who live on fixed incomes and have left their peak earning and taxpaying years behind. This age group will draw heavily on state health care and benefits and will spend less money, which will affect the state budgets and economies.

Sex Ratio

Learning Objectives
LO 3.10 Identify factors that may contribute to an imbalanced sex ratio in a human population.
LO 3.11 Discuss the consequences of an imbalanced sex ratio in a human population.

Geographers also study a population's **sex ratio** when evaluating population composition. This ratio represents the proportion of males to females in a population. In general, slightly more males than females are born. However, women tend to live five or more years longer than men do. Overall, the world has a ratio of men to women of 101:100. In Europe and North America, the ratio of men to women is 95:100. In the rest of the world, the ratio is 102:100. In peripheral countries, the number of women who die during childbirth contributes to a lower percentage of women. Societies with a high rate of emigration where men rather than women are more likely to migrate elsewhere will have more females than males. War can also create disparities because typically more men than women die in war.

Many of Asia's populations have long had a strong preference for sons. India's cultural preference for male children has led to many female children being neglected or abandoned, resulting in a high number of female deaths. This inclination can also be seen in India's sex ratio for its overall population, which was just over 108 males per 100 females in 2020. With

World Sex Ratios, 2022

Males per 100 Females
- More than 110
- 105–110
- 100–104.9
- 90–99.9
- Less than 90
- No data

Reading Maps Some countries in Southwest Asia, such as Saudi Arabia and the United Arab Emirates, have a highly unbalanced sex ratio due to a large male migrant population—men who come to work in mostly low-wage jobs. Consider Saudi Arabia, where there are 137 males for every 100 females. In 2022, fewer than half of foreigners in Saudi Arabia were female. ▌How might the high male population in Saudi Arabia impact the country's marriage rate, fertility rate, and the rate at which the population grows?

India's huge population, this means in that year there were just over 717 million males and not quite 663 million females, a difference of more than 54 million. Nearly as out of balance in 2020 was China, with a sex ratio for its population of 105.3 males per 100 females, representing 738 million males and 701 million females—a gap of 37 million in its world-leading total population of just under 1.4 billion in that year.

Many aspects of society, including political factors, have led to such an imbalance. For example, China's one-child policy that launched in 1980 and remained in effect for more than 35 years has had lasting effects on the country's population structure. Under this policy, Chinese couples were allowed to have only one child, so they often sex-selected their first pregnancy. The World Health Organization defines sex-selection as "the practice of using medical techniques to choose the sex of offspring," which can include choosing specific embryos to be implanted or terminating a pregnancy after the sex of the fetus has been identified.

In some rural areas and for most ethnic minorities, couples were allowed a second pregnancy if the first child was a female. If the second (or subsequent) pregnancies were female, the result was often sex-selective pregnancy termination, infanticide, or abandonment. In many cases, male births were reported, and female births were not. Even though the one-child policy was modified in 2015 to allow for two children—and then again in 2021 to allow for three children—its aftereffects continue to produce measurable impacts now and well into the future as China faces a rapidly aging population and shrinking workforce. Though its sex ratio at birth has fallen since 2010 when it was 117 boys for every 100 girls, China in 2020 still had a dramatically imbalanced sex ratio of about 113 boys for every 100 girls.

In both China and India, marriage and family determine social status and acceptance. The imbalanced sex ratio in these countries means greater competition for a bride and inevitably more single men who aren't fulfilling traditional social and economic roles. Additionally, when there is a shortage of women, the men who remain unmarried are generally those in the lower socioeconomic classes of society, who then become marginalized. In China, crime rates have exploded, with two-thirds of violent and property-related crimes in the country being committed by young unmarried men.

The sex ratio is an important social indicator as it affects marriage rates, women's labor-market participation rates, and the socialization of men and women within their cultural norms. It also impacts fertility rates, birth rates, and the natural rate at which a population grows.

Geographic Thinking

1. Describe the possible limitations of dependency ratios.
2. Identify three factors that may cause an imbalance in a country's sex ratio.
3. Explain why the sex ratio of a population is an important social indicator.

3.4 Measuring Growth and Decline

The number of births, deaths, and people moving in and out of a location all contribute to population growth and decline. But it is difficult to pinpoint and explain the reasons for these changes. By measuring specific population characteristics—including fertility and mortality—geographers can identify trends that help explain how environmental, economic, cultural, and political factors play a role in the patterns of population.

Fertility

Learning Objectives
LO 3.12 Explain how fertility measurements can help identify population trends.
LO 3.13 Define *total fertility rate* (TFR) and *crude birth rate* (CBR).

Demographics are data about the structures and characteristics of human populations. Though population groups can be identified and measured in different ways, geographers measure some key factors, such as fertility, to identify population trends. **Fertility**, or the ability to produce children, influences the birth rate of a population. Experts analyze the growth or decline of populations at various scales because the data impacts the natural environment and society, in the form of human health and the economy. Governments and public health officials study fertility rates to see if the population will grow, decline, or stay the same.

As world population rises, pressure increases on the agricultural and housing sectors to provide food and shelter for the added millions. Experts in public health use the data to predict and plan for health care. Fertility also affects budgets and policies surrounding education. International organizations, such as the United Nations, measure fertility to help solve international problems and make decisions about where to focus global resources.

Measures of Fertility The **crude birth rate** (CBR) is the number of births in a given year per 1,000 people in a given population. In order to compare CBR data among countries or specific regions, a standardized measurement for CBR is used, providing the rate of births in a population of 1,000. If a country or region has a population of 1 million, and 15,000 babies were born in that country last year, both numbers are divided by 1,000 to obtain the rate of 15 per

Total Fertility Rates, 2019

Number of children born per woman
- 6 or more
- 4–5.9
- 2–3.9
- 0–1.9
- No data

Reading Maps Over the past 50 years, the global total fertility rate (TFR)—the average number of children a woman will give birth to—has been cut in half due to factors including the rising cost of raising children and increased education and employment opportunities for women. ▌Choose a country on the map and explain how its TFR provides insight into the status of women in that country.

Patterns of Population 81

1,000. This means that 15 babies are born every year for every 1,000 people in that country or region. The global CBR in 2022 was estimated at 17.7 births per 1,000, with core regions having an average CBR of about 11 and peripheral regions having a much higher average CBR of 21.

Generally, the lowest CBR is in Europe and the highest is in Africa. If only 500 of 1,000 people are women and only a portion of the 500 are in their childbearing years, then only a certain percentage will give birth in a given year. Similarly, a country may be experiencing significant immigration of men of working age, which may cause a low CBR because the total population contains an unusually high number of men. The CBR does not account for these variables, but it is useful for geographers to calculate the rate of natural increase and population growth or decline in a country or at another scale.

A potentially more informative measure is the **total fertility rate** (TFR), which is the average number of children one woman in a given country or region will have during her childbearing years, which are calculated as between the ages of 15 and 49, according to the Population Reference Bureau. This rate relates births to an age and gender-specific group. It helps eliminate distortions that may arise because of different age and sex distributions among populations; therefore, the TFR provides a better comparison of fertility levels across countries and populations than CBR measurements. The TFR is also viewed as a more meaningful statistic than CBR because it helps reveal where the population is growing or shrinking.

In 2022, the global average fertility rate was just below 2.5 children per woman. But in semi-peripheral and peripheral countries, the rate can be 5.0 or higher, and in core countries, it is often 2.0 or less. A country's TFR can be impacted by government policy, war or conflict, increased urbanization, economic stability or instability, higher levels of education among women, higher rates of women participating in the labor force, and the influence of culture and society. The TFR often reflects the quality of a country's health-care system and a population's access to doctors, nurses, hospitals, and medicine.

A population's replacement level is the number of children per woman necessary to keep a country's population constant. A level of 2.1 is the point at which a population is neither growing nor declining but remains stable (with the replacement-level TFR being slightly above 2 to counteract infant mortality). If the total fertility rate is higher than 2.1, the population long term will grow. If the ratio is below 2.1, the population long term will decrease.

The lowest TFR is in Europe (1.6 in 2020), and the highest is in Africa (4.7 the same year). Fertility in all European countries is now below the level necessary to keep the population from declining in the long run. Better education and increased job opportunities for women have contributed to the low fertility rates in most European countries. Bulgaria, with a 2020 TFR of 1.6, is representative of the situation overall in Europe. Its population has declined by 11.5 percent in the

FERTILITY AND MORTALITY RATES (2020–2025)

World Region	Crude Birth Rate (births per 1,000 people)	Total Fertility Rate (children born per woman)	Crude Death Rate (deaths per 1,000 people)
Africa	31.6	4.16	7.6
Asia	15.2	2.09	7.2
Europe	9.8	1.62	11.3
Latin America and the Caribbean	15.3	1.96	6.6
Northern America (United States and Canada)	11.8	1.76	9.0
Oceania	15.8	2.30	6.8

Comparing fertility and mortality data provides valuable information about the demographics of a region. ▮What factors might contribute to the high CBR and TFR in Africa? What factors might contribute to the low CDR in Asia?

Source: United Nations

decade between national censuses in 2011 and 2021—from 7.3 million to under 6.5 million, a drop of 844,000 Bulgarians. Experts attribute Bulgaria's decline to a low CBR, a high death rate, and a steady migration of working-age people leaving to search for better work and education prospects in other European countries and beyond.

The shift from agriculture to manufacturing has greatly increased opportunities for women in the labor force in Europe and other industrialized countries around the world. In addition, the desire to have fewer children, partly due to a lack of affordable childcare, has been a growing trend in Europe and elsewhere. Although the United States is a core country and China is a semi-peripheral one, they had similar low TFRs in 2019 (1.71 and 1.69, respectively). China's recent TFR represents a significant change from its high TFR of the late 1960s, which was more than 6.0—an average of six children per woman.

Mortality

Learning Objectives

LO 3.14 Explain how mortality measurements can help identify population trends.

LO 3.15 Define crude *death rate* (CDR), *infant mortality rate* (IMR), *and life expectancy*.

Another key demographic factor affecting the growth of population is **mortality**, or deaths as a component of population change. Circumstances that affect the mortality rate include the availability and affordability of health care as well as clean water, adequate food, and shelter. Geographers and other researchers across many disciplines, as well as public health and governmental officials, measure and study mortality to explore solutions and policies that can improve the health and well-being among all populations, including children and older adults. Policymakers analyze and use data to help them promote equity and health-care access. The United Nations measures mortality in its efforts to protect human rights and to reduce hunger and deaths from disease. The UN also assists with economic development and provides disaster relief, both of which impact mortality rates.

Measures of Mortality The **crude death rate** (CDR) is the number of deaths of a given population per year per 1,000 people. Geographers consider CDRs of below 10 per 1,000 people as low and CDRs of higher than 20 per 1,000 as high. Core countries tend to have a higher CDR than peripheral countries do, but generalizations of—or conclusions based solely on—CDRs can be misleading. A high CDR may result from a multitude of conditions (unclean water, poor diet, or poor-quality housing) that lead to poor health. However, it can also result from a high number of natural deaths occurring within the oldest portion of the population, when that segment represents a large portion of the population.

For example, in 2019, Mexico had a CDR of 6, while the United States had a CDR of 8.7; this was largely because the United States had a much higher percentage of older people. The U.S. population in the age range of 65 and older is projected to almost double from 51 million in 2017 to 95 million by 2060. The CDR remains lowest in regions and countries with a high percentage of younger people. For example, 16 percent of Asia's population is 15 to 24 years old and 17 percent of Latin America and the Caribbean is 15 to 24 years old. Both regions have relatively low CDRs: Asia has a CDR of 7, and Latin America has a CDR of 6.

The global crude death rate is currently 7.5 and has been on the decline since 1960. Significant declines around the world—more dramatically in peripheral and semi-peripheral economies—are due to longer lifespans resulting from better food supply and distribution, better nutrition, higher quality and more widely available medical care, advancing medical technology, improvements in sanitation, and expanded clean water supplies. The CDR can spike temporarily due to events such as pandemics and natural disasters or can increase for longer periods of time due to the spread of diseases such as HIV/AIDS. Death rates are affected by other factors such as war, poverty, and education.

The **infant mortality rate** (IMR) is the number of deaths of children under the age of 1 per 1,000 live births. The IMR is often considered a better indicator than the CDR in terms of health and health care in a country. Infant mortality is high when medicine and maternal care are limited and also when access to clean water, sanitation, and food is inadequate. Currently, the highest IMRs are found in the peripheral countries of Africa and Asia. For example, in 2021, Afghanistan had an IMR of 110.6, and Somalia had an IMR of 94.8. In contrast, core countries have much lower IMRs. In 2021, both Finland and Norway had IMRs of 2.5, while the U.S. IMR was 5.8.

Life expectancy is the average number of years a person is expected to live. In core countries, the average life expectancy is more than 80 years; in peripheral countries, it's around 50 years. Life expectancy data summarizes the mortality pattern across all age groups in a given year. Although not a classic mortality measurement, life expectancy analyzed with IMRs can provide useful information regarding the health of the country. The highest life expectancies are generally in the wealthiest countries and lowest in the poorest, though there are exceptions.

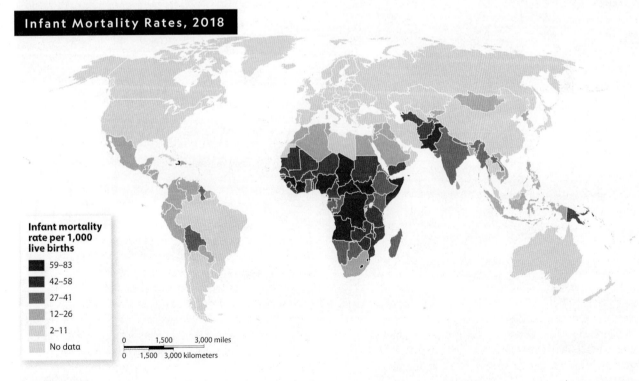

Infant Mortality Rates, 2018

Reading Maps This map reveals that the highest infant mortality rates in 2018 were found in peripheral countries. Notice that most of these countries are in Africa and Asia. ▍Compare this map to the earlier "Total Fertility Rates, 2019" map. Describe how the maps help you understand the long-term population growth of countries in different world regions.

Patterns of Population

In addition to access to health care, clean water, and adequate food and shelter, other factors also play a role. High rates of alcoholism in Russian men or higher rates of obesity across both genders in the United States are factors that decrease life expectancy in those countries. Climate, hygiene, crime rates, and genetics also affect life expectancy. In 2020, Japan had a life expectancy of 85.5 years, which can be attributed to accessible health care and lifestyle choices such as diet and exercise. Additionally, the Japanese government has invested heavily in public health for decades, which has resulted in a health- and hygiene-conscious culture.

In general, women live longer than men (see Japan's population pyramid on page 85), but the average in high-income areas is six years longer, while the average in low-income areas falls to three years. Much of the difference can be attributed to the investment high-income countries have made in combating disease such as heart disease and cancer. In 2018, women in the United States had a life expectancy of 82.3 years, while men had a life expectancy of 77.8. By contrast, in some African countries, the average life expectancy for women was 64, while the average life expectancy for men was 61. The smaller gap between the genders is largely the result of AIDS and poverty, which lower the life expectancy of both groups.

In regions where the life expectancies of men and women are close, the root cause is typically a systemic issue such as poverty or inadequate medical care—or sometimes a combination of both. India had life expectancies of 69.1 for men and 71.8 for women in 2020. This small gap — smaller than the global average — can largely be attributed to how gender inequality impacts the health and well-being of women. Many Indian girls and women are kept from receiving an education, are considered a drain on family finances, and are frequent victims of violent crimes.

By comparing CBRs, CDRs, and TFRs, geographers have concluded that the world has a large population of youth—the largest percentage of which is in countries in the periphery. This circumstance pressures education and health-care systems, as well as employment opportunities, within peripheral countries. People in core countries are growing older, and these populations add stresses to health care, retirement pensions, and social protections. The worldwide population in the oldest age group, age 60 and over, is growing faster than all younger age groups. Globally, the population is aging.

Bearing in mind that scale can be an issue, geographers study fertility and mortality rates across global, national, and local scales. Regional averages can disguise significant variation among countries, and country-level measurements can mask dramatic local variations, such as urban versus rural, male versus female, and wealthy versus poor. Ethnic and cultural variations that exist within some countries, such as the vast differences among states in the United States and India, must also be considered.

The exceptionally long life expectancy in Japan can be attributed, in part, to the government's emphasis on health. For example, the town of Kamikatsu keeps the elderly population active and employed in local gardens.

Visualizing Population Composition with Population Pyramids

Learning Objective
LO 3.16 Explain how population pyramids display and analyze population.

Geographers use **population pyramids** to interpret the implications of the changing structure of a population. These charts show the age–sex distribution of a given population (by groups of five-year increments called cohorts) which helps indicate whether the population is growing rapidly, growing slowly, or in decline. The population size of each age group affects the demand for goods and services in the region's economy. Projections and trends revealed in population pyramids inform policymakers about what goods and services are needed, such as schools, elder care, and health care, and what to plan for in the future.

Some countries have reached zero population growth or are experiencing negative growth, or decline, because of low birth rates and an old-age structure coupled with minimal net migration. Population growth rates are negative in many European countries—most notably in Russia (which lies in both Europe and Asia)—where the population growth rate was estimated to be –0.07 percent in 2022. Russia's trends of population growth and aging have been profoundly affected by high rates of alcoholism and suicide among its men as well as catastrophic events in the 20th century. The typical age distribution and male and female balance in the population became distorted as a result. The millions of losses incurred from World War II have caused Russia to have the lowest overall sex ratio in the world, at about 86 males per 100 females in 2018. Focusing on Russia's population age 65 and over shows an even lower male–female ratio of 46:100.

In 1992, Russia entered a period of decline, with the number of deaths exceeding the number of births combined with the number of immigrants. In 1999, the TFR dropped below 1.2 after the economic upheaval that followed the breakup of the USSR. Drops like these will continue to impact the number of births and the rate of population growth and aging for decades to come. In turn, this downturn also affects retirement, education, and employment. Russia's low TFR of 1.5 in 2020, combined with only 17 percent of its population being below age 15, has resulted in severe labor force gaps. If the growth rate continues to decline in Russia, the population size will slowly decrease as well.

Countries with rapid population growth, such as the Democratic Republic of Congo, have a large-based and narrow-topped population pyramid, indicating a high child dependency ratio. This shape can also signify a peripheral country as it results from high birth rates that supply increasingly more people into the lowest five-year cohorts of the pyramid and in turn shrink the relative number of people in the oldest stages. As the death rate declines, more people survive to reproductive ages and beyond. As these people reproduce, the pyramid base widens. Peripheral and semi-peripheral countries that have experienced comprovements in life expectancy but continue to have high birth rates tend

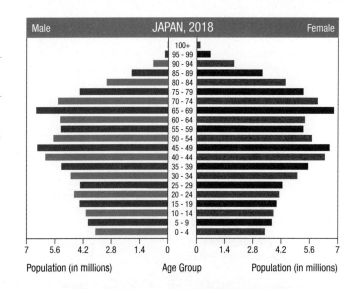

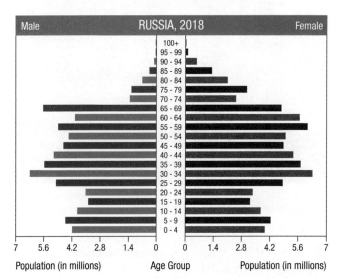

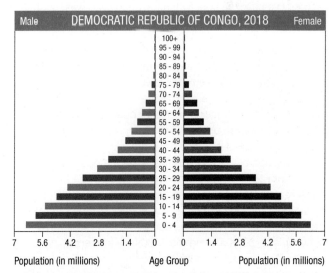

Each population pyramid reveals key factors about a country's population. The Democratic Republic of the Congo's wide-based pyramid shows rapid growth, with many children and fewer working-age and older adults. Japan's pyramid, with a narrowing base, shows slow or negative growth and an aging population. ■ What does each pyramid suggest about what the population of each country will look like in the future?

Source: CIA World Factbook

A group of young boys train at a fútbol school in Goma in the Democratic Republic of Congo (DRC). The DRC has one of the highest child dependency rates in the world. In 2020, 46 percent of the population was under the age of 15.

to have this shape. The pattern reflects a history of rapid population growth and the potential for future rapid growth. The narrow top is the result of a smaller number of births for many years in the past (when the population was much smaller) and higher death rates at older ages.

A slowly growing population, such as the United States, has had declining fertility and mortality rates for most of the past 100 years. Because of the lower fertility rates, fewer people have entered the base of the U.S. pyramid. As life expectancy has increased, a greater percentage of the births have survived until old age. The proportion of older people in the population has been growing. This trend of the population aging was interrupted by the postwar Baby Boom from 1946 to 1964 when birth rates climbed again. This break in the trend can be seen in the population pyramid for the United States later in this chapter. After the Baby Boom factor, birth rates continued to trend downward until the late 1970s. During the 1980s, as the remaining members of the Baby Boom approached their childbearing years, the number of births rose again and peaked in 1990. Notice the bulge in the 15-to-34-year-old population in the pyramid; this shows the Baby Boomers' children.

Despite the number of births per woman being at its lowest, the U.S. population continues to grow, if slowly, because of the children and grandchildren of the enormous Baby Boom generation. Immigration also plays a part in population growth in the United States; more people enter the country than leave, and many of them add children to their families. Recent data, for the July 2020–July 2021 period, show that the U.S. grew just 0.1 percent, adding a mere 393,000 people in a year (births over deaths plus net migration). Deaths from the COVID-19 pandemic clearly influenced this, as did slowly increasing mortality rates and curtailed immigration, but experts contend that the leading cause for this lowest percentage population growth in U.S. history was a decline in fertility rates—with a TFR of just 1.64 in this period.

Geographic Thinking

1. Explain what factors may cause TFRs to be lower in core countries.

2. Describe how China's population will change in the long term, according to its 2019 TFR of 1.69.

3. Identify what technological, economic, and social factors might cause levels of mortality to change.

4. Choose a population pyramid and explain Russia's, Japan's, or the Democratic Republic of Congo's current needs for goods and services and how these needs may change in the future.

Examining Population Pyramids at Different Scales

Learning Objective
LO 3.16 Explain how population pyramids display and analyze population.

When you analyze population pyramids, you have to consider scale. The national median age in the United States in 2018 (38.2) disguised the tremendous variations that exist at state and local levels. To understand the connection between population characteristics and the demand placed on any given region, the lens through which data is examined must adjust to national, state, and local levels. For example, Virginia had a median age of 38 in 2018—close to the national median age. If geographers dig deeper into specific cities and counties, however, they uncover drastic variances and trends that differ from this average. Virginia's Lexington City and Highland County are within 80 miles of each other, yet their population compositions in 2017 were radically different.

That year, Lexington had a population of 7,113 with a median age of 21.6. It is the home of Washington and Lee University and Virginia Military Institute. Over 75 percent of students enrolled at Virginia Military Institute are male. As evidenced by the population pyramid, the city attracts high numbers of college-age men and women. Almost 20 percent of the city's total population is 20- to 24-year-old males, and the overall sex ratio is 110:100. The pyramid also shows that these same college students graduate and move elsewhere rather than aging up into Lexington's population. The trend demonstrates that this cycle repeats, with more college students replacing those who left. The population percent change for Lexington is 1.09 percent growth, and a high percentage of the population, 69 percent, has never married.

By contrast, Highland County, with a population of 2,213, is one of the least-populated counties in Virginia and has a median age of 60.5. The population pyramid for Highland County shows a much larger percentage of aging population. This pyramid suggests a trend in the migration of retirees to the area. As you have read, people are less likely to live at high elevations and—at 3,000 feet—Highland County has one of the highest average elevations east of the Mississippi River. The Interstate Highway System doesn't pass through the county, and there are no colleges there. The population percent change is a 4.7 percent decline, and a high percentage of the population, 63 percent, is married. ∎

Geographic Thinking

Explain how the economic and health services needs of the two Virginia populations differ.

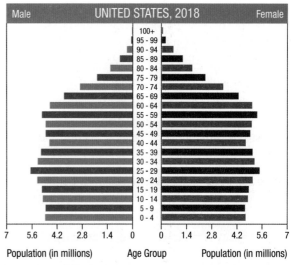
Source: CIA World Factbook

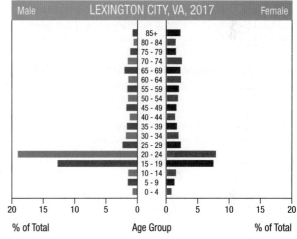
Source: U.S. Census Bureau

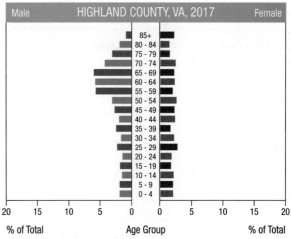
Source: U.S. Census Bureau

Chapter 3 Summary & Review

Chapter Summary

Geographers study the impact that physical, environmental, and human factors have on population distribution and density to analyze population patterns.

- Physical and environmental factors that influence population distribution include climate, landforms, and water accessibility.
- Human factors that influence population distribution include economics, politics, culture, and history.

Geographers use three methods for calculating population density to uncover different information about the pressure the population exerts on the land.

- Arithmetic density is the total number of people per unit area of land.
- Physiological density is the total number of people per unit of arable land.
- Agricultural density is the total number of farmers per unit of arable land.

Geographers study the impacts of population distribution and density on society and the environment.

- Societal challenges include pressures on the government to grow the economy, provide more services, and create more jobs. Societal benefits include localized and more efficient services in densely populated areas.
- Environmental challenges include carrying capacity and the depletion of resources. Benefits include efforts to expand arable land and to support ecosystems.

Geographers analyze composition data in various ways to evaluate present and future population needs.

- The dependency ratio is the number of people in a dependent age group (under age 15 or age 65 and older) divided by the number of people in the working-age group (15 to 64), multiplied by 100.
- The sex ratio represents the proportion of males to females in a population.

Geographers analyze data about fertility, mortality, and age–sex distribution to understand population trends.

- Fertility, the ability to produce children, is measured using the crude birth rate (CBR) and the total fertility rate (TFR).
- Mortality, deaths as a component of population change, is measured using the crude death rate (CDR) and the infant mortality rate (IMR).
- Life expectancy is another measure used to compare countries.
- A population pyramid is a graph that shows the age–sex distribution of a given population to indicate whether that population is growing rapidly, growing slowly, or in decline and to understand what this might mean for that population.

Review Questions

Use complete sentences to answer the questions.

1. **Apply Conceptual Vocabulary** Consider the terms *distribution* and *density*. Write a standard dictionary definition of each term. Then provide a conceptual definition—an explanation of how each term is used in the context of this chapter as it relates to population.

2. How do physical, environmental, and human factors affect population distribution? Provide at least one example of each factor.

3. Provide an example to support the theory that past population distributions influence present and future population concentrations.

4. Explain why geographers compare physiological density and agricultural density.

5. Why is arithmetic density also called crude density?

6. What are the characteristics of an area with a low agricultural density? Use a real country as an example.

7. Define *subsistence agriculture* and explain why it is common in some countries in the periphery.

8. Identify two demographic factors that geographers use to measure population groups and define both terms.

9. Define the term *carrying capacity*. Explain its origin and how the term may relate to human populations.

10. Compare and contrast the population densities of Egypt and Canada.

Interpret Graphs

Study the population pyramid and answer the following questions 14–17.

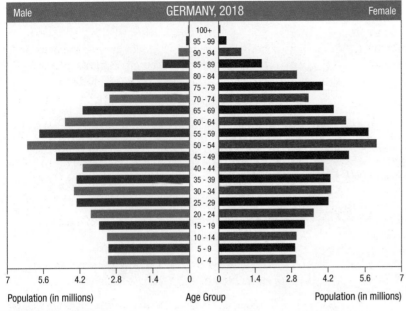

Source: CIA World Factbook

11. How can both the crude death rate and crude birth rate indicate population growth or decline?

12. Where are the highest infant mortality rates found?

13. Explain how a region's income level affects the gap in average life expectancies between men and women.

14. **Identify Data & Information** What type of CBR does this graph represent?

15. **Predict Outcomes** Describe Germany's likely dependency ratio over the next 20 years.

16. **Analyze Visuals** What economic and social challenges might Germany face in 20 years?

17. **Explain Patterns & Trends** How does Germany's population continue to grow despite its death rate exceeding its birth rate?

Chapter 4
Population Growth and Decline

Critical Viewing In Ankara, the capital city of Turkey, a crowd of people enjoy picnicking in a city park. Ankara is home to about 5 million people and has a high annual rate of population growth. ▪ Why might a country carefully monitor the size of its population?

Geographic Thinking What factors contribute to changes in population size?

4.1
Why Populations Grow and Decline

4.2
Theories of Population Change

National Geographic Explorer Pardis Sabeti

Case Study: Zika Virus in South and North America

4.3
Population Policies

Case Study: China's Population Policies

4.4
Consequences of Demographic Change

4.1 Why Populations Grow and Decline

The trend in human population can be summarized in one word: *growth*. However, the social, economic, and geographic factors underlying this growth are nuanced and complex. Geographers analyze these factors and how they affect fertility, mortality, and migration to understand population changes at a range of scales, from local to global.

Trends in Population

Learning Objectives
LO 4.1 Explain population growth rates at a range of scales.
LO 4.2 Apply the concepts of *natural increase* and population doubling time to understand characteristics of population.

Earth's population has been expanding since human life began. For many thousands of years, humanity's numbers grew relatively slowly, but the rate of growth has multiplied dramatically in the past 200 years. In 2020, the world's population numbered 7.8 billion, a nearly five-fold increase over the 1900 estimate of 1.6 billion. When plotted on a graph, the change in human population over time results in a curve that is relatively flat at the beginning and then undergoes a rapid increase in growth, producing a shape that somewhat resembles the letter *J*.

To analyze trends in population growth and decline, geographers look at the rate of natural increase and doubling time. The **rate of natural increase** (RNI) is the difference between the crude birth rate (CBR) and crude death rate (CDR) of a defined group of people. At the country or region scale, a high RNI can indicate rapid population growth, but because it does not take migration into account, RNI does not tell the whole story of an area's growth or decline. Considering Earth's population as a whole, however, RNI does render an accurate picture of trends. The global RNI peaked at 2.2 percent in 1963, and projections indicate that it will decline to 0.1 percent in 2100, while the actual number of humans will continue to increase. For the period of 2015 through 2020, the United Nations estimated the global RNI to be 1.1 percent.

Doubling time (DT) is the number of years in which a population growing at a certain rate will double. The formula for calculating DT is 70/RNI = DT. Using this formula, it is possible to calculate Ethiopia's projected DT in 2019 based on its RNI of 2.6 percent for that year: 70/2.6 = 26.9, or just under 27 years. The 2020 RNI for the United States was 0.3 percent, giving a DT of 233.3 years. Looking at DT can be a useful way to compare trends among countries or regions, but DTs can change from year to year. As a country's RNI changes, its estimated population DT must also be recalculated.

The recent striking increase in human population growth was sparked by developments beginning in the mid-1700s. Among the most prominent of these was the Industrial Revolution, which was launched by major technological innovations in manufacturing and profoundly affected where and how people lived and worked. The effects of the Industrial Revolution did not touch all parts of the world equally or at the same time. In general, industrialization occurred most quickly and had the greatest impact in Western Europe and North America. You learn about the impact of the Industrial Revolution in greater detail in Chapter 18.

At the same time that manufacturing was being transformed, new understandings, practices, and technologies were increasing agricultural productivity: people could grow more food with less labor and often on less land. In addition, scientific discoveries contributed to new and better medicines, and sanitation practices began to improve. Combined with technological developments, these advances in health care helped lower rates of disease and infant mortality and improve the overall health of populations in the countries most affected by the Industrial Revolution.

As new agricultural and health-care practices spread throughout the world, they helped increase population growth in nonindustrialized countries as well. The resulting higher birth rates and lower crude death rates led to a rapidly increasing worldwide RNI.

In the present day, as in the past, the world's population is not increasing at the same rate in all areas. Switching focus from the global scale to the national scale, differing population trends become apparent. Certain countries in the periphery are seeing the highest rates of natural increase. In Chapter 3, you read about the Democratic Republic of the Congo's (DRC's) population pyramid, which reflects its rapid population growth. In 2020, Niger, also in Africa, had the highest RNI—3.8 percent—followed by Mali (3.6 percent), Angola (3.5 percent), Uganda (3.2 percent), and the DRC (3.2 percent). Rapid population growth can challenge a country to meet its people's needs for food, housing, medicine, schools, jobs, and other services. Experts worry that the most rapid population growth is taking place in areas that may be least equipped to support larger numbers of people.

Meanwhile, many countries in Europe have a negative RNI, meaning that the population in those places is declining or will soon begin to decline. Looking to the future, experts predict that 48 countries will show a population decline by 2050, even as Earth's overall human population continues to increase.

The changes initiated during the Industrial Revolution had a powerful and unequal effect not just on the size of the population but also on its distribution and density in different areas. As a result of the new technologies, jobs in many home-based industries, such as sewing and weaving, moved into factories, which were largely located in cities. The result was a mass migration from farms to cities, where people could find work and where population densities were higher than in rural areas. This movement contributed to **urbanization**, the growth and development of cities, and took hold more strongly in the industrialized countries that form the core. Countries in the periphery, with lower rates of industrialization, have generally remained more rural. In the present day, however, many peripheral countries are experiencing rapid urban growth as populations move from rural areas to cities.

Past, Present, and Future By the turn of the 20th century, the changes brought about by industrialization and improvements in health and sanitation had contributed to unprecedented population growth. In just 100 years, from 1900 to 2000, the world population increased from 1.6 billion to more than 6 billion. Within another two decades, that number jumped by another 1.8 billion people. Though the global RNI has slowed substantially in recent decades and will likely remain low in the coming decades, projections by the United Nations and others indicate that the world population will still grow to nearly 10 billion by 2050, and perhaps by several hundred million more by the 2080s. After that, the planet's population will likely decline—but experts also say that the further into the future projections are made, the more uncertain they become.

Many researchers worry about the impact of this population growth, fearing the number of humans will exceed Earth's carrying capacity. Eminent scientist E.O. Wilson has calculated that the planet can support between 9 and 10 billion humans. Others point out that most populations cannot continue to grow rapidly and that limiting factors such as the available natural resources cause a population to level off over time. Differences in how we perceive the future depend in part on how we interpret and analyze the data. Varying scenarios can be used for forecasting population growth and total population. For example, different researchers might make different assumptions about future fertility, mortality, quality of life, and sustainability.

Factors That Influence Population Growth and Decline

Learning Objectives
LO 4.3 Explain how economic, political, and environmental factors influence population growth and decline.
LO 4.4 Explain the social and cultural factors that affect fertility rates.

Changes in population in a given place or region are driven by the balance among three factors: mortality, fertility, and migration. These are influenced by the interplay of economic, political, environmental, and cultural factors, which vary from region to region and even from place to place.

Economic Factors The strength of a country or region's economy can have a significant impact on both fertility and mortality. Birth rates tend to decline in times of economic hardship, particularly if people are concerned about having sufficient food and resources to support their children. Conversely, birth rates often rise during more prosperous times, when people are feeling optimistic about their future.

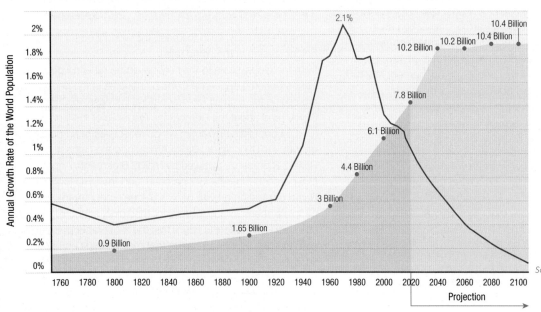

World Population Growth, 1760–2100 This graph shows the J-shaped curve of world population growth since the beginning of the Industrial Revolution. During the millennia prior to this time period, population growth was relatively flat. The "medium scenario" of the UN predicts that the 2100 global population will reach perhaps 10.4 billion, even as the population growth rate (in red) falls to 0.1 percent.

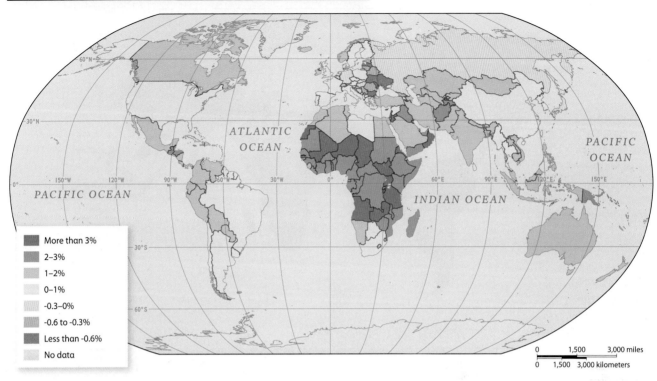

As of 2017, countries in Eastern Europe are seeing the greatest rates of population decrease. Countries in Africa and Southwest Asia have the highest rates of population increase.

Access to health care, a key factor in fertility and mortality, can also be largely dependent on the economy. In general, wealthier countries can provide better access to health care, advanced medical treatments, good nutrition, and clean water, all of which can help prevent or cure disease and lower the crude death rate. Good-quality pre- and postnatal health care and nutrition may also make it easier for women to carry their babies to term and lower a country's infant mortality rate. Both recessions (temporary periods of economic decline) and public health crises have historically been associated with reductions in births. The COVID-19 pandemic, which had both elements, thus had the potential to lead to a sizable decrease in births, which some term a "baby bust." Indeed, the Brookings Institution calculated 60,000 U.S. "missing births" in the period from October 2020 through February 2021, which correspond roughly to conceptions that would have happened from January through May 2020 but did not as the pandemic first appeared and portions of the economy shut down.

Types of economic activity also influence families' decisions about childbearing, and thus fertility rates. Families tend to be larger in agriculturally based economies, where children are considered essential for labor. In industrial and postindustrial economies, on the other hand, children may be viewed as an economic burden, leading to a lower crude birth rate.

Political Factors War, peace, and government policy play an important role in population trends. At different times, governments in different parts of the world have adopted measures attempting to control the rate of population growth in their countries. Some governments, concerned about rapid population growth, aim to dissuade families from having too many children by offering incentives or disincentives—measures intended to discourage a certain behavior. China went a step further with its policy permitting only one child per couple. China's total fertility rate, birth rate, and rate of natural increase all dropped during the years the law was in force.

On the other hand, countries with declining populations may enact policies that encourage people to have children. Singapore, for instance, offers tax breaks and extended maternity leave. South Korea, Sweden, France, and Iran have also passed policies that favor families with children. You learn more about China's and Singapore's policies on fertility later in this chapter.

War contributes to population decline in ways both obvious and more subtle. In addition to a higher mortality rate as the result of combat-related deaths and inevitable civilian casualties, war often results in food shortages or mass migration away from the conflict. At the end of 2018, for example, 5.7 million people had fled Syria's civil war to become registered refugees in neighboring countries. War also may influence the birth rate, as soldiers on the front lines are separated from their spouses or partners. Additionally, just as people tend to have fewer children in times of economic hardship, they also hesitate to start

Population Growth and Decline

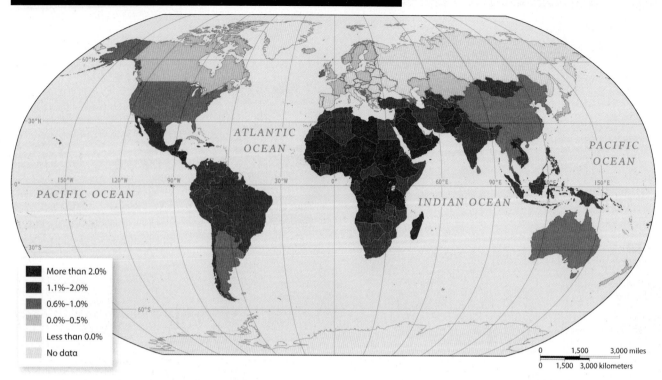

Average Annual Population Growth Rate, 1997–2015

Legend:
- More than 2.0%
- 1.1%–2.0%
- 0.6%–1.0%
- 0.0%–0.5%
- Less than 0.0%
- No data

One of the main drivers of population growth is fertility. According to the United Nations, global fertility is now 2.5 children per woman, but the rate varies considerably across the world. Singapore, a country in Southeast Asia not visible on this small-scale map, has the lowest total fertility rate (TFR), with just 0.82 children per woman. Eight countries—all in Africa—have TFRs of more than 5.5 children per woman.

families in times of conflict. A war can have a long-term impact on the demographics of a country or region by reducing the number of people of childbearing age. For example, France experienced a drastic reduction in the number of births immediately following World War I, a conflict that killed more than a million French soldiers.

In contrast, the onset of peace after a long war or political strife may cause a spike in population. For instance, many countries, including the United States, experienced an increase in the birth rate during the two decades following World War II—a period known as the Baby Boom.

Environmental Factors Natural disasters are just one environmental factor that can radically affect a region's population. Examples of such events abound: between 1 million and 4 million people lost their lives in floods in China in 1931; in 2004, an earthquake in the Indian Ocean caused a tsunami that killed almost 228,000 people; and earthquakes in Haiti in 2010 and 2018 resulted in 318,500 deaths.

Famine and the spread of deadly disease can have a similar impact. For instance, the world's population experienced a radical dip in the 1300s as the result of the bubonic plague, which swept through much of Europe and Asia. According to some estimates, perhaps 60 percent of Europe's population was wiped out. In a modern-day example,

the World Health Organization (WHO) confirmed nearly 550 million cases of COVID-19, including 6.34 million deaths, by early July 2022.

Famine is often caused by drought, a natural disaster, or by political factors such as war. It may affect a population not only through deaths by starvation but also by lowering fertility rates due to poor maternal health. In China, a famine that began in 1876 killed upward of 9 million people. Since the year 2000, famines in Sudan, Uganda, Somalia, and the DRC have cost more than 3 million lives. The United Nations World Food Programme reported in mid-2021 that 41 million people in 43 countries were "teetering on the very edge of famine" and that nearly 9 million Afghans were at risk of famine in the winter of 2021–2022, as the Taliban took control of Afghanistan's government and most humanitarian aid was disrupted. Both long-term drought and war that erupted in 2021 combined to threaten nearly 7 million Ethiopians with famine by early 2022. In Madagascar, a series of tropical storms in early 2022 coupled with the worst drought in decades caused a hunger crisis that authorities feared would become a famine.

Environmental factors also affect population numbers and distribution within regions and worldwide by encouraging migration. During the Great Famine of 1845–1849, for example, around 1 million people in Ireland died when a crop disease destroyed potato harvests, and nearly 2 million Irish

This female construction worker in China gives orders during work on the Three Gorges Dam, the world's largest hydroelectric dam, which was completed in 2006. Countering worldwide trends, the percentage of Chinese women working outside the home has been steadily declining in recent years, from 73 percent in 1990 to 61 percent in 2019.

people emigrated to the United States and other countries. Recent analysis finds that environmental factors explain migration patterns; for example, rural migrants fleeing their home region after natural events like hurricanes in Central America or the Caribbean but often staying within their country and returning home when conditions improve. The study concludes that while migration patterns vary among income groups and across world regions, environmental factors generally cause migration that is regional, short distance, and temporary—and therefore generally cause short-term changes in population distribution. You learn more about reasons for and patterns of migration in Chapter 5.

Cultural Factors Cultural expectations play an important role in fertility rates. In societies where women tend to marry at a relatively young age and large families are the norm, the birth rate can be expected to be higher. In Afghanistan, where girls often marry before they turn 18, the crude birth rate in 2018 was 37.5. (A CBR of greater than 30 is considered high.) Religion also has historically played an important role in a family's decisions to have children. Some conservative Islamic and Christian leaders, for example, have explicitly taught against using various forms of contraception.

Family planning—availability and access to contraceptive options—also has an impact on birth rates. Societies that discourage the use of contraception, often due to religious beliefs, tend to have higher birth rates than those that do not. As many people worldwide are becoming more secular (i.e., not following a religion), they are often ignoring religious rules about birth control. Catholic teaching still forbids contraception, for example, but birth rates in many traditionally Catholic countries such as Ireland and Italy remain low.

Another factor that may influence population growth is education, particularly as it relates to medicine, prenatal care, health care, and nutrition. More information results in fewer infant deaths (a lower infant mortality rate), a lower CDR, and greater life expectancy. While these factors might tend to increase population, this effect may be balanced by better education about contraception, which may result in a lower total fertility rate (TFR).

The Changing Role of Women Changing social, economic, and political roles for women have influenced patterns of fertility. In many core countries, women in previous generations tended to stay home and raise children upon marriage. Today in those countries, women are more

Population Growth and Decline

likely to remain in the workplace full time as they balance parenting and household responsibilities. Many women are postponing having children until after they have established their careers, thus reducing their childbearing years. In addition, some women may choose not to have a family at all—or to have fewer children—because of career concerns. In the United States, studies find a number of factors driving this trend: shifting expectations about gender roles, economic insecurities, a lack of family-friendly governmental support, and a lessening stigma around the choice not to have children. Though the COVID-19 pandemic showed just how little governmental or employer support most women in the United States receive relating to maternity leave and childcare, some women had already made up their minds before the disease disruptions. Combined with the availability of effective birth control options, greater levels of education and the improved status and decision-making power of women contribute to a lower birth rate.

In many cultures, however, strict gender division remains the norm. In patriarchal societies, male babies are still preferred over female babies, girls receive less education, and women have few rights. According to the United Nations Educational, Scientific and Cultural Organization (UNESCO), fewer than 40 percent of countries provide girls and boys with equal access to education. Studies indicate that women who receive more schooling have a lower TFR, at least in part because they tend to marry later. Nobel Laureate Amartya Sen stands behind the empowerment of women as a way to advance societies, seeing a variety of inequalities imposed upon girls and women. He points out that in some societies where male children are favored—such as China, South Korea, Singapore, and Taiwan—expectant mothers are pressured to use what he terms "high-tech sexism" to terminate pregnancies if it would result in a female birth. Sen also notes that the Taliban in Afghanistan specifically prohibits girls from attending school as one aspect of its strict enforcement of gender inequality. He is highly critical of such practices, particularly when they are justified by traditional cultural values.

In many countries, too, women are discouraged from working outside the home, and cultural expectations call for them to care for elderly members of the family, as well as their own husbands and children. Evidence suggests that the chronic stress associated with caring for elderly parents contributes to poor health outcomes for caregivers.

According to some recent evidence, increases in political power for women may also affect mortality. One study conducted in 2019 examined countries that in the 1990s had established gender quotas for their legislatures. In other words, each country's legislative body was required to have a certain minimum number of women. Researchers found that these countries had experienced an estimated 9 to 12 percent decrease in maternal mortality rates. Other studies have found that greater political power for women—as demonstrated by voting rights and the number of women serving in the legislature—correlates with a lower infant mortality rate.

Geographic Thinking

1. Identify which factors affected population growth beginning in the mid-1700s, and explain which factors had the greatest influence.

2. Explain whether population projections are more likely to be accurate over the short term or the long term.

3. Compare cultural perceptions of the role of women in different parts of the world; explain how they affect a country's RNI.

4.2 Theories of Population Change

How can we explain the history of Earth's population and predict its future? In the 1700s, scholars began to develop theories and models to analyze population growth and decline. Thinking on the topic continues to evolve even today.

Malthus's Theory of Population Growth

Learning Objective
LO 4.5 Explain Malthusian and Neo-Malthusian population theories in historical and contemporary contexts.

One of the earliest theories of population growth was developed by Thomas Malthus, a demographer who lived and worked in England in the late 1700s. This was the beginning of the Industrial Revolution, which, as you have read, was an era of new technologies and scientific advances that prolonged life and reduced the death rate in much of the world. Malthus interpreted the rising life expectancy and lower death rate as a sign that the world's population would grow exponentially—that is, that it would expand at an ever-increasing rate. At the same time, he observed that food productivity was growing more slowly at an arithmetic, or constant, rate. In his *Essay on Population*, Malthus speculated that Britain's accelerated population growth would contribute to a food shortage and famine by the late 1800s. He coined the term **overpopulation** to describe a population that exceeds its sustainable size, or carrying capacity.

Malthus envisioned a bleak future in which only war, famine, and the spread of disease would check the excessive population growth. He believed it was critical for people to prevent the impending population explosion by lowering the birth rate through what he called "moral restraint"—delaying marriage and childbirth.

Criticism of the Theory Malthus's prediction that the world's population would increase exponentially was accurate up to a point. As we have seen, the global RNI has decreased in recent years. Centuries ago, Malthus could not foresee the development of today's effective contraception methods or other factors that have slowed population growth. In addition, his theory did not envision the inventions and understandings that have dramatically increased food production, such as mechanized farming, hybrid seeds, and chemical fertilizers. These developments have led to greater yields and hardier crops, and have enabled agriculture to expand into marginal areas—places with soil less suited to farming.

Technologies that improved agricultural efficiency and increased productivity were coupled with other advances, most notably improved storage and refrigeration, that enabled food to last longer. Grain elevators and other storage structures better protected farmers' crops from moisture and pests. Refrigerated rail cars and trucks rolling on the better transportation network that was developed made it possible to distribute perishable goods farther from farms, while refrigerators provided a way for families and markets to preserve food for longer periods. Combined, these advances increased the food supply to a much greater degree than Malthus had predicted.

Numerous scholars have also countered the presumptions Malthus used in his forecasts. More than 100 years after Malthus introduced his theory, Danish economist Ester Boserup posited that the food supply was affected directly by population size. In other words, people would find ways to produce enough food to support a growing population. Another limitation of Malthus's ideas is scale. While the theory may be appropriate for analysis in particular regions where local agriculture cannot ramp up food production to meet an exponential population growth, geographers feel it is not useful at a broader scale.

Not everyone discounts the Malthus theory altogether, however. You have learned about the concept of sustainable development, which requires using natural resources in such a way that they will not be exhausted. A new school of thought—called **Neo-Malthusian**—raises concerns about sustainable use of the planet, claiming that Earth's resources can only support a finite population. (The prefix *neo-* means "new.") Neo-Malthusians argue that today's declining trends in fertility and crude birth rates will be reversed in some countries. They believe growing populations may bring about increasingly unsustainable development. Take, for example, the agricultural advances mentioned earlier that have helped feed ever larger numbers of people. In many regions, these technologies and practices have placed a serious strain on resources such as water and soil. Neo-Malthusians caution that future resource scarcity will lead to famine and war, and they advocate population control programs as a preventive measure.

Demographic Transition Model

Learning Objective
LO 4.6 Explain the five stages of the Demographic Transition Model (DTM).

The **demographic transition model (DTM)** represents the shifts in growth that the world's populations have undergone—and are still experiencing—over time. It is based on population trends related to birth rate and death rate, and each stage of the model is characterized by the relationship between these two factors.

Geographers use the DTM to better understand the relative stability of a population and, in particular, the factors that have contributed to population growth. Researchers proposed the model in the 1930s and 1940s to explain the changes they saw in Europe following the Industrial Revolution.

Initially, the DTM depicted four stages in the development of a population, but demographers have recently added a fifth one.

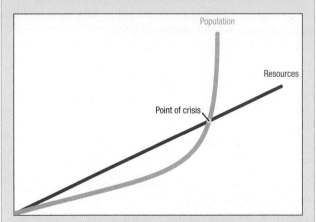

Malthus's theory is based on the premise that exponential population growth will outpace the increase in resources, including food supply. The point where the two lines intersect is where the food shortage would begin.

Population Growth and Decline 97

The first stage depicts the long period of human history before the improvements in health care and other changes at the time of the Industrial Revolution reduced the crude death rate. The population growth in Stage 2 corresponds with the continuing high birth rate and falling death rate that accompanied industrialization in Western Europe and the United States. Stages 3 and 4 reflect trends seen in societies as birth rates begin to slow due to a variety of economic and social factors. In Stage 5, population begins to decrease.

According to the model, all regions, countries, and societies go through the first four stages, and they rarely return to an earlier stage. In the present day, no country as a whole is in Stage 1, but some societies—usually indigenous groups—may be classified as such. *The Economist* cited findings in late 2021 that of 186 countries analyzed all but 11 had made the transition to lower death rates well below preindustrial levels. Seventy countries began the transition to lower fertility rates in the 30 years between 1960 and 1990, and 80 countries have shifted in the most recent period to lows in both fertility and mortality. Importantly, these findings show that the time frame during which countries undergo their demographic transitions has sped up, primarily where countries that are both geographically and culturally close have also undergone a decline in fertility—with the overall result being "a steady drop in global fertility and population growth." A few countries, like Japan and Germany, are thought to have reached Stage 5.

Evaluate the Model Although each stage seems distinct when observed on the chart, the divisions between them are not clear-cut; countries at the beginning of one stage may have more in common with countries at the end of the previous one. In addition, regions and countries may progress through stages at different paces and time scales.

It is important to note also that the DTM does not imply value judgments on societies. A society classified as being in a later stage is not "better" than one in an earlier stage. While the economies of countries at later stages tend to be wealthier than those in earlier stages, there are many other factors that determine economic success. This is particularly true with regard to Stage 5. The long-term prospects for countries in this stage are a matter of speculation. The effects of a declining population over time probably depend on the population's overall size and the nature of a country's economy. Changes in other factors, such as technology and immigration, may also play a key role in whether and how the population in a Stage 5 country will continue to decline.

The main criticism of the demographic transition model is that it is based solely on the experience of Western Europe and may not be applicable to other parts of the world, particularly those that do not undergo the dramatic changes of rapid industrialization. Critics suggest that peripheral countries today may not experience economic progress in the same way as Western Europe and the United States did, and thus may not undergo the same demographic transition.

In addition, some feel that the DTM contributes to misleading interpretations of the data, implying causes and effects that may not exist. Social and economic changes, such as the rise of industry, appear to cause the demographic transitions seen in the model, but some researchers suggest that the opposite is true—that changes in societies are the result of changes in population. Similarly, the fact that a reduction in fertility rates corresponds with economic development does not necessarily mean that economic development is the cause of the decline. Geographers and demographers must probe more deeply to seek the underlying causes of population change at each stage—and in different countries and societies.

Epidemiological Transition Model

Learning Objective
LO 4.7 Explain the five stages of the epidemiological transition model (ETM).

The **epidemiological transition model (ETM)** describes changes in fertility, mortality, life expectancy, and population age distribution, largely as the result of changes in causes of death. The stages of the ETM do not correspond to the stages of the DTM; the ETM is an independent model incorporating cause-of-death patterns to explain population growth and decline.

The ETM was introduced in the early 1970s by Abdel R. Omran, a professor of epidemiology at the University of North Carolina's School of Public Health. Essentially, epidemiological transition is the process by which a country or society goes from having high mortality rates as a result of infant mortality and episodes of famine or disease to having lower mortality rates that reflect longer lifespans and deaths due to diseases that largely affect older adults. Prior to the 20th century, the transition was related to rising standards of living, nutrition, and sanitation. In the countries of the core, the transition is often fueled by advances in medicine and better access to health care.

The first stage of the ETM is characterized by high and fluctuating mortality rates and low and variable life expectancy rates, resulting in short periods of population growth that are not sustained. Populations in this first stage are beginning to live closer together, increasing their exposure to human and animal waste and contaminated water and facilitating the rapid spread of infectious disease. Attacks by animals and other humans are another main cause of death. The deadliest epidemic of this phase in much of Europe and Asia was the bubonic plague, often called the Black Death, which killed about 25 million people in Europe in the 1340s.

Stage 2 is marked by an increased average life expectancy from about 30 to 50 years. The main causes of death are pandemics, or diseases that affect a significant percentage of the population over a large area. As populations in urban areas grow, people are at risk from diseases like cholera, which is contracted from drinking contaminated water. Over the course of this stage, efforts to fight such diseases

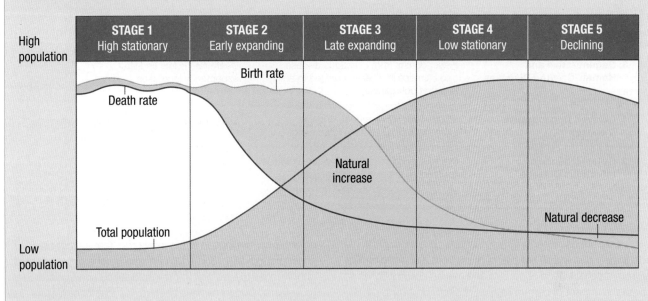

Model: Demographic Transition

The demographic transition model (DTM) is both a description and an explanation of the growth of the world's population over time. The graph illustrates how the global population has changed at the same time that birth rates and death rates have risen and fallen. Contemporary examples of countries in different stages of the DTM are provided below, but as time goes on, the list of examples for each stage will evolve. ▮ Do you think that all countries in the periphery and semi-periphery have followed or will follow the DTM and make the demographic transition? Why or why not?

	STAGE 1 High stationary	STAGE 2 Early expanding	STAGE 3 Late expanding	STAGE 4 Low stationary	STAGE 5 Declining
Description	Birth rates and death rates high; total population size low	Birth rates high; death rate falls; population grows	Birth rates gradually decrease; death rates low; population growth slows	Low birth rate; low death rate; constant population/zero population growth	Low birth rate; low death rate; elderly population; decreasing population
Examples	No entire countries	Nigeria, Guatemala, Afghanistan, Yemen	Mexico, India, South Africa	Canada, South Korea, the United States	Japan, Germany

become more successful and the number of deaths due to pandemics is reduced. Although cholera is still sometimes seen today in communities of the periphery and semi-periphery, outbreaks of cholera and other pandemics receded in Europe due to improved sanitation, sewer systems, nutrition, and medicine.

In Stage 3, the main causes of death are chronic disorders associated with aging, including cardiovascular diseases—illnesses of the heart and lungs—and cancers. Because such diseases affect people at a relatively late stage of life compared to infectious diseases or pandemics, Stage 3 is characterized by a longer life expectancy and lower death rate, translating to population growth.

Stages 4 and 5 are more recent additions to Omran's ETM model. Stage 4 reflects improvements in medicine that have extended life expectancy. New, more advanced methods for detecting cancer and cardiovascular diseases result in better patient outcomes. Some cancers are removed, and technological innovations such as new types of surgery and heart pacemakers address some cardiovascular disease.

Stage 5 suggests that life expectancy may decrease as a result of people living close together in urban environments and the development of bacteria that are resistant to antibiotics. Some medical professionals also point to deadly diseases such as AIDS and the Ebola virus as examples of what the future may hold. In 2020, the widespread outbreak of a form of coronavirus known as COVID-19 proved to the global community that new diseases can spread quickly across borders from their places of origin and that their progress is hard to stop.

Evaluate the Model One of the limitations of the epidemiological transition model is that it focuses only on health-related factors, and almost exclusively on disease. In addition, the ETM oversimplifies the causes and patterns of disease and mortality, which do not fit neatly into

Model: Epidemiological Transition

STAGES	DESCRIPTION	EFFECTS
1. Famine	Infectious and parasitic diseases mostly cause human death; animal attacks also cause deaths	Death rate is high and life expectancy is low
2. Receding Pandemic Disease	Improved sanitation, nutrition, and medicine lower spread of infection	Death rate decreases and life expectancy increases
3. Degenerative and Human-Created Diseases	Fewer deaths from infectious disease and increase in diseases related to aging (heart attack/cancer)	Death rate is low and life expectancy increases
4. Delayed Degenerative Diseases	Medical advances reduce or delay incidences of diseases related to aging	Life expectancy is at its highest
5. Re-emergence of Infectious Disease	Infectious and parasitic diseases become resistant to antibiotics and increase	Life expectancy decreases

historical periods or geographic locations. The model fails to adequately distinguish the risk of dying from a specific cause or from a combination of causes and makes assumptions about the impact of certain causes of death on overall mortality. In particular, the ETM overlooks the role that poverty plays in determining disease risk and mortality. Poverty is a leading cause of shortened lifespans and may account for differences among countries or regions more than disease.

The ETM also does not necessarily address changes that are occurring in how people live. New patterns in food consumption and other lifestyle choices often have unpredictable consequences for health and thus are not accounted for in the model. For example, people in core countries eat more processed foods than at any time in history, a change that may be linked with unprecedented and rising rates of obesity that some experts warn may shorten lifespans. Noncommunicable diseases such as type 2 diabetes, heart disease, and high blood pressure—which are potentially linked to lifestyle—are also on the rise in core countries. On the other hand, as the number of smokers has declined, so has mortality from lung cancer. Traffic accidents, a leading cause of death not only in Stages 4 and 5 countries but also in many Stages 2 and 3 countries, are also not accounted for in the ETM.

Another open question is how human-caused environmental factors might affect the causes of mortality described in the ETM. Air and water pollution are known health hazards. Gun violence and conflict are not addressed, either. Climate change, too, has been linked to the spread of communicable diseases, as warmer temperatures expand the habitats of disease-carrying mosquitoes.

Food production could also play a role in epidemiological transition. In the present day, many farmers use pesticides and other chemicals to keep weeds and pests from destroying their crops and to fight diseases such as blight. In addition to the potential contamination of the food crop, these practices may affect groundwater. Some farmers and ranchers give antibiotics or hormones to livestock to prevent disease and to make them grow faster. Concerns have been raised that these factors may have an impact on the health of those who consume meat.

Some chronic diseases may be caused by environmental factors that result from materials used in construction and technology. For instance, asbestos was a common building material before it was determined to be a cancer-causing agent. Similarly, lead was used in paint and as an additive in gasoline before it was determined to have deadly properties.

In addition, a highly mobile population and global economy increase the risk of deadly communicable diseases spreading rapidly. The 2010s, for example, witnessed the outbreak of Ebola, a deadly virus that was first identified in 1976 in tropical Africa. The first incidence occurred in March 2014 in a rural region of southeastern Guinea, but the disease soon spread to other parts of that country, as well as to neighboring Liberia and Sierra Leone. Ebola is very contagious through direct contact with an infected person or animal, so it has the capacity to spread quickly. Many people contracted Ebola by caring for an infected relative at home or preparing a body for burial according to tradition. At the same time, medical services in the affected countries were not sufficiently staffed or prepared for such an emergency.

By the time the outbreak was declared over in June 2016, there were 28,610 reported cases of Ebola, causing 11,308 deaths. In 2018, the second largest Ebola outbreak on record occurred in two northeastern provinces of the DRC. With much learned from the 2014 outbreak, the World Health Organization worked quickly with the country's Ministry of Health, the affected communities, and many other partners to train thousands of health-care workers to register, test, and vaccinate hundreds of thousands of residents and to

Critical Viewing In 2019, an outbreak of the Ebola virus had killed thousands of people in the Democratic Republic of the Congo. People crossing the border into Uganda had their temperatures taken as a way of checking for symptoms of the highly contagious disease. ▍ What are some of the challenges of preventing the spread of communicable diseases like Ebola?

care for all those who recovered. Made especially challenging because it occurred in an active conflict zone, the outbreak resulted in 3,470 cases with 2,287 deaths by the time it was declared over in June 2020. Knowing that it could have been much worse, the WHO committed to build on the lessons learned and the gains made in this outbreak to prepare for other public health challenges, including re-emergence of Ebola such as occurred in late 2022 in Uganda.

Many researchers and officials (including geographers, epidemiologists, health officials, medical doctors, and others) have long been concerned about the possibility of viruses or antibiotic-resistant strains of bacterial diseases spreading on a more global scale, traveling quickly along highway and rail networks and jetting across oceans with infected airline passengers. These concerns were confirmed with the arrival of what is now known by the medical and public health community as SARS-CoV-2 (severe acute respiratory syndrome coronavirus 2)—with the respiratory disease it causes known by everyone as COVID-19 (coronavirus disease 2019). Part of the coronavirus family, which includes both common viruses that cause usually mild diseases, such as head or chest colds, as well as less common but more severe diseases like Middle East respiratory syndrome (MERS) and other SARS illnesses, COVID-19 spreads quickly through droplets projected out of the mouth or nose when an infected person breathes, coughs, sneezes, or speaks.

First appearing in the city of Wuhan, China, in December 2019, COVID-19 was found in other Asian countries and Europe by early January 2020, and in the United States and other countries at least as soon as February 2020. Despite intensive efforts over the following months by governments across the globe to restrict travel and to order closures of or protections for businesses, schools, public facilities, and more, the virus continued to spread. Medical scientists in many countries raced to develop vaccines to protect against the most severe illness or death, with the first ready for use within a year; the first vaccine was approved for use in the United States in December 2020. Wealthy countries of the core were first to receive vaccines, with countries of the semi-periphery and periphery unable to procure the billions needed for months and years. By early July 2022, as the pandemic continued with evolving variants, the WHO reported nearly 550 million cases of COVID-19 worldwide, with 6.34 million deaths.

Geographic Thinking

1. Explain how Malthus's theory was affected by the time and place in which he lived.

2. Explain the degree to which the demographic transition model effectively explains population changes.

3. Describe limitations of the epidemiological transition model.

4. Explain how recent Ebola outbreaks and the COVID-19 pandemic fit into Stage 5 of the epidemiological transition model.

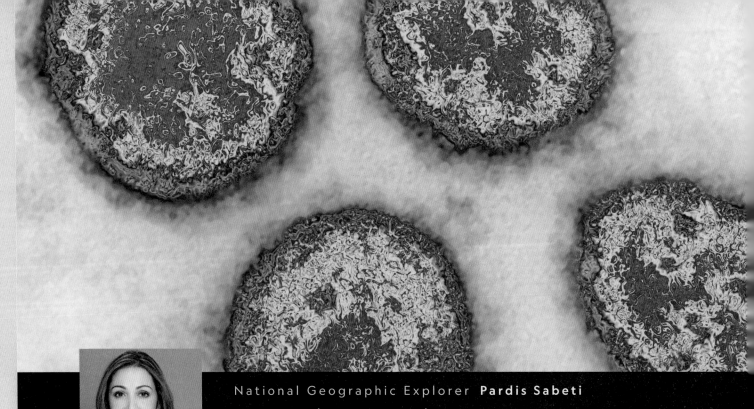

National Geographic Explorer **Pardis Sabeti**

Cracking the Genetic Code

When she's not making music with her indie rock band Thousand Days, Pardis Sabeti is a genetic researcher and professor at Harvard University.

Learning Objective
LO 4.7 Explain the five stages of the epidemiological transition model (ETM).

In her groundbreaking research on the evolution of disease, Pardis Sabeti wields the power of math in a quest to protect the human population.

"Humans and virulent microbes are both governed by genetic codes that allow them to evolve over time," Sabeti explains. "There's also a constant evolutionary arms race going on between the two. Humans develop genetic resistance to particular diseases, while microbes develop resistance to antibiotics and our immune defenses. Unlocking the genetic codes of humans and pathogens can help us understand how to intervene." Work like Sabeti's may someday help prevent other communicable diseases from making the comeback predicted in Stage 5 of the ETM.

To this end, Sabeti has developed methods for computers to analyze the DNA of groups of people in search of clues to disease resistance. Specifically, she is looking for genetic mutations that have become very common very quickly in certain populations. These mutations can help researchers identify genes associated with the ability to resist diseases such as malaria.

One of Sabeti's genome scans found evidence for a recent genetic adaptation that may be linked to resistance to the Lassa virus, an extremely dangerous pathogen. She has worked in Nigeria over the decade since, identifying individuals who have been exposed to the virus but have suffered few effects and analyzing their genomes to uncover the parts associated with this resistance. This decade-long project is not only exploring how certain people develop resistance but also helping develop improved diagnostics to quickly test for Lassa exposure and identify which strains of the virus are the most infectious. With the outbreak of COVID-19 in 2020, Sabeti's work found new avenues to help provide the global community with the tools and technologies needed to detect viruses and share data in an effort to improve human health.

Geographic Thinking

Explain how Pardis Sabeti is using quantitative data to have an impact on mortality.

Case Study

Zika Virus in South and North America

The Issue In 2015, hundreds of thousands of cases of Zika virus infection were reported in Brazil. The virus spread to other parts of the Americas and the Caribbean.

Learning Objective
LO 4.7 Explain the five stages of the epidemiological transition model (ETM).

By the Numbers

707,133
Zika virus cases reported in the Americas between May 2015 and December 2016

60
countries worldwide reporting cases of Zika virus in 2015–2016

2
U.S. states reporting cases of Zika in 2016: Florida and Texas

Source: Centers for Disease Control and Prevention

The Zika virus rapidly spread through Brazil in the years 2014–2016. A 2019 study by the journal *Nature Microbiology* predicted that in future years, warming global temperatures caused by climate change will foster conditions favorable to the mosquitoes that spread Zika and other diseases.

One Unpredictable Factor that influences population size is the spread of disease. Scary new viruses sometimes seem to emerge abruptly, and too often they re-emerge after being inactive for years or decades. One such virus is Zika, which was first discovered in 1947 in monkeys in Uganda. Five years later, the first human diagnoses of the Zika virus were made in Uganda and Tanzania. In May 2015, the virus made front-page news due to a major outbreak in Brazil, on the other side of the Atlantic Ocean. By November of that year, the Brazilian government had declared a national public health emergency and, by November 2016, more than 200,000 cases of Zika infection had been reported there. The virus soon spread to other parts of North, Central, and South America and the Caribbean, including the United States.

Researchers point to several factors that contributed to the rapid spread of the disease. The Zika virus is carried by the *Aedes aegypti* mosquito, a species that lives in tropical and subtropical regions. These mosquitoes reproduce in puddles and other standing water, and crowded living conditions in warm climates provide excellent breeding grounds. Although mosquitoes are the most common means of transmission, the virus can also be diffused through sexual activity. In addition, scientists say that the population of the Americas was susceptible to the disease because, unlike Africans, who had built up resistance to Zika, they had no previous exposure to the virus. Symptoms of Zika are often mild, so infected people may unknowingly spread the disease locally or travel to a new place after contracting it.

One of the reasons the Zika virus is so frightening is that it affects the brain development of a fetus and can cause microcephaly—a smaller-than-normal head size—and other malformations. Thus, Zika is especially damaging when a pregnant woman is infected. To address the crisis, governments engaged in public education campaigns and took measures to eradicate the *Aedes aegypti* mosquitoes. In the United States, the Centers for Disease Control and Prevention (CDC) cautioned pregnant women to avoid traveling to areas where Zika had been detected. Researchers later found higher risks for developmental, eye, and hearing abnormalities in children whose mothers had been exposed early in pregnancy. And while almost half of infants with abnormalities at birth actually had normal test results in the second or third year of life, 25 percent of children who tested normal at birth had below-average developmental levels or abnormal hearing or vision before their third birthday. The CDC has reported no local mosquito-borne Zika transmission in the continental U.S. since 2018.

Geographic Thinking

Explain whether the spread of the Zika virus challenges aspects of the epidemiological transition model.

Population Growth and Decline 103

4.3 Population Policies

Governments occasionally try to influence population trends through policy, in some cases trying to spur population growth and in other cases trying to limit it. Their goal is to avoid out-of-control increases in numbers that could stress resources, while at the same time maintaining a vigorous workforce and strong economy.

Types of Population Policies

Learning Objective
LO 4.8 Explain how governmental policies may affect population growth and decline.

Government policies enacted to influence the natural rate of increase fall into two categories. Those designed to curb population growth by discouraging citizens from having children are known as **antinatalist**, and they are usually a reaction to concerns about population growth exceeding resources. The goal of these policies is to reduce the risk of potential famine or disease due to overuse of natural resources and to ensure that there are sufficient schools, jobs, and services to support the future population.

In contrast, **pronatalist** policies encourage births and aim to accelerate population growth. Governments enact pronatalist policies for a variety of reasons, including to address concerns about an aging population. Low birth rates and fertility rates in some countries give rise to concerns about the size of the workforce and its ability to meet future economic needs as older workers retire. Some experts also argue that stronger birth rates promote a stronger economy, or that having a family promotes personal well-being.

Critical Viewing Countries monitor their population trends by looking principally at two areas: birth rate and flow of immigration. In 1979, China began a program to control its birth rate by enacting an antinatalist one-child-only policy. Billboards like these depicted happy one-child families. ▎How might this billboard have affected Chinese citizens' feelings about the country's One-Child Policy?

Case Study

China's Population Policies

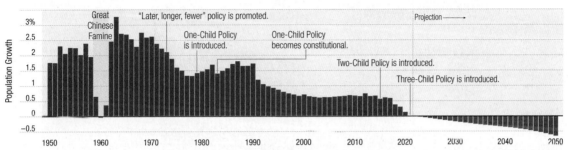

Annual Population Growth of China China's One-Child Policy resulted in a drop in the country's fertility rate. By the mid-2010s, with projections anticipating an alarming decline in population, the government ended the policy.

Source: U.S. Census International Data Base

The Issue China issued its One-Child Policy to avoid problems with population growth. Now it is experiencing unforeseen impacts of the measure's success.

Learning Objective
LO 4.8 Explain how governmental policies may affect population growth and decline.

By the Numbers

970 million
China's population in 1979

1.4 billion
China's population in 2017

2.8
China's fertility rate in 1979

1.7
China's fertility rate in 2017

Sources: Centers for Disease Control and Prevention; World Data Atlas

In 1979, concerned that the country's resources would be unable to support its growing population and that earlier measures to limit population growth had not shown sufficient results, the communist government of China enacted its One-Child Policy. Any couple who had a second child would be required to pay a fine that only the very rich could afford. In addition, families with more than one child were ineligible for government assistance. The only exemptions given were to members of some ethnic minorities or in cases where the firstborn child had a disability. The government also offered incentives to families who limited themselves to one child, including better employment opportunities and higher wages.

Even before enacting an official policy, China had worked to decrease its population by encouraging people to marry later and have smaller families. As a result, the fertility rate had fallen from a high of 6.1 children per woman in 1968 to 2.8 in 1979. By the 2010s, China's TFR was holding steady at about 1.6 children per woman. The policy is estimated to have prevented up to 400 million births.

China's One-Child Policy had unintended consequences, however. The elderly population grew more than anticipated because improvements in medicine and health care increased life expectancy. Eventually, government leaders began to worry that there were too few workers to support the aging population. In addition, the cultural preference for boys led to a gender imbalance. Infant girls were sometimes abandoned or given up for international adoption. Pregnancy terminations of female fetuses and female infanticide were also reported.

In 2015, China abandoned the policy and replaced it with one that allowed any couple to have two children, and after the 2021 census showed that RNI had slowed to 0.53 percent, authorities changed the policy again to allow three children, along with supportive measures to convince couples to have more kids. Some experts believe these changes will not affect the population growth rate any time soon, for several reasons. Many Chinese women are past their childbearing age. Many men are having trouble finding wives because of the gender imbalance. Young couples often live in cities, where the high cost of living may deter them from having children. And couples are getting married later in life, a cultural change the government itself had encouraged for decades. Immigration—often involuntary—may come to play a role in China's future population. In recent years, news organizations have reported that some criminal groups have begun luring or kidnapping women in neighboring countries for forced marriages or childbearing. China will be coping with the impacts of its One-Child Policy for decades to come. With a TFR that had fallen to between 1.2 and 1.3 by 2022, China faces a decline in its population to as low as 700 million by the year 2100, half its current population.

Geographic Thinking

Identify the ways in which China's One-Child Policy was both successful and unsuccessful.

In the East African country of Tanzania, family-planning information is distributed by community-based organizations as well as by mass media campaigns funded by the United States Agency for International Development (USAID). In 2018, Tanzanian president John Magufuli ordered USAID to stop running commercials, citing concerns about a shrinking labor force.

Results and Consequences

Learning Objective
LO 4.8 Explain how governmental policies may affect population growth and decline.

Both pronatalist and antinatalist policies have had mixed success. Results generally depend on how well the population accepts the fundamental principles proposed by its government. If citizens agree that it is to their advantage to have more or fewer children, they will more likely comply with government policies. It is impossible to force people to have more children; pronatalist policies are generally fueled by a sense of ethnic or national pride and tend to succeed where there is a strong sense of nationalism or anti-immigrant sentiment. Pronatalist policies may include financial support, such as "baby bonuses," tax credits, and free daycare, to make it less expensive to have and raise children. Antinatalist policies, on the other hand, generally include elements such as contraception education and family-planning support, as well as financial incentives. A few countries have attempted to control the population through inhumane policies including forced sterilization, but other than in China, these policies have tended to have little impact on the total population growth.

India's government was among the first to control its citizens' fertility in an effort to curb population growth. In the 1970s, India enacted aggressive mandatory sterilization campaigns, mostly targeting men. In many cases, men were sterilized without their consent; in others, the men consented in exchange for land or for not being cited for riding a train without a ticket. In 1976 alone, the Indian government sterilized 6.2 million men. In more recent years, India has focused its sterilization efforts on women, generally targeting very poor areas, where the payment for sterilization may amount to more than two weeks' wages. More than 4 million sterilizations were carried out from 2013 until 2014, with fewer than 100,000 done on men. India's government insists that its sterilization programs are voluntary and claims they are essential to limit population growth in the country, which is projected to become the most populous one on Earth by 2030. India has also implemented media campaigns to promote contraception and family planning, as well as financial aid for people seeking birth control from health-care centers. In 2021, the government of India's most populous state, Uttar Pradesh, proposed legislation to carry out a two-child policy and voluntary sterilization program. As proposed, the policy would bar couples with more than two children from applying for state government jobs and receiving public benefits.

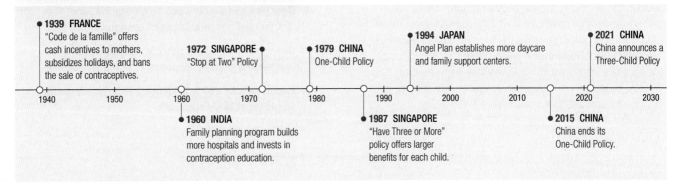

Pronatalist and antinatalist policies have been a part of population strategies in many parts of the world since the early 20th century. ▮ What does the time line illustrate about ways governments have tried to influence birth rates?.

Nigeria initiated an antinatalist policy in 1998 with the goal of reducing the total fertility rate from more than six to four births per woman. The government sought to encourage contraception and worked with international donors to increase family-planning education and the availability of contraceptives. The policy encountered opposition from Nigeria's Catholic and Muslim communities, which both frown on birth control. More than two decades later, the country's fertility rate is still above five children per woman—one of the highest in the world.

Many other countries of Africa have similarly tried to address population growth through antinatalist policies, with varying degrees of success. One country that may be seeing positive results is Rwanda, Africa's most densely populated nonisland state. With a population of 12.2 million growing at a rate of 2.3 percent, Rwanda's doubling time is 30 years. The government believes poverty and hunger are the results of the country's high fertility rate and has made family planning a top priority. Billboards promote the use of contraceptives, and community health workers go door to door to provide family-planning counseling to people in rural areas. Although it may be too early to draw conclusions about the long-term effects of these efforts, initial reports suggest that the strategy is working. For instance, family-planning use jumped from just 4 percent in 2000 to 51 percent in 2010, and the fertility rate had fallen to 3.8 in 2022 from a high of 8.4 in 1979.

Not all African countries are seeking to curb population growth. In 2018, Tanzania's president announced that the country needed to increase its population and encouraged women to stop taking birth control. Pronatalist policies are more common among European countries, some of which are experiencing a very low growth rate or an actual decrease in population. There, governments often provide economic incentives such as paid maternity and paternity leave, childcare, or free education to encourage families to have children. France, unlike most other European countries, has a relatively high and stable fertility rate, which some analysts attribute to effective pronatalist policies. France provides tax breaks to both married and unmarried parents, as well as social assistance and housing subsidies for poor families. Other benefits include generous parental leave, childcare services, and a family allowance. Among other advantages, families with three or more children can obtain a card that gets them discounts on train travel and other everyday expenses.

Hungary has put in place a number of measures to counter a downward trend in population that is due to both a low fertility rate—around 1.5 children per woman—and emigration. Concerned about labor shortages, the prime minister offered scholarships to university students who promised to stay in the country and gave citizenship to ethnic Hungarians living abroad. He then turned attention to increasing the fertility rate by exempting Hungarian women with four or more children from paying income tax. The prime minister has also recommended other incentives, such as reducing mortgage and automobile payments for parents, giving grandparents parental leave, and increasing daycare options. Hungary's pronatalist movement has strong nationalist roots. In speeches, the prime minister has explicitly refused to consider one obvious method for increasing the country's working-age population—immigration.

Singapore has engaged in policy experiments on both the antinatalist and pronatalist sides. For many years, Singapore, which is small in area and very densely populated, strived to decrease its population growth through a series of efforts that included making contraceptives available at low cost, creating family-planning clinics, and embarking on media campaigns that promoted the benefits of a smaller family. The policy appeared to work—perhaps too well. By the 1980s, the government began to reverse its antinatalist stance and promote larger families. Today, Singapore has comprehensive pronatalist policies that encourage marriage and provide support for families, including a baby bonus system to give cash to new mothers, financial benefits for female college graduates who have more than three children, and paternity leave for fathers. The government also has announced that it will provide more assistance for couples who have difficulty conceiving. Despite these efforts, Singapore's TFR remains low and declining; it dropped from just 1.25 in 2015 to 1.1 in 2022.

Geographic Thinking

1. Describe some of the positive, negative, and unintended consequences of pronatalist policies.

2. Describe some of the positive, negative, and unintended consequences of antinatalist policies.

3. Explain which policy or policies Neo-Malthusians would most likely support.

4.4 Consequences of Demographic Change

Although some countries still have high birth rates, the overall global fertility rate has steadily declined since 1950. In many countries, the demographic changes are bringing profound social alterations. Adapting to changes in population size and composition is a key challenge of the 21st century.

Changes in Size and Composition

Learning Objective
LO 4.9 Explain the implications of changes to a population's size and composition.

As you've learned, the growth or decline of a population can have economic, cultural, and political effects. On the economic front, a growing population creates greater demand for homes, goods, and services. As demand outstrips supply, the prices of goods and services increase. Rapid population growth may contribute to insufficient housing stock, crowded schools, difficulties with providing enough food, and inadequate health-care services. A decrease in population has the opposite effect: the prices of homes and other goods also often go down. Even with lower housing costs, however, cities and towns that have experienced population decreases confront problems with vacant homes and abandoned properties. In some cases, there may be too few qualified applicants for available jobs; in others, businesses may have too few customers to support them. Hospitals and other health-care services often cut back or close, unable to generate enough revenue or staff positions.

From the political perspective, governments must cope with a growing population's need for homes and infrastructure—essential networks and facilities such as schools, roads, power lines, and water and sewer systems. To pay for the increasing demands, governments typically raise taxes or fees. When they are unable to do so fast enough, the quality or quantity of services suffers. In less wealthy countries, this may translate into insufficient education, transportation, or water supply, sanitation, and health-care systems. An issue for local governments in wealthier countries is, again, often the reverse: a community's tax base shrinks along with its population, often resulting in a decrease in funding for schools, social services, and other basic needs. You will learn more about urban issues in Chapter 17.

Population size affects the environment as well. Rapid population growth can put groups at risk of exceeding the carrying capacity of their local or regional environment. In some cases, the use of limited resources has long-term repercussions such as **land degradation**, or long-term damage to the soil's ability to support life. Examples of land degradation include soil exhaustion—the depletion of nutrients in soil that has been farmed too long or too intensively—and deforestation caused by communities clearing wooded land for farming, homes, and infrastructure. The cutting of rain forests in West Africa, for example, has contributed to severe droughts that have caused widespread hunger and famine. In more densely inhabited areas, increases in solid surfaces such as roads, parking lots, and buildings prevent rain from soaking into the soil, creating problems with water runoff and increasing the risk of flooding. In view of these phenomena, some researchers believe global or regional declines in population could have their advantages, allowing for more sustainable use of Earth's resources.

Critical Viewing Vans in Idumota Market in Lagos, Nigeria, pick up workers returning home from work. Some worry that Lagos will be unable to meet the needs of its growing population. ▎What challenges resulting from Nigeria's growing population do you see in the photo?

Composition may be as important as size when examining a population's social, economic, and political impacts. As you have seen, countries with imbalanced gender ratios face cultural and economic difficulties. The average age of a population and the relative number of people in different age brackets can also have a profound impact on the economy. An excessively large working-age population may make it difficult for people to find jobs, driving down wages. Many government leaders, especially in core countries, are more worried about the opposite problem: a too-small working-age population means fewer people in the workplace, less economic output, and less money paid in income and other taxes. Government expenditures vary according to the age of the population as well. A country with a relatively young population may need to build more schools for its children, while an aging population may need greater investment in facilities to care for them and in long-term health care.

Nigeria, home to 215 million people in 2022, provides a useful example of the impacts of population size and composition. Already the most populous country in Africa, Nigeria is also one of the fastest-growing countries in the world, with a population growth rate of about 2.6 percent. Researchers worry this rapid growth will put a strain on the country's limited resources. Nigeria's unemployment rate in 2019 was 17 percent and its COVID-19-impacted unemployment rate in late 2020 was 33 percent—meaning that one of every three people of working age could not find a job at that time. The unemployment rate is even greater among young people—estimates suggest it may be as high as 60 percent. Rural parts of the country are also experiencing high levels of unemployment. The lack of opportunities in rural regions has contributed to migration to cities that are ill prepared to deal with the resulting rapid growth. Nigeria's health-care system has also been overwhelmed.

Not all the consequences of this rapid growth have been negative, however. High fertility rates of families working in the agricultural sector have increased the labor force and enhanced productivity, enabling the country to prevent famine and malnutrition. These factors, in turn, have lowered Nigeria's crude death rate. Predictions that rapid growth would contribute to economic decline have not borne out; in fact, Nigeria's gross domestic product grew at 7 percent per year from 2000 to 2014, though it has slowed since then, with a COVID-related contraction in 2020–2021. Nevertheless, some researchers say its population is one of Nigeria's greatest resources. At the same time, many middle-class and wealthier Nigerians with highly skilled jobs are moving to Europe, the United States, and Canada, a trend that causes some concern for the future of Nigeria's economy.

Population Growth and Decline

Consequences of an Aging Population

Learning Objective
LO 4.10 Explain the consequences of an aging population.

In Chapter 3, you read about the dependency ratio, which is the number of people in a dependent age group (under age 15 or older than 64) divided by the number of people between the ages of 15 and 64. In many countries, particularly core countries, the dependency ratio is increasing due to the number of older adults. Countries with aging populations generally have lower TFRs, often below 2.1 children per woman. When represented by a population pyramid, the higher tiers are larger than the bottom ones.

In 2020, 9.3 percent of the world's population was age 65 and older and projections suggest that this number will increase to 17 percent by 2050. Aging populations are not equally spread across the world, however. Presently, most periphery and semi-periphery countries do not have rapidly aging populations. But in 2020, 13 countries had populations in which 20 percent or more of the population was over 65, with some projections suggesting that a total of 82 countries will have a similar top-heavy population pyramid by mid-century. Globally, one in six people will be age 66 and older (16 percent) in 2050, up from one in 11 in 2019 (9 percent), and the number of persons 80 years and older is projected to triple, from 143 million in 2019 to 426 million in 2050.

The average global life expectancy of roughly 72 years for those born in 2019 represents a significant rise over the life expectancy of 66.5 years for those born in 2000. In some regions of the world, life expectancy has risen even more significantly. This change, coupled with decreased fertility rates, is the main reason for the aging population in many countries.

In general, people are living longer because they have better access to medical care, especially in core countries—but increasingly in semi-peripheral and some peripheral countries, too. A better understanding about healthy lifestyle habits also contributes to longer lifespans. Improved detection and treatment of infectious and degenerative diseases in countries at all levels of development is another factor in the overall life expectancy rise. As you have seen, many of these changes are captured in the epidemiological transition model.

Although living longer is a positive development, it is not without challenges. It has been noted that an aging society is literally and figuratively a "dying society." Without young people coming into the society, with their energy, work, and new ideas, the future cannot be bright. Like any other change in population size or structure, an aging population has both short- and long-term effects on the economy, culture, and politics of a community, country, or region. Societies cannot avoid an aging population; they must adapt by confronting the challenges and taking advantage of the benefits this demographic change can bring about.

Social Effects A graying population can challenge traditional family dynamics. In many cultures, for example, multigenerational families have long been the norm. It is expected in some societies for the bride to move in with her husband's family, allowing the couple to care for his parents as they age. In recent decades, however, fewer newlyweds are opting for this living arrangement. Social changes like

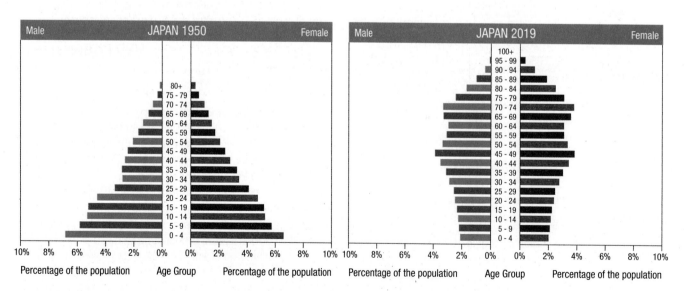

Population Pyramids, Japan, 1950 and 2019 It is projected that by the year 2060, 40 percent of Japan's population will be over the age of 65. Japan is feeling the impact of its rising elderly population in every area of modern life, from a reduced workforce in dire need of new workers to supermarkets needing more parking spaces near entrances to accommodate less mobile older customers.

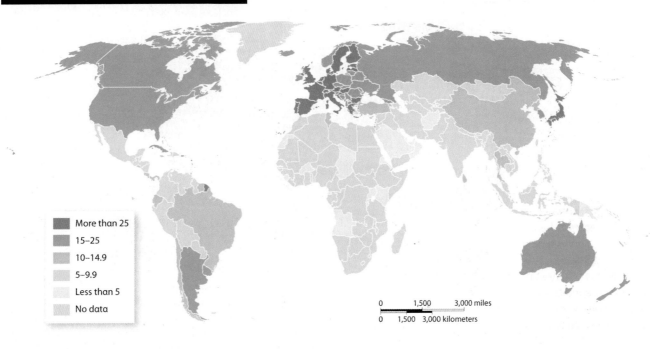

Reading Maps The elderly dependency ratio is the number of people age 65 and older relative to the working-age population.

this one, coupled with demographic change, mean that fewer young people are available to care for aging parents at home. In response, some countries have begun building living facilities for older adults, changing the social fabric of the culture.

India provides an example of this phenonenon. Traditionally structured Hindu families in India live in multigenerational households incorporating grandparents, aunts and uncles, and cousins. Brides move in with their husband's family, and elderly parents are cared for in the home. Recently, however, increasing numbers of young people have been moving away from their homes to seek better-paying jobs, and women are entering the workplace in greater numbers. These factors, as well as changes in cultural attitudes, have led many younger Indian families to form nuclear households—that is, homes consisting solely of parents and their children. Some social commentators in India have criticized this trend and the resulting increase in older adults moving into senior living facilities.

At the same time, the social benefits of a larger elderly population cannot be overlooked. Retired grandparents may be available to care for grandchildren, enabling both parents to return to work without the added cost of daycare. When parents are unable to raise their children, often it is the grandparents who step in. Older adults also play a key role in maintaining social networks that bind families and groups together and, through their engagement in community life, can help change negative societal attitudes toward aging.

Economic Effects Perhaps the largest economic challenge of an aging population comes from the fact that retirees pay less in income taxes, which in most countries is a major revenue source for the government. In many countries, including the United States, Germany, and Italy, record numbers of people are supported by government-sponsored retirement programs such as the United States' Social Security system. On average, retirees are living longer, adding to the costs of these programs, which are largely financed by taxes paid by younger people who are still in the workforce. As the cost of public retirement programs rises, taxes may also go up, or benefits for the retirees may be reduced.

People also rely on their private savings to maintain their lifestyles in retirement. Living 20 or 30 years beyond retirement is no longer unusual, and some retirees run out of their personal savings long before the end of their lives. In response, research suggests that in many countries people are postponing retirement and working longer. In some cases, retirees may return to work if their savings cannot support their retirement.

Health care is the other most notable cost associated with an aging population. Because people live longer, they are often ill or unable to fully care for themselves for a longer period of time. Statistics suggest that the cost of long-term care in the United States may double over the next 20 years.

The financial consequences of health care can span generations. When an aging parent requires round-the-clock

Population Growth and Decline

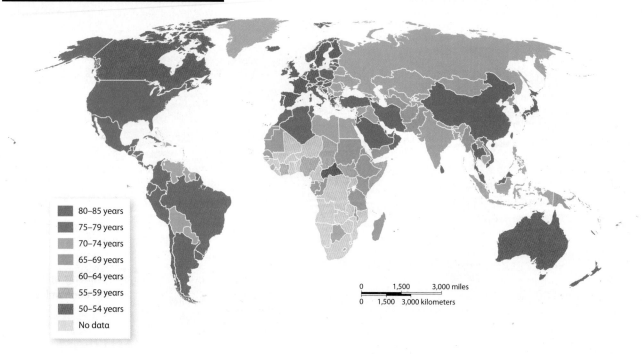

World Life Expectancy, 2020

Legend:
- 80–85 years
- 75–79 years
- 70–74 years
- 65–69 years
- 60–64 years
- 55–59 years
- 50–54 years
- No data

Reading Maps The elderly dependency ratio in the "Elderly Dependency Ratio, 2018" map can be compared with the life expectancy numbers in this map to better understand age-related demographics. For example, Nigeria has an elderly dependency ratio of 5–9.9 and an average life expectancy of 55–59 years. It makes sense that these numbers correspond because a low life expectancy would leave fewer older adults in the population in relation to number of people age 15 to 64. ▍ Compare the statistics for one or two other countries on this map and the "Elderly Dependency Ratio, 2018" map and explain how they might be related.

care for a serious physical or mental condition, an adult child may have to choose a less demanding job or leave the workforce altogether in order to care for that parent. In some places, the increased need for medical specialists who care for older adults has outstripped supply. Countries and communities may need to reconsider how services are provided. Sweden, for instance, has implemented an approach whereby mobile teams provide flexible home care for elderly residents.

These challenges can be partially offset by several economic benefits of an aging population. Older adults spend money on food, clothing, housing, and entertainment just as younger adults do, and they may stimulate growth in many economic sectors. For instance, older adults use services such as specialized health care and in-home caregivers. In recent decades, the number of geriatric physical therapists, or physical therapists who specialize in working with elderly patients, has grown considerably. One estimate placed the value of the elder care industry in the United States at $400 billion in 2018. Retirees also contribute countless hours in volunteer work, which may reduce the government's burden. In addition, older adults are less likely to commit crimes and do not attend public school, both of which represent major expenditures for local governments in many countries around the world.

Political Effects An aging population may also shape the political landscape of a country or region. Changes in the voting demographic may influence who is elected and what policies are enacted. In some countries, older citizens have a strong voice in politics. In the United States, for example, the percentage of people age 65 and older who vote in elections is consistently greater than that of any other age group. Some research has indicated that older voters see pensions and health care as their top priorities, with education and environmental concerns farther down the list. Yet older populations are diverse in their opinions. Researchers also concluded that as younger generations become seniors themselves, they will likely carry their own varying priorities into the voting booths.

Political effects of an aging population are often intertwined with economic effects. In the United States, Social Security is a regular topic of debate among citizens and lawmakers, as the Baby Boom generation leaves the workforce to enter retirement, adding to strains on the Social Security system. How to fund and maintain public retirement programs poses difficult policy questions for lawmakers worldwide. Germany and the United Kingdom have dealt with this problem by allowing larger numbers of immigrants of working age into the country to expand the workforce and relieve some of the tax burden.

Some countries, too, are attempting to control costs by raising the retirement age. The United Kingdom and the United States have both increased the age at which retirees can claim full benefits and plan to phase in more changes in the future. In some countries in the European Union (EU), including Finland, Denmark, Greece, and Italy, the retirement age is being linked to life expectancy. The general retirement age among EU member states is 65 years. Spain, Germany, and France are about to raise their retirement age from 65 to 67. Policy decisions relating to public retirement plans have economic impacts for both older and younger generations.

Immigration, mentioned previously as a way of adapting to the economic challenges of a shifting dependency ratio, also has a political dimension. Governments set immigration policies in response to a variety of economic and social pressures. In Germany, as in other European countries, increased immigration, peaking in 2015–2016 during the Syrian refugee crisis, has been unpopular among many citizens, contributing to divisions within the society. However, 2022 polling has shown a more moderating view among Germans, with majorities seeing immigration as a way to help offset its aging population, to compensate for a skilled labor shortage, and to generate additional revenue for retirement funds. In other countries, immigrants are also seen as welcome solutions to the problem of providing quality services to growing or aging populations. You explore the social, political, and economic dimensions of immigration in Chapter 5.

Geographic Thinking

1. Identify the stages in the demographic transition model where you expect to find countries with a high elderly dependency ratio. Explain what the DTM helps you understand about these countries.

2. Compare the positive and negative economic effects of an aging population.

3. Explain how changes in a country's working-age population can affect the country both economically and culturally.

Chapter 4 Summary & Review

Chapter Summary

The three main factors that cause changes in population size and structure are fertility, mortality, and migration. These factors are influenced by economic, political, environmental, and cultural factors.

- A wide range of changes that occurred during the Industrial Revolution caused the global population to start growing exponentially.
- The changing role of women is impacting the fertility rate in many places.
- An increase in life expectancy is contributing to population growth and influencing the composition of the population in many countries.
- Geographers use the rate of natural increase (RNI) and doubling time to explain changes in population size. The RNI is the difference between the crude birth rate and the crude death rate. Doubling time is calculated by using the formula 70/RNI to obtain the number of years a population will take to double in size.

Several models and theories have been developed to help explain and forecast changes in population.

- Malthus's theory suggests that population growth will outpace food production.
- The demographic transition model (DTM) portrays five stages of population development and is based on trends related to birth rates and death rates.
- The epidemiological transition model (ETM) seeks to explain trends in population by examining the principal causes of death at different stages.

Countries adopt policies to address needs related to the size of their population.

- Antinatalist policies are designed to address future problems due to overpopulation by discouraging people from having large families.
- Pronatalist policies are designed to address problems associated with a declining population by encouraging people to have children.
- Governments also use immigration policies to address population issues.

Although the population is growing rapidly in some countries (especially in Africa), other countries (especially in Europe) are experiencing declining population rates.

An increasing number of countries have aging populations, which leads to a higher dependency ratio.

- Social effects of an aging population include challenges to traditional family structures.
- Political and economic effects include changes to policies relating to public retirement funds.

Review Questions

Use complete sentences to answer the questions.

1. **Apply Conceptual Vocabulary** Consider the term *urbanization*. Write a standard dictionary definition of the term. Then provide a conceptual definition—an explanation of how the term is used in the context of the chapter.

2. What challenges might a country like Japan face in bringing up its rate of natural increase?

3. How does the environment affect the growth and distribution of population?

4. What is the significance to geographers and demographers of a population that follows the shape of the letter *J* when plotted on a graph?

5. Compare and contrast the demographic transition model and the epidemiological transition model.

6. What factors could cause a population's doubling time to increase or decrease?

7. What present-day trends might reinforce Malthus's concern about overpopulation?

8. Define and give examples of antinatalist policies.

9. Define and give examples of pronatalist policies.

10. What changes might a neo-Malthusian thinker project for a country with an aging population and an immigration policy such as Germany's?

11. Very few countries are thought to be in Stage 5 of the demographic transition model at present. Do you think many more countries will experience Stage 5? Why or why not? What would be the consequences?

12. What evidence is there that antinatalist or pronatalist policies can be difficult to reverse? Why do you think this is the case?

13. What does a country's elderly dependency ratio help explain about its politics, economy, and culture?

■ Interpret Graphs

Study the graph and then answer the following questions.

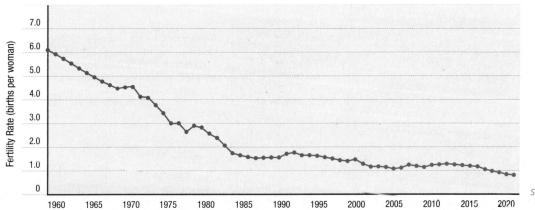

Total Fertility Rate in South Korea, 1960–2021

14. **Analyze Data & Information** Based on the graph, is South Korea experiencing positive or negative population growth? Explain.

15. **Ask Questions** What other information would you need to determine South Korea's overall population growth rate (or decline)?

16. **Predict Outcomes** What types of challenges might South Korea anticipate based on the population trends shown in the chart?

17. **Analyze Data** Do you think South Korea's natalist policies have changed between 1960 and today? Why or why not?

Chapter 5
Migration

Critical Viewing Rohingya refugees carry their belongings through a rice paddy in Bangladesh after fleeing from Myanmar. ▋ What cultural and economic effects might this migration have on Bangladesh?

Source: Jorge Silva/Reuters

Geographic Thinking How does migration shape our world?

5.1 Why Do People Migrate?	5.2 Types of Migration	5.3 Refugees and Internally Displaced Persons	5.4 Migration and Policy	5.5 Effects of Migration
	National Geographic Explorer Jason De León	**Case Study:** Surviving War in Syria	**Feature:** Walls That Divide Us	
	Case Study: Migration from Central America		**National Geographic Collaborator** Danielle Sanders	

5.1 Why Do People Migrate?

When people migrate, they take with them their customs, values, and beliefs—in short, their culture—creating a link between their origin and destination. Geographers study these links to better understand how migration affects populations and, in turn, changes the world culturally, politically, and economically.

What Is Migration?

Learning Objectives
LO 5.1 Compare the different types of human mobility.
LO 5.2 Define net migration rate.

Humans have been on the move for 2 million years. Explorers have journeyed across oceans in search of resources, and armies have crossed continents to establish empires. Human mobility has transformed the planet through the spread of cultures and ideas. **Mobility** includes all types of movement from one location to another, whether temporary or permanent or over short or long distances. Temporary, repetitive movements that recur on a regular basis are called **circulation**. This might involve short distances, like walking a few blocks or riding the bus a few miles to school each day, or it might mean long-distance movement like the seasonal routines of farm workers or retirees traveling to spend winter in a warmer place. The permanent movement of people from one place to another is called **human migration**.

Human movement affects the population of both the place of origin and the destination. **Emigration** is movement away from a location, and **immigration** is movement to a location. The difference between the number of emigrants and immigrants in a location such as a city or a country is the **net migration**. If more people emigrate from Japan than immigrate to Japan, the entire population decreases, creating a negative net migration. If fewer people emigrate from France than immigrate to France, the population rises and the net migration is positive.

Long-distance migrations can be dangerous, time consuming, and physically or mentally demanding. People—young, old, weak, strong, entire families—die trying to migrate. So why do people do it? Geographer Ernst Ravenstein studied the movements of migrants in England during the 19th century and proposed several principles or "laws" that describe trends in migration. Through his observations, Ravenstein concluded that economic conditions push and pull people in predictable directions. He also noticed that population size and distance affect migration.

Ravenstein's thinking laid the groundwork for the **gravity model**, which geographers derived from Newton's law of universal gravitation to predict the interaction between two or more places. (A diagram of the gravity model appears in the Maps and Models Archive at the end of this unit.) When used to describe migration patterns, the model suggests that as the population of a city increases, migration to the city increases, and as the distance to a city grows, migration to that city decreases. While economic circumstances play a major role in migratory decisions, environmental and political conditions, as well as demographic and cultural factors, also drive migration.

Geographic Thinking

1. Compare human migration and circulation.
2. Describe factors that might encourage young adults to migrate more often than families.

Model: Ravenstein's Laws of Migration

Ravenstein first published his laws of migration in a British journal in 1885.

1. Migration is typically over a short distance.
2. Migration occurs in steps, like from a rural area to a nearby city, and then perhaps on to a larger city.
3. Long-distance migrants often move to places of economic opportunity (urban areas).
4. Every migration generates a movement in the opposite direction, or a counterflow (not necessarily of the same number of migrants).
5. People in rural areas migrate more than people in cities.
6. Males migrate over longer distances than females.
7. Most migrants are young adult males.
8. Cities grow more by migration than by natural increase.
9. Migration increases with economic development.
10. Migration is mostly due to economic factors.

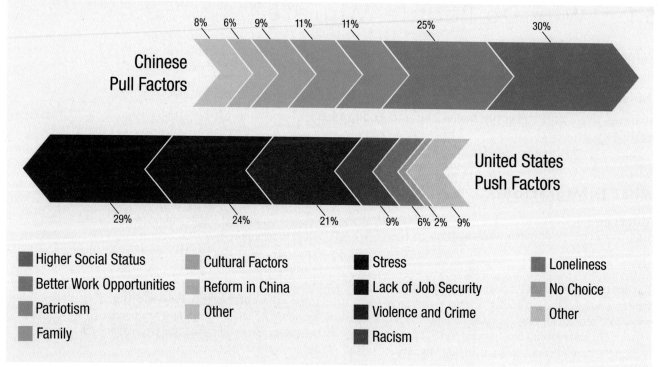

Push-Pull Factors for Chinese Students Deciding to Return to China, 1997 Many people migrate for a better quality of life, whether real or perceived. The arrows in this classic migration diagram show the pull factors (green) and push factors (purple) listed by Chinese students who chose to return to China after studying in the United States. ▌How do the push and pull factors relate to each other, and what does that say about the students' experiences in the United States? ▌Do you think that the factors might be similar or different today?

Source: David Zweig, Studies in Comparative International Development

Push and Pull Factors

Learning Objective
LO 5.3 Describe causal factors in migration decisions.

A combination of factors—rarely just one—typically drive people to migrate. For example, many changes are currently transforming rural communities worldwide and causing people to migrate. Declining agricultural productivity due to climate change; the use of farming equipment that reduces the need for laborers; and property-rights conflicts over boundaries, inheritance, and eviction impact rural populations and serve as reasons for migration. Such reasons are called push factors. A **push factor** is a negative cause that may compel someone to leave a location. Gang violence and hurricane damage are further examples of push factors. Sometimes push factors leave migrants no choice but to leave, as in the case of the millions who fled the civil war in Rwanda in 1994 that left an estimated 1 million dead and many more in fear of their lives. You will read more about types of involuntary migrations later in this chapter.

On the other hand, a **pull factor** is a positive aspect that attracts someone to a new location. A stable government, safe neighborhoods, and plentiful job opportunities are all pull factors. It makes sense that migrants perceive push factors more accurately than pull factors, since they are more familiar with their homeland than their potential destination. For instance, a community in a destination city may have a long-standing reputation as a safe, established neighborhood, but recently houses have been abandoned and crime is rising. In this case, a potential migrant's perception of the neighborhood as a pull factor is not entirely accurate. Perceptions about destinations are heavily influenced by social networks and media. Friends and family living in the destination country are important sources of information for prospective migrants, but images and stories from social media, television, and movies may have nearly as much impact.

Technology has transformed migration in recent years in both positive and negative ways. Though not all prospective migrants have access to the technology, smartphones and social media help many people decide when, where, and whether to migrate. For example, some Syrians who fled their country use cell phones to maintain contact with people back home and to convey what is happening in their destination country. Texts from family or friends and posts from trusted organizations provide information about increased conflict in a war-torn area, job opportunities, or the best time to move. Technology also offers migrants access to information about local resources, such as immigration lawyers and available health-care options. However, misinformation on the internet and the distrust that can result is problematic for finding safe passages and knowing who to rely on in a new community.

Economic Factors Most people migrate to support themselves and their families. Stable employment and higher wages are enticing economic pull factors that affect many populations, such as recent college graduates searching for career-starting opportunities and laborers looking for seasonal and permanent work in other countries. The Bracero Program, initiated in 1942, was created by the U.S. and Mexican governments to recruit much-needed Mexican laborers to replace American agricultural and transportation industry workers who were mobilizing for World War II. The expectation of earning the same wage as American farm workers pulled almost 4.6 million Mexican laborers to the United States over the course of the program, which ended in 1964.

Economic conditions are sometimes push factors. Job loss and the lack of employment opportunities can force people to look elsewhere for work. The global recession of 2008 greatly affected migration around the world. Ireland, for example, experienced negative net migration correlated with the downturn in its economy. High unemployment, poor career prospects, and low salaries pushed people away. As Ireland's economy improved by 2015, the country began to experience positive net migration.

Cultural Factors Cultural reasons for migration are often push factors involving discrimination, persecution, and political instability resulting from cultural diversity. Lack of religious freedom, racism, and unequal treatment have driven humans from their homelands for centuries. In the 1600s, the Pilgrims and Puritans broke from the Church of England and fled to North America to escape religious persecution. Three hundred years later in Turkey, authorities forcibly dispersed Kurdish people to different parts of the country in order to dilute the Kurdish culture. The president of Turkey in the 1920s, Mustafa Kemal Atatürk, saw the Kurds as a threat to his secular, modern vision for Turkey.

Gender roles and discrimination also affect migration. Since most people move for economic reasons, gender roles in terms of earning potential and responsibility for the family often determine who will migrate. When greater numbers of men are present in the workforce, more men relocate to provide for their families, as described by Ravenstein's laws. Recently, however, more women in many countries have taken on the role of provider or have chosen not to marry and start families. In some Southwest Asian and North African countries, legal barriers have been placed on female emigrants, lowering the rate of emigration from these gender-unequal states. When a traditional couple cannot migrate together because of restrictions on females leaving the country, emigration is limited for more than just females.

Demographic Factors Demographics—the characteristics of a population such as age, education level, and geographic location—often play a role in migration. For example, some people located in rural communities migrate because of lack of access to health care. Rapidly growing population and imbalanced gender ratios also spur migration. Large populations competing for limited employment opportunities and resources push people to move to less-populated regions. In Chapter 3, you read about China, which in 2018 had a ratio of 128 boys to 100 girls for children age 14 and younger. This imbalance between genders and the competition for wives result in men facilitating the migration of brides from Cambodia, Vietnam, or Russia. Thousands of women experiencing poverty or other difficulties in their home countries have migrated to China to get married.

Political Conditions and Conflict Historically, the promises of peace and freedom have served as compelling pull factors for countless migrants. For example, during World War II, millions of people left Europe to escape the Nazi regime and find security in the United States, Canada, Australia, and South America. In Syria, a civil war that began in 2011 pushed more than 5.7 million Syrians—including 2.7 million children—to flee their homeland by 2022. A case study about the Syrian crisis appears later in this chapter. As a result of multiple push factors—political upheaval, socioeconomic instability, human rights abuses, and violence—more than 5.6 million Venezuelans emigrated to Colombia, Brazil, and other neighboring countries in the

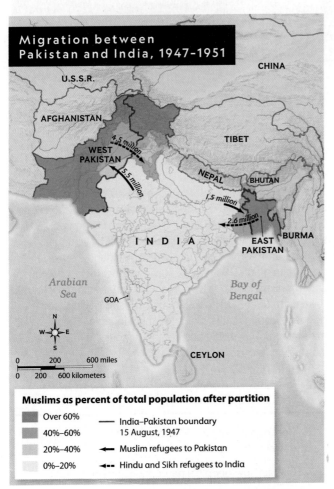

When the British left South Asia, it was divided into Hindu-majority India and Muslim-majority Pakistan. Cultural factors influenced millions of Hindus and Muslims to migrate.

Critical Viewing Freshwater flooding caused by Hurricane Florence, which struck the Carolinas in September 2018, left many homes in Lumberton, North Carolina, completely uninhabitable. Adverse physical conditions such as flooding can push people from impacted regions. ▮ What other factors related to Hurricane Florence might influence a Lumberton resident's decision to rebuild or relocate?

Source: JOHNNY MILANO/ The New York Times

years 2015–2020. The United Nations Refugee Agency calls this "one of the largest mass population movements in the history of Latin America."

Environmental Conditions A variety of environmental factors can also influence migration decisions, pushing or pulling people to relocate. A desirable climate or landscape can pull migrants to certain regions. Adverse physical conditions, including intense heat, drought, or substantial flooding, can push people from affected regions. Hurricane Katrina, the storm that battered the Gulf Coast of the United States in 2005, displaced hundreds of thousands of people and flooded much of the city of New Orleans. One year after the storm, almost 50 percent of residents had returned as the city continued to rebuild. For others, the damage to their homes was too extensive and costly to return and rebuild. Additionally, some residents with fewer ties to the city, including renters and people with no family members in Louisiana, simply relocated, many to Houston, Memphis, and Atlanta.

Scientific evidence indicates that climate change contributes to crop failure and water scarcity, intensifies storms, and causes sea levels to rise, displacing people from their homes around the world. The World Bank predicts that more than 200 million people worldwide will be displaced from their communities by 2050 because of problems related to climate change, and the figure likely will be higher. Some people will adapt to the new environmental conditions by voluntarily leaving home as what the UN Migration Agency terms "climate migrants," with most staying in their home countries. Others will be forced to become what other authorities call "climate refugees" as more areas, especially island, coastal, or desert communities, become uninhabitable. These migrants may have little choice about moving away, either internally or crossing international borders.

Geographic Thinking

3. Describe the types of factors that influence people's decision to migrate.

4. Explain how people's perceptions affect migration.

5. Explain why different communities have different push and pull factors.

5.2 Types of Migration

Think about why people might decide to leave the place they call home, and all that is involved in making such a move. How strong do the push and pull factors have to be to cause someone to endure the hardships and uncertainty of migration?

Voluntary Migration

Learning Objectives
LO 5.4 Describe types of voluntary migration including chain migration, step migration, and rural-to-urban migration.
LO 5.5 Compare transnational and internal migration.

Geographers distinguish between two types of migration: **voluntary migration**, in which people make the choice to move to a new place, and **forced migration**, in which people are compelled to move by economic, political, environmental, or cultural factors. In many cases throughout history, people have been enslaved through violent action by others and forced to move. But other instances blur the line between voluntary and forced migration.

People voluntarily migrate for many reasons. As you have read, push factors such as joblessness or drought might cause them to leave their current home; pull factors like economic opportunity or a mild climate might draw them to a new home. Whatever reasons migrants have for moving, these factors must outweigh the inconvenience of having to travel with everything they own or to leave possessions behind, as well as the uncertainty of what life will be like in the new location.

Crossing an international border brings an additional set of challenges. It often requires immigrants to travel long distances under difficult conditions, to seek permission to enter the destination country, and, once they've arrived, to attempt to adapt to the culture—the language, customs, and indeed, every aspect of life in a new country. The United States hosts the largest number of immigrants (51 million in 2020), followed by Germany (16 million), Saudi Arabia (13 million), Russia (12 million), and the United Kingdom (9 million). The countries with the most emigrants were India (with 18 million persons living outside of their country of birth), Mexico (11 million), Russia (11 million), and China (10 million). The single largest flow of immigrants from one country to another is from Mexico to the United States.

While about half of all foreign-born residents in the United States come from Latin America (primarily from Mexico, Central America, and the Caribbean), the fastest-growing group of immigrants comes from Asia. Between 2010 and 2017, Asians made up 41 percent of all U.S. immigrants. China and India contribute the largest proportion of Asian immigrants, with many settling in New York City. On the other side of the country, the city of Los Angeles is home to large groups of Filipino, Vietnamese, and Korean immigrants. Sizeable Asian American communities have sprung up in the South as well, in Atlanta, Georgia; Raleigh, North Carolina; and Austin, Texas. More and more, these new Asian residents are highly educated. Between 2000 and 2009, 30 percent of Asians arriving in the United States had a college degree. From 2010 to 2017, that number rose to 45 percent.

Overall, the tendency is for people to emigrate from countries of the periphery and semi-periphery and immigrate to countries with stronger economies. As of 2020, about 3.6 percent of the world's population—or 281 million people—were living outside the country in which they were born. The median age of international immigrants was 39 years, and more than 48 percent of immigrants were women.

In **transnational migration**, immigrants to a new country retain strong cultural, emotional, and financial ties to their country of origin and may regularly return for visits. In a sense, transnational migrants live in two cultures—and two countries—at once. Often, they send money to help support their families back home. They also use technologies such as social media to keep in close touch with the families and friends they left behind. Today there are three major flows of transnational migration in the world: from Latin America to North America, from Southwest Asia to Europe, and from Asia to North America. There will certainly continue to be changes in these major transnational flows, due to conflict, economic conditions, or climate change impacts.

Despite the large number of immigrants living around the world, it is more common for people to move from place to place within a particular country. Movement within a country's borders is called **internal migration**. Recall Ravenstein's laws, which state that most migration occurs over short distances. Moving a short distance is significantly easier than moving a long distance, both physically and psychologically. This is because of the **friction of distance**, a concept that states that the longer a journey is, the more time, effort, and cost it will involve. Consider the similarities and differences among cities around the world. How do towns and cities that are close together tend to be similar? How do towns and cities in different countries, regions, or hemispheres tend to differ? (See Unit 6 on cities and urban land use.) Factors such as language, customs, and climate can present difficult hurdles for international migrants. When moving from place to place within a country as internal migrants, however, they may face smaller changes—people likely speak the same language and may have similar social and cultural traditions.

Internal voluntary migrations often happen in waves or patterns. For instance, in the years following World War II (which ended in 1945), Americans moved in large numbers from the Northeast and Midwest to the South and Southwest, a broad region that became known as the Sunbelt. Think of the pull factors: taxes in the Sunbelt were low and land was cheap, which drew businesses and factories to the region, creating many job opportunities. The warm weather was a draw as well, and the invention of affordable, portable air conditioners made the summer's high heat bearable. So many people and companies made the migration to the Sunbelt that by 1962, California had become the most populous state in the country. And enough companies had left industrial cities in the Northeast and Midwest (an economic push factor) that the regions taken together were given a new, informal name—the Rustbelt—that highlights their abandoned factories, deindustrialization, and economic decline.

The COVID-19 pandemic caused extensive disruption to work and life for people across the planet. In the United States, the rise of prolonged remote working arrangements caused tens of millions to move temporarily and some to permanently relocate. Urban planner Richard Florida of the University of Toronto predicted in late 2021 that 14 to 23 million Americans would move, most of them as internal migrants: "People are asking deep questions about how and where they want to live." His analysis of U.S. Post Office data showed that these migrants are moving from large, densely populated urban centers to smaller cities and suburbs, with workers and even companies moving to low-tax, warmer-weather states in the South and Southwest. Florida found the three zip codes with the most new-arrival changes of address in 2021 were in areas outside Houston and Austin, Texas, and Jacksonville in the state of Florida.

Unlike the move to the Sunbelt, some internal voluntary migrations arise from long cultural tradition. **Transhumance** is a form of migration practiced by nomads who move herds between pastures at cooler, higher elevations during the summer and lower elevations during the winter. Transhumance is an international migration when nomads cross national borders during their seasonal movements.

Chain Migration Once immigrants have established themselves, they share information about their new home with people from their place of origin. Family members,

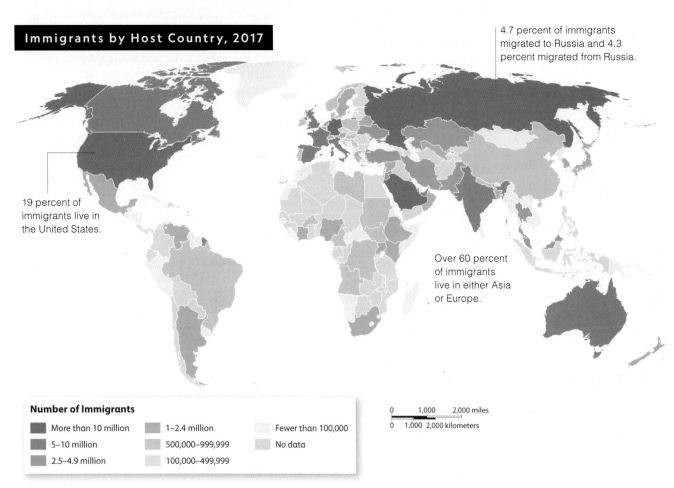

Immigrants by Host Country, 2017

4.7 percent of immigrants migrated to Russia and 4.3 percent migrated from Russia.

19 percent of immigrants live in the United States.

Over 60 percent of immigrants live in either Asia or Europe.

Number of Immigrants
- More than 10 million
- 5–10 million
- 2.5–4.9 million
- 1–2.4 million
- 500,000–999,999
- 100,000–499,999
- Fewer than 100,000
- No data

Reading Maps The number of immigrants in the world jumped by 85 million people between 2000 and 2017, and the majority of immigrants live in more economically developed countries. ▎What does this map suggest about the regions where the fewest immigrants reside?

friends, or other members of the community often follow in their footsteps. This type of migration, in which people move to a location because others from their community have previously migrated there, is called **chain migration**. Immigrants who follow others to a new location usually have similar reasons for leaving their home country, and they have an added pull factor—their way has been eased by those who arrived before.

Between 1820 and 1930, a large number of Irish immigrants—as many as 4.5 million—arrived in the United States. They were pushed out of Ireland by a series of factors, from oppression and religious persecution by the British government to starvation during the Potato Famine of 1845-1849. Chain migration helped determine where they ended up. The life they found in the United States was not easy. Irish immigrants left behind a rural lifestyle and tried to find work in industrialized cities such as Boston, New York, and Chicago, but the unskilled labor they could provide usually did not pay well. The Irish also experienced significant prejudice in their adopted country. To get by, they stuck together, sharing crowded housing with other Irish immigrants, saving money to send home, and bringing more family members and friends over from Ireland. In the 1840s, nearly half of all immigrants to the United States were from Ireland. Eventually, Irish immigrants and their descendants would dominate politics in many large U.S. cities in the 20th century. Indeed, the significance of this huge chain migration is apparent in the strong presence of Irish American culture throughout much of the United States today.

Step Migration Migration doesn't always occur along the most direct route. A migrant may set out to move from a rural farm to a big city but end up living for a time in villages, towns, or cities along the way. This series of smaller moves to get to the ultimate destination is called **step migration**. Whether a migration is voluntary or forced, most migrants have limited financial resources that make it necessary to take a long move in smaller steps. An occurrence that holds migrants back is called an **intervening obstacle**; an occurrence that causes migrants to pause their journey by choice is called an **intervening opportunity**.

For example, a family in India might set out from a farm in rural Bihar intending to migrate to a big city such as New Delhi but find that they've run out of money before they reach their destination and stop to look for work in a village along the way. The family might then move on to a larger village, town, or small city as they make their way toward their goal. Depending on their circumstances, the family might stay temporarily before they set out again for New Delhi, or they might remain in the village, town, or small city for good.

Rural-to-Urban Migration One of the most important migration trends at present is the movement of people from rural areas into cities. In 1950, 30 percent of the world's population lived in urban areas. Today that number is more than 57 percent, and it is projected to rise to 68 percent by 2050. Much of this urbanization is happening

Bridget Casey (center) sailed with nine of her children on the S.S. *Berlin* from County Cork, Ireland, to Ellis Island in New York City in 1929. They were following Casey's husband and two older chidren who had settled in Connecticut four years prior. One of the Casey children was left behind in Ireland after failing to pass an Immigration Department test.

Source: Bettmann/Getty Images

in Africa and Asia. People leave rural areas as the agricultural opportunities decline and head to cities where jobs may be easier to find.

Urban places offer a greater concentration of jobs than rural areas. In fact, towns and cities are responsible for more than 80 percent of all goods and services produced around the world. Cities also offer increased access to education and health care. Many people, especially women, find more personal freedom in cities than in rural areas. Greater access to reproductive health care promotes empowerment, and the urban lifestyle offers women some independence from their families, which might allow them to escape some of the traditional restrictions placed on them. In rural China, young women often quit school at age 16 or 17 to move to large cities and find work in textile or electronics factories. These jobs can be very difficult, with long hours, low pay, and miserable conditions. But they also offer the women freedom and independence from restrictive family life.

Some of the challenges brought about by urbanization include poverty and inequality, giving rise to difficult conditions for the poor. As cities grow more crowded, rural-to-urban migrants—especially those living in poverty—may face such issues as high crime rates, inadequate housing, and pollution.

Guest Workers Not all migration is intended to be permanent. **Guest workers** are migrants who travel to a new country as temporary laborers, like the Mexican workers who participated in the Bracero program. Guest workers are brought in to work jobs that cannot be filled by citizens of the host country, usually involving manual labor, agricultural work, or work in the service industry. For instance, after World War II, West Germany needed young workers to fill its factories and mines due to a postwar economic boom. The government brought in millions of Turkish workers to help rebuild the country's infrastructure. Guest workers are typically drawn by the opportunity to make more money than they could at home. Many of them send money home to family members or save money for when they return home. As in step migration, in which people change their original plans, guest workers often intend to remain in their host country for a short period of time but later decide to stay permanently. In fact, today, between 2.5 and 4 million people of Turkish descent live in Germany, making up the country's largest immigrant group. With globalization and improved transportation, however, **circular migration**—when migrant workers move back and forth between their country of origin and the destination country where they work temporary jobs—is becoming more common. The money workers earn in the destination countries may have greater purchasing power in their home countries, where prices are lower.

Model: Distance Decay

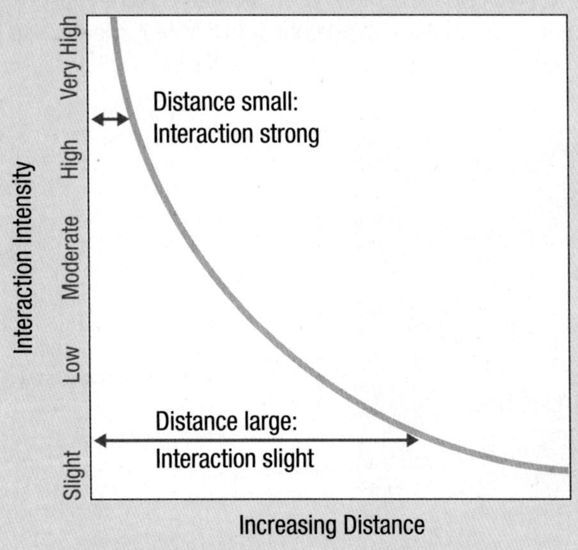

[Graph showing Interaction Intensity (y-axis: Slight, Low, Moderate, High, Very High) vs. Increasing Distance (x-axis). Curve shows: Distance small: Interaction strong; Distance large: Interaction slight.]

The farther away an object is, the harder it is to see clearly. This statement also applies to an important geographic concept and model called distance decay. The term is used to describe how distance affects interactions between locations. For example, people know more about their own town or a nearby city than they do about a city in another country. And the farther away a place is, the less people tend to know about it. Distance decay can affect where migrants decide to move. But today the effects of distance decay are lessened as technology connects us in new ways and puts knowledge about distant locations at the fingertips of anyone with an internet connection. ∎ How do you think the construction of new roads or railways and improvements in communication technology affect distance decay?

Forced Migration

Learning Objectives

LO 5.6 Define the terms internally displaced person, refugee, asylum, and human trafficking.

LO 5.7 Describe causal factors that lead to forced human migration.

For many, the concept of home brings to mind feelings of comfort, safety, and identity. People grow attached to the places they live, and it would take quite a lot of persuasion to convince them to move. But not everyone who leaves their home does so by choice.

As you have read, forced migration occurs when people are compelled to leave their homes by extreme push factors. People might be forced to move because of conflict, political upheaval, or natural disaster. They might be fleeing from persecution due to their ethnicity, religion, or social or political beliefs. They might be escaping the violence of war. **Refugees**, for instance, are people who are forced to leave their country for fear of persecution or death. Because of the danger they face at home, refugees may be granted special status when they attempt to enter a new country. They have the right to request **asylum**, or the right to protection, in the new country. Migrants who do not meet the definition of refugee established by the United Nations

High Commissioner for Refugees (UNHCR) are not eligible for asylum and must find another avenue to enter a foreign country. **Internally displaced persons** (IDPs) are people who have been forced to flee their homes but remain within their country's borders. IDPs include those running from conflict and those evacuating their homes because of a natural disaster such as a flood, earthquake, or hurricane.

The United Nations (UN) reported that the global number of forcibly displaced persons had rocketed from under 35 million in 2009 to more than 84 million by 2021. More than 26 million were refugees who had fled to escape conflict or persecution in places such as the Democratic Republic of the Congo, Ethiopia, Syria, and Myanmar. And 3.5 million were asylum seekers waiting to learn if they would be granted refugee status.

As is discussed further in Chapter 11, Russia invaded its neighbor Ukraine in early 2022, resulting in millions of refugees as well as millions more IDPs. The UNHCR found that within just three months, this conflict had caused an estimated 6.8 million Ukrainians to flee their country – the majority of these refugees crossing west into bordering Poland, Romania, and other European countries. The UNHCR describes a version of transnational migration here, too: more than 2 million of the 6.8 million Ukrainians who fled their homeland in the first weeks of the Russian invasion had later returned, in what the UN Refugee Agency called highly volatile and unpredictable "pendular" cross-border movements. Additionally, 8 million or more Ukrainians were displaced internally in the first half of 2022 – moving to a hoped-for safer area of their country.

For people forced from their homes, migration can be significantly more difficult than when the move is voluntary. Especially if they are fleeing from violence, people often do not have time to plan properly. They might have to leave with only the possessions they can carry, with no idea where their journey will take them. For many, like the millions of refugees who have fled from the civil war in Syria since 2011, the journey itself is fraught with danger, and they often end up living in harsh locations such as refugee camps for months or years.

Many Rohingya people, members of an ethnic minority in Myanmar, have experienced forced migration. The Rohingya are Muslims in a majority Buddhist country, and they have faced persecution for years. Their citizenship was denied by a law passed in 1982, leaving them "stateless" and subjecting them to extreme difficulties with employment, heath care, and more. In August 2017, Myanmar's military carried out widespread attacks on the Rohingya, burning villages and torturing and killing many individuals. Since then, hundreds of thousands of Rohingya have fled Myanmar for Bangladesh, making a dangerous journey through jungles and over mountains or risking a harrowing sea voyage across the Bay of Bengal. Most of those who have reached Bangladesh are women and children. Forty percent are under the age of 12. The UNHCR documented this humanitarian crisis, finding that 914,000 Rohingya had fled their homes by 2021, with most sheltered in a single refugee camp in Bangladesh called Kutapalong, the largest in the world. Indicative of the dangers that refugees face in housing built with flimsy bamboo and canvas shelters, a fire there in early 2021 displaced about 50,000 residents. The COVID-19 pandemic brought yet another threat to this and other refugee camps everywhere. Their future here is, like that of most refugees everywhere, uncertain.

The largest forced migration in history was the transatlantic slave trade. From the 16th to the 19th centuries, between 10 and 12 million men, women, and children were transported from Africa to the Americas. They were captured by their enemies or by European or African slave traders and forced onto crowded ships bound for North, Central, and South America and the Caribbean. Many died on the horrific voyage across the Atlantic, and those who survived were sold into lives of hard labor, often on sugar, tobacco, and cotton plantations. The slave trade had a devastating effect in Africa as well. People lived in fear of capture, and depopulation, especially of working-age adults, gutted families and societies.

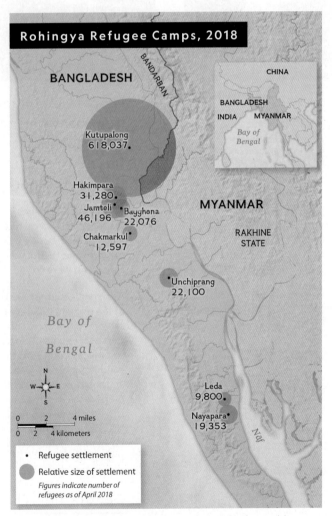

Kutupalong is the largest refugee settlement in the world, housing more than 600,000 Rohingya refugees in an area covering just 5 square miles.

Migration **125**

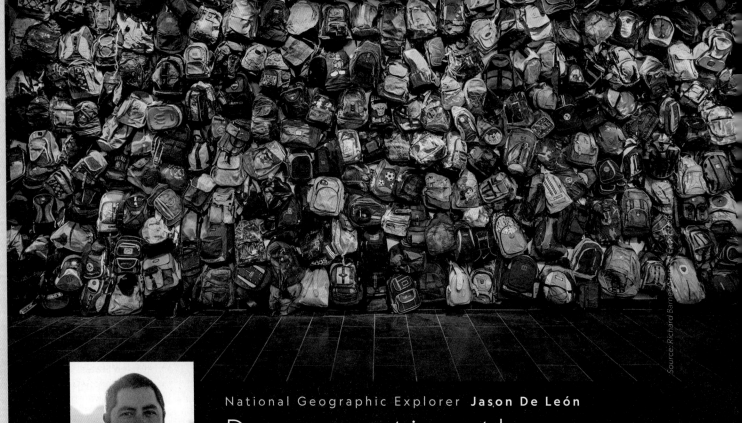

National Geographic Explorer **Jason De León**

Documenting the Stories of Migrants

De León collaborated with artist Amanda Krugliak on an exhibit called State of Exception/Estado de excepción, which includes a wall of discarded backpacks left by Mexican migrants in the Sonoran Desert.

Source: Michael Wells

Learning Objective
LO 5.3 Describe causal factors in migrations.

People view undocumented migration into the United States from many perspectives, often focusing on immigrants' effects on the politics, culture, and economy of the United States. Jason De León may be the first to study migration patterns from an unusual perspective—that of an archaeologist.

A spiky cactus bristles in the scorching Sonoran Desert sun. Caught on its spines—a baby's diaper. Miles away, a tattered backpack contains one roll of toilet paper, a love letter, and a prayer card. Beside it, a tiny child's shoe. Until Jason De León, a National Geographic Explorer and the founder of the Undocumented Migration Project, arrived on the scene, such items were considered garbage or never discovered at all.

As an archaeologist, De León applies scientific techniques to gain a deeper understanding of the migrants' experiences, paying particular attention to vulnerable groups such as women, children, and LGBTQIA+ individuals. De León studies artifacts he finds in the desert's harsh landscape, left behind by migrants from Mexico, Central America, and places farther south. In addition to collecting the objects they leave behind, he conducts interviews to help reveal the meaning of the items. De León explains, "We use archaeological surveys, linguistics, forensics, and ethnography to document how people prepare to cross the border, who profits from helping them, how they deal with physical and emotional trauma during their journey, and what happens to those who don't make it." His goal is to learn more about the impact of immigration on everyone affected. As of 2022, his Global Hostile Terrain exhibit had mobilized thousands of volunteers to document more than 3,400 migrants who had died crossing the Arizona desert.

Geographic Thinking

Describe how the migration data De León collects might help him identify patterns and trends among vulnerable groups of migrants.

Case Study

Migration from Central America

A mother and two young daughters, part of a caravan of thousands traveling from Central America to the United States in 2018, flee tear gas fired by U.S. Border Patrol agents over the security fencing that separates the United States from Mexico.
Sources: Kim Kyung-Hoon/Reuters

The Issue Since the U.S. Refugee Act became law in 1980, Central American immigrants, fleeing instability and violence in their countries of origin, have sought asylum in the United States.

Learning Objectives
LO 5.8 Compare immigrants, refugees, and asylum seekers.
LO 5.3 Describe causal factors as they relate to migration decisions.

By the Numbers

350,000
Central American immigrants living in the United States in 1980

3.8 million
Central American immigrants living in the United States in 2019

42%
of people apprehended at the U.S. southern border in 2016 came from the Northern Triangle

Source: The Brookings Institution

Beginning in the 1980s, the number of immigrants arriving in the United States from Central America rose dramatically. The reasons for this increase are numerous, including unstable political conditions and a lack of jobs, as well as high homicide rates and widespread gang violence in their home countries.

In 1980, immigrants had a new reason to consider coming to the United States. Congress passed the Refugee Act, which raised the ceiling for refugees allowed into the United States each year from 17,400 to 50,000. The act also affected who would be considered a refugee. Previously, the United States had granted refugee status only to those fleeing from a communist regime. The new law followed UN protocols, defining a refugee as anyone with a "well-founded fear of persecution on account of race, religion, nationality, membership in a particular social group, or political opinion."

Today, many of those seeking asylum in the United States come from El Salvador, Guatemala, and Honduras—the "Northern Triangle" of Central America. These countries, according to the aid organization Doctors Without Borders, are experiencing "unprecedented levels of violence outside of war zones." They have some of the highest homicide rates in the world, and kidnapping, extortion, and corruption are common. Sexual violence and violence against women are also widespread. Fear of these crimes has driven many from their homes, including families and unaccompanied children. Natural disasters, including a pair of devastating hurricanes in late 2020, have forced thousands more to flee Guatemala, Nicaragua, and Honduras. Between 1980 and 2019, the number of Central American immigrants living in the United States rose by more than a factor of 10, from 350,000 to 3.8 million. Movement restrictions as well as health and sanitation measures enacted by countries in the region as a response to the COVID-19 pandemic also impacted Central American migration flows.

In recent years, however, the United States decided that the difficult circumstances stemming from violence in the Northern Triangle do not rise to the level of persecution. In an effort to stem the flow of immigrants from this region, the U.S. government tightened restrictions on would-be immigrants. Many who arrive in the United States seeking asylum are deported back to their home countries. The irony is that the majority of the crime in these countries is committed by drug cartels and gangs, which is driven by demand for illegal drugs in the United States—and much of the violence is committed using weapons manufactured in the United States. The rising wave of Northern Triangle migration is reflected in numbers from 2021: the U.S. Border Patrol encountered 684,000 people from the three countries, including 309,000 Hondurans, 279,000 Guatemalans, and 96,000 Salvadorans. Outnumbering encounters with Mexican citizens during this time period, about 45 percent were single adults, 37 percent were traveling with family members, and 17 percent were unaccompanied minors. Nearly 60 percent of these migrants were expelled. ∎

Geographic Thinking

Explain whether the migration patterns of people coming from Central American countries to the United States are examples of forced or voluntary migration.

Slavery was outlawed in the United States more than 150 years ago, but a form of the institution still exists in the world today. **Human trafficking** is defined by the UN as "the recruitment, transportation, harboring, or receipt of persons by improper means (such as force, abduction, fraud, or coercion)." Victims are often illegally sold into forced labor, domestic servitude, and sexual exploitation. Human trafficking is a form of modern-day slavery, and according to the U.S. Department of Homeland Security, millions of people are trafficked around the world each year, including within the United States.

Traffickers prey on vulnerable people, such as homeless and runaway youths, poor immigrants, and people who have suffered physical and psychological trauma. They often smuggle their captives across international borders because victims find it more difficult to seek help in a foreign country. Nearly every country in the world is affected by human trafficking, either as a country of origin, a country of transit, or a destination country for victims.

Consequently, human trafficking frequently goes unreported and is considered a "hidden crime." Despite its invisibility, it is both a grave violation of human rights and a major issue. Experts estimate the crime brings its perpetrators billions of dollars each year. It is the second most profitable form of international crime, after drug trafficking.

The Gray Area Despite the terms' clear definitions, it can at times be difficult to distinguish between voluntary and forced migration. Consider the example of the Irish who immigrated to the United States in the 1800s. The emigrants who left Ireland during the Potato Famine of 1845–1849 were forced out by starvation, but the wave of Irish migration had actually begun earlier, in the 1820s. The country was under colonial rule by the British and, until 1829, Irish Catholics were prohibited from owning land, carrying weapons, and voting. In addition, the Industrial Revolution that swept England beginning in the mid-1700s left Ireland behind. The smaller country's agricultural economy and real estate values suffered, leaving the Irish even poorer.

For many, the conditions had become too difficult, and Irish citizens boarded ships bound for cities in Great Britain such as Liverpool and Glasgow, or they made the long sea journey to the United States to make a new life for themselves in cities such as New York and Boston. But many others remained behind in Ireland under the same conditions. The question for geographers, then, is this: When that wave of Irish immigrants boarded those ships, could their experience be described as voluntary or forced migration? Over the centuries and across the globe, myriad migrants have made their way—and continue to advance—in this "gray area."

Geographic Thinking

1. Explain how a region's climate, economy, politics, and culture might affect migration to and from the area.

2. Explain the degree to which friction of distance might explain the decisions of Irish people considering chain migration in the late 1800s.

5.3 Refugees and Internally Displaced Persons

Conflict and instability have driven millions from their homes, creating a widespread crisis that is one of the most challenging issues in the world today. For governments, it is a political and administrative issue. For the people forced to flee, it can be a matter of life or death.

Challenging Obstacles

Learning Objectives

LO 5.9 Compare the migration experiences of refugees and internally displaced persons.

LO 5.10 Explain how changes in environmental conditions affect migration patterns.

Migrants who seek asylum in another country are known as refugees and those who remain within their home country's borders are known as internally displaced persons. IDPs were estimated to number 55 million in 2021. In either case, they are likely fleeing into unfamiliar territory by any means available. They must rely on the help of others and may be forced to pay to be smuggled to safety or over borders. IDPs must also navigate the natural obstacles of unknown topography and weather conditions. International border restrictions cause many refugees to take dangerous alternate routes. If crossing a border illegally, refugees live in fear of being apprehended by authorities. Getting caught could mean being separated from family members or being deported back to the homeland they are trying to escape.

Obtaining Refugee Status Today, refugees can seek asylum in any of the 145 countries that have ratified the 1951 Refugee Convention. The document was created by the United Nations to address the many people displaced by World War II. Originally, the UN defined refugees as people who leave their home countries out of fear of persecution. In 1967, it expanded its definition to include people escaping any conflict or disaster, such as the famine in Ethiopia in 1984–1985 that forced hundreds of thousands to leave the country rather than face starvation.

Official refugee status must be granted by the country providing asylum or by an international agency. Obtaining refugee status can be a long and difficult process. This is especially true when applicants have little documentation to prove their situation complies with the official definition of a refugee. Factors such as illiteracy, severe trauma, and memory loss can make it difficult for refugees to tell a convincing story. Once an asylum seeker is approved for refugee status, the host country is expected to provide civil rights, the right to work, and access to social services.

The United Nations High Commissioner for Refugees, established in 1950, continues to be an international resource for refugees and for countries offering asylum. Organizations such as the UNHCR work to help displaced people find safety, health care, and shelter and apply for refugee status. In recent years, the number of refugees has reached record highs, and fewer refugees have been able to return to their home country—or **repatriate**—due to ongoing conflicts or postdisaster conditions. Consequently, organizations have needed to plan and construct more long-term camps where refugees can hopefully find support for the needed adaptations to a new location—and new hope for the future.

Top Refugee Countries of Origin A large portion of the refugees in the world come from just a few countries. In 2021, for instance, more than a quarter of the world's 26.6 million refugees were Syrians, and half of the remainder were from Venezuela, Afghanistan, South Sudan, and Myanmar.

These countries and their people have been ravaged by war, violence, and persecution. In South Sudan, for instance, a civil war that began in 2013 has embroiled most of the country in conflict and violence. As the war has reached the fertile land in the south, farmers have not been able to produce as much food, bringing famine and food insecurity to much of the country. By 2018 these issues had forced more than 4 million people, or 1 out of every 3 South Sudanese citizens, from their homes. More than 1.7 million of them remain within South Sudan's borders; over 2.4 million have managed to escape to neighboring countries.

The Hidden Crisis of IDPs News stories about boats full of refugees making dangerous sea crossings or caravans traveling long distances on foot draw attention to the plight of refugees. But the struggles of IDPs do not always receive as much coverage. Stories of IDPs are often subsumed by the narrative of the larger conflict taking place in their country. Therefore, despite the fact that there are about twice as many IDPs than refugees in the world—about 55 million total in 2021—the refugee crisis is more well known and better understood.

IDPs remain within their country's borders for a number of reasons. Some choose to stay close to the homes they were forced to leave, hoping the factors that pushed them out will improve. Others don't have the money or the physical means to make a long journey. Still others might be trapped in the area by the violence or conflict they're attempting to flee in the first place.

While refugees rely on the protection of a foreign government for their safety, IDPs remain under the laws—and therefore the protection—of their own government. Sometimes, however, the government is involved in the conflict the IDPs are attempting to flee and is unable or unwilling to provide protection. In addition, armed conflict or an unfriendly government often make it difficult for aid organizations to reach IDPs. All of this makes IDPs a particularly vulnerable group.

Climate's Role Changes in environmental conditions—floods, drought, volcanic eruptions—have always spurred migrations. But as climate change intensifies these effects, a new category of migrant has emerged: climate refugees. These hazards result from an increasing frequency and intensity of weather events and are already impacting millions, particularly the most vulnerable living in climate hotspots. These events include exceedingly heavy rainfall and flooding, prolonged droughts, environmental degradation, tropical cyclones, and rising sea levels. Scientists predict that in the coming years, potentially millions of people living in coastal areas will be forced to move due to rising sea levels. Increasing temperatures are already diminishing agricultural production in some areas, while rising sea levels are taking some farmland out of production.

While climate refugees haven't been given an official definition yet, their numbers are on the rise—since 2008, severe weather has driven an average of 22.5 million people from their homes each year. Though the term *climate*

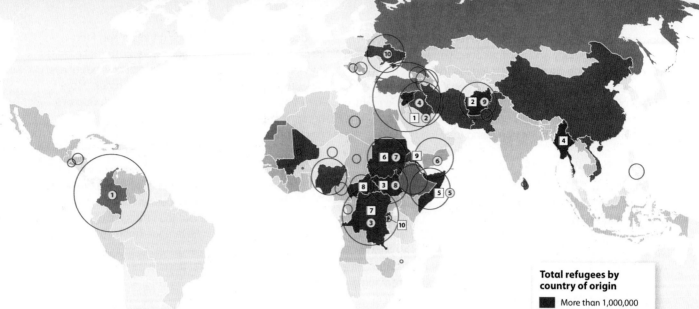

A Displaced World, 2017

How many people were displaced in 2017, and where were they from? Darker shaded countries produced the most refugees. (The top 10 countries of origin are numbered with white boxes.) Blue circles appear above countries with IDPs. (The top 10 countries with the most IDPs are numbered in blue below.)

Total refugees by country of origin
- More than 1,000,000
- 500,001–1,000,000
- 100,001–500,000
- 50,001–100,000
- 20,001–50,000
- 5,001–20,000
- 2,001–5,000
- 501–2,000
- 0–500

IDPs by country
- 2,000,000–7,700,000
- 1,000,000
- 100,000

Top 10 Countries with the Greatest Refugee Exodus, 2017

#	Country	Refugees
1	Syria	6,290,908
2	Afghanistan	2,621,091
3	South Sudan	2,439,848
4	Myanmar	1,106,555
5	Somalia	986,382
6	Sudan	691,430
7	D.R. Congo	611,875
8	Central African Republic	545,525
9	Eritrea	464,136
10	Burundi	439,329

Top 10 Countries with the Most IDPs, 2017

#	Country	IDPs
1	Colombia	7,677,609
2	Syria	6,150,005
3	D.R. Congo	4,351,376
4	Iraq	2,615,988
5	Somalia	2,116,705
6	Yemen	2,014,062
7	Sudan	1,997,022
8	South Sudan	1,903,953
9	Afghanistan	1,837,079
10	Ukraine	1,800,000

By the Numbers

55 million internally displaced people worldwide

26.6 million refugees worldwide

4.4 million asylum seekers

Sources: United Nations High Commissioner for Refugees, Internal Displacement Monitoring Centre; data from 2017

refugee seems to imply international movement, more than half of the people fleeing natural disasters each year become IDPs.

Climate change creates new patterns of displacement. For example, Bangladesh's southern coast was a traditional source of agriculture and trade for thousands of years. Within the last decade, however, unpredictable flooding and severe erosion have caused people to flee the area. Most have wound up in the urban slums of Dhaka, one of the world's fastest-growing cities. Bangladesh's southeast coast is also where nearly 1 million Rohingya refugees live in settlements, shown on the "Rohingya Refugee Camps, 2018" map earlier in this chapter, following Myanmar's acts of genocide. Deforestation and other environmental impacts caused by the presence of so many refugees have compounded the effects of erosion and other natural disasters in the area.

Geographic Thinking

1. Compare an internally displaced person and a refugee.

2. Identify three reasons—one political, one social, and one environmental—why refugees flee their homeland.

3. Describe the predicted impacts of climate change on migration patterns.

Critical Viewing Food is distributed to young victims of Hurricane Maria in Puerto Rico, where 86,000 residents were displaced by the storm in 2017. Economic and infrastructure problems on the island worsened exponentially after Maria hit, and in 2018, 123,000 more people left Puerto Rico than moved there. ▮ How can a storm's short- and long-term impacts on an area affect patterns of migration?

Sources: ERIC ROJAS/ The New York Times

Case Study

Surviving War in Syria

The Issue The civil war that broke out in Syria in 2011 has led to a humanitarian crisis, with more than 13 million Syrians forced to flee their homes.

Learning Objective
LO 5.9 Compare the migration experiences of refugees and internally displaced persons.

By the Numbers

6.7 million
Syrians internally displaced as of 2018

6.6 million
Syrians registered as refugees as of 2018

10%
of Syrian refugees live in refugee camps in Southwest Asia

80%
of Syrian refugees live in urban areas in Southwest Asia

Sources: Pew Research Center; United Nations High Commissioner for Refugees

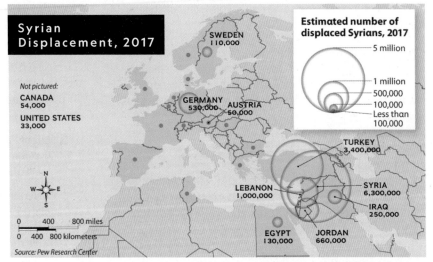

Numbers on the map estimate how many displaced Syrians were in each country, with the majority remaining in Syria as IDPs. Notice that most Syrian refugees stayed in neighboring countries, while a smaller fraction migrated to other regions.

In 2011, peaceful protest marches in Syria rapidly turned into a full-scale civil war. The marches were part of a wave of demonstrations, known as the Arab Spring, that occurred in Southwest Asia and North Africa beginning in 2010. Some earlier pro-democracy demonstrations, such as those in Tunisia, Egypt, and Libya, had been successful in overthrowing governments believed to be oppressive. Syrian pro-democracy activists were hopeful that peaceful protests calling for the resignation of President Bashar al-Assad would also succeed. By March 2011, demonstrations had spread throughout Syria. However, government troops responded by imprisoning, killing, and torturing demonstrators, including children.

What started as a demonstration by civilians quickly escalated into a complex war among religious groups within Syria. The violence against demonstrators spurred rebel groups to engage in combat with government troops, leaving civilians caught in the crossfire. Since 2011, the war has increased in complexity as more rebel factions formed and governments outside of Syria, such as Turkey, Russia, and the United States, backed different participants. Syrian neighborhoods became battlefields where families lived in fear of bombs, gunfire, imprisonment, and chemical weapons. It is estimated that the war has left more than 500,000 dead or missing.

By 2021, more than 13 million Syrian civilians had been forced from their homes. About 6.7 million were internally displaced. Of these, about 3 million fled to remote, besieged areas within the country. Most internally displaced Syrians live on the edge, struggling to find safe, affordable, healthy places to live. In 2019, U.S. president Donald Trump withdrew the majority of American troops from Syria, and the Turkish army advanced into the country. The Turkish government claimed it would establish a "safe zone" in Syria for the refugees Turkey had accepted during the war, but the invasion in reality unleashed another wave of displacements.

Since the uprising began, more than 6.6 million Syrians have fled the country and are registered as refugees. The majority of Syrian refugees went to urban areas in the region, including cities in Turkey, Jordan, Lebanon, Iraq, and Egypt. Just 10 percent settled in refugee camps in neighboring countries. In 2021, Jordan hosted two-thirds of a million Syrian refugees, with 20 percent in refugee camps—almost all with very limited means to cover even basic needs. About 1 million Syrian refugees fled to Europe. Even after obtaining asylum, refugees struggle for social acceptance and economic opportunities in their new country of residence.

Geographic Thinking

Explain the degree to which distance decay and the gravity model apply to the pattern of Syrian displacement shown in the map.

5.4 Migration and Policy

Since governments were first formed and national borders established, the movement of people has been affected by the border rules—or policies—of governments. Geographers analyze historical migrations to understand the effects migration patterns and policies have had on migrants and the places to which they move.

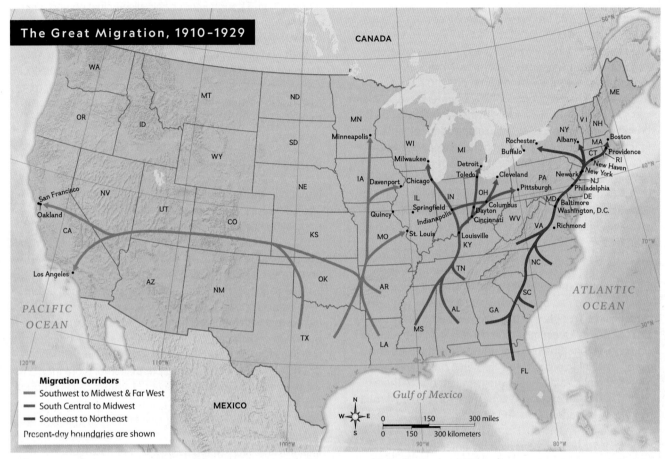

Reading Maps The map highlights the migration corridors that were established in the contiguous United States during the first part of the Great Migration. ▌ What social and economic impacts might these patterns of internal migration have had on the North, the South, and the West?

Historical U.S. Migrations

Learning Objective
LO 5.3 Describe causal factors in migrations.

U.S. policy and government action have influenced a number of important migrations in American history. One of the earliest and most infamous examples of this influence was the Trail of Tears, the forced internal migration of approximately 100,000 Cherokee, Creek, Chickasaw, Choctaw, and Seminole people from their land in the southeastern United States. The five tribes' lands, which were already attractive to European American settlers, became even more desirable to the United States when gold was discovered there. The takeover was finalized with the passing of the Indian Removal Act of 1830, which authorized the U.S. Army to force the tribes to move to one shared territory in Oklahoma.

Numerous voluntary migrations have also taken place throughout U.S history. In fact, internal migrations happen more frequently than international migrations, due largely to the friction of distance. The following historical examples illustrate voluntary internal migrations, some occurring after an international migration to the United States. They highlight both **interregional migration**—or movement from one region of the country to another—and **intraregional migration**, or movement within one region of the country. In each example, consider how U.S. policy played a role, what unintended effects may have resulted, and whether the goal behind the policy was achieved.

Migration 133

The Great Migration The Great Migration was a voluntary internal migration that occurred during the 20th century in the United States. Between 1916 and 1970, more than 6 million African Americans moved from the South to industrialized cities in the Northeast, Midwest, and West. The Great Migration altered the demographic makeup of the country. When it began, 90 percent of African Americans lived in the South. By the 1970s, 47 percent of African Americans lived in the Northeast, Midwest, and West.

Strong push factors propelled African Americans to leave the South. Racial discrimination and violence were rampant, and segregation laws known as Jim Crow laws severely restricted African American livelihoods. Moreover, a pest called the boll weevil wreaked havoc on cotton crops. Employed as sharecroppers or tenant farmers, most African Americans were ruined financially when a cotton crop failed. The main pull factor was economic opportunity, in the form of factory and mill jobs. Work opportunities opened up when the United States entered World War I and millions of men enlisted. Furthermore, U.S. immigration policy of the 1920s set **quotas**—or limits on the number of immigrants allowed into the country each year. This slowed the flow of Europeans into northern cities, leaving many jobs open in urban factories.

News of economic opportunities from growing African American communities in cities like Chicago, Detroit, and Baltimore spread among families and friends in southern communities. This caused migration waves to flow steadily from the South. Chain migration played a role in the Great Migration as **kinship links**—or networks of relatives and friends—led migrants to follow the same paths and settle in the same places as those who migrated before them. The move was an expensive struggle. Migrants often made the journey in stages, stopping to work in places on the way before continuing on. For many, this step migration to the final destination consisted of numerous stops and took years. For example, a sharecropping family from rural Alabama might have stopped in Nashville, Louisville, and Indianapolis on their years-long migration to Detroit. In each of these cities an intervening opportunity, in the form of a job or affordable place to live, would have held them for months or years. Intervening obstacles, such as a lack of funds or a sick family member, might also have paused the migrants' progress. Or the family might have stayed in one of those cities permanently without ever reaching their intended destination of Detroit.

Hispanic-Latino Migration Another massive, centuries-old and ongoing internal U.S. migration is that of Hispanics and Latin Americans. *Hispanic* is an overarching term that refers to U.S. residents whose cultural heritage traces back to a Spanish-speaking country, while *Latin American* (and derivatives Latino, Latina, and the gender-neutral term *Latinx*) refers to a native of Latin America. Hispanics and Latinos have inhabited parts of what is now the U.S. since the 16th century, and in the first half of the 19th century when Florida, the Louisiana Purchase, and the northern portion of Mexico were annexed, more than 100,000 Spanish-speaking residents became U.S. citizens. The 1850 census tallied more than 80,000 former Mexicans, 2,000 Puerto Ricans and Cubans, and another 20,000 people from Central and South America. The descendants of those 1850 residents are part of a Hispanic–Latin American population that has since grown enormously. By 1900, including immigrants from new U.S. possessions Puerto Rico and Cuba, these populations rose to almost a half million. Though xenophobic discrimination and even deportation campaigns in the early part of the 20th century temporarily reversed the trend, impressive Hispanic-Latino growth was augmented by decades of mostly voluntary immigration and high growth rates. The Hispanic-Latino population grew to 62.1 million (18.7 percent) in 2020; more than half (51.1 percent) of all growth in the United States between 2010 and 2020 came from this population.

Until recently, a large majority of the Hispanic-Latino population resided in the American Southwest—California, Texas, Arizona, Colorado, and New Mexico—but now about one-half of all Hispanics live outside these states. The geographical spread of America's Hispanic-Latino population has affected virtually all parts of the country; every state plus Washington, D.C., has seen growth in these populations between 2010 and 2020. Florida, California, and Texas have experienced increases of more than 1 million each in that decade. Both urban and rural areas have seen Hispanic-Latino population growth, much associated with industrial restructuring and recent job growth in construction, a wide range of service industries, agriculture, meat packing, and food processing. For example, the city of Chicago's population of more than 2.7 million in 2020 was 28 percent Hispanic-Latino. Though still among the smallest numbers in the country, North and South Dakota saw the highest Hispanic-Latino percentage growth in the 2010–2020 decade. In many rural counties, this growth has provided a "lifeline" that has offset White population declines. People of Mexican origin made up almost 62 percent of the country's overall Hispanic-Latino population as of 2019. The next largest group is of Puerto Rican origin, at about 10 percent. Through both immigration and widespread internal migration, Americans of Hispanic-Latino descent have made huge and ongoing impacts on the demographics, economies, cultures, and politics of the entire country.

Somali Migration Other internal migrations have occurred and still continue today in the United States. One contemporary example involves a refugee group from Somalia in East Africa. Somali refugees have been resettling in the United States since 1990. U.S. resettlement locations for this group are spread throughout the country in large cities such as Minneapolis, Houston, Atlanta, Boston, and San Diego. After making their international migration as refugees to U.S. resettlement locations, many make an internal secondary migration to other U.S. towns or cities where large Somali communities have emerged. This secondary migration has caused growing Somali

American Jacob Lawrence painted Panel 28 of his 1941 series, *The Migration of the Negro*. The 60-painting collection honors African Americans in the South, including his parents, who moved to cities in the North and West during the Great Migration.

Source: The Museum of Modern Art/Licensed by Scala/Art Resource, NY © 2019 The Jacob and Gwendolyn Knight Lawrence Foundation, Seattle/Artists Rights Society (ARS), New York

communities to form in Minneapolis, Minnesota; Columbus, Ohio; and Lewiston, Maine.

As Somali communities in the United States continue to grow, chain migration plays a stronger role. The increase in Somalis living in a given location causes more kinship links to form. Growing networks of Somali relatives and friends attract even more Somalis to the community. One such example of this can be seen in Minneapolis, which has the largest Somali community in North America. Not far from this urban center, a Somali community in the small city of St. Cloud, Minnesota, has grown. Some Somali refugees resettled directly in St. Cloud. Others migrated internally from within the United States, attracted to the small-town life and manufacturing jobs available there. Internal migration has trickled even farther as a few Somalis moved from St. Cloud to tiny, nearby agricultural towns such as Cold Spring, Minnesota.

Hmong Migration Another example of internal migration as a secondary migration occurred after Hmong refugees from Laos began moving to the United States in 1975. The Hmong had fought alongside U.S. soldiers in Laos during the Vietnam conflict. When the conflict ended, the new Laos government threatened to capture and kill Hmong soldiers and families who had sided with the United States. For the two decades that followed, more than 200,000 Hmong fled Laos to neighboring Thailand, where some spent up to 10 years in refugee camps awaiting resettlement. Most moved to the United States, with fewer moving to France, Canada, and Australia. Hmong refugees were widely dispersed in the United States. Many made internal secondary migrations to more populated Hmong communities in California, Minnesota, and Wisconsin. Leading factors that drove the Hmong resettlement pattern were kinship links and economic opportunity or sponsorship.

Numerous Southeast Asians, including Vietnamese and ethnic Chinese peoples, fled when the Vietnam conflict ended in 1975. As they sought asylum in the United States, it became apparent, given the circumstances, that American policy was too restrictive in its admission of refugees. The U.S. government quickly enacted the 1975 Indochina Migration and Refugee Assistance Act, which allowed 300,000 refugees from Southeast Asia into the country.

What Drives Policy?

Learning Objectives

LO 5.11 Define *xenophobia* as the term relates to migration.

LO 5.12 Explain how factors such as gender, age, asylum regulations, and other immigration-related legislation have driven government immigration policy.

Governments use immigration policy to achieve several purposes. The main goal behind the creation of most immigration policy is to meet labor market needs.

A secondary aim is to maintain or change current levels of immigration. Policies are also commonly structured to attract skilled workers, promote the well-being of immigrants and their integration into society, and address illegal immigration. Illegal immigration, in particular, has become an exceedingly complex issue to tackle. Related challenges such as age and gender discrimination, exploitation, and abuse of immigrants are difficult problems to uncover and address effectively.

Government policies intended to limit immigration have sometimes focused on the number of immigrants from a certain country or region. These limits may have as much to do with xenophobia—or fear and hatred of foreigners—as they do with the good of the country. In contrast, loosening quotas on immigrants from specific countries increases population diversity in the host country.

Contemporary and historical examples of factors that have driven government immigration policy include gender or age, asylum regulations, and other immigration-related legislation. By examining factors such as these, geographers can better identify how government immigration policies can foster positive patterns, limit negative patterns, and achieve a balance that works in conjunction with other existing policies.

Risks for Female Immigrants Gender and age play a role in the types of opportunities and risks that immigrants face. For instance, female immigrants face different opportunities and risks than their male counterparts. This is true even though the numbers of female and male immigrants are about equal. In 2020, a little more than 48 percent of all international immigrants were female. That year, female immigrants outnumbered male immigrants in Europe, North America, and Oceania.

If women are moving from a region with restrictive laws or traditions, their new home might offer more access to education, jobs, and status. These opportunities can increase as they further their education and grow more accustomed to their new culture. But in general, female immigrants are more vulnerable than males to violence, human trafficking, and sexual discrimination. Understanding the positives and negatives of female migration helps organizations like the United Nations Entity for Gender Equality and the Empowerment of Women push the international community to develop policies that reflect the needs of female immigrants.

Asylum in the European Union The early-21st century refugee crisis in the European Union (EU) was exacerbated by a policy requiring asylum seekers to remain in the first EU country entered and to apply for asylum there. Immigrants who traveled on to other EU countries risked being deported back to the first EU country they entered.

The policy posed a great strain on border countries along the Mediterranean Sea—such as Greece, Italy, Spain, and Malta—which experienced a mass influx of illegal entries by boat, most noticeably from 2014 to 2017. These dangerous and too-often deadly maritime crossings have continued: An estimated 23,000 people died or went missing between 2014 and 2021.

During the overflow, asylum seekers broke past the entry-point countries' borders and headed for Germany, Sweden, and other EU countries with more robust economies. These countries temporarily ignored the regulation to address the refugee crisis, accepting many asylum applications. The plight of asylum seekers was again highlighted in late 2021, when Belarus, hoping to embarrass Poland and the EU, brought hundreds of Iraqi refugees to the Belarus–Poland border with the promise of easy passage to the EU. Caught between Belarus security officers trying to compel them to cross the international border and the Polish army troops blocking their way, these unfortunate asylum seekers were pawns, treated inhumanely and eventually removed from the border and sent back home.

U.S. Restrictions The Chinese Exclusion Act was the first U.S. policy to broadly restrict immigration. Passed in 1882, it was meant to suspend Chinese immigration for a period of 10 years. However, its ongoing renewal by Congress kept the suspension constant for more than 60 years. By 1924, the act was expanded to include nearly all Asian groups. This greatly reduced the number of immigrants from an entire continent for decades. The act was finally repealed in 1943, but quotas enacted in the 1920s continued to severely limit the number of Asian, Arab, and African immigrants allowed into the United States. Not until the passage of the 1965 Immigration and Nationality Act was a more inclusive policy instituted.

In the aftermath of the 9/11 terrorist attacks on the United States, the U.S. government developed stricter border protection and immigration requirements. Tougher border patrols and immigration laws in the United States pushed immigrants crossing the U.S.–Mexico border, many of whom fled violence in Central America, to fall victim to smugglers and human traffickers. Afraid to seek help from U.S. authorities for fear of deportation, victims endured situations of forced labor or prostitution. In addition, certain U.S. visa programs have been criticized for not protecting against exploitation and abuse. Each year, more than 1 million temporary foreign workers—or guest workers—are employed in the United States. Once in the United States, temporary workers' visas almost always link them to a single employer, with no option to seek another employer if the immigrant is mistreated.

More recently, U.S. immigration policies have stirred debate about tougher restrictions that may infringe on the human rights of immigrants. For example, a 2017 policy meant to

This aerial view of the U.S.–Mexico border highlights the division between the residential area (right) stretching to the border fence in Tijuana, Mexico, and the open landscape outside San Diego, California. Parts of these homes and structures are slated for demolition as the border fence is replaced with a larger barrier.

Source: GUILLERMO ARIAS/AFP/Getty Images

Walls That Divide US

Learning Objectives

LO 5.11 Define *xenophobia* as the term relates to migration.

LO 5.9 Compare the migration experiences of refugees and internally displaced persons.

The recent unprecedented rise in the number of refugees and immigrants has prompted some world leaders to construct walls and fences along sections of their countries' borders. When the Berlin Wall was torn down in 1989, 15 other border walls still stood in various parts of the world. By 2018, more than 70 walls had been completed or were under construction. Most of the walls are in Europe, a continent that in the second half of the 20th century had seen a strong movement toward open borders, particularly within the member countries of the EU. That movement encountered some resistance in the beginning of the 21st century and was cited as one of the reasons Great Britain left the EU in 2020.

Though countries insist that border walls increase security and reduce illegal immigration, barriers are also about politics and can signal a nation's attitude regarding outsiders. For example, a 400-mile wall along the West Bank in Southwest Asia dividing Israelis from Palestinians reduces Palestine's territory and restricts Palestinians' movements. The wall is a constant reminder of Israel's occupation of the region and can provoke anger, resentment, and violence against the Israelis. A more secure wall built by Spain in Morocco in 2005 has significantly reduced the number of people entering Spain from the north coast of Africa illegally. But the wall hasn't stopped immigration completely. Immigrants find new, often riskier, routes around the barrier or enter Spain from neighboring countries. As described earlier, many immigrants now enter European countries by way of a dangerous journey across the Mediterranean Sea.

Throughout history, walls and fences have separated populations literally and figuratively, intensifying the divisions between people. Americans are greatly divided over the building of a wall spanning the 2,000-mile border with Mexico—an idea introduced by U.S. President Trump's administration to keep undocumented immigrants out of the United States. A Pew Research Center survey in 2018 revealed that younger generations were less likely to favor a wall, while non-Hispanic Whites were more than twice as likely to favor it as African Americans or Hispanics. Some people question whether a wall to stop illegal immigration is even feasible or necessary, or if it will instead put refugees escaping violence in greater danger.

Geographic Thinking

Identify the consequences that a border wall along the entire 2,000-mile U.S.–Mexico border could have on the United States and on Mexico.

National Geographic Collaborator **Danielle Sanders**

Journalism and the Great Migration

Danielle Sanders is the Managing Editor of the *Chicago Defender*. Her family has deep roots in the Bronzeville neighborhood, where her grandparents were professional jazz musicians and her great-uncle worked as a photographer for the *Defender*.

Learning Objective
LO 5.3 Describe causal factors in migrations.

The hardships of the Jim Crow South provided ample cause for African Americans to leave, but moving to a new place presented its own risks and uncertainties. The trusted voice of one Chicago newspaper encouraged many to migrate north.

Standing beneath a soaring sculpture that honors investigative journalist Ida B. Wells, journalist Danielle Sanders reflects on how she and her colleagues today "are being challenged to share their stories to diverse audiences using more diverse voices." The *Chicago Defender*'s role in the Great Migration exemplifies the power of such voices.

Sanders lives and works in the Bronzeville neighborhood of Chicago, Illinois, which she describes as a "mecca of Black history." In 1905, in the midst of Bronzeville's burgeoning jazz scene, publisher Robert Abbott founded the *Chicago Defender*. By 1916, the *Defender* had become the country's most influential Black newspaper, with a circulation of 50,000. Its contributors would include Langston Hughes, Gwendolyn Brooks, and Wells.

In the pages of the *Defender*, readers found not only honest reporting on life under Jim Crow, but also practical advice on places to live and shop in Chicago. In 1916, as the Great Migration was getting underway, the *Defender* began actively urging African Americans in the South to migrate north. The railroad system became a vital conduit of information as Pullman porters—Black men who worked on luxury trains—brought copies of the *Defender* to southern states. Sanders notes that the porters risked their lives in places where the Ku Klux Klan threatened to lynch anyone found with a copy of the newspaper. The information they distributed, including job opportunities and train schedules, helped bring at least 110,000 new Black residents to Chicago in just two years.

Geographic Thinking

Explain the impact of the *Chicago Defender* on the Great Migration.

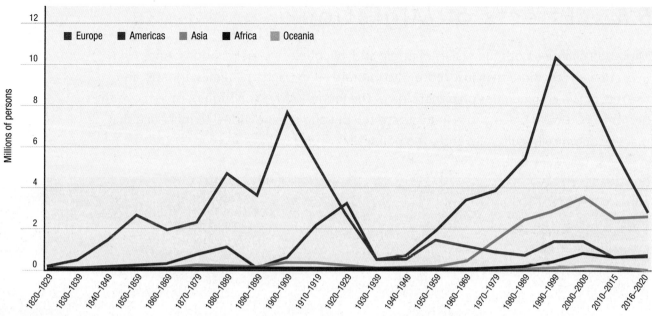

Volume of U.S. Immigration by Continent of Origin, 1820–2020 Spikes and dips in the immigrant populations coincide with the changes that took place to U.S. quota laws and immigration policies over time. Notice that the last time span shows data for just five years, whereas the other time spans represent data for 10 years.

Source: U.S. Department of Homeland Security

keep terrorists out of the country restricted entrance of foreign nationals from seven Muslim-majority countries and was criticized as being a "Muslim ban." Also, legislation known as Deferred Action for Childhood Arrivals (DACA) that protected immigrants who had entered the United States illegally as children was revoked in 2017. With its repeal, the program stopped accepting applications for work visas and for protection against deportation. Those who remained in the United States under DACA were expected to lose legal status within a few years. Equally controversial in 2018, a "zero-tolerance" immigration policy in the United States led to children being separated from their parents and held in shelters scattered across the country. The policy required that adults illegally crossing the U.S.–Mexico border be criminally prosecuted. Because it is illegal to send children to jail with their parents, families who were caught crossing the border were split up, with parents being sent to one facility and children to another. Another controversial Trump-era policy was called "Remain in Mexico," which, though enacted as a COVID-19 public health response, was seen as an attempt to close the U.S.–Mexico border. This resulted in as many as 70,000 people forced to stay south of the border in squalid, unsafe tent encampments, awaiting their U.S. immigration court hearings.

Restrictive immigration policies affect migration patterns. In addition to the immediate effects of deporting immigrants or turning them away at borders, policies like these can impact international migration flows. People who are planning to leave their home country but are not welcome in the country of their choice because of restrictive policies might decide to settle somewhere else, or not to move at all. Beginning in 2021, the Biden administration attempted to reverse many of these restrictive policies. These included executive actions to protect children and families and to stop the Remain in Mexico policy—as well as legislative proposals to allow more immigrants and refugees into the United States, and to provide the estimated 10.5 million unauthorized immigrants already in the country with a multiyear pathway to stay and earn legal status. While the question of long-term viability for the DACA program works its way through the U.S. court system, it remained in effect in 2022.

Geographic Thinking

1. Explain how migration policy created to meet labor market needs reflects Ravenstein's laws.

2. Identify what factors may cause a country to create a stricter immigration policy. Give examples.

3. Explain the degree to which changes in immigration policy affected the number of Asian immigrants entering the United States over time. Use the graph above to explain your answer.

4. Compare how step migration and chain migration play a role in the resettlement of refugees.

5.5 Effects of Migration

Migration changes both the places of origin and, of course, the destinations. And migration affects more than just those who migrate—it impacts the family, friends, and the rest of the communities on either side of the journey. When enough people move to or from a particular place, those impacts stretch beyond the individual to influence that place economically, culturally, and even politically.

Somali Muslims pray together during an evening soccer tournament at Hamline University in St. Paul, Minnesota. Cities such as Minneapolis and St. Cloud have become destinations for East African immigrants. Some Somalis came to Minnesota as refugees and were settled by nonprofits and religious groups. Others moved to the state to be part of these large Somali communities.

Source: The Washington Post/Getty Images

Economic Consequences

Learning Objective
LO 5.13 Explain the economic effects of migration on countries of destination and origin.

While a move drastically changes the life of an individual, a single migrant moving from one place to another has little effect on either location. But migration pathways and chain migration can push the number of migrants high enough for impacts to be felt in both the sending and the receiving communities. Geographers examine these effects in terms of the economic costs and benefits that migrants have on their destinations and their places of origin.

Destination Countries Recall that the overall trend in international migration is from less economically developed countries to more economically developed countries. One benefit to core countries receiving immigrants is that immigrants become a source of labor. They are often more willing than native-born citizens to accept less desirable, lower-paying work, such as agricultural, construction, and service sector jobs. This is true of immigrants in general, but especially true of those living in a country illegally. In addition, immigrants who are educated or skilled are often willing to work for less pay than their native-born counterparts. And qualified immigrants can reduce the **skills gap**—a shortage of people trained in a particular industry—in a country's workforce. The rate of immigrant entrepreneurship is high in core countries. Immigrants often open small businesses in commercial districts with high vacancy rates, revitalizing these areas.

Migration can have another important effect on destination countries—population growth. As you read in Chapter 4, countries that are in stages 4 and 5 of the Demographic Transition Model, which to date tend to be core countries, experience a decline in birth rates. Since the 1990s, migration has been the main source of population growth in core countries, and projections show that from 2020 on, migration has been and will be the only source of population growth in these regions.

Immigrants can profoundly affect population composition of their new homes as well. Since the majority of immigrants are of working age—in 2020, about 73 percent of all international immigrants were between age 20 and 64 compared to the overall population of 57 percent—migration lowers the dependency ratio of the destination country. Immigrants who move into rural areas can help offset the effects of urbanization, preventing further decline of rural communities.

Despite these positive effects, immigration has economic costs as well. A large influx of immigrants willing to work for lower wages might reduce the number of jobs available for native-born citizens. Industries that tend to hire immigrants, such as agriculture, can become too dependent on immigrant labor and suffer if immigration slows or laws change to make hiring immigrants more difficult. Additionally, immigrants often send **remittances**—money earned by emigrants abroad and sent back to home countries—to family members who remain in the country of origin. While helpful and often essential to those family members "back home," this means that a significant portion of their earnings is not being infused into the economy of the destination country. Receiving countries may also expend resources to meet the needs of immigrants, such as language lessons, assimilation classes, translators in medical facilities, and help finding employment.

Immigration's effects are readily observable in Sweden. The refugee crisis of 2015 brought a wave of more than a million asylum seekers to Europe from Syria, Afghanistan, and other Southwest Asian and African countries. Sweden, with a population of 10 million people, took in a record-breaking 162,000 asylum seekers—1.6 percent of its total population—in a single year. This exceptional influx has changed some people's attitude toward immigrants. A poll taken in 2015 revealed that 54 percent of Swedes would definitely help asylum seekers. In 2016 the number had dropped to 30 percent. In 2018, the country elected a government that promised more restrictions on immigration. But in the city of Malmö, refugees from Syria and other Southwest Asian countries have found a new home and become an integral part of the economy. In just a few years, the district of Möllevången transformed from a working-class neighborhood into a multicultural area full of Southwest Asian restaurants, cafes, and currency exchanges.

However, all has not been positive in the Swedish experience: A 2019 poll found that 51 percent of Swedes

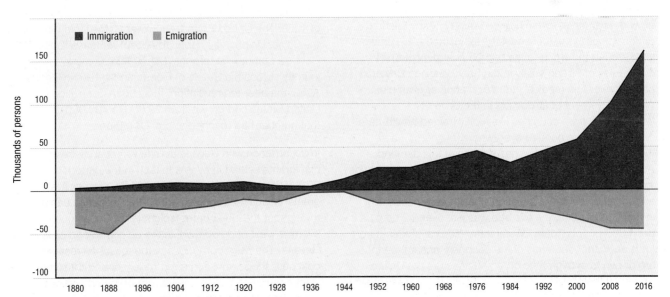

Swedish Immigration and Emigration Flows, 1880–2017 In the wake of the refugee crisis of 2015, when Sweden took in more asylum seekers than any other European country besides Germany, the country developed stricter immigration policies. ▎How does the graph help you understand why Sweden underwent a shift in attitude and policy toward asylum seekers?

Source: Statistics Sweden, "Befolkningsutveckling 1749–2017," accessed October 3, 2018

agreed that their country's borders should be closed to refugees, with doubts that additional newcomers can successfully integrate into the host society. Other European countries have seen similar recent change in public sentiment about accepting immigrants, including negative attitudes toward new immigrants in France, Serbia, and Hungary. Indeed, Hungary's government has established specific anti-refugee policies aimed at keeping asylum seekers out by building a barbed-wire fence along part of its border and forcibly removing 71,000 between 2016 and 2021.

Countries of Origin The economic effects on the countries left behind by immigrants are also profound. In the benefits column, the country of origin might see a decrease in unemployment as working-age immigrants leave, opening up jobs for those who stay. With fewer people, there is less strain on resources such as food and services such as health care. Immigrants returning home after a stint in a foreign country often bring new skills they learned abroad. And, significantly, the remittances that many immigrants send home are an important source of support for the economies of many countries. They help the families of immigrants as well as their communities. In the Philippines, for example, about 10 percent of the country's gross domestic product in recent years came from remittances. More than 10 million Filipinos were working in foreign countries, and they sent $35 billion back home in 2019. Overall, according to the World Bank, after a COVID-caused decline in 2020, global remittances grew to a total of $589 billion in 2021.

But there are costs to the countries of origin as well. The immigrants who leave are disproportionately young adults, and whether they are skilled workers, unskilled workers, or entrepreneurs, this loss can slow the home country's economy and ability to develop. The term **brain drain** refers to the loss of trained or educated people to the lure of work in another—often richer—country. Many talented professionals in the tech industry, for instance, are drawn from around the world to the computer and electronics companies of Silicon Valley in the San Francisco Bay area of California. The medical industry is another example. About 25 percent of all doctors in the United States are foreign born, from countries like India. Trained health-care professionals are needed everywhere, of course—and the coronavirus pandemic has deepened this demand. According to the World Health Organization, a region needs to average at least 4.5 doctors for every 1,000 people to adequately meet the population's health-care needs. But in India, as of 2019, the average was just over 0.9 doctors per 1,000 people. Many doctors and medical students from India and other countries of the periphery or semi-periphery are drawn to immigrate to core countries like the United States by the promise of higher salaries and a better quality of life.

Cultural and Social Consequences

Learning Objective
LO 5.14 Describe the social and cultural effects of migration on countries of destination and origin.

Countries of destination and origin are affected culturally and socially by migration as well. These effects, both positive and negative, are often intertwined with the economic consequences.

Destination Countries Immigrants bring aspects of their home culture with them, including music, literature, fashion, religion, and language. They typically add flavor to the destination country through the food and recipes of their homeland. Many immigrants are entrepreneurs, so not only do they bring these cultural features with them, they also share them with their destination country by establishing shops or restaurants. These businesses are often opened to meet the needs of immigrant communities, but they may end up becoming popular with native-born citizens as well. For instance, Indian restaurants opened by immigrants are found throughout Great Britain. The bestselling fast food in Germany is the doner kebab, brought to the country by the Turkish guest workers you read about in section 5.2. In the United States, Mexican food has been an integral part of mainstream American cuisine since the second half of the 20th century. Usually, though, as these foods gain popularity, the dishes are adapted to the tastes of the new country. One of the most popular dishes in Britain's Indian restaurants—chicken tikka masala—is not a traditional Indian dish. It is believed to have been invented in Great Britain to appease the British taste for meat served in a sauce.

The effects of migration can be seen, heard, and tasted in Miami. The southern Florida city saw a massive influx of Cuban immigrants in the years after the Cuban revolution of 1959. The United States, embroiled in the Cold War with the Soviet Union, was happy to accept refugees fleeing Fidel Castro's communist reforms. Many Cubans made the short journey across the Straits of Florida and settled in Florida. A decade after the revolution, the Cuban population in the United States had grown from 79,000 to 439,000. By 2018 that number had reached about 1.3 million.

In accordance with Ravenstein's laws, most of these immigrants settled in Miami. In fact, in the period 2014–2018, 63 percent of Cubans in the United States lived in the greater Miami metropolitan area. Their influence can be felt throughout the city, but the epicenter of Cuban culture in Miami is west of downtown, in the neighborhood called Little Havana. The neighborhood is home to Cuban restaurants, bakeries, shops, museums, art galleries, and theaters. Cuban music and dance styles can be heard and seen in the clubs that host Little Havana's lively nightlife. Spanish is

The vibrant colors of Cuban culture are on full display in Miami's Little Havana. Eighth Street—or *Calle Ocho* as it's known—is the neighborhood's main thoroughfare, lined with murals celebrating the country's rich culture and history, restaurants serving Cuban cuisine, and many shops, galleries, and museums. In Maximo Gomez Park, better known as Domino Park, locals and tourists alike come to witness the older generation of Cuban immigrants engage in an intense game of dominoes.

Source: LAIF/Redux

Migration 143

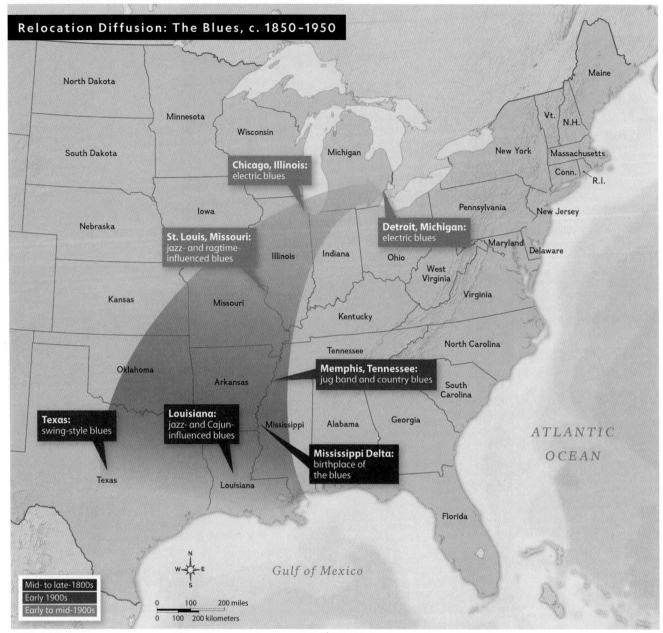

Reading Maps The music known as the blues originated on slave plantations in the Mississippi Delta and spread north and west as African Americans migrated during the Great Migration. As it spread, each region left its mark, combining the blues with jazz, gospel, country, and ragtime music. ▌ What cultural effects might this example of relocation diffusion have on the Midwestern United States?

spoken in the shops and on the streets, and in 1963 Miami opened the first bilingual elementary school in the United States.

Over the years there has been some backlash against the huge Cuban influence in Miami, with working-class Whites feeling left out of the job market, and some resistance to the prevalence of Spanish, but overall the city has prospered. Millions of tourists flock to Miami each year, and Little Havana accounts for a significant portion of the more than $20 billion that tourism pumps into the economy each year.

The spread of ideas and culture traits through migration is called **Relocation diffusion** (discussed in Chapter 7). Instances of relocation diffusion—caused both by immigrants and internal migrants—can be found throughout the United States. For example, during the Great Migration, African American migrants influenced northern culture by bringing blues music to cities like Chicago and Kansas City. With the Harlem Renaissance in New York City, they created an explosion of influential literature, music, and art. For more recent examples, think about the different types of religions represented by the churches, temples, mosques, and other religious buildings in a local town or city. Or consider the assortment of ethnic foods available in grocery stores or restaurants. Or how about music and fashion?

The cultural costs to the destination country are harder to gauge precisely because they are rarely concrete or measurable. In fact, they are often perceived as costs by only a portion of the receiving society. Some people believe a large influx of immigrants causes a society to change too much. They feel their country's traditions are endangered by the introduction of new customs. These attitudes can lead to prejudice against immigrants. Some countries have implemented policies to protect their culture from change. In the Netherlands, for instance, immigrants must take classes in Dutch culture to become citizens. The topic of immigration can also lead to strain within the receiving society as pro- and anti-immigration groups engage in public demonstrations, sometimes even becoming violent. In recent years, demonstrators on both sides of the issue have clashed in North America and Europe.

Because of the challenges involved in migrating to a foreign country, immigrants tend to cluster together, creating segregated neighborhoods like Little Havana in Miami or Chinatown in a number of U.S. cities. While these neighborhoods help immigrants retain their traditions and customs, they can also prevent them from fully acclimating to their new country. Conversely, many immigrants, as well as children of immigrants born in the receiving country, tend to lose some of their cultural identity as they adjust to the way of life in their new home.

Countries of Origin Immigration's social benefits for the source country are connected to the economic benefits. Remittances can improve the quality of life of families in the source country, which can in turn help support the economy. In Jalpan de Serra, a city in the State of Queretaro, Mexico, people flock from surrounding towns and villages to collect remittances sent from family members living in the United States. A worker in Mexico averages about $55 per week, which is less than an immigrant working construction or harvesting fruit in the United States can make in a single day. The extra money sent home as remittances pays for nicer clothes, better food, appliances, and other necessities and small luxuries for the immigrant's family in Mexico. Legal and illegal immigrants sent about $43 billion in remittances home to Mexico in 2020 and were projected to increase in succeeding years.

On the other hand, migration can have a disrupting effect on the social structure of a place of origin. For example, the rural villages surrounding Jalpan de Serra have lost many young men to migration. It's difficult to make a living on the area's small, overworked farms, so young men are drawn to the promise of higher wages in the United States. Older mothers often don't see their sons and daughters who emigrate for years, and young women who stay in their hometowns have trouble finding life partners with whom to start a family. Life in these rural villages has been completely altered. The effect on the dependency ratio is the opposite of the change immigrants bring to destination countries—in many of these villages, it is mostly older people and young children who remain.

Geographic Thinking

1. Explain how remittances act as both a benefit and a cost to countries of origin and destination.

2. Compare the positive and negative economic effects of migration for countries of origin and destination.

3. Describe how the ethnic diversity resulting from migration benefits the destination country.

4. Explain how Ravenstein's laws are represented by the large-scale immigration of Cubans to Miami.

Chapter 5 Summary & Review

Chapter Summary

Migration is the movement of people from one location to another, with the intent of the move being permanent. It is influenced by the interplay of environmental, economic, cultural, and political factors.

- Push factors encourage migration from a place; pull factors encourage migration to a place.
- Emigration is migration from a location; immigration is migration to a location.
- Net migration is the difference between emigration and immigration.

Geographers classify different types of migration. In voluntary migration, people choose to migrate, and in forced migration, they are compelled to move by extreme push factors.

- Internal migration—movement within a country's borders—can be interregional or intraregional.
- Transnational migration involves individuals who have emigrated but retain close ties to their country of origin and may regularly return for visits.
- Refugees are people who have been forced to migrate to another country, fearing persecution or death.
- Internally displaced persons are people forced to migrate within their country.

Most countries establish migration policy to control immigration. Some policies work to help immigrants whereas others work to limit the number of immigrants.

- The main purposes of most immigration policies are to meet labor market needs and maintain migration levels.
- Quotas set limits on the number of immigrants allowed to migrate to a particular country and may set preferences for who is allowed into the country.

Geographers study the economic and social effects of migration on the countries of both destination and origin.

- In destination countries, immigrants can fill labor needs and reduce the skills gap, as well as add diversity and cultural variety. Costs include a reduced number of jobs available for native-born citizens and possible discomfort with cultural changes brought by immigrants.
- In countries of origin, the gap left by emigrants can reduce unemployment and strain on resources, and remittances sent home by immigrants can help support the economy. Costs include brain drain—the loss of trained workers—and disruption of the social structure.

Review Questions

Use complete sentences to answer the questions.

1. **Apply Conceptual Vocabulary** Consider the terms *mobility* and *refugee*. Write a standard dictionary definition of each term. Then provide a conceptual definition—an explanation of how each term is used in the context of this chapter.

2. How are the terms *kinship links* and *chain migration* related?

3. Define the term *net migration* using the concepts *emigration* and *immigration*.

4. Describe how push factors and pull factors affect a person's decision to migrate. List examples of both push factors and pull factors.

5. Give examples of a demographic push factor and a demographic pull factor.

6. Compare the terms *voluntary migration* and *forced migration*. Cite examples of each.

7. How are the terms *internal migration* and *internally displaced person* related?

8. Explain how a refugee is different from a guest worker. Give an example of each.

9. Provide an example of an intervening obstacle. Then, provide an example of an intervening opportunity.

10. Explain how the terms *asylum* and *transnational migration* are related.

11. Define the term *brain drain* using the concept *skills gap*.

12. Explain the relationship between the terms *distance decay* and *friction of distance*.

13. Explain the degree to which the gravity model could impact relocation diffusion.

14. Define the term *quota* in relation to migration policy. Cite an example of when, how, and why a quota was used.

15. Describe how the terms *remittance* and *guest worker* are related.

Interpret Graphs

Study the graph and then answer the following questions.

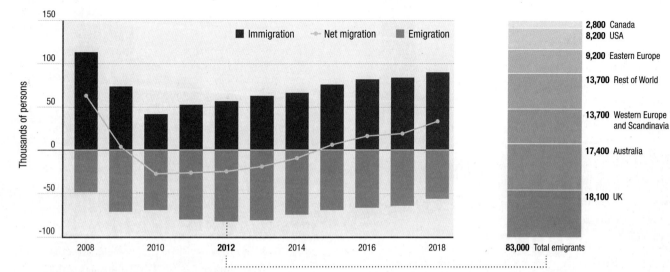

Ireland's Net Migration, 2008–2018
Source: Irish Central Statistics Office, Population and Migration Estimates

16. **Identify Data & Information** In what year was emigration highest in Ireland? In what year was immigration highest?

17. **Analyze Visuals** How did the rise in both immigration and emigration affect Ireland's net migration in the year 2011?

18. **Describe Spatial Relationships** Describe Ireland's pattern of migration between the years 2008 and 2018.

19. **Apply Models & Theories** How does the number and location of emigrants from Ireland in the year 2012 support the distance decay model of migration?

Unit 2 | Writing Across Units, Regions, & Scales

Climate Change Migrants
by Tim Mcdonnell

A teenage boy works at a brick field in Bhola, Bangladesh. Brick fields in urban areas are a common destination for coastal climate migrants, especially young men.

Golam Mostafa Sarder starts every day before dawn, rising from a thin reed mat in the shed that he shares with 15 roommates. Each has just enough space to lie flat. Outside the shed's open doorway, in the outskirts of Dhaka, the sprawling megacity capital of Bangladesh, is the brick factory where Golam and his neighbors work for 15 hours a day, seven days a week, at least six months a year. His home in Gabura, a remote village on the country's southwestern coast, is more than a day's journey from the city by bus, rickshaw, and ferry.

Golam's job is to push wheelbarrows of mud down the production line. Waist-high rows of drying bricks spiral off from a towering kiln that belches smoke over an area the size of a city block. Golam has never heard of global warming. But he says he knows one thing for sure: "If the river didn't take our land, I wouldn't need to be here."

Climate-Driven Displacement Fifty-five million people in Bangladesh live along the southern coast, a lush honeycomb of island villages, farms, and fish ponds linked by protective embankments. During the rainy season more than one-fifth of the country can be flooded at once. Climate change is disrupting traditional rain patterns—droughts in some areas, unexpected deluges in others—and boosting silt-heavy runoff from glaciers in the Himalaya Mountains, leading to an increase in flooding and riverbank erosion. Meanwhile, sea-level rise is pushing saltwater into coastal agricultural areas and promising to permanently submerge large swaths.

Over the last decade, nearly 700,000 Bangladeshis were displaced on average each year by natural disasters. As people flee coastal areas, most are arriving in urban slums, or informal housing, particularly in Dhaka, one of the world's fastest-growing megacities. The city is perceived as the country's bastion of economic opportunity, but it is also fraught with extreme poverty, public health hazards, human trafficking, and its own vulnerability to floods.

A City Altered by Climate Change For climate migrants who arrive in Dhaka, life is seldom easy. Men and boys work in brick factories, drive rickshaws, and build skyscrapers. Women and girls clean houses, stitch Western fashions, and raise families—often fending off sexual violence at multiple steps along the way. Education is a luxury; rent is preposterous. As a result, 40 percent of the city's residents live in slums, hundreds of which are spread across the city. According to the International Organization for Migration, up to 70 percent of the slums' residents moved there due to environmental challenges.

Slums emerge unplanned and unsanctioned in the backyards of glassy skyscrapers, straddling railroad tracks, on stilts above water-logged floodplains, on the fringes of construction sites. Single beds are frequently shared by five or more family members. Sewage runs freely. Structure fires spread easily. Most electricity, when it's working at all, is tapped illegally from the grid. Insect infestations are inescapable. Skin and gastrointestinal diseases transmitted by dirty water are routine, and the infant mortality rate is twice that of rural areas. Rent money flows into a real estate black market controlled by corrupt local officials and businessmen.

"It's very hard to get a living here," says Sahela Begum, who moved to Dhaka with her four young daughters after losing her home along the Padma River. "But my life is my children's life. If I can make a good future for them, that's the best thing I can hope for."

Cities for Climate Migrants When Bangladesh gained independence in 1971, the population was 91 percent rural. Today, nearly one-third of the population lives in cities. Throughout that process of growth, "low-income people were totally left out of the development framework of the city," says A.Q.M. Mahbub, an urban studies researcher at the University of Dhaka. Affordable housing and public transit connecting the city center to suburbs were never priorities.

Local officials tend to view slum dwellers as illegal squatters, rather than residents with a right to basic services, preferring to leave slum residents reliant on aid from local and international nongovernmental organizations. "If we invest money directly in slum areas, or give them an electricity supply, they will start to think, 'O.K., we have these facilities, so we have the ownership of this land,'" says a senior official at a government agency that manages the city's infrastructure. "Once we give them improved services, they become permanent."

Many of the country's leading public policy experts think that attitude—that climate migrants are a regrettable burden—is shortsighted. Mongla, a port town on the country's south-central coast, is one of several emerging "secondary cities," models of climate-savvy urban planning where investments in sea walls and other adaptive infrastructure are being paired with blue-collar job opportunities, as well as affordable housing, schools, and hospitals.

In the last five years, Mongla's population has jumped nearly 60 percent to 110,000, and the price of land has skyrocketed. The town's positive reputation is spreading. "Because of salinity and flooding, there's not much opportunity in my village. But here, I can make good money," says Mohammed Kabir Hossain, who drives a rickshaw. He came to Mongla a few years ago. "A lot of people are coming here from across southern Bangladesh, especially those who are unwilling to go to Dhaka."

Source: Adapted from "Climate Change Creates a New Migration Crisis for Bangladesh" by Tim McDonnell, nationalgeographic.com, January 2019

Write Across Units

Unit 2 explored factors that influence changes in population and how population changes and migration profoundly affect a country's development. This article reveals how one factor, climate change, is affecting Bangladesh, altering not only the lives of those experiencing a natural disaster, but also those who take in climate migrants. Use information from the article and this unit to write a response to the following questions.

Looking Back

1. What geographic questions does the article pose about the situation in Bangladesh? Unit 1

2. What economic impact can climate change migrants have on their destinations? Unit 2

Looking Forward

3. What aspects of Bangladesh's culture are depicted in the article? Explain how one human–environment interaction mentioned in the article is affecting the cultural landscape. Unit 3

4. How do political attitudes affect treatment of Bangladesh's climate migrants? Unit 4

5. What effects is climate change having on the agricultural areas of Bangladesh, and how might those effects be felt by the entire population of the country? Unit 5

6. Describe how climate migration is putting pressure on Dhaka's urban infrastructure. Unit 6

7. What can you infer about Bangladesh's level of development from the article? Unit 7

Write Across Regions & Scales

Research a community in a region outside of South Asia that has been directly affected by climate change or by the arrival of climate migrants. Write an essay comparing this community with either Dhaka or Mongla in Bangladesh. Drawing on evidence from this unit and the article, address the following topic:

Describe how a community can prepare for changes in population due to climate change. What might the ideal city for climate migrants look like?

Think About

- The factors that city officials must take into account as the population increases or decreases

- The similarities and differences in how each community handles the direct and indirect effects of climate and population change

| Unit 2 |

Chapter 3

Population Pyramids

A population pyramid illustrates the age and gender characteristics of a country's population and may provide insights about economic development and social circumstances, such as gender equality, ethnic tensions, or health conditions. The population is distributed along the horizontal axis. The male and female populations are represented as horizontal bars along the vertical axis according to age. The shape of the population pyramid may evolve based on fertility, mortality, and international migration trends. A country with a wide pyramid base likely has high fertility rates. A country with a narrow base likely has low fertility rates. ▌Compare the population pyramids shown here. Explain what they might indicate about the birth and death rates or about migration trends in each country.

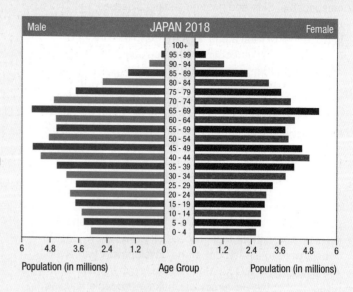

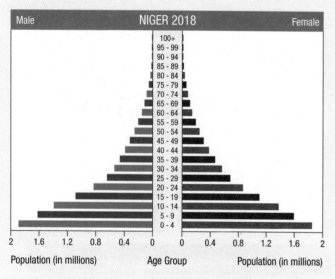

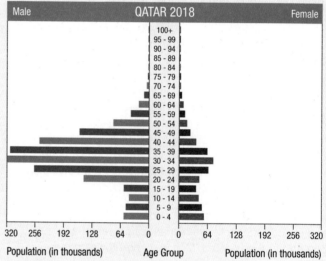

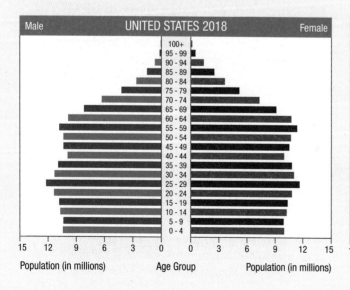

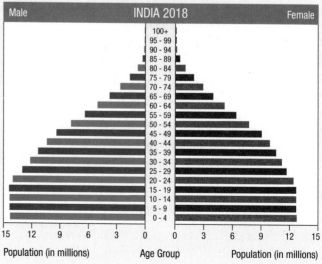

Source: The World Factbook

Maps & Models Archive

Chapter 4

Demographic Transition Model

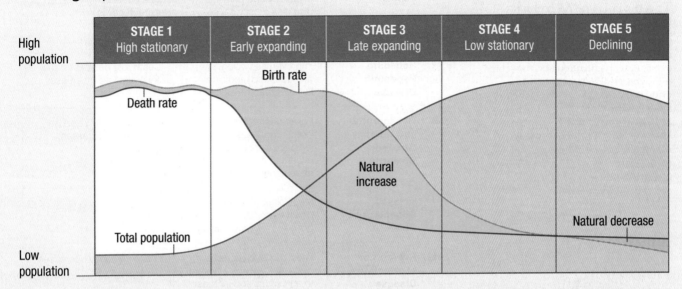

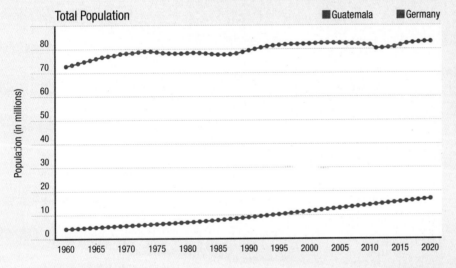

The demographic transition model (DTM) is a useful tool for analyzing a country's population and development. The model illustrates the interaction between a country's birth rate, death rate, and population. Historically, countries move through the stages of the DTM as they develop. ▪ Based on the total population and birth and death rates provided, identify the stages for Germany and Guatemala. Explain your answer.

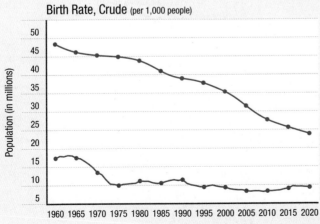

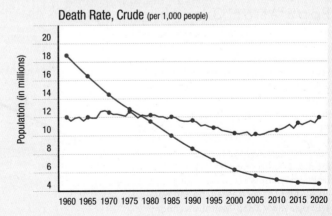

Source: Worldbank.org

Unit 2

Chapter 4

Epidemiological Transition Model

The epidemiological transition model (ETM) depicts patterns of diseases and causes of death and uses them to explain changes in population over time. Although the DTM and the ETM are often compared, the stages in the two models do not correspond. The ETM is an independent model for explaining how populations grow and, perhaps, eventually decline. ▌ Research online to find the life expectancy for Australia. Identify the country's stage in the ETM and explain your answer.

Stages	Description	Effects
1. Famine	Infectious and parasitic diseases mostly cause human death; animal attacks also cause deaths	Death rate is high and life expectancy is low
2. Receding Pandemic Disease	Improved sanitation, nutrition, and medicine lower spread of infection	Death rate decreases and life expectancy increases
3. Degenerative and Human-Created Diseases	Fewer deaths from infectious disease and increase in diseases related to aging (heart attack/cancer)	Death rate is low and life expectancy increases
4. Delayed Degenerative Diseases	Medical advances reduce or delay incidences of diseases related to aging	Life expectancy is at its highest
5. Reemergence of Infectious Disease	Infectious and parasitic diseases become resistant to antibiotics and increase	Life expectancy decreases

Chapter 4

Malthus's Theory of Population Growth

In 1798 Thomas Malthus argued that population grew exponentially while food production grew arithmetically. He predicted that population growth would surpass food production, leading to a hunger crisis. While agricultural advancements and globalization have caused food production and distribution to grow faster than Malthus predicted, neo-Malthusians have broadened the model, arguing that pressure on resources from overpopulation leads to dire consequences such as famine and war.

▌ Explain how a neo-Malthusian might apply Malthus's theory to climate change.

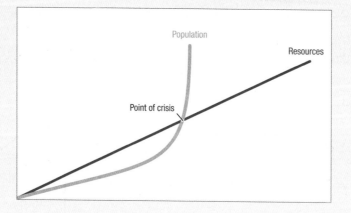

Maps & Models Archive

Chapter 4

Mapping Noncommunicable Diseases

The United States is considered to be in Stage 4 of the ETM. In this stage, noncommunicable diseases (NCDs), including heart disease, cancer, and diabetes, are the dominant cause of death. The number of people and communities affected by these diseases is growing due to insufficient physical activity, unhealthy diet, and the harmful use of tobacco and alcohol, among other factors. ▌ Explain how the information in the map below can be useful for reducing NCD death rates in the United States.

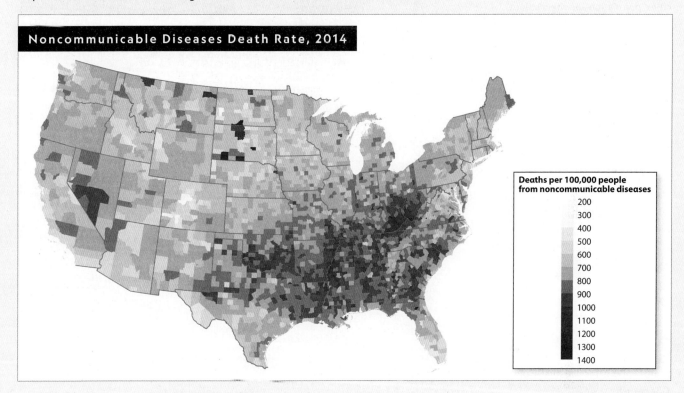

Noncommunicable Diseases Death Rate, 2014

Deaths per 100,000 people from noncommunicable diseases: 200, 300, 400, 500, 600, 700, 800, 900, 1000, 1100, 1200, 1300, 1400

Chapter 5

Ravenstein's Laws of Migration

Ravenstein's laws were developed by geographer Ernst Ravenstein in 1885 to describe trends in migration. Ravenstein concluded that factors such as economic conditions, population size, and distance affect whether people migrate and where they move to. While Ravenstein's laws are still relevant today, some are more relevant than others. For instance, Ravenstein stated that most migrants are young adult males. In 2020, however, a little more than 48 percent of international migrants were women, nearly equaling male migrants. ▌ Explain whether Ravenstein's laws apply equally to voluntary and forced migration.

1. Migration is typically over a short distance.
2. Migration occurs in steps, like from a rural area to a nearby city, and then perhaps on to a larger city.
3. Long-distance migrants often move to places of economic opportunity (urban areas).
4. Every migration generates a movement in the opposite direction, or counterflow (not necessarily of the same number of migrants).
5. People in rural areas migrate more than people in cities.
6. Males migrate over longer distances than females.
7. Most migrants are young adult males.
8. Cities grow more by migration than by natural increase.
9. Migration increases with economic development.
10. Migration is mostly due to economic factors.

Chapter 5

Gravity Model

The gravity model predicts the amount of spatial interaction, such as migration, tourism, or trade, that will occur between two or more places. According to the model, the level of interaction between two cities depends on the size of the cities' population and the distance between them. The larger the population, the more interaction a city will receive. And the shorter the distance between two cities, the more they will interact with each other. ▮ Describe the gravity model using Delhi, Kolkata, and Rajkot as examples.

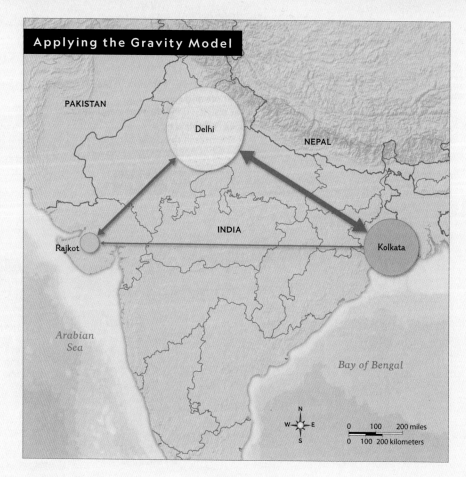

Applying the Gravity Model

Chapter 5

Remittances

Remittances are the money earned by emigrants abroad and sent back to their home countries, usually to family members. This money plays an important role in the economies of many countries. In some countries, such as those shown in the graph, remittances represent a significant percentage of the gross domestic product, which is the total value of all goods and services produced by a country in a year. ▮ Explain how the importance of remittances to the economies of the countries shown in the graph has changed since 1998.

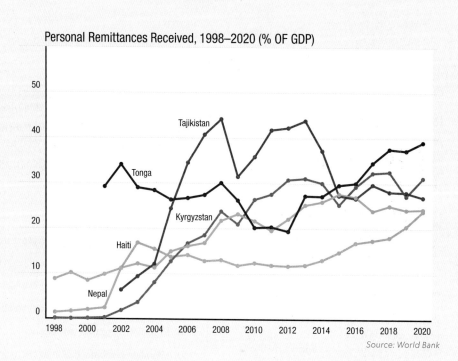

Personal Remittances Received, 1998–2020 (% OF GDP)

Source: World Bank

Maps & Models Archive

Chapter 5

Migration Data

Displaying geographic data in a graph or chart allows you to quickly compare two or more statistics or visualize how trends have developed over time. Study the following infographics and use the questions below to help you analyze the migration data. ▮ Explain what this graph tells you about immigrants in the U.S. labor force.

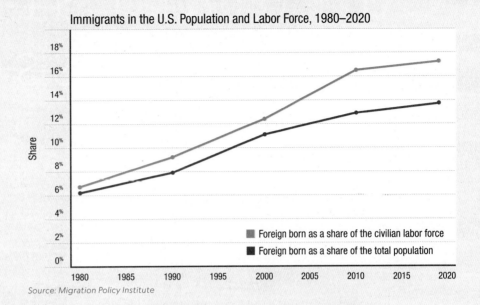

Immigrants in the U.S. Population and Labor Force, 1980–2020

- Foreign born as a share of the civilian labor force
- Foreign born as a share of the total population

Source: Migration Policy Institute

Global Displacement, 2007–2018

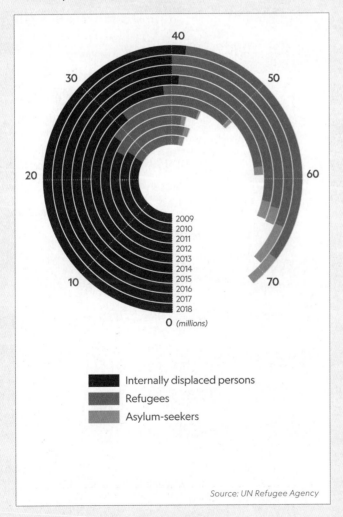

- Internally displaced persons
- Refugees
- Asylum-seekers

Source: UN Refugee Agency

▮ Describe the trend depicted in the graph.

Forced Migration Due to Natural Disaster, 2018

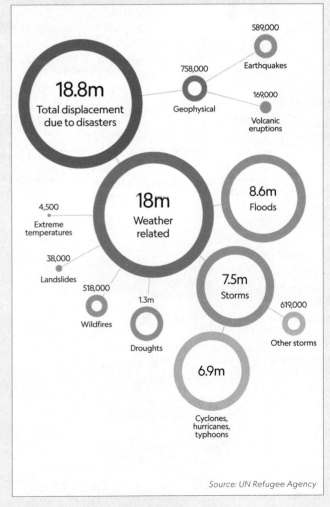

- 18.8m Total displacement due to disasters
- 758,000 Geophysical
  - 589,000 Earthquakes
  - 169,000 Volcanic eruptions
- 18m Weather related
  - 4,500 Extreme temperatures
  - 38,000 Landslides
  - 518,000 Wildfires
  - 1.3m Droughts
  - 8.6m Floods
  - 7.5m Storms
    - 619,000 Other storms
    - 6.9m Cyclones, hurricanes, typhoons

Source: UN Refugee Agency

▮ Describe in a sentence or two the role that weather plays in forced migration.

| Unit 3 |

Cultural Patterns and Processes

Who We Are

Village church in Machuca, Chile

Culture is a mix of tangible objects and intangible concepts—embodied in buildings, books, and paintings as well as in languages, religious ideals, and attitudes toward women. Human cultures are intimately tied to the places where they arise and where they spread.

People from diverse perspectives not only endow existing landscapes with lasting meanings but also create new landscapes that reflect their cultural visions. In Bolivia, participants in La Paz's Fiesta del Gran Poder, or the Festival of the Great Power (shown here), blend indigenous Aymaran dress and folklore with Catholic themes as they celebrate their belief in Jesus Christ.

Over time, cultures transform; as do the places they help mold. Rarely have cultural landscapes evolved as swiftly as they do today, with globalizing forces such as communication and transportation technology driving the pace.

Chapter 6
Concepts of Culture

Chapter 7
Cultural Change

Chapter 8
Spatial Patterns of Language and Religion

Unit 3 Writing across Units, Regions, & Scales

Unit 3 Maps & Models Archive

National Geographic Explorer **Fred Hiebert**

Discovery and Preservation

As a field archaeologist, National Geographic Archaeologist-in-Residence Fred Hiebert travels the world to unearth remains of ancient civilizations. As a museum curator, he looks for innovative ways to educate present-day audiences about these cultures. And as an advocate for preservation, he strives to make sure that irreplaceable artifacts will be available for future generations.

Learning Objective
LO 6.1 Explain the subject matter and focus of cultural geography as a sub-field of human geography.

Ancient Geography Hiebert is always conscious of the ways geography influenced the location and cultural practices of ancient civilizations. He defines geography as the ability to identify regions and put them in context in terms of conflict, history, and culture. He's applied that understanding to archaeological digs in locations such as Peru, the Black Sea, Afghanistan, and Jerusalem.

In 2014, Hiebert curated a landmark exhibition called "Peruvian Gold: Ancient Treasures Unearthed" at the National Geographic Museum in Washington, D.C. Hiebert linked Peru's geography to its archaeological richness and cultural diversity. He explains, "What makes Peru one of the great centers of civilization is its geography," describing the country's high mountains, valleys, and coastal deserts. Hiebert believes curating and displaying Peru's intriguing artifacts helps people connect the present with the distant past—which enriches the lives of all who come to see them.

Hiebert also curated "The Greeks: Agamemnon to Alexander," an international 2016 exhibition that explored 5,000 years of ancient Greece through more than 500 artifacts from 22 Greek museums. Greece's geographic diversity includes a mountainous mainland and thousands of islands dotting the Aegean, Adriatic, and Ionian seas. "Greece's challenging geography is also one of its greatest assets," Hiebert says. "The ancient Greeks adapted to living on islands, coastline, rugged terrain, and hard-to-farm land, which helped them establish a unique identity."

Past and Present "Culture—the past, the remains of our history—is nonrenewable," Hiebert told an audience at National Geographic headquarters in Washington, D.C. He was talking about the urgent need to protect archaeological sites from looters who remove artifacts and sell them to collectors. These thefts rob archaeologists of the chance to study such objects in their proper context and deprive the general public of the chance to learn from and enjoy them.

Hiebert also pointed out that archaeological treasures are sometimes at risk of deliberate destruction. Demolishing cultural pieces violates international conventions of war. Some conflicts seek to negate cultures and demoralize populations. He highlighted a 2001 incident in which the ruling Taliban party in Afghanistan dynamited a pair of monumental statues of Buddha dating to about 500 C.E. Taliban members also smashed artifacts in Afghanistan's National Museum. Fortunately, the museum's curators had the foresight to remove and hide many of its most valuable artifacts years before the Taliban's rampage. Hiebert was instrumental in rediscovering these objects. In 2004, he curated National Geographic's exhibition "Afghanistan: Hidden Treasures from the National Museum." Clearly, preserving artifacts from disappearance or destruction serves the needs of archaeologists in search of knowledge.

Museum exhibitions are not Hiebert's only method for sharing connections with antiquity. National Geographic reaches nearly 400 million individuals through print, video, and digital media. Hiebert uses all these 21st-century modes of communication to make the ancient world come alive for people who may never step inside a museum. His curation of *The Greeks* exhibition led to National Geographic's involvement in the 2016 restoration of the Edicule—believed by the faithful to be the tomb of Christ—in the Church of the Holy Sepulchre in Jerusalem as well as the 2017 "Tomb of Christ" 3-D exhibition in the National Geographic Museum.

Geographic Thinking

Explain how Fred Hiebert's definition of geography influences his perceptions of ancient cultures, and describe what his work might convey to others.

Top: Fred Hiebert's role as an archaeologist leads him to more than "lost" artifacts—he can also search for lost people. In 2019, Hiebert and anthropologist Jaime Bach inspected a site on Nikumororo Island, Kiribati, in the western Pacific Ocean, to look for evidence of Amelia Earhart's 1937 plane crash. *Bottom left:* The drinking vessel made by the Recuay culture in Peru around 300 c.e. whistled when liquid was poured from it. *Bottom right:* The original Mask of Agamemnon, named for the mythical Greek king, dates to the 16th century b.c.e.

Chapter 6
Concepts of Culture

Critical Viewing In 2016, the Standing Rock Sioux Tribe protested the construction of the Dakota Access Pipeline on the grounds that this oil pipeline would disturb their reservation in North Dakota. ▌For what reasons might a cultural group like the Standing Rock Sioux Tribe want to protect their homeland?

Geographic Thinking How do location and resources influence cultural practices?

6.1
An Introduction to Culture

Case Study: Wisconsin's American Indian Nations

6.2
Cultural Landscapes

Case Study: Tehrangeles

National Geographic Explorer
Jim Enote

6.3
Identity and Space

6.4
Cultural Patterns

National Geographic Explorer
Sandhya Narayanan

6.1 An Introduction to Culture

Evidence of people's culture, or way of life, can be found in the places where people live. Cultural geography is the study of the ways humans organize space and the varying characteristics and patterns of the places people create.

Culture

Learning Objectives

LO 6.1 Explain the subject matter and focus of cultural geography as a subfield of human geography.

LO 6.2 Compare the three categories of cultural traits (artifacts, sociofacts, and mentifacts).

As the heartbeat of human geography, **culture** refers to the beliefs, values, practices, behaviors, and technologies shared by a society and passed down from generation to generation. Some elements of culture, such as clothing, literature, food, and music, are relatively easy to identify and experience. The United Nations Educational, Scientific, and Cultural Organization (UNESCO) refers to these physical artifacts as "tangible cultural heritage" that are produced, maintained, invested with importance, and passed intergenerationally within a society. Other cultural attributes, however, are not as easily observed. For example, a group's festivals or holidays might be obvious, but their beliefs, attitudes, and values may not be so visible. This "intangible cultural heritage," as UNESCO explains, includes knowledge, skills, expressions, representations, and practices that cultures recognize as part of their heritage.

To illustrate the many types of cultural attributes, geographers often use the metaphor of a "cultural iceberg." The exposed tip of the iceberg represents everything fairly easy to observe about culture, such as language, the arts, land-use patterns, and personal behavior. The larger, submerged part of the iceberg represents the subconscious values and thought patterns in culture that are generally unseen, such as a group's beliefs, values, and rules. This invisible part of a culture strongly influences the visible part.

At the tip of the cultural iceberg are material objects shared by a group, such as food and clothing, as well as language and other shared cultural practices, which include activities that most group members do. Each attribute is considered to be a **cultural trait**. Cultural traits can vary widely across regions and even within societies. For example, a cultural trait many Muslim women follow is wearing a hijab in public, so their hair and ears are completely covered by a scarf outside of the home. However, not every Muslim woman wears a hijab. Hijabs are common for religious, cultural, and social reasons. In other places and among other groups, Muslim women will wear no head covering at all. Another cultural trait is how people greet one another. A handshake may be acceptable in one group but not in another.

British scholar Julian Huxley identified cultural traits as artifacts, sociofacts, or mentifacts. **Artifacts** are the visible objects and technologies that a culture creates, such as houses and buildings, clothing, tools, toys, and land-use practices. Artifacts change readily. Some objects are basic necessities (buildings), while others can be used as part of a culture's religious or societal expression (clothing) or for recreation (toys). **Sociofacts** are structures and organizations that influence social behavior, such as families, governments, educational systems, and religious organizations. Slower to change than artifacts, sociofacts define the way people act around others and establish rules

A *quinceañera* is a traditional celebration in Latino communities around the world during which a 15-year-old girl is said to move from girlhood to womanhood. Here, a young woman named Lacey celebrates her 15th birthday in Culiacán, Mexico.
A girl's *quinceañera* dress can be considered an artifact.

Concepts of Culture **161**

that govern behavior. **Mentifacts**—the central, enduring elements of a culture that reflect its shared ideas, values, knowledge, and beliefs—are the slowest to change. Examples include religious beliefs and language, which you will read about later in this chapter.

Cultural Dynamics

Learning Objectives

LO 6.3 Describe how cultural aspects change and spread in response to environmental, human, and technological forces.

LO 6.4 Compare the dynamics of popular culture and traditional culture.

Modern cultural geography examines the wide-ranging ways in which culture evolves and makes a difference to everyday life and places. Geographers study, for example, the cultural politics of different groups with respect to issues such as race, gender, ethnicity, sexuality, and disability. They also examine how the processes and practices of religion, colonialism, imperialism, and nationalism shape the lives of people in different places and contexts, fostering senses of belonging or of exclusion. Geographers have looked at how culture is reflected and influenced through representations such as music, art, film, and other mass media; material culture such as food, memorials, and monuments; and practices of creating knowledge and communicating through language.

Culture is socially constructed, which means it's created jointly by people rather than by individuals. Culture is also naturally dynamic: it's subject to change, and it transforms in response to countless environmental, human, and technological forces. The speed at which a culture transforms and the aspect of a culture that changes vary from society to society.

For example, the role of women in U.S. culture changed rapidly in the 1960s. A growing population and booming job market enticed women to pursue paying careers. They moved beyond the established social expectation for them to remain at home, working as unpaid housewives. As these societal values changed, education and job opportunities for women expanded. From 1964 to 1974, the number of women working outside the home grew by 43 percent. Simultaneously, the women's rights movement helped women take more prominent roles in the workplace and in society. Women's labor participation rate grew steadily until its plateau in the early 2000s in the mid–50 percent range; in 2019, the U.S. Bureau of Labor Statistics put the figure at 57.4 percent. Again, cultural expectations changed, which permanently altered the U.S. workforce. (You learn about labor-market participation in Chapter 19, including how the COVID-19 pandemic changed work for women—and men—in the United States and across the world.)

Sociofacts, such as the role of women in society, play out in regions all over the world. In Kenya, the *sepaade* tradition—which was developed during the mid-19th century—required that some women could not marry until all their brothers had married. Many women didn't marry until they were beyond childbearing age. By 1998, a variety of social, technological, and environmental factors had altered how Kenyans regarded this custom. A cultural value shifted and, as a result, the tradition ended.

When Cultures Mix Cultural change can occur when groups from one culture move to a new place that is home to a different culture. People may retain some or much of their original culture while taking on traits of the new-to-them culture. Over time, immigrants may lose some of their original cultural traits, and the cultural traits they maintain can eventually differ from those of their home countries or regions.

One example can be seen in the way Spanish is spoken throughout the United States. Many people who originally immigrated to the United States from Spanish-speaking countries continue to speak Spanish at home. However, while children born to these families in the United States often learn Spanish, they learn it as a heritage language, a minority language learned primarily at home in informal settings. A heritage language differs from the Spanish spoken in their family's country of origin and serves as an illustration of how a cultural trait—language—can become different as two cultures come together.

Cultural Norms and Differences As you have read, the effects of distance decay have been reduced through innovations in transportation and communication. Regardless of geographic location, people now can connect with other people and communities easily via the internet, social media, and smartphones. Also, global travel is easier than ever.

Buddhist monks at Fuguo Monastery in Yunnan Province, China, playing basketball—a sport more typical to the United States—is an example of cultural mixing.

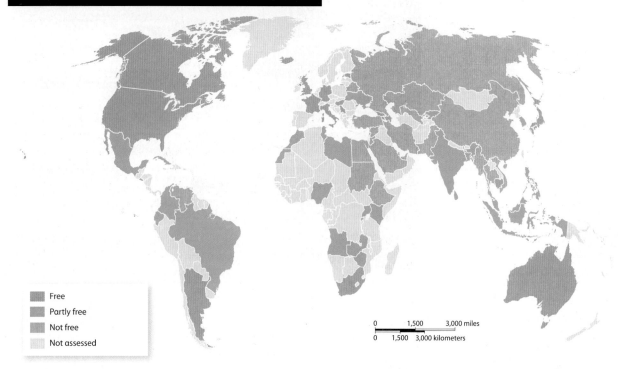

Internet Freedom around the World, 2018

Free
Partly free
Not free
Not assessed

Reading Maps An analysis of internet freedom conducted during 2017 and 2018 showed that government censorship and surveillance of the internet varied widely around the world. For example, China's Cyber Security Law makes it illegal to transmit banned content. ▌ How does this map support the claim that U.S. popular culture influences the world?

The increased use of these technological advances has led to rapid, far-reaching cultural changes. Trends in music, dance, fashion, and food now bounce around the world in a matter of seconds. These examples of **popular culture**—defined as the widespread behaviors, beliefs, and practices of ordinary people in society at a given point in time—tend to change quickly. And with increased communication and transportation technology, the rate of change continues to accelerate. Television, movies, streaming services, and social media have drastically expanded cultural reach across the world. For example, slang from U.S. movies and TV shows often weaves its way into the cultures of other countries, thanks to the power of global broadcasting.

In contrast to popular culture, **traditional culture** is composed of long-established behaviors, beliefs, and practices passed down from generation to generation, such as languages, food, ceremonies, and customs. **Cultural norms**, or the shared standards and patterns that guide the behavior of a group of people, play an important role in upholding traditions and keep traditional culture from changing. In some countries, cultural norms may even be tightly governed. Some societies requiring strong conformity may limit people's exposure to popular culture by forbidding them from accessing the internet and social media. Societies with more relaxed cultural norms are more likely to embrace the influences of popular culture.

People around the world have a wide range of attitudes toward cultural differences. One such attitude is **ethnocentrism**, the tendency of ethnic groups to evaluate other groups according to preconceived ideas originating from their own culture. Ethnocentrism exists in varying degrees. More profound ethnocentrism may include the belief that one's own cultural group is superior, which may result in overt or clear discrimination against other groups.

In contrast, the attitude known as **cultural relativism** is the evaluation of a culture solely by its unique standards. This approach requires putting aside one's own cultural criteria to understand the context behind the cultural practices of another culture. Critics of cultural relativism challenge the belief that it is appropriate to accept extreme cultural practices, such as violations of human rights, as long as the cultural context behind such practices is understood. Such detractors suggest that cultural relativism allows societies to transcend the limits of what is morally acceptable, justifying extreme cultural practices through cultural context.

Geographic Thinking

1. Define *artifacts*, *sociofacts*, and *mentifacts* and compare the differences among them.

2. Identify and describe an example of a cultural trait from your own culture that has changed over time.

3. Explain how cultural relativism might impede attempts to have international agreements on practices such as genocide or child labor.

Concepts of Culture

Case Study

Wisconsin's American Indian Nations

The Issue The state of Wisconsin officially recognizes the cultural significance of the Wisconsin American Indian Nations, which has affected the state's own cultural values and practices.

Learning Objectives
LO 6.5 Describe the characteristics of cultural landscapes.

By the Numbers

50,094
Native Americans and Alaska Natives in Wisconsin (2020)

11
federally recognized tribes in Wisconsin as of 2021

21
states that observe Indigenous Peoples' Day as of 2022, either by proclamation or as an official holiday; Washington, D.C., also commemorates this holiday via proclamation

Sources: 2013–2017 American Community Survey; Wisconsin Tribal Judges Association; Smithsonian magazine, October 2019

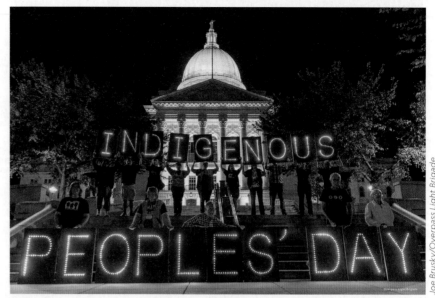

In 2019, Wisconsinites celebrated an executive order designating the second Monday in October as Indigenous Peoples' Day instead of Columbus Day. While the latter remains a federal holiday, states and localities can choose not to observe it. Indigenous Peoples' Day recognizes and honors the contributions of Native American nations.

In an Effort to Combat Misunderstandings and misinformation about American Indians, the state of Wisconsin and its tribal leaders passed Wisconsin Act 31 in 1989. (Other Indigenous peoples may refer to themselves as Native Americans, Alaska Natives, First Nations, First Peoples, or by their tribal or cultural names.) The statute was developed to ensure that Wisconsin's students had access to accurate information about the history, culture, and tribal sovereignty of the state's federally recognized American Indian nations and tribal communities. Required instruction under Wisconsin Act 31 includes educational programs that teach about terminology, value systems, and connecting cultures from the past to the present. The purpose of the act is to bridge cultural gaps and develop unity through mutual respect and understanding.

Higher educational institutions within Wisconsin have also taken steps to honor Native American heritage by using language that promotes unity and understanding among different communities within their respective regions. St. Norbert College in De Pere, Wisconsin, adopted a land acknowledgment statement in 2018 that reads, in part, "In the spirit of the Norbertine value of *stabilitas loci*, a deep commitment to the local community, we acknowledge this land as the ancestral home of the Menominee nation, which holds historical, cultural, and sacred significance to the community." This declaration serves as a first step in acknowledging both the historical and ongoing contributions of Native Americans, and similar statements are being implemented by colleges and universities globally.

To further honor Wisconsin's Native American history, Governor Tony Evers signed an executive order in 2019 to declare the state's first Indigenous Peoples' Day. By renaming and repurposing Columbus Day, Wisconsin and a growing number of other states—plus hundreds of U.S. cities as well—use Indigenous Peoples' Day to transform a traditional celebration of colonialism, honor the historical and contemporary cultures of Indigenous people, oppose current injustices against them, and recognize that original inhabitants of the Americas have shaped regional culture.

Geographic Thinking

Explain why the passing of Wisconsin Act 31 is culturally significant.

6.2 Cultural Landscapes

Every place and region has its own distinct, definable characteristics. These characteristics assemble to form the cultural landscape of a place or region. Made up of natural and cultural features, each cultural landscape is unique.

The terraced rice paddies in Shiroyone Senmaida, Japan, have been registered by the United Nations as Globally Important Agricultural Heritage Systems. The 20,000 solar-powered LED lights provide them with a modern look that attracts visitors and highlights the paddies' cultural importance.

Defining Cultural Landscapes

Learning Objectives
LO 6.5 Describe the characteristics of cultural landscapes.
LO 6.6 Define sequent occupance.

A natural landscape that has been modified by humans, reflecting their cultural beliefs and values, is a **cultural landscape**. It is the human imprint on the landscape, and it offers clues about cultural practices and priorities, both past and present. Geographers study cultural landscapes to answer questions such as these: What does the landscape communicate about the culture of the people who shaped it? How did they adapt the natural environment to fit their cultural practices and societal needs? And in terms of **identity**—how humans make sense of themselves and how they wish to be viewed by others—how do cultural landscapes influence and reflect a group's identity, and vice versa? A cultural landscape is not simply a material artifact or set of artifacts but is imbued with symbolic meaning that needs to be interpreted—"decoded"—with respect to its social and historical context, and with respect to the ways that differences and power relationships work in a society.

Indeed, every landscape is a cultural landscape, and each one has been shaped by a unique series of human ideas and activities over time. Humans have interacted with Earth's natural landscapes and resources for hundreds of thousands of years, creating varied cultural landscapes to meet their needs and express their values and beliefs.

Material expressions of culture found in cultural landscapes are evidence of how people live to meet their needs and aspirations. For example, geographers have examined the cultural landscape of the Shiroyone Senmaida rice paddies in Japan, shown in the photograph. Shiroyone is part of an agricultural tradition that has been in place for more than 1,300 years. The

Concepts of Culture 165

terraces built there support the local industry of rice production and create a habitat for wildlife. The population that gathered—and continues to gather—in this region has shaped a unique cultural landscape, defined by agricultural practices. The terraced rice paddies are visible manifestations of a coherent, organized culture that can maintain them.

Early-20th-century geographer Carl Ortwin Sauer saw cultures and societies as both developing out of and being affected by landscape. He famously stated, "The cultural landscape is fashioned from a natural landscape by a culture group. Culture is the agent, the natural area is the medium, and the cultural landscape is the result." Sauer perceived the human impact on landscapes to be a reflection of culture and asserted that geographers could only understand a culture by first learning to read its landscape. One example can be seen in the U.S.–Mexico borderlands, which have been studied extensively by geographer Daniel D. Arreola. Each side of the border shares the same physical environment of mountains, deserts, and grasslands, but Arreola has identified varied cultural landscapes as a result of the differences between the two cultures. Visually apparent contrasts are readily seen when comparing the land use, agricultural practices, and architecture that exist on each side of the border.

Sauer also suggested that people need to understand the past to understand the present. He believed that cultural landscapes offered clues about the cultural practices and activities of a region's past inhabitants. Geographers call this **sequent occupance**—the notion that successive societies leave behind their cultural imprint, a collection of evidence about human character and experiences within a geographic region, which shapes the cultural landscape.

Consider New Orleans, Louisiana, as an example of sequent occupance. Its landscape reflects numerous cultures that have shaped the city over time. As each group sequentially occupied this space, its members instilled some of their values, economic practices, languages, and customs into the landscape. Native Americans initially inhabited the region in which present-day New Orleans exists. After European colonization of the Americas, it was a French city and then a Spanish city before becoming part of the United States due to the land deal known as the Louisiana Purchase. African American culture had a tremendous influence on New Orleans and the city's cultural landscape. Enslaved Africans and African Americans made New Orleans's Congo Square (now Louis Armstrong Park) their own. It is not coincidental that this location in the historically African American Tremè neighborhood is the birthplace of jazz, a uniquely American art form created by African Americans.

Examples of Cultural Landscapes

Learning Objectives
LO 6.7 Explain how studying ethnic neighborhoods can illuminate the relationship between culture and landscape.

LO 6.8 Compare how traditional and postmodern architecture reflect cultural beliefs and identity.

Landscapes you encounter every day as you walk or drive through your community, as ordinary as they may seem, are still unique cultural landscapes. Geographers analyze such landscapes to understand the cultures of the people who create them. Cultural landscapes can undergo both slow and rapid changes over time. These changes are accelerated by human factors and processes—such as technology, industrialization, urbanization, globalization, and sustainability initiatives—all of which are influenced by culture. By studying examples of different cultural landscapes, including neighborhoods, building styles, and land-use patterns, geographers can better understand the relationship between culture and landscape.

Ethnic Neighborhoods Ethnicity is the state of belonging to a group of people who share common cultural characteristics. Large cities typically contain minority clusters, who may, if their numbers and duration of residence allow, form **ethnic neighborhoods**—cultural landscapes within communities of people outside of their areas of origin. These include Chinatowns, Koreatowns, Japantowns, and Little Saigons. Chinatown in Vancouver, British Columbia, is an example of how the history, tradition, and social practices of an ethnic group can transform a landscape into one that is responsive to a community's needs. But why do ethnic neighborhoods form? Processes of exclusion, segregation, racist politics, and discrimination certainly cause minority groups to band together to maintain their identity. The advantages of uniting in a certain area can help a cultural group set themselves apart to practice their customs in their own neighborhood schools, places of worship, stores, and businesses.

The Goutte d'Or neighborhood of northern Paris, also known as "Little Africa," is another example of an ethnic neighborhood. The city of Paris appeals to emigrants from French-speaking African countries such as Mali and Senegal for an obvious reason: the common language. For decades, African immigrants have settled in Goutte d'Or and, consequently, the cultural imprint of both North and West Africa is present everywhere there—from the foods served in restaurants to the rich fabrics sold at the local open-air market. Additionally, the exceptional textile-making skills of African immigrants have propelled this Parisian neighborhood into one of France's creative hubs.

Attitudes that exist within the surrounding communities may draw visitors and businesses into an ethnic neighborhood, or they may promote discrimination against the ethnic neighborhood's residents and businesses. Sometimes, discrimination and economic factors can cause the decline of an ethnic enclave. Washington, D.C., once had a vibrant Chinatown, which was established in the 19th century, with large grocery stores, shops of varying kinds, and dozens of traditional Chinese restaurants. Today, D.C.'s Chinatown has but a remnant of its previous rich character, with many chain restaurants and gift shops but only a handful of traditional Chinese restaurants and far fewer Asian American residents. This and other ethnic Asian neighborhoods have also struggled to survive in the face of more recent anti-Asian sentiment and violence fueled by the COVID-19 pandemic.

Traditional Architecture

Traditional architecture—established building styles of different cultures, religions, and places—was originally influenced by the environment and is based on localized needs and construction materials. It tends to reflect local traditions and usually evolves over time to reflect the environmental, cultural, and historical context in which it exists.

In the 1500s in what is now the U.S. Southwest, Spanish colonizers found mud homes built by the Pueblo peoples. The Spaniards later combined the Spanish-Moorish adobe building technique with the traditional, regional style of the original American dwellings. (The Spanish word *adobe* is derived from the Arabic term *al-tob*, which means "mud-brick," and can describe either the material or the method of construction.) The original materials of wood, stone, and puddled adobe—sun-dried mud—were readily available resources that kept homes cool during the day and warm at night. The construction materials have been adapted over time and, with added modern energy conveniences, this style of home remains prevalent in states such as Arizona and New Mexico, where adobe continues to provide an energy-efficient material in hot, dry climates. Traditional architecture often reflects a historical association, continuity, and ties with the origins of a culture or place.

The round, movable homes—called gers by Mongolian nomads and yurts by other groups in Central Asia—reflect a nomadic lifestyle, dependency on traditional herding practices, and interactions with nature. As of 2017, approximately 1 million nomads lived on the vast, open plains of Mongolia. They principally herd sheep and goats, but many nomads also have cows, horses, dogs, camels, and yaks. In addition to milk and meat, these animals provide wool for the felt sidings that keep the gers warm and dry. Nomadic Mongolians set up their mobile dwellings between summer pastures and winter-feeding lands where they can shelter from the powerful, icy winds. This confluence between humans and their environment illustrates how established practices and natural elements are woven together into a cultural landscape.

Postmodern Architecture

Postmodern architecture emerged in the 1960s as a reaction to "modern" designs, which emphasized form, structure, and materials. Some people thought modern architecture could solve urban social problems. During the 1950s and 1960s, many urban renewal projects destroyed historic buildings—and even entire neighborhoods—in cities large and small. Postmodernism was, as historian Mary McLeod wrote, "a desire to make architecture a vehicle of cultural expression."

In contrast to the restrictions of modernism, postmodern architecture values diversity in design, and public spaces that can be enjoyed by anyone are integral. Urban skylines

Critical Viewing A street market on Rue Dejean in the Goutte d'Or neighborhood of Paris bustles with activity. ▌Which details in the photo would you investigate to understand the ways in which this ethnic neighborhood is a unique cultural landscape?

Concepts of Culture

can reflect a city's robust economy. In recent decades, Dubai has become a global center for commerce. Highly visible from great distances, its skyscrapers—including Burj Khalifa, one of the tallest structures in the world—represent both rising power and economic strength. As part of the cultural landscape, skylines make the most effective use of space in a crowded city and reflect the local culture's admiration of the values of ambition and success.

However, geographers such as David Harvey criticize postmodern architecture and urban design as failing to deliver on promises of architectural diversity and equity in public spaces. In his influential book, *The Condition of Postmodernity*, Harvey finds serious shortcomings in postmodern architecture: Instead of "infinite variety," it "degenerates" into monotony; rather than providing relief to the poor and oppressed, it simply marginalizes these groups in new ways; by constructing extravagant spaces like San Francisco's Fisherman's Wharf as a "sanitized" version of urban life, it simply masks the realities; and postmodern construction and identity of space simply "pander" to consumer demands.

Landscapes of Religion and Language

Learning Objectives
LO 6.9 Describe how religion significantly impacts cultural landscapes.
LO 6.10 Explain how visible language, like toponyms, can leave a significant imprint on the cultural landscapes people inhabit.

From the markets of Goutte d'Or to the restaurants of Tehrangeles, each diverse cultural landscape you have read about contains evidence of how it has been constructed to reflect the cultures of its inhabitants. These cultural practices might be seen in the land-use patterns, the types of housing, the transportation systems, or other cultural landscape features. Two of the most deeply influential aspects of culture—religion and language—leave distinct and identifiable imprints on almost any cultural landscape.

Religion Religion is a system of spiritual beliefs that helps form cultural perceptions, attitudes, beliefs, and values—the motives behind observable cultural behaviors and practices. Although these deep cultural foundations may be less visible, religion significantly impacts cultural landscapes. As a mentifact, religion is the slowest type of cultural trait to change, and because it is central to the shared beliefs and ideas of a group, it is reflected in the sociofacts and artifacts of that group's culture.

While immigrants and their descendants often adopt the language and other cultural traits of their new country, they typically continue to uphold the religious beliefs and practices of their home country or region. People's religion reflects their core beliefs, which contributes to their cultural values and identity. And there is a strong connection between religion and ethnicity—so strong that the effects of both are evident in an ethnic group's lifestyle and value system. In fact, some major world religions are so closely connected to the cultural identities of certain countries that it is difficult to think of the country without thinking of the religion, too. Buddhism, for example, is closely linked with the cultural identity of Thailand, and Hinduism is a prominent part of India's cultural identity.

Each religion has a different spatial organization. The practices of the religion influence where its followers worship, which then affects the design and architecture of its places of worship. Members of some religions assemble for congregational worship or community assembly in churches, mosques, and synagogues. Others worship or observe religious practices individually in smaller pagodas, temples, or shrines.

The Kaaba, a square stone building in the center of the Great Mosque in Mecca, Hejaz, Saudi Arabia, is considered to be the most sacred Islamic shrine. Wherever they are in the world—even on airplanes—Muslims make sure to face the Kaaba during their five daily prayers. They also bury their dead facing Mecca. Adult Muslims are expected to make a **pilgrimage** to Mecca called the hajj at least once during their lifetime to walk around the Kaaba. A pilgrimage is a journey to a holy place for spiritual reasons. In contrast, Hinduism's temples are meant to house shrines devoted to certain gods, rather than provide places for congregational worship. Individual Hindus practice private worship, alone or with family members, performing rituals that include making offerings and repeating mantras—sounds or words—three times a day.

Language Language is the carrier of human thoughts and cultural identities. As you might expect, each distinct system of communication has a significant imprint on the cultural landscapes people inhabit, linking language to sequent occupance. Nearly all languages have a literary tradition, or some form of written communication. Visible language is a clue to the identities of the people who live in that area, and it threads its way into daily life within a cultural landscape through place names and street names—and even public murals. These uses of language often reflect the linguistic heritage of a particular geographic location.

Place names, or **toponyms**, help define what is unique about a place, such as its geographic features or history. Coconut Creek, Florida, for example, was named for the large number of coconut trees planted by developers before the area became a city. Big Bend National Park in Texas was established in 1944 alongside a large, natural curve of the Río Grande. A toponym such as Battle Creek, Michigan, provides insight into a city's history, while a French street called Avenue Charles de Gaulle pays homage to a famous individual.

Case Study

Tehrangeles

The Issue The concentration of Iranians living in the Los Angeles community known as Tehrangeles forms a cultural landscape that preserves some of their native language and traditions.

Learning Objective
LO 6.7 Explain how studying ethnic neighborhoods can illuminate the relationship between culture and landscape.

By the Numbers

40%+
of all people of Iranian ancestry in the United States live in California (2019 article)

224+
languages spoken in Los Angeles County (2019 estimate)

87,000
people of Iranian descent live in Los Angeles (2019 article)

Sources: Los Angeles Times, LA Tourism & Convention Board, U.S. Census Bureau

A U.S. flag flies near a Persian restaurant in Tehrangeles, an Iranian neighborhood in Los Angeles that makes up a unique cultural landscape. The name *Tehrangeles* is a blend of two place names: Tehran (the capital of Iran) and Los Angeles.

Los Angeles Is a Large, Diverse City filled with ethnic communities such as Chinatown, Koreatown, and Little Armenia. On the city's west side live a concentration of Iranians who form a neighborhood known as Tehrangeles. Tehrangeles started as a small community of immigrant students and entrepreneurs in the 1960s and has evolved into the largest community of Iranians outside of Iran. The main driver of the community's growth was the 1979 Iranian Revolution, when the last shah of Iran was overthrown and the government was transformed into an Islamic republic. Many Iranians fled the oppressive new regime, and thousands of them settled in Los Angeles. In the years that followed, chain migration drew more Iranians to the community to be with their families, and the Iran-Iraq war of 1988 pushed others to leave their native country.

Los Angeles is an appealing destination because many immigrants find that it reminds them of Iran: the mountainous landscape, the climate, and even the car culture. The community has offered a safe haven for Iranians during times of conflict between Iran and the United States, such as the hostage crisis of 1979, which exposed sharp feelings and subsequent actions of prejudice against Iranians. The Iranians who helped establish Tehrangeles are now in their 70s and 80s and are bonded to the community by their role in its history as well as their contributions to cultural preservation. Members of the younger generation inherit Iranian traditions from their elders and embrace their American identity at the same time.

What distinguishes this prosperous neighborhood of middle-class homes from any other in Los Angeles? Its Iranian character might not be readily apparent when traveling through Tehrangeles, but it's there; for example, children learn to play traditional Iranian instruments, and rug shops sell exquisite Persian rugs. Traditional Iranian restaurants exist near those that offer Iranian fast food such as "Persian pizza," which is made with Persian ingredients and spices. Iran's primary language, Farsi, is spoken throughout the community and appears on the signs of many shops and restaurants. Some businesses display only Farsi names, while others have signs in both English and Farsi. Similarly, some businesses fly only the Iranian flag while others display both U.S. and Iranian flags. In addition, Tehrangeles is a media hub, broadcasting several Farsi-language television stations and a Farsi radio station. ∎

Geographic Thinking

Describe how Tehrangeles illustrates the connections among language, ethnicity, and geography.

With improvements in transportation and crowd control and rising global wealth, more than 2 million Muslims make the hajj—or pilgrimage to Mecca—every year. As a result, the government of Saudi Arabia has expanded its facilities in Mecca to handle the large number of pilgrims.

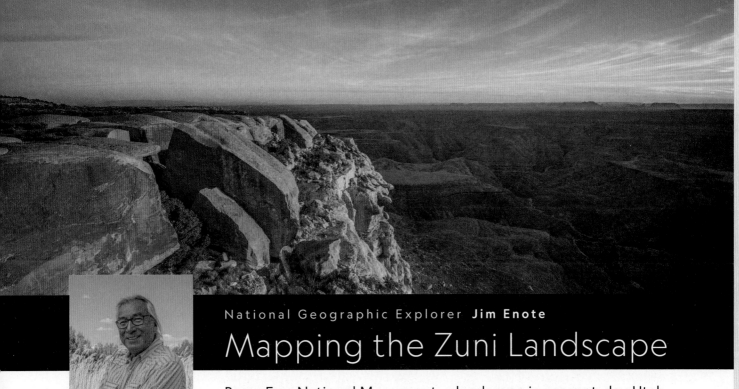

National Geographic Explorer Jim Enote

Mapping the Zuni Landscape

Jim Enote is a traditional Zuni farmer and CEO of the Colorado Plateau Foundation, which supports Native communities in protecting the region's water and sacred places, preserving Native languages, and promoting sustainable agriculture. He is the former director of the A:shiwi Awan Museum in Zuni, New Mexico.

Learning Objectives
LO 6.10 Explain how visible language, like toponyms, can leave a significant imprint on the cultural landscapes people inhabit.

Bears Ears National Monument, a landscape in present-day Utah that is sacred to the Hopi, Navajo, Ute, and Zuni peoples, was expanded in 2021, ensuring protection of its natural and cultural resources. Zuni tribal member Jim Enote views Bears Ears in the context of a much larger cultural landscape that his people have explored and inhabited since time began.

Cascading over the Grand Canyon's North Rim into a cool oasis, the spectacular Ribbon Falls is more than a popular backpacking destination to the Zuni people. Enote identifies it as the place of emergence—Chimik'yana'kya dey'a—where his ancestors began their life on Earth. From there, clans set out to explore the tributaries of the Colorado River. They sought their "center place," a place with plentiful water and good land for farming where they would eventually settle.

The story of this migration and exploration is told in petroglyphs, or rock carvings, found throughout the Colorado Plateau. One petroglyph depicts a spiral—a continuous line making a circular path toward the center. The spiral petroglyph represents the people's journey. Where it appears, it is a record showing that people journeyed and tested several places before moving on to find the center of their modern world. The presence of the spiral petroglyph in Zuni, New Mexico, in the Grand Canyon, at Bears Ears, and at many other sites throughout the region unites this sprawling cultural landscape across time and space.

Realizing that conventional maps do not adequately represent how Zuni people interpret their environments, Enote hired Zuni artists to represent the Zuni world with alternative maps. A Zuni map "may be something we heard from our grandmother about a place. There are maps in songs and in prayers. There are maps that are etched in stone and woven into textiles and painted on ceramics," he says. Incorporating petroglyphs and other images that convey the Zuni's lived experience in their environment, the maps give vibrant form to the abstract concept of a cultural landscape. Enote believes Zuni maps challenge mapping standards with different ways of knowing and a unique memory of place.

Geographic Thinking

How do petroglyphs throughout the Colorado Plateau help define the region as a cultural landscape?

Concepts of Culture

Toponyms can also be changed, with evolving or competing views of what a particular place name should or should not represent. For instance, more than 900 U.S. cities have renamed one of their major streets to memorialize civil rights leader Dr. Martin Luther King Jr. In late 2021, Chicago renamed 18 miles of its iconic Lake Shore Drive to honor Jean Baptiste Pointe DuSable, the Black owner of a trading post credited with being the city's founder. Shortened on some signage to DuSable Lake Shore Drive, the new toponym was criticized by some for taking away a name they liked as it was and forcing address changes for many.

Geographers have been reevaluating the study and meaning of toponyms in recent decades. Rather than being simply "linguistic indicators or artifacts of culture, history and environment" to be classified and mapped, cultural geographers Derek Alderman, Reuben Rose-Redwood, and Moaz Azaryahu see toponyms employed in the cultural naming, claiming, and contestation of places. They and others stress place naming as a tool used by the culture in control to shape society and life—that can also be employed by those resisting that culture. This contestation can be seen in what Alderman notes as the "racialized place naming patterns that have now become the target of challenge by anti-racist activists."

Geographic Thinking

1. Describe the characteristics of a cultural landscape.
2. Describe evidence that might help you identify the languages used in ethnic neighborhoods.
3. Explain the degree to which architecture can reflect cultural beliefs and identity.

6.3 Identity and Space

A single space in a cultural landscape—a person's workspace, for example—will likely reflect some part of that individual's identity. A person's age, ethnicity, and gender can often be "read" in that space. Geographers apply this same principle to help understand the ways that identities of groups are reflected in the cultural landscape.

Shaping Space through Identity

Learning Objectives
LO 6.11 Describe how landscape features can reflect the residents' cultural beliefs or identity.

LO 6.12 Explain the ways in which a culture's identity is connected to its use of land and resources.

So who are you, anyway? You read that identity is how humans make sense of themselves and how they wish to be viewed by others. Culture, ethnicity, and gender are major factors that establish a person's identity. And identity affects how people occupy space in society and shape the landscape. People's cultural beliefs and identities influence how they use the land they occupy and its resources.

The artifacts found in a person's workspace or home can tell you quite a bit about that person's life. Things like photographs, books, artwork, and technology can reveal what a person values and what that person finds less important. On a broader scale, neighborhoods reflect the cultural attitudes of their residents. For instance, the places of worship, restaurants, and shops of an ethnic neighborhood show how cultural attitudes and practices affect the way space is used. As you read, ethnic neighborhoods are examples of concentrated populations of an ethnic group that retain elements of the culture of origin while functioning within the new culture. In fact, however, all neighborhoods are landscapes that reflect a unique set of beliefs and attitudes.

Landscape Features and Identity Homes within neighborhoods may display features that reflect the residents' cultural beliefs or identity. People use these home landscapes to communicate their religious beliefs. For example, regardless of where they are located geographically, it is not unusual for people who practice Catholicism to have a statue of the Virgin Mary or a cross on a wall inside their home as a symbol or reminder of their faith. A Buddhist may build a garden that reflects the Buddhist principles of peace, serenity, goodness, and respect for all living things. A statue of Buddha in that garden would face the home, a position that Buddhists believe results in abundance.

Czech communities in the United States have created cultural landscapes that reflect both their religion and ethnicity. Religious monuments are significant components of Czechia's architectural heritage and identity, and Czech immigrants brought the tradition of erecting them along roads when they immigrated to Wisconsin in the mid-1800s. These stone or metal shrines were three to four feet high and were typically inscribed with a scripture verse in Czech and topped with a crucifix. Similarly, Czech immigrants and their descendants have left their mark on the landscape in Fayette County, Texas, where they hold traditional Czech festivals, operate bakeries specializing in Czech pastries, and have built the famous "Painted Churches of Texas"—churches that are unassuming from the outside but painted in vivid colors on the inside.

In the western United States, residents sometimes mark treeless hills and slopes with one or more gigantic capital letters to signify the names of towns or schools. Often, these mountain monograms or cultural signatures are constructed using whitewashed rocks, and they represent an invented regional tradition by which residents demonstrate their community or school pride. This practice of imprinting the landscape remains concentrated in the West and has become a means of regional identification over the past hundred years. Indeed, cultures across the world have stacked rocks, sometimes called cairns, for thousands of years—often to help people find their way safely through areas with little vegetation. Other cairns in other cultures have been created as memorials, tomb markers, symbols of faith—even lighthouses or simply works of art.

Land and Resource Use Amish immigrants first arrived in eastern Pennsylvania in the 18th century, and a large Amish settlement still thrives there today. Sizable Amish agricultural communities can also be found in Ohio, Indiana, and Missouri. The landscapes inhabited by these communities are distinct—large plots of farmland whose characteristics reflect the Amish identity. People who subscribe to the Old Order Amish refuse modern conveniences like cars, trucks, modern farming equipment, and even electricity, instead using horses and plows and gas- or battery-operated lights. They view their farming practices as coinciding with their values of discipline and hard work. Members of the community help each other farm, and Amish "barn-raisings" are an integral cultural tradition. Amish women are renowned for their baked goods that provide a source of income and are often made using milk, eggs, and produce from their own farms. Simple living, community cooperation, and acting as stewards of the land are values that the Amish identify with and are reflected in how they farm the land and use its resources.

The Inupiat people offer another example of the ways in which a culture's identity is connected to its use of land and resources. The Inupiat live on the North Slope, an Arctic coastal plain north of the Brooks Mountain Range in northeastern Alaska. The area stretches over 89,000 square miles from the Brooks Range to the coastline of the Arctic Ocean. They hunt marine and land mammals, fish, and migratory birds to feed their population—which totals fewer than 10,000 people in eight small communities—and sustain their economic livelihoods and cultural lifestyles. Hunting also serves as a vehicle to pass on tribal knowledge. The most culturally significant resource harvested on the North Slope is the bowhead whale, which the Inupiat have hunted for a thousand years. At an early age, children are taught about subsistence whaling as a culturally acceptable practice (as opposed to commercial whaling) and play a part in the harvest to ensure survival of this central part of the Inupiat culture. Subsistence whaling is also culturally important because sharing resources is a valued tradition for the North Slope people.

Women and Gendered Spaces

Learning Objective
LO 6.13 Explain how attitudes toward gender shape the use of space in a cultural landscape.

When a society has strict roles for men and women, certain spaces may be designed and deliberately incorporated into the landscape to accommodate gender roles. However, philosopher Judith Butler reminds us that both sex and gender are socially constructed: "gender is an identity tenuously constituted in time, instituted in an exterior space through a stylized repetition of acts." These spaces—exterior or interior—are called **gendered spaces**, and can exist in homes, workplaces, and public areas. Depending on the cultural and societal factors that establish them, gendered spaces can be supportive, positive places or restrictive places. Throughout history and across the world, gendered spaces have been more restrictive for women than for men. Traditional gender roles often keep women from playing a part in certain aspects of society and push them to fulfill cultural expectations that can be limiting.

In the rural areas of some countries, gender differences tied to laws or cultural beliefs about land ownership influence how men and women use land. For example, in Kenya, it's much more likely that a man will own land because women in this country don't have equal property rights, and property rights normally pass from father to son. In fact, since 2013, fewer than two percent of title deeds issued in

In the Indian city of Delhi, women-only train cars have been introduced to the city's metro system in an attempt to provide safer spaces for Indian women. Public awareness of sexual harassment and sexual violence against women in India has risen sharply since the 2010s.

Kenya have been granted to women. However, according to the World Bank, more than three-quarters of the farms in Kenya are run by women. They do much of the country's agricultural work while the men own and control the vast majority of the farmland.

For both men and women, some spaces are perceived to be inclusive whereas others are perceived to be exclusive. Urban planners all over the world recognize that women experience city life differently than men. In the United States, 10 million of the 18 million women who own homes live alone, and demand has increased for housing that provides increased safety and affordability and requires little maintenance. In general, women place more value on safety than men do, preferring urban spaces with well-lit parks and parking areas. Some urban places have responded to that preference by introducing spaces where men are not allowed. In Mexico City, for instance, the public transportation system includes bright pink buses that are available to women and children only.

Gendered spaces can offer women more than just physical safety; they can be places where women feel emotionally safe as well. In some regions of India, for example, women are discouraged by the culture from openly sharing their viewpoints in the company of men. To address this issue, the Zenana Bagh, or "women's park," was created in Delhi as a safe place for Indian women to gather for prayer and to openly discuss religious, environmental, and political issues.

Until recently, restaurants in Saudi Arabia were required to have separate entrances for men and women, and screens were set up to divide men-only areas from areas where families and tables with only women could eat. In December 2019, however, these restrictions were eased by the government and restaurants were no longer required to segregate by gender.

What happens when societal conventions related to gendered spaces are broken? In Egypt, during the political revolution known as the Arab Spring in 2011, women joined demonstrators in Tahrir Square, a public space they had once been prohibited from entering. Many of these women were threatened and physically attacked while demonstrating. In Egyptian culture, men and women are often expected to follow strict gender roles, with a woman's role mostly limited to serving men in domestic capacities

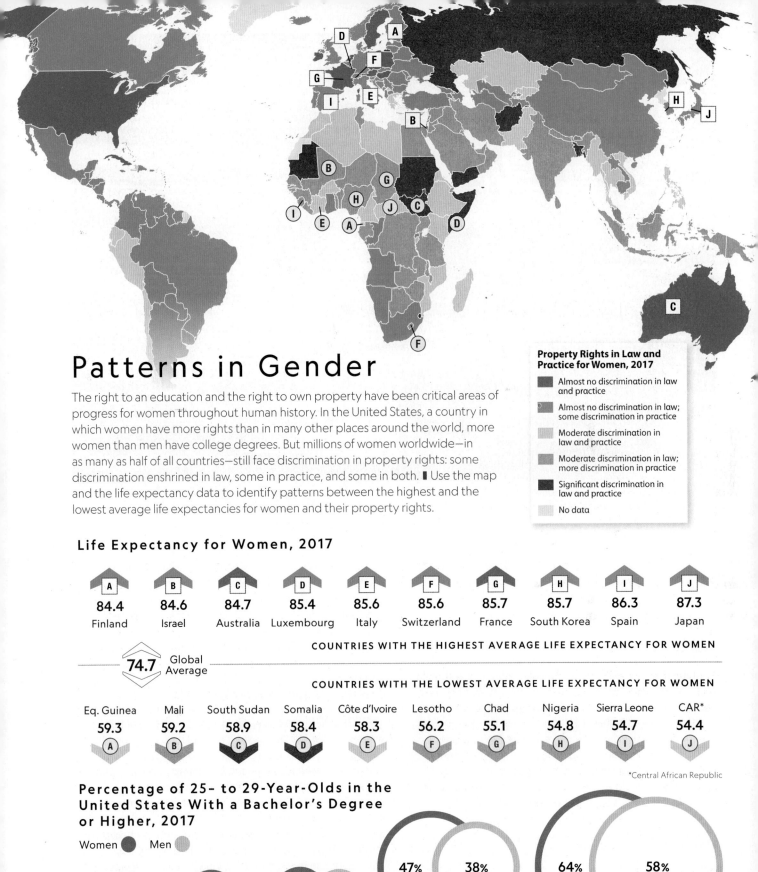

Patterns in Gender

The right to an education and the right to own property have been critical areas of progress for women throughout human history. In the United States, a country in which women have more rights than in many other places around the world, more women than men have college degrees. But millions of women worldwide—in as many as half of all countries—still face discrimination in property rights: some discrimination enshrined in law, some in practice, and some in both. ■ Use the map and the life expectancy data to identify patterns between the highest and the lowest average life expectancies for women and their property rights.

Property Rights in Law and Practice for Women, 2017
- Almost no discrimination in law and practice
- Almost no discrimination in law; some discrimination in practice
- Moderate discrimination in law and practice
- Moderate discrimination in law; more discrimination in practice
- Significant discrimination in law and practice
- No data

Life Expectancy for Women, 2017

COUNTRIES WITH THE HIGHEST AVERAGE LIFE EXPECTANCY FOR WOMEN

A	B	C	D	E	F	G	H	I	J
84.4	84.6	84.7	85.4	85.6	85.6	85.7	85.7	86.3	87.3
Finland	Israel	Australia	Luxembourg	Italy	Switzerland	France	South Korea	Spain	Japan

Global Average: 74.7

COUNTRIES WITH THE LOWEST AVERAGE LIFE EXPECTANCY FOR WOMEN

Eq. Guinea	Mali	South Sudan	Somalia	Côte d'Ivoire	Lesotho	Chad	Nigeria	Sierra Leone	CAR*
59.3	59.2	58.9	58.4	58.3	56.2	55.1	54.8	54.7	54.4
A	B	C	D	E	F	G	H	I	J

*Central African Republic

Percentage of 25- to 29-Year-Olds in the United States With a Bachelor's Degree or Higher, 2017

Women ● Men ○

- American Indian/Alaska Native: Women 19%, Men 8%
- Hispanic: Women 22%, Men 15%
- Black: Women 24%, Men 22%
- White: Women 47%, Men 38%
- Asian/Pacific Islander: Women 64%, Men 58%

Sources: The Women's Atlas, World Bank, National Center for Educational Statistics

Concepts of Culture 175

In Judaism, a bar mitzvah celebration takes place on a boy's 13th birthday, which marks the day when a boy becomes accountable for his own actions and sins. The corresponding ritual for girls is called a bat mitzvah and takes place when a girl is 12 or 13.

within a household. The events of Tahrir Square demonstrate how men can carry over dominant behaviors from the domestic sphere into the public sphere.

Geographers can use spatial analysis and maps, particularly mental and sketch maps, to identify and highlight gendered spaces. As geographer Jack Jen Gieseking writes, "The mental mapping method affords a lens into the way people produce and experience space, forms of spatial intelligence, and dynamics of human–environment relations." Mapping gendered spaces in this or other ways can reveal patterns that might otherwise be overlooked. Geographers can also help identify where gendered spaces are compartmentalized in society and determine the accessibility of resources. For example, they could use land ownership mapping to show discrimination in places that is based on gender.

LGBTQIA+ Spaces

Learning Objective
LO 6.14 Explain the importance of creating safe spaces in the cultural landscape for members of the LGBTQIA+ community who are sometimes marginalized by society.

According to the Human Rights Campaign, **gender identity** is "one's innermost concept of self as male, female, a blend of both or neither—how individuals perceive themselves and what they call themselves. One's gender identity can be the same or different from their sex assigned at birth." Many factors contribute to a person's gender identity and can include a culture's mentifacts. The gender with which a person identifies may also differ from the gender perceived by society, and this can profoundly affect how certain individuals participate in society and shape the landscape.

The LGBTQIA+ community is an inclusive group of people whose gender identity, sexuality, or both do not fall within "traditional" cultural norms. This community generally celebrates diversity, pride, and individuality and provides a place for members to feel safe and accepted. Over the years, the LGBTQIA+ acronym has evolved to be more inclusive, but its most recent form stands for lesbian, gay, bisexual, transgender, queer or questioning, intersexual, and allied or asexual. The plus sign at the end represents anyone not included in the abbreviation whose gender identity differs from traditional expectations.

Safe Spaces and Neighborhoods Social media, school-based groups, sports leagues, and performance groups have created **safe spaces**, or spaces of acceptance for people such as members of the LGBTQIA+ community

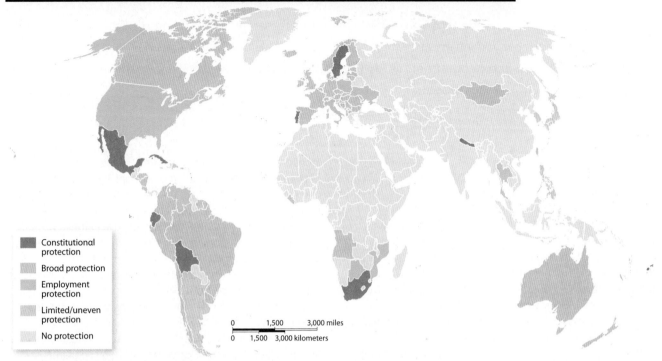

Protection against Discrimination Based on Sexual Orientation, 2019

- Constitutional protection
- Broad protection
- Employment protection
- Limited/uneven protection
- No protection

Reading Maps Every year, the International Lesbian, Gay, Bisexual, Trans and Intersex Association publishes a report on sexual orientation laws across the world. As of 2019, consensual same-sex acts between adults are punishable by death in 11 countries. ▌ What conclusions can you draw about regions that discriminate against sexual orientation?

who are sometimes marginalized by society. Safe spaces exist within larger, more traditional landscapes where LGBTQIA+ people can share interests and experiences with others who identify with the community. LGBTQIA+ community centers throughout the world, such as The Center in New York City, established in 1983, offer advocacy, health and wellness support, and access to cultural events.

A number of urban neighborhoods, such as Midtown in Atlanta and Schöneberg in Berlin, have become cultural landscapes with many safe spaces for LGBTQIA+ people. Such communities formed years ago when gays and lesbians moved into and renovated urban neighborhoods to form places free of discrimination. However, areas such as Boystown in Chicago and Shoreditch in the East End of London are now attracting residents from outside the LGBTQIA+ community. **Gentrification**, or the renovations and improvements conforming to middle-class preferences, has driven up the demand for housing and the cost of living in these neighborhoods, making it difficult for less affluent, more vulnerable LGBTQIA+ populations to live there. Some LGBTQIA+ businesses have also been forced out of their neighborhoods, changing the cultural landscape.

The term **third place**, while originally coined within postcolonial studies to refer to a space between the colonizer and the colonized, was repurposed in the late 1980s to refer to a communal space such as a coffee shop, fitness center, or bookstore that is separate from home (first place) or work (second place). In all communities, having a physical third place is important; individuals need social spaces where they can develop a sense of self, let their guard down, and form relationships with others. The need for accessible, safe, welcoming, and inclusionary third spaces is especially important for groups such as members of the LGBTQIA+ community, who feel marginalized by mainstream society.

More than 20 U.S. states and the District of Columbia have adopted legislation to prevent the bullying of LGBTQIA+ students in grades K–12. Various organizations help schools create safe environments by offering educator guides and "safe space kits" to help teachers and students. In similar fashion, colleges and universities across the United States have instituted programs for LGBTQIA+ students, offering resources, support, and safe spaces to foster a campus environment where all students are welcomed, respected, valued, and safe. Organizations that rank colleges and universities produce "best of" and "LGBTQIA+-friendly" lists for prospective students.

Geographic Thinking

1. Describe an example of a gendered space for women.

2. Explain how educational institutions might further develop or create safe spaces for LGBTQIA+ students.

6.4 Cultural Patterns

Think about a place that is special to you. What makes it so remarkable? Good memories of a place can affect how you interpret that geographic space. All humans associate feelings with locations, and geographers study these emotions because they help explain patterns found in cultural landscapes.

Sense of Place

Learning Objective
LO 6.15 Define sense of place.
LO 6.16 Describe how placemaking facilitates creative patterns of use in a landscape.

Geographers claim that when people develop a **sense of place**, they fill a geographic location with meaning by connecting memories and feelings to it. Or as geographer Yi-Fu Tuan terms it, "Topophilia"—or love of place—in his ground-breaking book of the same name, published in 1974, in which he explored the attachment to place and how people relate on an emotional and perceptual level to the places they inhabit.

As memories and stories of a place accumulate, transform, and build or even fade over time, an individual's sense of place will continue to adjust, which impacts that person's understanding of that place. Hearing new stories about your childhood home may add new dimensions to your sense of place for that home. Even fictional stories associated with a location can contribute to a sense of place. Consider works of literature, poetry, music, and film that evoke a sense of place and geography—and that both reflect and shape what might be called "placeness." The setting, including place, of most short stories and novels is central to these works. Musical offerings ranging from classical and folk to hip-hop and rap are often grounded in specific places, or types of places. A classic film example would include the sequence that shows King Kong climbing the Empire State Building and having that episode shape your sense of place about the famed New York City skyscraper. And it's difficult to think of *The Sound of Music* without pastoral Austria for its setting, or the Harry Potter films without Scotland as a mystical backdrop. About his novel, set in a place he knows intimately and describes in great detail, James Joyce noted, "If Dublin were to be destroyed, *Ulysses* could be used to rebuild it brick by brick."

Placemaking is a community-driven process in which people collaborate to create a place where they can live, work, play, and learn. Placemaking facilitates creative patterns of use in a landscape, and in turn the landscape

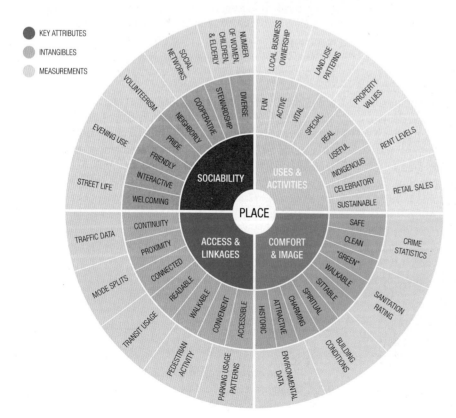

Placemaking The Project for Public Spaces (PPS) is an organization that helps people build public spaces that support local communities. This PPS infographic is a tool designed to help communities evaluate public places. The inner circle shows a place's most important qualities, the middle circle describes intangible attributes, and the outer circle shows measurable data.
▌ With a partner, use the intangible attributes in the middle circle to evaluate your college or university in each of the four qualities shown in the center circle. Then use the outer circle to identify one piece of data you would like to measure as part of your evaluation.

178 Chapter 6

reflects the culture, feelings, experiences, and perceptions of the people who use the place. Placemaking is a dynamic process that adapts to the needs of the physical, cultural, and social identities that participate in defining and using a specific place.

As introduced in Chapter 1, people's sense of place shapes their identities. Residents of South Boston also identify as New Englanders, just as residents of Colorado identify themselves as residents of the Mountain West. Examples of the perceptual or vernacular regions (defined by a person's perception and sense of place) in the United States are the West, the Midwest, the East Coast, and the South. Each perceptual region has distinct environmental, cultural, and economic features that cause people to perceive it as unique from the others. The West, for example, is commonly viewed as a place of adventure and opportunity where the outdoors represents a way of life, whereas the East Coast is known for its large cities, industry, and diversity.

Recent research in "maroon geographies" provides an example of placemaking processes instituted by Black, Indigenous, and other communities of color in the Americas, when previously enslaved peoples created communities separate and distant from dominant societies. Such places, often in locations considered marginal, inaccessible, or without value by the dominant society, provided community members an opportunity to shape what geographer Celeste Winston termed "zones of liberation." This process continues today in many urban areas of the United States where African Americans live in sites she characterizes as places of "radical transformation."

The Valley of the Gods is an area within the Bears Ears National Monument in Utah. Its topography provides a classic example of the landscape of the American West.

Concepts of Culture **179**

Language, Religion, and Ethnicity

Learning Objective

LO 6.17 Explain how patterns of language, religion, and ethnicity contribute to a sense of place and shape the global cultural landscape.

You already know the importance of language, religion, and ethnicity to culture and the cultural landscape. So it should come as no surprise that these factors contribute to a person's sense of place. The distribution of cultural traits creates unique landscapes with which people identify. Language, religion, and ethnicity all work together to form regions. For instance, the southwestern United States is shaped by the Spanish language, Catholicism, and Hispanic population. Geographers identify the patterns and landscapes that exist for individual languages, religions, and ethnicities by examining various regions.

Linguistic Patterns As we discuss in greater detail later in chapter 8, languages and words play a crucial role in the establishment of landscape. The speech patterns of members in a group within a geographical area distinguish these people from other groups and also shape their sense of place. Within a single language there may be wide variations in grammar, vocabulary, and the pronunciation of words. These variations are often regional and provide clues to migration and settlement patterns and the historical development of each region.

Early European settlers to the United States established themselves in different areas along the East Coast. Eventually, settlers moved, and more migrants came to America contributing to the growth of American **dialects**. A dialect is a variation of a standard language with a distinct but changing geography. It is distinguished by differences in pronunciation, degree of rapidity in speech, word choice, and spelling. As groups settled in different regions, their unique ways of speaking contributed to the cultural landscape and became part of the regional identity. Vocabulary is one way that groups distinguish themselves. The use of the second person plural varies among people living in the South, the Midwest, New England, and other U.S. regions: *you*, *y'all*, *you guys*, *youse guys*, or *you'uns*.

In addition to geographic location, social factors such as class, race, or ethnicity can also impact language patterns. In New England, for example, a distinction is sometimes made between people in the upper and working classes based on dialect. The so-called Boston Brahmin dialect is historically associated with people who shaped the New England cultural landscape, including the descendants of the American founders and the powerful and wealthy. In contrast, the traditional Boston accent, with its "lost *r*'s" ("pahk the cah" instead of "park the car") and other unique grammatical constructs, is traditionally linked to the working class. Another example of societal factors influencing language patterns is African American English (AAE), a distinct dialect that developed because many African Americans were socially isolated, even after emancipation.

The southern United States uses a regional variety of English that distinctly sets it apart from other regions. The dialects and accents of the South have become part of the region's identity. Despite stereotypes, there are linguistic variances in Southern American English (SAE), some of which can be attributed to geographical location and many of which are determined by whether a person lives in a city or a rural area, as well as that individual's family history. For example, a person from Georgia has a different southern accent than a person from Texas. SAE garners a great deal of attention because it is so widely recognized. While some southerners may work to modify their accents, most maintain them as part of their cultural identity. Regional dialects aside, the United States acknowledges the existence and use of Standard English (SE) as an idealized norm, but to what degree SE is used or adopted depends largely on location.

Certain places in the world have more involved language structures in which language use, identity construction, and place formation are intertwined. Complex regional patterns such as these can be found in Hualien County, Taiwan, an agricultural township that spans the island's Coastal Mountain Range and East Rift Valley. Rich in indigenous diversity, this region is packed with various ethnic groups, including Taroko, Amis, and Han.

The Hualien County government reports that there are at least nine languages spoken in the region. In Hualien, different ethnic groups name places based on how each group views and uses the surrounding landscape as well as the group's historical legacy. Mandarin Chinese is the most common language, but different ethnic groups will switch from Mandarin to an indigenous language when referring to place names. Groups do this because places have different cultural meanings to different groups and switching languages emphasizes the identity of the speaker. In this example, indigenous place naming coincides with indigenous knowledge and experience, much of which centers around a relationship with a certain place.

The Basque are an ethnic group with a population of approximately 1 million people who live in the western Pyrenees and along the Bay of Biscay in Spain and France in an area comprising about 3,900 square miles. The Basque language (also called Euskara) is spoken by the Basque people and is a unique contributor to placemaking and the Basque landscape. This language has eight different dialects and is known as an isolate, as it has no common link to any modern languages. The rugged Pyrenees have long provided a natural fortress of safety and seclusion for the Basque homeland. This isolated existence helped preserve the language in the face of heavy language competition. As one of the oldest living languages, derived from the pre-Indo-European group (which you will learn in Chapter 8), the Basque language has become symbolic of the Basque people's cultural identity as they are one of the oldest ethnic groups in Europe. In an effort to help protect this language, knowing how to speak and write it is a requirement to work for the region's Spanish Basque government, and the language is being taught in some schools across the region today.

Religious Patterns As we discuss later in chapter 8, religions tend to organize space in distinctive ways, based on how the religion is organized, its beliefs, and how it is practiced. These three factors also impact the distribution of **adherents**, or the people who follow a belief, religion, or organization. Some religions appeal to a wide variety of people across the globe, no matter their race, ethnicity, or class, and other religions appeal to a specific ethnic group living in particular regions of the world.

Religions are often organized into smaller groups including branches, denominations, and sects. A branch is a large fundamental division within a religion. Protestant, Catholic, and Eastern Orthodox are the major branches of Christianity. Sunni and Shiite are two branches of Islam. A branch is sometimes divided into **denominations**, which are separate organizations that unite a number of local congregations. Lutheran, Methodist, and Baptist are a few denominations of the Protestant branch of Christianity. A **sect** is a relatively small group that has separated from an established denomination.

Sometimes the terms *branch*, *denomination*, or *sect* are used interchangeably, but it is important to understand that geographers use these terms to define these types of organizations to study all religions' influence on the landscape. It is also relevant to note that religions transition between these types of organizations—for instance, from sect to major religious branch—as their popularity or acceptance grows.

In the United States, the distribution of adherents of various religions and denominations follows a regional pattern, which is also related to ethnicity. According to the Public Religion Research Institute, a U.S. nonprofit organization, the number of White evangelical Protestants is twice as large in the South (22 percent) and Midwest (20 percent) as it is in the Northeast (8 percent). White evangelical Protestants account for 12 percent of residents in the West. In contrast, Catholics represent a much larger share of the northeastern region: 29 percent of residents of the Northeast identify as Catholic, compared to 21 percent of Westerners, 19 percent of Midwesterners, and 17 percent of Southerners. In Chapter 8, you learn in detail about the regional distribution of adherents in the United States.

Places of worship are closely related to how a religion is practiced, and they also represent a physical connection between religion and the cultural landscape. Places of worship, including churches, synagogues, temples, and other sacred places, are sites where worshipers sometimes gather or have spiritual significance or power. Mosques are community spaces where Muslim worshipers assemble for prayer or to celebrate holy festivals. Each mosque has a minaret, or tower, where a man summons people to worship. In contrast, pagodas in China and Japan are designed to hold relics of the Buddhist religion, not as sites of community worship. Pagodas are tall, ornate structures with balconies and slanting roofs.

Critical Viewing The location of the Basque homeland in a remote and mountainous area of Spain has insulated its language from external influences. As a result, certain dialects of the Basque language have remained largely unchanged since the Roman era in the 2nd century B.C.E. ▌Explain how the Basque language contributes to a sense of place.

Concepts of Culture

National Geographic Explorer **Sandhya Narayanan**

Studying Indigenous Languages

Sandhya Narayanan studies indigenous multilingualism in the Puno region of Peru. The community gathers at the foot of Allin Qhapaq annually to watch performances of traditional dances. Her newest work focuses on Indigenous language reclamation and revitalization with Native American communities from the North Atlantic coast.

Learning Objective
LO 6.17 Explain how patterns of language, religion, and ethnicity, contribute to a sense of place and shape the global cultural landscape.

Growing up in a multilingual home in multilingual communities led Sandhya Narayanan to research indigenous languages, focusing on how multilingualism affects a culture and the future of language.

The high-altitude plain in the Andes Mountains in South America has been home to speakers of Quechua, Aymara, and other indigenous languages for centuries. Spanish is also a vital part of the linguistic mix in this region. The Puno region of Peru, on the border of the Quechua and Aymara regions, is the perfect place to study the interactions among cultural landscapes. The Puno region drew linguistic anthropologist and National Geographic Explorer Sandhya Narayanan to its local marketplaces. There, women have traditionally sold produce and craft goods to consumers from the area's diverse linguistic communities, making it a living laboratory for the study of the historical effects of multiple languages on the region.

Through fieldwork that includes interviews, recordings, observations, and conversations with local consultants, Naranayan has noted a generation gap, as members of the younger generation tend to employ Spanish exclusively to communicate. Interestingly, this is not youthful rebellion against the traditional languages but a practice created by parents who discourage learning indigenous languages. Naranayan says, "There's an idea that children won't advance if they don't speak Spanish, and that speaking an indigenous language jeopardizes their ability to learn Spanish." Members of that younger generation see this practice as a loss that disconnects them from their traditional culture and heritage.

Are these language dynamics at play in other multilingual areas? Narayanan may explore this question in future research, as she intends to do similar fieldwork in India.

Geographic Thinking

Compare how and why different generations in the Puno region of Peru may have differing senses of place.

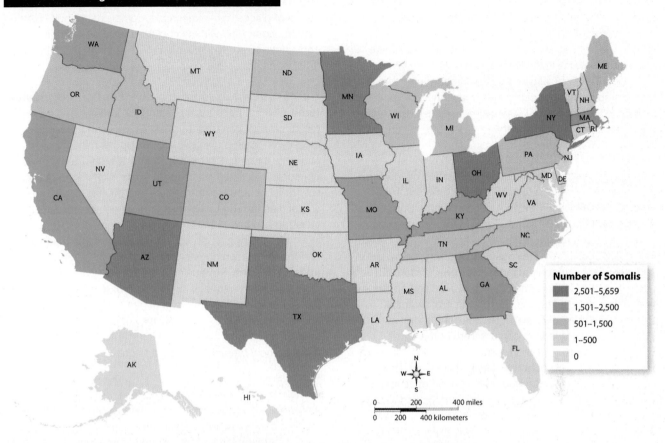

Reading Maps Minnesota remains the top state of primary resettlement for Somali refugees, with almost 12 percent of total Somali refugees settling there. New York State is second with 8 percent.
❚ How might this map be used to attract more Somalis to the United States?

Patterns related to the selection of sacred places contribute to a regional cultural landscape and can be expressed through interactions with the natural environment, human-made features or design, and even through culture and lifestyle. For example, some religions choose sacred sites based on places that were significant in the lives of various religious founders or leaders. Lumbini, Nepal, where Buddha was born; Bodh Gaya, India, where he received enlightenment; and Sarnath, where he first preached, are all considered sacred sites and are heavily visited by Buddhist pilgrims from around the world. The Maya Devi Temple in Lumbini is considered Buddha's birthplace and is a site of pilgrimage. Sacred Buddhist spaces, including pagodas and stone stupas (dome-shaped Buddhist shrines), and Buddhist texts written in Pali and Sanskrit languages and other ancient languages of India contribute to a sense of identity and form the regions where Buddhism is prevalent in East and Southeast Asia. Ethnicity is another factor that helps define a region.

Ethnic Patterns Patterns of ethnicity are deeply linked to religion and language. The Somali in Minnesota exemplify how ethnicity, religion, and language contribute to a sense of place within a specific region. As you read in Chapter 5, the largest community of Somalis in North America lives in Minneapolis, where Somalis continue to shape the cultural landscape both in their ethnic neighborhoods and throughout the state. In 2020, Somalis represented the second-largest group of foreign-born Minnesotans, with a population of about 32,000. The Cedar-Riverside neighborhood of Minneapolis, which houses the largest concentration of Somalis in the city, is also home to Dar Al-Hijrah, the oldest mosque in Minnesota. The mosque has become part of the cultural landscape, signifying an important component of Somali identity and providing a space for religious worship, an Islamic school, and community gatherings. More than 30 mosques now exist in the metropolitan area of Minneapolis and St. Paul.

Overcoming the language barrier has been challenging for Somalis. Many of them arrived in Minnesota proficient in multiple languages, like Somali and Arabic, but gaining English proficiency made their transitions difficult. Somali families have started charter schools to support Somali language preservation and make it easier for students to adhere to religious practices. Demographics reveal that half of the Somali population in Minnesota is age 22 years or younger. As this young population ages, its economic and cultural influence in the city and state will likely continue to grow. The first Somali American (and first of two Muslim women) secured a seat in Congress in 2018, and immigrant entrepreneurs, in general, currently contribute $489 million annually to Minnesota's economy.

Concepts of Culture

Regionally, why Minnesota instead of other states? Though Minnesota may not seem like an obvious destination for Somali immigrants, highly active volunteer agencies there have contracted with the U.S. State Department to help refugees learn English, find housing, access health care, and start a new life in the United States. For similar reasons, Minnesota is a population center for Hmong refugees. When ethnic groups find that public and private institutions are welcoming and supportive, the news spreads back to the home country or region and influences where second waves of relatives and friends choose to settle.

Historically, three major ethnic groups formed clusters in particular regions in the United States: Hispanics in the West and Southwest, African Americans in the South, and Asian Americans in the West. Over time, these patterns have changed through migration and new immigrant flows. The U.S. population overall has been moving southward and westward in recent decades. However, certain ethnic patterns in the United States have prevailed and continue to enhance the placemaking of regions. For example, the main urban ethnic neighborhoods established in U.S. cities during the 19th century were home to the Catholic Irish, Italians, Polish, and East European Jews. In Chicago specifically, patterns of Irish, Italian, and Polish neighborhoods have changed over a period of almost 60 years. African American and Hispanic immigration to these neighborhoods has changed the landscape. Today, all of these groups have a strong presence in the form of places of worship, restaurants, stores, and visible manifestations of culture such as murals and other artistic expressions—many of which reflect their native languages and all of which contribute to a sense of place within a culturally diverse city.

Africa is made up of more countries than any other continent, and its cultural landscape is a mosaic of hundreds of ethnic groups. Some groups formed before the colonial period, and others emerged because European colonial governments categorized groups by differences that often held little meaning to the people themselves. Colonialism forced Africa into a period of intense environmental, political, social, and religious change as European powers created new political realms and divided existing ones. Established cultural groups that had long been in existence became split, and other groups—different from one another—were forced to live together. The cultural tensions created by this politically motivated restructuring created deep strains and tumultuous conflict that still exists today and has become part of the global cultural landscape.

Patterns of Unity and Division

Learning Objective
LO 6.18 Explain how religion, language, and ethnicity can act as both centripetal and centrifugal forces under different circumstances.

A century ago, Sauer taught us the need to interpret the landscape to understand the culture that shaped it. Over time, geographers have broadened their understanding of cultural landscapes to recognize the contested and dynamic nature of places and landscapes, complex and multifaceted constructions filled with symbolic power. Some of this contestation and dynamism can be seen in the unifying and divisive forces at work on cultures across the globe.

A force that unites a group of people is called a **centripetal force**. Cultural traits act as centripetal forces when they create solidarity among a group of people and provide stability. A common language or a popular national sport can unite citizens. Conversely, a force that divides groups of people is called a **centrifugal force.** Cultural traits that sow division between the people of a country or region—sometimes leading to violence, civil unrest, or war—act as centrifugal forces. A state with two or more ethnic groups aiming for their own political status and wishing to separate and form their own country can be a dominant centrifugal force. The former Soviet Union, which broke up into 15 independent countries, is an example of a multinational state that experienced the effects of centrifugal forces. Centripetal and centrifugal forces and how they affect patterns of political geography are covered in Chapter 11. Here, you focus on how these forces relate to culture.

Centripetal Forces Shared religion, language, or ethnicity can create a shared sense of identity—which can act as a centripetal force. Even very diverse cultures may have centripetal forces that help hold them together. A dominant religion, such as Roman Catholicism in Mexico or Brazil, can be an extremely strong centripetal force. Similarly, a shared language helps unify people by facilitating communication. As a flood of immigrants came to the United States in the 19th century, many learned English, which helped unify these people from diverse cultures. A shared ethnicity can be a strong centripetal force as well. Although China has 55 ethnic minorities, more than 91 percent of the country's population is Han Chinese. This shared ethnicity serves as a unifying force.

Centrifugal Forces Religion, language, and ethnicity—which were just identified as centripetal forces—can also act as centrifugal forces under different circumstances. For example, the official "language of the Union" of India is Hindi. Hindi and English are both used in parliament, and India's constitution includes 22 official regional languages as well. Because language is so critical to communication, having several languages within a country can be a dividing force. Ethnicities are often extremely divisive as well. Conflict between ethnicities in many parts of the world has escalated to genocide or ethnic cleansing.

For example, the country of Azerbaijan in western Asia has seen ongoing disputes between ethnic Azerbaijanis and Armenians. Disagreements over territorial issues, specifically the Nagorno-Karabakh region along the border between Armenia and Azerbaijan, have fueled the continuing

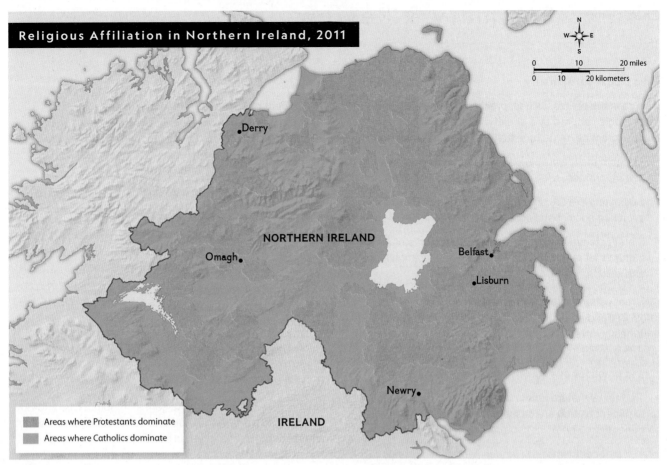

Reading Maps A research project mapped where Protestants and Catholics resided in Northern Ireland in 2011. ▪ Explain how the map supports the idea that religion has been a centrifugal force in Northern Ireland.

clashes since the late 1980s, resulting in bloodshed and political and ethnic tension. The Nagorno-Karabakh region was established within Azerbaijan's borders by the Soviet government in the 1920s. Approximately 95 percent of the population living within this region is Armenian. When the Soviet Union collapsed in 1991, the region declared independence, and the Armenians and Azerbaijanis have been fighting for control in the more than three decades since. Repeated mediation efforts have failed, ceasefire attempts have been violated, and militarization has increased throughout this serious ethnic conflict.

Religion has been a centrifugal force in Northern Ireland for many decades, as Catholics and Protestants have fought for control. Much of the conflict has been grounded in politics and cultural identity. Protestants, who have largely identified as British because of their British ancestors, have desired to remain part of the United Kingdom, while the majority of Catholics, who have seen themselves as Irish, have wanted a united Ireland free from British rule.

Violence between the two groups ceased, thanks to the 1998 Good Friday Agreement between the British and Irish governments and most of the political parties in Northern Ireland, about how Northern Ireland should be governed. A significant demographic shift—a rapid increase of the Catholic minority—is occurring within Northern Ireland.

The 2011 Census showed that the Protestant population had decreased to 48 percent while the Catholic share had risen to 45 percent, with Protestants also comprising an older demographic than Catholics. Geographers cataloged the impact of these changes by using census data to map areas where Protestants and Catholics live. The data reveals the deep divisions that remain in the country. Most Catholic and Protestant children are taught in separate schools, and their families live in segregated neighborhoods. Rivers and other natural features often serve as borders between Catholic and Protestant areas, but there are also human-made borders called peace walls. It is speculated that the 2021 Census results—to be available in 2022 or 2023—might show a Catholic majority.

Geographic Thinking

1. Explain the linguistic patterns of the United States.
2. Explain how placemaking impacted the Somalis in Minnesota and shaped the cultural landscape.
3. Describe an example of how language, religion, or ethnicity work together to form regions.
4. Compare how one factor can be both a centripetal force and a centrifugal force.

Chapter 6 Summary & Review

Chapter Summary

Culture is the beliefs, values, practices, behaviors, and technologies shared by a society. Elements of a shared cultural practice, or cultural trait, fall into three categories.

- Artifacts are the visible objects and technologies a culture creates.
- Sociofacts are the structures and organizations that influence social behaviors.
- Mentifacts are the central, enduring elements of a culture that reflect its shared ideas, values, knowledge, and beliefs.

Different cultural aspects change and spread at different rates, depending on the degrees of technology and cultural attitudes involved.

- Popular culture reflects widespread behaviors, beliefs, and practices of a society at a given point in time.
- Traditional culture reflects long-established, shared experiences that are passed from generation to generation. There are two prominent cultural attitudes:
  - Ethnocentrism is the tendency of ethnic groups to evaluate other groups according to preconceived ideas originating from their own culture.
  - Cultural relativism is the evaluation of a culture by its own standards.

A cultural landscape is a natural landscape modified by humans, reflecting their cultural beliefs and values.

- Sequent occupance is the notion that successive societies leave behind a collection of evidence about human character and experiences within a geographic region, which shapes the cultural landscape.
- Large cities typically contain many ethnic neighborhoods, which are cultural landscapes within communities of people outside of their areas of origin.
- Traditional architecture and postmodern architecture are two building styles that reflect different cultures, religions, and places in cultural landscapes.
- Religion, a system of spiritual beliefs, greatly impacts cultural landscapes because it drives much of what is hidden beneath the observable surface of culture.
- Language, the carrier of human thoughts and cultural identities, is a distinct system of communication and usually has a literary tradition, or some written form.

Identity is how humans make sense of themselves and how they wish to be viewed by others. Landscape features, land and resource use, and attitudes toward gender shape the use of space in a cultural landscape.

The subjective feelings people associate with a geographic location are known as sense of place. Placemaking is how people collaborate to create a place where they can live, work, play, and learn. Patterns of language, religion, ethnicity, unity, division, centripetal forces, and centrifugal forces contribute to a sense of place, enhance placemaking, and shape the global cultural landscape.

Review Questions

Use complete sentences to answer the questions.

1. **Apply Conceptual Vocabulary** Consider the terms *centripetal force* and *centrifugal force*. Write a standard dictionary definition for each term. Then provide conceptual definitions—an explanation of how each term is used in the context of this chapter.

2. Explain why geographers often use the metaphor of an iceberg to illustrate the concept of *culture*.

3. Provide examples of how artifacts, sociofacts, and mentifacts can be both similar and different around the world.

4. Explain why some places have stricter cultural norms than other places.

5. Explain the existence of ethnocentrism with the Wisconsin American Indian Nations Case Study.

6. Define the term *sequent occupance* and provide an example of it.

7. Describe the characteristics of an ethnic neighborhood using an example.

8. Explain why places of worship are often featured prominently in cultural landscapes. Provide an example.

9. How can a person's identity shape a place? Give two examples in your response.

10. Explain the origin of the term *gendered space* and tell how cities are creating safer gendered spaces today.

11. Compare the building styles of traditional architecture and postmodern architecture and how they shape cultural landscapes.

Interpret Graphs

Study the graph and then answer the following questions.

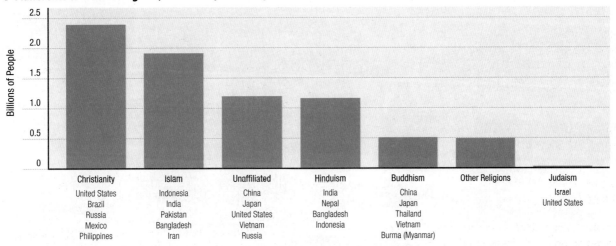

Adherents of World Religions, in Billions (2020 EST.)

Source: The Pew-Templeton Global Religious Futures Project

12. **Identify Data & Information** What information about religion does the bar graph show?

13. **Describe Spatial Patterns** What is spatially significant about the countries with the most adherents?

14. **Analyze Visuals** What generalizations can you make about Hinduism based on the bar graph?

15. **Make Inferences** What conclusions can you draw about the influence of Christianity on the global cultural landscape?

Chapter 7
Cultural Change

Critical Viewing A woman sells goods at the Siti Khadijah market in Malaysia. ▌ How might a region's cuisine reveal details about its culture?

Geographic Thinking What are the causes and consequences of cultural change?

7.1
Cultural Diffusion

Feature: A Portrait of Relocation Diffusion

Case Study: African Culture in Brazil

National Geographic Explorer
Justin Dunnavant

7.2
Processes of Cultural Change

Case Study: Fútbol—A Globalizing Force

National Geographic Photographer
William Allard

7.3
Consequences of Cultural Change

7.1 Cultural Diffusion

Cultures are not set in stone. Cultural ideas, practices, and innovations change or disappear over time. Cultural change results from both internal pressures within a group of people and through external contact with others. By studying the processes that cause cultural change, geographers can gain a better understanding of culture and its impact on human societies.

Relocation Diffusion

Learning Objectives
LO 7.1 Define relocation diffusion.
LO 7.2 Describe the cultural impacts of relocation diffusion.

The process by which a cultural trait spreads from one place to another over time is called **diffusion**. A trait often originates in a **cultural hearth** and initially diffuses, or spreads, from that source (which cultural geographers often term its "hearth," analogous to the source of a fireplace's heat). As you've read, flow is an important concept in human geography. And just as geographers study the flow of humans in migration, they also study how cultural ideas, innovations, and trends flow in spatial patterns. Historically, diffusion occurred through imperialism and colonialism, which drove exploration, military conquest, missionary work, trade, and migration. Today, diffusion continues through some of the same historical ways, but it also occurs through mass media and the internet. Diffusion has introduced customs, traditions, language, technology, means of communication, consumer products, and other aspects of culture to people across the planet. There are two broad categories of diffusion: relocation diffusion and expansion diffusion.

Relocation diffusion, as you read in Chapter 5, is the spread of cultural traits or ideas through the movement, or relocation, of people—either individuals or groups. Throughout history, people migrating to new places have brought artifacts and mentifacts of their culture with them. Religions, for instance, often spread through relocation diffusion as people bring their deeply held beliefs with them and continue to practice the rites and rituals of their faith in their new homes.

Beginning in the late 15th century, as Europeans moved to the Americas, they brought aspects of their culture with them. The languages spoken by most people in North and South America today—Spanish, English, Portuguese, and French—are the result first of relocation diffusion, then expansion diffusion. Europeans also brought Christianity and Judaism, changing the religious beliefs and practices of many with whom they came into contact, often through force and conquest.

Relocation diffusion is also illustrated in the example of the African diaspora, the spread of people of African descent from their ancestral continent to other parts of the world as a result of forced migration. The huge number of people—10 to 12 million—taken from their homes in Africa during the transatlantic slave trade inevitably left its mark on the cultures of the Americas. The men, women, and children brought over on the slave ships carried their cultures with them, and despite the brutal conditions under which they lived, many of their cultural traits and practices survived. They made instruments for playing traditional African music, which is community based, with everyone participating by playing, singing, or dancing to the rhythmic beats. Aspects of African agricultural practices spread to the Americas through farming techniques, food preparation methods, and specific ingredients. Enslaved Africans popularized new foods such as okra, melons, and bananas and contributed new techniques for farming rice. They made clay pottery, baskets, and clothing with African designs. Over time, some of these cultural traits were adopted by the greater population. Traditional African music influenced the blues and jazz, which went on to influence rock music. The farming techniques the Africans brought with them greatly improved rice production in the American South.

Another migration—this one occurring in the aftermath of American slavery—provides an example of relocation diffusion on a regional scale, within a country. As you know, the Great Migration was the internal migration of millions of African Americans from the rural South due to the push factors of racial discrimination, segregation laws, and widespread financial difficulties. Large numbers of African Americans moved to industrialized cities in the Northeast, Midwest, and West between 1916 and 1970. The culture they brought with them—the literature, music, theater, and visual art they created—spread to these new regions as people living in these places adopted some of the cultural traits introduced to them by the migrants.

Relocation diffusion can occur at an even smaller scale as well. When a Chief Executive Officer (CEO) leaves one company and goes to work for another company, for instance, the person will often take ideas, tools, or systems that made the original company successful and bring them to the new company. In fact, observers of the business world describe the aspects of a company that affect employees' daily work environment as a "company culture." A positive culture can lead to a more successful company than a negative culture, and a new CEO might be hired specifically to make positive changes to the company culture. These changes occur through a combination of relocation and expansion diffusion.

Critical Viewing *Left:* Paintings of Charlie Lewis (*top*) and Ossa Keeby, two of the formerly enslaved people who helped found Africatown. *Right:* Lewis's great-great-granddaughter Lorna Gail Woods (*top*) and Keeby's descendant Karliss Hinton. ▌Explain the role that relocation diffusion played in the lives of the Africatown residents pictured in the paintings and photos.

A Portrait of Relocation Diffusion

The effects of relocation diffusion as a result of the African diaspora can be seen today in Africatown, Alabama. The town was founded by formerly enslaved people who had been brought to the United States on the *Clotilda*—the last known slave ship to arrive in the country. The *Clotilda* landed illegally near Mobile, Alabama, in 1860, 52 years after the importation of slaves had been outlawed. Transported across the Atlantic Ocean in a clear example of forced migration (the Middle Passage), the enslaved Africans on the ship formed tight bonds with one another.

After slavery was abolished in 1865, the formerly enslaved people worked and saved and pooled their money together to buy land to establish their own community. The inhabitants planted gardens and fruit trees, built a church that faced east toward Africa, and passed their cultural traditions down to their children and their children's children. Today, Africatown is still inhabited by the descendants of these formerly enslaved people, and the sunken remains of the *Clotilda*, which evaded researchers for decades, were discovered in 2019 along the Mobile River.

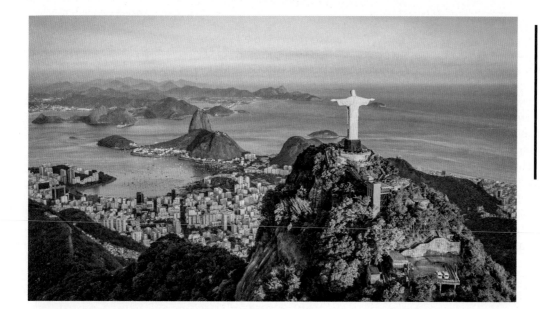

The Christ the Redeemer statue in Rio de Janeiro signifies the relocation diffusion that took place when Portuguese colonists brought their language and religion with them to Brazil in the 1500s. Today, Brazil has the largest Roman Catholic population in the world, with an estimated 126 million Catholic Brazilians.

Expansion Diffusion

Learning Objectives
LO 7.3 Define three types of expansion diffusion.
LO 7.4 Describe the cultural impacts of expansion diffusion.

Expansion diffusion occurs when an aspect of culture spreads outward from where it originated. As it spreads, the trait also remains in its place of origin—that is, it expands outward from its source (or cultural hearth). The main difference between relocation diffusion and expansion diffusion is that with expansion diffusion, cultural traits move even though the people who are a part of that culture do not relocate. There are three types of expansion diffusion: contagious diffusion, hierarchical diffusion, and stimulus diffusion. The three types differ in the spatial patterns by which an idea or trait spreads and in the mechanisms that trigger the flow.

Contagious Diffusion **Contagious diffusion** occurs when an idea or cultural trait spreads adjacently, or to people or places that are next to or adjoining one another. Contagious diffusion occurs among people of all social classes and levels of power. An outbreak of a disease, when it spreads through direct person-to-person contact and expands outward to people living nearby, is an example of contagious diffusion. Another example is when a new slang word spreads through a school or workplace as more and more students or workers hear their friends and acquaintances say it and begin using it themselves.

Originally, contagious diffusion referred only to the spread of a trait outward from person to person through direct contact. However, technology has changed the ways in which people connect with one another. The internet or social media can cause a song, fad, or idea to spread independent of proximity to its point of origin. The contact between people online is virtual, but a meme or Tik Tok can be created by anyone who has access to the internet and spread to everyone that person has contact with online. They, in turn, can spread it to all of their contacts, and on and on.

Because traits that spread through contagious diffusion are available to people without regard to social status, wealth, or power, they can spread quickly and widely. Contagious diffusion is sometimes compared to a wave spreading through a population.

Hierarchical Diffusion A hierarchy is a system through which people, groups, or places are ranked one above the other according to their size or status. **Hierarchical diffusion** is the spread of an idea or trait from a person or place of power or authority — at the top of the hierarchy — to other people or places. A cultural trait that becomes popular in a large urban center will spread down the urban hierarchy to midsize cities and then on to small cities, towns, and villages. Traits can diffuse hierarchically through people as well. A new fashion trend, for instance, might start when a popular celebrity or public figure introduces a clothing style, hairstyle, or other fad, prompting others to adopt it. Such trends are often then reinterpreted and recreated to be sold more affordably in department stores to the general public.

Professional innovations can spread hierarchically as well. For instance, a farmer might innovate or adopt a new planting or plowing method or machine that is more effective—and then lets neighboring farmers know about the new practice. Perhaps all would hope to adopt the practice or equipment, but only the neighbors with the ability to implement the innovation—including purchase of new equipment and also, importantly, the time and effort to learn the new agricultural practice—would be able to do so. In this way, it's likely the innovation would spread across the region first among wealthier farmers with the resources and time to adopt. If it becomes popular enough that the equipment becomes less expensive and perhaps easier to use, it might then diffuse to farmers with fewer resources.

Hierarchical diffusion can also occur from the bottom of the hierarchy to the top. This is sometimes referred to as reverse hierarchical diffusion. Blue jeans, for example, were initially designed for miners during the 1849 California

Cultural Change

Gold Rush. Soon, cowboys and other laborers were wearing jeans because they could stand up to hard work. The durable garment became a symbol of the American West and the working class, and as such it traveled up the hierarchy to be embraced in the 1950s by actors like Marlon Brando and James Dean. From there, blue jeans traveled back down the hierarchy, becoming popular with teens and college students, and eventually becoming a staple of the American middle-class wardrobe. Today, blue jeans are worn by people around the world. Original blue jeans are still appreciated for their rugged wear, but designer brands have also adopted this aspect of the culture.

Stimulus Diffusion The third type of expansion diffusion—**stimulus diffusion**—occurs when the fundamental idea behind a cultural trait stimulates a new innovation. In stimulus diffusion, the trait itself does not spread, but it triggers an idea that leads to a new cultural trait. In the early 19th century, blacksmith John Deere found that wood and cast-iron plows, common in the eastern United States at that time, didn't work well in the heavier soils of the Illinois prairies. By 1838, in an example of stimulus diffusion and experimentation, he successfully modified the older design into a steel-bladed plow that did the trick—and made his name synonymous with American agriculture. More recently, in 2014, Elon Musk, the CEO of Tesla, a company that makes electric cars, announced that anyone could freely use the company's patents to advance electric car technology. Musk made the decision in hopes that his technology will trigger others to come up with innovations to make electric cars better and more popular, fighting climate change as a result.

A Mix of Diffusion Types It is useful to learn about the different types of diffusion individually, but in the real world, culture usually spreads through a combination of diffusion types. A cultural trend might start out diffusing hierarchically and then continue to spread through contagious diffusion. For instance, when the major networks first started broadcasting regular television programming in 1947, fewer than two percent of U.S. households owned a TV set. It was an expensive luxury that spread through hierarchical diffusion, beginning with larger cities that had some of the first TV stations and eventually spreading to smaller and smaller cities and towns. When a television station began broadcasting in a community, people would gather to watch in the homes of those who had TVs or in public places like bars. As the medium became more popular and TVs became more affordable, television spread through contagious diffusion, from household to household. By 1956, 70 percent of American households had a TV.

Cultural Diffusion Online

How does culture diffuse online? Is the spread of an idea over the internet an example of contagious or hierarchical diffusion? The answer is: it depends.

On one level, all online activity is hierarchical because not everyone has access to the internet or speaks the language in which a meme or trend is shared. Those who do have an internet connection, however, can access a seemingly infinite amount of information, so the internet is often thought of as a place where everyone is equal. When an idea, trend, or meme is passed between contacts online without regard to status or wealth, it is an example of contagious diffusion. Indeed, the phrase "going viral" refers to an idea spreading over the internet through virtual contact in the same way that a contagious disease spreads through physical contact.

Even online, however, hierarchies exist. A popular social media star has much more influence than the average user. A fashion trend posted to an account with millions of followers can spread around the world quickly, and the opinions of popular influencers might be accepted more readily than opinions shared by acquaintances on social media. This type of expansion—the spread of an idea through an influencer—is an example of hierarchical diffusion.

The Diffusion of Misinformation In recent years, with the enormous expansion and popularity of the internet and online social media, there has been a growing availability of "user-generated" content. This has resulted in what might be seen as a differentiated pattern of contagious diffusion, with aggregations of like-minded people holding common interests, views, and opinions communicating with each other—but not with others who share differing worldviews. In this way, though the online access among users might be the same, the acceptance of this information differs. Such clusters of users tend to send and receive content that only they think is important, while refusing to accept any ideas from others, in what has metaphorically been called an "echo chamber." This can also result in the rapid contagious diffusion of misinformation, unsubstantiated rumors, and "conspiracy theories"—explanations that events or situations occur as a result of conspiracies between sinister, powerful groups—that may bring huge social and often illogical responses. False information can gain broader acceptance, and can then result in false beliefs that, once adopted by an individual or group, can be quite resistant to correction. This kind of misinformation has become so pervasive in online social media that the World Economic Forum has identified it as a major threat to human society.

Geographic Thinking

1. Explain how relocation diffusion through forced migration might differ from relocation diffusion through voluntary migration.

2. Introduced by Italian immigrants, pizza spread from the Italian neighborhoods of New York City outward to every corner of America. Explain how this spread represents both relocation and expansion diffusion.

3. Compare the three types of expansion diffusion and explain how each is affected by wealth and power.

RELOCATION DIFFUSION
Cultural traits spread through the movement of individuals or groups.

- Relocating Population
- Destination Population

EXPANSION DIFFUSION
Cultural traits spread independently of the movement of people.

1. CONTAGIOUS DIFFUSION
Traits spread from person-to-person contact regardless of social class or level of power.

2. HIERARCHICAL DIFFUSION
Traits jump from powerful places or people to other powerful places or people, then spread down the hierarchy (or up from the bottom of a hierarchy to the top).

- Hearth
- Early Diffusion
- Later Diffusion
- Non-powerful Place or Person
- Powerful Place or Person

3. STIMULUS DIFFUSION
Traits spread to another culture or region but are modified to adapt to the new culture.

- Original Trait
- Early Adaptations
- Later Adaptations

Types of Cultural Diffusion Cultural traits and ideas can spread across space and time in different ways. In general, culture originates from a cultural hearth, such as an urban area, and spreads to smaller cities and towns. ▮ Explain which type of cultural diffusion results in the greatest change to a cultural idea or trait, and why.

Cultural Change 193

Case Study

African Culture in Brazil

An African drumming band performs during Carnival in Pelourinho, the historical center of Salvador, Brazil. West African culture remains a dynamic part of Carnival celebrations in the city, as well as in everyday life in Brazil.

The Issue Descendants of the Africans brought to Brazil as enslaved labor from the 16th to 19th centuries have left an indelible mark on the country's vibrant culture, including its Carnival celebrations.

Learning Objective
LO 7.2 Describe the cultural impacts of relocation diffusion.

By the Numbers

1888
the year Brazil became the last country in the Americas to ban slavery

40%
of all Africans enslaved during the Middle Passage were in Brazil

MORE THAN 50%
of Brazilians identify as having African heritage

Source: World Population Review

The City of Salvador knows how to throw a party. It's the capital of Bahia, a state in northeastern Brazil. The city has about 3 million residents, but each year over a million more are drawn to its Carnival festival, which goes on for the six days leading up to the Christian observance of Lent. The celebrations include parades, shows, street parties, and the pounding beat of drumming bands that can be heard for miles around. Brazil's diverse culture is on full display during the festivities, and while its history as a Portuguese colony is reflected in the country's language and main religion, many other cultural traits have roots that trace back to West and Southwest Africa—an example of cultural diffusion through forced migration.

Portuguese settlers in Brazil began the brutal process of importing Africans as enslaved labor in the 16th century and continued the practice for approximately 300 years. In that time, nearly 5 million Africans were stolen from their homes, chained and brought to Brazil in slave ships, and forced to work on sugar plantations and in the mining industry. Brazil enslaved more people than any other country— an estimated 40 percent of all Africans enslaved in that time period—and did not abolish slavery until 1888, the last country in the Americas to do so. Today, more than half of Brazilians identify themselves as having African heritage.

This heritage is evident in the country's culture, for even when migration is forced and people leave their homes with none of their belongings, they carry their culture in their hearts and minds. And when they arrive at their destination, they continue, as they are able, to play the music, make the food, recreate the fashion, and tell the stories of their homeland. The cultural ideas and traits they bring with them change as they interact with the people in the destination country, combining into a new form of cultural expression with roots in the old. Samba, an African-influenced style of music and dance that is one of the defining features of Carnival, developed in the country's predominantly Black neighborhoods. Candomblé, a religion that mixes Christian and African rituals and traditions, originated in Bahia and today has about 2 million followers in South America. The elaborate costumes and decorations of Carnival, with their vibrant colors, are influenced by African cultures as well.

Geographic Thinking

Explain how aspects of Salvador's Carnival celebration are an example of cultural diffusion through forced migration.

Dr. Justin Dunnavant is a professor in the Department of Anthropology at the University of California, Los Angeles. A certified SCUBA diver, he also conducts maritime archaeology with the nonprofit group Diving With a Purpose.

Learning Objective
LO 7.2 Describe the cultural impacts of relocation diffusion.

National Geographic Explorer **Justin Dunnavant**

Retracing Afro-Caribbean Maritime Routes

Millions of Africans brought their culture to the Americas as a result of forced migration. Archaeologist Justin Dunnavant is interested in the next chapter in this story, in which some of these people used maritime routes to escape slavery and carry their cultural traditions to new places.

The town of Loíza, located just outside Puerto Rico's capital of San Juan, is known for its thriving Afro–Puerto Rican culture. When Dunnavant visited Loíza in 2021, its residents proudly introduced him to art, music, dance, food, and poetry reflecting their African heritage as well as the influences of Spanish and Taíno culture. African drums provide the rhythms to which Loízans dance Bomba. Crabs from the coastal mangrove forest are a key ingredient in empanadas de yuca y jueyes, a local delicacy.

Some ancestors of Loíza's Black residents were brought to Puerto Rico from West Africa on slave ships. However, others landed on the island actively seeking refuge from slavery. Black people who had escaped slavery, called maroons, established communities throughout the Caribbean. On the island of St. Croix, a community called Maroon Ridge developed in the remote mountains where the people found relative safety. Eventually, however, Danish plantations began to encroach on the area, and raids to recapture the maroons became more frequent. People looked to the sea as their next escape route.

Using GIS software, Dunnavant has identified ocean currents that would have carried boats from St. Croix to Puerto Rico, a distance of about 100 nautical miles. "The same ocean and wind currents that brought enslaved Africans to the Americas," he notes, "are the ones self-liberated Africans would use to reclaim their freedom." Maroons, many of whom were skilled sailors, navigators, and fishermen and women, either stole boats from the shore or made their own canoes. The government of Puerto Rico encouraged the migration as a means to increase their labor force. Dunnavant explains that "maroons who converted to Catholicism, pledged allegiance to Spain, and worked for a year . . . were freed and given land to cultivate." Those who settled in Loíza created a rich cultural legacy that endures to this day.

Geographic Thinking

What types of cultural diffusion have contributed to the unique culture of Loíza?

7.2 Processes of Cultural Change

Cultural patterns are shaped by processes occurring at global, regional, and local scales. Historically, cultures came into contact with—and altered—one another through colonialism, imperialism, and trade. While these processes are still at work, other processes, such as urbanization and globalization, are currently causing changes to cultural ideas and practices through the media, technology, politics, economics, and social interaction.

Historical Causes of Diffusion

Learning Objective
LO 7.5 Compare the historical causes of diffusion, including trade, migration, military conquest, and imperialism.

The cultural patterns seen in the world today have their roots in the past. Countries with strong military and economic power have always imposed elements of their cultures on countries and regions with less power. Trade between different cultures has also facilitated the spread of ideas. Migrations of all kinds have impacted cultures worldwide. The effects of these interactions are observable in cultures around the world.

Colonialism and Imperialism Many countries have sought to gain power, wealth, or other advantages through policies of colonialism and imperialism. As you know, colonialism is when a powerful country establishes settlements in a less powerful territory for economic or political gain. Imperialism is a related concept, occurring when a country enacts policies to extend its influence over other countries or territories through diplomacy or force. These two processes have had a strong influence on cultural patterns in the world. The dominance of European countries on the world stage in the 17th, 18th, and 19th centuries, and those countries' policies of colonization and imperialism, reshaped the world map and spread European cultures in ways that still resonate today.

For instance, in the 17th century the Netherlands established a colony on the southern tip of Africa as a supply port for Dutch East India Company ships trading with Asia. Settlers moved there to farm the land in order to provide food for the Dutch ships. Because there were no sources of local labor, they brought enslaved people with them. The Dutch introduced cruel policies against the enslaved and native peoples, as well as new land ownership laws and farming techniques that were advantageous to Dutch settlers. Over the next several decades, Dutch influence spread, giving rise to a new language, Afrikaans, which integrates the languages of the native groups of southern Africa into the Dutch language. The culture of the area was forever changed. Descendants of these original Dutch colonists (along with some of German and French Huguenot descent) established apartheid in South Africa in 1948, a racist policy of segregation that wasn't repealed until 1994.

In the 19th century, European countries increased their imperialist activity, racing to establish colonies across Africa that would give them access to the raw materials found there as well as control over shipping routes and ports. At the Berlin Conference in 1884, the European empires met, with no African representatives present, and split up Africa to avoid fighting each other for control. The European countries drew borders through African territories and then took control of the areas within those boundaries, forever changing the cultures of people throughout Africa. The effect on culture is particularly evident in the languages spoken in varied African countries.

The cultural impact of European imperialism can be seen in Asia as well. The British East India Company, for instance, had taken control of parts of India in the 1700s and ruled most of India by the 1820s. The British government took control of India in 1858 and held power there until it granted the territory independence in 1947, dividing the land into separate countries that are today India, Pakistan, and Bangladesh. The effects of British rule on India's culture are evident in many ways, such as the language (125 million Indians speak English), religion (Christianity is the third most practiced religion), and recreation (cricket, soccer, and field hockey, all introduced by the British, are extremely popular).

Spain colonized the islands known as the Philippines, securing control of Manila with its excellent harbor by 1565. Within several decades, Manila had become a leading commercial port, carrying on a prosperous trade with China, India, and the East Indies as well as the Spanish colonies in Central and South America. Spain controlled the Philippines until 1898 when the United States acquired it at the close of the Spanish-American War. The Philippines did not gain independence until after the Second World War. The effects of centuries of rule by both Spain and the United States on the cultures of the Philippines are evident even today: in religion (most Filipinos are Roman Catholics), in language (many people speak English), in sports (basketball is the most popular), and in other aspects of culture from architecture to music.

The Dutch also colonized what is today Indonesia. Traces of Dutch influences there include Dutch-origin loanwords (words adopted from another language with little or no modification) in Indonesian and cuisine (from casseroles

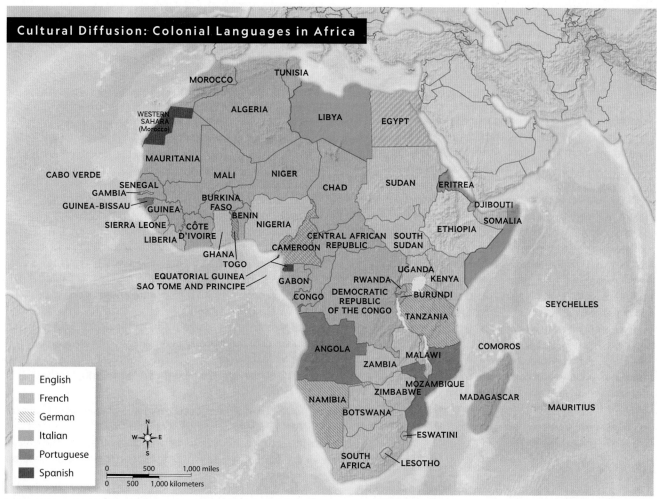

Reading Maps When Africa was colonized by Europe, European languages became the official language in those African colonies. Today, most Africans speak both colonial languages and indigenous African languages, or a combination of the two. ▌Explain how the map illustrates cultural diffusion and its impact on Africa during European colonization.

and soups to pastries, cakes, and cookies). And, of course, some Indonesian dishes have been adopted and, in turn, have influenced Dutch cuisine in the Netherlands.

Military Conquest Some of the most significant cultural changes have occurred when powerful military leaders took over new lands. Military conquest can cause cultural change to take place more rapidly than through processes like trade or migration because the conquerors impose their way of life on the conquered people. In the 300s B.C.E., Alexander the Great spread Greek culture throughout his empire, which at its height extended from Macedonia and Greece through modern-day Turkey, south into Egypt, and across Southwest Asia into northern India. He established cities in the conquered lands, including Alexandria, Egypt, which became a cosmopolitan center of learning. He also encouraged the men of his army to marry the local women. Greek culture became entwined with those local cultures, bringing about changes so significant that historians have a name for it: the Hellenistic Age.

Hundreds of years later, Spaniards conquered the Aztec of Mexico and the Inca of Peru. Spanish conquistadors,

or conquerors, had many goals, including the discovery of fabled riches, but their main stated goal was to spread Christianity. They also spread other aspects of their culture. Today, Spanish is the main language of most of Central and South America, and Catholicism is the most practiced religion.

As global forces—such as the Spanish conquest of much of Central and South America—come to bear on a society, interactions between different cultures can lead to new forms of cultural expression. In this case, Spanish became the region's **lingua franca**, or common language used among speakers of different languages, as the conquered peoples adopted the language of the conqueror. A lingua franca might also develop through more peaceful means, however. Through trade, speakers of different languages might develop a simplified combination of two languages or a third language altogether. And while a lingua franca might become the official language of a country, as Spanish has in many South American countries and French in former colonies of West Africa, many people still speak their indigenous languages as well. Today, English is a lingua franca around the world. The historical expansion of the

British Empire, and later the United States' dominance on the world stage, has made English the dominant language in business, political, and cultural transactions around the globe. It is also the predominant language of the internet, and throughout the world, all airline pilots must communicate in English with air traffic controllers.

Sometimes, rather than leading to the adoption of a common cultural trait, interactions between cultures can result in two or more cultural elements blending together. This process is called **creolization**. It occurs when people incorporate particular elements from an incoming culture and endow them with new meanings, forming a hybridized culture. A creole language, for instance, results from the blending of two or more languages that may not include the features of either original language. Haitian Creole is the blend of French and several African languages. (You learn more about the spread of language and religion in Chapter 8.) The Creole culture that developed in Louisiana in the 1700s blended West African, French, Spanish, and Native American cultural elements. Controlled first by France and then Spain (before acquisition by the United States in 1803), Louisiana became home to a hybrid culture with both traditional African practices as well as European social and political formations surviving but recognizably different. The variant of the French language spoken had many African and Indigenous words mixed in and, while Roman Catholicism was practiced by most, Voodoo was brought with enslaved people from West Africa and also was followed by some. Many people followed both faiths. The Black, White, Indigenous, and mixed-race Creoles of Louisiana forged a new identity, with their food, literature, music, and values also reflecting this hybridization of culture.

Trade Trade brings people together. They interact in order to buy and sell goods and, as you've seen, interaction between people from different places leads to an exchange of ideas, beliefs, values, technologies, and practices. Geographers examine trade patterns from the past to determine how historical processes impact current cultural patterns.

The Silk Road, an ancient network of trade routes that stretched from East Asia to the Mediterranean, brought about an unprecedented exchange of ideas that transformed cultures from China to Europe in ways that can still be observed today. For instance, along with luxury items such as Chinese silk, Roman glass, and Arabian spices, Buddhism also spread via the Silk Road. The process of religious diffusion was facilitated by the towns that sprang up along the Silk Road, as Buddhist temples were built to attract Buddhist merchants. And as Buddhism spread westward along these trade routes, Christianity spread eastward along the Silk Road through Asia to China.

Silk Road traders from many cultures along the way also helped spread new ideas related to medicine, mathematics, astronomy, and technology. Paper-making methods and printing technology spread from Asia to Europe and gave rise to Gutenberg's printing press, leading to the spread of knowledge and religion. New ship designs and tools such as the astrolabe enabled traders and explorers to travel farther. Seeking new trade routes, European explorers happened upon the Americas, lands previously unknown to them. As European countries scrambled for control of these lands, trade between Europe and the Americas continued to transform cultures on both sides of the Atlantic Ocean.

Today, trade and culture sometimes go hand in hand. For instance, American entertainment, such as movies, television, and music, is popular around the world. And while people in other countries enjoy the boost to the economy that trade with the entertainment industry brings, they sometimes resent the effect that foreign media has on their culture. In order to stop their culture from becoming too Americanized, many countries take measures to ensure that there is room in the marketplace for their own entertainment. Some restrict the number of American films that can be exhibited each year. Others use government funds to subsidize their own film and music industries.

Migration As discussed, migration has helped to shape the patterns and practices of culture through the spread of ideas and cultural traits between countries and regions. Global patterns of language developed in large part because of migration.

The Romance languages, which include French, Italian, Spanish, Portuguese, and Romanian, evolved from Latin, which spread across much of Europe with the expansion of the Roman Empire. As Europeans settled in the Americas, Africa, and Asia, the Romance languages spread throughout the world. Today more than 900 million people speak a Romance language as their first language.

Like other aspects of culture, language is constantly changing. People who migrate to new areas introduce words for unfamiliar things they see or experience. And as they come into contact with people who speak different languages, words or phrases from the foreign language may be folded into the dominant language. The result is that people in different places may speak different dialects of the same language. Isolation can also influence vocabulary and spelling. People in different places may speak the same language but with different pronunciation and speed. People in England, the United States, Canada, Australia, New Zealand, and South Africa all speak English, for instance, but they have very different accents. In addition, some words are spelled differently in American English than in British English and there are different vocabulary words for the same thing: What Brits call a lift, Americans call an elevator. Americans order fries with their burger while people in the United Kingdom order chips.

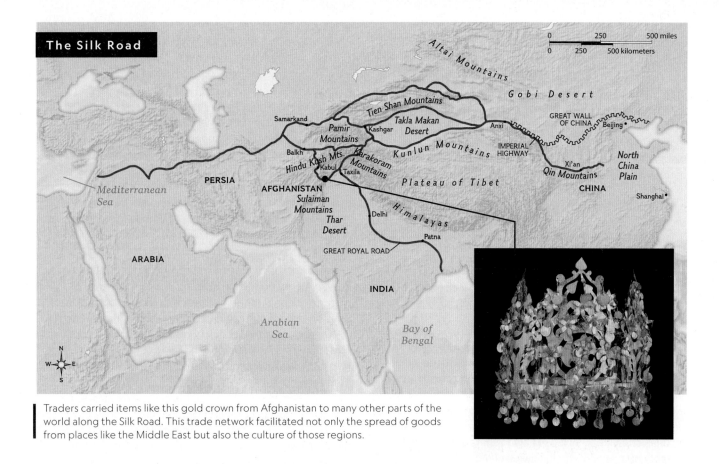

Traders carried items like this gold crown from Afghanistan to many other parts of the world along the Silk Road. This trade network facilitated not only the spread of goods from places like the Middle East but also the culture of those regions.

The snack called potato chips in New York is called crisps in London.

Some things that people view as integral parts of American culture are in reality a result of the process of cultural interaction that leads to new forms of cultural expression. Take cowboys, for example. Geographer Terry Jordan traced the cattle industry in the United States to three Old World sources: Spain in the Iberian Peninsula, West Africa, and Great Britain. Together these three traditions shaped the cowboy culture that was then brought to the United States by Mexican vaqueros in the 19th century. Vaqueros had been taming wild horses, herding cattle, and leading them on cattle drives for hundreds of years before they introduced these practices to Texas ranges. The vaqueros introduced the 10-gallon hat, spurs, and the lasso—all of which are today associated with the all-American cowboy. Many quintessential American foods also were introduced by immigrants. The hot dog was introduced to the United States by immigrants from Frankfurt, Germany, and the hamburger came from Hamburg, Germany. Flatbreads with toppings were consumed by ancient civilizations of the Mediterranean, but the modern pizza was developed in the southern Italian region of Campania, and—via relocation diffusion—came to New York City and other U.S. cities with immigrants from the port of Naples. Migration also influences styles of music and art, sometimes giving rise to new genres.

Geographic Thinking

1. Compare the diffusion of culture through trade, migration, military conquest, and imperialism.

2. Explain whether you think global trade has as much influence today as it had in the past, and why.

Contemporary Drivers of Change

Learning Objective
LO 7.6 Compare the contemporary causes of diffusion, including globalization, urbanization, migration, and technological advances.

Cultural ideas and practices are socially constructed; in other words, they are created by a group of people. Interactions between different places and populations expose people to new artifacts, sociofacts, and mentifacts, and ultimately, through the process of cultural diffusion, some of these traits are picked up by the populations exposed to them. As you've read, artifacts change frequently, sociofacts like marriage customs change slowly, and mentifacts like language and religion change extremely slowly.

Contemporary cultural change—change happening around you today—is driven by large-scale and small-scale processes. The main processes driving cultural change, such as globalization, urbanization, migration, and technological advances, affect culture through media, politics, economics, and social relationships.

Globalization As you read in Chapter 1, globalization is the process by which people across the world have become increasingly connected through travel, trade, and technology. As transportation and communication become faster, more affordable, and more accessible, people can more easily travel to, move to, or communicate with others in locations far away. Globalization also means that people have access to a variety of goods from diverse places around the world. The impact of globalization is often discussed in terms of economics, but it also has a profound effect on culture.

On a small scale, individual travelers experience the art, architecture, food, and other cultural traits of the places they visit and bring these experiences home, where they may influence the travelers' values, beliefs, behaviors, and diets. Immigrants, too, bring their culture with them as they settle in new lands. By holding onto the cultural traits that are familiar and comfortable to them, they share something that might be new and different to their new neighbors. On a broader scale, the internet is available to many (but not all) people around the globe. The internet is critical in the spread of ideas today, making it possible to share or learn about aspects of a culture with the swipe of a touchscreen or the click of a mouse.

Because it appeals to a diverse group of people and tends to spread quickly, popular culture has the potential to spread globally. Compared to aspects of traditional culture, such as customs or beliefs, pop culture can often be traced to a specific person or group in a particular place. For example, hip-hop, which began in the Bronx in New York City in the 1970s, influenced Korean artists who introduced a new genre called Korean pop, or K-pop for short. Emerging in the early 2000s, K-pop became a global phenomenon in the 2010s. Catchy melodies, synchronized dance routines, and distinctive style made the genre a hit in its native South Korea. The music spread from there, as viral videos were shared over social media, and K-pop bands were featured in the opening and closing ceremonies of the 2018 Winter Olympics, held in Pyeongchang, South Korea. Today K-pop has fans in places as far-flung as Japan, Bangladesh, India, Latin America, and the United States—truly a global audience.

Urbanization Urbanization is another significant driver of cultural change. You've read that 57 percent of the world's population live in urban areas and that this number is growing. The flow of people from rural to urban areas means that more and more of us are living in large, diverse settlements where the mix of people brings many cultural ideas and practices together. Cities are centers of innovation and they often attract people who adhere to less traditional beliefs than non-urban dwellers. In addition, people living in a city are exposed to a wider variety of cultural elements than people living in rural areas. City dwellers interact routinely with others of different races, religions, ethnicities, and cultures. Urbanization can drive cultural change on a small scale, such as when people move to a city and hear a genre of music they've never heard before. Or it can happen on a larger scale, such as when cultural interactions between large urban centers spread to the surrounding areas. Fashion that starts in Paris or New York, or a new food trend that starts in Tokyo, might spread hierarchically to other large cities and then make their way to clothing stores, or restaurants in smaller cities and towns.

Migration When they move, migrants bring with them the language, traditions, customs, political values, and cultural norms from home. Over time, these cultural traits have an impact on the culture of the migrants' new home. In recent years, for instance, conflict, drought, and poverty have caused many migrants from Muslim majority countries in Southwest Asia and northern Africa to seek asylum in European countries. These immigrants bring their religious faith and practices with them. Many Muslims gather for traditional daily prayers and women may choose to wear a hijab or a burqa. These aspects of Muslim culture are highly visible, and the reaction of some people to the immigrants illustrates how people sometimes resist cultural change. Some European countries have put bans on burqas or other full-face veils, and many Muslim immigrants report feeling discriminated against in their new European countries.

Migration is connected to urbanization as a driver of cultural change. As you've learned, Ravenstein's laws state that long-distance migrants tend to settle in urban areas. As we've seen in examples of ethnic neighborhoods, immigrants from the same country or region often cluster together in their adopted city. As a group, they have more of an impact on a city's culture and more of a voice in the political and social spheres within a community.

Technology You've already read about how technology has a profound impact on culture. The internet, smartphones, and social media allow for great connectivity between people and provide access to a wider variety of cultural ideas and practices. Global news channels and the internet spread news about faraway events as they are still unfolding. Entertainment streaming services such as Netflix, Hulu, and Britbox feature movies and television series from around the world, introducing people to new cultures and languages. Often, however, having access to the music, movies, TV shows, recipes, and clothing styles of any country in the world leads to less variety rather than more. Teens in Hong Kong and Quebec listen to the same music; families stream the same television programs from laptops in Mozambique and Brazil; the same movies

The music group BTS performs onstage in Los Angeles, California. Korean pop, or K-pop, music groups originated in South Korea, but their popularity has spread across the world through the internet and other media. Today, K-pop is a $5 billion industry and a global phenomenon.

play in theaters on different continents—often with the same opening day. The result is a more uniform world perspective and homogeneous culture. Some people worry that places and peoples are losing what makes them unique.

Cultural Convergence

Learning Objectives
LO 7.7 Explain the factors that drive cultural convergence.
LO 7.8 Compare Tobler's First Law of Geography to the distance decay model.

As cultures interact with one another, they become more similar, sharing and adopting one another's ideas, innovations, and other cultural traits. This phenomenon is called **cultural convergence**. Historically, the spatial relationship between cultures had a major impact on the amount of interaction they had. The farther away one group was from another, the less likely the two groups were to interact. As a result, cultures in close proximity were more alike than those that were far apart. Groups within a region often spoke the same language, ate the same food, and practiced the same traditions. Natural, cultural, and political barriers stopped or slowed the spread of ideas and cultural traits to faraway groups. These patterns followed geographer Walter Tobler's first law of geography, proposed in 1970, which states that while all things on Earth are related to all other things, the closer things are to one another, the more they are related.

Tobler's law is related to the distance decay model, which suggests that the farther people are from the hearth of an idea or cultural trait, the less likely they are to be affected by it. The model also suggests that as an idea or trait travels, the friction of distance takes effect and the idea "decays," or changes. As a cultural trait makes its way from location to location, the people who adopt it might alter it—intentionally or unintentionally. The trait might even die out and disappear altogether. Distance decay explains why people in different parts of the world who speak the same language speak with different accents or why people have different rites and rituals for the same religion. It also explains why people in one region might have a practice or tradition that seems very foreign to those in another part of the world. Arranged marriages, for example, are part of the tradition of many people in South Asia but are highly unusual in the United States today.

Cultural Change **201**

Because of time–space compression, distance decay no longer has the same impact on cultural interaction that it once did. Cultural convergence, therefore, occurs much faster today than in the past. Remember that time–space compression describes the shrinking of the world due to improvements in communication and transportation technologies. It shows how technology has overcome the friction of distance by shortening the time it takes to travel or communicate between two places, making places seem closer together. With each progression in technology, ideas flow faster, and people and goods move more quickly. Their values, ideologies, behaviors, arts, and customs begin to reflect these interactions, driving cultural convergence. As globalization, communication, and transportation technologies make interactions easier and more frequent, cultures often share traits and mimic one another.

Cultural convergence can be observed in the ways in which aspects of Hispanic culture have spread to the United States. Quinceañeras, which you read about in Chapter 6, are popular among Hispanics living in the United States. During the Christmas season, the soft glow of luminarias—traditional Mexican lanterns—illuminate pathways, squares, and churches in many places in the Southwest. And Tex-Mex food, which evolved along the border of the United States and Mexico, is an Americanized version of Mexican cuisine, adapting spicy Spanish, Mexican, and Indigenous flavors.

Cultural Divergence

Learning Objective
LO 7.9 Explain the factors that drive cultural divergence.

Just as increased interaction can cause cultures to become more similar, conflicting beliefs or other barriers can cause two cultures to become less similar. This process is called **cultural divergence**. Divergence can happen when a person or group moves away from their core culture and is exposed to new cultural traits.

Contemporary drivers of change, most notably communication technologies, can cause cultural divergence. Access to technology may vary considerably from one culture to another. As technology speeds up cultural change, the divide will grow between people who have smartphones, the internet, and other technologies and those who lack these technologies. In addition, different cultures may view new technologies differently. Some cultures may reject new technologies, contributing to cultural divergence. The Amish people, a traditional Christian group in North America, do not use many conveniences of modern technology, such as cars, telephones, and even electricity, because they believe these technologies have a negative effect on community and family life.

Divergence can also happen due to physical barriers, such as mountains or rivers, which might separate groups of people living in different parts of a cultural region. As time passes, geographers observe that each group may develop different cultural traits. For example, the Indigenous peoples of the islands of Polynesia are separated by the vast expanses of the Pacific Ocean and by the varying politics of the countries that control particular islands or groups of islands. The culture of Polynesians in New Zealand, known as Maoris, differs greatly from that of Polynesians living in Hawaii or Samoa. The cultures of the different islands evolved differently in response to their particular physical geographies and natural resources, as well as colonizing forces, particularly Britain's annexation of New Zealand in 1840.

Barriers to Diffusion Besides physical barriers like mountains, rivers, oceans, deserts, swamps, and rain forests, cultural and political barriers can slow or stop the spread of culture as well. Cultural barriers include taboos or bans on certain practices. In the Hindu religion, cows are sacred, so it is a taboo to eat beef. McDonalds found a way around this barrier in India by developing a special menu with vegetarian items. Muslims are prohibited from drinking alcohol, and some Muslim-majority countries including Saudi Arabia and Kuwait have complete bans on its sale or consumption. Language can act as a cultural barrier to diffusion as well. People cannot adopt a cultural idea or trait if they don't understand the language in which it is communicated.

Political barriers include policies and borders. For instance, a country that allows free expression and does not place limits on communication technologies like the internet will be much more open to the diffusion of new cultural traits than a country that places restrictions on speech and technology. Iceland, Estonia, and Canada ranked as having the highest degrees of internet freedom among countries in 2021, while the People's Republic of China, Iran, and Myanmar had the lowest degrees of internet freedom. China's Cyber Security Law makes it illegal to transmit government-banned internet content. Travel, business, trade, and border policies vary, sometimes dramatically, between neighboring countries. Some countries work together to open their common borders. You will learn more about international boundaries and border policies in Chapter 9.

Geographic Thinking

3. Explain how communication and transportation technology have affected the distance decay model.

4. Describe how religion can be a strong force of cultural divergence.

Case Study

Fútbol—A Globalizing Force

The Issue The popularity of soccer transcends geopolitical boundaries.

Learning Objective
LO 7.7 Explain the factors that drive cultural convergence.

Megan Rapinoe (*far right*) helped lead the United States to victory in the 2019 FIFA Women's World Cup in Lyon, France—a match watched by viewers around the world.

By the Numbers

3.5 Billion
people watched the 2018 Men's World Cup

22 Percent
more U.S. viewers watched the 2019 Women's Final than watched the 2018 Men's Final

4 Percent
of the world's population actively play soccer in organized leagues

Two hundred eleven member nations belong to FIFA.

Source: FIFA

Soccer . . . Football . . . Fútbol . . . It's called different things in different parts of the world, but this sport is loved around the globe. According to the Federation Internationale de Football Association (FIFA), soccer's official governing body, over 265 million people play the sport on clubs in organized leagues. Recent surveys show that more than 4 in 10 people consider themselves fans, making soccer the world's most popular sport. In some countries, more than 75 percent of survey respondents say they follow soccer.

International rules for soccer were officially set in 1863, but it was already being played in countries across the globe. It spread from its hearth in England through trade and colonialism. By the time the colonial era ended, the sport had become an integral part of the culture in many of England's former colonies. One of the reasons for soccer's popularity is that it can be adapted to local circumstances. In many peripheral countries, children make balls out of whatever materials are available and play on abandoned fields or lots. The game can be learned and played for free, and yet the pros play in stadiums that hold as many as 100,000 fans. In 2022, three of the four highest paid athletes in the world were soccer players: Lionel Messi, Cristiano Ronaldo, and Neymar.

FIFA has 211 member nations, each of which has leagues at all levels. The result is a common culture grounded in the sport. Refugees and other migrants who move from one part of the world to another take comfort in soccer as a common bond. Elite players might migrate to play in a league a country—or even a continent—away. Many professional teams have fans who live halfway around the world.

Every four years, players from all over the world join national teams to participate in the World Cup. The Men's World Cup has been held since 1930, and the Women's World Cup since 1991. Countries compete to host the tournaments, which are attended by hundreds of thousands of fans and watched by billions. Worldwide, more than 3.5 billion people watched at least some part of the 2018 Men's World Cup, with 1 billion tuning in for the final match between France and Croatia. The 2019 Women's World Cup Final had 22 percent more U.S. viewers than the 2018 men's final, with 14.3 million U.S. viewers watching the United States defeat the Netherlands. Soccer, football, or fútbol—whichever name you prefer—is a globalizing force that has created a common culture that crosses boundaries. ∎

Geographic Thinking

Identify a question that you might pose to determine whether the popularity of soccer will continue to grow.

National Geographic Photographer William Allard

William Allard's photos put people front and center, allowing viewers to make their own connections across cultures.

Learning Objective
LO 7.9 Explain the factors that drive cultural divergence.

Introducing People Across Cultures

For more than 50 years as a National Geographic photographer, William Allard has been known for his vivid, memorable portrayals of people in their daily lives. "I really believe if you can make that picture well enough, you can introduce someone who might be a continent removed, a language removed—somehow that viewer can get the feeling that they know something about the person in the picture," he says.

On assignment in 1969 to photograph Surprise Creek, a Hutterite community in central Montana, Allard connected deeply with the families he met. The Hutterite religious group arose from the same movement as the Amish and, like the Amish they live apart from mainstream U.S. culture. The first Hutterites arrived in North America in the 1870s, and today hundreds of Hutterite communities, called colonies, dot the northwestern United States and southwestern Canada. Hutterites are guided by a strong, Bible-based faith and adhere to traditional ways of dress and behavior, including firmly defined roles for men and women. At the same time, they do not avoid technology to the same extent as the Amish. They use the latest in agricultural machinery on their farms, have televisions in some colonies' common areas, and some colony members even maintain a website about Hutterite life.

In 2005, Allard photographed and wrote another article about the Hutterites of Surprise Creek. It was a personal essay about his friendship with colony members, and it recounted the heartfelt support they offered when Allard learned his son had died after a battle with cancer. Sharing his own emotions in words and pictures, Allard once again introduced viewers to the people of a unique American culture.

Critical Viewing According to Allard, Hutterite children are great at entertaining themselves, despite the fact that TVs and computers aren't allowed in homes. ▍ Explain the degree to which the Hutterite community represents an example of cultural divergence.

Critical Viewing *Top:* No matter the activity, Hutterite girls wear traditional dresses or skirts and head scarves. *Bottom:* Guests blow bubbles at a Hutterite wedding reception. Weddings are celebrated with a shivaree—a big, joyous meal with singing. Explain how these photographs illustrate the cultural processes at work in the Hutterite community.

7.3 Consequences of Cultural Change

By now you should understand how colonialism, imperialism, globalization, and urbanization have impacted—and continue to impact—cultural patterns. Geographers also identify ways in which cultural diffusion affects societies as people and groups adopt or reject the cultural artifacts, sociofacts, and mentifacts they come into contact with.

Acculturation

Learning Objective
LO 7.10 Describe the processes of acculturation.

When cultures come together, either through migration or some form of expansion diffusion, one effect is **acculturation**, where people within one culture adopt some of the traits of another while still retaining their own distinct culture. It often occurs as the result of prolonged contact between two or more cultures and can happen at a group or individual level.

Acculturation can occur with almost any cultural trait. It is often discussed in terms of a minority culture that adopts elements of a majority culture. Immigrants to a new country, for instance, go through a process of adapting to their new cultural context—they often learn the language, wear the fashions, and follow some of the customs of their adopted country. In the process of taking on these new traits, some aspects of the original culture can be lost over time. As the pie charts show, 97 percent of Hispanic immigrants speak Spanish to their children at home, while 71 percent of second-generation Hispanic immigrants share their native language with their children. The number falls to less than half for third-generation immigrants and beyond.

Fashion offers a clear example of acculturation. Muslim women who immigrate to a Western country from a Muslim-majority country might adopt aspects of their new culture by wearing many of the popular fashion trends of their adopted country but still retain their own culture by continuing to wear the traditional hijab.

Destination countries and cultures experience acculturation as well. Currywurst, a popular food sold by street vendors in Berlin and elsewhere in Germany, combines a traditional German sausage with curry powder—a spice from India—and ketchup, which originated in the United States.

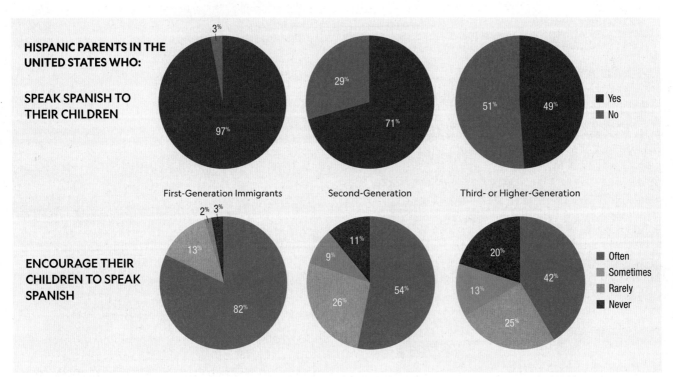

Source: Pew Research Center

Critical Viewing Acculturation has caused the number of Spanish-speaking Hispanics in the United States to decline. Notice that the pie charts show declining numbers of both Hispanic parents who speak Spanish to their children (*top*) and Hispanic parents who encourage their children to speak Spanish (*bottom*). ▮ Describe how the two sets of charts relate to each other, and explain the overall trends depicted in the charts.

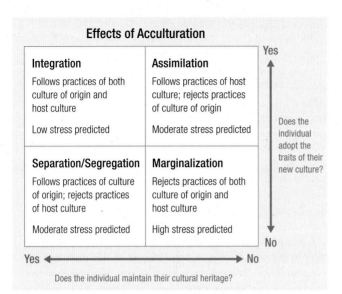

The table shows the results of cultural psychologist John W. Berry's research on acculturation.

Geographers studying acculturation point out the complicated nature of the process. Factors such as age, personality, upbringing, education, and many others determine how comprehensively immigrants incorporate a new culture into their life. Some immigrants adopt the receiving culture wholeheartedly and drop as much of their heritage culture as possible, while others go in the opposite direction, holding onto as much of the culture they grew up with as possible. Some adopt elements of the new and retain elements of the old, and some marginalized immigrants reject both the receiving and heritage cultures. Because of **intersectionality**—the idea that there is a cumulative way in which the effects of multiple forms of discrimination like racism and sexism may overlap, combine, or intersect in the experiences of some marginalized groups or individuals—different immigrants will assimilate in different ways, based on their ethnicity, race, gender, and class. Cultural psychologist John W. Berry studied the ways in which individuals choose to acculturate and the effects of these choices. Berry found that no matter how people attempt to handle the collision of two or more cultures in their lives, they will experience some stress. Those who integrated elements of both their original and adopted cultures into their lives experienced the least amount of stress, and those who rejected both cultures experienced the most. People who accepted the adopted culture and rejected their original culture, and people who retained their original culture and rejected the new culture, both experienced moderate amounts of stress.

Assimilation

Learning Objective
LO 7.11 Describe the process and results of assimilation.

Assimilation is a category of acculturation in which the interaction of two cultures results in one culture adopting almost all of the customs, traditions, language, and other cultural traits of the other. An individual or group that has fully assimilated into a culture will be indistinguishable socially and culturally from others in the culture. It is the most comprehensive form of acculturation, and it might happen voluntarily or be forced upon a group by a dominant culture. But since it involves replacing deeply held beliefs and cultural practices, complete assimilation is rare.

In North America, the governments of Canada and the United States removed thousands of American Indian, First Nations, and other Indigenous children from their families and forced them to attend boarding schools where they had to speak English and take on English names. This is a very clear case of forced assimilation. Sometimes, however, the line between forced and encouraged assimilation can be blurry. To encourage immigrants to assimilate, Germany requires them to pass an oral and written language test for permanent residency and citizenship, though the new government planned changes to the policies in 2022.

Voluntary assimilation can occur when it is advantageous for an immigrant group to fit in with its adopted culture. When Europeans immigrated to the United States in the 1800s, they often faced prejudice from anti-immigrant groups. By assimilating into the culture, they were more likely to find employment and avoid persecution.

The downside of assimilation is the loss of cultural identity—the traditions, languages, holidays, and practices that make a culture unique. Historically, there are many examples in which colonial forces and imperialism have minimized or banned cultural traditions. Whole languages have faded out completely as people from colonized groups have been forced to learn the colonizer's language. But assimilation can happen in a less extreme manner as well. Traditions might be lost as generations die off and children of immigrants naturally absorb the practices of the adopted culture.

Syncretism

Learning Objective
LO 7.12 Describe the process and results of syncretism.

Sometimes traits from two or more cultures blend together to form a new custom, idea, value, or practice. This process of innovation combining different cultural features into something new is called **syncretism**. Cultural syncretism is an effect of diffusion that can occur through circumstances such as immigration, military conquest, marriages between groups, and others. The result is a combination of two cultures to create new ideas, values, or practices.

Religious syncretism is common around the world. Indeed, as many religions found a foothold in new areas, they incorporated traditional customs into their rites, rituals, and beliefs. Santería, for instance, is a religion that developed in Cuba as a syncretic hybrid of a traditional African religion and Roman Catholicism. Ethnic Yorubans of West Africa who were enslaved and transported to Cuba brought with them a

Cultural Change

religion in which practitioners worship deities called *orishas*, who are thought to be messengers between humans and the divine. The religion was influenced by Roman Catholicism, which was already present in Cuba, and developed into Santería, which blends the *orishas* with Roman Catholic saints. Since its development in Cuba, the syncretic religion has spread through relocation and expansion diffusion to groups of people in Latin America and the United States.

Cultural syncretism is often evident in celebrations. Halloween, for instance, has its roots in both Christian and pagan practices. In Mexico, the Day of the Dead is celebrated from October 31 to November 2 to coincide with two Catholic feast days, but the festival's origins are rooted in traditional celebrations of Indigenous peoples.

Syncretism occurs in music as well, as new musical styles often combine the musical traditions of several different cultures. Bluegrass music, which was developed in the Appalachian region of the United States, fuses Irish and Scottish musical traditions with African American influences. As the popularity of bluegrass spread across the United States and to other parts of the world, younger bluegrass musicians incorporated influences from rock music.

Multiculturalism

Learning Objective
LO 7.13 Explain the consequences of multiculturalism.

Sometimes diverse cultures coexist within a shared space. People in these spaces do not belong solely to one culture or another. They may share some cultural features with others around them while retaining some of their original cultural traits or practices. This **multiculturalism** often occurs in large cities, where people from many different cultures live in close proximity. The United States is a classic example of a multicultural country. Although Americans share a dominant culture based on democratic ideals, there are many different ethnicities that retain aspects of their own cultures, as evident in the varied cuisines, religions, arts, and languages found within the country.

Multiculturalism can create an atmosphere of acceptance and lead to a rich, vibrant blend of cultural traits. In Canada, for instance, multiculturalism is an official policy. In 1971, the government adopted the stance that multiculturalism was a fundamental right of all citizens, and that by keeping their cultural identities and taking pride in their ancestry, all Canadians would become more open and accepting of diverse cultures. This climate of tolerance has contributed to an influx of skilled workers and foreign investment that has bolstered Canada's economy. In addition, the marriage rate between native-born Canadians and immigrants has risen and immigrants report approximately the same level of satisfaction with their lives in Canada as native-born citizens.

The coexistence of cultures is not always easy. Sometimes people can feel caught between two cultures. These difficulties are typical of cultural changes caused by diffusion, and their severity depends on a range of factors, such as the degree of cultural difference. In Australia, a recent study of immigrant families from Burma, Nepal, India, Afghanistan, Bangladesh, Iraq, and countries of Africa, found that many immigrants struggled as they attempted to balance the need to acculturate with the desire to retain their home culture. In particular, researchers noted challenges that relate to the clash between Western ideals and the traditional values of the countries of origin. The cultures of Africa and Asia are **collectivist cultures**, where people are expected to conform to collective

Young women celebrate *Día de los Muertos*, or Day of the Dead, by dressing up and painting their faces. Throughout Mexico, and in U.S. locations where people of Hispanic descent live, people honor family members who have passed away by singing, dancing, and holding parades and parties during this holiday.

In London, Ontario, people from diverse cultures take part in a Canadian citizenship ceremony to celebrate Canada Day. Multiculturalism is an accepted and celebrated part of life in Canada.

Cultural Appropriation

Cultural appropriation is the act of adopting elements of another culture. It is usually used to describe the adoption by a dominant culture of one or more elements of a minority culture. The term was coined in the 1960s to critique the effects of colonialism, and it is used today to describe any intentional adoption of cultural traits of another culture. The term is most often used to describe instances when such adoption is inappropriate or out of context.

Cultural appropriation is controversial. Some people complain that it is disrespectful to adopt cultural traits that have meaning to a group of people without fully understanding or embracing the meaning behind these traits. In some cases, participating in a custom, tradition, or celebration has sparked cultural debate, as has wearing the clothing or hairstyles of another culture. The cultural appropriation controversy often gets attention at Halloween, as people's choice of costumes may reinforce racial, ethnic, or cultural stereotypes.

Sometimes there is a backlash against a group or a person—often a celebrity or a public figure—for appropriating an aspect of another culture, or otherwise offending people of that culture. This outcry is typically amplified through social media, with the perceived offender subject to "mass shaming." In turn, there may then be a backlash, with others contending—again, via social media—that this represents "cancel culture," and that these attacks against cultural appropriation demonstrate an oversensitivity. People on this side of the argument claim that if cultural traits of a particular group become something that only that single group has a right to have or use, then all types of cultural diffusion would be wrong and restricted. Indeed, not everyone agrees that cultural appropriation should be avoided. They argue that the appropriation of cultural traits is a natural aspect of positive change, as cultures "try on" the traits of others.

responsibility within the family and to be obedient to and respectful of elder family members. These expectations are at odds with the spirit of individualism and independence of the culture of Australia. The disparity between these sets of values can both make it difficult for young people to assimilate into the new culture and also can lead to conflict at home. The fact that these issues are less obvious than other challenges of acculturation, such as learning a new language, means that they can be even more difficult to overcome.

Geographic Thinking

1. Compare acculturation and assimilation.
2. Identify and describe consequences of multiculturalism.
3. Explain whether Australia's problems with multiculturalism would be solved by adding a government policy of acceptance and tolerance like Canada's. Why or why not?
4. Explain how cultural appropriation is different from acculturation.

Cultural Change

Chapter 7 Summary & Review

Chapter Summary

Cultural traits spread from one place to another over time through diffusion. Historically, diffusion occurred through imperialism, colonialism, and other ways like urbanization. Today, diffusion occurs through some of the same historical ways, but also through mass media, social media, and the internet.

- The spread of traits through the migration of people is called relocation diffusion.
- Expansion diffusion occurs when an idea or cultural trait spreads outward from its hearth through contact. There are three types of expansion diffusion: contagious diffusion, hierarchical diffusion, and stimulus diffusion.

Drivers of cultural change include colonialism and imperialism, military conquest, trade, migration, globalization, and urbanization.

- European countries spread their cultures around the world through colonialism and imperialism, beginning in the 15th century.
- Military conquest sometimes causes rapid cultural change because the conquerors impose their way of life on the conquered.
- Trade brings together people from different places. Buying and selling goods and services requires interaction, which results in the spread of culture.
- Migration brings new languages, religion, fashion, music, and food to the places of destination.

- Contemporary cultural change is driven by processes such as globalization, urbanization, migration, and technological innovation.
- Cultural convergence is the process by which cultures begin to share traits and mimic one another.
- Cultural divergence is the process by which conflicting beliefs or other barriers cause two cultures to become less similar.

The effects of cultural change range from disagreement and conflict to greater understanding and sense of connection among individuals and groups.

- Acculturation is when people within one culture adopt some of the traits of another while still retaining their own distinct culture.
- Assimilation is a category of acculturation in which the interaction of two cultures results in one culture adopting almost all the customs, traditions, language, and other cultural traits of the other.
- In multiculturalism, which often occurs in large cities, people share some cultural traits with those around them while retaining some of their original cultural traits.
- Syncretism is the process by which traits from two or more cultures blend together to form a new custom, idea, value, or behavior.

Review Questions

Use complete sentences to answer the questions.

1. **Apply Conceptual Vocabulary** Consider the terms *convergence* and *divergence*. Write a standard dictionary definition of each term. Then provide a conceptual definition—an explanation of how each term is used in the context of this chapter.

2. What does it mean for cultural ideas or practices to be socially constructed?

3. Explain the role of a cultural hearth in the spread of cultural traits through diffusion.

4. Define and give an example of relocation diffusion.

5. Describe how the African diaspora is an example of relocation diffusion.

6. How have colonialism and imperialism contributed to relocation diffusion?

7. How are contagious diffusion, hierarchical diffusion, and stimulus diffusion similar? How are they different?

8. How and why has English become a lingua franca?

9. What is cultural convergence? Give an example.

10. Explain how migration contributes to cultural divergence.

11. How does distance decay affect cultural diffusion?

12. How has time–space compression affected cultural convergence?

13. Compare the concepts of acculturation, assimilation, and cultural appropriation.

14. What would you expect to find in a multicultural country or community?

15. How is the term *syncretism* different from the term *multiculturalism*?

■ **Interpret Graphs**

Study the graph and then answer the following questions.

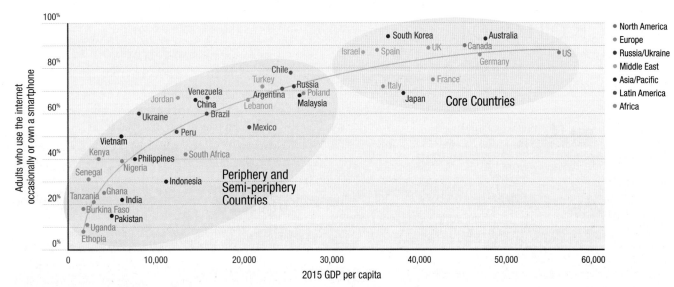

Internet Use and GDP A strong correlation exists between the wealth of a country as measured by its Gross Domestic Product (GDP) and internet access. Wealthier core countries such as Australia, Germany, and the United States have much higher rates of internet use than periphery countries in Africa and Asia/Pacific regions. ■ Do you think that the trends shown in 2015 have changed?

Source: Pew Research Center

16. **Explain Patterns & Trends** According to the graph, what is the relationship between per capita gross domestic product (GDP) and internet use around the world?

17. **Synthesize** Based on what you know about how culture spreads, would you expect the cultures of the core countries or the periphery and semi-periphery countries to be more alike? Why?

18. **Evaluate Geographic Concepts** Choose two countries and, using their GDP and percentage of internet access, describe what type of cultural diffusion you would expect to see between them.

Chapter 8
Spatial Patterns of Language and Religion

Critical Viewing In 2016, the Tunisian-French artist known as eL Seed painted a mural across a group of 50 buildings in one of the poorest neighborhoods in Cairo, Egypt. The mural is Arabic calligraphy and it quotes a Coptic Christian bishop: "If one wants to see the light of the sun, he must first wipe his eyes." In what ways do both language and religion play a role in the mural?

Geographic Thinking How do language and religion affect each other's spread?

8.1 Patterns of Language

8.2 The Diffusion of Language

Case Study: French or English in Quebec?

National Geographic Explorer Maria Fadiman

8.3 Patterns of Religion

Case Study: Shared Sacred Sites

8.4 Universalizing and Ethnic Religions

8.1 Patterns of Language

More than 7,000 distinct languages are spoken in the world today. Geographers group these languages according to various factors, including their shared origins and how they have evolved. Comparing languages and studying their history and diffusion is a window into the interactions of people over space and time.

Language and Culture

Learning Objective
LO 8.1 Explain how language reflects and transmits culture.

Language is an important vehicle by which culture is transmitted and preserved. Through language, whether oral or written, a culture's values, beliefs, traditions, and norms are shared and passed down. Language can identify and differentiate a culture, unifying the people who speak a common tongue and separating them from those who speak other languages. For example, the hundreds of millions of people in the world who speak Arabic share a common bond, whether they come from Saudi Arabia, Nigeria, the United States, or New Zealand. The differences among languages reflect the differences among cultures, and language is sometimes used to highlight those differences. In 2019, Ukraine's parliament enacted a law making Ukrainian the official state language. The law called for any print media published in Russian (but not English) to also be published in an equal number of copies in Ukrainian. The law was aimed at diminishing the use of Russian and weakening the spread of Russian media in Ukraine, while strengthening Ukrainian culture and identity by protecting the use of its language. More people speak Ukrainian now than 20 years ago and a strong majority supports further expansion of the use of Ukrainian in the country, especially in the face of continued aggression from Russia.

Language and culture both reflect and shape our ways of life. The Kuuk Thaayorre, an Aboriginal people living in northern Australia, use their words for north, south, east, and west to describe the location of people and objects. They do not have words such as *left*, *right*, *in front*, or *behind*. While English speakers might refer to their left hand, Kuuk Thaayorre individuals refer to the north, south, east, or west hand—depending on the direction they face. These two approaches of spatial thinking are affected by both cultural differences and the language English speakers and the Kuuk Thaayorre have available to them.

As the needs of a culture change, so does its language. Throughout history, people have created and borrowed words to describe new experiences and encounters that are made possible through migration, trade opportunities, and technological advances. For example, the term *text* has evolved to refer to the way people share messages by cell phone. As people migrate, their native languages intermingle, and new languages evolve. Occasionally, the language brought by migrants to a different region even replaces the original language of that region entirely.

Linguists and geographers study the methods and sequences of the diffusion of different languages, looking for connections between languages that will help describe human movement and development. Mapping the locations of where different languages are spoken is an important step in seeking connections among them. A system for comparing and categorizing languages aids this process.

Categorizing Languages

Learning Objectives
LO 8.2 Compare the hierarchical categories of language including family, branch, group, and dialect.
LO 8.3 Identify the language families with the largest number of languages.

While languages like Spanish and Portuguese are distinct, speakers of these two languages can usually understand each other because the languages are closely related. Linguists use the similarities and differences among languages to organize them into categories with increasing degrees of differentiation from a common language. From closest to the common language to farthest from the common language, the levels of organization are families, branches, groups, and dialects.

Language Families A **language family** is the largest grouping of related languages and includes those languages that share a common ancestral language from a particular hearth or origin. Ancestral languages are not extant, meaning they are not being actively used anymore. In fact, they are not even identifiable from written records because they developed in ancient times before any culture developed a writing system. Therefore, linguists infer the existence of these ancient languages based on similarities in grammar and root words in existing languages, rather than based on clear written evidence. The first language family to be proposed was Indo-European—a diverse language family that includes most of the languages of Europe and many languages of South and Southwest Asia, ranging from Hindi to German, for example. Indo-European was described by a British scholar in the 1780s based on similarities he found between Greek, Latin, and Sanskrit, an ancient language of South Asia.

Spatial Patterns of Language and Religion

Language is closely tied to culture, as evidenced by this signboard of ads for retail stores written in Hangul, the language of South Korea. The name of one American business, Baskin-Robbins, appears in English.

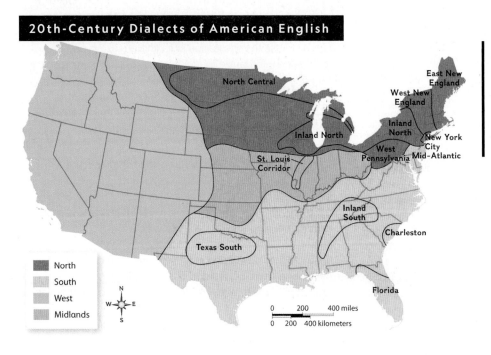

20th-Century Dialects of American English

Reading Maps The contiguous United States has four main dialect regions: North, South, West, and Midlands. Subdialect regions are areas within a larger dialect region that share certain dialect characteristics. ▌Describe two patterns of language development that you can infer from the map.

Ethnologue—a respected organization that catalogs languages—has identified 142 language families. Six of those are major language families, and together they account for 63 percent of all living languages and 85 percent of the world's population. Nearly 6 billion people speak a language that is included in the six major language families. Indo-European languages, with more than 3.2 billion speakers, make up the largest language family with the widest distribution, including speakers in North America, South America, Europe, and South Asia. The Sino-Tibetan family, with nearly 1.4 billion speakers in East and Southeast Asia, is the second largest.

Some languages are not assigned a language family. An isolated language, or **isolate**, has no known historic or linguistic relationship with any other known language. Isolates may occur because related languages have gone extinct. As discussed in Chapter 6, Basque—the native language of about 1 million people living in the Basque region of northeastern Spain and southwestern France—is one such example. Huave, an Indigenous language of the Oaxaca state in Mexico, is another. The rugged mountainous terrain of each of these regions has helped to isolate each culture and preserve its language.

Language Branches The next level in the language hierarchy is the **language branch**. This collection of languages within a family share a common origin and were separated from other branches in the same family thousands of years ago. Languages among various branches may show some similarities in grammatical structure, but are so distinct that speakers of languages from two branches cannot understand each other. For example, in the Indo-European family, the Romance branch—which derives from Latin and includes French, Italian, Romanian, Spanish, and Portuguese—is very different from the Germanic branch, which includes German, English, Dutch, as well as the Nordic languages such as Swedish, Icelandic, Danish, and Norwegian. Both of those branches differ significantly from the Slavic branch, which includes Russian, Polish, Czech, and Bulgarian. Each of these three branches has a distinct geographic distribution, with the Germanic languages found mainly in northern Europe, the Romance languages largely in southern Europe, and the Slavic languages mostly in eastern Europe.

Language Groups **Language groups** are languages within a branch that share a common ancestor in the relatively recent past and have vocabularies with a high degree of overlap. Speakers of different languages within a language group will recognize many similar words in each other's languages. Using the example of Spanish and Portuguese again, speakers of these two languages typically can understand each other because Spanish and Portuguese belong to the same language group within the Romance language branch.

Dialects Language branches and groups are divided further into individual languages and dialects. A dialect is a variation of a standard language distinguished by differences in pronunciation, degree of rapidity in speech, word choice, and spelling. Speakers of the same language sometimes talk with a unique dialect. Various Spanish dialects are spoken throughout regions of Spain and in Latin American countries. So a native resident of Spain will understand a native resident of Mexico but may use distinct pronunciation for some of the same words. For example, the Spanish word *guagua* refers either to a baby or a bus depending on the speaker's dialect. Similar variations in words and meanings exist in all countries and across all languages.

English has numerous dialects, a result of the large number of primarily English-speaking countries distributed around the world, including Australia, the United States, those in the Caribbean, and, of course, those in the United Kingdom. Both migration and isolation play a role in diversifying the English dialect spoken in each of these countries and within regions in the countries.

The migration of colonists from Europe to North America during the 17th century contributed to the American English dialects. The drawl, or long, drawn-out vowels, characteristic of the southern dialect reflects the speech of southern England, source of the first southern settlers. African Americans formed their own dialect by mixing English and their native West African languages. Other groups who came to the English colonies, like the Scots-Irish settlers of Appalachia or Germans who settled in Pennsylvania, contributed to the development of regional dialects as well. These people, of course, originally brought their native languages but often learned English as the language of government and the majority population.

As people—both English speakers and others—moved west, the American English dialects that had developed in the original colonies blended to create a less distinct western accent. Immigrants of the 19th and 20th centuries, including the Irish, Scandinavians, Italians, and Eastern European Jews, influenced the dialects that developed both along the Atlantic coast and inland. Expansion north from Mexico into areas from Texas to California that had originally been settled by Spanish speakers influenced the dialects of the Southwest. The separation of English speakers across oceans works to preserve the differences in the English language, but that may be short-lived. Communication technologies and social media continue to transform the English language by allowing constant interaction among people who speak a multitude of different languages.

Geographic Thinking

1. Describe how and why language differentiates one culture from another.

2. Explain why languages change as cultures change. Give two examples as part of your answer.

3. Explain how the categorization of languages provides insights into human history.

Spatial Patterns of Language and Religion

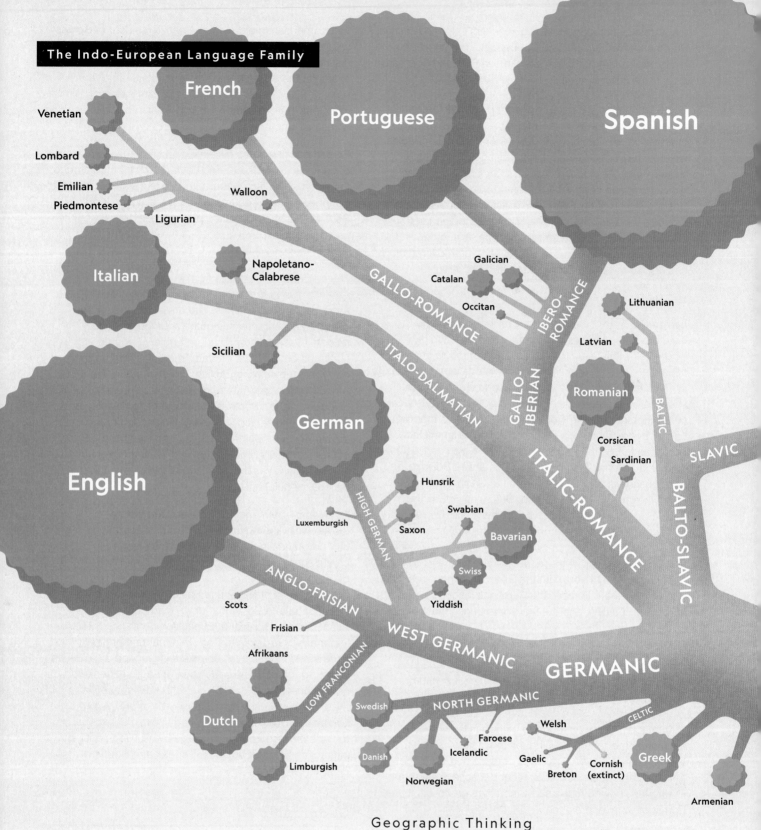

Geographic Thinking

4. Identify the five largest Romance languages on the diagram.

5. Explain how Hindi and Urdu are related.

6. Explain how the diagram illustrates the increasing differentiation of languages as one moves from ancestral languages to dialects.

There are 142 language families in the world, and Indo-European is the largest with more than 3.2 billion speakers. The size of the leaves on the diagram approximates how many people speak that language, and not every Indo-European language is represented.

216 Chapter 8

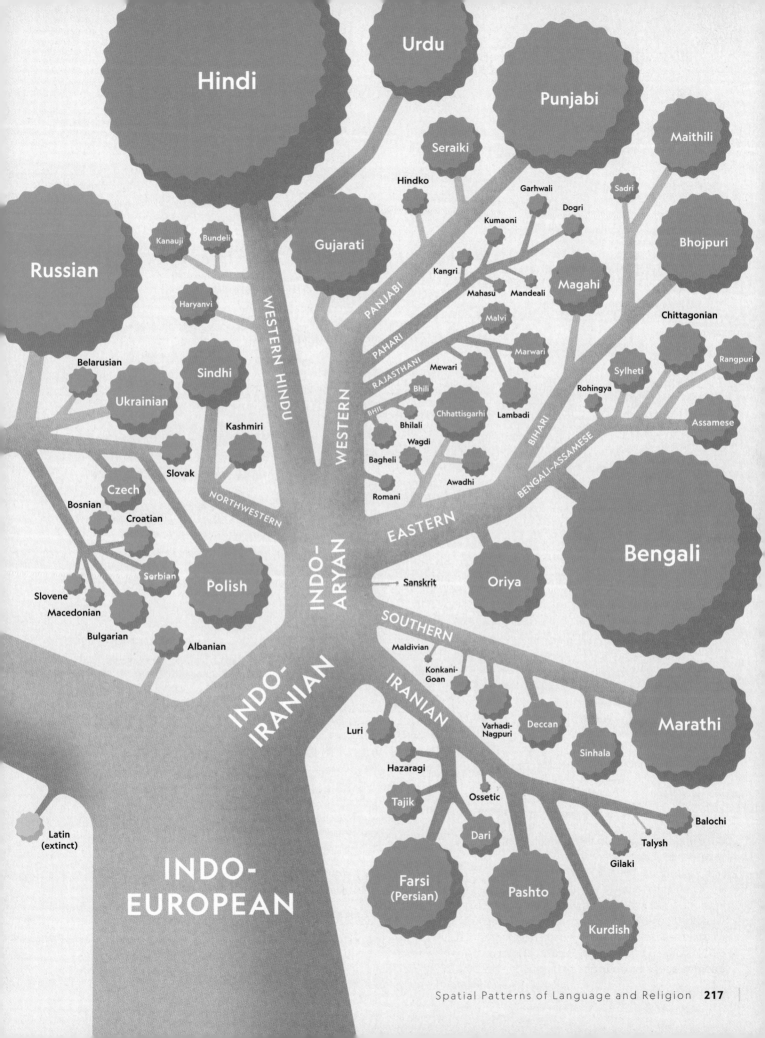

Spatial Patterns of Language and Religion 217

8.2 The Diffusion of Language

Languages have distinctive distribution patterns, sometimes contiguous and sometimes not, tying their points of origin and diffusion pathways. Maps and charts depict the extent to which languages have spread from their hearths. Language distribution is also reflected in the names given to regions and places throughout the world.

Distribution of Languages

Learning Objective
LO 8.4 Describe the geographic distribution of world languages.

Global patterns of language have changed over time. The "Distribution of Language Families" map documents the location of various language speakers today. It looks different than a map from centuries ago. For instance, English was not the predominant language in North America until British colonization of eastern North America in the 17th and 18th centuries sparked immigration by English-speaking people. English became the dominant tongue of these colonies and, subsequently, of the United States. When others immigrated to the region, they generally adopted English. As cultures interact, maps documenting the location of language families will continue to change.

Today, five languages—Arabic, English, Portuguese, Russian, and Spanish—show wide transcontinental or intercontinental distribution as a result of trade, migration patterns, and language adoption practices. For instance, the dominance of Spanish in South and Central America and Portuguese in Brazil is a function of Spain and Portugal forming colonial empires in the region beginning in the 16th century, the subsequent imposition of the colonizers' language on Indigenous peoples, as well as both forced migration from Africa and voluntary migration from Europe. A map of language distribution in 1400, before systematic colonization and migration, would show this same region dominated by various Indigenous language families.

Five other major languages—Mandarin, Hindi, Bengali, Japanese, and Western Punjabi—have grown largely as a result of population growth. East Asia and South Asia are two of the most densely populated regions in the world. Mandarin has also grown because of an effort to expand Han Chinese control to more remote places like Tibet and Xinjiang. Modern maps of world language distribution can represent the relative "size" of different languages geographically; however, the land area a language covers may differ from the actual size of the population speaking the language. Compare the nearly 400 million speakers of Hindi in India and neighboring South Asian countries, who considerably outnumber the 137 million speakers of Russian across the sprawling land area of the Russian Federation. The speakers of Hindi occupy a much smaller geographic area than the speakers of Russian.

How Languages and Language Families Spread and Change

Learning Objective
LO 8.5 Describe the geographic distribution of language families.

Changes in languages offer clues about historical events such as migration, trade, exploration, and imperialism and colonialism—or any combination of these factors. Using historical documents, archaeological artifacts, and genetic research, scholars study the evolution of languages, in work termed *glottochronology*.

Indo-European Hearth and Diffusion Almost half of the world's population speaks one of the 445 living languages that are part of the Indo-European language family, either as their native tongue or to conduct business as a lingua franca. Scholars speculate about the existence of a common ancestral language of the Indo-European family, which they call Proto-Indo-European. This hypothesized language is considered by some to be the predecessor to languages such as Greek, Latin, and Sanskrit. As noted earlier, it is difficult to confirm an ancestral language like Proto-Indo-European since it doesn't exist in any written form.

Clues to the origins of a language can be discovered in cognates, which are words found in different languages that have similar meaning and a shared descent from a common ancestor. Just because two words sound similar or have a common meaning doesn't mean they share a common cognate. A word in one language—the English *mountain*, for example—could simply be borrowed from another language—the French *montagne*. The Romance languages derived from Latin share cognates for their words for *water*, such as *acqua* (Italian), *apă* (Romanian), *eau* (French), *agua* (Spanish), and *água* (Portuguese), because they all derive from the Latin original, *aqua*. As these examples show, cognates can have significant differences in spelling and pronunciation, making the identification of cognates difficult.

Scholars argue about where the Indo-European language family originated, identifying two leading areas. One idea, called the Kurgan hearth theory, places the origins on the steppes, a flat grassland area of modern Russia or Ukraine. Another theory proposes the hearth is in Anatolia, the Asian section of Turkey.

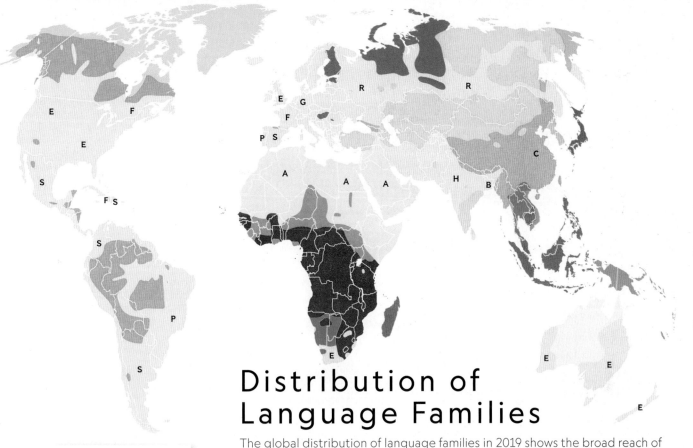

Distribution of Language Families

The global distribution of language families in 2019 shows the broad reach of the Indo-European language family and other major language families such as Niger-Congo or Altaic. Individual languages are often distributed regionally, such as Arabic across North Africa or Russian in northern Eurasia. The majority of languages that have more than 50 million speakers belong to the Indo-European family. However, the number of speakers of Mandarin Chinese, part of the Sino-Tibetan language family, far exceeds the number of speakers of any other language worldwide. ▮ Explain how the data in the chart and the map work together to provide a current picture of global languages.

Language Families
- Afro-Asiatic
- Altaic
- American Indian
- Australian Aborigine
- Austro-Asiatic
- Austronesian
- Basque
- Dravidian
- Eskimo-Aleut
- Indo-European
- Japanese
- Khoisan
- Korean
- Niger-Congo
- Nilo-Saharan
- North Caucasian
- Sino-Tibetan
- Tai-Kadai
- Trans-New Guinea
- Uralic
- Vietnamese
- Yeniseian

Major Languages
- A Arabic
- B Bengali
- C Chinese
- E English
- F French
- G German
- H Hindi
- P Portuguese
- R Russian
- S Spanish

Sources: Goode's World Atlas, Ethnologue

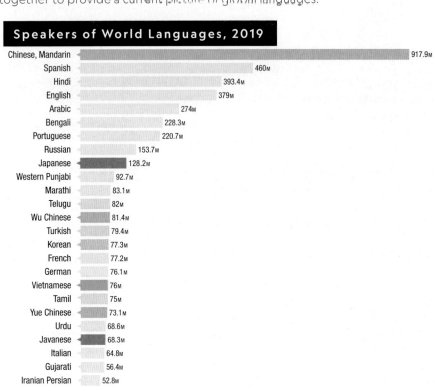

Speakers of World Languages, 2019

Chinese, Mandarin	917.9M
Spanish	460M
Hindi	393.4M
English	379M
Arabic	274M
Bengali	228.3M
Portuguese	220.7M
Russian	153.7M
Japanese	128.2M
Western Punjabi	92.7M
Marathi	83.1M
Telugu	82M
Wu Chinese	81.4M
Turkish	79.4M
Korean	77.3M
French	77.2M
German	76.1M
Vietnamese	76M
Tamil	75M
Yue Chinese	73.1M
Urdu	68.6M
Javanese	68.3M
Italian	64.8M
Gujarati	56.4M
Iranian Persian	52.8M

Spatial Patterns of Language and Religion

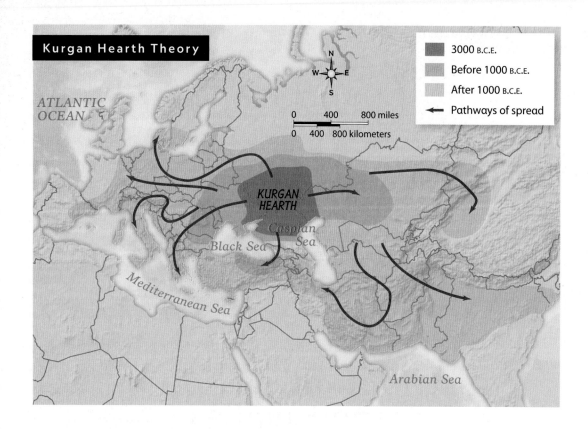

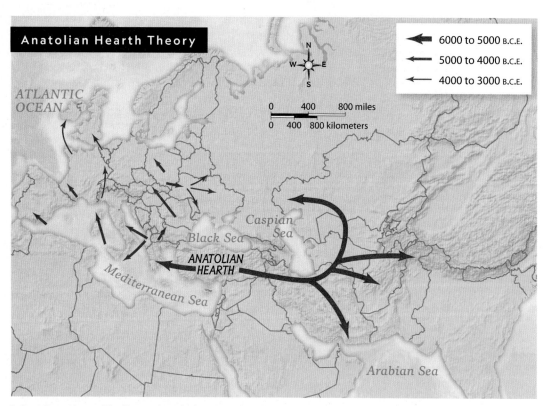

Reading Maps The Kurgan hearth theory asserts that the Kurgans, a nomadic-warrior people, migrated from an area around the present-day border between Kazakhstan and Russia into Europe, Siberia, Iran, and South Asia. The Anatolian hearth theory argues for an earlier diffusion and claims the first speakers of Proto-Indo-European were farmers who moved into Europe and South Asia in conjunction with agricultural practices. ▪ Compare the Kurgan hearth theory and the Anatolian hearth theory using elements from the maps.

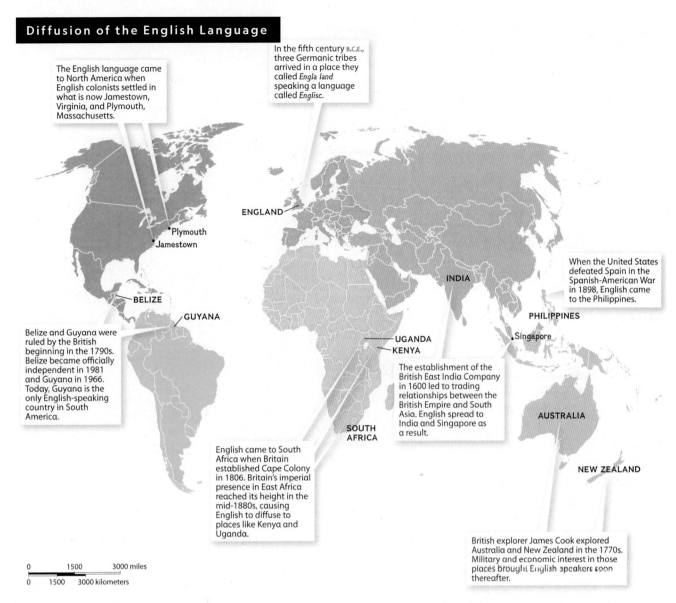

Reading Maps The history of the diffusion of the English language reaches back thousands of years. Identify two historical factors that contributed to the diffusion of the English language across the globe.

Cognates found among the languages of the Indo-European family point to the Kurgan hearth, since the words that would be used to describe climate and landscape, like *winter* and *snow*, apply to the Kurgan region's climate, animals, and trees. They suggest that the people of the region conquered surrounding lands and advanced to Europe and South Asia, thus diffusing the Proto-Indo-European ancestral language. Those who favor the Anatolian hearth theory suggest the diffusion of Proto-Indo-European relates to agricultural practices. The adoption of agriculture promoted population growth, causing migration from the hearth to new regions.

Diffusion of the English Language The English language formed as a mixture of several linguistic influences—Celtic, Germanic languages, French, Latin, and borrowings from other languages. English originated on the island of Great Britain in the northwest corner of Europe and has spread to become the dominant language in areas as far away as Australia and North America. With nearly 400 million native speakers and more than 1 billion second-language speakers, English is one of the most spoken languages in the world and has become the dominant language of politics, business, finance, and technology in much of the world. The spread of the English language exemplifies both main types of diffusion—relocation diffusion and expansion diffusion. (You learned about these types of diffusion in Chapter 7.) The hierarchical type of expansion diffusion, in particular, impacted the spread of the English language, as leaders from various kingdoms and regions established English as the language of their expanding domains.

Spatial Patterns of Language and Religion

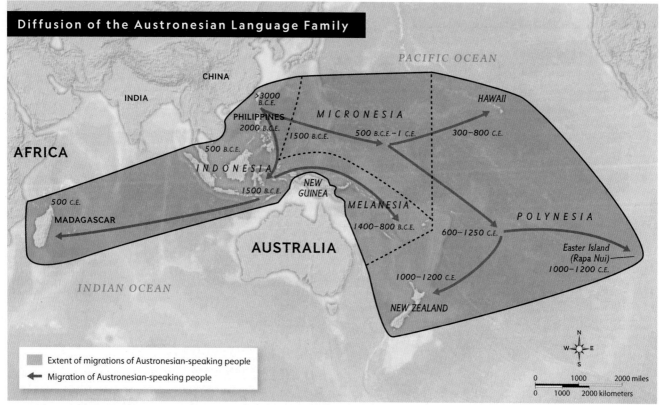

Reading Maps The Austronesian language family has spread among many islands of different regions. ▎How does the map support the idea that geographic proximity facilitates the diffusion of language? In what way does the Austronesian language family example also offer a counterargument?

Little is known of the language of the first inhabitants of the British Isles, but recent research using DNA from the remains of 400 ancient Europeans documents a previously unknown major migration into Britain from 1,300 B.C.E. to 800 B.C.E. The new inhabitants from Europe may have brought Celtic languages. The Angles, Jutes, and Saxons, who spoke Germanic languages, invaded from continental Europe during the fifth century C.E., pushing Celtic-speaking people to more remote locations in Ireland, Scotland, and Wales. As a result of this relocation diffusion, both the Celtic and Germanic languages have influenced the English language.

In 1066, Normans from northern France invaded Great Britain. Their leader installed himself as king of England, and the Normans settled in many parts of the island, introducing a Romance-based vocabulary and pronunciation of words through relocation and hierarchical diffusion. French, the language of the Normans, became the language of the British aristocracy for 300 years, while most people continued to speak English. Latin—the language of the church and law—contributed words to English as well. After years of conflict with France, French fell out of favor and English—with its combination of Celtic, Germanic, French, and Latin influences—became the official language of the court and law in the 14th century. Through hierarchical diffusion, this new form of English effectively became the official language of the English realm.

Beginning in the 17th century, English spread by relocation diffusion to a new continent with the establishment of colonies in North America and the Caribbean. In the 18th and 19th centuries, the British expanded their worldwide empire to Australia and New Zealand, South and Southeast Asia, and parts of Africa, leading to hierarchical and relocation diffusion to these regions. In addition, English-speaking Americans and Canadians carried the language from the Atlantic coasts of North America to more central areas and the Pacific coast. American traders, settlers, missionaries, and government officials also spread English to Hawaii and the Philippines.

Austronesian Language Family Relocation diffusion and expansion diffusion have both contributed to the unique distribution pattern of the Austronesian language family. The people of Madagascar, an island off the eastern coast of Africa, don't speak languages similar to people living in Africa. Rather, they speak languages belonging to the same language family as the languages of those living on the distant Indonesian islands of Southeast

Asia. But why? Evidence points to Indonesian sailors crossing the Indian Ocean and colonizing Madagascar in the first century c.e. Africans didn't sail to Madagascar until centuries later. The Austronesian languages had been firmly established on the island by the time Africans arrived, and that language family still prevails in Madagascar today.

Impacts of Language Diffusion

Learning Objectives
LO 8.6 Explain how political and cultural forces affect the diffusion of languages.
LO 8.7 Explain how language diffusion can result in convergence and divergence.

The spread of language impacts cultures in different ways. Earlier in this unit, you read about the interaction of cultures and cultural convergence or divergence, which causes the cultures to become more similar or dissimilar. One culture either adopts another's language, the languages blend to form a new language, or the languages become isolated from each other and are preserved.

Sometimes, as a result of diffusion, cultures borrow from another language. Such words as *kindergarten*, *coffee*, and *karate* are now common English words that originated in the German, Turkish, and Japanese languages, respectively. Today, the existence of the internet, plus widespread access to movies in English and to American, British, and Australian TV shows, affects worldwide language use, causing language and thus cultural convergence through these lines of communication.

Convergence is evident in the development of blended languages like "Spanglish," a form of expression spoken by some Hispanics living in the United States that combines English and Spanish grammar and vocabulary. For example, one might ask *"Are you ready?"* in English, or *"¿Estás listo?"* in Spanish. The Spanglish version of this question is a combination of both languages: *"¿Estás ready?"* Sometimes English words are transformed into Spanglish, taking on spelling, construction, or pronunciation from Spanish. The English verb "to click," as in clicking on a website, would become in Spanglish *"clickear."* The *-ear* at the end of the word is a common verb ending in Spanish. Many Spanglish words come from business or technology, reflecting the dominance of English in these fields.

Language divergence is also a result of the formation of group boundaries, or barriers that separate people into groups. Causes for these boundaries include topography and cultural distinctions. Both are at play in the world's most linguistically diverse country, Papua New Guinea, which is less than one-twentieth the size of the United States but with more than 800 languages. The mountainous terrain and multitude of islands result in populations occupying small, isolated pockets of land and promote the development of distinct languages. Strong in-group identification leads these populations to protect their linguistic autonomy and reject the language of other communities when there is contact.

Another example includes the people of North and South Korea, who come from the same ethnic group and both speak Korean but have been separated politically since 1945. During this relatively short period of isolation, significant language differences have developed between the two countries. Since South Korea is open to trade and political contact, many foreign words have been incorporated into South Korea's version of the Korean language. The communist government in North Korea has resisted the introduction of foreign interactions—and consequently, foreign words. Those that did enter the language were often Russian, reflecting the influence of the long-time North Korean political ally. The Korean word for *friend*, *tongmu*, was once used widely across the Korean peninsula. However, when the North Koreans began using *tongmu* to mean "a fellow socialist citizen" just like the Russian word *comrade*, the South Koreans stopped using the word *tongmu*, replacing it with the word *chingu*.

The people of some countries or cultures have assembled formal bodies and written decrees aimed at preserving their language in the face of language diffusion. The Arab Academy of Damascus, founded in 1919, has the goal of setting standards for the Arabic language, creating Arabic equivalents for scientific terms and promoting the growth of Arabic content on the internet. The French—who established their language academy in 1635—resist the promotion of English, establishing several official organizations with the purpose of protecting the French language. They even recommend French equivalents for English terms used commonly, such as *la fin de semaine* instead of "le weekend." In Israel, Hebrew was established as the official language when the state formed in 1948 to differentiate from other languages spoken in the region.

The Influence of Power You have read that power, whether economic, political, or military, impacts cultural change. Power influences which languages become dominant and subordinate. Dominant languages tend to be distributed widely or spoken by a large number of people, and subordinate languages are in danger of disappearing.

Linguists disagree over whether the subordinate or dominant group, or a combination of the two, is the primary force behind the development of creole languages. Recall that a creole language is a blended language that arises to facilitate communication among the dominant and subordinate groups with diverse languages in a colonial system. Some scholars believe population size and duration of interaction between the cultures influence how languages develop. In the case of Haitian Creole, which has a strong African influence, Africans outnumbered the French colonists in Haiti, and the French controlled the island for less than 200 years. Over time, the influence one language has over the other lessens, and creole languages develop their own vocabularies and more complex grammar.

While commerce and trade often necessitate the development of a lingua franca, a common language between speakers of different languages also develops for diplomacy. A lingua franca can help governments and organizations build international relationships. Speakers of local languages sometimes adopt the imperial language, either through contact with native speakers of that language or due to the prestige the language carries. Arabic became a lingua franca as the Islamic Empire spread throughout Southwest Asia and North Africa, not only to facilitate commerce and diplomacy but as part of the diffusion of the Islamic religion.

Today, English is the dominant language used in the fields of science, technology, and diplomacy. The economic power of the United States influences the use of English in global commerce as well. English is also the lingua franca of numerous university programs around the world. Non-native English speakers learn English to increase their career opportunities and improve their ability to attend an international university. English is the indispensable language used worldwide between aircraft pilots and air traffic controllers and between ship captains and harbor authorities.

Power also influences toponyms, or place names. As regimes change throughout the world, leaders sometimes change the names of the cities or streets under their control. The Russian city of St. Petersburg was founded

The city of St. Petersburg is an example of how changes in political power impact toponyms. The name of the city changed each time power in the region shifted, either to pay homage to a leader or to restore a name some believe is more appropriate.

in the 18th century by Czar Peter the Great. After the Russian Revolution, which put communists in control of what became the Soviet Union in 1918, the city's name was changed to Petrograd—a Russian form of the name. That name remained until the death of the communist leader Vladimir Lenin, when the city was renamed as Leningrad. When the Soviet Union collapsed in 1991, the city was once again named St. Petersburg.

Many cities in India have been renamed over the centuries, each name revealing something about the culture of those who chose the name. In 2018, the government of the state of Uttar Pradesh changed the 500-year-old name of the city Allahabad, which means "place where Allah lives," back to its original name of Prayagraj, which means "place of sacrifice." These toponym changes reflect the religious beliefs of those controlling the city; the Muslim emperor Akbar gave the city its Muslim name in 1580, and Hindu political leaders made the change in 2018.

Indigenous peoples around the world are working to restore both their languages and place names, after centuries of colonial policies and laws often sought to eradicate them. The surveyors, cartographers, and political leaders of colonial powers generally assigned toponyms that reflected their own languages and cultures—place names from their home countries, religious figures, royalty, even naming places after themselves or family members, thus erasing the deeply meaningful and often spiritual naming practices of indigenous cultures. Today, only a small fraction of these original place names is found on "official" maps, documents, and signs. Efforts by Indigenous groups to rename places is part of the difficult process of reclaiming their language, heritage, knowledge, and lands—to inscribe, once more, deeper meanings in the cultural landscape. For example, the iconic Australian landform known as Uluru to its original inhabitants, the Anangu people, was renamed Ayers Rock by a British explorer in 1873 after a South Australian leader. Efforts by Indigenous people and others led first to a dual name in 1993—Ayers Rock/Uluru—and then to reverse the name order to Uluru/Ayers Rock in 2002. However, Parks Australia, the agency that manages national parks and conservation areas in the country, refers only to the original indigenous name in its Uluru-Kata Tjuta National Park.

Endangered Languages Language diffusion sometimes results in endangered languages—the languages of small groups of people, often of indigenous cultures, that become in danger of disappearing due to declining populations and outside cultural pressures.

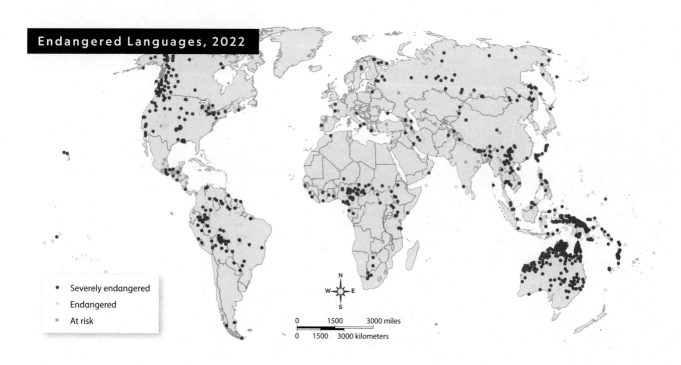

Reading Maps A group of geographers, linguists, academics, and other experts have joined together to form the Endangered Languages Project, which tracks languages in danger of disappearing and encourages the sharing of information on how to preserve them. This map from the Endangered Languages Project identifies three categories of endangered languages. ▌What do you notice about the pattern of languages that are under threat across the world?

Experts identify language hotspots that have a high degree of linguistic diversity—determined by the number of language families present in an area—and several levels of endangerment, based on the number and age of native speakers. Some languages succumb to these pressures. According to the United Nations Educational, Scientific and Cultural Organization (UNESCO), about 230 languages became extinct from 1950 to 2010, and a language dies, on average, every two weeks.

Social scientists, linguists, and native speakers are making efforts to preserve endangered languages, and some with considerable success. In 1983, only about 50 children spoke Hawaiian. An effort to teach the language in the state's schools has made it far more widespread. In Wales, part of the United Kingdom, champions of the Welsh language won the establishment of a Welsh-only public television channel that broadcasts Welsh-language programming. In Japan, members of the Ainu are taking steps to preserve their ancient language, which likely originated more than 8,000 years ago. They include free Ainu courses at various sites on the island of Hokkaido; publication of books and stories from the Ainu oral tradition in both Ainu and Japanese; and creation of an Ainu radio station. Groups like the Endangered Language Fund, Cultural Survival, and UNESCO's Languages in Danger program seek support to preserve endangered languages.

Geographic Thinking

1. Identify an example of hierarchical diffusion affecting language.
2. Describe how the diffusion of language creates unity.
3. Explain how changes in regime can affect toponyms.

Case Study

French or English in Quebec?

In 2013, protesters demonstrated against a language bill in Montreal. The bill strengthening the use of French in Quebec failed, facing opposition by those who favored bilingualism. However, new legislation in 2021 made French the official language of Quebec.

The Issue The French-speaking majority in the Canadian province of Quebec wants to preserve French language and culture in the predominantly English-speaking country.

Learning Objective
LO 8.6 Explain how political and cultural forces affect the diffusion of languages.

By the Numbers

7.9 million
French speakers in Canada

6.9 million
French speakers in Quebec

26.2 million
English speakers in Canada

1.1 million
English speakers in Quebec

6.9 million
bilingual speakers in Canada

3.6 million
bilingual speakers in Quebec

Source: Statistics Canada

Canada Is a Bilingual Country with two official languages: French and English. In the past, cultural and political conflict between French Canadians and English Canadians has led Quebec's majority French-speaking citizens to call for the separation of the province from the rest of Canada. Quebec is the second most-populous province in the country. A large portion of Francophones (people who speak French) in Quebec believe the Canadian government will never respect the French language and culture that is so prominent there. To address their concerns, Bill 101, the Charter of the French Language was passed in 1977. This language legislation of the Parti Quebecois government made French the official language of the Government and the courts of Quebec. French became the "normal, everyday language of work, instruction, communication, commerce and business."

However, still unsettled is the tension over French dominance and bilingualism, driven by demographic changes. Quebec has grown more multicultural, particularly in and around Montreal, which holds two-fifths of Quebec's population. As more French speakers moved from the city to its suburbs, the provincial government passed a law unifying the city and its suburbs into one municipal unit to maintain Montreal as a French-speaking city. The provincial government answered its Francophone majority's continued concerns by passing a nonbinding resolution that called for shop employees to replace the bilingual, tourist-friendly Montreal greeting of "bonjour hi"—with just "bonjour."

Most recently, Bill 96, or officially, "An Act respecting French, the official and common language of Québec," is the first major revision of the 1977 law. It is designed, according to the Coalition Avenir Québec government, to combat a perceived decline in the use of French in the workplace, educational institutions, and in the overall culture of Quebec. The controversial bill was adopted in principle in November 2021, confirming French as the official language of Quebec. Among its provisions ensuring fundamental French-language rights in business and commerce is a requirement to protect the right of employees to work in French and for customers to be informed and served in French. English speakers have chafed at this new law, irritated that the government went to such lengths to diminish bilingualism, which they perceive as a strength in Quebec. The law also shows that efforts to negotiate language and cultural differences remain a struggle. ∎

Geographic Thinking

Explain how non-French-speaking immigration might affect the goals of Quebec's French-speaking citizens.

National Geographic Explorer **Maria Fadiman**
Preserving Indigenous Knowledge

Dr. Maria Fadiman, whose TEDx Talks demonstrate her gift for lively storytelling, is a professor in the Department of Geosciences at Florida Atlantic University.

Learning Objective
LO 8.6 Explain how political and cultural forces affect the diffusion of languages.

Kiha, a Bantu language spoken in western Tanzania, preserves the cultural memory of those who speak it. As an ethnobotanist interested in the relationships between people and plants, Maria Fadiman has helped Kiha speakers record their own ancient knowledge about the plants in their environment.

The village of Bubango in Tanzania sits just outside Gombe Stream National Park, where Jane Goodall conducted her groundbreaking research on chimpanzees. Like most forest environments, the area is threatened by deforestation and habitat loss, even though the people of Bubango avoid cutting too many trees to plant their crops. Fadiman and Grace Gobbo, a fellow ethnobotanist, knew that the villagers were interested in a project to record their traditional knowledge about plants in the forest. They also believed that preserving this knowledge would strengthen the people's sense of connection with their environment, giving them an even greater incentive to protect it.

In the fall of 2014, Fadiman and Gobbo began by interviewing elders in the community. They carefully recorded information about how various plants could be used—for food, medicine, in religious rituals, to build houses, weave baskets, and make musical instruments. The activity soon piqued the curiosity of the village's children, who the elders worried were losing interest in their traditional ecological knowledge. The researchers invited the children to create illustrations of the plants, labeled with their Kiha names.

The contributions of both young and old members of the community were collected in a trilingual book, with text in Kiha, Kiswahili, and finally English. The people of Bubango loved the book and appreciated seeing their own language in the most prominent position on each page. Fadiman hopes that the book will also promote the cause of conservation: "People value what they use, and use what they know. If they are able to utilize plants in local forests, people are more likely to see the value of habitat protection." Follow-up studies will determine whether the book has had a lasting impact on the community's relationship to their natural environment.

Geographic Thinking

How can indigenous languages be used to help protect vulnerable ecosystems?

8.3 Patterns of Religion

Interaction among people over the centuries has contributed to the creation, evolution, or splintering of thousands of religions that attract different groups of people. The distribution of religions today varies, based on their different hearths, teachings, values, and patterns of diffusion.

Religion and Culture

Learning Objective
LO 8.8 Explain how religious beliefs influence individuals and culture.

Religions deeply impact individuals and cultures. They shape cultural beliefs and traditions and serve as a reflection of how people think about the world. As an integral part of culture, religious beliefs can cause political divisions within a country and can impact the environment. According to the Dalai Lama, a Buddhist leader, the Buddhist view is that humans have a responsibility to protect nature. Some Christians hold the view that God gave humans authority over all creation, to use it as they wish. Other Christians argue that humans have a moral responsibility to exercise stewardship over the entire natural world, which echoes Buddhist values.

The beliefs and traditions shaped by different faiths have played an important role throughout the history of the world because religions are so closely tied to identity and ethnicity. This has impacted how religions have spread and where certain religions are practiced today.

Patterns of Distribution

Learning Objectives
LO 8.9 Describe the contemporary geographic distributions of the religions with the largest number of adherents.
LO 8.10 Explain the factors that influence the distribution of religions worldwide.

To understand the geographic distribution of religions worldwide, geographers consider the places of origin, extent and methods of diffusion, and contemporary cultural processes of different religions. Some religious groups are concentrated within a region, while others are dispersed with a presence on several continents. Practices and belief systems attract different people and impact a religion's distribution. Historically, for example, many Hindus perceived to be of a lower caste, or social status, were attracted to Islam because its teachings value all humans equally. Some Europeans became Protestants because of disagreements with Roman Catholic principles and to break from oppression.

Like the previous world languages map, the "Estimated Majority Religions" map is a snapshot in time, documenting current patterns of distribution—not those of earlier centuries. Christianity is the dominant religion in most of Europe, the Americas, central and southern Africa, and Australia and New Zealand. With origins in Southwest Asia, the spread of Christianity has occurred as a result of relocation diffusion (migration) and expansion diffusion (conquest and missionary work). Islam, which also originated in Southwest Asia, is the main religion in Central and Southwest Asia, North Africa, as well as Indonesia, Bangladesh, Malaysia, plus portions of southeastern Europe and India. Trade, conquest, missionaries, and leaders adopting Islam as the religion of their people contributed to its current distribution. Hinduism is distributed near its hearth in South Asia and hasn't spread, in great numbers, far beyond the borders of India. While Buddhism has spread across countries, fewer adherents remain in and around its hearth of India, now dominated by

Voodoo is an ancient religion with roots in the West African nation of Benin, which officially acknowledged it as a religion in 1989. It is practiced by about 60 million people worldwide. Practitioners of Voodoo gather in Ouidah, Benin, for an annual religious festival, which includes elaborate ceremonial costumes.

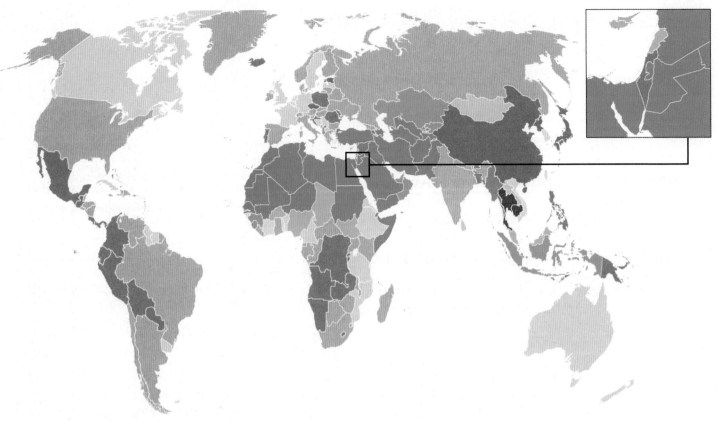

Estimated Majority Religions, 2020

The coloring of each country corresponds to its majority religion, and the darker the shade, the greater the prevalence of that religion. No folk religion is the majority in any country. Guinea Bissau, Ivory Coast, Nigeria, Singapore, South Korea, Taiwan, Togo, Vietnam, and Macau (autonomous region) have no clear majority religion. Explain how the map and chart work together to provide a deeper understanding of the distribution of religion in the world.

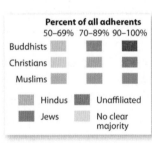

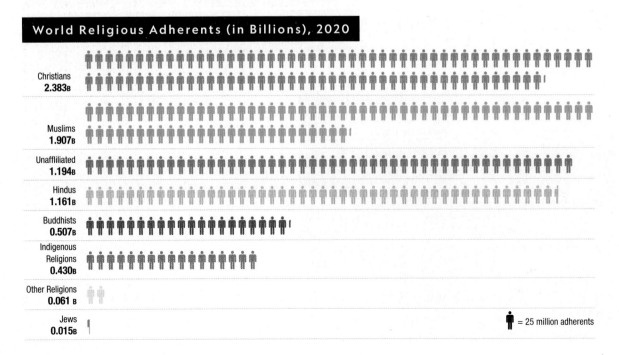

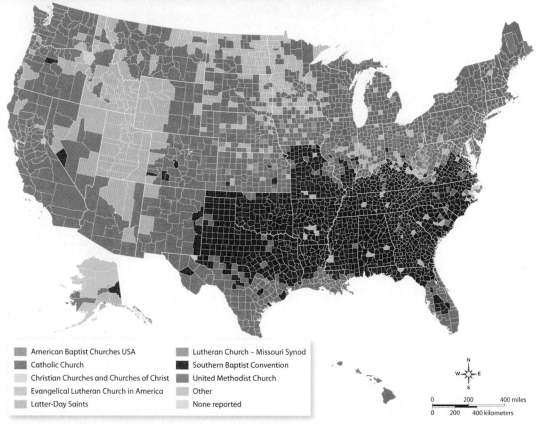

Majority Religious Adherents by U.S. County, 2010

Legend:
- American Baptist Churches USA
- Catholic Church
- Christian Churches and Churches of Christ
- Evangelical Lutheran Church in America
- Latter-Day Saints
- Lutheran Church – Missouri Synod
- Southern Baptist Convention
- United Methodist Church
- Other
- None reported

Reading Maps Every ten years, a study is conducted of religious adherents in the United States, but COVID delayed the 2020 count. You might notice that some major religions, such as Judaism, aren't on the map. Judaism has many adherents in large metropolitan areas such as Los Angeles, Miami, and New York City, but it is not the majority religion in any U.S. county. ▌ Does the map support the idea that historical settlement patterns and religious patterns are aligned in this country? Explain your thinking.

Hinduism. Indigenous religions are strong in many areas of rural Africa, the interior Amazon Basin, and northern North America, locations with large Indigenous populations. Though they display significant differences, these traditional religions are sometimes grouped under the name *animism*, which denotes a belief that humans share the world with numerous spirits whose actions can help or hurt humans.

The global scale of the "Estimated Majority Religions" map makes interpreting the locations of where different religions are practiced a bit misleading. The map displays the religion that is most widely practiced in a country or region, not accounting for the relative mix of faiths in a given area. For example, the large majority of religious adherents in the Philippines is Christian, yet followers of Islam dominate in southern parts of the island country. And the prevalence of Islam found in North Africa decreases when moving to the south, but it is unclear if the remaining population in the Muslim-majority areas is Christian, follows traditional religions, or practices some other faith. Additionally, some individuals embrace more than one belief system. Recall that some Roman Catholics in Latin America and the United States practice rituals that belong to traditional African-based beliefs called Santería. In some countries, including China, the largest number of people are not affiliated with any religion.

Regional Patterns Patterns revealing greater religious differentiation are evident with different scale maps. Consider the "Majority Religious Adherents by U.S. County" map, depicting Roman Catholicism as the chief religion across much of the country, though Southern Baptists dominate the South. Historical settlement patterns help explain present-day locations of religious groups. For example, in the 1850s, many Germans and Scandinavians who identified as Lutheran settled in Pennsylvania and eastern Ohio and moved throughout the upper Midwest, attracted by employment opportunities and available farmland. Their settlement is still reflected in the distribution of Lutheranism today. The Church of Jesus Christ of Latter-day Saints, which originated in New York State, dominates in Utah and is significantly represented throughout the West. Early followers of the faith were not welcome in certain established parts of the United States, so they migrated west to largely unsettled territories where they could practice their religion freely. Members of this church have remained in this region for more than 150 years.

Geographic Thinking

1. Compare the spatial distribution of Christianity, Islam, and Buddhism.

2. Identify where Christianity, Islam, and Hinduism originated.

Spatial Patterns of Language and Religion

Case Study

Shared Sacred Sites

The Issue Disputed land in India sacred to Hindus and Muslims has caused tensions between both religious groups for decades.

Learning Objective
LO 8.10 Explain the factors that influence the distribution of religions worldwide.

India is home not only to 94 percent of the world's Hindus but also to the second-largest Muslim population living in a single country (after Indonesia). These Hindu and Muslim children are praying for peace at their school in the city of Ahmedabad.

By the Numbers

228.9 million
people living in Uttar Pradesh (2022)

79.7%
Hindu

19.3%
Muslim

1%
Other

Source: World Population Review, Census Organization of India

The Birth and Spread of Religions spanning centuries has resulted in examples where one geographic location is significant to more than one faith. Angkor Wat, located in Cambodia, was first a Hindu temple and later turned into a Buddhist temple. Jerusalem's Old City remains a sacred place to three major religions—Judaism, Christianity, and Islam. Hindus and Muslims both consider a three-acre plot of land in Ayodhya, a city in the state of Uttar Pradesh, in India, sacred to both of their religions.

Hindus believe the site was the birthplace of their revered god Rama and that a Hindu temple was once located there. Some believe the temple was demolished by the Mughal empire, a Muslim empire that ruled northern India from the early 16th century to the mid-18th century, to build an Islamic mosque called Babri Masjid. The mosque stood on the disputed site until 1992 when Hindu extremists destroyed it, sparking nationwide riots.

Ownership of the land was debated in Indian courts for decades. A group that oversees Sunni Islamic properties argued that there is no proof of the Hindu temple ever existing. Hindus presented the courts with ancient texts that purportedly describe Ayodhya as a holy Hindu city. Finally, in 2019, the Supreme Court of India unanimously ruled in favor of the Hindus, giving them permission to construct a temple at the site. The court also ruled that Muslims will be given land to build a mosque at a prominent alternative site in Ayodhya.

India's Muslims were divided about the ruling. Some would like to contest the decision, viewing it as a statement about Muslims' place in India as second-class citizens. Some have called for action to contest the decision, fearing it has emboldened Hindu extremists to target and attempt to halt the construction of mosques in the future. Others are accepting of the fact that five acres will be provided to build a new mosque in another location. They believe accepting the decision is the first step toward easing tensions between Hindus and Muslims regarding the dispute. The prime minister of India, Narendra Modi, who was re-elected in 2019, has further inflamed sentiment against the country's 14 percent Muslim minority by enacting anti-democratic policies to make India more Hindu and less welcoming to its minorities.

Geographic Thinking

Describe how historical events have impacted the interactions between some Hindus and Muslims in Uttar Pradesh.

8.4 Universalizing and Ethnic Religions

Geographers rely on their own knowledge of history and the scholarship of historians when they examine how religion affects patterns of culture. An in-depth account of religious history is beyond the scope of this textbook, but knowing a few key milestones for each of the major world religions will help you put events and concepts in context.

Universalizing Religions

Learning Objectives
LO 8.11 Identify the characteristics of a universalizing religion.
LO 8.12 Explain how universalizing religions spread through expansion and relocation diffusion.

Universalizing religions attempt to appeal to a wide variety of people and are open to membership by all, regardless of a person's location, language, or ethnicity. Examples of universalizing religions are Christianity, Islam, and Buddhism, as well as faiths with fewer adherents including Sikhism, Baha'i, and the Church of Jesus Christ of Latter-day Saints.

By nature, universalizing religions are open to diffusion, and two in particular—Christianity and Islam—have become the largest religions in the world today. At different periods in history, different processes of diffusion have spread religions. Often more than one type of diffusion is reflected as different elements of human interaction occur in different places. Universalizing religions have commonly spread through relocation and expansion diffusion.

Christianity Started in what is now the West Bank and Israel around the beginning of the common era, **Christianity** has spread to all continents. The religion is based on the teachings of Jesus, a man believed by the faithful to be God's son. Jesus taught his followers that they should love and care for their fellow humans.

The Christian religion began to spread through relocation and expansion diffusion as a small group of followers called disciples carried the Christian message throughout the Roman Empire and Mediterranean region. In the 300s, the emperor Theodosius declared Christianity the official religion of the Roman Empire, and the religion spread hierarchically from Rome to other cities and regions of the Empire.

For the next 1,000 years, Christian missionaries traveled to promote the religion in new regions, spreading Christianity from Rome throughout Europe. Church officials made their faith more welcoming by absorbing some of the beliefs and practices of local groups of people. These actions are an example of the acculturation discussed in Chapter 7. For example, pagan Europeans converted readily when Christianity adapted and included their beliefs and practices, such as the Celtic holiday Samhain marking the end of the harvest season. By the 10th century, Christianity had spread throughout most of Europe as far east as Russia, though some areas—such as Sweden in the 12th century—did not become Christian until later. Some Europeans were forced to adhere when their monarchs converted to Christianity, such as occurred in Lithuania. Throughout the Crusades (1096–1204), Christians in Europe fought non-Christians, including Muslims, to conquer Jerusalem and retake Spain. During the wars, the rulers of Spain used a powerful court called the Spanish Inquisition to punish non-Christians who would not convert or who secretly practiced their former religion. Christianity became strong in northern Africa from Roman times and in Ethiopia from the fourth century when, through hierarchical diffusion, the king converted and adopted it as the state religion, also forcing the people to adhere.

In the 11th century, differences over the organization and doctrine (or teachings) within the church led to a split that resulted in the divergence of religious groups and the formation of two distinct Christian branches called Roman Catholicism and the Eastern Orthodox tradition.

The Roman Catholic Church grew in wealth and power, raising concerns of corruption and eventually causing the Catholic Church to fracture in the 1500s, spurring a third Christian branch called Protestantism. European geopolitics affected the Protestant branch, and the development of denominations, or separate church organizations, was shaped not only by differences in doctrine and ritual but also nationalism, or a strong sense of national identity. A recent estimate placed the total number of Christian denominations across the globe at more than 30,000. Independent nation-based organizations such as the Greek Orthodox Church, the Bulgarian Orthodox Church, and the Russian Orthodox Church are all examples of denominations of the Eastern Orthodox branch of Christianity.

The next significant period of Christian diffusion began in the 15th century and brought Christianity to areas beyond Europe through imperialism and colonialism. Emigrants left Europe to settle in new colonies in the Americas, and later in Australia, New Zealand, and southern Africa. In addition to relocation, Christianity continued to spread through expansion diffusion. Power dynamics and persuasion by missionaries led to the conversion of many Native Americans and enslaved Africans to Christianity while incorporating their previous spiritual beliefs. Missionaries also brought Christianity to Asia, and it took strong hold in the Philippines in the 16th century. Christianity gained less of a stronghold in places like China and Japan during this age of European expansion. Rulers and their advisors worried that the spread of Christianity would threaten traditional culture and their own power, and the Japanese moved to close all churches and to bar Europeans from the islands in the 17th century. Only in the 19th century did renewed missionary activity lead to larger Christian populations in Asia.

Similarly, hierarchical diffusion in Africa impacted areas in immediate contact with European settlers. Settler-colonies were established in Algeria, Tunisia, South Africa, Namibia, Angola, and Mozambique and were among the first African populations to become Christian. The king of Kongo (present-day Angola) adopted Christianity to gain favor with Portuguese traders, but the religion did not take hold there until long afterward. Others converted to Christianity as part of the assimilation to the imperial power structure.

Sometimes convergence came into play, as evidenced by the development of the syncretic faiths like Voodoo and Santería in the Caribbean, which blended traditional West African and native Caribbean beliefs and practices with Christian ones. Today, Christianity is the largest and most widespread religion in the world.

Islam The world's second-largest religion, Islam, originated in the cities of Mecca and Medina on the Arabian Peninsula in the seventh century. Muhammad, who is considered by adherents the last messenger of Allah (God), introduced Islam to the Arab people. It has deep ties to Judaism and Christianity. Like those religions, Islam is a monotheistic religion, meaning it is based on belief in one God. Muslims, followers of Islam, believe that Muhammad is a direct descendant of Abraham, who is also the founder of Judaism. Islam emphasizes moral behavior based on, first, the text of the Quran (the holy book of Islam) and, second, on the traditions of Sunnah, the "well-trodden path." Sunnah is based on teachings of Muhammad and the writings of Muslim scholars. Muslims also believe that God is all-powerful and all-knowing and that the events in one's life are predetermined by God. However, this doesn't exclude humans' free will to make good or bad decisions for which they will be judged at the end of their life.

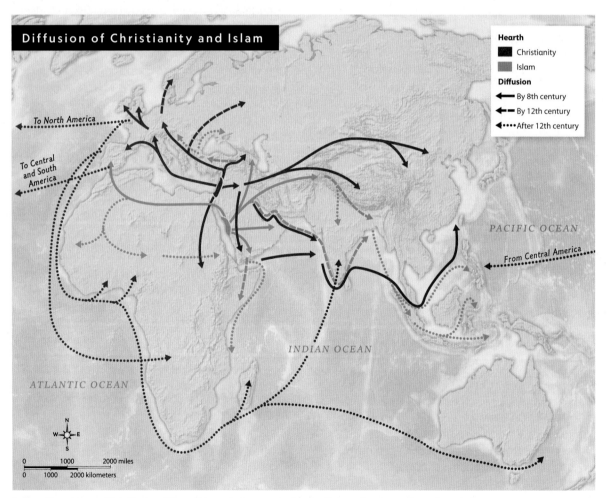

Reading Maps Christianity diffused through Europe during the Roman Empire and continued via the work of missionaries to other regions. Islam spread through both hierarchical and contagious diffusion across Africa and into Asia. ▮ Using map details, describe the similarities among the diffusion of Christianity and Islam.

After the death of Muhammad, Muslims spread their religion through conquest, trade, and missionary work. Arab armies took control of a vast region from the Indus River Valley in the east, across North Africa to the Iberian Peninsula in the west. As Arab soldiers settled in the region and assumed political control, Islam spread throughout this area. Over time, millions of locals converted to Islam in a hierarchical pattern, from the largest urban settlements to more remote rural areas. Muslim traders also helped spread the faith into new areas, including African kingdoms that lay just south of the Sahara and via sea routes as far east as Indonesia, Malaya, and the Philippines. Hierarchical and contagious diffusion influenced these areas as scholars settled in Muslim-dominated cities and taught the principles of the faith and of Islamic law.

Similar to the religious divergence that occurred with Christianity, the early period of Islam saw the religion break into two major branches consisting of the Shiites and Sunnis. Shiism remains a strong but minority branch of Islam to this day. Shiites form majorities in Iran, Iraq, and Bahrain. But the great majority of Muslims—an estimated 87 to 90 percent—are Sunni, the branch that follows the traditional and consensus, or unified, view of what it means to be a Muslim. While there are a small number of other branches of Islam, the faith has far fewer distinct denominations than are found in Christianity. Political fragmentation broke the Muslim empire into several competing states, but the Muslim world continued to expand. Trade, conquest, and the appeal of the faith brought Islam deeper into South Asia and, by sea, to the eastern coasts of Africa and to Southeast Asia. Today, Indonesia has the largest number of Muslims in the world—more than 230 million, far more than the Arabian Peninsula.

In 2022 an estimated 3.5 million Muslims lived in the United States, many of them as a result of relocation as immigrants or refugees. Islam in America dates back much further, however. Some 15 percent of the Africans brought to the United States as enslaved people are thought to have been Muslims, though many found it difficult to continue practicing their faith. Beginning in the late 19th century, immigrants from Southwest Asia began arriving in the United States in substantial numbers, many settling in the Midwest. Refugees from war-torn Muslim areas such as Afghanistan, Iraq, and Somalia have led to an influx of Muslims in recent decades. Most Muslim immigrants, though, have come from Iran or South Asia (India, Pakistan, and Bangladesh). They have come seeking economic opportunity, looking for political freedom, or hoping to flee religious persecution.

Buddhism The oldest universalizing religion is **Buddhism**, which arose from a hearth in northeastern India between the mid-sixth and mid-fourth centuries B.C.E. Buddhism is based on the teachings of Siddhartha Gautama, known as Buddha, or "the Enlightened One." At the time Buddhism arose, the people of India were seeking spiritual alternatives to the practices of early Hinduism, an ethnic religion also practiced in India. The Buddha traveled the region and preached, gaining a set of followers.

While Buddhism now has fewer adherents than Christianity or Islam, it has a large number of followers found around the world. The highest concentration of Buddhists live in Japan, China, Sri Lanka, and South Korea and countries in Southeast Asia. At first, Buddhism's growth was through contagious diffusion, as the Buddha's direct teaching inspired many followers. After his death, these followers recorded his teachings and spread his views. Some traveled across Asia; others established monasteries, which attracted visitors who wished to learn about the Buddha's message.

After becoming a Buddhist, Ashoka, the emperor who ruled much of South Asia from 268 to 232 B.C.E., promoted the religion both inside and beyond his empire. He sent missionaries to carry the teachings of Buddha as far west as Greece and Egypt and south to Sri Lanka. As with Islam, traders carried Buddhism to new areas; by the first century of the common era, the religion was established in Central Asia and had reached China.

Changes in the ruling families throughout Asia had a negative effect on Buddhism, as fewer people practiced the religion. By the 12th century, Buddhism had become a minority religion in India, and it lost its dominant position in Korea in the 14th century. Buddhism nevertheless has remained a presence throughout South, East, and Southeast Asia to the present day.

Buddhism gained attention in the Western world beginning in the 19th century and grew in popularity in the 20th. Some Buddhist groups adapted their teachings to make them more accessible or appealing to westerners. Growth of the religion was also aided by the moral and spiritual example of such figures as the Dalai Lama, a Buddhist leader who has lived in exile in India since China's takeover of Tibet in 1959. Buddhism grew in the United States partly as a result of relocation diffusion through immigration, but only about a third of the estimated 3 million American Buddhists are of Asian heritage, and not all of them are immigrants. Most American Buddhists are converts, not born into the faith.

Sikhism The newest universalizing religion is **Sikhism**, founded by Guru Nanak, who lived from 1469 to 1539. Its hearth is the Punjab region of northwestern India. Sikhs, followers of Sikhism, identify ten gurus, or religious teachers, who guided the community in its first century and a half.

Most of the 25 million Sikhs today live in the Punjab region that borders Pakistan. Sikhs believe there is one God and that devotion to that God will bring liberation from the cycle of death and rebirth. They believe that all humankind are one people and that men and women are equal.

While gurus preached throughout India, most Sikhs remained concentrated in Punjab, which became the center of Sikh culture and home to the Sikhs' most sacred site, the Golden Temple. When the British took control of much of South Asia, many Sikhs served in the colonial army and were transferred to other British colonies, such as Hong Kong and Malaya, sparking the diffusion of Sikhism. Later, Sikhs relocated to other English-speaking areas including the United Kingdom,

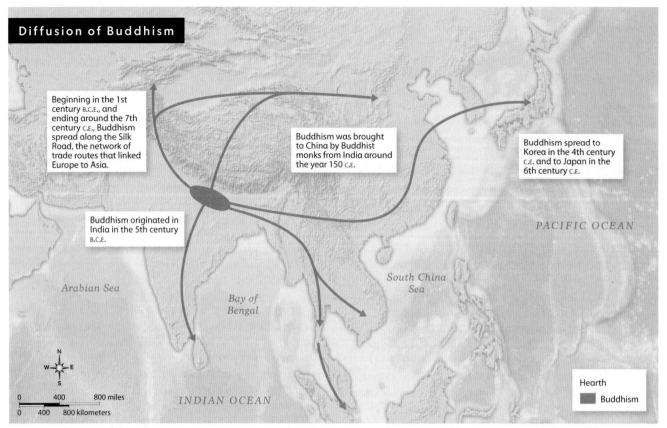

Reading Maps Buddhism's initial diffusion was mainly to East and Southeast Asia, but it eventually spread all over the world, including to the United States. ❙ How was the spread of Buddhism related to trade?

In the fall of 2018, Sikhs began a year-long celebration of the 550th birthday of Guru Nanak, the founder of Sikhism. These Sikhs are participating in a religious procession outside the Golden Temple in Amritsar, India. The holy book of the Sikh religion is inside the golden palanquin they are carrying.

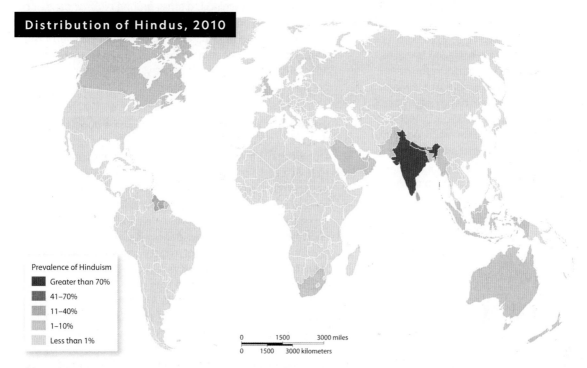

Distribution of Hindus, 2010

Prevalence of Hinduism
- Greater than 70%
- 41–70%
- 11–40%
- 1–10%
- Less than 1%

Reading Maps The world's Hindus are largely concentrated in India and Nepal. Guyana, in the northern part of South America, has the highest percentage of Hindus in the Western Hemisphere.
▌ How does the map support the idea that the diffusion of Hinduism was related, at least in part, to the expansion of the British Empire?

Australia, parts of Africa, and the United States. Many Sikhs in South Asia were frustrated when the British left the region and partitioned British India into Muslim and Hindu states. They had hoped for an independent Sikh state in Punjab, but it was placed within India. Many Sikhs in Muslim Pakistan moved to India at the time of the partition in 1947.

Today, about 500,000 Sikhs live in the United States, with the majority living on the east and west coasts. Many American gurdwaras, where Sikhs gather and worship, have Punjabi language programs to preserve the language and the Sikh faith.

Geographic Thinking

1. Explain how Christianity spread to Africa.
2. Describe how Buddhism exemplifies both contagious diffusion and relocation diffusion.
3. Explain why many Sikhs migrated from Pakistan to India in 1947.

Ethnic Religions

Learning Objectives
LO 8.13 Describe the characteristics of ethnic religions.
LO 8.14 Explain how ethnic religions spread through relocation diffusion.

Ethnic religions are closely tied with a particular ethnic group generally in a particular region. Examples include Hinduism and Judaism, as well as Shinto, found in Japan; Confucianism and Taoism, which originated in China; and the indigenous religions of Africa, North and South America, and Oceania. The world's two major ethnic religions—Hinduism and Judaism—are ancient religions. The diffusion of these ethnic religions has been limited, but still substantial, and is largely due to relocation diffusion, as neither religion has actively sought to recruit new believers.

Hinduism The earliest evidence of the origin of **Hinduism** dates from about 1500 B.C.E. in South Asia where, today, it is practiced by some 80 percent of India's population, approximately 1 billion people. Most Hindus live in India and Nepal. Hinduism is specifically tied to India and has left its mark on various natural and human features of that South Asian region, including the Ganges River, which is considered sacred to Hindus.

Hinduism is not a single set of beliefs or practices, but in general, Hindus believe in one eternal spirit called Brahman. Hindu deities, or gods and goddesses, are different expressions of Brahman. Deities or spirits from outside Hinduism were sometimes incorporated into the Hindu faith, which facilitated the diffusion of the religion as it appealed to more people familiar with the other gods. Hindus also believe in reincarnation, or the idea that souls are constantly being reborn. Karma, the negative or positive effects of one's actions, determines if the soul moves to a higher or lower state of being.

Hinduism developed over centuries and grew out of the beliefs and practices of the Aryans, who lived around 1500 B.C.E. The Aryans belonged to the Indo-European people who had

Spatial Patterns of Language and Religion

populated central Asia. They developed a social hierarchy, known as the caste system, that was hereditary. People could not change the social class into which they were born. For centuries, some cultural practices were shaped by caste, but for decades many have been re-examining the historic interpretation of caste and its controversial role in shaping Indian society. The spread of Hinduism has mostly remained limited to South Asia. Many Indians are regionally isolated and only experience the type of Hinduism practiced in their area, creating variations in Hindu thought and practice. The popularity of certain deities, temple design, and the approaches to celebrations like Holi vary among regions. Hinduism absorbed many of these assorted local beliefs, practices, and deities—making it a highly diverse religion.

Over time, the travel of priests, traders, and teachers carried Hinduism from South Asia to Southeast Asia through relocation and expansion diffusion. Hinduism reached Indonesia first on the island of Bali in the seventh century but took firmer hold in the 16th century when Islam became the chief religion of nearby Sumatra, and Hindu believers relocated to Bali to continue practicing their faith. Bali remains strongly Hindu to this day. The Khmer kingdom of Cambodia, which flourished from the 9th to the 12th centuries, adopted Hinduism and built the magnificent temple complex at Angkor Wat to celebrate the god Vishnu, although the complex was later rededicated to Buddhism.

After the British began to assert political power in India, Hinduism reached parts of the Americas through relocation diffusion. As part of the British Empire, Trinidad and Tobago saw the arrival of indentured servants from India in the 18th century and in what is now Guyana in the mid-19th century.

Hinduism also spread to Africa, including Uganda and South Africa, and to the Fiji Islands in the Pacific in the 19th century because of British imperialist policies inviting Indians to work in these parts of the empire. Teachers who came to the United States in the late 1800s and early 1900s brought Hindu teachings to small groups of Americans and introduced the practice of yoga. More recently, many Hindus have relocated to the United Kingdom and the United States for employment opportunities. By 2010, more than a million Hindus lived in each country, making up the largest communities of Hindus outside Asia.

Judaism Another ancient ethnic religion is **Judaism**. According to the Hebrew Bible, the Jewish prophet Abraham founded the monotheistic religion, which developed among the Hebrew people of Southwest Asia in present-day Israel and Lebanon about 4,000 years ago. While Judaism is currently much smaller than both Christianity and Islam in terms of numbers of adherents—approximately 14.6 million people—its importance and impact on the two religions and society go beyond numbers. In terms of religion, both Christianity and Islam have ties to Judaism. Jewish prophets such as Abraham are honored in Christianity and Islam, and the scriptures of Judaism appear in Christian texts. The city of Jerusalem is also home to important sacred sites for all three religions. Secular society has also been impacted by Judaism. For example, the religion introduced the idea of the seven-day week with one day of Shabbat, or Sabbath, on which no work is to be done.

The Jewish faith largely spread through relocation diffusion. Deteriorating conditions in the region of the Roman Empire inhabited by Jews led them to disperse to other areas of the Empire. They settled across the Mediterranean and farther into Europe. The expansion of Islam in the seventh and eighth centuries put many Jews under Muslim control, but the similarities between Islam and Judaism allowed Jewish communities to flourish in what is now Iraq, Iran, and Spain.

The relatively peaceful existence of Jews was shattered beginning in the 11th century. The Crusades put pressure on Jews who found themselves in lands under Christian control, as Christian kingdoms tried to oust non-Christians from Spain. Jews were forced to convert to Christianity, leave the

Israelis celebrate Jerusalem Day every year to commemorate their victory in the Six-Day War. In June 1967, Israel defeated its Arab neighbors in that conflict, resulting in the reunification of Jerusalem under Israeli control as well as territorial gains that remain in dispute.

kingdom, or be killed. Jews were also expelled from England in 1290, France in 1306, and Spain in 1492. Jews elsewhere in Europe faced persecution as well.

A general period of tolerance for the Jewish faith followed during the 18th century Enlightenment. However, in the late 19th century, Russian authorities allowed destructive attacks on Jewish communities. Hundreds of thousands of Jews from this region emigrated to the United States, Canada, and other parts of the world over decades, seeking an end to persecution and better economic opportunity. Later, in the 1930s and 1940s, the Nazi regime in Germany conquered much of Europe and acted on ethnic hostility toward Jewish people. Some 6 million Jews were killed during this period of extended genocide called the Holocaust.

The harrowing tragedy of the Holocaust gave new impetus to the Zionist movement. This movement had arisen in the late 19th century among European Jews as a reaction against anti-Semitism, or hostility against Jews, by launching a renewed expression of the millennia-old Jewish yearning for a return to the homeland. This movement resulted in the settlement of approximately 90,000 Jews in Palestine by 1914. The dream of a Jewish state seemed to receive a boost by the Balfour Declaration of 1917, a public statement in which the British government promised to support creation of a Jewish state in Palestine. In 1948, Jews living in Palestine proclaimed the creation of the state of Israel. More Jewish people moved to that state, and it became one of three principal centers of Judaism today, the other two being the United States and areas of the former Soviet Union.

Over decades, Judaism divided into different branches. Orthodox Jews strictly adhere to ancient laws, and Reformed Jews are part of a more evolved branch of Judaism allowing for greater diversity while preserving some traditions. Many Jewish people are more **secularized**, or not religious, identifying as Jews through ethnicity and culture and not religious practice. This is particularly the case in the United States, Canada, and Europe.

Unfortunately, anti-Semitism has not disappeared. Jewish communities in Europe, the United States, and elsewhere have faced rising instances of harassment, vandalism, and violence, directed at both adherents and rabbis (Jewish leaders) and at synagogues (Jewish places of worship). In recent years, worshipers have been threatened and physically attacked—even killed in synagogues—in increasing numbers.

Geographic Thinking

4. Explain how the map of Hinduism supports its definition as an ethnic religion.

5. Explain the degree to which diffusion has impacted Judaism's influence in the world.

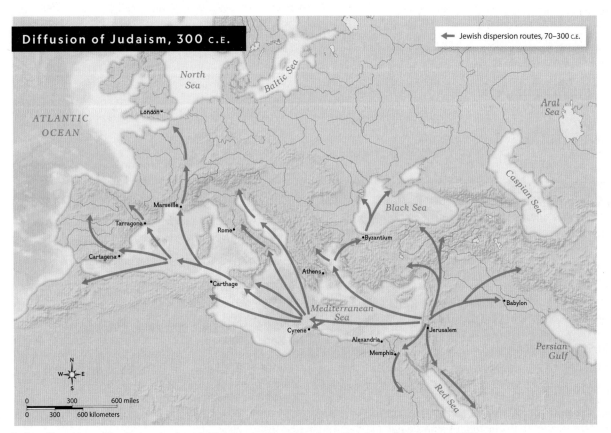

Reading Maps The first major Jewish Diaspora, or dispersion from their homeland, began in 587 B.C.E. when the Babylonians conquered the Kingdom of Judah. Enslavement and exile of Jews by the Romans in subsequent centuries accelerated the Diaspora. ▍ What pattern do you notice on the map relating to the Jewish dispersion routes?

Spatial Patterns of Language and Religion

Chapter 8 Summary & Review

Chapter Summary

Culture and language are linked and shape people's identities. Studying language provides clues to the history and migration of humans.

- To categorize languages, linguists study the similarities and differences among them.
- Languages are categorized according to family, branch, group, and dialect.

The distribution of world languages and world religions reflects their diffusion from a hearth, or place of origin, to other locations.

- Diffusion can take place by relocation (moving to a new area) and expansion (adoption by new individuals or groups).
- Languages show convergence (blending or mixing) and divergence (increasing differentiation).
- Practices and belief systems impact how widely a religion diffused.

Universalizing religions like Christianity, Islam, Buddhism, and Sikhism are open to membership by all. They spread through expansion and relocation diffusion.

- Christianity and Islam both began in Southwest Asia. Buddhism and Sikhism both began in South Asia.
- Christianity has the widest distribution and in recent years has more adherents than any other religion. Christianity has three major branches: Roman Catholicism, Eastern Orthodox tradition, and Protestantism.
- Islam, the second-largest religion, is the dominant faith from North Africa to Southeast Asia. It has split into two major groups: the Shiites and the Sunnis.
- Buddhism spread from South Asia around the world, but the highest concentration of Buddhists is in East and Southeast Asia.
- Sikhism is concentrated in its hearth, the Punjab region of northwestern India.

Ethnic religions like Hinduism and Judaism are closely tied to a particular ethnic group living in a particular place. They are generally practiced near the hearth or spread through relocation diffusion.

- Hinduism, with about one billion adherents, is found mainly near its hearth in South Asia but has also spread to other areas where adherents have relocated.
- Judaism has its largest communities in Israel (its hearth) and in the United States.

Review Questions

Use complete sentences to answer the questions.

1. **Apply Conceptual Vocabulary** Consider the term *isolate*. Write a standard dictionary definition of the term. Then provide a conceptual definition—an explanation of how the term is used in the context of this chapter.

2. Describe how the terms *language families*, *language branches*, and *language groups* are related. Give an example of two branches and two language groups that come from the same family.

3. Define the term *dialect* and provide an example within the United States.

4. Explain how both relocation diffusion and expansion diffusion are related to the history of a particular language.

5. Explain how cognates can provide clues to the origin of a language. As part of your answer, discuss why identifying cognates can be problematic.

6. Give an example of a lingua franca and explain how that language became a lingua franca.

7. Explain how political power can influence toponyms.

8. How do Spanglish and creole languages demonstrate convergence?

9. Explain in what ways Sikhism is a universalizing religion.

10. Describe the different types of diffusion that caused early Christianity to spread throughout the Roman Empire and Mediterranean region.

11. Explain how Judaism, Islam, and Christianity are related.

12. How are denominations related to divergence in the development of a religion?

13. Explain why Buddhism is considered a universalizing religion and Hinduism an ethnic religion.

14. Explain what it means that many people in the United States and Europe are secularized.

Interpret Charts

Study the chart, which shows the rank of the languages spoken in U.S. homes, as determined by the number of speakers. Then answer the following questions.

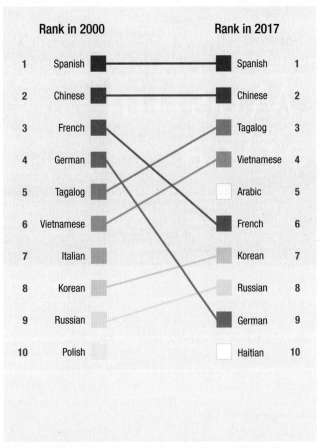

Languages spoken at home in the United States (other than English)

15. **Identify Data & Information** What do the data in the chart show?

16. **Identify Data & Information** Which languages had the number of speakers who spoke the language at home increase between the years 2000 and 2017?

17. **Analyze Data** What factors may contribute to a language's change in rank from the year 2000 to 2017?

18. **Describe Spatial Patterns** What spatial patterns can you infer from the data?

19. **Apply Models & Theories** What geographic model most likely accounts for this data? Explain your thinking.

Spatial Patterns of Language and Religion 241

Being Muslim in America
by Leila Fadel

In Hamtramck, Michigan, children flock to an ice-cream truck to buy cool treats. In this neighborhood most of the children have parents who came from Yemen. The city (population 28,433), which is surrounded by Detroit, has a mostly Muslim population, and Muslims hold the majority on the city council.

Today an estimated 3.45 million Muslims in America are living in a climate of hostility, their faith distorted by violent extremists on one end and an anti-Muslim movement on the other. The rise in animosity was stoked by the fiery anti-Muslim rhetoric from conservative commentators and politicians, including U.S. president Donald Trump.

And yet Muslim communities in America are thriving. Modest clothes for women who cover their hair are being created by Muslims in the United States under labels like Haute Hijab and Austere Attire, and Macy's is now selling fashion for Muslim women. Halal food products, which follow rules of Islamic law as defined in the Quran, are available at Costco and Whole Foods. Mattel has even debuted a Muslim Barbie. The doll, complete with a head scarf, is modeled on Olympic fencer Ibtihaj Muhammad.

Uniquely American Islam Relatively new immigrants and their children make up the largest portion of the faith's adherents. Many prayer leaders across the United States are foreign born. But there's also a growing number of Muslim leaders and scholars speaking in the language of a U.S.-raised generation of Muslims—black, white, brown—to translate a faith often seen as foreign into culturally relevant lingo. Almost half of Muslims in America were born here. The face of the faith in the United States is young: nearly half are millennials and came of age after the terrorist attacks of September 11, 2001.

One of the leaders who speaks to that generation is Usama Canon, a biracial preacher from California. He converted in 1996 and studied with Islamic scholars in the United States and abroad.

"One of the great scholars said that Islam was kind of like a pure and clear water that takes the color of whatever riverbed it flows over," Canon says, talking to me after an all-night lecture at a mosque in Houston. "And so I'm hoping that Muslims in America can kind of color that bedrock in a beautiful way and can contribute to what is the American project in a way that when that water flows over it, it has a uniquely and distinctly American color and flavor but is authentic to itself as a faith tradition."

A Diverse Faith Community That color and flavor are coming from a mosaic of Muslim communities across the country like nothing I saw in over a decade of writing about the Muslim world. Islam is one of the most diverse faiths in the United States, with adherents from some 75 countries bringing distinct ways of worshipping.

I met Bosnian Muslims in Chicago listening to a girls' choir sing hymns praising the Prophet Muhammad. I went to the mosque my father attends, a largely South Asian Shiite congregation, next to a strip mall in a Chicago suburb. And I visited the mostly Arab Muslims in Dearborn, Michigan, where much of my father's family lives. There, people mix Arabic and English in local slang, signs are in both languages, and Middle Eastern cuisine is more common than pizza.

In Houston, I spent time at the country's first Spanish-language mosque, Centro Islamico. It opened its doors in 2016, and on the night I visited, more than 100 people gathered to pray. Latinos are one of the fastest-growing Muslim populations in the country.

Islam and Black America Any mapping of U.S. Muslims must include Black U.S. adherents, many of whom call themselves indigenous Muslims. They make up at least one-fifth of the country's Muslims. Islam is often referred to as hip-hop's unofficial religion. Praises to Allah, the Arabic word for God, and other phrases such as *assalamu alaikum*, or peace be upon you, are in the lyrics of celebrated artists from the Roots and Rakim to a Tribe Called Quest. New York City's Harlem is still home to historic Black Muslim communities with signs in their restaurants that say "no Pork on my Fork." And in Philadelphia, Arabic words from the faith such as *salaam* (peace) and *zakat* (alms) are uttered by Muslims and non-Muslims alike. In South Los Angeles, Jihad Saafir, an imam and son of a Black Muslim cleric, turned his father's storefront mosque into a vibrant community center and school that teaches the history of Islam through the lens of empowerment and African and African American history.

It's at Saafir's mosque that Hamidah Ali sits next to me as we break our fast during Ramadan on halal Chinese food. I ask her what Islam in America is to her. She doesn't hesitate. "Islam is cool," the 27-year-old African American second-generation Muslim says. Her hoop earrings peek out from her turban-wrapped head scarf, and she lists her idols, from hip-hop artists such as Yasiin Bey (formerly known as Mos Def) to America's most famous Muslim, the late Muhammad Ali, who once said, "I am America. Only I'm the part you won't recognize. But get used to me. Black, confident, cocky; my name, not yours; my religion, not yours; my goals my own—get used to me!" There's even a book on the topic—*Muslim Cool*, by anthropologist Su'ad Abdul Khabeer. Khabeer writes on the Muslim identity formed at the intersection of Black culture, hip-hop, and Islam.

All of this is Islam in America. Canon, the preacher from California, says the United States is one of the only places where pluralism openly flourishes because Americans have the freedom to practice the religion in all its forms.

Source: Adapted from "Being Muslim in America," by Leila Fadel, *National Geographic*, May 2018.

Write Across Units

Unit 3 explained that culture and geography are inextricably linked. Through language, religion, technology, and many other practices, humans create distinctive and evolving cultural landscapes within the physical landscape. This article explores how one religion—Islam—has become part of the American cultural landscape. Use information from the article and this unit to write a response to the following questions.

Looking Back

1. How would using different scales of analysis help you understand the influence of Islam in America? Unit 1

2. What pull factors might encourage Muslim immigration to the United States? Unit 2

3. Find examples in the article of acculturation, syncretism, and multiculturalism among American Muslims. Unit 3

Looking Forward

4. How do you think political boundaries affect the spread of Islam in the United States? Unit 4

5. How might a growing Muslim population affect U.S. agricultural patterns? Unit 5

6. In what ways can the diffusion of religions such as Islam affect the U.S. urban landscape? Unit 6

7. How might changes in the world economy affect the diffusion of Islam away from its hearth? Unit 7

Write Across Regions & Scales

Research the present-day Muslim communities in a European country. Write an essay comparing Muslim diffusion in that country and in the United States. Drawing on your research, this unit, and the article, address the following topic:

In what ways have Muslim citizens and immigrants changed the cultural landscape in different locations? In what ways have Muslims assimilated or acculturated as Islam has diffused beyond its hearth?

Think About

- Examples of both similar and unique cultural practices that have arisen in Muslim communities in Europe and the United States

- Challenges that Muslim communities face in Europe and the United States

Unit 3

Chapter 6

Cultural Iceberg

The cultural iceberg metaphor is a useful way to visualize the many obvious, not-so-obvious, and seemingly invisible cultural attributes that unite a group of people. This graphic divides cultural attributes into three levels: surface, deep, and unconscious. ▌ Consider your culture and other cultures you have read about, and identify cultural attributes that could be categorized at each level. Then explain why geographers assigned emotional levels to each part of the iceberg.

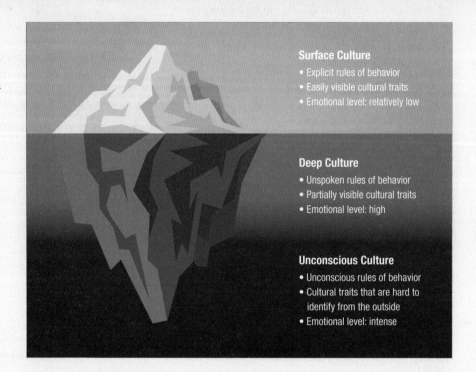

Chapter 6

Neighborhood Map

A neighborhood map reflects multiple landscapes of a city: cultural, ethnic, religious, geographic, and historical, among others. The toponyms, or place names, on such a map often reveal something about the people who live in those neighborhoods or people who migrated to those locations in the past. ▌ Identify San Francisco neighborhoods on the map that reflect the ethnicity of their residents, their geographic location, or California's Spanish and indigenous histories.

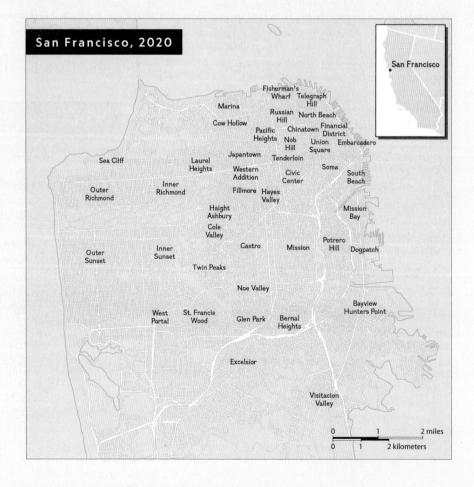

Maps & Models Archive

Chapter 6

LGBTQIA+ Spaces for Youth

Members of the LGBTQIA+ community often face discrimination and harassment at work, in school, and in public spaces. In the United States, elementary, middle, and high schools have undertaken efforts to create safe spaces for LGBTQIA+ youth, but protections from harassment are not always encoded into local or state laws. This map shows that in 2022, more than 20 states as well as the District of Columbia enacted anti-bullying laws to protect LGBTQIA+ students from being intimidated by fellow students, teachers, or school staff because of their sexual orientation or gender identity. ▍ Identify the spatial patterns shown on the map. Explain what the map might reveal about attitudes toward LGBTQIA+ people in the United States.

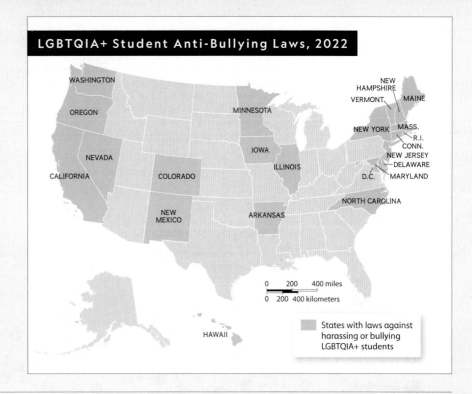

Chapter 7

Languages on the Web and IRL (In Real Life)

Languages and cultures diffuse through a variety of means. In recent decades, the globalization of some languages has been accelerated by the internet, which hosts innumerable cultural landscapes for the billions of people who log on to their devices every day—even though the internet itself is not a physical space. ▍ Compare the number of people who speak English in real life with the percentage of website content written in English. Identify possible reasons for the present-day domination of English on the web.

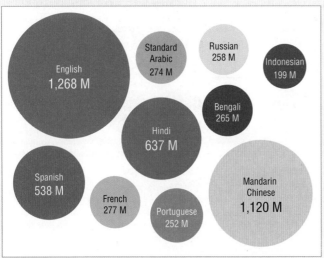

Languages Spoken by the Largest Numbers of People in Real Life, 2020

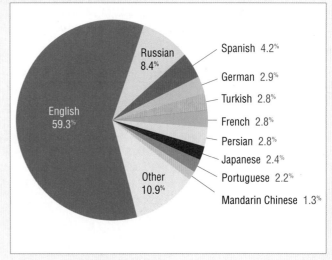

Most Common Languages Used in Websites, 2020

Sources: Ethnologue, W3Techs

Unit 3

Chapter 7

Cultural Diffusion

Cultural traits spread through various types of diffusion. Relocation diffusion occurs when people move from place to place, and expansion diffusion describes the movement of cultural traits through communication rather than migration. There are three separate types of expansion diffusion: contagious, hierarchical, and stimulus. In many cases, culture spreads through a combination of types of diffusion. ▌Identify examples of relocation diffusion and expansion diffusion in your own culture or daily life. Use the diffusion visuals to describe how these traits spread and changed your community's cultural landscape.

Expansion Diffusion
Cultural traits spread independently of the movement of people.

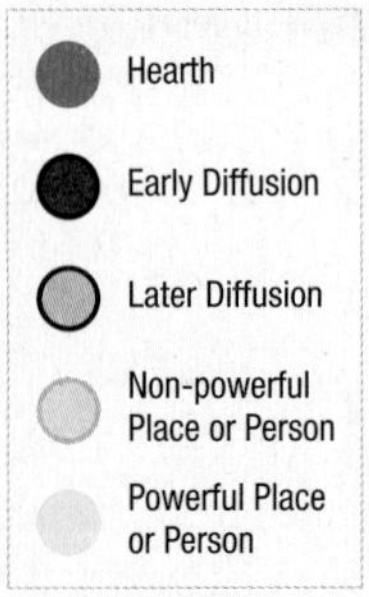

1. CONTAGIOUS DIFFUSION
Traits spread from person-to-person contact regardless of social class or level of power.

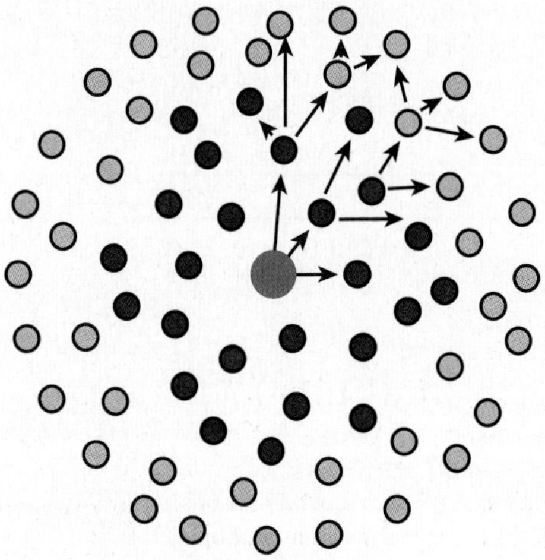

2. HIERARCHICAL DIFFUSION
Traits jump from powerful places or people to other powerful places or people, then spread down the hierarchy (or up from the bottom of a hierarchy to the top).

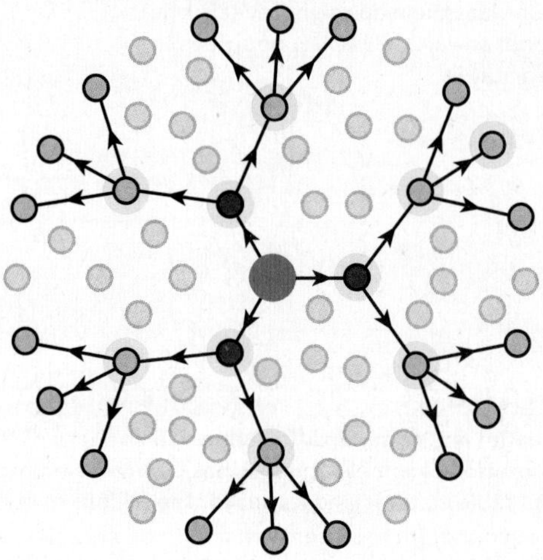

Legend:
- Hearth
- Early Diffusion
- Later Diffusion
- Non-powerful Place or Person
- Powerful Place or Person

3. STIMULUS DIFFUSION
Traits spread to another culture or region but are modified to adapt to the new culture.

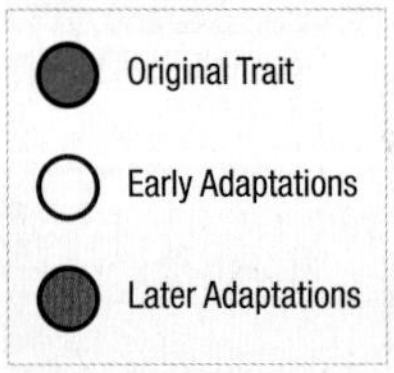

Legend:
- Original Trait
- Early Adaptations
- Later Adaptations

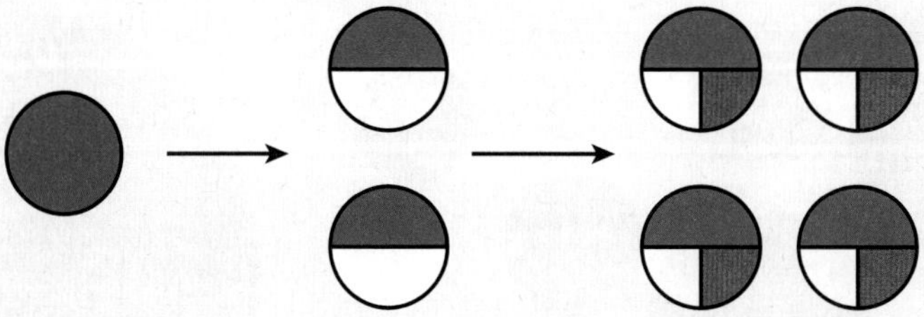

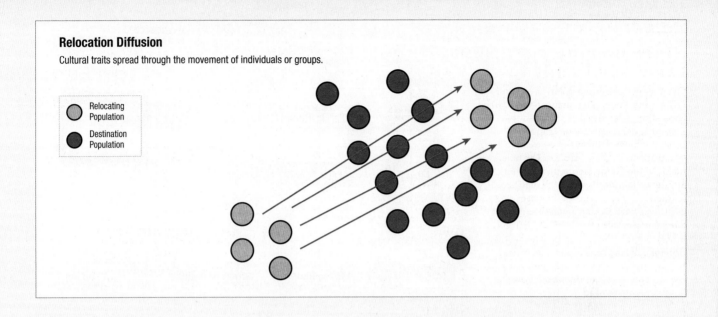

Relocation Diffusion
Cultural traits spread through the movement of individuals or groups.

- Relocating Population
- Destination Population

Chapter 7

Fashion Diffusion

Although people who study fashion trends use different terms than human geographers do, the processes are the same. Fashion experts have identified three theories to explain how fashion trends diffuse through different types of cultural diffusion: trickle-down, trickle-across, and trickle-up. The trickle-down theory portrays fashion that originates with the social elite and flows to people who want to gain upward mobility by imitating celebrities or wealthy people. The trickle-across theory demonstrates how fashion can move between groups who occupy the same social and economic levels. The trickle-up theory describes fashion innovations that begin in lower-income groups and eventually ascend social and income scales. ▌Identify which type of diffusion each theory of fashion represents. Explain the degree to which trickle-down, trickle-across, or trickle-up diffusion explains a national or international fashion trend you are aware of.

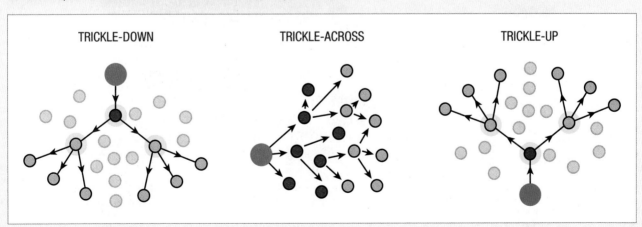

TRICKLE-DOWN | TRICKLE-ACROSS | TRICKLE-UP

| Unit 3 |

Chapter 8

Niger-Congo Language Family

The majority of people in Africa—around 85 percent—speak a language in the Niger-Congo family. It is estimated that around 1,400 Niger-Congo languages exist, but most of them can be classified into distinct language groups. The nine major Niger-Congo language branches are shown in green ovals in the diagram. (Linguists do not agree on the exact classification of Dogon, hence the dotted line.) ▌Identify the types of information in this diagram that would be unavailable on a map of languages spoken in Africa today. Describe how linguists may have reached their conclusions about how the languages are related.

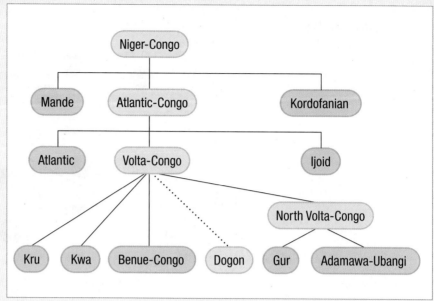

Source: Encyclopedia Britannica

Chapter 8

Religious Influence

The "Estimated Majority Religions, 2020" map in Chapter 8 shows the majority religion in each country but does not include data about the number of people who practice that religion. This map shows the percentage of people in each country who replied "very important" in a survey about the importance of religion in respondents' lives. ▌Compare the two maps and explain how data from both of them can help researchers consider the extent to which religion influences culture.

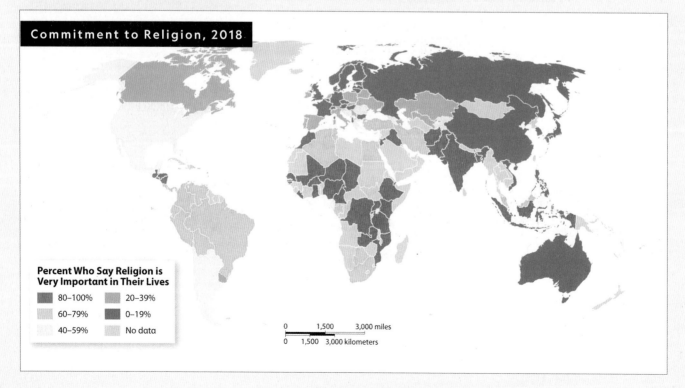

Chapter 8

Denominations of Christianity

Christianity is a widespread religion with numerous denominations, or separate church organizations that unite a number of local congregations. This diagram traces the principal organizations in the development of Christianity, although many additional denominations exist. ▌Identify the two church organizations that have the highest number of connected denominations. Then explain how diffusion may have led to the creation of Christianity's multiple denominations.

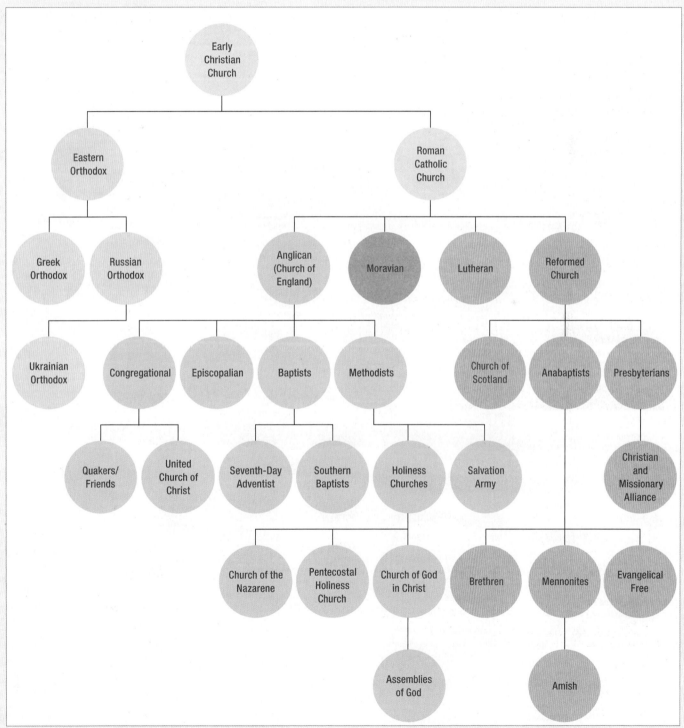

Source: Thinking Through Christianity

| Unit 4 |

Political
Patterns and Processes

Dividing Lines

Imjingak Park, South Korea

What could be simpler or more definitive than boundary lines on a map? In fact, the reality represented by those lines is complex and almost never clear cut. International borders are a reflection of political power, and they result from an interplay of cultural, historical, economic, and other forces.

Some borders, like the 2.5-mile-wide strip of land that separates North Korea from South Korea, act as buffer zones between hostile countries. This border is known as the Demilitarized Zone (shown here) and it's one of the most heavily guarded and fortified places on Earth. Still, tourists—like these visitors from South Korea—are often drawn to borders, the symbolic and literal dividing lines between peoples and cultures.

For geographers, borders on a map are just the starting point for understanding the many ways that groups define themselves and each other and lay claim to their spaces on the globe.

Chapter 9
The Contemporary Political Map

Chapter 10
Spatial Patterns of Political Power

Chapter 11
Political Challenges and Changes

Unit 4 Writing across Units, Regions, & Scales

Unit 4 Maps & Models Archive

National Geographic Explorer **Aziz Abu Sarah**

Reconciliation Through Narrative

National Geographic Explorer Aziz Abu Sarah calls the Israeli–Palestinian conflict "one of the oldest, most complicated conflicts in the world." Both Jews and Palestinians, most of whom are Arab Muslims, claim centuries-old cultural, religious, and national ties to a small but historically vital strip of land between the Mediterranean Sea and the Jordan River. After the formation of the modern state of Israel in 1948, a series of wars with its Arab neighbors culminated in the country's occupation, beginning in 1967, of territories beyond its original borders: the Golan Heights (which had been part of Syria), the West Bank (which had been part of Jordan), and the Gaza Strip and Sinai Peninsula (which had been part of Egypt). The Sinai was returned to Egypt in 1982, but Israel, primarily citing security concerns, has maintained varying degrees of control over the other regions to this day. Among the Palestinian residents of Gaza and the West Bank, opposition to the Israeli occupation twice erupted into major uprisings, or *intifadas* (1987–1993, 2000–2005), which became increasingly violent. From within Gaza, it turned repeatedly into outright war, as indiscriminate rocket barrages in the direction of Israeli cities were answered by overwhelming military force. Even during quieter times, tensions in the region continue to simmer, and a political resolution to the conflict seems to many beyond reach.

Learning Objective
LO 9.3 Define territoriality in the context of political geography.

Growing up Palestinian When Abu Sarah was a young boy attending a Palestinian elementary school near Jerusalem, conflict was an everyday fact of life. His mother made certain he had an onion with him every day when he went to school, because an onion's fumes were thought to counteract the effects of teargas, which Israeli soldiers sometimes use when clashes with Palestinian protesters become violent. Later, Abu Sarah became a dedicated participant in the fight against the Israeli occupation, motivated by an urge toward revenge for the death of his older brother, who perished after spending a year in an Israeli prison. As a teenager, Abu Sarah wrote prolific articles for anti-Israeli publications and took part in violent confrontations, throwing stones at Israeli soldiers.

When Abu Sarah was 18, he decided to learn Hebrew because it is difficult to work and live in Jerusalem without knowing the language. He found himself the only Palestinian in a classroom full of Jews from other countries who had recently immigrated to Israel. Bonding with his classmates over relatively frivolous connections such as a shared love of country music—which Abu Sarah claims is practically unheard of among his fellow Palestinians—led to deeper conversations about personal experiences and values. These exchanges humanized Israeli Jews in Abu Sarah's eyes. "I knew after that class where the problem is," he says. "When you don't know somebody, you're going to hate them, and you're going to fantasize about their horns and tails. . . . You start thinking they're less of a human than you."

Changing Narratives When describing his work as a peace activist and cultural educator, Abu Sarah often uses the words *narrative* and *reconciliation*. He seeks out ways to foster meaningful, in-person encounters between Israelis, Palestinians, and others involved in the conflict because sharing personal narratives helps individuals find commonalities and perceive one another as fully human. By building empathy, Abu Sarah hopes to foster a spirit not only of tolerance but of reconciliation—a true coming together.

To this end, Abu Sarah speaks frequently to groups in churches, synagogues, and mosques, as well as working to share his message across all media. In a series of videos for National Geographic, he stood on both sides of skirmishes between Israeli soldiers and Palestinian protesters to gather their stories firsthand. Abu Sarah has also taken on the role of social entrepreneur, or one who founds a business that aims to solve a social problem. His company organizes tours of Israel and the Palestinian territories narrated by diverse guides who share multiple points of view—Palestinian and Israeli; Jewish, Muslim, and Christian. While Abu Sarah is optimistic about the power of narrative and personal connections, he knows the wall of ignorance separating Jews and Palestinians is an imposing barrier. "What needs to be done is put cracks in that wall," he says. "One of my colleagues says what we [as peace activists] do is bang our heads against that wall until we break it down."

Geographic Thinking

Describe how people's narratives can affect the expression of territoriality in their communities.

Top: In Jerusalem, people pray at the Western Wall, all that remains of the Second Temple of Jerusalem, an ancient Jewish holy site. The gold dome in the background is the Dome of the Rock, an Islamic shrine built on a site that is sacred to Muslims, Jews, and Christians.
Bottom: Aziz Abu Sarah leads a tour group.

Chapter 9
The Contemporary Political Map

Critical Viewing A military parade through the Arc de Triomphe is part of the Bastille Day celebrations in Paris, France, on July 14, 2019. Demonstrations of national pride such as this one help define and unite nations. What might this celebration suggest about the political situation in France?

Geographic Thinking How do maps reflect political borders around the world?

9.1
The Complex World Political Map

9.2
Political Power and Geography

9.3
Political Processes over Time

Case Study: The Kurds

9.4
The Nature and Function of Boundaries

Case Study: The DMZ in Korea

National Geographic Photographer
David Guttenfelder

9.1 The Complex World Political Map

Take a close look at a political map of the world. Notice how lines—straight and crooked, solid and dashed—divide the land into countries. Who created these spaces and gave them their particular boundaries? On a world political map, geographers see the impact of people's need to control territory and exert power.

Organizing Space

Learning Objectives

LO 9.1 Describe the different ways that areas are organized into political entities.

LO 9.2 Differentiate between nation, state, and nation-state, and explain how these concepts apply in varying political contexts.

When studying a map or globe, human geographers are interested in understanding how and why countries and regions of the world came to be organized politically. As you have learned, the world political map has changed enormously since the first civilizations began to mark their territory and establish governments. Some early civilizations had vague, loosely defined borders, but others did demarcate boundary lines: the Babylonians erected boundary stones while the Chinese built frontier walls. Sometimes these territories encompassed just a single city and its hinterland; in other instances, much larger areas were organized through loose political alliances and governments. Over the centuries, however, as groups established themselves and claimed land through means both peaceful and not, these lines—the borders between groups—often became more clearly defined. **Political geography** is the study of the ways in which the world is organized as a reflection of the power that different groups hold over territory.

Although we tend to view maps as representations of settled facts, political maps can express particular interpretations of the world. For example, the government of the People's Republic of China has never recognized the island of Taiwan, off its east coast, as a separate country. From 1949 until 1979, the United States did consider Taiwan a sovereign state, but this status was then downgraded based on the Taiwan Relations Act—after U.S. ties with the People's Republic of China were strengthened. However, the United States still, along with the nearly 24 million Taiwanese, do consider Taiwan to be an independent entity; the United States treats Taiwan in most ways as if it was a country. So, a map of China approved by the Chinese government looks different from one published by a mapmaker in the United States. Travelers might experience the same phenomenon when using online map applications in different countries. If a boundary dispute exists, the border could appear in a

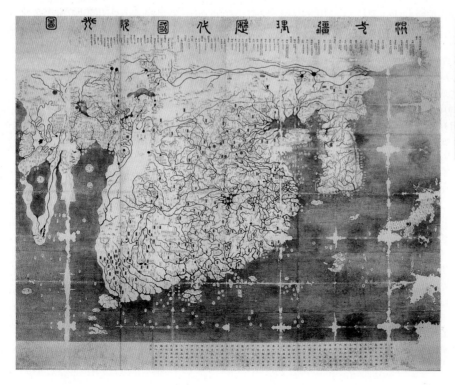

This map of the Eastern Hemisphere, known as the Kangnido Map, was created by a Korean geographer in the 15th century. The mapmaker's inaccurate size of Korea (the peninsula to the right of the large landmass) in relation to the rest of Asia is an example of how maps can express skewed interpretations of the world. The Kangnido Map also shows Africa (*far left*) to be much smaller than its actual size and a barely visible Europe.

The Contemporary Political Map **255**

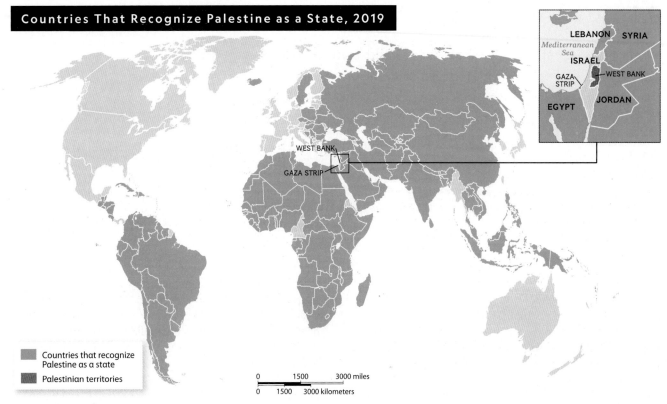

Countries That Recognize Palestine as a State, 2019

- Countries that recognize Palestine as a state
- Palestinian territories

Reading Maps Since the 20th century, the Palestine region has been a source of persistent conflict as Jewish Israelis and Arab Palestinians fight for control of territory. One hundred thirty-eight members of the United Nations have recognized Palestine as a state, while 55 have not. ▎ Why do you think a larger number of UN member countries recognize Palestine as an independent state?

different place depending on which country a person is standing in when accessing a smartphone map.

States States are created by humans as a way to organize and manage themselves. A **state** is a politically organized independent territory with a government, defined borders, and a permanent population—in short, a country. State governments have power over a population that works together to contribute to an economy and is connected by networks of transportation and communication systems. A state has **sovereignty**, which is the right of a government to control and defend its territory and determine what happens within its borders. If a state is not recognized as an independent country by other states, it is not considered sovereign. However, as geographer John Agnew points out, sovereignty is not always associated with territory: "Effective sovereignty is not necessarily predicated on and defined by the strict and fixed boundaries of individual states." His research in political geography recognizes the increasingly important roles that globalization and supranationalism (an alliance of three or more countries to pursue common goals; discussed in Chapter 11) have played in changing notions of sovereignty—and differences in its meaning over space and time. Agnew argues that there are unique "sovereignty regimes," which he describes as combinations of state authority, strong or weak, and territory relationships, consolidated or open. The use of the term *state* can sometimes be confusing for students in the United States: an independent state such as Sweden is not the same as a U.S. state such as Nebraska.

The world is organized into a number of diverse sovereign states. Some, like China and Canada, cover vast territories, whereas others, such as Belize and Togo, are very small in area. The number of independent states around the globe often changes in response to pressures from political circumstances. In 1991, when the Soviet Union dissolved, 15 republics that had been part of that massive state, including Estonia, declared their independence. Following Estonia's declaration, other countries recognized it as a sovereign state. Among its first steps, the new Estonian government reestablished its military and its currency, the kroon, which immediately became a symbol of the country's independence. Other former Soviet republics, including Ukraine and Belarus, also took steps to establish their sovereignty and were recognized by the international community.

There is no general agreement on the number of independent states around the globe. The United Nations recognizes 195 countries, but not every member state of the UN agrees on which countries are independent and which are not. For example, the UN as an organization recognizes Palestine as an independent state, but 55 of its member countries do not. On the other hand, Kosovo, which had been part of Serbia but declared its independence in 2008, is not recognized as an independent state by the United Nations. However, 111 UN members do consider Kosovo to be sovereign.

Nations States should not be confused with **nations**. Whereas states are political entities, nations are cultural entities, meaning that they are made up of individuals who have forged a common identity through a shared language, religion, ethnicity, or heritage—often all four of these. Some define a nation as including a "reasonably large population," whereas others argue that the size of the population does not matter. The people of a nation share a common vision of the future, which produces an undeniable feeling of togetherness.

Several examples of nations exist within the United States. Native Hawaiians, for example, are descendents of Polynesian people who landed on the Hawaiian islands more than a thousand years ago and are united by their common history, culture, and language. In the southwestern United States, the Navajo people are also a nation bound by tradition, history, and a common language. In fact, the U.S. government recognizes 573 native, or tribal, nations living on United States territory.

Nation-States In a **nation-state**, the territory occupied by a group who view themselves as a nation is the same as the politically recognized boundaries of the state they call their own. The concept of a nation-state is an ideal; no existing country can be described as a pure nation-state because all are home to at least small ethnic or cultural minorities who might consider themselves as a nation within their country. Still, some countries come closer than others to the definition of a nation-state. Estonia, following its independence from the Soviet Union, is often viewed as a nation-state because most of its people share a common identity in terms of ethnicity and language. In Asia, Japan is commonly used as an example of a nation-state because nearly all its people share a common culture. Examples of other countries often identified as nation-states include Iceland, Iran, Albania, Croatia, Poland, and France. However, some critics challenge this categorization. France, for example, is home to regional nations with distinctive identities such as Breton and Provence, and also diverse minority ethnic communities originating from North Africa, Africa south of the Sahara, and Southeast Asia. Although the French government banned censuses by race and ethnicity in 1978, an independent study in 2017 found that around 5 percent of the French population was non-White and non-European.

The concept of the nation-state first emerged in Europe, which until the 20th century was ruled by a small group of monarchs. A map of Europe from the early 19th century looks similar to a modern map in some places, but very different in others. Italy was a loose confederation of separate states until united as the Kingdom of Italy in 1861, while the German Empire brought together a number of German states finally in 1871. France, however, mostly looks as it does today.

Toward the end of the 19th century, however, the idea of drawing state borders to match national identities was taking hold, and when the victorious Allied leaders met in Versailles, France, in 1919 and 1920 after the end of World War I to redraw national boundaries in Europe, they had this ideal in mind. Poland, for example, which had been divided among Russia, Germany, and Austria-Hungary, was reunited. The rest of Europe, too, was largely composed of countries intended to be nation-states. But even today, some European countries lack the unity that defines most of the continent's nation-states. In Spain, both the Basque people and the Catalans, ethnic groups living in parts of northern Spain, have strong national identities, and each group has agitated for independence from Spanish rule. As political geographer Alexander B. Murphy writes, ". . .the nation-state concept was a fiction of sorts from the beginning, and it became even more so as the modern global political map emerged in the wake of the dissolution of Europe's colonial empires. It may be common practice to call the units shown on the world political map nation-states, but the typical state today encompasses multiple cultural-historical communities."

Multistate Nations and Multinational States A **multistate nation** consists of people who share a cultural or ethnic background but live in more than one country. Ethnic Russians are considered to be a multistate nation, because sizeable numbers of them live outside of Russia. They form substantial minorities in several countries that once belonged to the Soviet Union, including Kazakhstan, Moldova, Belarus, and Ukraine. Some consider the two Koreas as one nation but two states, while others disagree with this view.

Multistate nations can pose challenges to political borders because people may feel a stronger affinity for a neighboring state that is home to others in their ethnic group than to their own state. Surveys of ethnic Russians living in Estonia, Latvia, and Ukraine, for example, show strong pro-Russia attitudes. Sometimes this situation leads governments to establish a policy of **irredentism**, attempting to acquire territories in neighboring states inhabited by people of the same nation. Russia was accused of irredentism when it annexed territory in Ukraine in 2014 that had a significant Russian population. Russia repeated this reasoning as cause for its invasion of Ukraine in 2022, targeting the Donbas region of eastern Ukraine. You learn more about irredentism and the case of Ukraine in Chapter 11.

A country with various ethnicities and cultures living inside its borders is termed a **multinational state**.

Multinational states sometimes struggle to create a sense of unity among different peoples. Iraq, for example, has long suffered from internal conflict because of a lack of shared identity among its Sunni, Shia, and Kurdish populations. Sometimes multinational states split up because the differences are unbridgeable, like Yugoslavia after the breakup of the Soviet Union. Other times, multinational states are able to forge a national identity despite the presence of many different ethnicities, cultures, and religions. Although culture and ethnicity have sometimes been a source of conflict, the United States has been broadly successful in integrating and assimilating different groups after they immigrate. In reality, because of global migration and the diverse nature of boundaries, most countries today are multinational states.

Autonomous and Semiautonomous Regions Some countries contain regions that are either **autonomous** or **semiautonomous**, meaning they are given some authority to govern their own territories independently from the national government. Greenland is a part of the Kingdom of Denmark, but has gained a great measure of autonomy, beginning with "home rule" in 1979 and then a self-governing arrangement in 2009. Though still supported financially by Denmark, Greenland's government has taken over most tasks from the Danish government. In China, the territory of Hong Kong was autonomous, using a system of government and currency that differed from the ones used throughout the rest of the country. However, since 2020, China's growing influence and control of Hong Kong's internal affairs has diminished the territory's autonomy over local issues. In the United States, American Indian reservations are semiautonomous places with the authority to operate under certain different laws. This is the reason gambling casinos are permitted on Indian reservations even when they are not legal in the states where the reservations are located. Despite their semiautonomy, the view of the U.S. government is that Native American nations are not sovereign because their power is limited. Some tribal nations disagree with this interpretation, and the nature of tribal sovereignty in the United States is a topic of serious ongoing debate. This is particularly pressing in criminal matters, where tribal police have limited powers. There are also thorny jurisdictional issues among tribal, local, state, and federal law enforcement agencies, which has made tackling crime on reservations challenging. This has proven to be especially difficult in cases involving non-Native people on indigenous-owned lands. Additionally, it is estimated that more than 5,000 American Indian and Alaska Native women and girls are missing—believed to have been murdered. Indigenous advocates cite the uncertain status of semiautonomous territories as contributing to this tragic issue, as responsibility for investigations is often unclear.

Stateless Nations The term **stateless nation** describes a people united by culture, language, history, and tradition but not possessing a state. Tribal nations in the United States are stateless nations. Similarly, the Basque

Reading Maps State boundaries in Europe today primarily reflect the ideal of nation-states. Prior to World War I, however, this was not always the case. Identify the boundary changes that have taken place in Germany between 1914 and today. ▌ How do these changes reflect the trend of drawing state borders to match national identities?

people in Spain have a unique culture and language, but despite the formation of secessionist organizations in the past, they do not have a separate, independent state. The Palestinians are considered a stateless nation because much of the world does not recognize Palestine as an independent state. Before Israel was established in 1948, the Jewish people were considered to be a stateless nation. The Kurds, yet another stateless nation scattered across several states in Southwest Asia, are examined in a case study later in this chapter.

Geographic Thinking

1. Identify how maps can be a tool used to express power.
2. Explain why Kosovo seeks recognition as a sovereign state by the United Nations as a whole.
3. Explain your argument to answer the following question: Is the United States a nation-state?

9.2 Political Power and Geography

Territory is a powerful word describing a concept that's meaningful on many scales. At the individual level, your family might view your home as personal territory, protected by a fence around the yard or apartment doors that lock against intruders. States, too, place tremendous value on claiming, controlling, and defending the land that they consider their territory.

Issues of Space and Power

Learning Objective
LO 9.3 Define *territoriality* in the context of political geography.

Looking beyond your personal space, you can quickly perceive the countless ways different groups or entities claim their territories. For example, schools at all levels typically have home courts or fields, mascots, logos, and slogans that give them a unique identity. Communities of all sizes also define themselves using markers such as signs, slogans, and sometimes nicknames like "The Big Apple" or "City of Brotherly Love." In many cities and towns, gated communities are neighborhoods surrounded by actual fences to ensure that only residents or people invited by residents can enter. At the national scale, countries control their land by forming borders, and they establish a national identity in a variety of ways, including through their names, flags, anthems, and citizenship requirements.

These are all examples of **territoriality**, a concept that has multiple dimensions. Geographer Robert Sack defines territoriality as the "attempt by an individual or group to affect, influence, or control people, phenomena, and relationships by delimiting and asserting control over a geographic area." It is also an expression of a group's historic and personal links to a place—the connection of people, their culture, and their economic systems to the land. According to Sack, territoriality is the basis for the power that people try to exert and the political spaces they create. Governments form around these spaces, build political power, and establish sovereignty, which allows them to control their territory and protect it from outside interference. Sovereign countries, under international law, or the set of rules and standards that are generally recognized as binding between states, are permitted to defend their borders militarily and establish the laws that govern the people who live there.

Controlling People, Land, and Resources

Learning Objective
LO 9.4 Explain the ways in which states impose control over people, land, and resources to maintain or assert political power.

To assert and maintain political power, states impose control over the people, land, and resources in their territories. At times, states also attempt to control resources outside of their territories using tools of trade, diplomacy, or war.

Neocolonialism As you have learned, European countries began to establish colonies throughout the world starting in the 16th century, gaining control over lands and resources in Africa, Asia, the Americas, and elsewhere. The term **colonialism** describes this practice of claiming and dominating overseas territories. Colonial relations of domination can also happen internally, within countries. Russia expanded across Siberian Asia beginning in the 1600s, colonizing disparate areas and peoples. Many

Along some parts of its border with Mexico, the United Sates erected fences to outline its territory. A photograph of the eyes of a young undocumented immigrant, known as a "Dreamer," spreads across this fence in Tecate, Mexico. The border has been the site of tension between migrants, asylum seekers, and border patrol agents, but the artwork is meant to be used as a giant shared picnic table among residents on both sides of the wall.

residents of Scotland still perceive their position in the United Kingdom as an example of colonization, and nearby Ireland is considered to be England's first colony, and remained so for hundreds of years.

Although most former colonies have declared independence and claimed their sovereignty, **neocolonialism** endures in the use of economic, political, cultural, or other pressures to control or influence other countries, especially former dependencies that are strategically important. Neocolonialism, a term coined by the first president of Ghana, Kwame Nkrumah, is seen in many former African colonies that are free states but have economies that rely on outside investment and are therefore vulnerable to excess influence by outside powers. To cite one example, Kenya, in eastern Africa, needed to replace aging railroad infrastructure to transport cargo across the country. In 2014, the Kenyan government agreed to pay a company owned by the Chinese government to build a railroad line from the Indian Ocean port of Mombasa inland to Nairobi, the capital, and on to neighboring Uganda. The cost of the project was $3.8 billion, an amount that critics say will place Kenya in debt to China for many years. It also leaves China in control of decisions about when and how to build the railroad, which has not yet been completed.

Choke Points A **choke point** is a narrow, strategic passageway to another place through which it is difficult to pass. Because they are limited in size and because there is a great deal of competition for their use, choke points can be sources of power, influence, and wealth for the countries that control them. Waterway choke points can be straits, canals, or other restricted passages.

Choke points have historically played a significant role in military campaigns, as large armies or navies have difficulty moving through narrow passages. A classic example of a land-based choke point is Thermopylae, a mountain pass in Greece where a Greek force estimated at 7,000 men was able to hold off an invading Persian army of between 70,000 and 300,000 soldiers for three days in 480 B.C.E.

Global Oil Choke Points, 2022

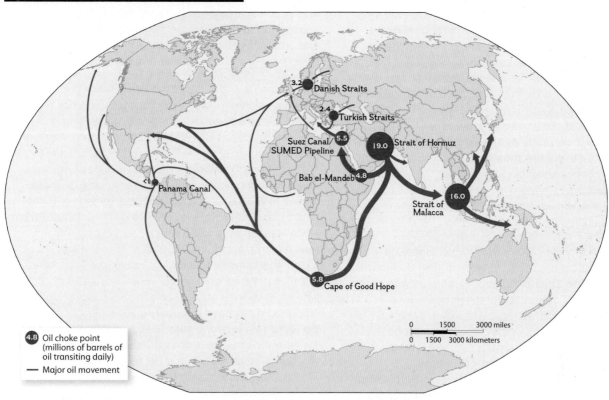

Reading Maps Oil transit choke points are a critical part of the global economy, as millions of barrels of oil are shipped by tankers around the world each day. The Cape of Good Hope is not technically a choke point, but is considered a critical route. ▍ Explain how oil prices in the United States would be affected if the choke points in Asia and Africa were blocked.

In 2021 the giant container ship, the Ever Given, bound for Rotterdam from China, blocked Egypt's strategic Suez Canal for six days, after being wedged across it by high winds. Hundreds of other ships were stuck in the jam, disrupting global trade.

The Contemporary Political Map

Today, waterway choke points command the most attention and are a cause for international concern because high volumes of crucial commodities, such as oil and food, pass through them. The Strait of Malacca—between Malaysia, Indonesia, and Singapore—is a choke point for all goods shipped by sea between Europe, Africa, Southwest Asia, and South Asia to East Asia. For example, more than one-quarter of all soybean exports move through the Strait of Malacca annually, which is only 1.7 miles wide at its narrowest point. About 54 percent of internationally traded grain and fertilizers passed through at least one maritime choke point. Blocked choke points can put the global food supply at risk. In 2021, a massive container ship blocked the Suez Canal, through which 12 percent of global trade passes, for nearly a week—with serious repercussions for global supply chains and crucial oil supplies that lasted for months.

Countries that control choke points sometimes use them to expand their global influence or gain political advantages. The Strait of Hormuz, located between Iran and Oman at the intersection of the Persian Gulf and the Gulf of Oman and leading to the Indian Ocean, is a significant choke point in the global oil trade and the subject of rising international tensions. Iran, struggling with economic sanctions imposed on it by Western powers since 1979, has used a hybrid warfare strategy to disrupt trade through the strait, threatening the stability of that vital transit route.

Shatterbelts Territoriality and the quest for political power sometimes lead to instability in regions known as **shatterbelts**. In a shatterbelt, states form, join, and break up because of ongoing, sometimes violent, conflicts among parties and because they are caught between the interests of more powerful outside states. Often, shatterbelts exist in areas that have been beset by violence for many years due to long-lasting antagonism between the religious, ethnic, and linguistic groups who live there. These hostilities may be exacerbated by alliances between countries or by outside powers seeking to expand their own influence over the region.

The Balkan Peninsula in southeastern Europe is a well-known shatterbelt. Over the centuries, conflict has repeatedly broken out in this region as the influence of outside powers has intensified internal divisions. Although relatively small in area, the Balkan Peninsula is home to Muslims, Roman Catholics, Eastern Orthodox Christians, Slavic-speaking people, and non-Slavic peoples. These groups have often been governed as a single unit. Before World War I, the region was claimed by the Ottoman Empire, which was resisting pressure from neighboring Austria-Hungary and other European states. After the war, the region was formed into the single country of Yugoslavia. In the post–World War II years, Yugoslavia's government was firmly under Soviet authority. After the Soviet Union collapsed in 1991 and its military could no longer control the Balkan Peninsula, armed conflict broke out. Today, the former Yugoslavia is divided into independent states: Slovenia, Croatia, Bosnia and Herzegovina, Serbia, Montenegro, Kosovo, and North Macedonia.

Geographic Thinking

1. Describe the relationship between territoriality and the open display of local pride.
2. Explain why choke points are important to the global economy.

9.3 Political Processes Over Time

You've learned about the right of sovereign countries to represent their interests on the world stage and determine what happens inside their own boundaries. Although the theory of sovereignty may seem clear cut, its practice in an interdependent world can become complex.

The Complicated Nature of Sovereignty

Learning Objectives

LO 9.5 Explain how sovereignty is impacted by power imbalances between states.

LO 9.6 Define *self-determination* as a concept in political geography.

The concepts of sovereignty, nation-states, and self-determination shape the contemporary world. As you have learned, sovereignty is a state's authority to control its territory and govern itself, and the ideal of a nation-state is a country that defines itself as belonging principally to a single nation. **Self-determination** is the right of all people to choose their own political status.

The Balkan Peninsula shatterbelt is a consequence of nations of people fighting for self-determination against outside powers and against their neighbors. In the 18th and 19th centuries, the different ethnic groups living on the peninsula formed nation-states, including Montenegro, Bulgaria, Romania, and Serbia. Because of the two world wars and the influence of the Soviet Union, boundaries between the countries changed, and the newly formed states such as Yugoslavia often failed to respect national and ethnic identities. Following the collapse of the Soviet Union in 1991, Yugoslavia began to splinter as groups within the country asserted their self-determination and formed independent states. Croatia and Slovenia were the first to declare their independence, followed by Macedonia (now North Macedonia) and the state called Bosnia and Herzegovina. Years of conflict among ethnic groups followed, but eventually an agreement was reached to dissolve Yugoslavia entirely. By 2003, what had once been one country made up of different groups who did not have shared values or cultures became seven separate countries.

States can sometimes be independent but not entirely sovereign. For several decades during the latter half of the 20th century, Cuba depended on the Soviet Union because the island's economy had been cut off from the economies of democratic countries such as the United States, leading to shortages of food, fuel, and other necessities. The Soviet Union, looking to support a communist country in the Western Hemisphere, provided aid in the form of loans, oil, military supplies, and technical expertise. This aid put the Soviet Union in a position to strongly influence Cuba's internal and external affairs.

Sometimes a country's right to self-determination is violated, such as when other countries interfere with its natural development. In the Central American country of Guatemala, for example, the United States gave support to a coup in 1954 that ousted a democratically elected president whom the U.S. government believed to have communist leanings. The new ruler, Carlos Castillo Armas, opposed communism but was himself a ruthless dictator. Once he took control, Castillo Armas's regime rounded up thousands of suspected communists and executed hundreds of prisoners. Because of his repressive regime, he was assassinated three years after taking office. The political turbulence caused by the coup fostered a civil war that began in 1960 and lasted 36 years, costing 200,000 lives. Today, while trying to address the atrocities committed during the long civil war with special courts called tribunals conducted through an impartial and independent justice system, Guatemala remains plagued by violence and organized crime.

Legacies of Colonialism and Imperialism

Learning Objective
LO 9.7 Identify the effects of colonialism and imperialism that endure in contemporary political geographies.

The driving force behind European colonial efforts was imperialism, which may be defined as a country's policy or practice of extending power and ownership, especially by gaining political or economic control of other territory or people.

In this quest for empire, Spain, Portugal, Britain, France, the Netherlands, Belgium, and later Germany took control of distant territories already claimed by other peoples, sent colonists who imposed their cultural values on the inhabitants, and exploited the lands and people for economic advantage.

Two of the largest empires were controlled by Spain and Britain. Spanish territories spanned the globe, including land in North and South America, Europe, Africa, Southeast Asia, and even Oceania. The British said theirs was "the empire on which the sun never sets" because they controlled territory across the globe. The impacts of this imperialist wave endure today, with some of the most obvious colonial exports being language and religion. People speak English in the United States and Australia because these lands were controlled and settled by British colonists; by the same token, imperialism is the reason Spanish is spoken throughout much of Central and South America. Latin America has the world's largest percentage of Catholics, due in large part to the Portuguese and Spanish colonists and missionaries who landed on its shores centuries ago.

Case Study

The Kurds

The Issue Even though the Kurdish people in Southwest Asia have a long history as a distinct ethnic group and view themselves as a nation, they do not have their own state, despite years of trying.

Learning Objective
LO 9.5 Explain how sovereignty is impacted by power imbalances between states.

By the Numbers

19 percent

Kurdish population of Turkey

10 percent

Kurdish population of Syria

15–20 percent

Kurdish population of Iraq

10 percent

Kurdish population of Iran

Source: CIA World Factbook

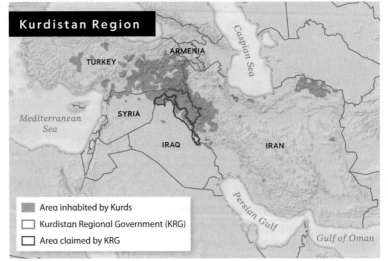

An estimated 30 million Kurds live throughout Southwest Asia in an area covering about 74,000 square miles.

A Distinct Ethnic Group of People called the Kurds lives in a region that covers parts of Iraq, Iran, Syria, and Turkey, unofficially called Kurdistan ("Land of the Kurds"). However, the estimated 30 million Kurds do not have their own state, despite having a history that dates back more than a thousand years, if not longer. The dream of an independent country was born in the late 19th century, and it came close to being realized after the breakup of the Ottoman Empire following World War I. At that time, a treaty was proposed that would define the borders of the region's new countries—including the option for a Kurdish state. However, the treaty was never ratified, and the one that replaced it did not include a provision to carve out a state for the Kurds. The treaty was negotiated between Turkey, Greece, and the Allied powers; the Kurds were not represented.

Since then, the Kurdish people have continued to work for a sovereign state, at the same time that their minority status in the four countries gives them minimal power and has led to persecution and worse. Under the rule of President Saddam Hussein in Iraq during the last decades of the 20th century, thousands of Kurdish men, women, and children were killed with poison gas.

Political obstacles have stalled any momentum toward a Kurdish state. To allow for the creation of a new state, the four countries where the Kurds reside would need to give up some of their territory. Not surprisingly, the countries' governments find this prospect very unappealing. Within Kurdish territory, infighting among leadership groups has proved to be another obstacle, as the lack of unity makes progress difficult. Pressure from external powers, such as the United States or European countries, in favor of an independent Kurdistan could build enough political will for a new country to be created. That pressure has not yet materialized, however, even though Kurdish fighters allied with U.S. forces in the complex conflict that engulfed Syria beginning in 2011, and led the ultimately successful fight against the Islamic State (a violent Sunni group intent on controlling territory across Iraq and Syria). Though they gained respect and some degree of autonomy in these two countries for a time, the Kurds again have found themselves under increasing pressure from Syria and also Turkey to abandon territories previously claimed for Kurdistan. At present, the Kurds can be described as a stateless nation, because they do not have an independent country to call their own, and a multistate nation, because they live in a region that spans multiple countries.

Geographic Thinking

Describe what the map above illustrates about the Kurds' challenges as a stateless nation.

A later wave of imperialism hit Africa in the late 19th century, when European empires, looking to extend their economic, political, and social power, began to take African territories by force. In 1884, European leaders met at the Berlin Conference to arbitrarily define the boundaries between their conquered African possessions—boundaries that still exist between states today. Africans were not even present at the meeting, and no consideration was given for traditional ethnic boundaries or the governance structures that had existed for centuries.

Many believe the economic and social problems affecting Africa today can be traced back to imperialism and the 1884 Berlin Conference. By extracting wealth, establishing export-driven economies, and creating the conditions for conflict, European imperial powers laid the groundwork for events such as the Rwandan genocide of 1994, in which one ethnic group, the Hutus, slaughtered hundreds of thousands of Tutsis. Differences between the two groups had been profoundly exacerbated during colonial times, when the Belgian rulers had greatly favored the Tutsis over the Hutus, granting them better jobs and educational opportunities. Decades later, Rwanda still wrestles with governance and human rights issues, rooted, in part, in its colonial experience.

In countries or regions affected by imperialism, peoples have sought self-determination through independence movements. When World War II began in 1939, for example, much of Africa and Asia were still under European control, although independence movements were growing in numerous places. Following World War II, many groups fought for and gained independence from their colonial rulers. Among them was India, a British colony for 200 years that won its independence in 1947 and was split into the countries of India and Pakistan. The present-day boundary between Pakistan and India, which is still disputed in places, resulted from an agreement among leaders of the independence movement and the departing British authorities that the majority-Muslim and majority-Hindu regions should form separate states. In North Africa and Southeast Asia, the colonies of Morocco, Algeria, Vietnam, Cambodia, and Laos also exemplified the postwar movement to throw off colonial powers, gaining their independence from France during the 1950s after intense fighting. The contemporary political boundaries of all these countries are the result of their successful struggles to achieve independence from their former colonial rulers.

Sometimes related to independence movements, the process of **devolution** occurs when the central power in a state is broken up among regional authorities within its borders. Devolution tends to happen along national lines, allowing members of a nation to claim greater authority over their territory. In the case of the former Soviet Union, devolution led to the creation of 15 independent states. You learn about devolution in greater detail in Chapter 11.

Geographic Thinking

1. Explain how the concept of self-determination might challenge the political structure of a multinational state.

2. Describe the cultural and economic aftereffects of imperialism.

9.4 The Nature and Function of Boundaries

The states you've learned about would not exist without boundaries. Geographers recognize many different types of boundaries, but they all serve the same goal—to define political spaces and territories.

Defining Political Boundaries

Learning Objective
LO 9.8 Explain the ways in which countries establish and administer their boundaries.

The amount of territory that falls within a state is defined by the boundaries that surround it. International boundaries, or borders, are the outcome of geopolitical relationships and expressions of territoriality—people's sense of connection to a place and their drive to control it. As such, boundaries are subject to change when relationships among countries change, or when people assert a claim to territory. As you have read, boundaries in Europe have been contested and redrawn many times as a result of conflicts, negotiations, and independence movements. Recall how the breakup of the Soviet Union resulted in the creation, or re-creation, of independent states such as Latvia, Lithuania, and Estonia. Each of these states had to establish its borders.

The Contemporary Political Map **265**

Even boundaries based on physical features can fluctuate. Natural features such as rivers, in fact, make notoriously poor borders because they often change course. This has repeatedly happened along portions of the southern border of the United States, where the Rio Grande, which has changed its course more than once, was defined as the boundary with Mexico when the two countries signed a treaty in 1848.

Countries establish boundaries by defining, delimiting, demarcating, and defending them. When **defining** boundaries, countries explicitly state in legally binding documentation such as a treaty where their borders are located, using reference points such as natural features or lines of latitude and longitude. Definitional boundaries are typically straightforward and all interested parties agree on them, but there are sometimes exceptions, such as in the case of Belize and Guatemala. In 1859, Guatemala had achieved its independence from Spain, but Belize was still a British colony called British Honduras. That year, the British and Guatemalan governments signed a treaty establishing the boundary between the two states. In the 20th century, however, Guatemala declared the treaty was invalid and staked a claim to more than half of Belize's territory. The citizens of both countries have agreed to submit the border dispute to the UN's International Court of Justice, and both governments have agreed to abide by the court's decision, which may take several years to be concluded.

Countries **delimit** their boundaries by drawing them on a map in accordance with a legal agreement, as the United States did in its 1848 treaty with Mexico. Sometimes boundaries are **demarcated** with physical objects such as stones, pillars, walls, or fences. However, many long stretches of border between countries have no demarcation at all because physical markers or barriers are thought to be impractical or unnecessary—or just too difficult to construct. In the 21st century, the United States government debated whether its entire border with Mexico could or should be demarcated with a wall, and the difficulties of building such a physical barrier were very much part of the discussion. The border is more than 1,900 miles long, and much of it crosses rugged, isolated landscapes. Estimates for building the barrier ranged from $17.3 million per mile to $36.3 million per mile. American citizens and lawmakers disagreed over whether the potential benefits of completing a border wall would exceed the effort and expense of building it. Construction begun in 2016 was halted by presidential order in 2021. However, funding was provided to clean up construction sites and to close some gaps in California, Arizona, and Texas.

To defend their borders, countries must take steps to **administer** them, or manage the way they are maintained and how goods and people will cross them. Most of the world's borders are, to some extent, restricted, or closed. This means that people cannot freely cross the border from one country to the other. Instead, one must have official government permission to enter a country unless one is a citizen of that country, or the countries have agreements to allow entry to one another's citizens. Permission to enter a country typically comes in the form of documentation such as a visa. In rare cases, where borders are completely restricted, people are not permitted to cross at all. An example of this is the demilitarized zone between North Korea and South Korea. Some countries, such as the 26 found in what is known as the Schengen Area in Europe, which includes France, Germany, and Sweden, have decided to allow their borders to be largely open to members of the European Union (EU). Citizens of participating countries are allowed to freely move about the Schengen Area in order to work, travel, and live, as there are no passport or other types of border control. Tourists and other travelers are also able to freely cross borders within the Schengen Area. Today, the Schengen Area comprises most EU countries, with the exception of Bulgaria, Croatia, Cyprus, Ireland, and Romania—though Bulgaria, Croatia, and Romania are currently in the process of joining, and already apply Schengen rules to a large extent. In addition, the non-EU states Iceland, Norway, Switzerland, and Liechtenstein have joined the Schengen Area. As a way to slow disease transmission, some border protocols, or sets of rules, were reinstated during the COVID-19 pandemic.

While the United States government debated the building of a border wall with Mexico, large sections of wall or fence were contructed in South Texas with considerable environmental damage. Openings allow residents to access the land just south of the barrier. Notice that the Rio Grande—the river that forms the natural boundary between the United States and Mexico—lies just beyond the fence.

Case Study

The DMZ in Korea

The Issue One of the most tightly closed borders in the world exists along the Demilitarized Zone on the Korean Peninsula, but glimmers of greater openness are appearing.

Learning Objective
LO 9.8 Explain the ways in which countries establish and administer their boundaries.

South Korean soldiers stand guard by the fence that marks the Demilitarized Zone (DMZ) between North Korea and South Korea. Even though the DMZ is heavily guarded, the two countries have worked to reconnect parts of the region in recent years.

By the Numbers

150 miles
Length of Demilitarized Zone

51.7 million
Population of South Korea

25.8 million
Population of North Korea

Sources: Encyclopedia Britannica; CIA World Factbook

Between 1950 and 1953, North Korea and South Korea fought a war that ended with no victor when an armistice agreement was signed. The Armistice ended the fighting and established a boundary called the Demilitarized Zone (DMZ) that divides the Korean Peninsula roughly in half. At 150 miles long and 2.5 miles wide, it's a buffer between the two countries that has kept hostilities mostly at bay for seven decades.

The war, fought between communist forces in the north and anticommunist capitalists in the south, was a fight that was part of the larger Cold War. The north received aid and training from China and the Soviet Union, while the south was supported by the United Nations and principally the United States. Because the two countries remain divided along democratic capitalist and authoritarian communist lines, South Korea is a strong ally of the United States while North Korea is an adversary.

Despite the differences between the two countries' economies and approach to governing, the north and south have found a way to coexist. The DMZ is heavily fortified with large numbers of troops from both countries on both sides, who are not allowed to cross the Military Demarcation Line that runs through the center of the DMZ. At the Joint Security Area in the village of Panmunjom, where the armistice agreement was signed, soldiers from North and South Korea stand face to face. The two countries still use the site for meetings and negotiations.

Although this boundary is seemingly impenetrable, North and South Korea do have some connectivity and, in recent years, more links have formed. The two countries have reconnected their railways and roads and, in late 2018, a train crossed the border for the first time in more than a decade. Other discussions about connecting the two countries have taken place, and there is some hope that they will host the 2032 Olympic Games together.

Geographic Thinking

Explain why it is important that the border between South and North Korea is clearly demarcated.

The Contemporary Political Map

National Geographic Photographer **David Guttenfelder**

Revealing Mysteries

Guttenfelder's photographs of everyday life in North Korea have made him a seven-time finalist for the Pulitzer Prize.

Learning Objective
LO 9.8 Explain the ways in which countries establish and administer their boundaries.

One of the most mysterious places on the planet is North Korea. Since the country was established in 1948, the outside world has had little opportunity to see what happens inside its borders and how its people live. Recently, National Geographic photographer David Guttenfelder has helped to lift the veil of secrecy that shrouds North Korea.

In 2011, the Associated Press opened its first bureau in North Korea, and Guttenfelder became the first Western photographer to cover the rarely photographed country. His purpose, as he sees it, is to show the world how normal, everyday North Koreans live. "My job was really to get as far out there as I could go and as close and intimate as I could be with people and tell their story."

Guttenfelder's photographs add up to a picture that is larger than the sum of its parts. Although each individual shot might seem ordinary, when taken collectively they tell a relatable story of life in North Korea beyond the seemingly impenetrable boundary of the DMZ. These pictures of people going to work, waiting for the bus, hugging loved ones, or riding an escalator show that there's much more to North Korea than what we learn in the news about its government and its role in global politics.

"No one believes there's real people there," Guttenfelder says. "I find myself arguing with people who would say, 'Oh that's not true. There's not a father with his daughter in a supermarket. That's all just fake.' It was very surprising to me that I was having to defend that there was a real life and real people living lives in North Korea."

Critical Viewing Shoppers go about their daily lives, comparing goods in a supermarket in Pyongyang, North Korea's capital. Describe how the images of North Korea and its cultural landscape at this particular scale alter your perspective on the country.

Critical Viewing *Top:* Bicyclists ride past portraits of North Korean leaders Kim Il Sung and Kim Jong Il in Pyongyang. Portraits of the leaders are displayed prominently throughout the country. *Bottom:* Commuters in Pyongyang head home from work on the subway. What do these photographs suggest about how North Korea uses its leaders as a centripetal force?

Types of Boundaries

Learning Objective
LO 9.9 Compare the different types of political boundaries that separate geographic areas.

Geographers define many different types of boundaries by considering not just their physical features but also how, when, and why they were created. **Antecedent boundaries** are established before many people settle into an area. An example of this is the boundary between the United States and Canada, established at the 49th parallel in 1846, before most European American settlers moved into the territories that became Minnesota, North Dakota, Montana, and Washington. The boundary between Malaysia and Indonesia on the island of Borneo is also considered to be antecedent because the Dutch and British colonists established it when the area was lightly inhabited. Like those drawn in Africa during the Berlin Conference, this boundary is also a legacy of imperialism.

Unlike antecedent boundaries, **subsequent boundaries** are drawn in areas that have been settled by people and where cultural landscapes already exist or are in the process of being established. These types of boundaries are the most common, since the process of establishing them is lengthy and related to territoriality. Many of the boundaries in Europe are subsequent, having evolved over centuries between neighboring states. France and Germany, for example, are delineated with subsequent boundaries.

A **consequent boundary** is a type of subsequent boundary. Consequent boundaries take into account the differences that exist within a cultural landscape, separating groups that have distinct languages, religions, ethnicities, or other traits. Recall what you have learned about the formation of new states in the Balkan region after the breakup of the Soviet Union and the devolution of Yugoslavia that followed. The borders between many of the newly formed countries in the former Yugoslavia may be called consequent boundaries because they follow ethnic and cultural divisions in the region. Serbia, for example, encompasses territory inhabited in large part by a single ethnic group, the Serbs. Croatia's boundaries enclose a population that is more than 90 percent Croat. When India gained its independence from Britain in 1947 and the territory was divided into modern-day India and Pakistan, the consequent boundary that was defined, delimited, and partially demarcated between the two states ran along religious lines: Pakistan's population is mostly Muslim, whereas India is largely Hindu.

While subsequent and consequent boundaries could be said to arise naturally from patterns of human settlement and the growth of cultures, other boundaries are **superimposed**, or drawn over existing accepted borders, by an outside or conquering force. This occurred in Africa when European colonial powers met at the 1884 Berlin Conference. The European empires drew up the boundaries of the new countries they conquered with no regard to the culture or ethnicity of the people who lived on the land. In many cases, ethnic groups were split by the superimposed boundaries established by the 1884 Berlin Conference. When groups from a variety of cultural and ethnic backgrounds are forced to live alongside one another inside boundaries superimposed from the outside, the possibility of conflict within countries, or even civil wars, increases.

Looking at the "Africa: Political Boundaries and Cultural Groups" map, it's impossible not to notice that some boundaries feature many curves and squiggles, whereas others are perfectly straight. These **geometric boundaries** are mathematical and typically follow lines of latitude and longitude, or are straight-line arcs between two points, instead of following the irregular lines of physical and cultural features. Many states in the western United States, such as Colorado, Wyoming, New Mexico, and Utah, have geometric boundaries. Geometric boundaries may be superimposed, as in Africa, or they may be antecedent.

Like boundaries based on natural features, geometric boundaries can be flawed and cause conflict when they are applied without thought for the people living on the lands being delimited. Consider major portions of the Kenya–Tanzania and Somalia–Ethiopia borders created by the 1884 Berlin Conference, both of which are straight-line borders that cut across ethnic lines. Ill-considered superimposed geometric boundaries are one reason armed conflict has occurred in parts of Africa.

Africa: Political Boundaries and Cultural Groups

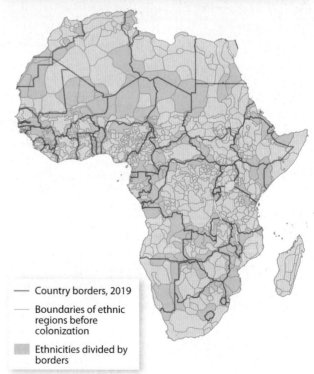

— Country borders, 2019
— Boundaries of ethnic regions before colonization
▨ Ethnicities divided by borders

This map highlights how the political boundaries created at the 1884 Berlin Conference were drawn with no regard to the diverse nations already living in Africa.

Former boundaries that once existed but no longer have an official function are considered to be **relics**. These borders illustrate how the control and management of geographic space changes over time as a result of different circumstances. The boundary between the former East and West Germany, for instance, is a relic whose presence continues to be felt. After its reunification in 1990, Germany became a single, unified country. Still, the impact of the former boundary lingers. The capitalist economy of West Germany created a very different physical and cultural landscape than the communist economy of East Germany during the 45 years they were separated. The drab, utilitarian housing of communist East Germany can still be seen in some neighborhoods of Berlin which were previously in East Germany, while areas that were previously in West Germany generally have more varied and attractive housing that was designed to appeal to consumers.

In Southeast Asia, Vietnam also was divided into two countries as part of the Cold War conflict between the United States, Soviet Union, and China. Unlike Korea, Vietnam was reunited under a single government in 1976, making the former border between North and South Vietnam a relic. Many hope the boundary between North and South Korea will also cease to exist someday. If that comes to pass, the former border will be a relic with a lasting and profound impact because of the enormous differences between South Korea, a capitalist democracy with a core economy, and North Korea, a totalitarian, peripheral state.

Sea Boundaries Not all boundaries exist on land. In fact, many are miles out to sea. Also called maritime boundaries, sea boundaries allow countries access to offshore resources such as coastal sites for wind farms and oil and other minerals beneath the sea floor. Countries with sea boundaries are typically more economically developed than those that are landlocked because having maritime ports of entry makes it significantly easier to conduct trade with other countries. Bolivia is an example of a landlocked country that is poorer than its maritime neighbors, who have little incentive to help an economic rival reach their ports. Landlocked countries have also suffered from not receiving the flow of people and ideas that have generally made maritime countries with ports—and thus greater access to the rest of the world—more dynamic through interaction and innovation.

The 1982 **United Nations Convention on the Law of the Sea (UNCLOS)** established the structure of maritime boundaries, stating that a country's territorial seas extend 12 nautical miles off its coast and that its **exclusive**

This cement barrier is the longest surviving stretch of the Berlin Wall. After West Germany and East Germany were reunified, artists began painting colorful murals on the once-drab surface.

The Contemporary Political Map

World Maritime Boundaries

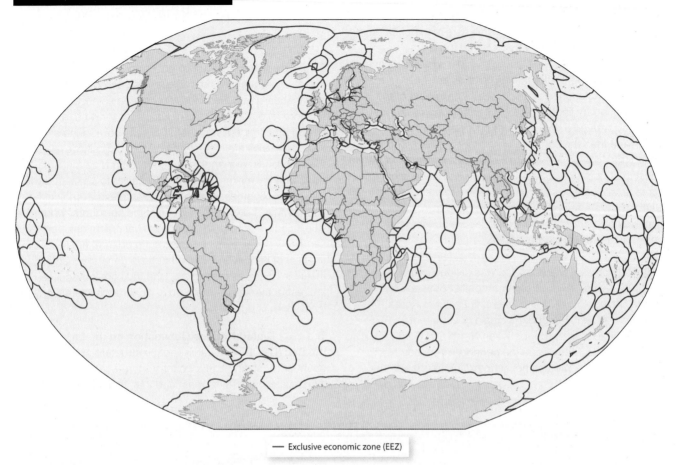

— Exclusive economic zone (EEZ)

Reading Maps The map shows the location of the world's exclusive economic zones (EEZs), which extend 200 nautical miles off the coasts of the countries that control them. ▌Describe how maritime boundaries like the EEZs might influence foreign policy among the coastal states of the world.

economic zone (EEZ) extends 200 nautical miles from its coast. (A nautical mile is 1.1508 land miles, or 6,076 feet.) UNCLOS also specifies rules for determining how territorial seas and EEZs should be measured and delimited.

Countries exert different levels of control over their territorial seas and their EEZs. States have complete sovereignty over their territorial seas, covering not only the surface but reaching down to the layers beneath the seabed and up into the airspace above the water. The principal restriction on this sovereignty is that countries must permit "innocent passage" of foreign ships through their territorial waters. Innocent passage is defined as nonstop direct travel through territorial waters between two points outside of a country's borders or from a point outside the country's borders to one of its ports. States do not have full sovereignty over their EEZ, but they do have sole access to resources found within the waters or beneath the sea floor of the EEZ, such as fish, oil, and natural gas. Countries also have the exclusive right to generate energy from the waves, wind, or currents inside their EEZ.

Along with the rights governing territorial waters and EEZs come certain responsibilities. For example, UNCLOS specifies that within its EEZ, "the coastal State, taking into account the best scientific evidence available to it, shall ensure through proper conservation and management measures that the maintenance of the living resources in the exclusive economic zone is not endangered by over-exploitation." In other words, UNCLOS requires coastal countries to employ sound environmental practices in the waters they control, such as avoiding overfishing and depleting ocean species. In addition, states are required to make public any dangers to navigation that they know of within their territorial waters.

Disputes over maritime boundaries sometimes arise, as has happened in the strategically important South China Sea. There, China has laid claim to territory that, according to the 200-nautical-mile rule, falls within international waters. China argues that it has historical claims to the sea, dating back to naval expeditions in the 15th century, and after Japan was defeated in World War II, China laid claim to 90 percent of the sea. To solidify its claim, China has built artificial islands on top of a number of reefs it claims in the South China Sea because the 200-mile line can be extended outward from a country's islands.

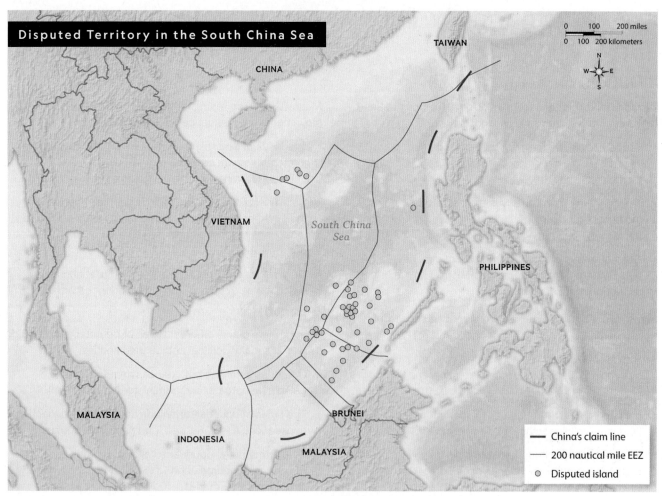

Reading Maps More than $3 trillion in trade passes through the South China Sea every year. As tensions rise over the disputed ocean territory, countries including China, Vietnam, and the Philippines have increased military spending. ▌ Describe how the unresolved dispute over maritime territory in the South China Sea likely impacts each of the states that asserts claims to the region.

China has continued to behave aggressively in the region. South China Sea experts with the Center for Strategic and International Studies (CSIS) report an ". . .expansion of China's maritime militia—a force of vessels ostensibly engaged in commercial fishing but which in fact operate alongside Chinese law enforcement and military to achieve political objectives in disputed waters." The 2021 CSIS study used remote sensing data and Chinese language sources to identify more than 120 such vessels, whose operations discourage commercial fishing, challenge oil and gas operations, and work purposely to disrupt other maritime activities by neighbors Vietnam, Malaysia, and Indonesia—violating the EEZs of these countries. Though China is negotiating with the affected countries, the disputes remain unresolved. Many are concerned that a military conflict could one day break out between China and these neighbors, who assert their own claims in the South China Sea.

Why Do Boundaries Matter? Political boundaries are the result and the reflection of the ways humans divide space. As you have seen, some boundaries are products of balanced negotiation between groups, whereas others demonstrate the power imposed by one group over another. Sometimes boundaries follow ethnic or cultural lines in an attempt to delimit nation-states, and sometimes they divide nations among multiple countries. As sources of both conflict and harmony, change and permanence, political boundaries shape the contemporary world.

Geographic Thinking

1. Compare subsequent and superimposed boundaries.

2. Describe how the nature of a country's boundaries is linked with its history and political situation.

3. Explain why countries find it necessary to delimit their boundaries.

4. Explain why coastal countries seek to expand and defend their maritime boundaries.

Chapter 9 Summary & Review

Chapter Summary

Territoriality is the connection of people, their culture, and their economic systems to the land.

- Territoriality is the basis for the power that peoples assert over the geographic areas they call their own and the political spaces they create.
- The concept of territoriality is the basis for sovereignty, the right of governments to control their territory and decide what happens inside their borders.
- Coastal waters are considered part of a maritime country's territory, and control of them is an important aspect of this sovereignty.

Maps reflect the ways people organize their spaces into political entities.

- States, or countries, are independent political units created by people to organize and manage territory and themselves.
- Nations are cultural entities comprising people who have a shared identity through traits in common, such as language, religion, ethnicity, and heritage.
- Nation-states are countries with political boundaries that match the cultural boundaries of a people who consider themselves a nation.

The contemporary world has been shaped by a variety of factors, including the concepts of sovereignty and self-determination, and attempts to form nation-states.

Meanwhile, colonialism, imperialism, and independence movements have influenced political boundaries.

- The issue of sovereignty has been complicated by the interconnectedness of the modern world, as well as by imbalances in power in relationships between states.
- Self-determination is the right of people to choose their own political status. Some peoples have launched independence movements in the name of self-determination.
- Colonialism and imperialism have left a lasting mark on countries that have won independence from their colonial rulers.

Geographers define different types of boundaries that mark the world's many political and cultural entities.

- Countries establish boundaries when they define, delimit, demarcate, and administer them.
- The types of boundaries include antecedent, subsequent, consequent, superimposed, and geometric boundaries.
- Sea, or maritime, boundaries allow countries access to offshore resources and to exercise their sovereignty over offshore territories.
- Cultural boundaries differ from political boundaries and run along lines that separate people of different ethnicities or cultural backgrounds.

Review Questions

Use complete sentences to answer the questions.

1. **Apply Conceptual Vocabulary** Consider the terms *political geography* and *nation-state*. Write a standard dictionary definition of each term. Then provide a conceptual definition—an explanation of how each term is used in the context of this chapter.

2. Provide an example of how a country uses its sovereignty to control what happens within its borders.

3. Why could the Panama Canal be considered a choke point?

4. Describe how a government might manipulate a map to tell a particular story.

5. Explain why the country of Bolivia is considered an independent state, while the Canadian province of Manitoba is not.

6. Identify a nation that does not have its own state.

7. What is the difference between a multistate nation and a multinational state?

8. Define the term *stateless nation*.

9. Explain why the 1884 Berlin Conference was so devastating to the continent of Africa.

10. Describe the difference between a delimited boundary and a demarcated boundary.

11. What type of boundary typically features straight lines?

12. How do the terms of the United Nations Convention on the Law of the Sea benefit countries with maritime boundaries?

■ **Interpret Maps**

Study the map and then answer the following questions.

13. **Identify Data & Information** What types of information does this map give about the countries of South America? What information does it not include?

14. **Analyze Maps** What feature defines the border between Chile and Argentina?

15. **Explain Patterns & Trends** What evidence of colonialism does this map show?

16. **Analyze Geographic Concepts** In the early 1800s, Colombia, Ecuador, Panama, Venezuela, and parts of other countries all formed the state called Gran Colombia. For what reasons might Gran Colombia have broken up?

17. **Analyze Maps** Why is Paraguay at a possible disadvantage compared to most of its neighbors? Explain your answer.

18. **Explain Geographic Concepts** How does a map like this one illustrate the concepts of territoriality and sovereignty?

Chapter 10
Spatial Patterns of Political Power

Critical Viewing In 2019, protesters demonstrated in front of the U.S. Supreme Court against the controversial process of gerrymandering. The Court was hearing two cases that impacted legislative districts in Maryland and North Carolina. ▮ How do political boundaries affect people's lives?

Geographic Thinking What role do boundaries play as symbols of political power?

10.1
Organization of States

Case Study: Political Control and Nunavut

10.2
Electoral Geography

Case Study: Gerrymandering and Race

10.1 Organization of States

States are divided into smaller regional units to make governance more efficient. (As introduced in Chapter 9, in this context, *state* means "country.") The ways in which a state's internal boundaries are set and its regional units administered reflect the balance of power between the central, or national, government and its internal political units.

Unitary States

Learning Objectives
LO 10.1 Identify the advantages and disadvantages of unitary states.
LO 10.2 Compare the political geographies of unitary systems and federal systems.

Different forms of governance, or how a state is organized politically and spatially, affect a country's economic and social affairs. Most state governments are organized in one of two ways, either as a **federal state** or as a **unitary state**. Where power is held within a country affects the amount of authority governments have at national and regional levels. In federal states, considerable power is held by regional units, such as the states of the United States or the provinces of Canada. These political units typically have their own governments that maintain some autonomy and hold substantial power. In unitary states, more power is held by a central government that maintains authority over all of the state's territory, its regional units, and its people.

A unitary state has a form of government that follows a top-down approach in which policies are conveyed by the central government and funneled down to regional units to be carried out. The central government creates its internal units, such as provinces, states, or other regional and local divisions, which are sometimes given a degree of power, perhaps even to make regional laws. But a unitary state always maintains supreme authority at the top.

The vast majority of the world's countries are unitary states. The United Kingdom is one example. The UK Parliament is the central governing authority over its four constituent units: England, Wales, Scotland, and Northern Ireland. Parliament controls a range of national affairs, including military defense, foreign relations with other countries, and immigration for the entire United Kingdom. To provide for each unit's own particular needs and concerns, the central government allows Scotland, Wales, and Northern Ireland to have their own national assemblies and to administer systems of local governments called districts, boroughs, and councils. (England does not have its own assembly.) Local governments oversee issues such as health services, housing, education, and the environment.

Other unitary states include Poland, France, Spain, China, Indonesia, Bangladesh, Algeria, and Japan. The constitution of Japan establishes a democratically elected central government with a prime minister, while also giving power to local units. The central government in Tokyo holds most of Japan's authority, with some autonomy given to a system of 47 local units called prefectures. While each prefecture has sizable administrative power within its territory, these local units are overseen by a central ministry and grouped into eight regions. Lower levels of government lie in each prefecture's cities, towns, and villages. In France's unitary system, the central government has supreme power over its major subunits, called provinces. The provinces do not have the power to act independently. France is also a republic, a state in which power is held by the people and their elected representatives; French citizens vote in democratic elections and choose a president, who then appoints a prime minister.

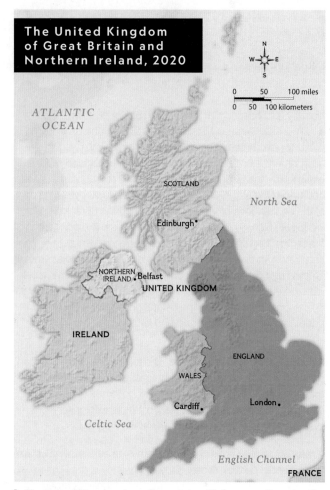

The United Kingdom is a unitary state with four constituent units: England, Wales, Scotland, and Northern Ireland. Though each unit has its own government (except for England), Parliament, the national legislature in London, holds central power over them.

Spatial Patterns of Political Power **277**

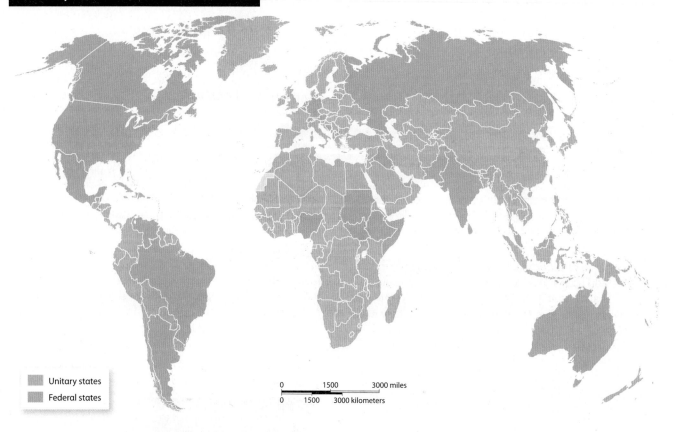

Unitary and Federal States, 2020

- Unitary states
- Federal states

Reading Maps Of the 193 countries that are United Nations members, the vast majority are unitary states. ▌Explain one possible reason why there are so many unitary states in Africa.

Norway also has a unitary, top-down power structure. County and municipal authority is solely and totally granted by the national government. For instance, the central Norwegian government is in charge of devising health policy and establishing the budget for health services throughout the country. In the early 2000s, the Ministry of Health, which is part of the central government, coordinated efforts to tackle health inequity in Norway (equity, relating to fairness, is discussed further in Chapter 19). Through comprehensive legislation and oversight by the central government, health policies were created and distributed to county and municipal governments who are each responsible for enacting different health services. This coordinated effort by the central government ensured coherence across the regional and local levels.

Advantages of Unitary States The top-down nature of unitary systems gives them several advantages. One advantage is that unitary systems tend to have fewer government agencies, especially those dealing with taxation. States with unitary systems also tend to be less corrupt at the local level.

A major advantage of a unitary system is its efficiency: laws are implemented quickly, evenly, fairly, and with less duplication. For example, the efficiency of Singapore's highly regarded transportation system is credited to that city-state's unitary system of government. Since the 1970s, spatial planning has been done according to a nonstatutory framework called the Concept Plan, reviewed every 10 years, and by which all public transportation decisions are made. In 1995, a governmental agency called the Land Transport Authority (LTA) was established to implement, make policy for, and regulate all urban transportation systems throughout Singapore—all based on the Concept Plan. The LTA also builds and maintains the roads and transit infrastructures, both local and regional, providing a range of connected options for millions of daily commuters. Many take advantage of the extensive road–rail–bus network, or cycle and walk to destinations. The fact that one organization holds so many responsibilities related to transportation allows decisions and actions to be taken more deliberately. Furthermore, because the LTA's decisions are always connected to the original Concept Plan, upgrades and expansions are implemented quickly and fairly.

Another example of unitary efficiency is found in the Netherlands, where decentralization has been applied to social policy. In 2011, the Dutch central government decided to shift certain tasks and responsibilities to regional and local units. Regional governments took on spatial planning, environmental concerns, regional economic issues, and transportation. Local authorities began to manage disabled and elder care, as well as support for youth. A law called the Child and Youth Act went into effect January 1, 2015. It empowered local municipalities to handle youth-related issues and tend to specific needs in a timely manner through the use of local social teams, which are groups of like-minded experts, in each municipality. While a federal agency, the Ministry of Interior Affairs, supports the municipalities' work and fosters good governance, the country's 42 youth care regions (and further subregions) do the actual work.

Disadvantages of Unitary States A unitary system of government often has negative characteristics. The overarching issue is that highly centralized governments can become disconnected from local areas and lose touch with the issues that concern people living there. Unitary systems tend to favor the politically or culturally dominant group, resulting in governments that often issue one-sided policies that ignore the concerns of minority groups and local cultures. Also, the policies tend to serve the needs of the region adjacent to the capital or where the ruling elites reside. As a result, unitary governments can be slower than federal governments in responding to local issues. They may also fail to equitably distribute goods and services to peripheral areas or even have difficulty providing services to localities at all.

An incident in the People's Republic of China illustrates how a unitary government must balance centrality and efficiency with the concerns of local areas. When a massive earthquake struck an area of Sichuan Province in 2008, the government quickly sent 130,000 workers to remote areas to help with relief efforts. Yet the event also exposed the need for stronger building codes and improved early warning systems in those places. The centralization of the government had actually been creating inefficiency: the responsibilities of several agencies and ministries overlapped to such a degree that the main issue—disaster management—was not properly tended to. In 2018, the central government in Beijing created an agency devoted entirely to improved emergency management. In the face of COVID-19, China implemented a unified health code for the entire country, following this trend to centralize policies. Supported by digital health records and other advanced surveillance technologies, China has used contact tracing and other strategies to monitor the pandemic. However, some fear this centralization has been used to solidify the authoritarian power of the central government, deepening its ability to intrude upon the lives of its 1.4 billion citizens.

Federal States

Learning Objectives
LO 10.3 Identify the advantages and disadvantages of federal states.
LO 10.2 Compare the political geographies of unitary systems and federal systems.

A federal state has a form of government in which the country's power is more broadly shared between a federal—sometimes called national—government and its regional units. These regional units, such as provinces or states, maintain greater

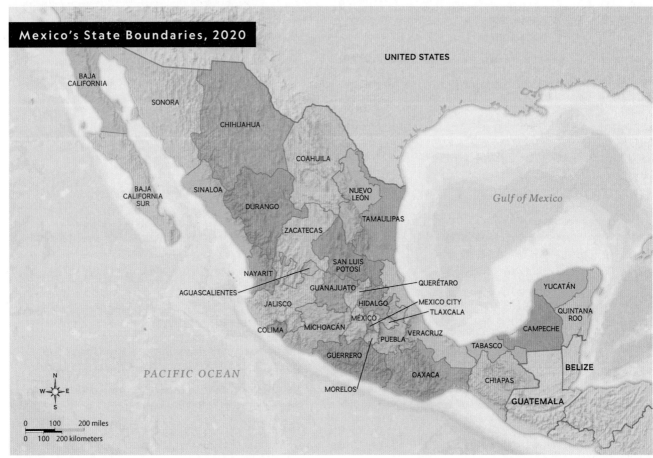

Reading Maps Mexico comprises 31 states and a federal district in Mexico City. The geography, size, and population vary among each state.
▍ Based on what you observe in this map, what factors might impact the way internal state boundaries are drawn under a federal system?

Spatial Patterns of Political Power **279**

The Sharing of Powers

The federal system in the United States results in numerous concurrent, or shared, powers between the federal government, state governments, and local governments. ▌Why do you think states decide school curricula?

Federal

All U.S. citizens are subject to federal laws, regulations, and policies and use the same currency no matter what state they live in. State and local governments help to enforce federal laws and policies.

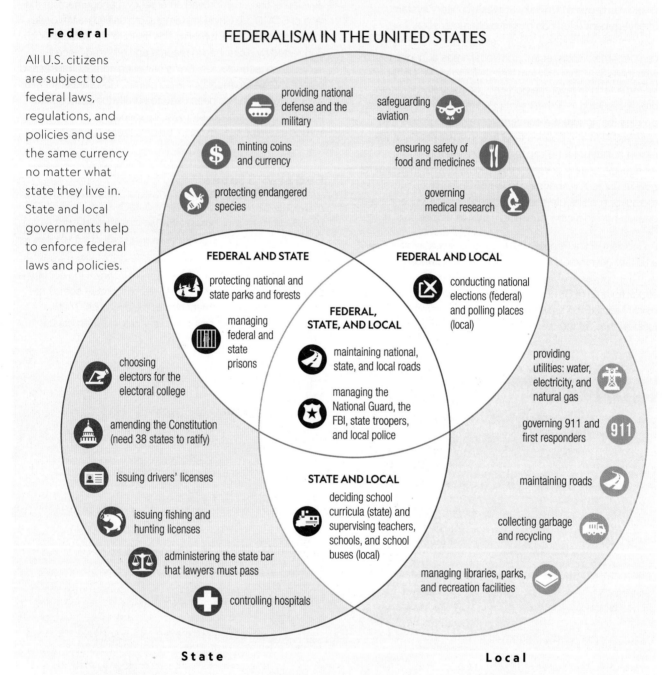

FEDERALISM IN THE UNITED STATES

- providing national defense and the military
- safeguarding aviation
- minting coins and currency
- ensuring safety of food and medicines
- protecting endangered species
- governing medical research

FEDERAL AND STATE
- protecting national and state parks and forests
- managing federal and state prisons

FEDERAL AND LOCAL
- conducting national elections (federal) and polling places (local)

FEDERAL, STATE, AND LOCAL
- maintaining national, state, and local roads
- managing the National Guard, the FBI, state troopers, and local police

- choosing electors for the electoral college
- amending the Constitution (need 38 states to ratify)
- issuing drivers' licenses
- issuing fishing and hunting licenses
- administering the state bar that lawyers must pass
- controlling hospitals

STATE AND LOCAL
- deciding school curricula (state) and supervising teachers, schools, and school buses (local)

- providing utilities: water, electricity, and natural gas
- governing 911 and first responders
- maintaining roads
- collecting garbage and recycling
- managing libraries, parks, and recreation facilities

State

Local

All citizens of a particular state are subject to its state laws, regulations, and policies. Moving from one state to another, citizens are still subject to federal laws and policies, but a different set of state laws and policies. What is legal in Nevada may be illegal in Alabama, and state policies and services in New York may be different from those in Arizona.

At the local level, citizens are still subject to federal and state laws, but the regulations and policies in one city may differ even from the neighboring city. The services provided by each city and the local taxes residents pay also differ. Aspen, Colorado, utilizes 100 percent renewable energy to provide electricity to the city, while Pueblo, Colorado, still used coal as of 2022.

Patriotism and Nationalism

Patriotism, the love that one feels for one's country, can be found in both unitary and federal systems. In a unitary system, patriotism is often bolstered by the sense of uniformity that results from the central system itself. In a federal system, patriotism tends to be more complex. It is not unusual for someone to feel a sense of belonging to both their local town or region and their country simultaneously. For example, a long-time resident of New York City may feel like a New Yorker as much as she feels like an American.

An increase in patriotism, however, can have both negative and positive results. On the positive side, patriotism causes citizens to have more pride in their country's history, culture, and accomplishments, leading to a stronger desire to find common good. Ruling governments may discover more popular support for their policy initiatives when patriotism is high. The development of a national identity can also help combat, or at least offset, certain forces that divide a state, such as political or economic inequality. These forces, called devolutionary forces, are discussed in Chapter 11.

Patriotism is sometimes problematic, too, particularly when it leads to nationalism. When individuals feel greater devotion to the state than they do to other group interests, nationalism can occur. Nationalism as an ideology can be both positive and negative. The desire of a people to form their own nation and determine their own destiny drove the creation of the United States of America, for example. However, taken to an extreme, nationalist sentiments can lead to intolerance and exclusion of groups deemed outside of the national norm, such as racial, ethnic, or religious minorities. Indeed, history is littered with examples of toxic nationalism resulting in national, international, and intranational tensions and conflicts, including the Holocaust in Europe, Japanese internment in the United States during World War II, and the attempted insurrection at the United States Capitol on January 6, 2021.

autonomy, have their own governments, and have more authority to administer their regional territories to meet the needs of diverse groups. As a result, power is shared between central, regional, and local governments much more broadly than in unitary states. Federal states have often been formed where populations are very large, highly dispersed, or both. Examples of federal states include Russia, India, Brazil, Malaysia, the United States, Australia, Germany, Sudan, and Mexico.

The Mexican Federal Constitution provides for a central government that shares power with 31 organized political districts, called states, and a federal district, each with their own government. Similar to the United States, the federal government in Mexico (centered in its capital, Mexico City) has specific powers, while its state governments possess other powers. Mexican states hold the right to pass laws related to issues of state importance, such as raising local taxes and conducting state elections.

Mexico also has more than 2,000 municipalities, which are organized by location and centered in cities, towns, villages, and rural areas. Municipal governments enact local policies and oversee issues like public parks, public services, public safety and traffic, and urban planning. The municipalities manage their local concerns, the states manage their regional interests, while the central government focuses on issues affecting the entire country.

Germany has a federal system in which the central government shares its political power with its subunits, called Länder, but holds ultimate authority over them. Germany is a republic as well, and German citizens vote to elect representatives to the Bundestag, one of two legislative bodies. Members of the Bundestag, in turn, elect a chancellor who heads the government.

The United States is also a federal state. Its central government shares power with the governments of its 50 states (here, the term *state* refers to one of the United States, such as Ohio) and their numerous counties, cities, and towns. The federal structure has given state governments the sole power to conduct elections, issue marriage licenses, regulate intrastate commerce (meaning business within the state), administer driving regulations, and other activities. In fact, the Tenth Amendment to the U.S. Constitution "reserves to the States" any powers not explicitly "delegated to" the federal government. The federal government and state governments also have some **concurrent**, or shared, powers. Those include levying taxes, making and enforcing laws, establishing courts, and borrowing money.

The power structure of federal and state governments impacts groups of people across the United States in different ways. One area where this is exemplified is transportation, specifically the nation's highway system. The federal government initiated a national system of interstate highways, called the Interstate System, in the 1950s. While the federal government funded the entire system, each individual state constructed the actual highway (or highways) that ran through it.

Today, the Interstate highways are owned by the state in which they were built. Individual state transportation agencies set and enforce their own speed limits, fix the highways when they need repair, and some collect tolls that help pay for those repairs. Each state also has its own system of state highways. State highways are funded by a combination of state-determined tolls, user fees, and taxes (including gas taxes). U.S. interstates and state highways are recognized by their differing signage. Both are shield shaped, but interstate signs are blue and red, while state highway signs vary from state to state.

State governments also have the power to establish local governments and distribute certain powers to them. In the United States, governments of counties and municipalities are responsible for decisions made regarding local parks and recreation, emergency services, police and fire departments, public transportation, and public services such as sewers and snow removal.

Advantages of Federal States One positive result of decentralizing power in a federal system is the reduction of conflict between regions that differ on civil and political issues. A regional unit can pass a law that applies to it and not to the rest of the country. In this way, a federal state allows for the diversity of opinions, as reflected in its laws. A good example of this is the death penalty in the United States. The U.S. Supreme Court has ruled that the death penalty is constitutional, except in cases where the offender is mentally disabled or a juvenile. However, each of the U.S. states, like Pennsylvania or Texas, is permitted to ban the death penalty if it wishes.

Federalism also allows room for diversity. Multiple political parties can be in power in different areas of a country, and this pluralism, or coexistence of more than one party, helps keep oppression by one authority at bay. It also pushes against divisive forces that result from economic or cultural differences within a state. Attention to local issues within a federal system also boosts political participation among its citizens who want to make a difference in their local community. There are more than half a million elected officials in the United States, the vast majority (96 percent) of them in local offices, as in county, city, town, and village posts—plus school boards and other special districts. In a federal system, government efficiency occurs when local governments can tend to local needs.

Disadvantages of Federal States A federal system is not perfect. Many disadvantages are the downsides of its perceived advantages. For example, the focus of a "state" in a federal system (like Minnesota or Mississippi in the United States) on regional and local issues can allow regional and local leaders to stymie, or prevent, progress on issues that may impact the whole country. Policy areas like civil rights, energy, poverty, and pollution have all experienced roadblocks at the state or local level. Federalism can give undue, or improper, power to localized special interests. During the civil rights era in the United States, those in favor of racial segregation claimed that U.S. states' rights trumped national powers, which allowed them to dodge federal discrimination laws. In fact, decentralization contributed greatly to the spread of segregation. When civil rights for African Americans were mandated by Supreme Court rulings, including *Brown* v. *Board of Education*, in the 1950s and 1960s, some progress was made to protect rights for all.

Another example of the disadvantages of a federal system is illustrated by a 2022 U.S. Supreme Court decision, in which the Court struck down *Roe* v. *Wade*, a 1973 ruling by the high court that gave women the constitutional right to abortion. The court thus shifted the fight over a woman's right to choose from the federal level to the individual state level. Within days of the decision, more than 20 states with conservative governments enacted bans on terminating pregnancy, while states with more liberal governments worked to protect medical providers from prosecution and to ensure access to contraception and other women's health services—even for out-of-state women in need of assistance. Women's rights advocates decried the negative impacts on women's health of the Court's decision—particularly for low-income and minority women. Across the country, both women's health providers and women in need of pregnancy services had to navigate a new and varied patchwork of states' laws that replaced the "law of the land"—a federal right to choose. To combat total bans in some states, a federal (Executive Branch) action came within two weeks of the Court ruling: the Department of Health and Human Services issued guidelines asserting that "federal law requiring emergency medical treatment supersedes any state restrictions" on the termination of pregnancies "in cases where the pregnant patient's life or health is at risk." Political and legal battles at multiple levels of United States government had just begun in mid-2022, exposing a major difficulty with its federal system.

Another negative aspect of federalism is that the costs and benefits of federal policy and aid are often distributed unevenly among the country's regional or local governments. Political motivations impact policies and affect where money is directed. Therefore, poorer communities can suffer in a federal system when they receive relatively fewer or lesser-quality services in areas like health and welfare, police, and environmental protection because they aren't being represented as robustly as other, wealthier communities.

An example of spatial inequality resulting from a wide disparity in tax revenue can be found in the funding of public education in the United States. Public schools are supported by a combination of federal, state, and local funds, with federal money accounting for only about 10 percent of the total figure. Money raised by property taxes at the local (and sometimes state) level is the major source of funds for a locality's school districts, while the monies are directed and supervised at the state level (in all but five states). As a result, a school district that is home to high-value properties generates higher property taxes, and its schools are therefore more well funded. By contrast, a less-populated or poorer district would generate lower property taxes, and its schools would be more reliant on state funding, which is often scarce.

COMPARING TWO FORMS OF GOVERNANCE

Federal	Unitary
Power is divided between the national government and state and local governments.	Power is held in one central, national government.
Promotes diversity.	Promotes strong sense of national identity.
Power may be diffused.	Very little power is diffused.
State or provincial governments have some degree of self-rule and have their own legislatures.	Laws are standardized and implemented across the country.
Change can come slowly; conflicts between governments occur; abuse of power is prevented.	Change can come quickly; there is less intergovernmental conflict; abuse of power is more likely.

Federal States versus Unitary States There are many distinctions between federal and unitary systems of government that can impact a state's unity. Sometimes tensions occur between different groups living within a state.

Case Study

Political Control and Nunavut

The Issue Though federalism tends to support diversity—whether cultural or political—many countries with federal systems struggle to adequately serve their minority populations. Canada provides a good example of a federal system that accommodates multiple nationalities, ethnicities, and cultures.

Learning Objective
LO 10.3 Identify the advantages and disadvantages of federal states.

By the Numbers

1,296 Miles
Distance from Ottawa to Iqaluit, capital of Nunavut

39,589
Population of Nunavut (2021)

38,436,447
Population of Canada (2021)

Sources: Google Maps; The Canadian Encyclopedia; Statistics Canada

Nunavut is homeland to the Inuit, an Indigenous people of Canada who keep their traditions alive by wearing traditional clothes and engaging in activities like dog sledding. Nunavut's government works to meet the needs of its majority Inuit population using resources provided by Canada's federal government.

Canada Has a Federal System with a central government seated in Ottawa, its capital city. Canada's federal government holds authority over 10 provinces and three large territories: Yukon, the Northwest Territories, and Nunavut. Each province has its own legislature, and the territories are governed by the federal government while also having their own territorial governments.

The Inuit, an Indigenous people of the Arctic, make up four-fifths of Nunavut's population. Nunavut was part of the vast, remote Northwest Territories until the 1990s. The Inuit had pushed for independence at a time when their lifestyle was transitioning from one based on hunting, fishing, and fur trapping to one in which they were more permanently settled in small communities; much of this was a result of the federal government's forced settlement policy of these migratory hunters and fishers. Eventually, the Inuit wanted more control of their own local affairs, and they began to pressure the federal government to grant them more autonomy. In 1993, an act of Parliament recognized Inuit land rights and established Nunavut. The Nunavut Land Claims Agreement gave the Inuit political control over these lands, which became an official Canadian territory in 1999. The Inuit, who were once marginalized and discriminated against, finally had their status elevated through this spatial political reorganization.

The federal distribution of power has been beneficial for the Inuit because the territorial government, located in the capital city of Iqaluit, better attends to their needs. For example, most of Nunavut's population speak the Inuktitut language. Accordingly, Inuktitut is used in the day-to-day running of the territorial government, and Inuktitut is officially recognized alongside English, French, and Inuinnaqtun (a dialect of Inuktitut). In fact, the territorial government is working to preserve the Inuktitut language in different ways, including ensuring Inuit schoolchildren are taught in their native language.

Canada's federal government has also played a role in helping the Inuit community by redistributing resources for the development of culturally sensitive policies and programs that benefit the Inuit. In 2019, the federal government and Inuit partners implemented an indigenous skills and job training program that provides more funding and improved training for the Inuit and other First Nations peoples. However, many Inuit are still discontent with what they see as their continued marginalization and seek complete self-government through negotiations with the central government in Ottawa. Nonetheless, Nunavut provides a useful example of how the structure of political power can impact the representation and recognition of groups within a federal system.

Geographic Thinking

Describe how Canada's federal system serves Nunavut.

A federal state like the United States can also experience conflict within its constituent units. Minimum wage laws provide an everyday example: the municipal government in Birmingham, Alabama, passed a local law in 2016 raising the city's minimum wage from $7.25 an hour to $10.10 an hour. Before the law could take effect, however, the Alabama state legislature intervened to address business concerns, passing a law preventing localities from setting their own minimum wage. A group of Birmingham citizens and state officials then sued the state to force it to comply with the local law. A series of federal court decisions culminated in the dismissal of the lawsuit. By the end of 2019, the state had won the battle over the minimum wage. As of 2022, the federal minimum wage is $7.25—set by the U.S. Congress in 2009. Alabama, Iowa, and 18 other states simply use this long-established federal minimum, while others set their own minimums above the federal wage floor. By early 2022, only California had established a $15/hour minimum wage, though a number of other state legislatures have passed laws that will eventually bring their rates to that level (while the District of Columbia reached the $15/hour level in 2021, climbing to $16.10 by 2022). Cities that have established the highest wage minimums above their states' bases include four cities in California's Bay Area plus Seattle, Washington—which has the highest hourly minimum in 2022, at $17.27. Arguments for and against higher minimum wages fuel ongoing political discussions and decisions at federal, state, and local levels.

Geographic Thinking

1. Compare key spatial elements in unitary and federal systems.

2. Explain how the Chinese government's response to the Sichuan earthquake in 2008 highlights a disadvantage of unitary states.

3. Describe how the U.S. government is spatially organized and how power is distributed among its units.

10.2 Electoral Geography

When geographers study electoral geography, they look at the spatial aspects of voting: how the boundaries of voting districts are set, how that process influences voting results, and what those outcomes mean for candidates and voters.

Representing the People

Learning Objective
LO 10.4 Describe how the boundaries of voting districts influence election outcomes and the distribution of political power.

International boundaries are drawn to define and organize states. Boundaries are also drawn *within* states to divide areas into manageable spaces that are governed by different authorities. These internal boundaries define provinces (in Canada); prefectures (in Japan); states, counties, and county-equivalents (in the United States); and municipalities in many countries. Internal boundaries also form voting districts, the defining and drawing of which are sometimes manipulated to influence elections, and therefore political power. Congressional districts—voting districts in the United States—offer a prime example of the controversy that can surround the drawing of internal boundaries.

U.S. Congressional Districts In the United States, one measure of a state's political power is the number of members it has in the House of Representatives. Here again, *state* refers to one of the United States, like Virginia or Utah. A state's number of representatives depends on its population. A highly populated state like California, with 53 congresspeople, has more power in the House than a less populated state like Wyoming, with one congressperson.

As you learned in Chapter 2, every 10 years in the United States, a census is conducted to determine the number of people living in each state. These numbers are used to reconfigure each state's congressional district map. Each congressional district elects one congressperson. The number of districts a state has equals the state's number of congresspeople. The total number of U.S. representatives is always 435, so the state's population growth or decline and migration rates are key pieces of data. When one state loses people and another gains or a state's population doesn't grow as fast as others, a process called **reapportionment** takes place, in which seats in the House of Representatives are reallocated to different states. The 2020 census revealed that Pennsylvania's population growth slowed and, as a result, it lost one congressional

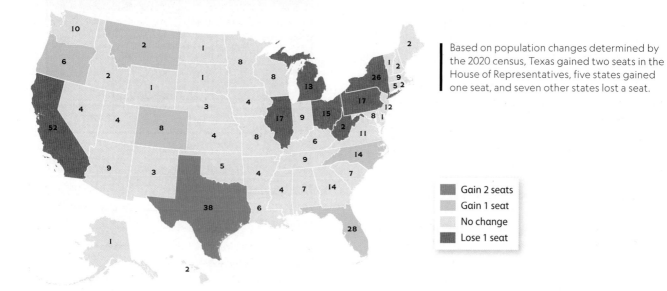

Congressional Reapportionment, 2020

Based on population changes determined by the 2020 census, Texas gained two seats in the House of Representatives, five states gained one seat, and seven other states lost a seat.

Legend: Gain 2 seats; Gain 1 seat; No change; Lose 1 seat

seat. Conversely, Texas saw high rates of population growth and gained two congressional seats. Overall, seven states lost a congressional seat in 2020, while Oregon, Montana, Colorado, North Carolina, and Florida each gained one. Reapportionment ensures that a state's population is accurately represented in the House of Representatives.

A shift in population from one geographic region to another has implications for the whole country as well. The president of the United States is not elected directly by the people, but rather by the **electoral college**, a set of people—called electors—who are chosen to elect the president. The total number of available electors (and, therefore, electoral votes) is 538, which is the same as the total membership of Congress—435 members of the House of Representatives and 100 senators—plus three electoral votes for Washington, D.C., whose citizens vote in federal elections but who do not have a voting member of Congress. Consequently, to be elected president, a candidate must win a combination of states whose electoral votes add up to a minimum of 270—one more than half. Therefore, when a state loses a congressional seat, it loses an electoral vote as well. The loss of an electoral vote is also a loss in electoral and political power for the state.

The U.S. Senate has its own representation issue—it represents states, not people. Every state has two U.S. senators in Congress, which does not result in proportional representation. North Dakota, with about 779,000 people, has the same number of senators as Florida, with more than 21.5 million people. The Senate has significant powers that the House does not, and some argue that gives small population states unfair power. For example, it now takes only 50 votes in the Senate to confirm a justice to a lifetime appointment on the U.S. Supreme Court. In 2018, Judge Brett Kavanaugh was confirmed receiving 50 votes (plus the deciding vote from then Vice President Mike Pence) that came from states holding only 45 percent of the U.S. population. In 2021, Judge Amy Coney Barrett was confirmed to the same court by a vote of 52 to 48 without a single vote from a Democrat, the minority party at the time.

Electoral District Boundaries

Learning Objectives
LO 10.5 Explain how electoral district boundaries are established.
LO 10.6 Describe how the practice of gerrymandering affects electoral outcomes.

After each census is complete, reapportionment takes place along with a process called **redistricting**. During redistricting, a state's internal political boundaries that determine voting districts for the U.S. House of Representatives and the state's legislature are redrawn to accurately reflect the new census data. Redistricting is both a geographic and political process. It is a geographic process because the boundaries of districts must be redrawn to reflect any changes in population. It is a political process because those boundaries are often drawn by the state's legislature—a political entity. In principle, voting districts should meet three criteria: they should be spatially compact, they should be physically connected or contiguous, and they should include communities with similar interests, that is, they should respect existing communities such as ethnic neighborhoods, school districts, or tribal lands. Districts for state governments are also redistricted after every census. Changing congressional or state legislative district boundaries can have a radical effect on who is elected to represent a geographic area.

Gerrymandering The party that controls a majority of seats in the state legislature typically draws legislative maps with a partisan advantage that favors their party over any

Spatial Patterns of Political Power

other. This is called partisan **gerrymandering**, named for Governor Elbridge Gerry of Massachusetts, who in 1812 signed a law approving a map of state legislative districts that were drawn to favor his own political party. One of the districts resembled a lizard-like salamander, and a Boston newspaper coined the term "Gerry-mander."

Legislators can gerrymander a district by either "packing" the district or "cracking" the district. Packing a district is when local population data is used to draw a district that is full of the opposing party's voters. Concentrating opposition voters into a single district allows a greater number of the surrounding districts the opportunity to be won by the party that is in power. By contrast, cracking a district is the practice of splitting up the opposition party's voters across many districts, thereby diluting their electoral strength. Either way, partisan gerrymandering is often used by majority parties to tip the electoral scales in their favor.

A crucial aspect of partisan gerrymandering is the role that race and ethnicity play in drawing the district's boundaries. The Voting Rights Act of 1965 was established to remedy the disenfranchisement of minorities attributed to the common practice of cracking, which spread African American voters into multiple districts to diminish their impact on elections. The act created new **majority-minority districts**. These gerrymandered districts, in which minorities made up the majority of voters, were designed to help ensure, for example, that African American voters could elect their candidates of choice. Gerrymandered districts have resulted in an increase in the number of minority representatives in Congress.

In recent electoral history, African Americans—and to a lesser extent, Latinos—have favored Democratic candidates. So, Republican lawmakers in some states have packed African American voters into a single district (or small number of districts), thereby creating majority Republican districts in the rest of the state. Consequently, the Voting Rights Act is now being manipulated to again disenfranchise minority voters. The distinction between partisan and racial gerrymandering is so slight that many who study the gerrymandering issue believe they have essentially become the same practice.

Opposition and Remedies Gerrymandering is considered by many to be unfair because voters who favor the opposition party in a gerrymandered district are essentially disenfranchised, which means they are prevented from having the right to vote. The argument is that their vote doesn't count because the district has been drawn so their party cannot win. Another argument points out that gerrymandering prevents an accurate representation of a state's partisan makeup. For example, in Ohio's 2020 congressional election, Democrats won 42 percent of the state's popular vote, but the party won just four of the state's 16 congressional districts because the districts had been heavily gerrymandered to favor Republicans.

Gerrymandering Tactics The top diagram shows what happens when one party (blue) packs voters from the opposite party (red) into one district. The bottom diagram shows cracking, when voters from an opposition party (blue) are spread into several districts. ▮ How do the graphics help you understand arguments against gerrymandering?

Legal challenges to gerrymandering have had mixed results. For Democrats, one bright spot was in Pennsylvania, when, in early 2018, the state Supreme Court ruled that the state's congressional map, drawn by the Republican-held state legislature in 2011, was a gerrymander in violation of the state's constitution. As a remedy, the court drew its own map and ordered both parties to comply with it. The effect was marked. Pennsylvania voters are pretty evenly divided between Democrats and Republicans, but in 2016, Democrats held just five of the state's 18 congressional districts. In late 2018 elections, with the court-ordered map, the two parties evenly represented the districts—nine seats each, a pattern that prevailed in 2020.

In 2019, two key judicial rulings on gerrymandering were handed down. First, the U.S. Supreme Court ruled that gerrymandering was a political issue, not a judicial one, and therefore not a matter for federal courts to decide. That ruling appeared to keep gerrymandering firmly in the hands of state legislatures. However, just a few months later, North Carolina's state court ruled that the Republican-drawn map of that state's 13 congressional districts was an "extreme partisan" gerrymander that violated not the U.S. Constitution but North Carolina's state constitution. That ruling raises the possibility that legal challenges may continue at the state level. By the end of 2019, a new Republican-drawn (and state court–approved) congressional map had somewhat eased the Republican gerrymander. Since the 2020 census, there have been a number of court challenges to maps drawn by both state legislatures and by independent commissions.

Case Study

Gerrymandering and Race

The Issue Electoral districts are sometimes gerrymandered along racial lines, a practice that often triggers legal challenges.

Learning Objective
LO 10.6 Describe how the practice of gerrymandering affects electoral outcomes.

By the Numbers

20 percent
of North Carolina's congressional districts are majority-minority (3 of 13 districts in 2019)

13 percent
of Wisconsin's congressional districts are majority-minority (1 of 8 districts in 2019)

28 percent
of U.S. congressional districts are majority-minority (122 of 435 districts in 2019)

Source: Ballotpedia

For Decades, Republican lawmakers around the country have packed African American voters into majority-minority districts in hopes that the remaining districts would lean in their favor. This is because African Americans have been a reliably Democratic-voting constituency. Racial gerrymandering has affected North Carolina for decades. After the 1990 census, population growth resulted in the creation of the 12th Congressional District. The then-Democratic-controlled state legislature packed the district with African Americans so that it became North Carolina's second majority-minority district. In 1993, the U.S. Supreme Court took notice of this majority-black district, and most of the justices objected to it, calling it a clear "racial gerrymander." District lines were later redrawn in North Carolina four more times as the constitutionality of majority-minority districts was examined.

By 2010, Republicans had secured control of North Carolina's state legislature. Lawmakers drew a map in which the state's 1st and 12th Congressional Districts were majority African American in order to favor the Republican Party, but in 2016, a federal court declared (and in 2017, the U.S. Supreme Court agreed) that the districts were unconstitutional gerrymanders. In response to the ruling, Republicans drew a new map that was used during the 2016 federal elections. The battle over political maps continued following the 2020 census when the Republican-dominated state legislature approved political maps that granted 11 of 14 seats in Congress to Republicans despite the fact that North Carolina's voters are almost evenly divided between political parties.

The challenge to North Carolina's 12th Congressional District is a good illustration of how racial gerrymandering and partisan gerrymandering can be difficult to separate, a notion that you have already learned about. A similar instance occurred in Wisconsin in 2016, when a legal challenge to the state's legislative districts was treated as a partisan gerrymandering issue, but racial issues were also at play.

Beginning in 2011, the Republican party controlled both houses of the Wisconsin state legislature as well as the governor's mansion—a situation called a *trifecta*. The state legislature drew a state legislative map that favored their party. The map, called Act 43, was legally challenged in court by 12 Democratic voters who claimed Republicans had both packed and cracked districts to give their party an advantage.

There was a racial component as well, which was revealed in another court challenge to Act 43. In Milwaukee, Wisconsin's most populous and diverse city, arguments were made that one district had been cracked so that non-White voters, and especially Latino voters, were spread out. Milwaukee is infamous for having a stark divide between its largely non-White central city and its largely White outer neighborhoods. After redistricting, neighborhoods that were 60 percent non-White were tossed into a district that became 87 percent majority White. As a result, it was unlikely that a non-White representative would be elected.

While a U.S. District Court panel agreed that Act 43 was a statewide partisan gerrymander, the U.S. Supreme Court essentially agreed with the Republican-led state legislature, saying that the 12 Democratic plaintiffs had no right to sue in the first place and that gerrymandering is a political issue and not a judicial one. The Princeton Gerrymandering Project, a nonpartisan, interdisciplinary group of scholars that works to bring more fairness into redistricting maps, gave the Republican-drawn Wisconsin map an "F" rating, finding it "very uncompetitive relative to other maps that could have been drawn," and Governor Tony Evers vetoed the map in late 2021. But in 2022, after the Wisconsin Supreme Court initially chose Governor Evers' map from among eight competing legislative and congressional voting maps, the U.S. Supreme Court struck down the Governor's map, based on concerns related to the Voting Rights Act. Ultimately, the Wisconsin Supreme Court ruled in favor of a new map drawn by the Republican-controlled legislature, with few alterations to their original 2011 map—a major win for the Republicans. A redistricting expert termed this final map "a very strong partisan gerrymander, the kind you would expect from a legislature that was trying to tilt the playing field in its favor."

Geographic Thinking

Explain the role race played in the legal challenges to either North Carolina's or Wisconsin's gerrymandered congressional maps.

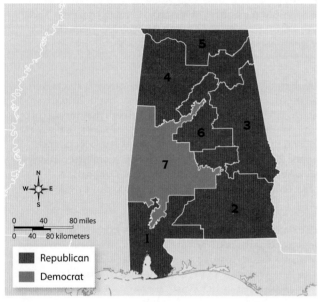

Alabama's 7th Congressional District is an example of a packed district. Republican state lawmakers constructed several oddly shaped districts so that Democratic voters would be packed into the 7th, while Republican voters would outnumber Democrats in the other six districts.

The states themselves determine their own redistricting methods. These methods vary from state to state and, sometimes, within a state (e.g., different methods may apply to congressional redistricting than to state legislative redistricting). State legislatures play the dominant role in congressional redistricting in 33 states. Commissions draw congressional district lines in eight states, and in two states, legislatures share redistricting authority with commissions. The remaining seven states comprise just one congressional district each, so redistricting is unnecessary. Some redistricting commissions are bipartisan, meaning they are made up of members of both parties, and others are independent. For example, voters in Michigan approved a plan in 2018 to form a commission made up of 13 citizens—four Republicans, four Democrats, and five independents. The districts the commission draws must avoid any partisan advantage and must, in an effort to avoid oddly shaped districts, "reflect the state's diverse population and communities of interest." The state of Iowa has an even more straightforward remedy—a nonpartisan body called the Legislative Services Agency draws districts that must be made up of "convenient contiguous territory." In all cases, the goal is to reduce the acute partisanship of states' maps.

At-Large versus Single-Member Districts Are the results of an election truly representative of a geographical area? The answer largely depends on the election method that area uses. In most local municipalities around the United States, for example, voters elect their representatives in one of three ways: by at-large elections, by district elections, or by a mixture of both. In an at-large election, the entire population of a geographical area, such as a city, town, or school district, elects someone to represent them as a whole. In a district election, a single individual is elected to represent the population of a smaller geographical area (in most local municipalities these subunits are called wards). A good example of a mixed system is seen in the school board of Minneapolis, Minnesota. The board has nine members, three of whom are elected at large, and six represent certain individual school districts.

Political geographers debate which election type achieves better representation. Proponents of at-large districts argue that at-large representatives keep the interests of the entire community in mind and tend to be less partisan. Opponents argue that minority groups in at-large districts are underrepresented because those groups tend to be concentrated in certain areas. By contrast, single-member districts allow for greater representation of all groups and their representatives can be more attentive to the particular needs of a local community.

Geographic Thinking

1. Explain how reapportionment can impact the political power of a state.

2. Describe the arguments against gerrymandering.

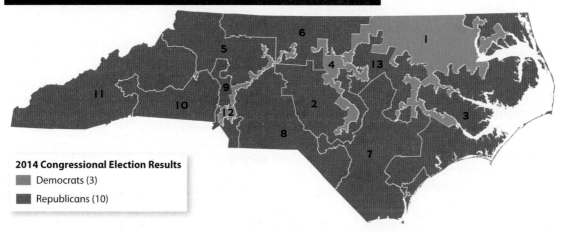

After the 2010 census, Republican state lawmakers drew this map of North Carolina's congressional districts. Notice the unusual shape of the 1st, 4th, and 12th districts. Both a U.S. federal court and the U.S. Supreme Court ruled that the 1st and 12th districts were unconstitutional racial gerrymanders.

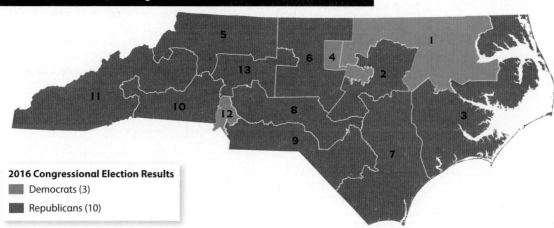

This map of North Carolina's congressional districts was drawn by Republican state lawmakers in 2016 and was used in the 2016 and 2018 elections. Describe how the boundaries of the 1st, 4th, and 12th districts changed between 2014 and 2016.
▌ How did this impact the congressional election results between the two years?

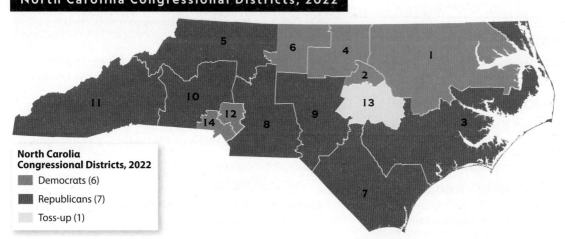

This 2022 map of North Carolina's congressional districts was drawn by a special panel of judges following a ruling that previous maps drawn by Republican state lawmakers were partisan gerrymanders. Note that, as a result of the 2020 Census, there are now 14 congressional districts, with district 13 seen as a toss-up, reflecting the state's continued population growth. ▌ What are several districts that changed shapes the most in the years 2012–2020?

Spatial Patterns of Political Power **289**

Chapter 10 Summary & Review

Chapter Summary

State governments—country governments, in this context—are organized based on how political power is divided. There are two main approaches to organizing state governments: a unitary state or a federal state.

- In a unitary system of government, power is concentrated in the central government, which decides how much power to distribute to its regional units.

- Advantages of a unitary state include efficient implementation of laws and services across the state, less potential for corruption of local government, and fewer government agencies.

- Disadvantages of a unitary state include disconnection between the central government and local regions, favoritism for the dominant political or cultural group, and slowness to respond to local issues.

- In a federal system of government, power is shared between the central government and its regional units.

- Advantages of a federal state include reduced regional conflict, political diversity, and attention to local issues.

- Disadvantages of a federal state include undue power given to localized special interests that sometimes block national issues and uneven distribution of costs and benefits of government policies.

Electoral geography is the study of the spatial organization of voting districts.

- The United States' 435 congressional districts are reapportioned and redistricted every 10 years following the census.

- Changes due to reapportionment can result in shifts in political and electoral power.

- Gerrymandering is the process of drawing internal legislative districts to secure an advantage for one party or another by either "packing" or "cracking" the districts.

- Gerrymandering can take place on the basis of partisanship or race, and therefore many see it as an unfair practice.

Review Questions

Use complete sentences to answer the questions.

1. **Apply Conceptual Vocabulary** Consider the terms *unitary* and *federal*. Write a standard dictionary definition of each term. Then provide a conceptual definition—an explanation of how *unitary state* and *federal state* are used in the context of the entire chapter.

2. Give an example of the efficiency that can result from the organization of government in a unitary state.

3. Explain how Norway's approach to public health exemplifies its unitary system.

4. Why is a federal system a good fit for a large, diverse country?

5. Explain how spatial inequality can occur in a federal state.

6. Summarize how federal (country wide) and state (regional) governments share power by giving an example from the federal system of the United States.

7. Define *concurrent powers* and include an example of them in your response.

8. Describe the difference between patriotism and nationalism.

9. Explain how the term *reapportionment* is related to a census.

10. Describe the difference between reapportionment and redistricting.

11. Explain how the electoral college works, and identify who is elected using this process.

12. Define *gerrymandering* and give two examples of strategies used to gerrymander a district.

13. Identify and describe one remedy for gerrymandering.

14. Explain how majority-minority districts are related to gerrymandering.

15. Describe the difference between an at-large district and a single-member district.

Interpret Maps

The 2011 map outlines Pennsylvania's 15th Congressional District as it was drawn as part of a statewide Republican gerrymander. The 2019 map shows the 7th Congressional District that was drawn by the Pennsylvania State Supreme Court after it struck down the former gerrymandered map. The former 15th Congressional District was broken up to form parts of other districts including the 7th Congressional District. The map data showing Democratic and Republican voting behavior is based on the results of the 2016 presidential election between Hillary Clinton and Donald J. Trump. Study the maps and then answer the questions.

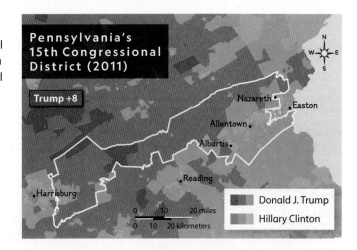

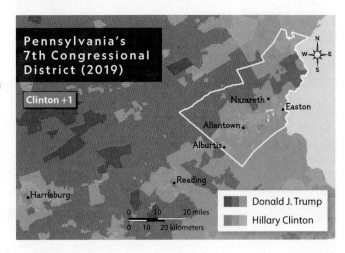

16. **Analyze Models & Theories** Identify aspects of the 2011 15th district that could be cited to argue that it was gerrymandered.

17. **Analyze Geographic Concepts** What can be inferred about the Republicans' strategy in drawing the 2011 15th district by looking closely at the cities of Allentown and Easton?

18. **Explain Geographic Concepts** The 2019 7th district appears to have more area that is red than is blue. Does that necessarily predict that the district will elect a Republican representative? Explain.

19. **Identify Data & Information** Identify data that supports that districts in 2019 were less gerrymandered than districts in 2011.

Chapter 11
Political Challenges and Changes

Critical Viewing Demonstrators in London gathered in March 2019 in support of Brexit, the United Kingdom's decision to leave the European Union. ■ What cultural and economic impacts might these demonstrations have had on Europe?

Geographic Thinking What causes states to unify and to divide?

11.1
Devolution: Challenges to State Sovereignty

Case Study: Irredentism in Ukraine

National Geographic Explorer
Michael Wesch

11.2
Supranationalism: Transcending State Boundaries

Case Study: Brexit

11.3
Forces That Unify and Forces That Divide

National Geographic Explorer
Anna Antoniou

11.1 Devolution: Challenges to State Sovereignty

A state's sovereignty is not absolute. Different processes and forces may place stress on a government's ability to control the land and people within its borders. Under certain conditions, these stresses can cause a state to destabilize or even disintegrate. Geographers study these processes to understand the causes and effects of shifts in political power at all scales, local to global.

The Process of Devolution

Learning Objectives
LO 11.1 Explain the factors that can lead to devolution of a state.
LO 11.2 Define irredentism and ethnic cleansing in the context of political geography.

As you learned in Chapter 9, the world political map contains nearly 200 states of various sizes, population densities, and government structures. Each of these states, whether they're centuries old or nearly new, has experienced an evolution to arrive at its present situation. This may have involved a civil war, colonial rule, a revolution, conflict with other states, or more. But all states, even the oldest and strongest, face a range of divisive pressures that stress their existing structure. This process, termed *devolution*, can destabilize a state. Geographers study the forces that drive devolution, as well as the ways in which governments respond to this process.

A number of factors can challenge state sovereignty and may lead to the devolution of a state: the division of groups of people by physical geography, ethnic separatism, the practice of ethnic cleansing or terrorism, and the policy of irredentism, as well as economic and social problems. It is useful to look at each of these factors separately, but in many cases, there are multiple factors that lead to challenges to a state's sovereignty.

Physical Geography One factor that can pose great challenges to the sovereignty of a state is physical geography, including distance and topography. Devolutionary forces are most often prompted by the distance that exists within a state from its center of power. As you know, friction of distance describes the cost and difficulty of overcoming distance between two locations. As this distance increases, the quantity and quality of interaction between the locations will decline, which illustrates distance decay. Although the challenges of physical geography and actual or travel-time distance are less of an issue in today's world due to advances in transportation and communications technologies, the friction of distance and resulting distance decay still matter. States with fragmented physical geography, such as a country that is spread out over a group of islands, or those disrupted by major topographic features such as mountain ranges, can have challenges with unity. It is more difficult to build a cohesive state when division—in the form of mountains, deserts, or other physical impediments—is a factor in the state's physical geography.

Think about the difficulties involved in governing the Philippines, a country comprising more than 7,600 islands. The Philippines had the 13th-largest population in the world in 2020 (about 110 million), more than half of which lives in rural areas, in smaller villages in the rugged, mountainous interiors of its many islands. To travel to cities and towns on the coastline, people from the interior use roads that for the most part are unpaved. The climate of the Philippines is tropical and monsoonal (with a long period of rains from May to September), which creates additional challenges with natural disasters such as floods and landslides, as well as hurricanes, which are called typhoons in Southeast Asia.

In the 1900s, the Philippine government in Manila—located in the northern part of the country—made a decision that led to destabilization. The government attempted to develop the island of Mindanao in the south, more than 500 miles away from Manila. Mindanao was rich in natural resources and had fewer destructive typhoons than the northern islands. In order to encourage economic development there, the Philippine government resettled Christian people from other parts of the country on Mindanao, which was inhabited by native Filipinos mostly of Muslim heritage known as Moros. Part of this resettlement provided the Christian migrants with the best lands so they could prosper from the natural resources. They were also given government services that were not given to the Moros in 1968. Conflicts broke out between the new settlers and the Moros. In turn, the Moros started a movement to secede from the country, leading to violence and leaving the region one of the poorest in the country. The sheer distance between this southern region and the Manila-based government was an important factor contributing to devolutionary pressures on the Philippines.

Ethnic Separatism Another factor that can lead to devolution is **ethnic separatism**, which occurs when people of a particular ethnicity in a multinational state identify more strongly as members of their ethnic group than as citizens of the state. As we have seen, the Basques of Basque Country in northern Spain are a primary example. Basque nationalism is rooted in the region's history, culture, and unique language, cultivated and molded in its region separate from the rest of Spain. As Spain's efforts to control the region increased, and as Spanish immigrants were coming into the area at an alarming rate, the Basques wanted to salvage their autonomy and be recognized as their own nation. Basque Country was declared an autonomous region in 1979, which meant that it was given independence in some areas of governance, but the push and pull between the Basques and the Spanish government continues.

The root of many cases of ethnic separatism is disparity, or difference, in how an ethnicity is treated, both culturally and by the laws of the state. The people of an ethnic group may feel like they do not have the power and autonomy that they deserve. This is especially common in the case of stateless nations. As you read in Chapter 9, a stateless nation is a group of people united by culture, language, history, and tradition but not possessing their own state. In extreme cases, the state government may attack the ethnic group and try to eliminate it through expulsion, imprisonment, or killing. This is known as **ethnic cleansing**.

Consider the discussion of the Rohingya people in Chapter 5. The Rohingya, a mostly Muslim group, faced multiple military crackdowns in their homeland within Myanmar and were driven into neighboring countries, particularly Bangladesh, in 2017. The Myanmar government's official stance on the Rohingya is that they are not an Indigenous people but rather are illegal immigrants from Myanmar. Another example is the National Socialist German Workers' Party, or the Nazi Party, led by Adolf Hitler beginning in the 1920s. Its members believed in the supremacy of the Aryan people and blamed Jews and other ethnic groups for the economic and social problems in Germany that followed World War I. The Nazis had an ultimate plan for the systematic murder of all European Jews, which resulted in a staggering 6 million deaths.

Another example of ethnic cleansing occurred during the Bosnian War in the 1990s, in the wake of the devolution and breakup of the country of Yugoslavia. Both Bosnian Muslims known as Bosniaks and Bosnian Croats, who were largely Catholics, were subject to brutal killings and the destruction of their communities and cultural heritage by Serbian forces. The Srebrenica Massacre, part of this ethnic cleansing campaign, led to the deaths of more than 8,000 Bosniak men and boys in 1995. Ethnic tensions remain high today in Bosnia (officially, Bosnia and Herzegovina), where Muslim Bosniaks, Catholic Croats, and Eastern Orthodox Serbs govern the country in a tenuous relationship that requires consensus from all three groups to implement policies. These are three of many examples across the world and down through history that show that the goal of ethnic cleansing is to achieve ethnic homogeneity by eliminating those who are ethnically different.

Destabilization can also occur when a region shares ethnic, cultural, or historical traits with the people of a neighboring state. This can lead to irredentism, which was introduced in Chapter 9. Irredentism occurs when the majority ethnicity on one side of a boundary wants to claim territory from a neighboring state in order to bring in a minority group of the same ethnicity or other commonality—like language—who resides across the border. In the 1930s, for example, a region known as the Sudetenland in western Czechoslovakia was populated predominantly by Germans who felt they were being discriminated against by the Czech government. Propaganda by the Nazi Party was both anti-Czech and anti-Semitic (hatred of and hostility toward Jews), which drew these Sudetenland Germans to support annexation, and in 1938 the region was seized by Germany until after World War II.

Advances in communication technology have greatly impacted ethnic identity, sometimes fueling separatist movements. The internet and smartphones have made it easier for groups to organize, advertise, and recruit for their causes. Social media can be used as a low-cost outlet for sharing and voicing concerns regarding treatment of a group. This in turn gives a group that has been disenfranchised, or deprived of rights, power that they wouldn't have had even several decades ago. For example, in Afghanistan, despite its poor infrastructure connections and limited freedom of speech, more than 22 million people had telecommunications devices in 2020, and there were an estimated 4.4 million social media users in 2021. The Taliban, a political and military organization that long waged war within the country, once banned the internet because they thought it did not align with Islamic practices. But their perspective changed when they realized they could use it to promote their ethnic identity, spread their message, communicate with other members, and gather supporters. The Taliban now rules Afghanistan and aggressively employs social media to strengthen and legitimize its government.

Economic and Social Problems Economic divisions within a state often work in tandem with ethnic and geographic pressures to cause devolutionary forces. These divisions can result from variations in economic productivity or development between regions within a state. Economic divisions may also arise because of the way funds are allocated by the government to different regions. The people of a given region might feel that the central government is misspending the taxes they pay or that the central government is not providing enough funds or services to their region.

Canada provides an example of this regional division. Over the past several decades, the rich mineral and agricultural resources of provinces like Alberta and Saskatchewan have resulted in booming provincial economies and, while Ontario is the mid-country manufacturing powerhouse, some eastern provinces like Quebec and Nova Scotia have, for varying reasons, seen declining economies. To address these imbalances, since 1957 Canada's central government has conducted an annual procedure called "equalization," in which some federal tax revenues are divided among provinces to provide roughly equal public services for all. Authorities in the capital of Ottawa use revenues from the "haves" (Alberta, British Columbia, Saskatchewan, Ontario, plus Newfoundland and Labrador) to help out the "have-nots" (Quebec, Manitoba, Nova Scotia, New Brunswick, and Prince Edward Island; Canada's three territories are funded in a separate process). Though written into the Canadian constitution, this practice has upset some in "have" western provinces. For some years, one manifestation of this resentment was the Prairie Freedom Movement that argued western provinces should have more control over this revenue to invest in areas that generate more wealth, rather than using it to invest in the public services of the "have-nots." While late 2021 polling showed that a majority of Canadians favor the policy, nearly 62 percent of Albertans, the largest contributors on average to the equalization program, voted in favor of removing it from the Canadian constitution.

Economic issues fuel devolutionary forces in Catalonia, Spain, as well. Many citizens of this region, located in the northeastern portion of the country and focused on the city of Barcelona, feel they are victims of funding discrimination by the Spanish government, located in Madrid in the center of the country. As with some in western Canada, the Catalans believe the government takes too much of their tax money and unfairly redistributes it to the poorer regions of Spain. Governments like Canada and Spain distribute revenue to reduce the gap between rich and poor regions, but Catalans think the redistribution is dividing the country, as it provides more resources per capita to the poorer regions than what comes back to Catalonia for its essential services. They also point out that the more autonomous Basque Country doesn't have to send all of its tax revenue to Madrid.

Social issues can also destabilize a state. Discrimination against a minority group, for instance, can cause rifts between people living within a state's borders that act as devolutionary forces. The small nation of Wales, one of the four principal subdivisions within (the state of) the United Kingdom, has a markedly different culture and language than the rest of the country. And while the people of Wales—called the Welsh—are citizens of the United Kingdom every bit as much as the English, Scottish, and those of Northern Ireland, anti-Welsh sentiment is common in other areas of the country. Examples of public officials and personalities insulting or making discriminatory jokes about Wales and its people are commonplace. This casual discrimination among minority groups living within a single country may prevent them from developing a feeling that they are all one people.

Case Study

Irredentism in Ukraine

The Issue In 2014, Russian president Vladimir Putin incorporated Crimea into the Russian Federation, causing ongoing territorial disputes in eastern Ukraine.

Learning Objectives
LO 11.1 Explain the factors that can lead to devolution of a state.

LO 11.2 Define irredentism and ethnic cleansing in the context of political geography.

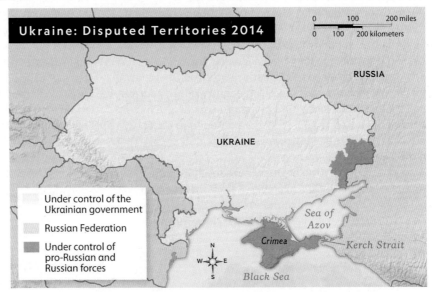

The territories of Crimea and eastern Ukraine have had a long and tumultuous relationship with Russia. The region remains disputed to this day.

An Example of Irredentism occurred in Eastern Europe in March 2014, when Russia sent troops into Crimea, an autonomous republic in southern Ukraine. Russia annexed the region and connected it to the Russian Federation by a 12-mile bridge spanning the Kerch Strait.

Crimea is a peninsula on the Black Sea that became a part of Ukraine in 1954, when both Ukraine and Russia were part of the Soviet Union. After World War II, Crimea was a part of the *Ukrainian Soviet Socialist Republic*. Prior to this, Crimea's identity had flip-flopped. Joseph Stalin's regime suppressed many of Crimea's minorities (primarily the Tatars, a native Turkish-speaking people) and forcibly deported them.

The ethnic Russian majority in Crimea made for strained relations with Kiev (Kyiv), the capital of Ukraine. In addition, Moscow signed a lease agreement allowing it to base the Russian navy's Black Sea Fleet in the port at Sevastopol in Crimea, making the region strategically important to Russia.

Over the years, Ukraine tried to balance its ties between Europe (with which it hoped to connect) and Russia (with which it was historically tied). However, Crimea continued to lean toward pro-Moscow involvement. In 2013, the pro-Russian president of Ukraine, Viktor Yanukovych, backed out of a deal with the European Union in favor of stronger ties with Russia. Protests erupted in Kiev. The Ukraine Parliament voted to remove the president from power and to ban Russian as the second official language of Ukraine, angering the Russian-speaking inhabitants of Crimea.

In 2014, Russian gunmen seized government buildings in Crimea's capital. The Crimean Parliament called a referendum and while the results have been disputed, voters decided to join Russia. Ukrainian troops pulled out of Crimea. Since then, Russia has supported Ukrainian separatists occupying key cities in eastern Ukraine, a region known as Donbas. Fighting in the Donbas continued for eight years, with more than 14,000 casualties. Desperate to keep Crimea out of Russian control, Western countries responded by imposing sanctions. In 2019, the United States' relationship with Ukraine was scrutinized because of the Trump administration's delay of financial aid to support Ukraine's military defense. Such aid had been provided regularly by the United States since 2014. In early 2022, the Russian military launched a massive invasion of much of the eastern portion of Ukraine, but eventually focused on attempting to seize and control the resource-rich Donbas region and to connect it to Crimea. The United States along with European and other allies imposed more drastic economic sanctions against Russia, and began to provide weapons and other military hardware to Ukraine. In the first six months of the conflict, tens of thousands of Russian and Ukrainian troops had been killed or injured, and the United Nations verified that at least 5,500 Ukrainian civilians had been killed—and estimated

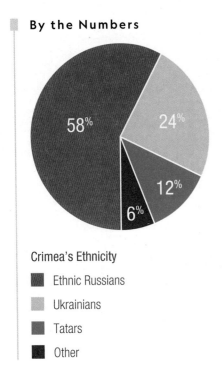

By the Numbers

Crimea's Ethnicity
- Ethnic Russians — 58%
- Ukrainians — 24%
- Tatars — 12%
- Other — 6%

Source: Ukraine Census, 2001

that many more likely had perished. By late 2022, Russia annexed more regions of eastern Ukraine. The outcome of this aggression, portrayed by Russia as an effort to reclaim lands occupied by ethnic Russians, was uncertain, but most military experts expected a long and deadly conflict.

Geographic Thinking

Describe how Russia's annexation of Crimea is an example of irredentism.

Responses to Devolutionary Forces

Learning Objective
LO 11.3 Describe the different responses by states to devolutionary forces and destabilization.

Just as causes of devolutionary forces are varied, so are responses. States respond differently to processes of destabilization depending on their particular mixture of ethnic, cultural, economic, and internal territorial divisions. For instance, a state may address devolutionary forces by sharing more power with subnational or regional units.

Canada has used this method to address various challenges to its state sovereignty. In response to demands by people in the Yukon Territory that a greater percentage of the money from mining and other resources be controlled within the territory, Canada shifted land- and resource-management responsibilities to the Yukon government. To appease French-speaking citizens in Quebec, Canada has decreed French to be the province's official language.

Devolutionary forces can also lead to sovereignty and self-determination for the people of a region within a country. For example, in the late 1970s, many people in Scotland pushed to form a Scottish Parliament within the United Kingdom that would allow the Scottish people to have greater control over their own affairs. It took 20 years, but after a series of referendums, the Scottish Parliament was formed in 1999. The Scots continued to push for more independence, and a separatist movement gained traction. This movement is an example of a devolutionary force since it destabilized the United Kingdom through disagreement, disunity, and a desire for more autonomy. Brexit has also renewed calls for Scottish freedom. In 2016, the UK government gave Scotland more power as a compromise to counter the Scots' desire for complete independence.

Devolutionary forces may result in a shift in a state's form of governance. The European country of Belgium, for instance, transitioned from a unitary to a federal state. Belgium's citizenry consists of French-speaking people called the Walloons, concentrated in the southern

The Scottish Parliament Building, completed in 2004, houses Scotland's legislative body. The building has been called a "poetic union" between the Scottish landscape, its people and their culture, and the traditional capital of Edinburgh. The Parliament now holds more power, as a result of devolutionary forces and the achievement of greater autonomy.

Political Challenges and Changes

Reading Maps The collapse of the Soviet Union is an example of devolution. This map shows the many Soviet Socialist Republics (S.S.R.) that composed the Union of Soviet Socialist Republics until 1991.
▌ Use the map to describe how friction of distance may have contributed to devolution in the U.S.S.R.

provinces, and Flemish- or Dutch-speaking people called the Flemings, concentrated in the north. Until 1970 the country was a unitary state, but tensions between the two groups caused the Belgian Parliament to gradually shift to a federal government. Changes to the constitution created independent administrations within the Walloon and Flemish regions, giving each region control over economic, educational, and cultural decisions. A third region was created in the late 1980s that consists of the metropolitan area of Brussels, the country's bilingual capital, which acts as a bridge between the two regions.

The oil-rich country of Nigeria, in West Africa, responded to devolutionary forces by breaking the country up into subnational political units. In the 1960s, a different major ethnic group dominated each of three regions of Nigeria, and conflict between the groups threatened to tear the country apart as leaders of each group fought for control. The government decided to break the country up into 12 states to lessen the power held by each of the three main regions. The decision led to a bloody civil war as one region attempted to declare independence, but the government was able to hold the country together, partly with the help of money generated by an oil boom in the aftermath of the war. This decades-long response to devolution, from the government centrally located in the capital of Abuja, has continued, and there are now 36 Nigerian states.

Eritrea is a small African country on the Red Sea that was once a province of its more powerful neighbor, Ethiopia. The region's inhabitants are extremely diverse: they are Christian and Muslim, they come from as many as nine different ethnic groups, and they speak at least seven indigenous languages. This diversity resulted in tensions that acted as devolutionary forces that the Ethiopian government attempted to counteract by banning political parties and trade unions and weakening the Eritrean provincial government. As Eritreans started to lose their autonomy, their discontent with the Ethiopian government grew, eventually leading them to declare their independence. Eritrea became a sovereign state in 1993. The unrest did not end there, though. A series of border conflicts began in 1998 and tens of thousands of lives were lost, resulting in only minor border changes. In 2018 Ethiopia and Eritrea reopened trade and diplomatic relations, declaring an end to 20 years of war. Ethnic tensions again erupted in Ethiopia in 2021, leading to a bloody civil war between the Tigrayan People's Liberation Front from the northern province of Tigray and the government of Ethiopia centered in the capital, Addis Ababa. At the heart of the conflict is debate over the political organization of the country and the autonomy of its 10 ethnically based states versus centralized power.

In Sudan, in North Africa, a civil war erupted in 1955. The attempts to form a democratic government failed, and

After the Soviet Union's Collapse

Reading Maps On December 31, 1991, the Soviet Union was formally dissolved. This map shows the independent states that were formed as a result. ▌ What challenges might the newly autonomous states that were part of the former Soviet Union face?

military forces pushed Islamic ideals in the name of national unity. Years of fighting took place between the north and the south—civil strife that acted as a powerful devolutionary force on Sudan. In 2011, the people of the south voted for independence, and the country was split into two: Sudan and South Sudan. The split was a direct result of devolution.

When a state cannot resolve issues causing destabilization, these devolutionary forces may result in the disintegration of the state. The Soviet Union, for example, broke apart in 1991 as a result of stresses that challenged state sovereignty and led to devolution. Its republics faced many challenges, including physical geography—the country's vast size, the distance between regions, and the severe climate made unity difficult. Additional factors contributed to the destabilization: a failing economy, a weakened military, ethnic separatism among at least 100 groups, a series of unpopular social and political reforms, and public dissatisfaction with Soviet President Mikhail Gorbachev. These forces destabilized the state to such an extent that, one by one, its republics declared independence, and the Soviet Union dissolved.

The dominant Soviet republic was always Russia, with its capital Moscow. Its president for most of the years since 1999, Vladimir Putin, has sought to reintegrate a number of the post-Soviet independent republics, sometimes through economic or political destabilization and sometimes through direct military invasion, as with the Caucasus country of Georgia in 2008. By 2022, Belarus had become a puppet state controlled by Moscow and, as you've learned, Russia annexed Crimea in 2014, backed ethnic Russian forces in eastern Ukraine, and invaded Ukraine in 2022. However, the majority of the Ukrainian people rejected reintegration with Russia and fiercely fought for their independence. Observers fear that the Russian leader, who once stated that the breakup of the Soviet Union was the "greatest geopolitical tragedy of the 20th century," is attempting to rebuild the Soviet Union.

Geographic Thinking

1. Describe how sovereignty is related to devolution.
2. Describe how the physical geography of the Philippines acts as a devolutionary force.
3. Identify devolutionary forces in Canada and the Soviet Union, and describe the results of each.

Political Challenges and Changes

National Geographic Explorer **Michael Wesch**

Technology's Impact on Society

Wesch, an anthropology professor at Kansas State University, helps others understand the effects of the internet. *Above:* A long-exposure photograph reveals light trails from thousands of smartphones used by demonstrators in Hong Kong during a protest.

Learning Objective
LO 11.1 Explain the factors that can lead to devolution of a state.

Advances in communication technology and social media continue to infiltrate societies all over the world. Geographers are working with other scientists who study culture—such as anthropologists and ethnographers—to examine the effects of the digital revolution. Michael Wesch is one of those scientists.

In 2005, there were just over 1 billion internet users. By 2022, that number had grown to over 4.95 billion users worldwide. This exponential growth has affected the world in ways we do not yet fully understand. Wesch studies these effects by examining how new digital technologies are altering human interaction. In one of his studies, he looks at the implications of an environment where information is available to us, anywhere, 24/7.

Some of Wesch's most important work is his study of the effects of YouTube, the popular video-sharing website, on viewers. Wesch observed that when YouTube vloggers (video bloggers) share videos, "It's a gateway to anyone, anywhere, throughout all time. This inspires some to feel a profound connection with the entire world. But that's not the same connection felt with a close family member. It's a relationship without any real responsibility, one you can turn off at any moment."

The political ramifications of the constant connection that Wesch studies are widespread. Information and news stories are now broadcast across borders. People have access to knowledge and services from all over the world. They are able to communicate with large groups instantly, which has fueled devolutionary forces such as large-scale protest movements around the world in recent years. Propaganda and disinformation can be communicated just as quickly and can be used to sow disunity within a state. And some governments restrict access to the internet, which allows them to control the information that their citizens have access to. Geographers are discovering that these effects are felt by both the governments that control states and the people who live within them.

Geographic Thinking

Describe how technology could make devolutionary pressures increase or decrease.

11.2 Supranationalism: Transcending State Boundaries

Despite the importance of state sovereignty, sometimes the advantages of states working together outweigh the disadvantages. Geographers study how challenges that transcend—or reach across—states' boundaries may lead them to form cooperative alliances. These organizations can help member countries, but they can also challenge state sovereignty by limiting the power of the individual states.

What Is Supranationalism?

Learning Objectives
LO 11.4 Define supranationalism.
LO 11.5 Describe the goals of various supranational organizations in modern history.

A **supranational organization** is an alliance of three or more states that work together in pursuit of common goals or to address an issue or challenge that these countries share. The goals of supranational organizations might be economic, political, military, cultural—or a combination. Some of these organizations have formed under treaties, while others are considered general alliances. States join supranational organizations because they see an advantage in working with the other member countries, or they want to avoid the disadvantages of being left out of such a group.

The first example of supranationalism in modern history was the formation of the League of Nations. In 1920, at the end of World War I, the Allied Forces came together and, spurred by public demand for a lasting peace, formed an organization for international cooperation. The organization's guiding principle was the belief that war is a crime against more than just those who are attacked. Rather, it is a crime against humanity, and it is the right and duty of all states to work together to prevent it.

The League was disbanded in 1946 and its powers were transferred to another newly formed group, the United Nations (UN). The UN was established in October 1945. The group's initial design was based on the concepts and guidelines of the League of Nations, but it developed into a more complex organization. The UN added a focus on respecting and protecting human rights; working to solve international economic, social, cultural, humanitarian, and environmental problems; and committing to promote economic and social development throughout the world.

Another supranational organization, the North Atlantic Treaty Organization (NATO), was established in 1949, primarily as a military collaboration in response to the occupation of parts of central and eastern Europe by the Soviet Union after World War II. NATO had 12 founding member states in 1949 (most of them in Europe), and 70 years later the organization had 30 members. In reaction to Russia's 2022 invasion of Ukraine, NATO invited Finland and Sweden, strategically-important countries that had previously maintained neutrality, to join the organization. Each member state commits to the collective defense of the entire alliance, with Article 5 of the NATO agreement stating that "an attack against one Ally is considered as an attack against all Allies." Article 5 was invoked for the first time in NATO's history when its members joined the fight against terrorism after the attacks on the World Trade Center in New York City on September 11, 2001.

The European Union (EU) formed after a series of smaller organizations came together to build economic and security alliances in the wake of World War II. The EU was officially created in 1993 by the Maastricht Treaty. Its original intent was to address the economic, social, and security issues of Western Europe, but in the early 21st century it expanded into Central and Eastern Europe. The 27 members of the EU share common trade and foreign policies, citizenship rights, environmental rules, and judicial cooperation. A unified economic and monetary system is a goal of the EU, though only 19 of its member states use the euro as currency. The EU has seen ups and downs. Several countries have expressed concern about losing state sovereignty when joining the EU. Some of its prosperous members worry about struggling members hindering the EU's overall progress. Issues like immigration, asylum, and open borders have led to disagreements among EU members. In 2020, the United Kingdom made a high-profile and complicated withdrawal from the EU, known as "Brexit."

The Arab League, also known as the League of Arab States (LAS), formed in 1945 as a loose coalition of Arab countries in North Africa and Southwest Asia to strengthen and regulate political, cultural, economic, and social programs and to settle disputes among its members. There have been some political disagreements over time, and the Arab League has had to adjust and respond to social and political changes among its member states. The League took action during the 2011 revolution in Libya, but it has failed to achieve diplomatic success in other significant regional disputes, including in the civil war in Syria or against the so-called Islamic State.

ASEAN, or the Association of Southeast Asian Nations, formed in 1967 by the governments of Indonesia, Malaysia, the Philippines, Singapore, and Thailand. Other countries have joined since then. The goals of ASEAN are to advance economic growth through trade in the region, encourage social progress, and bolster peace and security. ASEAN's peace and security goals were extremely important through the Vietnam War and its

Nato Members and Partners, 2022

- Member countries
- Euro-Atlantic Partnership Council
- NATO's Mediterranean Dialogue
- Istanbul Cooperation Initiative
- Partners across the globe

Reading Maps The North Atlantic Treaty Organization (NATO) is one example of a supranational organization. NATO includes 30 member countries from Europe and North America, as well as numerous partners across the globe that cooperate with NATO and share the common goal of maintaining worldwide security. Sweden and Finland were invited to join NATO in 2022 while Bosnia and Herzegovina, Georgia, and Ukraine are aspiring members as part of NATO's Open Doors enlargement policy. ▌ How does the map contribute to your understanding of the benefits of supranationalism?

Supranational organizations decide policy through negotiations and meetings. At this gathering of leaders of the African Union in 2018, member states signed a free trade deal designed to create a unified market for goods and services across the African continent.

aftermath, during the Cold War between the United States and the Soviet Union, and still today amid growing tension between the U.S. and China. ASEAN also worked to resolve the conflict in East Timor, caused by Indonesia's unpopular occupation and takeover of the small island country. East Timor has applied to become the 11th member of ASEAN. For a number of years, ASEAN has been in diplomatic negotiations with China to develop a "Code of Conduct" for activities in the highly contested South China Sea (that you read about in Chapter 9). These efforts have so far had little success, with China failing to respect the maritime claims of ASEAN member states.

The Arctic Council is a supranational organization that promotes cooperation and interaction among the countries with territory in the Arctic—Canada, Denmark (which owns Greenland), Iceland, Norway, Sweden, Finland, Russia, and the United States—and also works with Indigenous groups. The Council, whose member states serve as "stewards of the region," is primarily concerned with issues of sustainable development and environmental protection, including biodiversity, pollution, and climate change.

The African Union (AU) formed in 2002 to advocate for unity and solidarity of African states and to endorse international cooperation. All 55 states on the continent are represented in this truly pan-African organization, divided into five regional subdivisions. Its predecessor, the Organization of African Unity, was successful in mediating several border disputes, including between Algeria and Morocco and between Kenya and Somalia. It also pushed for international sanctions against South Africa when apartheid—a set of policies of racial segregation enacted by the White minority government—was in place there. In recent years, the AU has been involved in attempts to peacefully resolve conflicts across the continent, including those in Ethiopia, Libya, and South Sudan. A major AU focus during the COVID-19 pandemic was to help its member countries secure vaccines for Africa's huge population. The Union's Agenda 2063 envisions an African continent in which there is "a universal culture of good governance, democratic values, gender equality, and respect for human rights, justice and the rule of law."

The Key Benefits of Supranationalism

Learning Objective
LO 11.6 Describe the key benefits of supranationalism for the member states.

The collective power of supranational organizations works in several ways to benefit member states. One way it can benefit is economically. Through supranationalism, countries can increase trade and bargaining power and create **economies of scale**, where more goods and services can be produced for less money on average. Some ASEAN member states, for instance, are not economic powerhouses. Brunei, one of the group's members, has a small gross domestic product (GDP; about $14 billion). Collectively, however, the GDP of ASEAN is approaching $3.6 trillion. The organization has been successful in reducing trade barriers, or government-controlled restraints on the import and export of international goods. These restraints are usually controlled through embargoes (blockades) or tariffs (taxes) on imports.

ASEAN has also worked to eliminate nontariff barriers, which includes customs surcharges (fees on imported goods) and monopolistic measures. The elimination of these barriers has improved trade between the countries of the organization. However, this trade accounts for only about one-third of its members' exports. It is the trade outside of the bloc that has boosted ASEAN members' profits. Another example of the economic benefits from trade within a supranational organization is the free trade system within the EU. There are no borders for trade between the EU countries, which means no customs and much easier and faster trade operations.

Supranational organizations may also benefit through heightened military power. The combined forces of the member states of NATO, for instance, carry much more military might than the individual countries. NATO has the military capacity to react to a global crisis at any time. NATO maintains an active schedule of exercises involving thousands of military personal from its member countries to remain prepared. With Russia's invasion of Ukraine in 2022, NATO increased its mobilization of troops, particularly in member states bordering Russia, Belarus, and Ukraine.

States also benefit from supranationalism through other types of collaboration. In addition to the eight countries with sovereignty over lands within the Arctic Circle, the Arctic Council includes representation from six other Indigenous groups called "permanent participants," including the Aleut, Inuit, and Saami. The council conducts and supports research on general topics such as climate change, emergency preparedness, and wildlife and flora conservation, as well as more specific topics such as sea ice loss, mercury and microplastics in food chains, and ocean acidification. The member states and permanent participant groups of the Arctic Council bring distinct perspectives, unique solutions, and combined monetary support to issues faced in this remote yet valuable and vulnerable region, which is also home to more than four million people.

In some supranational organizations, ease of travel provides a benefit popular with citizens. In many places, travelers who want to cross from one country to another have to go through customs, show their passport, and abide by customs laws when entering a different country. As you learned in Chapter 9, travel is simplified in much of the EU—and in several other European but non-EU countries. Citizens of one of the member states of the Schengen Area can travel freely to other member states with no customs or passport control.

The Key Drawbacks to Supranationalism

Learning Objectives
LO 11.7 Describe the key drawbacks of supranationalism for the member states.
LO 11.8 Explain how supranationalism limits the power and sovereignty of individual states.

SUPRANATIONAL ORGANIZATIONS

ORGANIZATION	PURPOSE	MEMBER STATES
United Nations (UN)	International; maintain peace and security; promote economic and social development	193 member states (global)
North Atlantic Treaty Organization (NATO)	Military alliance	30 member states in Europe and North America, with partners across the globe
European Union (EU)	European organization governing common economic, social, and security issues	27 member states (Europe)
Association of Southeast Asian Nations (ASEAN)	International; accelerate economic growth; maintain peace and security in SE Asia	10 member states: Brunei, Cambodia, Indonesia, Laos, Malaysia, Myanmar, Philippines, Singapore, Thailand, Vietnam (Southeast Asia)
Arctic Council	High-level Intergovernmental; issues related to the Arctic, such as sustainable development	8 member states: Canada, Kingdom of Denmark, Finland, Iceland, Norway, Russia, Sweden, and the United States
Arab League	Political, cultural, economic, and social pursuits; mediate disputes among members	22 member states (North Africa, Southwest Asia)
African Union	Intergovernmental; promote unity and solidarity in Africa; spur economic and social development	55 member states (Africa)
United States-Mexico-Canada Agreement (USMCA)	Trading bloc among countries of North America	3 member states: Canada, Mexico, United States

Supranationalism involves making commitments that can challenge the sovereignty of member states. The degree to which sovereignty is compromised varies greatly depending on the scope of the agreement, but the responsibilities associated with membership in any supranational organization have the effect of limiting the political and economic actions of member states to some degree.

The countries that belong to the EU, for instance, give up a great deal of sovereignty. The difficulties involved in balancing out the needs of individual states for the good of the group became apparent as more countries joined the union in the late 1990s and early 2000s. Starting in 2002, the EU began the process of drafting a constitution for the growing union. The drafters wanted to incorporate the ideal of deeper integration between member states while protecting members' individual traditions. They also needed to figure out how to distribute power between member states with populations and economies of vastly different sizes and adapt the organization's institutions to meet the needs of a larger than expected membership. When the constitution was completed, some countries had reservations about ratifying it. Because it created an office of EU president, gave the EU power over some foreign policy decisions, and gave more power to the EU courts and parliament, some countries felt the organization's constitution threatened their own sovereignty. After negotiations lasting through 2009, however, an agreement was reached and the constitution was ratified.

Some still feel a threat to their sovereignty. One of the biggest challenges is the financial distress of certain member states. Greece, which fell into debt and borrowed from the EU, realized it could not repay what it owed. To help the country avoid default, the EU loaned Greece enough money to continue making payments. The EU's bailout came with strict austerity measures, or reductions in government spending, such as cuts to government employees' benefits. The bailout program ended in 2018, but Greece is still economically unstable. Because of this, according to some, the EU is not as strong as it could be.

Another potential challenge to the sovereignty of EU countries is the group's consolidated immigration policies and border security. For years in the United Kingdom, for instance, the free movement of labor caused overcrowding in some cities. Housing prices rose, and greater demand on resources strained cities. These and other challenges contributed to the United Kingdom's decision to leave the EU.

In addition, the refugee crisis that you learned about in Chapter 5 has led to the largest influx of refugees, asylum-seekers, and other migrants in Europe's history. In 2015, more than 1 million people migrated to Europe from war-torn countries like Syria, Iraq, and Afghanistan, compared to only 280,000 the year before. As a result, some European countries have tightened their borders, leaving many migrants stranded in eastern and southern European countries. Hungary, an EU member state, has strict anti-immigrant policies and has built a fence to keep out asylum seekers traveling through Serbia and Croatia from Southwest Asia. Many asylum seekers attempt to make their way to Germany, due to its willingness to take people in. Germany, in turn, has criticized other countries for their lack of commitment to an open borders policy. The EU has struggled for years to harmonize an asylum policy among its member states, but it has proven difficult with different judiciary systems and attitudes about immigration.

Geographic Thinking

1. Describe the benefits of a supranational organization for a state with a small GDP.

2. Explain how a supranational organization might challenge the sovereignty of a state.

Case Study

Brexit

The Issue Economic stagnation, the refugee crisis, and threats to sovereignty led the citizens of the United Kingdom to vote to leave the European Union in 2016.

Learning Objective
LO 11.7 Describe the key drawbacks of supranationalism for the member states.

Political cartoons are often used as a way to find humor in contentious political times. This cartoon illustrates some people's feelings that the United Kingdom's decision to leave the European Union was made for the wrong reasons.

On June 23, 2016, a national referendum was held in the United Kingdom. On the ballot: whether to remain in the European Union or to leave—an option known as "Brexit." The decision was a close one, but when the results were counted, nearly 52 percent had voted to leave and about 48 percent to remain, making the United Kingdom the first country to vote to leave the supranational organization.

The Brexit results came as a surprise to many, but a lot of people in the United Kingdom had long been skeptical of a union of European countries. There was some opposition when the United Kingdom signed the treaty joining the EU in 1993, and the British remained protective of their sovereignty in the following years, refusing to adopt the euro as their official currency. Still, supranationalism had its advantages, such as economic growth from free trade and the ability of its citizens to travel freely throughout the EU.

But the global economic crisis of 2008 hit Europe hard, resulting in high unemployment in parts of the continent. Conflict in Southwest Asia, North Africa, and other regions has caused many people to seek asylum in Europe. The EU has had to decide how to take care of this new influx of migrants. Europe has also become the target of terrorist attacks, which brings up issues of safety, a decline in tourism, and other economic impacts on this huge supranational organization.

These issues resulted in the Brexit vote in the United Kingdom. Arguments backing the proposal to leave the EU revolved around Britain's control over its borders, the large amount of money it paid to be a part of the EU, and the perceived burdens Britain faced from EU regulation. The underlying issues illustrate the potential downsides of supranationalism. While countries in a supranational organization can benefit from enhanced collective power, a downturn may impact EU countries negatively—and may be more burdensome for the strongest countries of the organization.

The United Kingdom officially withdrew from the EU on January 31, 2020, and untangling the affairs of the United Kingdom from those of the EU has proven to be difficult. The two entities struck a trade deal in late 2020, avoiding tariffs and quotas on goods but introducing difficulties and burdensome paperwork on British imports and exports, especially food products—which led to shortages in the United Kingdom. Some European businesses have decided that exporting to the United Kingdom is not worth the effort. Furthermore, labor shortages in the United Kingdom, especially in trucking, food processing, and services such as hospitality, have made many who voted for Brexit regret their decision. Overall, Brexit has resulted in a shrinkage of the UK economy by about 4 percent. ∎

Geographic Thinking

Explain how Brexit can be thought of as an example of devolution.

By the Numbers

513 million
citizens of the European Union in 2019

22%
of the world's economic output generated by the EU

51.9%
of UK voters voted to leave the EU in 2016

Sources: Eurostat, BBC

11.3 Forces That Unify and Forces That Divide

All political states face varying forces that unite or divide their citizens. These forces can create cohesion, prosperity, and national harmony, or they can create dysfunction, distrust, and dissolution. Studying these forces helps geographers understand how states' decisions about how to govern affect the people who live within their borders.

Centripetal Forces

Learning Objectives
LO 11.9 Identify social and cultural phenomena that contribute to centripetal political forces.
LO 11.10 Explain common consequences of centripetal political forces in a state.

Recall that a centripetal force is one that unites groups of people. It is named after the force studied in physics that draws an object in orbit toward the center. As it relates to political geography, a centripetal force can be thought of as a force that draws people together to support the sovereignty of the state. For example, when an attack by an enemy causes the people of a country to unite against a common foe, that is a centripetal force. When education and mass media promote a shared culture and language within a state, that is a centripetal force. When government policies allow for expressions of differences and acceptance of others within a country, that is a centripetal force. These forces can strengthen, stabilize, and coalesce a state.

Centripetal forces can be temporary, occurring in the aftermath of an event such as a disaster or fleeting threat. For example, the coordinated attacks by al Qaeda on the United States on September 11, 2001, shocked the country. One of the largest terrorist attacks in history, 9/11 resulted in a massive outpouring of humanity and patriotism that brought Americans together in their grief and in a battle against international terrorism.

On April 15, 2019, fire broke out on the roof of the Notre Dame Cathedral in Paris, France. The disaster came at a time when the French people were deeply divided by the "yellow vests" movement—a months-long series of protests against rising fuel prices and high taxes on the working and middle class. The movement involved large demonstrations that shut down roads and led to violent riots and police action. The unrest was ongoing at the time of the Notre Dame fire, but while the cathedral burned, Parisians gathered and watched the destruction of their beloved monument and symbol of Paris. The dying embers united Parisians in mourning, and immediately after the fire, President Emmanuel Macron announced that Notre Dame would be restored within five years. Just two days after the fire, French citizens and companies had donated the vast majority of the more than $1 billion raised for this important rebuilding effort—a collective act that appeared to be at least a temporary centripetal force.

Reading Maps Centripetal forces can sometimes politically unify multiple states, as in the confederation that was formed among 39 German states in 1815. The goal of the German Confederation was to strengthen political and economic bonds and defend itself against neighboring powers France and Russia. This political confederation went beyond what a modern supranational organization sets out to do. ▌Describe how the formation of the German Confederation is an example of a centripetal force.

Critical Viewing The burning of the Notre Dame Cathedral in Paris in April 2019 helped to unify the citizens of Paris and all of France as they responded together to the tragic fire. Many people donated to the iconic church's rebuilding. ▌Explain why the destruction of this particular building caused the French people to unite.

Centripetal forces can also be longstanding and help to increase cultural cohesion within a state. Factors such as a singular ethnicity, religion, or language; common social and economic standards; a strong infrastructure; a strong, patriotic leader; and a fair and just legal system can aid in the formation of a strong national identity. For example, many citizens in Nepal practice Hinduism, which contributes to a sense of unity in the country. In India, the connections formed by the country's infrastructure, such as its excellent railroad system, tie its remotest villages to its major cities, providing an economic centripetal force. In states that don't share unifying commonalities like a single ethnicity or religion, centripetal forces can still be fostered over time. Celebrating national holidays, like the Fourth of July in the United States and Mexican Independence Day, September 16, in Mexico, can unify generations of people countrywide.

Symbols can act as a display of unity within a state and can also be a cause of that unity. A symbol like a national flag often incorporates elements that represent the principles upon which a state has been founded. The Union Jack—the national flag of the United Kingdom—consists of intersecting red and white crosses representing the patron saints of England, Scotland, and Ireland. Americans often rally around their distinct symbols. The American flag, the Statue of Liberty, the Liberty Bell, and the bald eagle are all symbols of the United States that inspire national pride. Some countries may have more figurative symbols, though. In 1995, South African President Nelson Mandela used the country's national rugby team to help unify a very divided country that had suffered under more than three centuries of racial injustice. He brought attention to the team and celebrated it. South Africans banded together to support the team, and this led to a more unified state.

Sometimes centripetal forces can transcend a single state and unify multiple states. This kind of unification goes beyond what a supranational organization is designed to accomplish. States might voluntarily join together to work toward a common political, economic, or military purpose. A federation, for instance, is a group of states with a central government but some independence in internal affairs. For years, six East African countries—South Sudan, Tanzania, Uganda, Kenya, Rwanda, and Burundi—have been working to create the East African Federation to gain more political and economic power. While still in its planning stages, if it becomes official, the federation will have an official

Political Challenges and Changes

Ethnonationalism

Nationalism can take different forms. Some, such as civic or liberal nationalism, are based on the values of equality, inclusion, and the separation of government and religion—and act as centripetal forces within a state. Others, such as populist nationalism or right-wing extremism, are exclusionary. **Ethnonationalism** is when the people of a country identify as having one common ethnicity, religion, and language. This commonality creates a sense of pride and identity that is tied to the territory. This form of nationalism, also referred to as **ethnic nationalism**, puts up barriers for participation in the life of the state because identifying a country by a single ethnicity, religion, or language excludes people who don't fit within those parameters and is considered a form of racism.

Ethnonationalism is typically thought of as a centrifugal force, since it divides the people of a country based on their ethnicity. An ethnonationalist movement is an attempt to create or change a state based on a shared ethnicity. Such movements are often not spontaneous or natural but the result of social, political, historical, and geographic forces. Sometimes such forces are manipulated by individuals seeking power by promoting the interests of certain groups over others. The Nazi Party in Germany, led by Adolf Hitler beginning in the 1920s, was an example of an ethnonationalist movement. The Nazis believed that the only true Germans were people of Germanic or Nordic heritage, referred to as Aryans. Hitler's attempts to rid the country of non-Aryans, which culminated in the Holocaust, were an extreme version of ethnonationalism.

In recent years, ethnonationalism has become more common, with hate speech and intolerance accepted for political gain. There are many examples around the world, but particularly in Europe and North America, of anxiety about national identity leading to both clear and coded calls for ethnic purity and the preservation of "traditional" cultural practices. For example, Islamophobia, the irrational fear of Muslims and a form of ethnonationalism, led to immigration policies banning Muslims' entry into the United States during the Trump administration.

While these examples feed centrifugal forces, ethnonationalism can result from centripetal forces as well. In recent years, for instance, Russia has threatened the sovereignty of neighboring Belarus by taking steps to annex the smaller country. These actions have brought Belarusians together in opposition against Russia, giving rise to ethnonationalism. They have a renewed sense of pride in their heritage and history, and while Russian is the main language spoken in Belarus, there has been an increased interest in recent decades among many citizens for learning Belarusian, the country's historical language. However, despite this Belarusian ethnonationalism, by 2022 the country was considered to be under political control by Russia, merely a puppet state. In contrast, Ukrainian ethnonationalism has not only persisted but has grown markedly since Russia's seizure of Crimea and its backing of Russian separatists in the Donbas region in 2014. The national pride among its majority, coupled with strong leadership by President Volodymyr Zelenskyy, led to a tenacious defense of the Ukrainian territory following the Russian invasion in 2022.

In Belarus, events like this traditional harvest festival became more common as Russia challenged the sovereignty of its neighbor.

language (English), a lingua franca (Swahili), and a single currency.

Similar to a federation is a confederation, which is also a group of sovereign states that have banded together. Unlike a federation, however, in a confederation the autonomy of each member state takes precedence over the common government. Over time, a confederate arrangement often leads to the more organized structure of a federation. The German Confederation, formed in 1815 to replace the destroyed Holy Roman Empire, consisted of 39 German-speaking states. Germany's modern federal government can be traced back to this earlier confederation.

Centrifugal Forces

Learning Objectives

LO 11.11 Identify social and cultural phenomena that contribute to centrifugal political forces.

LO 11.12 Explain how centrifugal forces can lead to conflict, rebellion, and civil war in a state.

A centrifugal force, as you've learned, is one that divides groups of people. In physics, a centrifugal force is one that appears to pull an object in orbit away from the center. In political geography, it can be thought of as a force that acts to pull a state apart or diminishes the ability of the state to govern. When multiple nationalities within a state compete for control and allegiances, this is a centrifugal force. When economic or social inequality is rampant, this is a centrifugal force. When a state's infrastructure weakens connections between regions or localities, centrifugal forces are also at work.

Centrifugal forces can lead to uneven development within a state. A country with a weak infrastructure, for instance, has poor connections between regions and cities. Places that are difficult to reach will be less likely to develop at the same rate as areas with stronger transportation and communication connections. The populations in such isolated areas might rightly feel disconnected from – "left behind" by the rest of the country and the central government.

Centrifugal forces can also lead to separatist movements. Separatists may want complete secession, or they may just want greater autonomy or recognition as a national minority. A separatist movement has been simmering in Italy for several decades. A political party formed in northern Italy in 1991 to formulate a plan to separate from the southern half of Italy. This region, known as Padania, is Italy's industrial and economic powerhouse, and many in the region feel that their southern neighbors are taking advantage of the northern region's financial success.

An example where a complete secession occurred is the Civil War in the United States. The acrimonious issue of slavery heightened when new states were forming in the western territories. Many people in the South, where slavery was legal, wanted these new states also to allow slavery, while many in the North did not want slavery to spread and wanted the new states to be free states. Political leaders in the South used the argument of "states' rights" to defend their practices, worrying that if states where slavery was illegal outnumbered states where it was legal, it would greatly impact their economic system, which depended on enslaved labor. Eleven states eventually seceded and formed the Confederate States of America. They even nominated their own president, Jefferson Davis. This issue of slavery culminated in the bloody Civil War.

In another example, the Southeast Asian country of Indonesia experienced two decades of violent separatism after invading neighboring East Timor in 1976. Indonesia declared it was reunifying the island (the western portion of the island already belonged to Indonesia), but the East Timorese resisted the occupation and fought back, ultimately losing tens of thousands of their countrymen from fighting, famine, and disease. Worried about its continued ability to govern East Timor, and due to international pressures, Indonesia authorized a referendum in which the East Timorese could vote to determine the future of their country. Nearly four-fifths of the voters supported East Timorese independence from Indonesia and, after a period of transition supervised by the United Nations, East Timor became an independent sovereign state in 2002.

While cultural diversity can be a great strength in a state, it can also act as a centrifugal force under certain circumstances. The differences between groups living within the same country are sometimes so deeply rooted that conflict arises despite efforts to encourage unification. Leaders must find a way to balance the interests of the different groups. This is often accomplished with a federal system of government, but it still may be challenging to establish a unified identity when diverse groups within a state clash. Insurgents, or rebels, in the Sahel, the region south of the Sahara Desert, loosely affiliated with al-Qaeda and the Islamic State, have targeted Burkina Faso, Mali, Chad, and Niger, causing major disruptions, large movements of

CENTRIPETAL FORCES	CENTRIFUGAL FORCES
Ethnic unity and tolerance	Ethnic conflict
Social/economic equity	Social/economic injustice
Charismatic leadership	Dictatorial leadership
Strong infrastructure connections	Weak connections (transportation and communication)
Religious acceptance	Religious intolerance
Nationalism/symbols	Populist nationalism
Fair and just legal system	Loss of rights
Common language	Multiple languages

Centripetal Versus Centrifugal Forces Centripetal forces support a state's sovereignty, while centrifugal forces challenge a state's sovereignty.

At a Peace Day rally in Istanbul, a protester holds a pro-Kurdish Peoples' Democratic Party flag. The oppression of Kurds in Turkey is an example of a centrifugal force that has long worked to divide diverse groups in the Middle East.

internally displaced people, and the deaths of more than 7,000 people. The region is poor and politically unstable. Military coups took place in each of these countries in 2021–2022. The jihadists, or religious fighters, exploit ethnic differences in the region to recruit new members, destabilize already weak governments, and build their power base. The United Nations along with France, the former colonizer of the region, have troops engaged with maintaining peace but they have been largely unsuccessful. Russia and the United States have also been involved in efforts to help, but the centrifugal forces are strong across the states in this volatile area.

These clashes can be especially polarizing in countries that have a stateless nation living within their borders. The Kurds, the fourth-largest ethnicity in Southwest Asia, are one of the most persecuted minorities in modern times. In Turkey, where they represent about 20 percent of the population, they have been oppressed and assaulted by different factions for their religious and cultural diversity. This has caused a significant division within the country.

When centrifugal forces are strong enough to threaten a state's sovereignty and the government can no longer provide the services essential to governing, it is said to be a failed state. A sovereign state should provide for its citizens: security from invasion, a judicial system to uphold the rule of law, the opportunity to participate in the political process, essential freedoms such as human rights and tolerance of dissent, access to health care and education, and access to transportation and communication infrastructures, among other things. Strong successful states rate highly across these categories. Weak states perform poorly in some or all of these areas. Countries weak enough to be considered failed states usually break down into factions. They often become dangerous places that are full of conflict and might even descend into civil war.

Geographic Thinking

1. Explain how symbols such as the Union Jack may act as a centripetal force.

2. Describe how infrastructure, such as a quality transportation system, can act as a centripetal force.

3. Explain how nationalism can act as either a centripetal or a centrifugal force.

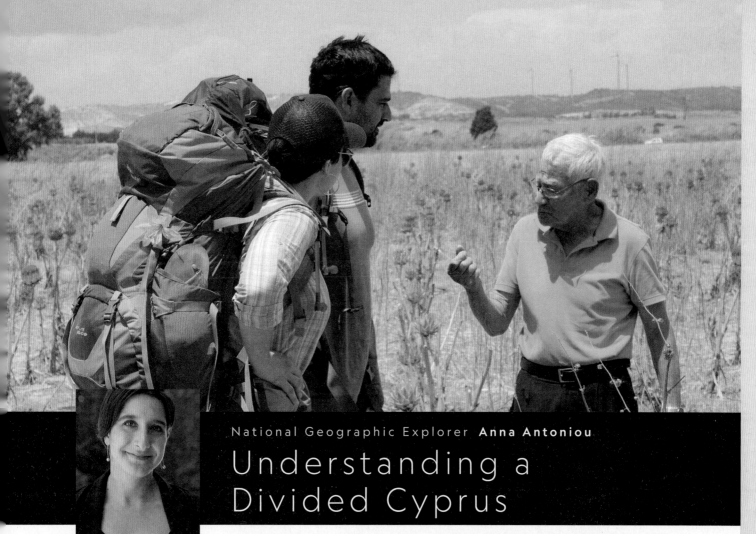

Anna Antoniou mixes archaeology and storytelling to discover more about the cultural heritage of communities around the world. Talking to residents during her exploration of Cyprus with other team members (*above*) led to a discovery that the long-segregated nation is not as divided as previously thought.

Learning Objective
LO 11.11 Identify social and cultural phenomena that contribute to centrifugal political forces.

National Geographic Explorer **Anna Antoniou**

Understanding a Divided Cyprus

Anna Antoniou circumnavigated the Mediterranean island of Cyprus for 60 days to understand the effects of the internal boundaries that divide the landscape geographically, politically, and ethnically.

Cyprus is home to a majority ethnic Greek population and a small population of ethnic Turkish residents. An ongoing dispute between the two groups over settlement issues of the island has existed since the 19th century. Consequently, Cyprus has been partitioned since 1974, resulting in separate governments and territories. The southern two-thirds of the island form an independent country with majority Greek ethnic ties, called the Republic of Cyprus. The northern portion remains a Turkish-occupied territory called the Turkish Republic of Northern Cyprus.

Anna Antoniou, an anthropological archaeologist and Cypriot-American, set out to explore the cultural and political dynamics of Cyprus's division. She wanted to know, "What unites this divided island?" Antoniou talked to real people about their lives to hear if there was more commonality among the people of Cyprus beyond the issues of the political divisions that split the land.

From her discussions, Antoniou found that though the boundaries divide the economies and governments of the people of Cyprus, most residents feel like they are part of the same culture and community and desire to be a unified island again. Antoniou told National Geographic, "Although they do not share a language, a religion, and now a territory, Greek Cypriots and Turkish Cypriots believe they possess a single ethnic and cultural identity, with 'Greek' and 'Turkish' being a secondary affiliation.... Greek and Turkish Cypriots have a stronger bond with each other than they do with their mainland counterparts."

Geographic Thinking

Identify centripetal and centrifugal forces that impact Greek and Turkish Cypriots.

Political Challenges and Changes

Chapter 11 Summary & Review

Chapter Summary

Devolution is the process by which a state destabilizes as a result of factors that divide groups. It occurs when states fragment into autonomous regions or subnational units, or when they disintegrate. Several different factors can lead to devolution of a state.

- Physical geography is a factor that can lead to devolution, as distance and geographic features can make a state more difficult to govern.
- Ethnic differences can lead to devolution. Disparity in treatment, identification with a neighboring state, and advances in technology are all contributors to ethnic separatism.
- Economic and social problems can lead to devolution. Inconsistent productivity, development, or allocation of tax funds can cause dividing pressure on a state.
- States respond differently to devolutionary forces depending on their unique internal divisions. States may respond by sharing more power with subnational units.
- Devolutionary forces can lead to sovereignty of a region within a state or to the disintegration of the state.

Supranationalism is an alliance of three or more states that come together in pursuit of common economic, political, military, or cultural goals.

- Examples of supranationalism in modern history include the League of Nations, the UN, the EU, NATO, ASEAN, the African Union, the Arab League and the Arctic Council.
- A key benefit of supranationalism is that it creates a collective power that can benefit member states.
- A key drawback of supranationalism is that it challenges state sovereignty and can cause states to make compromises and lose some independence.

Geographers study centripetal and centrifugal forces to understand how governments' decisions affect their citizens.

- Centripetal forces are both temporary and longstanding forces that unify a state's power across space. They can also transcend states and unify multiple states.
- Centrifugal forces challenge state sovereignty and divide a state's power. They can lead to separatist and nationalist movements and even the breakup of a state.

Review Questions

Use complete sentences to answer the questions.

1. **Apply Conceptual Vocabulary** The prefix *supra-* means "above" or "over." Break *supranationalism* into its prefix and root and write a standard dictionary definition of the term. Then provide an explanation of how the term is used in the context of this chapter.

2. What is devolution? What factors lead to devolution, and what are some potential consequences?

3. Differentiate between a federation and a confederation.

4. Is irredentism a political, economic, cultural, or technological challenge? Explain.

5. Define the term *ethnonationalism* using the concept of *centrifugal force*.

6. Use an example to show how ethnic differences can act as devolutionary forces.

7. Explain the relationship between terrorism and devolution.

8. Explain the relationship between physical geography and devolutionary forces using the related concepts of *friction of distance* and *distance decay*.

9. Give one example of a supranational organization that the United States is a member of and one that it is not a member of.

10. What is one reason a state would want to join a supranational organization? Explain.

11. What is one reason a state would not want to join a supranational organization? Explain.

12. Describe processes that might lead to ethnic cleansing.

13. Explain the ways ethnonationalism can create barriers for participation in the life of a state.

14. What is the Arctic Council and what transnational challenges does it face?

15. Explain the concepts of *centripetal* and *centrifugal forces* and how they apply to human geography. Why are these physics terms used in human geography?

Interpret Graphs

Study the graphs and then answer the following questions.

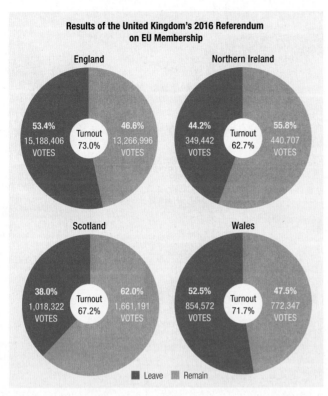

Source: BBC

16. **Identify Data & Information** Which political subdivision of the United Kingdom had the highest percentage of voters that wanted to remain in the European Union? Which had the highest percentage that wanted to leave?

17. **Analyze Visuals** Compare England's voting results with those of Wales. What conclusion can you draw about the relationship between England and Wales?

18. **Compare Patterns & Trends** Compare voter turnout in Northern Ireland to the rest of the United Kingdom. What conclusions can you draw about the feelings of the people of Northern Ireland about Brexit?

Unit 4 | Writing Across Units, Regions, & Scales

The Healing of Colombia

by Alma Guillermoprieto

Pedestrians watch a wedding party's arrival in the historic district of Cartagena, a coastal resort city popular with tourists. According to the Colombian government, tourism in the country has risen 250 percent since 2006.

María Magdalena, a 10-year-old known as Mayito, could see the black smoke coming from farther down the hill as the *paramilitares*, criminals with a right-wing ideological bent, advanced on the town of El Salado, setting fire to her neighbors' houses as they approached. Mayito's mother emptied all the corn out of a burlap bag so the chickens would have enough to eat, threw some clothing into it, and climbed on the back of the family donkey with Mayito, as her two older brothers walked alongside. For a full week they hid, with little water and almost no food, in the shacks that *campesinos*, or rural peasants, keep in their fields here.

From a distance the terrified family could not have guessed the full extent of what was taking place in El Salado, at the center of territory disputed between left-wing guerrillas and their paramilitary adversaries. The assault produced one of the most horrifying episodes in Colombia's five decades of brutal ideological warfare.

Death and Life in El Salado Villagers who hadn't had time to flee were rounded up in front of the church, in a field normally used for pickup soccer games. The victims, accused of sympathizing with guerrillas, were taken into the center of the field one by one, tortured, mocked, knifed, and then strangled or shot.

The killing in El Salado and nearby towns lasted six days, from February 16 to 21, 2000. By the end of it, 66 people were dead. Within a week, all of El Salado's 4,000 residents had fled, joining more than 2 million other internally displaced Colombians at that time who were robbed of their families, their homes, their livelihoods, and their peace.

What makes this story different from other episodes of horror and heartbreak in Colombia is that the people of El Salado came back. In a stubborn return to this most unlikely promised land, the Saladeros took back their town two years after the killings.

Today, El Salado and Colombia are transforming their grim heritage. After half a century in which the war circled repeatedly in on itself, the country's oldest guerilla group, the *Fuerzas Armadas Revolucionarias de Colombia*—or FARC, by its Spanish initials—turned over the last of its weapons in June 2017 to a United Nations team. By then the entire country had been reshaped by violence. Now a lasting peace will have to be won, inch by inch.

Colombia's War The reality is that in the two centuries since it gained independence from Spain, Colombia has rarely been without violent conflict. Some would argue that the latest cycle of bloodshed began on April 9, 1948, with the assassination of an overwhelmingly popular leader of the traditional Liberal Party, Jorge Eliécer Gaitán. The murder sparked deadly riots in the capital, Bogotá, and a 10-year wave of partisan killings—La Violencia—in the countryside. But long before that, members of the Conservative Party had been killing Liberals, and often enough, vice versa.

In 1957 an agreement to end the violence by rotating power between both parties led to a decade or so of relative peace, and in the cities not many people took notice of a few dozen Liberal *campesino* families who'd been radicalized by a forceful communist organizer. Among those who did were the army, the sitting president, and the archconservative senator who accused the campesinos of wanting to create "independent republics" inside Colombia. In 1964 a military operation involving thousands of troops overran the Liberal group's small, precipitous holdings in Colombia's Andean foothills. Further radicalized by being bombed, the campesinos adopted the FARC name and embarked on a guerilla war against the state that was to last 52 years.

It was the FARC that the paramilitary killers in El Salado accused the villagers of sympathizing with, and it was the FARC that, backed into a corner militarily, finally signed a peace accord with the Colombian government on November 24, 2016.

An Uncertain Future How will the 23 percent of Colombians who live in the countryside fare now that there's peace? Seventeen years after the massacre that emptied out the town, El Salado is a good place to look for first results.

Luis Torres led the return campaign 17 years ago, and when the first 130 people agreed to come back to El Salado, he raised the funds to hire the trucks that brought them home. Now he glowed with a sense of accomplishment as he showed me the sights of his hometown: a cell phone tower that at last allows Saladeros to communicate with the outside world, a preschool, a hundred new houses for the community's poorest families, a couple of storefront groceries, an evangelical church, a street lively again with scampering children, neighbors waving hello.

Depending on who was taking stock of the improvements—Torres or, say, me—one could see either heroic achievements against all the odds or modest recovery to the tune of millions of donated dollars, without solving many of the town's most basic problems, including water, jobs, and education. And El Salado is just one small town out of thousands in similar straits.

Luis Torres has an ultimate dream: He sees himself standing in the crowd and applauding as the ribbon is cut on a technical school in his hometown, one that will train the kids who now zip around so aimlessly on their motorbikes for something better than a dirt-poor life. "Once I see that ribbon being cut, I'll die in peace," he said.

Source: Adapted from "The Healing of Colombia," by Alma Guillermoprieto, *National Geographic*, January 2018.

Write Across Units

Unit 4 delved into the complexities and nuances behind the boundaries on the world map, revealing the political processes and other factors that support the creation of countries. This article discusses political forces that have remade the country of Colombia over the past five decades. Use information from the article and this unit to write a response to the following questions.

Looking Back

1. What are the advantages and disadvantages of using the small-town example of El Salado to analyze Colombia's present situation? Unit 1

2. What effect do you think the civil war has had on Colombia's population pyramid? Unit 2

3. What effects do you think the war in Colombia might have on the country's cultural landscape? Unit 3

4. What centrifugal and centripetal forces have been at work on Colombia's government? Unit 4

Looking Forward

5. How do you think the civil war affected Colombia's agricultural patterns? What do you expect for agriculture in Colombia over the next few years? Unit 5

6. In what ways would you expect the events in rural Colombia during the war to affect its cities? Unit 6

7. How could you use measures such as GDP, GNP, and income distribution to analyze how the civil war has affected Colombia's economic development? Unit 7

Write Across Regions & Scales

Research a country in a region outside South America that has experienced a civil war sometime during the past five decades and conduct additional research on Colombia's civil war. Write an essay comparing the causes and results of the conflicts in both countries. Drawing on your research, this unit, and the article, address the following topic:

What centrifugal forces push countries toward civil war? How can countries seek to counter these forces?

Think About

- Historical, economic, cultural, and political factors that cause internal conflict in a country

- Centripetal forces that may lead to greater cohesion within a society

Unit 4

Chapter 9

One Map, Different Perspectives

In 2014, Russia annexed Crimea, a region that belonged to Ukraine, in a move that the Ukrainian government does not recognize to this day. Officially, Russia and Ukraine draw the border around Crimea in different places. Online mapping applications chose to show the border differently depending on where users were located when they accessed the map. When viewed from a browser in Russia (*right*), this 2014 map marks Crimea with a solid border that shows it to be part of Russia. When viewed from a browser in Ukraine (*left*) there is no border, indicating that Crimea is part of Ukraine. Those viewing the map from outside the region see a dotted line that indicates a disputed border. ▌ Describe how these maps portray political power in the Crimea region.

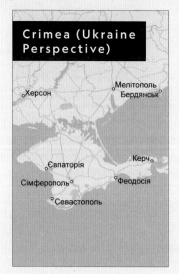
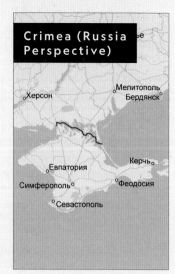

Chapter 9

The Former Yugoslavia

The section of the Balkan shatterbelt that once formed the country of Yugoslavia is commonly referred to as "the former Yugoslavia." This map shows the countries that now exist in the former Yugoslavia and the distribution of ethnic groups in the region in 2015. ▌ Compare the ethnic distribution of people in the former Yugoslavia with the national borders. Identify which countries are nation-states. Explain whether you think the current map of the former Yugoslavia will remain stable in the future.

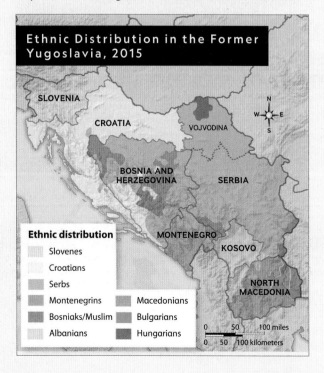

Chapter 9

Provinces of Canada

Canada's provinces share several types of borders. Some are legacies of European settlement, while others are determined by the country's physical geography. ▌ Identify borders on the Canada map that you believe are superimposed, subsequent, antecedent, geometric, or consequent. Explain your reasons.

316

Chapter 9

The Future of Choke Points

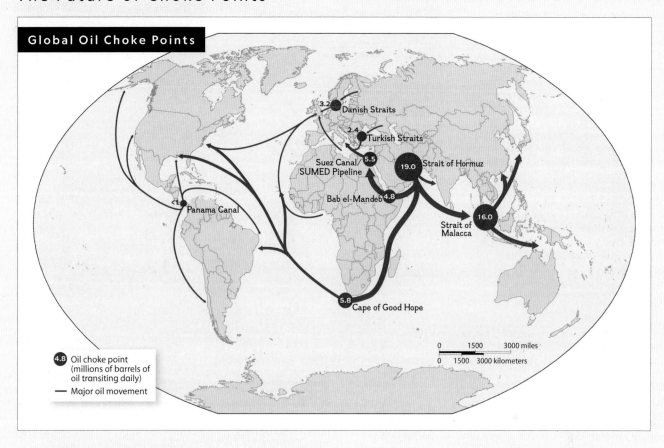

Global Oil Choke Points

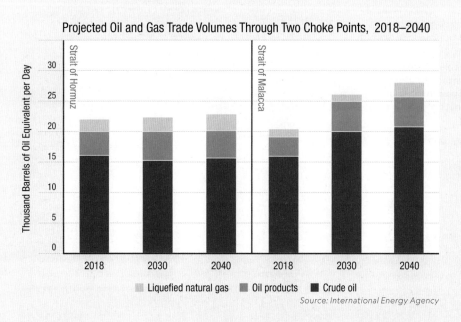

Projected Oil and Gas Trade Volumes Through Two Choke Points, 2018–2040

Liquefied natural gas | Oil products | Crude oil

Source: International Energy Agency

Most countries depend on oil for a large proportion of their energy needs, and much of the world's oil is transported by oceangoing freighters. Thus, countries controlling waterway choke points have the potential to wield political and economic power over countries that need to import oil. The International Energy Agency (IEA) produces an annual report on the present state of energy production and consumption. The report includes projections of possible energy trends. This chart from the 2019 IEA report projects the quantities of oil likely to pass through two choke points through 2040, assuming that the countries of the world maintain the energy policies they have stated. ▍ Describe the trends predicted for each choke point. Explain how these trends will affect countries that control the waterway choke points on the map.

Chapter 9

Conflict Over Borders

After about a century of domination by the British East India Company, India officially came under British rule in 1858. In 1947, Britain granted the territory's independence while at the same time dividing it into two separate states: Muslim-majority Pakistan and Hindu-majority India. Partition, as this act was called, did not result in peaceful borders, however. India and Pakistan have remained in conflict over the province of Kashmir since 1947, and each country claims the Siachen Glacier for its own. India and China also have border disagreements, with both claiming the regions of Aksai Chin and Arunachal Pradesh (the dotted line on the map representing their disputed borders). ▌ Identify the effects of colonialism evident on the map of India.

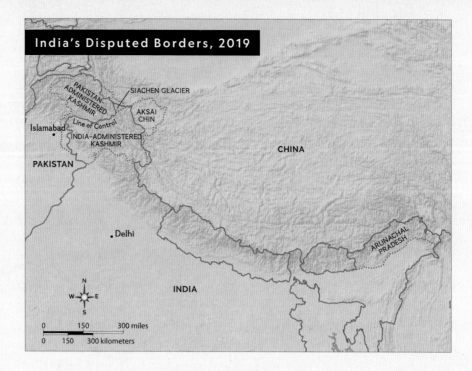

Chapter 9

Nations and States

Catalonia is a semiautonomous region of Spain with a distinct culture and language—Catalan—as well as its own parliament. In recent years, Catalan separatist movements have gained strength and visibility. After a majority of Catalan people voted in favor of independence in a 2017 referendum, Spain revoked Catalonia's autonomy, only to restore it in 2018. Tensions between the Spanish and Catalan governments continue to be high. ▌ Compare the maps of the semiautonomous region of Catalonia and the extent of Catalan speakers. Explain whether or not an independent Catalonia would be a nation-state. Describe Spain in terms of nation-state, multistate nation, or multinational state.

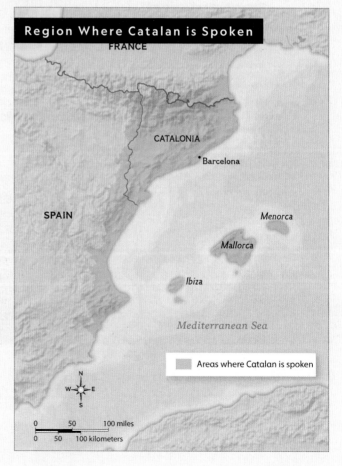

Maps & Models Archive

Chapter 10

Are You Satisfied with Your Government?

As part of its annual end-of-year survey for 2018, the polling organization Gallup International asked people in 58 countries to rate their satisfaction with their government. Respondents lived in countries with federal and unitarian governments, in democracies, republics, and dictatorships. ▮ Explain the value of studying people's attitudes toward government at a range of scales, national and regional (aggregated data used for regions). Describe the level of satisfaction in the United States with other countries and world regions.

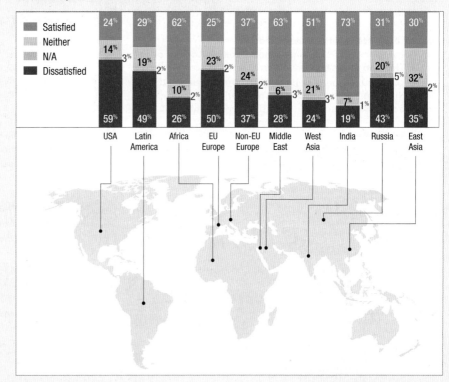

Source: Gallup International

Chapter 10

Who Creates Electoral Boundaries?

In most states, the legislature is in charge of drawing congressional district borders, but increasing numbers of states have begun employing commissions. Not all commissions function in the same way, however. Advisory commissions participate with legislators in the process of drafting districting plans. Backup commissions step in to assist when a legislature does not successfully pass a districting plan. Politician commissions are smaller groups of elected officials who are in charge of drawing district boundaries. Independent commissions consist of members who are not state legislators. ▮ Identify who is in charge of congressional districts in the majority of states. Describe the possible effects, at the national scale and the state scale, if more states move to using independent commissions.

Chapter 10

Leaving the EU

In 2016, the citizens of the United Kingdom voted to leave the European Union—a political and economic coalition of European states. This map shows how the different districts voted in the referendum. ▌ Describe, in general, how the legislature of the United Kingdom might employ gerrymandering if it wanted to propose the referendum again and obtain a different result.

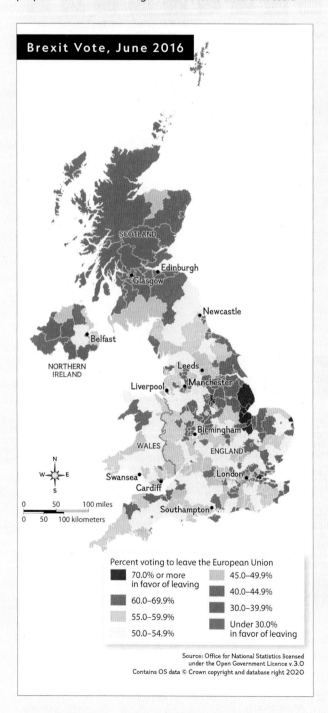

Chapter 11

Distance Decay and Devolution

Manila, the capital city of the Philippines, is on the island of Luzon in the north of the archipelago. It is separated from Mindanao, the second-largest population center, by a stretch of mountainous islands and ocean hundreds of miles long. At the same time, Mindanao is home to the majority of the country's Muslim population, as well as its largest concentration of ethnic minorities. ▌ Explain the degree to which the distance decay model could explain pressures on the Philippines' central government. Describe how distance decay could combine with cultural forces to push the Philippines toward devolution.

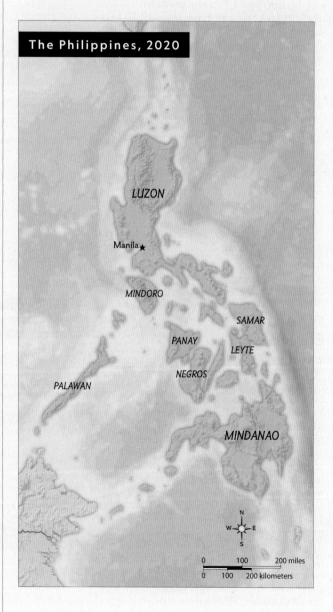

Maps & Models Archive

Chapter 11

Measuring Governmental Stability

The Fund for Peace (FFP), an international nongovernmental organization, publishes an annual report on fragile states, or countries in danger of becoming failed states. The map represents the report's findings for 2019. When determining each state's risk for failure, the FFP examines several centripetal and centrifugal forces such as infrastructure and public services, tensions among societal groups, a country's economic health, and the presence of refugees and internally displaced people. ▌Choose a country or region you have read about or that you know about from other sources. Identify its fragility status and describe the centripetal and centrifugal forces that explain the rating.

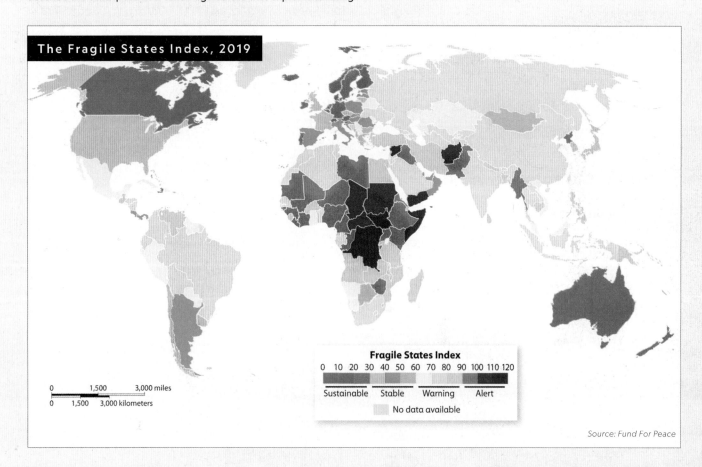

Source: Fund For Peace

State Fragility Trends, 2009–2019

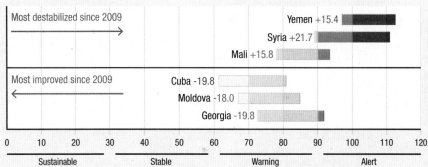

▌Based on what you've learned about devolutionary forces, identify some factors that may have caused the six countries in the graph to become more fragile or more stable.

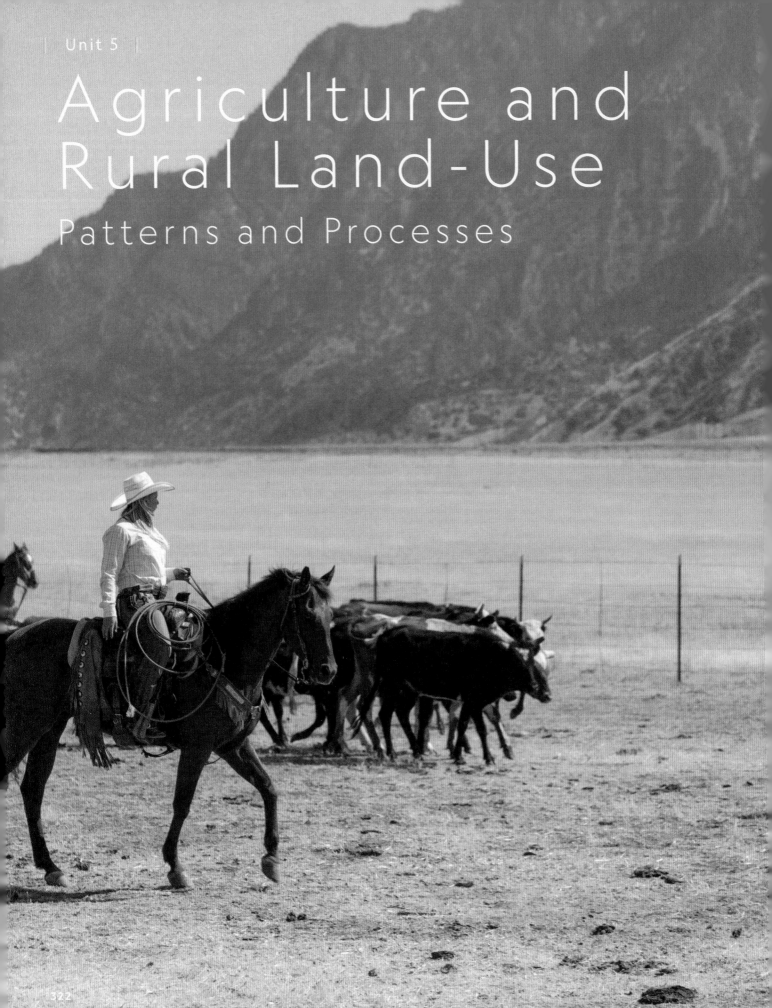

Unit 5

Agriculture and Rural Land-Use
Patterns and Processes

Agriculture's Impact

Zanskar Valley, northern India

Every person is affected by agriculture. It provides the food we eat and some of the fuel and products we trade and consume. The agricultural practices used to produce crops and livestock vary as widely as the physical geography of different regions, including Utah's grasslands on which cattle (shown here) thrive.

Agriculture (commercial or subsistence) is impacted by geography, economics, politics, and technology. How people obtain and distribute food differs among regions and countries, creating many agricultural opportunities and challenges. Global networks of trade and transportation allow millions of people to access agricultural products that are grown in diverse climates across the planet. The challenge for agriculture lies in developing sustainable farming practices that feed a growing population.

Chapter 12
Agriculture: Human-Environment Interaction

Chapter 13
Patterns and Practices of Agricultural Production

Chapter 14
Agricultural Sustainability in a Global Market

Writing Across Units, Regions & Scales

Unit 5 Maps & Models Archive

National Geographic Explorer **Jerry Glover**

Agriculture for a Hungry Future

Earth's most extensive type of land cover wasn't set by nature but by human activity. According to National Geographic Explorer Jerry Glover, agriculture takes up more acreage than any type of natural ecosystem. Glover is one of many researchers grappling with problems caused by farming practices that have replaced natural ecosystems and diminished biodiversity worldwide.

Learning Objective
LO 12.1 Explain how agricultural practices have environmental and societal connections.

Perennial Solutions As an agricultural ecologist studying farming systems around the world, Jerry Glover regularly sees the harmful consequences of farming techniques. "We've done more damage with agriculture in the past 50 years than I think were done in the previous thousands of years," he observes. In Glover's view, present-day farming practices are firmly tied to a 20th-century vision of agriculture, as farmers continue to plant the same crops that humans have been raising for centuries. These crops result in high rates of soil loss through erosion and require large amounts of fertilizers and pesticides to maintain them. Glover says, "Farmers have the thinnest and most nutrient-depleted soil in history. Given the new challenges we face, we need a uniquely 21st-century vision for food security."

Glover advocates for innovative solutions to the problem of depleted soil, including a strategy called perennation—the use of perennial plants in farm fields. Nearly all agricultural crops are annuals, or plants that must be replanted every year after harvesting. Perennial plants survive from year to year and are more resilient during droughts and other climate-related stresses. Glover works to expand the use of perennation in regions of Africa where the soil is especially thin and starved for nutrients. In Malawi, for example, he studied how farmers greatly improve their yields of maize by planting "fertilizer trees" with roots that draw water and nutrients from deep within the soil. When the trees drop their leaves, they return the nutrients to the shallower layer of soil, where the maize has its roots. Glover also promotes the development of food crops that are perennials themselves, but while some promising strains of perennial wheat and rice have been created, none have entered large-scale cultivation as yet.

Local Conversations While some have claimed that Earth is capable of producing enough food for all its human citizens, Glover points out that the best farmlands are unequally distributed, explaining, "An adequate global food supply does not solve regional problems." At the same time, barriers to improving crop yields may arise from social and economic concerns, as well as from geographic challenges. For this reason, Glover strives to adapt agricultural solutions to a variety of local needs. This may mean breeding perennial crops that are adapted to specific soil and climate conditions, and it means working with farming communities and forming an understanding of their social structures. Glover has learned that women farmers in Malawi and elsewhere in Africa are best at identifying the most resilient and nutritious solutions for their families. To communicate effectively with women in some communities, however, scientists must consider cultural and family dynamics because women's social position may affect the ways in which researchers can interact with them.

Glover envisions future generations feeding themselves with food sources that have been developed in only limited ways at present—such as seaweed and insects raised for protein. He also points to advanced technology that allows researchers to more directly manipulate GMOs (genetically modified organisms) to better adapt them to local growing conditions. As the human population grows, Glover believes a new agricultural revolution is needed. "Farming is what we ultimately rely on to survive," he says. "You know we can't go back to low-yielding traditional practices. We can't go forward with . . . just pouring the chemicals onto the landscapes and hoping for the best."

Geographic Thinking

Explain the long-term environmental and societal effects Glover's work might have on communities and regions.

Top: National Geographic Explorer Jerry Glover (left) learns about local farming practices as he talks with a family in Malawi that grows corn and tree crops, such as mangos. *Bottom:* Grains are harvested on the lands of the Agroscope research center in Switzerland in order to study GMOs and improve agricultural practices.

Chapter 12
Agriculture: Human-Environment Interaction

Critical Viewing The Chianti region of Italy is famous for its production of red wine, and its Mediterranean climate strongly influences its agricultural practices. Initially covered in dense forest, the land was transformed during the Middle Ages into a place where wine grapes could be cultivated. ▌ What interactions between people and the environment are evident in the photo?

Geographic Thinking How is agriculture a human-environment interaction?

12.1
Agriculture and the Environment

National Geographic Explorer
Stephanie Pau

12.2
Agricultural Practices

Feature: Rural Survey Methods

Feature: Rural Settlement Patterns

National Geographic Photographer
George Steinmetz

12.3
Agricultural Origins and Diffusions

12.4
Advances in Agriculture

Case Study: Women and Africa's Green Revolution

12.1 Agriculture and the Environment

Agriculture is a major way of life around the world—and, of course, it feeds everyone across the globe. The ways in which agriculture is practiced, the number of people involved, and the resulting cultural landscapes all vary among Earth's environments.

Introduction to Agriculture

Learning Objective
LO 12.2 Explain the connections between agriculture and the physical environment.

Agriculture is the purposeful cultivation of plants or raising of animals to produce food and other goods for survival. The first crops to be harvested through agriculture were food crops, such as fruits, vegetables, and grains, the most widespread being corn (or maize), wheat, and rice. Other crops, such as oats and alfalfa, are important for feeding livestock. Agriculture is more than growing food—though that is the primary purpose of farming and livestock raising in many parts of the world. Fiber crops, such as cotton, are used for textile and paper products. And oil crops can be used for consumption or for industrial purposes. For example, olives, corn, and soybeans may be harvested and processed into oils used for cooking, for machinery lubrication, or as biofuel.

Geographers study agriculture to understand how humans have modified the environment to sustain themselves. The types and patterns of agricultural production and the processes that affect these patterns exist at a range of scales. Observing these global, regional, and local patterns informs geographers about the sustainability of agricultural practices, especially in food production.

Environmental Factors Cultivating plants or raising animals requires adaptation to environmental limitations. Sunlight, water, and nutrients are all factors that affect plant growth. Agriculture is bound to the physical environment, and four factors have a profound effect on the agriculture that can take place there: climate, which includes temperature and precipitation; elevation; soil; and topography.

You've read that climate is the long-term patterns of weather in a particular area. (Weather is what happens today: sunny

Global Cropland, 2020

Cropland covers 18 percent of the United States, which contains 8.9 percent of the world's cropland.

India has the highest percentage of the world's cropland: 9.6 percent.

China has 8.8 percent of the world's cropland within its borders. Cropland covers 18 percent of the entire country.

Reading Maps A recent study revised what geographers had previously thought about Earth's cropland. India, rather than China or the United States, ranks first, with 9.6 percent of the world's net cropland area. ▪ Describe the general distribution of croplands across the continents.

or cloudy, rainy or snowy, hot or cold.) Climate is a major influencer of agricultural choices and practices because it provides precipitation and temperatures needed for seeds to germinate, plants to grow, or livestock to have the food needed to survive. Water, from natural precipitation or irrigation, critically provides the moisture plants and animals require. Temperature is the key factor in determining the growing season—the length of the year during which plant life can grow.

Generally, the greater the distance from the Equator, the shorter the growing season. At the Equator and in the tropics, the growing season can be year-round. In the temperate and subarctic zones, however, the colder temperatures of winter prevent plant growth for a varying number of weeks or months. In those regions, the growing season is measured in the number of frost-free days, as frost can kill plants.

Elevation also affects the growing season and what plants can be grown. Each increase of 1,000 feet above sea level means a corresponding decrease of about 3.6° F in average temperature. As a result, the higher the elevation, the shorter the growing season. Elevation can create different cultivation opportunities in all mountainous regions. For example, in tropical regions in Central America and South America, the hotter lowlands are used to grow tropical crops such as bananas and sugarcane. In the next highest zone, farmers can grow coffee, corn, and other vegetables. At higher elevations, crops must be hardier, like barley and potatoes. Higher still, the land can only be used for grazing livestock because only grasses can grow there.

Soil, a vital factor in determining the agricultural potential of a given area, is the biologically active coating of Earth's surface. This layer can range from a few inches to several feet in depth. It is formed by the weathering of rock by wind, water, and other factors, which break the rock into increasingly smaller pieces over an extremely long period. It can take thousands of years to form an inch of soil. Soil has four constituent parts: mineral particles, water, air, and organic matter like decaying plant material. The key characteristics of soil are its fertility, texture, and structure.

Topography, or an area's land features, includes the slope of that land, which affects the ability of the soil to stay in place and retain water. The steeper the slope, the more likely the soil will be affected by runoff. Slope can also be a factor in land productivity due to the position of the land toward or away from the sun, which affects how much of the sun's energy the land receives.

The most favorable land for growing crops has ideal temperatures, precipitation, soils, and slope. Sometimes landscapes are modified for better environmental factors. Terrace building for farming protects soil on steep slopes, while irrigation or drainage schemes influence water availability. (You learn about these agricultural practices in depth in Chapter 14.) Adding fertilizers enhances the soil fertility of cropland. In general, it is not feasible to modify the other environmental factors—soil texture, soil depth, soil mineral content, temperature, and terrain—at large scales.

Climate Climate varies greatly across the globe and is based on four key factors: distance from the Equator, wind and ocean currents, proximity to large bodies of water, and elevation and topography. These components interact with one another in different ways to create different climate regions.

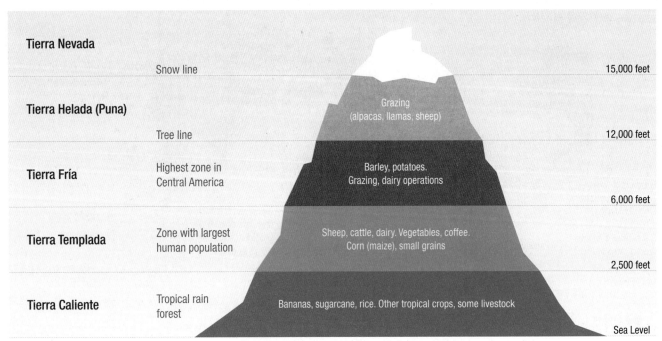

Altitudinal Zonation Central America and South America both show how elevation can affect agriculture (remember that the average temperature decreases by about 3.6° F with each 1,000 feet increase in elevation). ▮ Describe the temperature ranges for the agriculture that can be practiced in the five elevation zones in the diagram; start with 80°F at sea level. Where would you like to live?

Agricultural practices are impacted by climatic conditions. For example, in the tropical climate of Malawi, tea fields are built along the contours of the sloping landscape to slow erosion. Terrace farming is another way to reduce soil erosion and water loss.

Distance from the Equator determines the length of daylight and average temperatures. The Equator and the regions near it to the north and south, called the tropics, receive direct rays from the sun year-round. These regions have nearly equal periods of daylight and nighttime every day and are warm throughout the year. The northern and southern limits of these warmer regions are called the Tropic of Cancer and the Tropic of Capricorn, respectively.

Just to the north and south of the tropics are huge dry bands containing deserts including the Sahara, Kalahari, Sonora, and Atacama. Farther north and south of the desert bands are the temperate zones—large bands between the tropics and the frigid polar circles that have hot summers, cold or cool winters, and transitional springs and autumns.

Ocean and wind currents circulate cold or warm water and air masses over Earth's surface. The circulation of water and air masses in turn affects patterns of temperature and precipitation. Ocean currents that flow north and south transfer heat between lower latitudes (close to the Equator) and higher latitudes (farther from the Equator). When these currents flow away from the Equator, they bring warmth to temperate zones. For example, in January, the Gulf Stream carries warm water from the Gulf of Mexico into the Atlantic Ocean, where the current travels north along the eastern coast of North America before moving east toward Europe. Currents flowing from a temperate zone toward the Equator carry colder water in that direction. Winds also carry air at different temperatures from one region to another. Because warmer air tends to hold more moisture and colder air tends to be drier, these wind patterns affect precipitation patterns.

Location relative to large bodies of water affects climate in two ways. First, these bodies of water warm and cool more slowly than land. As a result, most coastal areas tend to have milder climates than regions farther inland. Second, coastal areas are cooled during the day by cool winds displacing warmer air rising from the land. These regions are warmed at night by warm air blowing onshore as cooler night air pushes down toward the water. Location also affects precipitation, which tends to be heavier near coastlines and lighter farther inland, with some exceptions.

You read that elevation can affect temperature even in the tropical zone, but topography impacts climate in other ways. Coastal mountains have much heavier precipitation on the side facing the wind and very light precipitation on the side away from the wind. This effect can be seen in the

Agriculture: Human–Environment Interaction

Köppen Climate Classifications

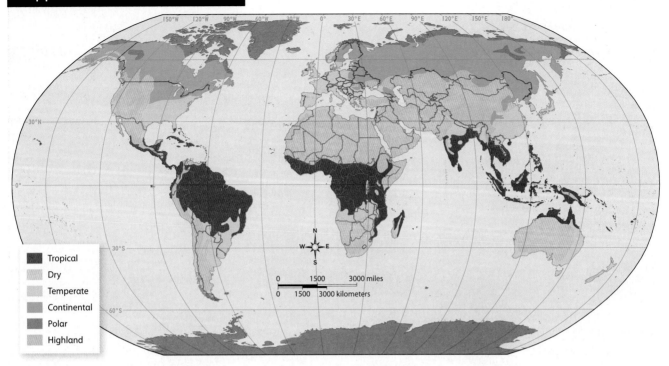

To classify the main climates on Earth, Vladimir Köppen considered global patterns of average temperatures, average precipitation, and natural vegetation. Climate in highland areas, represented by gray on the map, cannot be easily classified because both elevation and topography can vary widely within these regions.

Pacific Northwest of the United States, which produces temperate rain forests along the Pacific Coast and dry conditions farther inland in areas such as the Yakima Valley of Washington State.

Climate Regions

Learning Objectives

LO 12.3 Describe the factors that influence patterns of global climates.

LO 12.4 Compare the temperature and precipitation characteristics of different climates.

The combination of temperature, precipitation, wind patterns, and topography produces different **climate regions**, or areas that have similar climate patterns generally based on their latitude and their location on coasts or continental interiors. Vladimir Köppen, a Russian-German scientist, developed a system for classifying the world's climates. He identified five broad climate types: tropical, dry, temperate, continental, and polar, which are represented in the "Köppen Climate Classifications" map in this section. Each of these broad climate types can then be subdivided into more specific climate types.

Areas with tropical climates all have warm temperatures year-round but vary in their amounts of precipitation. The wet tropical climate has plentiful precipitation, which fosters the growth of tropical rain forests. The tropical monsoon climate—found in South Asia and West Africa—has extremely heavy summer rains and dry winters. The tropical wet and dry climate also has distinct rainy seasons but less precipitation than the tropical monsoon climate. One example of the vegetation associated with this climate is the savanna grasslands in East Africa. As noted, agriculture in tropical climates is wide-ranging, and varies with precipitation patterns and elevation.

With some exceptions, dry climates are commonly found in continental interiors and are either arid (very dry) or semiarid (minimal precipitation). Semiarid climates receive enough precipitation to allow the growth of grasslands. Agriculture in arid climates is typically only possible with irrigation, while semiarid agriculture often features the grazing of cattle or goats.

There are three basic types of temperate climates, or climates with moderate temperatures and adequate precipitation amounts. All three tend to have long, warm summers and short winters. Humid temperate climates, typically found on the eastern sides of continents, have colder winters and year-round precipitation. The marine west coast climate (on the west coast of continents at higher latitudes) has cooler winters and plentiful rain, supporting the growth of temperate rain forests. Mediterranean climates are commonly found on the west coasts of continents near deserts and around the Mediterranean Sea in countries such as Italy, Greece, and Tunisia.

The province of Khövsgöl in northern Mongolia is a mix of tundra and taiga, which are evergreen forests found in the far Northern Hemisphere. Because temperatures are so cold there, little farming occurs. Instead, the Tsaatan (or Dhukha), a local people, herd reindeer.

The milder, wet winters of the Mediterranean climate are conducive to agriculture, as you will read about in detail in Section 12.2, and the hot, dry summers are ideal for cultivating certain vines and trees. **Mediterranean agriculture** consists of growing hardy trees (such as olive, fruit, and nut trees) and shrubs (like grape vines) and raising sheep and goats. These animals forage in the sparse, scrubby summer growth and maneuver around the region's steep landscape.

Continental climates are found in the interior of continents in the Northern Hemisphere and are characterized by distinct seasons that include cold winters and snow. Summers are of varying length and temperature averages, with shorter growing seasons than humid temperate areas. The two polar climate types, tundra and ice cap, are found near the North and South Poles and are both extremely cold. The tundra climate has a short, mild summer but is too cold to allow farming, although some of the Sami people of northern Europe engage in reindeer herding. The temperatures in the ice cap climates of the Arctic and Antarctic rarely rise above freezing.

The warm temperatures of the tropics allow for year-round agriculture, which can also permit multiple harvests of crops, like rice, in a year. The type of tropical climate limits that potential, however. Areas with a tropical wet and dry climate may not receive sufficient precipitation in the dry months to allow crop production. Even the plentiful precipitation of the tropical wet climate poses challenges. The heavy year-round rains allow for the flourishing growth of rain forests, but the soils in these areas tend to be poor. The main source of nutrients needed to produce crops in rain forests is plant matter.

The temperate zones, with their long growing seasons, are home to major grain-producing regions in the United States and Canada. Hardier grains like wheat, which can thrive in shorter growing seasons and drier conditions, grow well toward the north of the temperate zones. Corn, which requires a higher average temperature to germinate than wheat, generally grows farther south. Rice, which needs an even longer growing season, grows in the southernmost parts of the temperate zone, near the tropics. Rice requires more water than wheat and corn and is grown along the warm, wet southern portion of the Mississippi. The humid tropical and subtropical climates on Earth are well suited to wet-rice farming. And in some other areas, such as California, rice growing needs irrigation. Climate change, however, may be altering many of these generalizations.

Geographic Thinking

1. Identify the four elements that make land favorable for growing crops.

2. Describe how the agricultural practices of certain regions are influenced by the Mediterranean climate.

3. Explain why areas on different continents—such as Western Europe and coastal East Asia—have similar climates at different latitudes.

Agriculture: Human-Environment Interaction

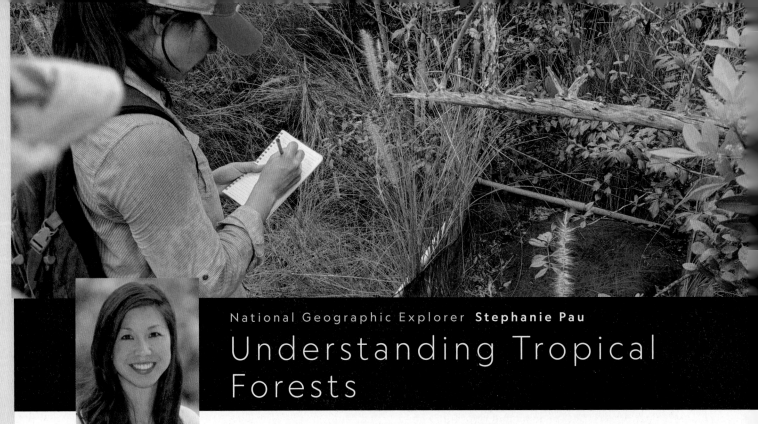

National Geographic Explorer **Stephanie Pau**

Understanding Tropical Forests

Stephanie Pau is a professor in the Department of Geography at Florida State University. Her research explores how climate change impacts ecosystem functions in forests, savannahs, and grasslands.

Learning Objective
LO 12.2 Explain the connections between agriculture and the physical environment.

Seasonal patterns of temperature and precipitation dictate which crops will thrive in a region. In wild environments, plants have developed strategies to survive in very specific climates. Geographer Stephanie Pau wants to find out which plants in tropical forests are likely to survive rising global temperatures and preserve biodiversity in these rich ecosystems.

Pau's research focuses on plant phenology as a potential indicator of a species' resiliency. Phenology, sometimes called "nature's calendar," refers to the timing of periodic biological events, such as the flowering of plants. Research in temperate and high-latitude ecosystems indicates that climate change has a significant impact on plant phenology. One example is "spring advancement," when warmer temperatures trigger plants to bud and flower earlier in the year. However, less evidence has been gathered in tropical ecosystems.

Pau is targeting two tropical forest habitats on the island of Hawaii that are part of the Polynesia/Micronesia biodiversity hotspot. The island's large climatic gradients enable her to compare one forest that is seasonally dry with another that is consistently wet. Pau's fieldwork involves monthly collection of litterfall—seeds and leaves that fall to the forest floor as part of plants' life cycles. Analysis of the litterfall will allow her to identify different reproductive strategies. "Classifying species with different reproductive strategies may help us predict the potential winners and losers of climate change," Pau says. "For example, species that continuously flower throughout the year may be less sensitive to climatic variation and may thus be less sensitive to future climate change."

Pau combines her field data collection with remote sensing to gain a better perspective on what is happening in the tropical forests. Satellite observations reveal the "greening" and "browning" of areas on Earth's surface as plants go through their life cycles. By comparing the timing of satellite observations to the changes in leaf litterfall, Pau's research can lead to a more accurate interpretation of the greenness that appears in satellite images. In turn, it can clarify how site-specific processes give rise to larger global patterns and provide insights into the future impact of climate change on vulnerable ecosystems.

Geographic Thinking

Explain why tropical forests on the island of Hawaii provide an ideal context for Pau's research on plant phenology.

12.2 Agricultural Practices

The physical geography, available resources, and cultural traditions in an area influence rural land-use patterns and agricultural practices such as the choice of crops to grow and livestock to raise. Such farming practices fall into two overall categories: subsistence agriculture and commercial agriculture.

Subsistence Agriculture and Commercial Agriculture

Learning Objectives
LO 12.5 Compare subsistence agriculture and commercial agriculture.
LO 12.6 Describe how *bid-rent theory* affects agricultural practices.

For some farmers, growing food is a matter of survival. They grow and raise a diverse range of crops and livestock for their family's consumption. This form of farming is called **subsistence agriculture**, which you briefly read about in Chapter 3. Occasionally, subsistence farmers enjoy a plentiful harvest and produce more than meets their needs. They may barter or sell their excess products for cash. Ultimately however, subsistence farming is simply about obtaining enough yield to feed one's family and close community using fewer mechanical resources and more hand labor to care for the crops and livestock.

Other farmers grow crops and raise livestock for profit to sell to customers, who buy these goods in a form of agriculture known as **commercial agriculture**. The goods commercial farmers produce depend on a range of geographic and economic factors, including the comparative advantages of their farmland and environment, market demands for particular products, and their agricultural practices. Subsistence and commercial agricultural practices help define the agricultural production regions across the world, as conveyed in the "Agricultural Regions Around the World" map later in Section 12.2.

Both subsistence and commercial agricultures are practiced at different scales known as intensive and extensive scales. The **bid-rent theory** explains how land value determines how a farmer will use the land—either intensively or extensively. Where land value is high, farmers will buy less land and use it *intensively* to produce the most agricultural yield per unit of land. Where land has a lower value or is farther from the market, farmers will buy more land and use it less intensively, or *extensively*. You learn more about intensive and extensive agriculture later in this section.

According to this theory, dairy and produce farmers, concerned with issues of freshness, perishability, and transportation, for example, are willing to pay higher costs—or "rent"—for land close to the market. However, farmers growing grains and cereal crops, which are easily stored and transported, will not pay as much for land

Model: Bid-Rent Theory

The bid-rent theory, which is used to describe how land costs are determined, was developed in 1964 by American economic geographer William Alonso. This theory explains the relationships between land value, commercial location, and transportation (primarily in urban areas) using a bid-rent gradient, or slope. The gradient is based on the practice of land users bidding against one another for land. The most desirable land, which is also the most accessible, receives the highest bids. The majority of consumer services are located in the center of a city because the accessibility of the location attracts these services. This central location is called the **central business district (CBD)**. The demand for central locations is translated into high land values. As the distance away from the CBD increases, land value decreases. The bid-rent theory assumes there is one CBD.

In terms of agriculture, the bid-rent theory explains how land costs determine how intensively the land is farmed.

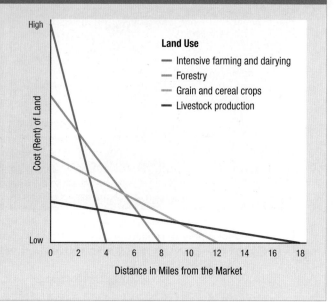

Rural Survey Methods

Learning Objective
LO 12.7 Describe the three methods used to draw land boundaries in North America.

Land ownership requires owning a title to the land, but for clear title to exist, the boundaries of one owner's land must be clearly differentiated from those of another's land. The scientific technique called land surveying is used to determine a property's three-dimensional position of points and the distances and angles between them. Land boundaries are drawn according to land survey methods, and the North American landscape is shaped by the use of three different methods.

In the 17th century, a system called *metes and bounds* spread from Great Britain to its North American colonies through relocation diffusion. This system describes property boundaries in terms of lines drawn in a certain direction for a specific distance from clear points of reference. Those points of reference were typically a natural feature, such as the crest of a hill or even a particular tree. This process resulted in unusually shaped land parcels that can still be seen in aerial photographs of land in New England and the mid-Atlantic region.

French and Spanish colonies from the same period used the *long-lot survey system*. Property was divided into a series of adjacent long strips of land stretching back from frontage along a river or lake. Land divisions like these can still be seen in Louisiana, parts of Missouri, some Great Lakes states, French Canada, and along the Río Grande. This system allowed equal access to the waterway and a mix of soils: richer soils near the river to the woodlands and less fertile areas farther away. Over time, lots became narrower when property was divided, per French tradition, equally among all children at the owner's death. That issue did not arise in British colonies, where an entire estate went to the eldest male child.

With the arrival of improved scientific surveying practices in the 18th century, people began to use rectangular grid systems to divide land. Some areas of the original 13 colonies as well as parts of Ohio that were surveyed by private land companies followed this system. When the U.S. government organized the Old Northwest—the area from Ohio to Minnesota—it adopted a cadastral, or land ownership, system that created rectangular land lots. This survey system was called the *township and range system*, and it was designed to create survey townships of 6 miles by 6 miles, giving a total of 36 square miles. Each square mile contained 640 acres and was called a section. Land was sold by the full, half, or quarter section. This system was used to survey and sell land controlled by the U.S. government as the country acquired new territories. Most land west of the Mississippi is surveyed using this system, which can be seen in the largely rectangular agricultural landscape visible when flying over areas like the Great Plains.

Exceptions to this rule show the results of sequent occupance—the cultural imprint left behind by successive societies. Parts of modern Missouri settled by the French have property lines that reflect the long-lot system. Areas controlled by Spain, such as present-day Florida and the region from Texas to California, show a mix of the metes and bounds and long-lot systems. But areas of these states that had not been surveyed under those systems divide land according to the township and range system, creating a mix of property types that demonstrates historical changes written on the land.

Geographic Thinking

Explain how geographic features would affect the grid pattern of the township and range system.

Farmland along the Richelieu River in central Quebec, Canada (left), shows the influence of the long-lot system. By contrast, farmland along the Pacific coast of California (right) shows a rectangular agricultural landscape, a result of the township and range system.

Intensive Agriculture

Learning Objectives
LO 12.8 Describe the characteristics of intensive subsistence agriculture.
LO 12.9 Describe the characteristics of intensive commercial agriculture.

With **intensive agriculture**, farmers expend a great deal of effort to produce as much yield as possible from an area of land. To achieve high productivity, they rely on high levels of "inputs" and energy. In some regions, commonly used inputs include chemical fertilizers, herbicides and other pesticides, and growth regulators. In other regions, the inputs may be human or animal labor, natural fertilizers, and thoughtful care of the soil. Regardless of the inputs, large amounts of energy are always needed in intensive agriculture—to run machines, to work the land by hand, and to utilize various degrees of technology. Technology and greater amounts of energy speed up the essential steps of farming—plowing, planting, and harvesting. These inputs and expenditures of energy maximize crop yields in intensive agriculture.

Intensive Subsistence Agriculture When people work the land intensively, putting forth a large amount of human labor to generate high crop yields on small plots of land to support their family and local community, they are practicing intensive subsistence agriculture. This type of farming feeds more than half the people living in densely populated semi-peripheral and peripheral countries.

In addition to heavy human labor, intensive subsistence agriculture relies upon careful land stewardship practices and sometimes fertilizers—mainly animal manure—to maintain soil productivity.

As with any type of agriculture, the yields from intensive subsistence farms are impacted by weather, seed quality, and the use or lack of fertilizers and pesticides. Yet when farms are primarily being used to feed a family or community (the definition of subsistence agriculture), yields reduced by weather, disease, or pests can be devastating and can lead to undernutrition or even widespread starvation in a community versus a reduction in profits, as with commercial agriculture. In 2019, for example, Cyclone Idai hit southern Africa with heavy rains, flooding, and damaging winds that destroyed crops throughout the region, greatly reducing crop yields and contributing to widespread hunger. Fortunately, new technology is helping intensive subsistence farmers by providing weather information and data that will help them to better strategize on fertilizer use and harvest times and by improving seed quality and farming tools.

As populations increase in intensive subsistence agriculture regions, many farmers maximize food production by modifying their local environment. One such example of intensive subsistence farming is the wet-rice agriculture of Asia. In South, East, and Southeast Asia, growers make large investments in productive seed types and fertilizer and use human labor rather than mechanized equipment

Women farmers plant rice by hand in a field in Mandalay, Myanmar. This wet-rice intensive subsistence agriculture is labor intensive and supports large populations with its high yields.

Rural Settlement Patterns

Learning Objective
LO 12.10 Describe the three forms of rural settlements: clustered, dispersed, and linear.

In general, people all over the world tend to cluster together in villages, towns, and communities. Collective living is the norm, and isolated living is rare. Rural settlement patterns can take several forms, each with advantages and disadvantages.

The most common form of settlement is a **clustered settlement** (also known as a nucleated settlement), in which residents live in close proximity. Houses and farm buildings are near one another, with farmland and pasture land surrounding the settlement. This settlement pattern promotes social unity. Its physical closeness also allows residents to share common resources and expand their land outward. However, this immediacy can be too much of a good thing and may lead to social friction or to a family's fields extending too far away from the settlement.

In **dispersed settlements**, houses and buildings are isolated from one another and all the homes in a settlement are distributed over a relatively large area. Dispersed settlements often exist in areas with difficult terrain, such as places where resources like water and fertile land are scarce. These settlements promote independence and self-sufficiency but lack social interaction; access to shared institutions, such as schools; and the ability to easily defend their residents.

In a **linear settlement** pattern, houses and buildings extend in a long line that usually follows a land feature, such as a riverfront, coast, or hill, or aligns along a transportation route. The linear features—including railroads and roads—generally predate the settlement, and people settle along these features because they provide access to water or transportation. Fields stretch out from the line of settlement, making the far end of those fields distant from homes.

Geographic Thinking

Compare the advantages and disadvantages of clustered, dispersed, and linear settlements.

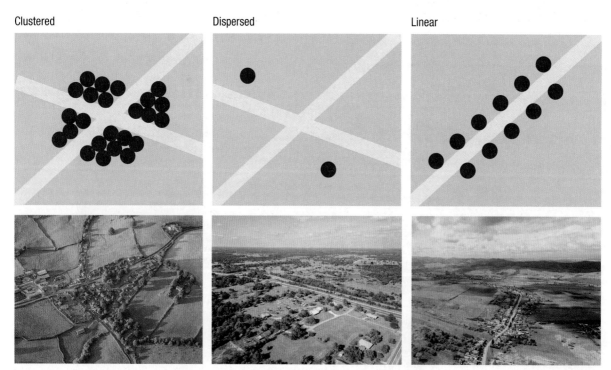

Most settlements are clustered, such as this one outside Usk, Wales (left). The town of Somerville, Texas (middle), is a dispersed settlement. This community in Hungary (right) is a linear settlement.

U.S. Corn Production, 2018

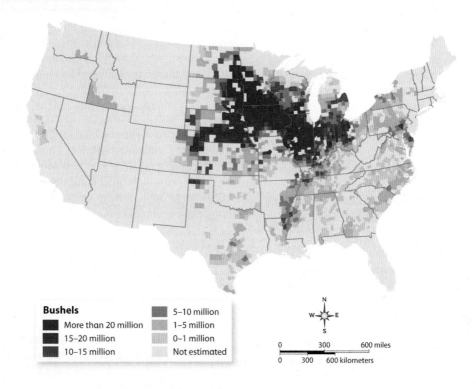

U.S. Soybean Production, 2018

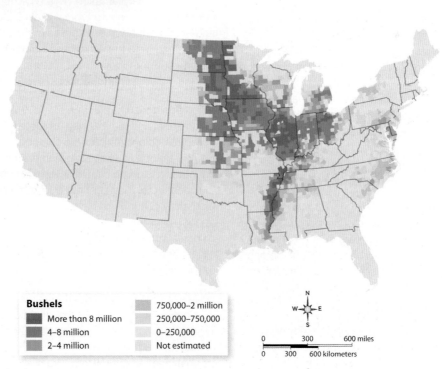

Reading Maps According to the U.S. Department of Agriculture (USDA), in 2018, farmers in the Corn Belt of the Midwest planted close to 90 million acres each of corn and soybeans. Commercial farmers sometimes rotate what they plant—one year corn fills their fields and the next year soybeans. Rotating crops strips fewer nutrients from the soil. ▌ Use the maps to describe the agricultural production regions of the midwestern United States. (These USDA data are gathered and shared every 5 years.)

to carry out the planting, weeding, and harvesting. Rice farmers in India modify their environment by leveling and flooding rice paddies, creating additional suitable land for growing a grain they depend on. In the mountainous terrain of China and Southeast Asia, farmers terrace the fields to effectively grow wet-rice crops. In dry-climate regions, farmers build irrigation systems to provide water to their crops as well as manage environmental degradation. These practices are indicative of intensive agriculture because of their reliance on heavy labor and the high crop yields they aim to generate.

Intensive Commercial Agriculture Farmers in some core and semi-periphery countries engage in intensive commercial agriculture, which involves heavy investments in labor and capital and results in high yields for profit—its products are commodities intended for sale at market. Intensive commercial agriculture often incorporates chemical fertilizers and machines, instead of relying mainly upon human and animal labor. It can be carried out close to or far from the market, or the place where the products are sold, processed, or consumed. For example, dairy farms in northern Europe are relatively close to their urban-dwelling consumers. However, the palm oil plantations of Malaysia and Indonesia are far from the factories that turn their product into cosmetics, soaps, ice cream, and chocolate bars. Regardless of the proximity to market, what all intensive commercial producers have in common is the use of intensive methods, whether these are capital- or labor-intensive, and a high yield.

The characteristics of intensive commercial agriculture can be observed in several specific types of farming including monoculture systems, plantation agriculture, market gardening, Mediterranean agriculture, and mixed crops and livestock systems. Many people who participate in intensive commercial agriculture in the United States focus on **monocropping**, the cultivation of one or two crops that are rotated seasonally—commonly corn, soybeans, wheat, or cotton. These crops are usually what the market demands and therefore can be very profitable for plantations as well as large corporate farms.

Similarly, **monoculture** refers to the agricultural system of planting one crop or raising one type of animal annually. In the years following World War II, farmers in the American Midwest shifted from farming many crops to focus on one or two crops—typically corn and soybeans, depending on market prices—or on raising hogs or cattle.

Monocropping allows for specialization, simplifies cultivation, and maximizes efficiency. Farmers can purchase seeds in bulk, as well as fertilizers and pesticides specific to a single crop, and they don't need as wide a variety of equipment or other inputs that may be necessary to cultivate several different crops. Growers choose a crop that is especially suitable for the environment available, which also increases efficiency. For example, the soil and average climate conditions in the midwestern United States are particularly suitable for growing corn and soybeans, and therefore farmers invest in fewer inputs to establish a favorable crop environment. Monocropping also leads to higher yields, which means increased sales and a strong return on investment.

However, monocropping can strip nutrients—like nitrogen, phosphorus, and other essential elements—from the soil; for instance, intensive cotton or corn production leads to soil exhaustion. One way to prevent this depletion of nutrients is through **crop rotation**, the varying of crops from year to year to allow for the restoration of valuable nutrients and the continuing productivity of the soil.

The shift to monoculture also raises the stakes for farmers. As yields go up, crop prices can decrease. Falling prices can force many farmers out of business and may encourage successful operators to purchase the land of failing farms. To compete, farmers must invest heavily in high-yield seeds, fertilizers, and pesticides to maintain the soil and maximize plant growth. Synthetic fertilizers—made with fossil fuels, while helping plants temporarily, can harm the soil long term, and herbicides and other pesticides can contaminate water supplies and hurt air quality.

Plantation agriculture involves large-scale commercial farming of one particular crop grown for markets often distant from the plantation. This type of intensive commercial agriculture typically takes place in peripheral and semi-peripheral economies in the tropical regions of Asia, Africa, and the Americas. Major plantation crops include cotton, tobacco, tea, coffee, sugarcane, bananas, oranges, palm oil, and rubber.

Plantation agriculture is one of the oldest forms of intensive commercial agriculture, with its roots going back to the European colonization of the Caribbean and Central and South America as well as Asia and Africa. As tropical countries like Jamaica, Brazil, and Sri Lanka cut ties from European rule, they continued to take advantage of their plantation culture and have become well known for certain goods—Sri Lanka for tea, for instance. These former colonies continue to rely on their production of specialty crops, and their economies depend on these neocolonial relationships.

However, many larger plantations are owned by European or American individuals and multinational corporations. These businesses wish to ensure a steady stream of revenues and profits, so they invest heavily in such inputs as pesticides and fertilizers. Brazil—a semi-peripheral country and leading producer of crops like sugarcane, oranges, and coffee—is one of the world's leading consumers of pesticides. Plantations tend to be labor-intensive operations, although since many are located in peripheral countries, the cost of that labor is relatively low.

Market gardening is farming that produces fruits, vegetables, and flowers and typically serves a specific market, or urban area, where farmers can conveniently sell to local grocery stores, restaurants, farmers' markets, and road stands. The practice of market gardening is driven by the perishability of the products, or their likeliness to spoil, and the demand by local consumers for fresh fruits and vegetables. Market gardens can be found near most large cities in the United States, from the Northeast to Texas to California. The farm-to-table movement, which emphasizes fresh, locally grown ingredients, is giving a new momentum to market gardening. Still, in the winter months, most fruits and vegetables consumed in the United States are grown in Chile or Mexico.

Operators of market gardens often need to invest in technology such as greenhouses—which also entail high energy costs—to germinate seeds before the growing season so that the operators can stagger or spread out crop production to provide a constant supply of products throughout the year. They also need to use costly high-quality seeds, fertilizers, and pesticides to ensure harvests. Some market garden operators practice organic farming, which involves using natural methods to fertilize the land and prevent pest damage.

Successful market gardens cultivate high yields on small tracts of land using intensive production methods. Selling directly to public markets, including community farmers' markets, increases a farmer's profits by eliminating the need to bring in another person or company to sell the produce. In 2019, the U.S. Department of Agriculture listed approximately 8,700 farmers' markets in the United States. They are popular with consumers who, for the most part, value locally produced food over processed and packaged items from grocery retailers.

Truck farming, once synonymous with market gardening, now serves markets that are often very distant from the farm. Large commercial farms in Mexico and the western and southern United States, where the climate is conducive to locally growing seasonal produce in high quantity, transport specialty crops to distant markets using large-capacity refrigerated trucks. The demand in markets like the Upper Midwest and Northeastern States is high because the cultivation in these areas is limited by climate. Truck farms produce crops such as tomatoes, lettuce, melons, beets, broccoli, celery, radishes, onions, citrus, and strawberries.

As you have read, agriculture thrives in the Mediterranean climate of southern Europe and North Africa, southwest Asia, South Africa, Australia, Chile, and California. In some Mediterranean agriculture regions, subsistence farmers may cultivate beans, lentils, onions, tomatoes, carrots, and leafy vegetables year-round for their family's consumption. Just down the road might be commercial orchards and vineyards producing citrus fruits, olives, figs, and grapes, which are mostly exported for market. Much of the land in these climate regions is used for this type of intensive commercial agriculture.

Another type of intensive commercial agriculture practiced frequently is **mixed crop and livestock systems**, in which both crops and livestock are raised for profit. There are two types of mixed farming: on-farm and between-farm. In on-farm mixed farming, the crops and livestock are raised on the same farm. In between-farm mixing, two farmers share resources, with one growing crops and the other raising livestock. One example of this is livestock ranchers in the United States who buy feed from grain-producing farms. This practice is also carried out in such areas as West Africa and India, where crop producers exchange their crops with livestock raisers for manure to fertilize the land or for milk.

On-farm mixing effectively combines a farm's focus on one or two crops, such as corn and soybeans, with the raising of animals to meet the demand for high-quality meat. Mixing provides a farmer with certain advantages: part of the crop can be fed to the livestock and the animals' waste can be used to fertilize the crops. This type of farming boosts labor needs, which makes it a form of intensive commercial agriculture. The diversity of mixing also may provide some protection from a bad crop year or low market value.

Extensive Agriculture

Learning Objectives
LO 12.11 Describe the different types of extensive subsistence agriculture.
LO 12.12 Describe the characteristics of extensive commercial agriculture.

With relatively few inputs and little investment in labor and capital, farmers who participate in **extensive agriculture** typically have lower outputs per unit area than farmers who employ intensive practices. Like intensive agriculture, extensive agriculture can be practiced in subsistence agricultural regions as well as in commercial agricultural regions. Extensive agriculture is found in countries in the periphery and semi-periphery, as well as in ranching enterprises in core countries.

Extensive Subsistence Agriculture Extensive subsistence agriculture is often found in regions in which intensive subsistence agriculture is not feasible because the environment is marginal—that is, too wet, too dry, or too cold—and thus the carrying capacity (the maximum population size an environment can sustain) is low. One type of extensive subsistence agriculture that uses relatively simple technology requiring little capital investment is shifting cultivation. **Shifting cultivation** is the practice of growing crops or grazing animals on a piece of land for a year or two, then abandoning that land when the nutrients have been depleted from the soil (or the grazing vegetation has been depleted) and moving to a new piece of land where the process is repeated. Although the size of the piece of land being used short term is generally not large, shifting cultivation requires a relatively large area in which to operate over time. This type of agriculture is practiced worldwide in marginal agricultural areas of the tropics, particularly in areas with high rainfall, such as in the rain forests of South America, Central and West Africa, and Southeast Asia.

Some farmers, including those in Colombia and Brazil in South America and Papua New Guinea in Oceania, use traditional subsistence farming techniques, such as **slash and burn**—a type of shifting cultivation—to maintain the land. They clear the land by cutting down the trees and brush and, after the vegetation dries, burning this "slash," resulting in a nutrient-rich ash fertilizer. The cleared land is then cultivated for several years until the soil becomes infertile. The process is then repeated on a new patch of land. While slash and burn has long been practiced, it is becoming unsustainable as more farmers engage in this practice. The cleared land can become severely degraded and open to erosion. In slash-and-burn agriculture—and in all agricultural methods—clearing forests for cropland leads to loss of habitat for local species and increases air pollution and the amount of carbon released into the atmosphere, contributing to global climate change.

Another example of extensive subsistence farming is **nomadic herding**, also called pastoral nomadism. People who practice this type of agriculture move their animals seasonally or as needed to allow the best grazing. It requires far-reaching areas of land to prevent overgrazing—the destruction of feed plants that results from livestock overpopulation or overfeeding.

Some nomads engage in **transhumance**, the movement of herds between pastures at cooler, higher elevations during the summer months and lower elevations during the winter. For example, the Kohistani people of eastern Afghanistan are nomads who move their herds of livestock among five different altitude levels from 2,000 to 14,000 feet above sea level over the course of a year. Families have five different homes—one at each level—to provide shelter during the seasonal stay at each level. Transhumance may be practiced by non-nomads, too, who move their herds upslope or downslope but live in only one home.

Extensive Commercial Agriculture Ranching is an extensive commercial farming practice. It takes place in semiarid grassland areas around the world in which crop production is difficult or impossible, including in the American and Canadian West; Brazil, Argentina, and Uruguay in South America; Australia and New Zealand in Oceania; and Botswana and South Africa in Africa. Ranching is not as labor intensive as other forms of agriculture. A rancher can rely on as little labor investment as one cowhand for every 800 to 1,200 head of cattle.

In the United States, livestock ranching is mostly found in the western states, where there are large, open tracts of land for livestock such as cattle and sheep to roam and graze. The arid grasslands of this region are suitable only for extensive agriculture, and therefore the price of this marginal land is low. Ranchers take advantage of the low land costs, including the leasing of lands managed by the federal government, and the fact that less labor and capital are required to prepare and use the land for grazing. Ranching is typically carried out in sparsely populated areas farther away from markets or city centers, and ranchers must transport their livestock to markets for sale.

Extensive commercial ranching, while a common agricultural practice of the western United States, is in direct contrast to the increasingly common intensive commercial practice of Concentrated Animal Feeding Operations, or CAFOs. With CAFOs, farmers on small tracts of land raise cattle, pigs, or other livestock in limited spaces called feedlots so they can maximize the potential of their land. This practice makes it easier to manage the animals and there are fewer costs involved in the operation, though concerns have been raised over animal welfare. Some cattle raised on the range may also be "finished" in feedlots for some weeks or months to reach desired market weight.

Geographic Thinking

1. Identify the environmental impacts of slash-and-burn farming techniques.

2. Explain why nomadic herding is the most extensive type of agriculture.

Comparing Intensive and Extensive Agriculture

Learning Objective
LO 12.13 Compare the geographies of land intensive and land extensive agriculture.

A society's agricultural practices depend on several factors, including climate, culture, the availability of capital, the quality of the land, the supply of labor, global markets, and the societal needs and demands for agricultural output.

National Geographic photographer John Stanmeyer photographs a child playing in an irrigation water tank on a rice farm in Punjab, India. Intensive subsistence farmers rely on inputs such as this rudimentary irrigation system to make dry soils more productive for farming.

Many countries will participate in a mix of practices. Kenya, for example, has both coffee and tea plantations—which reflect intensive commercial agriculture—and nomadic herding, a type of extensive subsistence agriculture.

Physical geography is a major determinant of the agricultural practices used. Large expanses of land with less nutrient-rich soil available for growing crops, such as in the rain forests of the Amazon Basin or in the semiarid grasslands of Central Asia and the American West, call for the use of extensive agricultural practices. The only way to make up for the low yields that such land produces is to work a more extensive expanse of land. In contrast, rich soils—such as those in eastern China and the American Midwest—are better suited to intensive agricultural practices because they can produce high yields.

Of course, inputs such as fertilizer or irrigation can be used to make deficient land more productive. However, there are limits to what inputs can achieve. The heavy rains of the tropical rain forest decompose organic matter quickly and leach or draw nutrients from the soil. Even if fertilizers are added, they can also be washed away in the region's regular rains. Irrigation to water crops in arid or semiarid areas can be expensive, may cause long-term environmental damage, and may prove unsustainable (you learn more about irrigation in Chapter 14).

Areas with marginal agricultural potential are generally only able to support small populations. In contrast, areas with highly productive agriculture are able to support large populations. And a populous society needs intensive agriculture. This situation generates a continued cycle. Intensive agriculture generates high crop yields, which can support a large population. A large population, in turn, requires high yields, which means emphasizing intensive agriculture.

Geographic Thinking

3. Compare the similarities and differences between subsistence and commercial farming practices.

4. Describe how intensive and extensive farming practices are determined in part by the bid-rent theory.

5. Identify an example where an agricultural practice is influenced by the availability of natural resources and the climate.

Agriculture: Human–Environment Interaction

Agricultural Regions Around the World

Agriculture takes many forms. In the late 1930s, American geographer Derwent S. Whittlesey developed a world agriculture map that identified 11 separate agricultural regions (as well as areas where agriculture is not practiced).

The large map shows the global distribution of agriculture today. For example, the intensive commercial agriculture practices of dairy farming and grain farming—represented by the colors brown and yellow, respectively—can be found in North America, South America, Europe, Asia, and Australia. Rice, produced through intensive subsistence farming practices and represented as dark green, mainly grows in Asia.

The text annotations on the large map provide geographic context for many of the types of farming, rural survey methods, and settlement patterns that have been represented in this chapter.

The smaller topographical map shows the elevation of global lands. Land elevation and topography play major roles in the kind of agriculture that can take place in a specific location. Some regions cannot sustain certain crops or animals—even with human intervention.

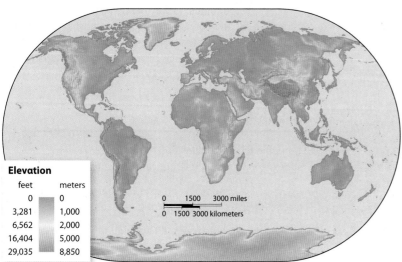

Land Elevations of the World

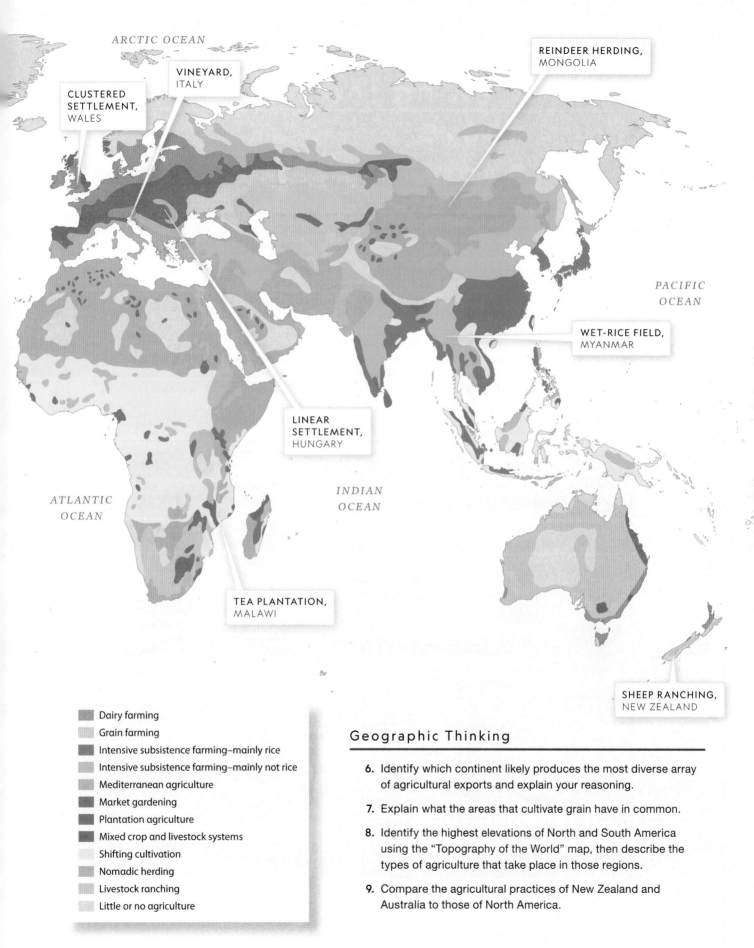

Geographic Thinking

6. Identify which continent likely produces the most diverse array of agricultural exports and explain your reasoning.

7. Explain what the areas that cultivate grain have in common.

8. Identify the highest elevations of North and South America using the "Topography of the World" map, then describe the types of agriculture that take place in those regions.

9. Compare the agricultural practices of New Zealand and Australia to those of North America.

National Geographic Photographer George Steinmetz

Viewing the World from Above

George Steinmetz specializes in remote deserts, little-known cultures, and new developments in science and technology.

Learning Objective
LO 12.1 Explain how agricultural practices have environmental and societal connections.

National Geographic Photographer George Steinmetz seems to think that a bird's-eye view of Earth is the best way to see the planet. This belief might explain why he's willing to strap himself into a motorized paraglider and soar into the air with his legs dangling below to photograph a lettuce farm in California. He calls his aerial photos "street photography from the sky." His images portray the varied and beautiful agricultural landscapes that societies create, from mechanized commercial operations to manicured subsistence farms and gardens.

Trained as a geophysicist, Steinmetz took up photography during a hitchhiking trip through Africa when he was in his 20s. He has returned to that continent many times to capture its land and people. Steinmetz has photographed deserts, dunes, cityscapes, and farmland—including the growing of crops in countries such as Brazil, Japan, and the United States.

While he has flown as high as 6,000 feet to take his photos, Steinmetz prefers a much lower altitude—around 100 to 500 feet. "You are seeing [the land] more obliquely, so you see the 3-D relationship," he explains. "It's the perfect mix where I can see the gross pattern—the infinite skyline or the mountains in the distance but also people and what they are doing."

Critical Viewing The Naxi (also called Nakhi or Nasi), an ethnic group in China, use terrace farming to cultivate millets, cabbage, corn, and wheat. Steinmetz's photo of a Naxi farm shows piles of manure dotting the landscape. The manure is mixed with straw before being folded into the soil. Identify human-environment interactions that are easily seen in the aerial photo.

Critical Viewing *Top:* The Adjder oasis in Algeria features small family gardens nestled into scalloped depressions created by the powerful Harmattan winds. *Bottom:* Organic lettuce is farmed near Hollister, California, by a machine that can harvest 10,000 pounds of lettuce per hour. ▮ Explain how farmers could use aerial photographs like Steinmetz's in their work.

Agriculture: Human-Environment Interaction 345

12.3 Agricultural Origins and Diffusions

When you eat pizza, you're enjoying an intercontinental food. Wheat and cattle—which provide grain for the crust and the milk that goes into the cheese—were first grown and raised in ancient Southwest Asia. South America was the ancient home of tomatoes. If you like onions on your pizza, thank Central Asia. And don't forget the people of Naples, who first put the ingredients together to make pizza and who then brought it to the United States when they migrated from Italy. The story of many foods is the story of people spreading plants and animals from their original homes.

Agricultural Hearths

Learning Objectives
LO 12.14 Explain the processes of domestication of plant and animal species.
LO 12.15 Compare characteristics of the world's major agricultural hearths.

Answering when, where, and why agriculture started is tricky. Unfortunately, no one at the time could post social media messages. In fact, people weren't even recording the event on cave walls, clay tablets, or papyrus scrolls. Finding answers to these questions might be more difficult than figuring out when dinosaurs lived. The fossils of seeds and pollen are much smaller than most dinosaur fossils. To compound the difficulty, modern plants don't look exactly like their ancient ancestors. Still, archaeologists and paleontologists have discovered fossils of ancient plants and early domesticated animals, and some conclusions have been drawn. Those answers can be summarized in this statement: People living in many different places domesticated different plants and animals at different times from about 11,000 to 1000 B.C.E. **Domestication** is the deliberate effort to grow plants and raise animals, making plants and animals adapt to human demands, and using selective breeding to develop desirable characteristics.

From Foraging to Farming Why did domestication take place? For thousands of years, humans lived as **foragers**, small nomadic groups who had primarily plant-based diets and ate small animals or fish for protein. Plant foods included fruits and vegetables, seeds and nuts, tubers, and other plants. Foragers fished in rivers and lakes or gathered shellfish and used traps, stones, or projectile weapons to hunt small game. Small bands sometimes formed within the group to hunt for larger animals. These humans lived by ranging over the land to exploit the food resources that were in season, often returning to the same areas each year.

Between 12,000 and 11,000 years ago, with the advent of the geological epoch called the Holocene, Earth entered a period of increased warming. The impact on the environment was substantial; this warming melted almost all the massive glaciers that had covered much of the Northern Hemisphere. Sea levels rose as a result, and climate regions changed, with more extreme changes in temperate regions than in the tropics. About 11,000 years ago, average rainfall in Southwest Asia dropped significantly for an extended time—perhaps as long as 1,000 years. People there adapted to this environmental stress by domesticating animals and plants to ensure a steady food supply, making them the first humans to do so. People also began migrating to and settling in warmer environments where more foods were available for foraging. Population growth increased pressure on the environment, and wild foods became scarce, which required the cultivation of a stable surplus.

The first animals to be domesticated were sheep and goats, which supplied hides, milk, and meat. With the changing climate, people in Southwest Asia began to plant seeds to secure a plant food supply in the new climate conditions. They selected plants such as cereal grains that produced plentiful seeds that could be ground or boiled to be eaten. In the switch to cultivation, people collected wild seeds, selected for their maturity and size, and then planted them. They tended the crops by ensuring that the plants had enough water and by removing weeds. Through selective breeding—accomplished by cultivating the plants that produced the most seeds—people gradually improved the plants to increase both their yields and nutrition.

Ancient Hearths Each area where different groups began to domesticate plants and animals is called an **agricultural hearth**. Scientists have identified several major agricultural hearths of domestication, and they continue to adjust their understanding of the timing of the emergence of these hearths as archaeologists and other experts unearth new information.

Domestication first took place in Southwest Asia. This hearth is called the **Fertile Crescent** because it forms an arc from the eastern Mediterranean coast up into what is now western Turkey and then south and east along the Tigris and Euphrates rivers through present-day Syria and Iraq to western parts of modern Iran. The people of this region grew wheat, barley, rye, and legumes (peas or beans) and domesticated sheep, goats, cattle, and pigs.

Another agricultural hearth arose in Southeast Asia, where people raised pigs and grew sugarcane and root vegetables. Domestication began there about 7000 B.C.E. Ancient North Central China had two hearths with distinct crop types, both developing between 8000 and 7000 B.C.E. To the north, in the somewhat dry valley of the Huang He, people used the river's fertile soil to grow millets, hemp, Chinese cabbage, and wheat. In the warmer, wetter south, they grew rice and harvested lemons and limes.

In South Asia, an agricultural civilization thrived in the Indus River Valley from 2500 to 1700 B.C.E. This group, known as the Harappan civilization, was based primarily in two large cities—Harappa and Mohenjo-Daro—as well as in other towns and villages. These people took advantage of the fertile valley and farmed wheat, barley, peas, sesame, and possibly cotton. Their domesticated animals included cattle, fowl, pigs, camels, and buffalo.

In northern Central America and into southern Mexico, an agricultural hearth began about 8000 B.C.E. People there grew sweet potatoes, beans, and other crops. They also domesticated the turkey. Maize (corn) later became the staple, or basic crop, of the region. About the same time, another agricultural hearth in the Andean highlands of South America began. Its chief crops were beans, tomatoes, and potatoes. The people there also domesticated llamas, alpaca, and guinea pigs.

Hearths were also located in Africa. In East Africa, crops including coffee, olives, peas, and sesame originated around 8000 B.C.E. Sorghum, a type of grain, was likely domesticated in Ethiopia around 4000 to 3000 B.C.E. In West Africa, by about 3000 B.C.E., people began to raise millets, sorghum, and yams.

Scientists distinguish such hearths as the Fertile Crescent from other ancient areas that adopted agriculture later through diffusion, which as you know is the spread of an idea or cultural trait from one place to another over time. For example, the ancient Nile River Valley—home to the Egyptian civilization that lasted for several thousand years and that was a hearth for other innovations—adopted farming and raising animals from Southwest Asia. The people who lived in the Nile River Valley did not develop these practices on their own. Sometimes it is difficult to know if a region was a hearth itself or if it was along a pathway of diffusion from that hearth. For example, rice may have been domesticated independently in the Ganges River Valley of India, or it may have been introduced by travelers from its home in China.

Shared Characteristics While ancient hearths have different physical characteristics, they share some features. Agriculture flourished in these regions because of fertile soil, the availability of water, moderate climates, and the organizational skills of the residents. These agricultural hearths were areas of independent innovations that people were able to develop over time, through trial and error, and with luck.

Because rains in these regions were not uniform throughout the year, the people in some areas developed methods of irrigation. In the valley of the Tigris and Euphrates rivers, for example, snow in the mountains that fed water to rivers melted in the spring, causing flooding. The floodwaters deposited rich soil that helped make the valleys fertile. By using low dams and reservoirs, the people made catch basins to hold the water to irrigate during dry months.

Many of the societies that developed in these hearths relied on the collective work of most of their members to tend the fields and harvest crops. This cooperation helped ensure success and encouraged settled life—and also put more pressure on the farmers to produce bountiful crop yields. Good harvests promoted population growth, which supplied more workers. Another factor that promoted agriculture was the development of efficient methods of storing seeds and harvests, as with using gourds and baskets.

The Diffusion of Agriculture

Learning Objective
LO 12.16 Explain how agricultural products and practices have spread through diffusion.

Agriculture depends on the land, but like all human activities, it is mobile because people are mobile. Diffusion of agricultural practices has produced patterns of flow throughout history, from ancient times to the present.

First Global Diffusions In ancient times, agriculture evolved independently and separately in several hearths. People migrated for various reasons, including population pressures, different opportunities, or conflict. These travelers introduced agriculture to new areas by relocation diffusion, which, as you know, is the spread of culture traits through the movement of people. As people migrated to different regions, the distribution of their crops and animals expanded. One study of archaeological sites in Southwest Asia and Europe suggests that agriculture spread steadily from the former region to the latter at a rate of about 0.6 miles a year over a period of about 3,000 years. Of course, the seeds these people brought did not flourish in all of the new areas. But through stimulus diffusion, humans could use their cultivation knowledge to adapt the practice to other, more resilient plants. In addition, some people already living in these areas recognized the value of the migrants' successful innovations and adopted them.

Over hundreds of years, many important crops diffused throughout Asia, Europe, and Africa as a result of relocation diffusion or through trade. People also created new breeds or varieties of animals that had been domesticated earlier. The spread of crops and domesticated animals was limited by the climate and resources that each plant or animal needed. By the time of the Roman Empire, in the first centuries of the common era, Egypt and other areas of North Africa were a major source of wheat. Rye, which can grow in cooler climates than wheat, spread to northern and Eastern Europe from its origin in Southwest Asia. Oranges

Hearths, Civilizations, and Domestication

Agricultural Hearths

1 Central America
Cassava, chiles and peppers, cocoa beans, cottonseed oil, maize, palm oil, sweet potatoes

2 Andean Highlands
Beans, potatoes, tomatoes

3 West Africa
Coffee, cowpeas, millets, palm oil, rice, sorghum, yams

4 East Africa
Bambara beans, coffee, cottonseed oil, cowpeas, millets, olives, peas, sesame, sorghum

5 Fertile Crescent
Barley, beans, peas, rye, wheat

6 Indus River Valley
Barley, cotton, peas, sesame, wheat

7 North Central China
Apples, grapefruit, grapes, lemons and limes, millets, oranges and mandarins, rice, soybeans, tea

8 Southeast Asia
Bananas, cloves, coconuts, grapefruit, rice, sugarcane, taro, tea, yams

Source: International Center for Tropical Agriculture

Agriculture developed in several early hearths of domestication. The first map shows eight regional hearths and their major crops. Some hearths raised the same animals and cultivated the same crops; for example, palm oil was produced in both Central America and West Africa.

spread from their native habitat in Malaysia to India to East Africa and then to the eastern Mediterranean. Bananas also reached Africa from their original home in Southeast Asia. Millets and sorghum, on the other hand, diffused from Central Africa to South Asia.

More centers of domestication are also being discovered. Recent research has found that wild potatoes were not only found and domesticated in South America but were being used by people in the western United States as long ago as 10,000 years and may have been domesticated there as well. Small-seeded plants, dogs, and turkeys were also domesticated in North America. In some cases, diffusion simply introduced new crops—and the knowledge of their specific needs—rather than the practice of agriculture itself. The indigenous peoples of

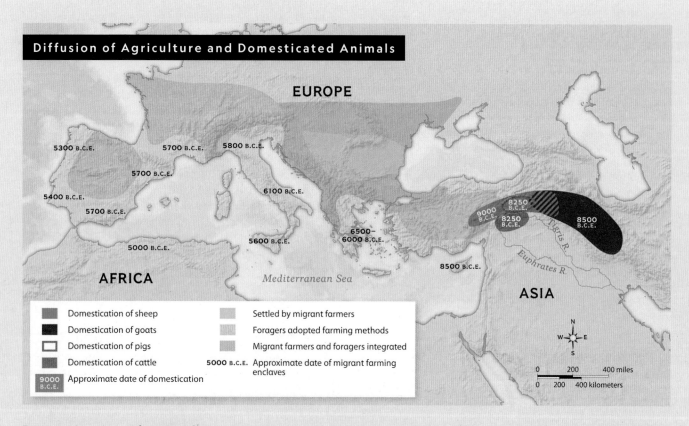

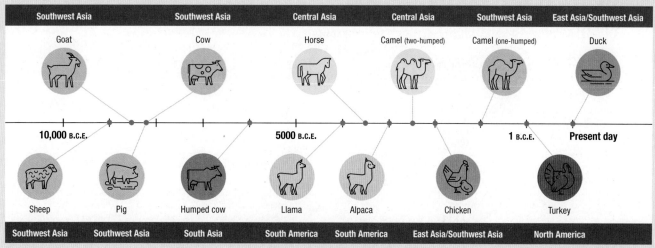

Source: Science News

The second map shows the spread of domesticated animals and farming methods from Southwest Asia. Farming enclaves, or groups, developed along the Mediterranean Sea more than 11,000 years ago. The information in the time line is based on archaeological remains. ▌ What crops and animals do you think the farmers in western Europe most likely cultivated? Explain using details from the visuals.

Agriculture: Human-Environment Interaction 349

eastern North America already knew how to farm when maize was introduced to the region in its diffusion from Central America. Similarly, tomatoes and quinoa diffused to Central America from South America. Of course, local domestication and crop variety development also continued in areas across the globe.

By the 15th century, many domesticated plants and animals had spread throughout Africa, Europe, and Asia. Other crops and animals had diffused widely between North and South America. The peoples of the Eastern and Western Hemispheres, however, had minimal contact with each other. That separation set the stage for the transformation that marked the Columbian Exchange.

The Columbian Exchange The exchange of goods and ideas between the Americas, Europe, and Africa, known as the **Columbian Exchange**, began after Christopher Columbus landed in the Americas in 1492. It had a huge impact on people, plants, and animals around the world and offers many examples of agricultural diffusion.

Columbus's arrival directly affected the human population. Isolation proved disastrous for the Indigenous peoples of the Americas, as they had no immunities to diseases from the Eastern Hemisphere, including smallpox and malaria. These diseases devastated the Americas' native populations. And while not considered part of the Columbian Exchange, the global diffusion of millions of people—including the forced migration of enslaved people from Africa (the Middle Passage) to provide labor for sugarcane and cotton production—throughout the Americas, along with their interactions with Indigenous populations, gave rise to new cultures.

In terms of agriculture, the Columbian Exchange had other momentous consequences on both sides of the Atlantic Ocean. Crops from the Americas like maize and potatoes packed a powerful nutritional punch. They were quickly adopted in regions of Europe, Asia, and Africa that had climate conditions similar to areas where the crops grew well in the Americas. Significantly, these crops often thrived in places where native domesticated plants could not flourish. One result was population explosions in Europe and Asia.

Over time, some crops from the Americas dominated the diets of many people in the Eastern Hemisphere. The results can still be seen today. Seven of the 10 countries that rely most on maize for daily caloric intake are in Africa and Asia. People who live in European, Asian, and African countries are the top 10 consumers of cassava and potatoes, and 8 out of the top 10 for sweet potatoes. The Columbian Exchange also transformed the Americas, with European crops and animals spreading widely throughout the Western Hemisphere. The fields of wheat covering much of the Great Plains in the present-day United States replaced native grasses. The ranches of the American West and the South American Pampas are home to cattle,

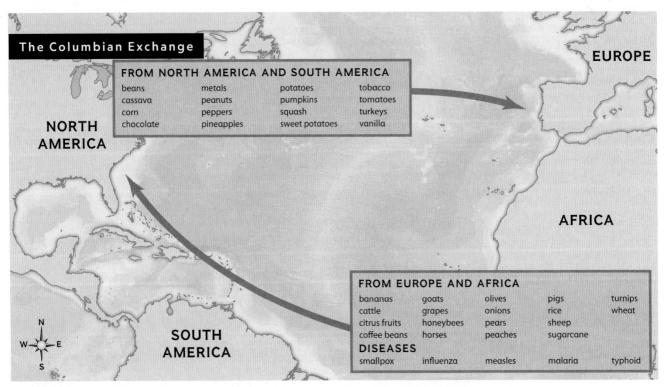

Reading Maps The Columbian Exchange was a massive global event. A wide variety of animals, plants, and diseases—and of course, people—moved between Europe, Africa, and the Americas. ▮ Explain how the Columbian Exchange contributed to the global diffusion of agricultural products and practices.

sheep, and horses introduced by Europeans. Sugarcane from the Southeast Asian agricultural hearth replaced native plants to become a plantation crop in the islands of the Caribbean and in Brazil. Coffee, which originated in the East African highlands of Ethiopia, became a dominant cash crop in some temperate highland areas of both Central and South America.

Diffusion in Modern Times Agricultural diffusion continues today. People have developed a worldwide system of agriculture with global markets and expanding tastes due to scientific advances and focused marketing. Producers seek new consumers for their products year-round. One recent example is the kiwi fruit. Native to China, it was transplanted to New Zealand in the early 20th century, where it grew well and became a popular food. During World War II, members of the U.S. armed forces stationed in New Zealand enjoyed the fruit and introduced it to the United States after the war. Today's top producers of kiwi fruit include countries as diverse—and far from China—as Italy, Chile, Turkey, and the United States.

There are other examples as well. Ostrich farms were first established in what is now South Africa in the mid-19th century to supply the market for the bird's exotic feathers. In the late 20th century, ostrich farming became popular in the United States and Europe, mostly for meat. Tilapia—freshwater fish from Africa—have become a major food in the 21st century because they are easy to raise and feed and mature fairly quickly. Tilapia are raised in more than 80 countries, with China producing half the world's output.

Geographic Thinking

1. Describe how environmental changes contributed to the development of agriculture.

2. Identify an environmental characteristic the ancient agricultural hearths in the Fertile Crescent and North Central China have in common and explain how it likely influenced the development of both hearths.

3. Identify the earliest animals to be domesticated and explain when and where the domestication occurred.

4. Compare agricultural diffusion today with diffusion during the Columbian Exchange.

12.4 Advances in Agriculture

Over millennia, humans have invented new technologies and practices that have had far-reaching impacts on life, society, and cultures. Geographers call these sweeping changes *revolutions* because they cause such radical transformations. There have been three—and some experts would say four—such revolutions in agriculture.

The First Agricultural Revolution

Learning Objective
LO 12.17 Describe the societal impacts of the First/Neolithic Agricultural Revolution

The first **agricultural revolution**, which occurred about 11,000 years ago and lasted for several thousand more, was the shift from foraging—or searching for food—to farming, which marked the beginning of agriculture. You read that this revolution occurred independently in different hearths across several continents. In other words, it did not occur simultaneously. Rather, the revolution occurred in different hearths at different times. From those points of origin, the first agricultural revolution diffused into other areas as groups of people transitioned from foraging to agriculture. Some advancements were independently developed, while others may have been borrowed from other hearths.

By some estimates, more than 80 percent of the world's diet comes from a dozen or so staple crops, such as rice, wheat, and maize. These staples were originally cultivated in the first agricultural revolution. In addition, this revolution included domestication of the most common animals that humans raise today, including sheep, goats, cattle, pigs, chickens, horses, and camels.

Çatalhöyük was a Neolithic settlement in what is now Turkey. An excavation begun in the 1960s revealed that Çatalhöyük was an early agricultural society; seeds, nuts, and edible grains were cultivated there. This reconstruction of Çatalhöyük shows what typical homes may have looked like.

The first agricultural revolution is sometimes called the Neolithic Revolution because it took place during a time in history now known as the Neolithic Age, or New Stone Age. In this period, people used tools made of stone or bone.

Societal Changes The first agricultural revolution profoundly changed the lives of the people who experienced it. For one thing, they went from being nomadic to being sedentary or semi-sedentary. *Sedentary* means settling in one place and making that place a permanent home. People built more durable housing and possessed goods that didn't have to be light enough to transport to a new area.

Living a settled life also meant increased reliance on one place rather than the variety of places exploited by foragers. While farming and herding promised a steadier food supply, these activities also decreased the variety of foods consumed. The focus of the human diet became the multiple staple crops that people produced. People often supplemented those crops with nuts, berries, and other foods gathered near the settlement when those resources were in season. Still, the overall diet was less diverse.

Farming practices in agricultural societies improved over time. Farmers learned to plant seeds from their strongest plants to generate more productive crops. New tools and practices made farming tasks easier. For example, some societies began to use domesticated oxen to pull plows, which made farming more efficient and improved yields. People in some areas began to practice irrigation, expanding the areas that could be farmed and ensuring a ready supply of water during the growing season.

Increased efficiency meant more food. Having more food supported a growing population, which provided more workers. Another impact of having more food was the need to store the surpluses for future use. People made clay pots and other containers for carrying and setting aside food. These methods of storage could not have been created without skilled artisans to make them, which presents another feature of the first agricultural revolution—the development of specialization of labor.

As farm fields became more productive, some members of society were not needed to cultivate food. They could live off the surplus produced by the farmers. Instead of food production, they focused on skills such as pottery-making or woodworking. Eventually, people began to work with metals, and skilled metal workers produced stronger tools and weapons or made luxury goods like jewelry.

In addition, farmers produced some nonfood crops. Cotton was grown in the Nile River Valley in Egypt, which also produced flax that was used to make linen. Cotton grew in the Indus River Valley as well. Sheep and goats were not kept solely for their meat; their wool could be sheared each year. Fibers such as cotton, flax, and wool could be woven into clothing, blankets, and other goods, so weaving became another specialized task.

As societies became more productive, they also became more complex. Larger settlements led to new forms of social organization. Ruling classes arose in many societies, as certain individuals or groups took charge of making laws, organizing productive activities, distributing resources, and settling disputes. Food surpluses created new dangers—they became potential targets for raids by other groups. The need to defend a society and its resources led to the development of a fighting class, which was typically under the control of the rulers. Some experts argue that members of certain societies established the idea that certain ritual practices could ensure good harvests. Some individuals assumed the role of priests to conduct these rituals and thereby gained higher status in the society.

Population growth meant larger and larger villages. Eventually, the first cities developed. Nearby settlements traded with one another and with distant communities as well, which provided locally available raw materials such as metals and luxury goods. Over time, the first ancient civilizations developed. They were characterized by large urban centers, complex societies, and advances in knowledge and the arts. You learn more about cities and their origins in Unit 6.

Mohenjo-Daro was one of the two main city centers of ancient Harappa, a civilization that developed in the Indus River Valley in the third millennium B.C.E. The Indus River Valley was extremely fertile, and barley, field peas, and sesame were some of the crops cultivated there. Harappa was also known for its terra-cotta pottery, which was used for cooking and storage.

The Second Agricultural Revolution

Learning Objective
LO 12.18 Explain the advances and impacts of the Second Agricultural Revolution.

For the next several millennia, people around the world continued to make breakthroughs that made agriculture more productive. The ancient Romans were systematic about improvements, taking notes on farming practices they saw in the lands they conquered and using methods that could be applied in other parts of their empire. In the 11th century C.E., farmers in southern China planted a faster-growing rice native to Vietnam. The new variety allowed Chinese farmers to produce two crops of rice a year. Farmers in northern Europe in the Middle Ages developed a wheeled plow that improved cultivation. These advances increased productivity but fell short of the widespread, deep impact of the Neolithic Revolution. Essentially, societies continued to grow food largely for themselves.

That changed in the early 1700s, when new practices and tools launched the **second agricultural revolution**, which began in Britain and the Low Countries (Belgium, Luxembourg, and the Netherlands) and diffused from those regions. This revolution saw dramatic improvements in crop yields, innovations like more effective yokes for oxen and later the replacement of oxen by horses, as well as advancements in fertilizers and field drainage systems.

A change in the way farms were organized was a major driver of the revolution. Whereas most agriculture was previously done by peasants who were, for the most part, growing food for themselves on communal, or shared, land, Britain gradually switched to an **enclosure system**. In this system, communal lands—lands owned by a community rather than by an individual—were replaced by farms owned by individuals, and use of the land was restricted to the owner or tenants who rented the land from the owner. This change gave owners more control over their farms and therefore led to more effective farming practices. However, it also pushed peasants off the land and created a labor surplus—a factor that contributed to the Industrial Revolution, which began in Britain in the 1700s and spread to Western Europe and the United States in the 1800s. The second agricultural revolution continued into the late 1800s, and much of it coincided with the Industrial Revolution.

One of the most important new technologies of the second agricultural revolution was a horse-drawn seed drill invented by Jethro Tull in England around 1701. Tull's machine meant that farmers no longer had to scatter seeds by hand, a time-consuming process that was potentially wasteful if seeds did not land in the plowed furrows or were not evenly distributed. Tull's drill was widely adopted across Europe.

New tools of this agricultural revolution were invented in the United States, too. Cyrus McCormick designed and produced a mechanical reaper in the 1830s that mechanized harvesting grain. As you learned, John Deere

Agriculture: Human-Environment Interaction 353

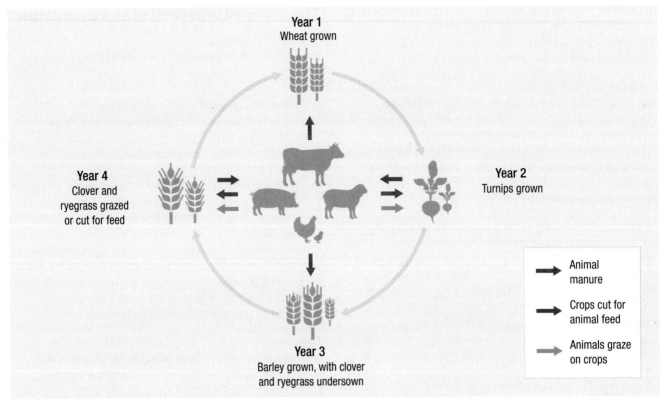

Norfolk Four-Field System The Norfolk four-field system was a key innovation of the second agricultural revolution. Soil benefited from richer manure because animals were better fed with crops produced by the improved soil. ▌Describe what happens during the fourth year of this field system.

invented a steel plow that by 1838 made it easier to plow in deep, tough soil, such as that found in the American Midwest. The steel plow made it possible to farm new areas that had not been arable—or able to grow crops—with earlier technology.

Farmers also adopted new methods of crop rotation that prevented soil exhaustion, cut down on insect pests, and increased yields. One system developed in England called for the rotation of several crops from year to year. Called the Norfolk four-field system, it involved the yearly rotation of several crops, including wheat, turnips, barley, clover, and ryegrass. The rotating of crops added nutrients to the soil, making it unnecessary to leave a field fallow—or unplowed— for a year, which had been the earlier practice. Crop rotation also fed livestock and promoted greater yields.

These changes resulted in another population boom, similar to the one that had accompanied the first agricultural revolution. People had more food, more nutritious diets, and longer life expectancies. As farms became more productive through mechanization and pushed peasant farmers off the land, these former laborers—and the overall growth of the population—provided a ready workforce for the new factories of the Industrial Revolution. These events all coincided with advancements in transportation made during the Industrial Revolution. The railroad allowed food to be transported greater distances, creating a larger market for the higher yields of the second agricultural revolution.

The Third Agricultural Revolution

Learning Objectives

LO 12.19 Discuss the technological advancements that characterized the Third Agricultural Revolution.

LO 12.20 Explain positive and negative aspects of the Green Revolution.

The **third agricultural revolution** began in the early 20th century and continues to the present day. It features further mechanization and the development of new technology, changes brought about by scientific and technological advances outside agriculture. The first shift was the move to mechanical and then electrical power, breaking away from the millennia-old reliance on animal power. Scientists also developed synthetic, or human-made rather than natural, fertilizers and pesticides. The third agricultural revolution occurred in core countries such as the United States before mid-20th-century scientists brought some of the revolution's advancements to countries in the periphery and semi-periphery.

Late in the 20th century into the early 21st, scientists added two more tools to improve agriculture. First, they applied advances in scientific understanding to manipulate the genetic makeup of plants and animals. The resulting **genetically modified organisms (GMOs)** can enhance the ability of the new strains to resist disease or drought or to have more nutritional impact or consumer appeal. For example, scientists added a gene from a bacterium that is a natural insecticide to such crops as potatoes, cotton, and corn. By making these plants genetically resistant to pests, the scientists reduced the need for growers to use chemical insecticides. Another innovation was to add a daffodil gene to rice to boost the amount of beta-carotene—needed for human consumption of vitamin A—in the grain. Producers are also using GIS and other information technology to monitor their fields for water and nutrient levels, allowing targeted delivery to meet crops' needs.

Some scientists refer to this use of information technology and data analytics as a fourth agricultural revolution. This revolution involves intensive data collection and manipulation that relies on global positioning system technology, smart technology farm equipment with sensors and wireless connections (precision agriculture), computer databases, and information-processing power. This recent period has been characterized by efficiency driven by data. For example, producers now use drones to deliver fertilizer and pesticide in measured quantities along precisely controlled routes to fields that were difficult to reach in the past.

Productivity through Technology The increased mechanization of farming initially took the form of the invention and use of motorized tractors. These multipurpose pieces of equipment had attachments that could be used for plowing, harvesting, and other functions. Tractors replaced horses, oxen, and other beasts of burden, speeding a farmer's work and facilitating cultivation of larger plots of land. Later, inventors developed special machines that were larger and more suited to the huge farms that came to typify intensive commercial agriculture in many core countries. Mechanical combines cut grain and separate the seed from the stalk, expediting the harvest. Similarly, corn picker-shellers pick ears of corn and strip them of kernels in one operation.

Adopting electricity greatly aided crop storage and preservation and enhanced livestock raising and dairy farming. Electric-powered livestock facilities allowed for the efficient rationing of feed and the maintenance of controlled growing conditions. Electric-powered milking and storage facilities radically changed dairy operations.

Another 20th-century innovation was the development and widespread use of synthetic fertilizers and herbicides and other pesticides. The fertilizers helped make fields more productive. Herbicides and other pesticides combated weeds and the potentially destructive effects of insects, other pests, and diseases—and boosted yields by preventing damage to crops. Crop dusting, in which low-flying airplanes spray pesticides over fields, made the delivery of these products more efficient.

The Green Revolution During the 1950s and 1960s, scientists used increased knowledge of genetics to develop new high-yield strains of grain crops, particularly wheat and rice. This movement, known as the **Green Revolution**, was an offshoot of the third agricultural revolution. The new crop strains—already in use in the United States—were introduced in areas with low yields and large populations, including Mexico, India, and Indonesia.

Spearheading this movement was Norman Borlaug, an American scientist from Iowa who was dedicated to the idea of transferring the advances of the third agricultural revolution to peripheral and semi-peripheral regions of the world. Borlaug first became involved in this work in Mexico in the 1940s, where he found a population of subsistence farmers plagued by chronically poor harvests. He worked for more than a decade to develop a disease-resistant strain of wheat and to convince Mexican farmers to plant these seeds. By the mid-1950s, Mexican farmers were able to produce enough wheat to meet their country's demand.

Borlaug turned next to South Asia, where his improved seeds and methods helped increase wheat harvests in both India and Pakistan fourfold. A crucial part of his vision of the Green Revolution was to train local agricultural scientists so that they could continue to make advances. For his work, Borlaug was awarded the 1970 Nobel Peace Prize and has been called the "father of the Green Revolution."

In the early 21st century, several groups began work on spreading the Green Revolution to Africa. A key to some of these efforts was the goal—set by former United Nations secretary-general Kofi Annan, of Ghana—to make the next part of the revolution environmentally friendly.

Impact and Response While it saved many lives and nourished millions, the third agricultural revolution came at a steep cost as well. First, increased mechanization reduced the need for human labor. Agricultural workers became displaced—a continuation of the trend begun in the second agricultural revolution. When human labor was required, many growers came to rely increasingly on migrant workers. Second, some technology of this revolution is dominated by multinational corporations, making producers vulnerable to the companies' marketing and sales practices.

For example, one multinational company sells crop seeds that resist one of its herbicides. That encourages farmers to buy the seeds and the herbicide, used to kill weeds. However, the corporation does not allow farmers to preserve seeds from one year's crop for the next year's planting—the growers must purchase new seeds each year. The expense of some of these practices forced many individual farmers out of business, leading to the replacement of small family farms with large-scale commercial agricultural operations.

The third agricultural revolution has also had environmental impacts. Green Revolution crops have increased growers' demand for water, causing regional inequities and the need for more water development projects. Some of those projects have had disastrous consequences. In the 1960s, the Soviet Union enacted an agricultural policy that diverted water from two rivers to irrigate cotton fields in Central Asia. As a result, the water no longer reached the Aral Sea, which shrank dramatically in size due to the semiarid climate of the region. Fish populations were devastated. The former seabed also had high amounts of salt, which resulted in health problems among the people of the surrounding area.

Widespread use of synthetic insecticides and other pesticides consisting of powerful chemicals can harm both pests as well as helpful insects and animals. Research in North America and Europe has shown that honeybees and other pollinators are weakened or killed by a widely used type of insecticide containing nicotine—and that its impact is made worse by concurrent use of other pesticides. The buildup of chemicals used in agriculture can pollute water supplies and cause human health problems. The concentration of livestock production in huge facilities can lead to difficulties related to the great amounts of waste products the animals generate. The spread of livestock production means diversion of more crop output to animal feed, which some critics view as both an economically and environmentally high-cost way of meeting nutritional needs.

Large agricultural facilities run by corporations require huge amounts of energy and other natural resources. Another cost is the loss of biodiversity as producers focus on heavily marketed, high-yield seed strains. Industrial-scale, market-oriented agriculture has also led to monocropping, which, as you know, is when a farmer or company produces the same crop year after year. This practice has contributed to a decline in soil fertility and biodiversity as well.

Many producers have reacted against these various drawbacks by focusing on alternative approaches to farming. One such method is sustainable agriculture, which is based on protecting the environment, ensuring profitability, and promoting greater social equality. Producers work toward ensuring the health of soils by avoiding the use of synthetic fertilizers and minimizing water use, among other practices.

Some people interested in sustainable agriculture practice organic farming. This method of growing crops completely eliminates the use of chemical fertilizers and pesticides, relying instead on natural products, like animal manure, and simple-but-proven farming practices, like crop rotation. These methods produce crops that are better for the environment but often have lower yields.

Geographic Thinking

1. Describe how the three agricultural revolutions were similar and how they were different.

2. Explain the advances and impacts of the second agricultural revolution.

3. Compare advantages and disadvantages of the Green Revolution for the food supply and environment in the periphery and semi-periphery.

4. Explain how the changes that some call the fourth agricultural revolution use geographic principles, skills, and technologies and have geographic consequences.

Critical Viewing A crop dusting plane sprays fungicide, a type of pesticide, to protect the crops on a banana plantation in Tagum in Davao del Norte province, which is located in Mindanao, a southern island in the Philippines. ▌Describe some of the possible negative impacts caused by the use of these types of chemicals.

Case Study

Women and Africa's Green Revolution

The Issue Women are important crop producers in areas of Africa South of the Sahara, but they suffer discriminatory barriers that impair their ability to benefit from their labor.

Learning Objective
LO 12.1 Explain how agricultural practices have environmental and societal connections.

Factors for Change

The Pathways program focuses on five "change levers" that help women gain success and increase agricultural output:

- Capacity—building skills, relationships, and self-confidence
- Access—technical advice, inputs such as seeds, and markets
- Productivity—improving output and crop diversification
- Household influence—women having more control over household decisions and assets
- Enabling environment—working to change attitudes toward women's roles

Source: CARE.org

Women work in a field in a rural farming community in Nkwandu, Arusha, Tanzania. Tanzania is one of four sub-Saharan countries in Africa where a program has been developed to help women farmers gain success and improve agricultural output.

For Millennia, women have worked in African farm fields. However, they have not benefited from their crucial position in the agricultural workforce. Historically, women were barred from owning property and denied access to financial backing, better seeds, water, tools, and agricultural extension services such as training or business planning. Patriarchal attitudes, which place greater importance on men's role in society than women's, left women as second-class citizens.

In recent years, international food agencies have developed programs to encourage women practicing agriculture, introducing some aspects of the Green Revolution. Pathways to Empowerment is a program for women in South Asia and Africa developed by nongovernmental organizations (NGOs) like CARE and the Bill and Melinda Gates Foundation.

The Pathways program was implemented in more than 400 villages in four African countries—Ghana, Malawi, Mali, and Tanzania—with more than 34,000 women receiving support. Special schools were set up to improve women's farming skills and help them believe they could succeed. The program encouraged greater connections to extension services and used innovative approaches to improve access to seeds and other inputs. Coaching helped women gain more access to markets, which not only allowed them to sell more of their output but to secure better prices as well.

Pathways participants in Malawi saw increased yields at a time when national output was falling by 30 to 50 percent due to adverse weather patterns. In Mali, women growers produced nearly 4,500 more metric tons of millets and more than 2,120 metric tons of rice than in the past, which was enough to feed 31,000 more families in that country for a year. Improvements to women's income were seen as well, including a 50 percent increase in Malawi villages over a few years.

While the gains are impressive, equally important is the societal change. Increasing numbers of women are more involved with their husbands in making decisions involving farmwork. In Ghana, the number of communities with laws that protected a woman's right to own and work land doubled. Attitudes about violence against women also changed, which substantially improved women's lives.

Geographic Thinking

Explain how the Pathways program has changed both agriculture and society in the four target African countries.

Chapter 12   Summary & Review

Chapter Summary

Agriculture is the purposeful cultivation of plants or raising of animals to produce food and other goods for survival.

- Four aspects of the physical environment have a profound effect on agriculture: climate (temperature and precipitation), elevation, soil, and topography.
- Temperature, precipitation, wind patterns, plus elevation and topography combine to produce climate regions.
- An area's physical geography, available resources, and cultural practices influence the area's agricultural practices.
- Subsistence agriculture provides crops or livestock to feed one's family and close community.
- Commercial agriculture focuses on producing crops and raising animals for others to purchase.
- The bid-rent theory explains how land value determines how the land will be used—intensively or extensively.
- Intensive agriculture involves a large amount of effort to produce as much yield from an area as possible (often a smaller amount of land).
- Extensive agriculture uses land extensively (generally a larger amount of land), with less investment in labor and capital, and results in relatively low levels of output per unit of land.
- Land boundaries in North America are drawn using three systems: metes and bounds, long-lot survey, and township and range.

- Rural settlements take three forms: clustered, dispersed, and linear.
- As humans moved from foraging to farming, their knowledge of domesticating plants and animals diffused throughout the world from various hearths.
- Agriculture was first employed from 13,000 to 3,000 years ago in hearths scattered across four continents.
- Agricultural products and practices spread through migration and contact with new peoples.
- Agricultural products and practices diffused globally in the Columbian Exchange, which connected the Americas to Europe and Africa, and still spread today.
- Advances in agriculture led to several revolutions that radically changed society and continue to resonate today.
- The first agricultural revolution resulted in population growth and major social changes.
- The second agricultural revolution resulted from new practices and tools that made agriculture more efficient.
- The third agricultural revolution featured further mechanization and the development of new technology.
- The Green Revolution was characterized by the development of high-yield strains of crops.

Review Questions

Use complete sentences to answer the questions.

1. **Apply Conceptual Vocabulary** Consider the terms *intensive* and *extensive*. Write a standard dictionary definition of each term. Then provide a conceptual definition for each—an explanation of how the term is used in the context of this chapter.

2. Compare the characteristics of subsistence agriculture and commercial agriculture.

3. Why might shifting cultivation take place in tropical rain forests?

4. Explain how transhumance is an example of extensive agriculture.

5. Compare the three types of rural settlement patterns.

6. Describe the role climate played in the shift from foraging to agriculture.

7. Identify where the original agricultural hearths were located.

8. Based on the "Köppen Climate Classifications" map and the "Agricultural Regions Around the World" map, compare how the climate regions of the original agricultural hearths are similar and different.

9. Describe how agricultural diffusion in modern times differs from how it occurred in ancient times.

10. Should the Columbian Exchange be classified as an agricultural revolution? Why or why not?

11. Identify what the three agricultural revolutions have in common.

12. Explain the technological advances of the second agricultural revolution and the impact of that event.

13. Are the three agricultural revolutions examples of intensive agriculture or extensive agriculture? Explain.

14. Describe the consequences of the Green Revolution on the food supply and environment in the periphery and semi-periphery.

Interpret Graphs

Irrigation and fertilizer are two inputs that determine higher or lower crop yields. Study the graphs and then answer the following questions.

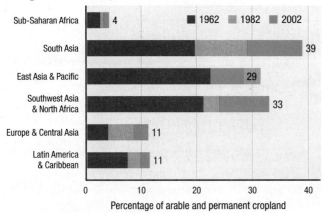

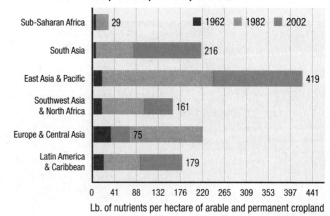

Source: World Development Report, 2008

15. **Analyze Data** Which areas showed the greatest growth in use of irrigation, and when? How about fertilizer use?

16. **Connect Visuals & Ideas** How are the data in both graphs related to the description in the text about the Green Revolution?

17. **Compare Geographic Concepts** Compare the graph on the use of irrigation with the "Köppen Climate Classifications" map, and then explain the greater use of irrigation in South Asia, Southwest Asia, and North Africa compared to Latin America and the Caribbean.

Chapter 13
Patterns and Practices of Agricultural Production

Critical Viewing On a large pineapple plantation in Oahu, Hawaii, workers use machinery to move pineapples from the fields onto a conveyor. ▌Compare how advanced machinery and technology might affect agricultural production on small-scale and large-scale farms.

Geographic Thinking What factors influence agricultural practices?

13.1
Agricultural Production Regions

Feature: The Changing Dairying and Ranching Industries

13.2
The Spatial Organization of Agriculture

National Geographic Explorer Tristram Stuart

13.3
The Von Thünen Model

13.4
Agriculture as a Global System

Case Study: Coffee Production and Consumption

13.1 Agricultural Production Regions

Agricultural practices help define the agricultural production regions around the world. Distinguishing regions of agriculture allows geographers to determine how economics, politics, culture, and the environmental characteristics of an area impact farming.

Economic Forces and Agriculture

Learning Objectives
LO 13.1 Describe how economic factors impact agricultural practices.
LO 13.2 Explain how technology, government policies, and consumer preferences influence agricultural practices.

In addition to environmental and cultural factors, why people farm, where people farm, and how people farm are also affected by the availability of resources and economic forces. Agricultural practices are influenced by such economic forces as the costs of materials, land, and labor; the availability of capital (money or other assets); the impacts of government policies; and ultimately, consumer preferences, or what people want to eat or consume, also known as market demands.

Economic forces help to distinguish subsistence agriculture from commercial agriculture. Most subsistence agriculture occurs in rural Africa and parts of Asia and Latin America, where connections to the global market are limited and farmers have less access to credit and financial capital. Many subsistence farmers live in poverty and do not have the economic resources to pay for labor or expensive machinery. Most often labor costs, either the farmer's own time spent tending to the land or wages for hired workers, are low relative to the costs of machinery.

Most commercial farming takes place in core and semi-peripheral countries with the **infrastructure** in place to access and supply the global market. Modern farm equipment, advanced technologies, and large plots of land are all characteristic of commercial agriculture. Commercial farmers maximize their income by purchasing a high level of external inputs. The impact of all of these costs can make commercial agriculture an expensive business. Therefore, it is important to have access to capital, which is easier to obtain in core countries than peripheral countries.

A **dual agricultural economy** refers to two agricultural sectors in the same country or region that have different levels of technology and different patterns of demand. In these areas, subsistence farms where food is grown for farmers to consume exist next to commercial operations that cultivate a crop to sell and often export to core countries where demand for the crop is high. South Africa and Zimbabwe are examples of dual agricultural economies. South Africa has both subsistence farms and well-developed commercial operations. Farmers who have the resources to invest in equipment, land, and materials participate in commercial agriculture. Farmers with fewer resources tend to be limited to providing food for their families.

In most instances, the costs of materials and labor are relative to the size of the farm. However, large-scale farming can be more cost-effective when fixed costs are spread over a greater area and lower bulk prices are negotiated for inputs, like seed or fertilizer. For example, farmers often receive a bulk discount when buying a large quantity of weed-controlling herbicides to spread over several hundred acres of land. As a result, the cost of the herbicide per acre is reduced. When production increases, expenses are lower per unit of output. This is an example of economies of scale, which have always existed in farming but have increased considerably since the introduction of modern farming technology and the innovations of the third agricultural revolution that helped to make agriculture a business.

The term **agribusiness** refers to the large-scale system that includes the production, processing, and distribution of agricultural products and equipment. Commercial farmers, large and small, are just one part of the agribusiness system, which has grown substantially over the last century. This growth has caused a major change in the nature of farming. Before the 20th century, farmers were typically self-sufficient small businesses, but as farm machinery, fertilizers, herbicides and other pesticides, genetically modified organisms, and smart technologies have made agriculture more efficient and specialized, farmers have become much more dependent on food manufacturers, distributors, and marketers. In fact, many farms are controlled by the producers, processors, and retailers who are part of the agribusiness system, which is described in more detail later in this chapter.

Technology and Increased Production

Modern equipment, improved fertilizers and pesticides, and new types of seeds all allow farmers to create higher yields. In fact, farmers today produce 262 percent more food with 2 percent fewer inputs than farmers in 1950. Technology has changed the growing season for many crops and improved production. For instance, scientists have taken the natural process of hybridization to intentionally create **hybrid** grains, fruits, and vegetables, in which different varieties of plants are bred to enhance desired characteristics and improve disease resistance. Some hybrids can grow in extreme temperatures or wet or dry conditions. For instance, seed technologies have helped corn and wheat become more frost tolerant, allowing growers to plant seeds sooner in cooler temperatures and harvest crops later. Additionally, farmers have new tillage practices, used to prepare the land for planting crops, that allow them to grow corn in areas that have historically not successfully produced corn.

One problem farmers all over the world continuously face is the reduction of crop yields due to pests. Weeds and other unwanted plants compete for resources and crowd out desired crops, and insects can destroy an entire field if not controlled properly. Since the 1960s, the use of pesticides, including herbicides and insecticides, around the world has increased considerably and led to higher crop yields. Herbicides are chemicals that are toxic to weeds and unwanted plants. Insecticides are chemicals that are toxic to insects. In the last six decades, the average yield of wheat and rice has more than doubled due in large part to the use of these and other pesticides. While pesticides can harm human and environmental health, without them crop production would drop and could result in soaring food prices.

Advanced farm machinery also increases production by improving efficiency. Farmers can plant and harvest more acres of land in a shorter period of time with modern tractors and combines, for example. Advances in irrigation have also provided better agricultural conditions in particular areas, which increase yields.

It is more difficult for subsistence farmers and small, family-operated commercial farms to adopt expensive technologies, like machinery. They generally don't have the capital input needed that large commercial farms have. Some smaller farms benefit from jointly owning machinery with neighboring farms or by joining cooperatives, but the economies of scale are much greater for large corporate-owned farms. A cooperative is a system in which farmers pool their resources to produce, market, and sell their crops.

The amount of capital farmers have access to affects their ability to operate and expand their farms. Capital, in turn, can impact a farm's productivity. Capital includes not only the money to purchase materials and equipment to make improvements but also a farmer's buildings, equipment, and animals. Farmers who can afford better (often expensive) machinery, fertilizers, and pesticides will increase their output and profitability. Large-scale farms often have more cash capital than small-scale farms, whose capital is found in their equipment, land, buildings, and for some, livestock.

Policies and Preferences Government policies and consumer demand both greatly influence agricultural practices. Most governments around the world intervene in agricultural markets in a few ways. They provide payments to farmers for growing certain crops or for not growing others, place regulations on agricultural imports and exports, or establish price supports in the form of crop purchases made by the government at a guaranteed price. All of these actions have been utilized by the U.S. government over the last century to protect farmers who have struggled at various times, including during the Dust Bowl of the 1930s, the farm debt crisis of the 1980s, and the volatile agricultural landscape of the 21st century.

In the mid-1930s, farmers limited their production and began receiving government payments to stabilize agricultural prices. This lasted until the late 1970s, when attitudes in the government changed. Farmers were no longer paid to produce "nothing," as some argued, but instead were incentivized to produce as much as possible, resulting in surplus grains and low prices. Today excess corn, in particular, is used as feed for animals, for ethanol fuel, and for producing high-fructose corn syrup—an inexpensive sweetener made from corn found in most processed foods. As a result, processed foods have become less expensive than fresh fruits and vegetables, directly impacting the food choices available to consumers.

Governments continue to control the supply of certain crops by enacting quotas, or the amount farmers can produce and sell. For instance, quotas are used in Canada in the dairy and poultry industries to keep prices of these goods stable and guarantee farmers a steady income. All of these policies affect pricing and therefore the quantity of crops farmers grow and harvest. In the United States, a farm bill is passed by the U.S. Congress about every five years to authorize policies regarding commodity programs and crop insurance, conservation, agricultural trade, nutrition, farm credit, rural economic development, and organic agriculture. These policies set guidelines for participating in agricultural and food programs such as the standards required to farm organically or the actions farmers must take to receive payments from the government.

Dietary preference and the kinds of agricultural products consumers choose to purchase also affect agriculture. Farmers will produce more of the products that are in demand. For instance, ahead of annual Super Bowl viewing parties across the United States, demand for avocados jumps. In 2020, almost 75 million pounds of avocados crossed the border from Mexico the week before America's largest sporting and television event. Farmers in Mexico employ methods to increase yields to meet the booming demand for avocados that has increased not just in the United States but around the world.

While government incentives and price controls help farmers, it is ultimately consumers who drive the market in commercial agriculture. In general, farmers will look to the market to determine what and how much to plant of the most sought-after crops each season.

Geographic Thinking

1. Compare the influence of economic factors in subsistence and commercial farming practices.

2. Explain how government policies have impacted farm practices and the foods available to consumers.

This farm near Goodland, Kansas, uses multiple combines to harvest wheat. Large-scale farming operations take advantage of economies of scale and technology to increase the amount of crops they produce. The number of farms larger than 2,000 acres continues to rise, as they are poised to take advantage of the latest machinery and agricultural practices.

The owner of an 800-acre ranch tends to her cattle in Woodland Park, Colorado. Changes in agriculture have impacted U.S. family farms and ranches such as this one for years.

The Changing Dairying and Ranching Industries

Learning Objective
LO 13.2 Explain how technology, government policies, and consumer preferences influence agricultural practices.

Commercial dairying is the production and selling of milk and related food products. In the United States, dairy farms are most prevalent in the upper Midwest and Northeast and are primarily run by families, but dairying is changing. Family-owned commercial operations are going out of business or changing their practices. According to the U.S. Department of Agriculture (USDA), the United States lost nearly 20,000 licensed dairy farms, a roughly 30 percent decline, between 2010 and 2020. Wisconsin, long known as "America's Dairyland," has experienced a decline from more than 10,000 herds in 2015 to fewer than 6,700 herds in 2022. New York and New England have also seen dramatic reductions in milk processors. Many of these farmers are selling their dairies to larger operations who have the capital to withstand a volatile market in which prices for milk fluctuate. Some dairy farmers are participating in cooperatives in which they pool their money together to purchase supplies and services. Making purchases collectively keeps costs down for farmers. Others are specializing in organic dairy products that are sold for a higher price than conventional ones. However, that strategy may no longer be successful in the face of competition from very large-scale organic dairy farms, located largely in the western United States.

Several factors affect the dairy market. Pricing is one of them. The price of milk is determined partly by market demand and partly by government pricing regulations. The price of milk had been generally falling since 2014, however this is no longer the case. The market for cow's milk has declined as alternative milk products made from soy or nuts have become popular.

Meat production has also experienced change in recent decades, as the demand for meat has increased in some places around the world. Ranching operations have grown in regions where the land is open and plentiful for cattle to graze, like in the United States and South America. As U.S. cattle mature, most are sent to feedlots, primarily located in Texas, Colorado, Kansas, and Nebraska, where they are fed corn to fatten them up before being processed into meat. The size of feedlots has increased in order to reduce costs and maximize profits. Animals are confined to smaller spaces than they have on the ranch and are fattened in a short period of time, increasing the number of animals that can be processed. Feedlots then sell to meatpackers where, because of recent mergers, only four conglomerates dominate and control 85 percent of the industry. These huge, multinational meatpackers impose low prices so both feedlot owners and ranchers receive little return on the cattle they sell. In 2022, the Biden administration and Congress sought to diminish the dominance of these four meatpacking agribusinesses who are squelching competition and intensifying inflation concerns. The USDA reported that more than 500,000 of the country's employees worked in the meatpacking industry in 2021 and that in 56 counties, mostly rural, meatpacking jobs were estimated to represent more than 20 percent of all county employment. Meanwhile, smaller-scale farmers in both the dairy and ranching industries are facing dire futures and further consolidation of agriculture.

While the number of meat producers who provide their animals access to open pastures throughout their lives is but a tiny percentage compared to feedlot-raised beef, about 3 percent of the total in recent years, this number has grown despite the fact that the price of grass-fed beef—whether strictly organic or not—is higher than conventionally raised beef.

The COVID-19 pandemic severely disrupted the global meatpacking industry, presenting a number of supply-chain and employee issues. The Centers for Disease Control and Prevention (CDC) found that processing lines as well as other areas of busy U.S. meatpacking facilities where workers have close contact with each other would and did contribute to coronavirus exposures. Designated as "essential workers," meatpacking employees stayed on the job as most of U.S. workers stayed home, and suffered high rates of coronavirus illness, hospitalizations, and deaths as a result. In 2021, the CDC recommended that meatpacking facilities pay particular attention to social distancing, duration of contact with others, shared spaces like break rooms, and shared transportation and provided optimal recommendations to keep this critical food industry operating.

Geographic Thinking

Describe the factors that impact commercial dairying and ranching.

Dietary preferences and prices alike guide what farmers grow for the market. Here, a father and his children shop for produce at a neighborhood store, which is well-stocked with many choices including organic.

13.2 The Spatial Organization of Agriculture

The rising demand for food and our changing appetites are fueling growth and change in agribusiness. Large-scale corporate operations are replacing more and more small family farms. Such changes impact all who are part of the complex global food supply system that includes not just farmers but governments, businesses, and consumers.

Family vs. Corporate Control

Learning Objective
LO 13.3 Compare the factors that impact small family farms and large commercial farms.

Family farms represent the vast majority of farms worldwide. However, these small family-owned enterprises account for less of the share of the world's total farmland. According to the Food and Agriculture Organization (FAO) of the United Nations, approximately 84 percent of farms worldwide are smaller than five acres, and these small farms operate about 12 percent of the total farmland. Meanwhile, approximately 16 percent of farms worldwide are larger than five acres and represent 88 percent of the world's farmland. Indeed, the top one percent, farms larger than 125 acres, accounts for 70 percent of all the world's farmland. The vast majority of the share of farmland controlled by larger farms is in core countries, such as the United States, Canada, Australia, and countries in Europe where commercial agriculture is more prevalent. Most family subsistence farms are located in the periphery, including countries in Asia and Africa; while they are small, collectively they produce food for a large portion of the world's population.

There are limitations to the estimated data gathered about agriculture worldwide, especially at the regional and local levels. These limitations include the number of participants in the agricultural census and variations in the manner in which the data is collected. The latest information available from the FAO reported that globally, between the years 1960 and 2021, the total number of farms increased to about 608 million. The average farm size decreased between the years 1960 and 2000 and remained about the same between the years 2000 and 2021. Of course, the figures vary country by country, and the difference is striking between countries in the core and those in the periphery.

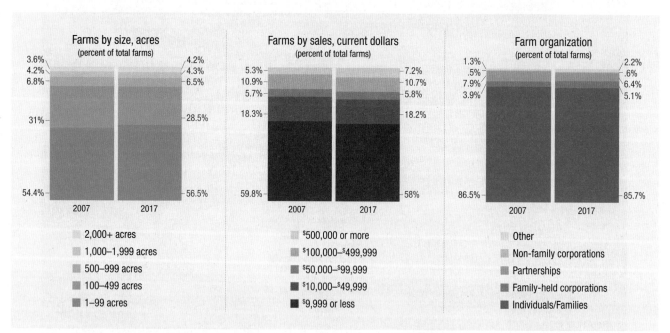

U.S. Agricultural Production, 2007 and 2017 Today, farming in the United States is made up of mainly family farms, but the number of family farms has decreased since 2007. The graphs give a closer look at the distribution of farms according to size, sales in dollars, and ownership.

Source: U.S. Department of Agriculture

Data gathered strictly about U.S. agriculture differs in some respects from the global data. Since the 1960s, the number of farms in the United States has decreased—from approximately 3.7 million farms to just over 2 million, a number that has held steady for the last few decades. Of these 2 million farms, more than 90 percent are classified as small and are mostly family owned and operated. In 2019, the average acreage of a U.S. farm was 444 acres, according to the USDA, down an acre from the previous year. Total U.S. land in farms was almost 900 million acres, or about 1.4 million square miles, representing nearly 37 percent of the entire country's area.

Recent trends are hurting the family farm and causing a shift in the spatial organization of agriculture. The overall population across the world is shifting away from rural areas and into urban areas. Many from younger generations see the amount of time and hard work established farmers devote to rural agriculture just to earn a very small profit. Therefore, fewer people are interested in or willing to take on the challenges that come with farming. Additionally, the farming population is aging. Farmers who were once prosperous are retiring or dying without successors in place.

Another challenge many farmers face is rising costs. When farming costs are greater than the income generated for many years in a row, farmers struggle to remain in agriculture. At times, a supply-heavy market causes the price of goods to fall. For example, when corn production is high, the price of corn decreases. Farmers make little, if any, profit when agricultural prices fall drastically. If another farmer or a corporation offers a struggling farm owner a tremendous amount of money for the land, it might be an offer that is hard to resist. Over time, such offers reduce the number of individual farm operations. The result is a shift from small family-operated farms to large corporate-controlled, vertically integrated agribusiness operations.

Vertical integration occurs when a company controls more than one stage of the production process. When a company manages all aspects of their business operations, from production to processing to shipping and then to selling, it helps reduce costs, improve efficiencies, and increase profits. It is difficult for small family-owned and -operated farms to practice vertical integration, but large corporate agribusinesses have the capital and systems in place to do so. Examples of vertically integrated agribusinesses include orange juice, chickens, cereal, and French fries. Some fast-food restaurant companies use vertical integration, which has allowed them to offer products at prices lower than the competition. McDonald's has complete control over its agricultural sources; its own processing facilities, distribution centers, and transportation systems; and the land that the restaurants occupy.

Commodity Chains

Learning Objective
LO 13.4 Identify different linkages that constitute commodity chains in agriculture.

The rise of agribusiness has led to the establishment of a complex network that connects places of production with distribution to consumers. This network is called a **commodity chain**. Numerous people in many regions have a hand in producing and distributing agricultural products. Agricultural commodity chains begin with inputs such as land, seeds, fertilizers, and animals all tended to by farmers to produce a crop or other agricultural product. After cultivation and harvest, the crop is processed, packaged, and then transported to wholesalers and retailers. Other agricultural commodity chains result in products like milk, cheese, meat, wool, and more. Eventually, the end result is a finished commodity that is marketed to consumers.

Successfully delivering a product from the farm to the consumer involves many exchanges that must be considered and planned. Factors that influence the agricultural process include the the physical environment (like weather and soils), financial markets, labor relations, government policies, and trade. Businesses sign labor contracts and create marketing plans, while governments negotiate trade agreements, both establishing relationships and policies to ensure the successful delivery of an orange from the grove, for example, to a consumer's table.

The commodity chain for orange juice reveals that the product changes hands several times during the process. The citrus industry operates on long-term contracts and the futures market (an obligation for a buyer to purchase—and the seller to sell—a crop at a set price at a future time). This creates an arrangement where parties in the production chain work together but remain independent entities.

First, orange growers manage, harvest, and sell the fruit. Some growers are small farms that sell their fruit to a fruit handler or a cooperative, where farmers pool their resources to produce, market, and sell their produce. Other growers are part of a larger fruit processing company. Fruit processors take the fruit and produce either packaged juice or concentrate. Processors that produce packaged juice then market and sell it. Bulk processors that produce frozen concentrated orange juice work on distributing it and marketing it to stores and consumers at the end of the commodity chain. Consumers often have a choice among multiple, competing orange juice products.

Pricing and Policies

Learning Objectives

LO 13.5 Describe how government policies can influence pricing and production costs in agriculture.

LO 13.6 Define the terms subsidy and tariff in the context of agricultural production.

The dramatic growth in the production of crops due to technology leads to a greater supply. When supply is high, prices go down—this is the function of the law of supply and demand. At times of high production success, prices can drop so low that production costs are higher than the value of the product. This can be catastrophic for farmers. Either by itself or if coupled with a devastating loss of crops due to weather events, farmers can go bankrupt and lose their farms.

Subsidies Some solutions to rising costs of production have been for the U.S. federal government to provide low-cost loans, insurance, and payments called **farm subsidies** to some farmers and agribusinesses. Farm subsidies originally started during the Great Depression of the 1930s to help struggling farmers and to make sure there was enough food being produced for American consumption.

In theory, subsidies continue to this day to protect all farmers and owners of farmland. But studies have shown that this isn't the case. Small family operations, the ones that need the help often more than large farms, aren't seeing the money. Instead, the highest quantity producers of commodities like corn, soybeans, wheat, cotton, rice, and sugar are benefiting the most. In fact, one study revealed that the largest 15 percent of farm operations receive 85 percent of the subsidies. The benefits of these farm subsidies are debatable. There are many proponents on both sides of the argument. Many believe that the subsidies help the agricultural economy of the United States, even though they are directed to large producers. Others believe that the money is just a bonus to operations that are already successful and don't need the assistance.

The U.S. government currently pays out about $20 billion each year in farm subsidies. In 2020, bailouts rocketed that number to over $50 billion with pandemic and also tariff relief programs (see "Tariffs") to help U.S. farmers. About every five years, with the help of the USDA, Congress evaluates and approves a farm bill that allocates the subsidies among other important regulations for farmers.

Tariffs Another factor affecting prices of crops is the use of government **tariffs**. A tariff is a tax or duty to be paid on a particular import or export. Tariffs can affect trade between countries and can bring about a trade war, in which countries try to negatively impact each other's trade. Tariffs are used to raise government revenue, or income, but they are also used to protect domestic industries or agricultural production against foreign competition. Tariffs raise the price of imported goods, making them more expensive to purchase than goods made within the country. Therefore, domestic producers of soy, for example, are given a price advantage, making soy imports less competitive. Consumers, however, face higher prices for products that contain soy as a result of the tariffs.

One example of tariffs leading to a trade war and impacting U.S. agriculture is when President Donald Trump, in 2018, imposed a 25 percent tariff on foreign steel and a 10 percent tariff on aluminum imports. This tariff affected more than 800 types of products—worth $34 billion—entering the United States from China. These products included industrial machinery, medical devices, and auto parts. Trump imposed the tariff for multiple reasons: to raise revenue for the federal government, to raise the price of Chinese products so Americans would not buy them, to promote steel and aluminum products made in the United States so Americans would buy those instead, and to force China to stop

THE FOOD SUPPLY SYSTEM

1 AGRICULTURAL INPUTS

Inputs are the items used by farmers to successfully produce agriculture.

- seeds
- fertilizers
- pesticides
- feed
- utilities

2 PRODUCERS

The producers grow, harvest, and raise the agricultural products based on farm size and type.

- farmers
- fishers
- ranchers

3 PROCESSORS

Agriculture is transformed into ingredients and packaged food after it is sorted, cleaned, milled, and prepared.

- manufacturers
- factories
- packagers
- storage facilities

Critical Viewing The production of food, from farm to consumer, is a highly complex system that goes through many exchanges. ▎How might external factors, such as physical environment and financial markets, impact agricultural products as they move through the food supply system?

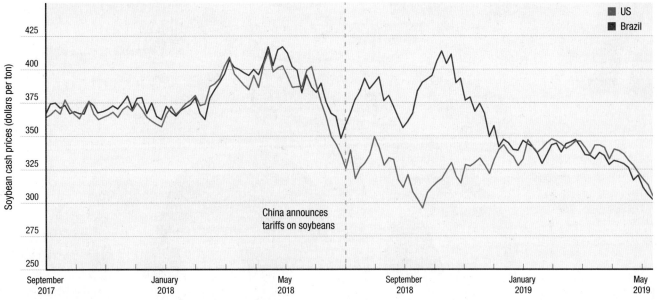

Impact of Trade War on U.S. and Brazil Soybean Prices Brazil and the United States are the top exporters of soybeans in the world. China is the top importer. After China implemented tariffs on U.S. products in response to President Trump's 2018 tariffs, there was less demand for U.S. soybeans. The U.S. price fell lower than the price of Brazilian soybeans, which fluctuated at a higher price. Prices converged in early 2019 when a limited number of soybean exports to China from the United States resumed and U.S. soybean exports reached additional countries.

Source: Bloomberg L.P.

participating in unfair business practices. China responded by imposing its own tariff on U.S. products worth $34 billion. These products included many American manufactured goods, as well as agricultural products like soybeans, beef, and pork. This was especially challenging for hog farmers, as the Chinese are the world's top consumers of pork. Imposing tariffs and engaging in such a trade war can greatly disrupt established commodity chains, lower the price of farm products, and cause farmers to lose business. Even with the change in presidential administrations, most of these tariffs remained in place in 2021; President Joe Biden announced some changes to tariff policies in 2022.

Geographic Thinking

1. Explain why the number of small farms is decreasing and the number of large commercial farms is increasing.

2. Describe how the closure of several food processing plants would disrupt agriculture commodity chains.

3. Explain how farm subsidies combat the issue of rising costs of production.

4. Identify an impact of the trade war between the United States and China on agriculture.

DISTRIBUTORS
Food and ingredients are distributed locally, regionally, and globally.
- importers
- exporters
- wholesalers

RETAILERS AND FOOD SERVICES
Retailers make food available for consumers.
- restaurants
- supermarkets
- convenience stores
- food banks

CONSUMERS
Consumers cook and/or eat the food.
- individuals
- families

Patterns and Practices of Agricultural Production

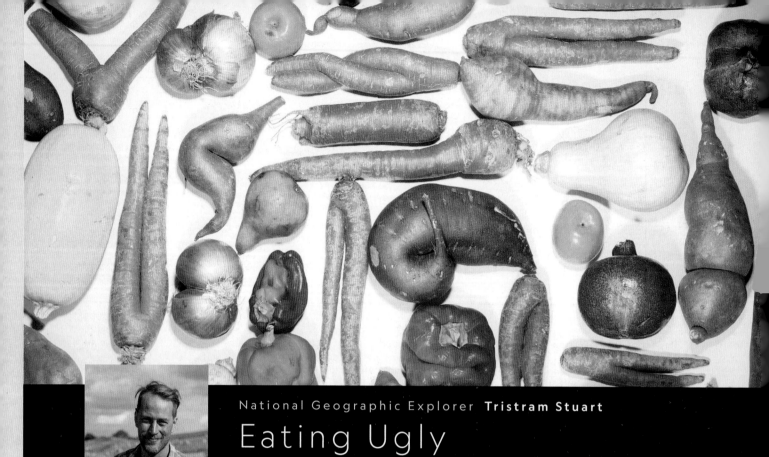

Stuart works to combat issues of food waste, including throwing away vegetables that are cosmetically imperfect but entirely edible.

Learning Objective
LO 13.7 Describe the relationship between production and consumption as it relates to agriculture and food supply.

National Geographic Explorer **Tristram Stuart**
Eating Ugly

Every year some 2.9 trillion pounds of food—nearly a third of all that the world produces—is discarded. National Geographic Explorer Tristram Stuart is on a crusade to reduce waste in the food supply system by encouraging the consumption of "ugly food" that is often thrown out only for aesthetic reasons.

Stuart's campaign started when he was young, collecting unconsumed food from his school kitchens and a bakery to feed hogs he was raising in the English countryside. He realized that the food many people waste is in fact fit for human consumption. Stuart then began studying the issue of food waste on a global scale. To support his research, he looked at quantitative data about the food supply of every single country and compared it to what was likely actually being consumed. Stuart's study considered obesity levels, diet intake surveys, and other factors that determine what people eat.

The study resulted in a chart with a line that shows the normal level of consumption that a country should have assuming every resident has a stable, secure, and nutritious diet. Countries above the line represent those with an unnecessary surplus of food, meaning food availability in its communities is much larger than consumption. So, what happens to the surplus food? Simply, food is being wasted, thrown in dumpsters or given to animals, often because the produce is considered subpar. Stuart notes, shocked, "Potatoes that are cosmetically imperfect, all going for pigs. Parsnips that are too small for supermarket specifications, tomatoes in Tenerife [an island in Spain], oranges in Florida, bananas in Ecuador, all being discarded ... perfectly edible, because they're the wrong shape or size."

Stuart's goal is to confront the business community, raise public awareness, and find solutions to these wasteful practices. "If we make noise about it, tell corporations about it, tell governments we want to see an end to food waste, we do have the power to bring about that change."

Geographic Thinking

Describe the impact Stuart's research could have on food consumption patterns.

13.3 The Von Thünen Model

Commercial agriculture is the large-scale production of crops and livestock to sell at market. The market may be either a wholesaler (which then sells the products to retailers) or the retail store itself. The distance and transportation costs to market can influence what the farmer grows or raises and create specific patterns of land use.

Rural Land-Use Patterns

Learning Objective
LO 13.8 Explain how the Von Thünen Model predicts the geography of agricultural production.

Johann Heinrich von Thünen was a German farm owner with an interest in the geography and economics of farming. In 1826, he wrote a book about observations he had made regarding the spatial patterns of farming practices in his community. He found that specific types of agriculture took place in different locations surrounding the market, or center where business took place in a city or town. A pattern of intensive rural practices close to the market and extensive rural practices farther from the market emerged. In his book, von Thünen suggested that a farmer decides to cultivate certain crops or raise certain animals depending on the distance between the farm and the market. Based on this principle, the **von Thünen model** hypothesizes that perishability of the product and transportation costs to the market each factor into a farmer's decisions regarding agricultural practices.

The model has four distinct concentric rings representing different agricultural practices. In the center is a core representing the market. The first ring outside the core represents intensive farming and dairying. The perishability of milk products and produce, like berries, lettuce, or tomatoes, makes it critical that they are produced close to the market and transported and sold within a limited time frame. These products cost more to transport, so having them produced closer to the market is a cost-saving measure. And even though this land is more expensive (remember bid-rent theory from Chapter 12), the products have a greater value and consumers will pay a higher price for them. For example, a farmer with land close to the market might choose to raise chickens and produce eggs for sale because the transportation costs would be low and the fragile, perishable eggs would arrive at the market in a shorter amount of time while still fresh.

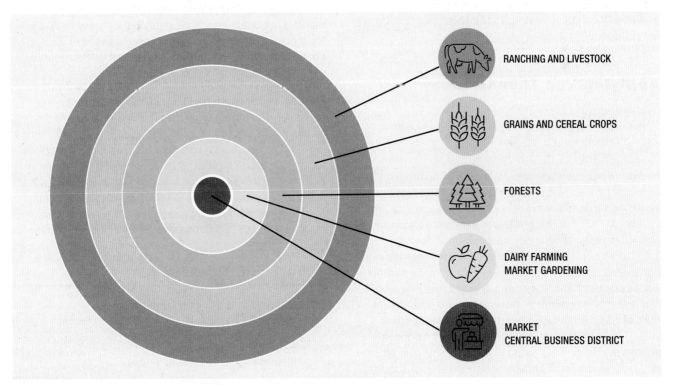

Critical Viewing The von Thünen model suggests that farmers make decisions about which crops and animals to raise based on their proximity to markets. The model was created before industrialization. ▮ Why was dairy farming placed so close to the market city in the early 1800s? Explain.

Patterns and Practices of Agricultural Production

Assumptions about the Von Thünen Model

The von Thünen model is based on an ideal environment that he specified back in 1826. These assumptions, when present, make this model work. They include:

- The market is located in an isolated, self-sufficient state without external influences.
- A commercial agricultural system exists where farmers will seek to maximize profits.
- A single, centrally located market is the destination of a farmer's produce.
- The land is *isotropic*, or flat and featureless, not containing mountains or rivers.
- There is only one means of transportation (oxen pulling a wooden cart over land).

The next ring represents forests. In the 19th century, timber and firewood were important commodities used for heating, cooking, and building. Wood is heavy and bulky, so its weight would make it expensive to transport. Producing wood close to the market reduces transportation costs.

The third ring is devoted to grains and cereal crops like rye, wheat, and barley—favored grains in 19th-century Germany. These crops are less perishable and not too bulky or heavy, so they can be grown efficiently farther from the market. The grain or cereal farmer could profitably raise and transport these crops to market. The land here, farther from the market, is less expensive and allows for less intensive agriculture.

Finally, the ring farthest away from the market in the von Thünen model is where livestock production occurs. Land is less desirable here and therefore less expensive. Farmers can buy or rent large pieces of land for extensive agricultural activities such as ranching. In the early 1800s, animals were walked to the market for sale, so transportation costs were not high.

Applying Von Thünen Today

Learning Objective
LO 13.9 Describe how technology and government policies have altered agricultural systems since the Von Thünen Model was proposed.

All models are based on assumptions, and no model accounts for every exception or deviation that takes place in real-life situations. Von Thünen made assumptions (listed in the box) based on the realities of the early 19th century. Most of his assumptions don't apply to today's world. Many cities have multiple centers of business, not just one. An isolated state that has not been influenced by outside cultures or events no longer exists. In fact, the influence of modern industrialization, technology, and government policies has dramatically altered agricultural systems around the world. Some governments have introduced policies that favor the growing of certain crops over others, which influences what farmers grow. However, despite the enormous changes over time, the model can still be loosely applied to contemporary agricultural production, especially when it comes to the role of transportation influencing patterns of production.

Perhaps the most innovative technology that has improved transportation of agricultural products is the refrigerated container that can be transported on trucks, trains, ships, and cargo planes. The development of this technology has permitted perishable items like eggs and dairy to be produced much farther from markets. At even greater distances, fruits and vegetables that are only grown in certain parts of the world can be flown to grocery stores in another hemisphere. Von Thünen could never have imagined the movement of produce over thousands of miles—like blueberries from Chile to stores in the United States.

Time-space compression—the ability to quickly exchange goods across distances, as a result of more efficient transportation systems—as well as growing demands for food worldwide have expanded the markets available to most producers. Whereas in the past, farmers sold all or nearly all of their goods at the nearest market, today's farmers can reach an abundance of hungry markets hundreds and thousands of miles away thanks to technology, including refrigerated containers, and a complex network of roads, rails, shipping channels, and air routes.

In today's global agricultural system, specialty farming thrives in regions with particular climates and soil types, unlike von Thünen's assumption that the land was all the same. Truck farming is common in regions like South Florida and California's Central Valley, where farmers ship produce nationwide and internationally. "Out of season" produce is essentially a thing of the past, as growers in tropical Central Mexico ship avocados and Chilean vineyards airfreight their Southern Hemisphere summer grapes all winter long to markets in the United States and other northern destinations.

Geographic Thinking

1. Explain the spatial relationship between land use and market areas, based on the von Thünen model.

2. Explain the von Thünen model assumptions that are no longer present in agriculture today.

3. Use the "Agricultural Regions Around the World" map in Chapter 12 to describe how the von Thünen model could be used to generally describe agricultural production across the continental United States.

13.4 Agriculture as a Global System

Importing and exporting agricultural products has been important to countries for centuries, but never before have so many regions been integrated into the global agricultural system. To manage the flow of food around the world, a large, complex network exists. Political relationships, trade patterns, and transportation networks all affect the distribution of food, which consequently impacts what appears on your plate.

Agricultural Interdependence

Learning Objective
LO 13.10 Explain the advantages and disadvantages for countries that are heavily dependent on commodity production.

In today's global economy, consumers purchase foods from all over the world—including bananas from Costa Rica and teas from India and China. Agriculture, like other economic activities, has become globally integrated and organized, often connecting peripheral countries with core countries. No single country produces all of the food its population consumes. Either a country's climate isn't conducive to growing certain foods, like bananas that require a tropical climate, or it is less expensive to import foods from a country that specializes in growing certain foods efficiently and in a high-quality manner. **Global supply chains**, which are the same as commodity chains but on a global scale, enable the delivery of a product between two different countries. For example, consumers in a store in Canada who purchase vegetables that originated from a farm in the Netherlands are each part of the global supply chain. As these networks have grown more complex, many regions of agricultural production and consumption have become increasingly interdependent.

Commodity agricultural products, including wheat, corn, soybeans, and cotton along with coffee, tea, cacao, and vanilla, are traded through global supply chains. Commodities are highly sought after in global markets. The supply chains for some agricultural commodities start with production in a peripheral country where crops are grown and harvested using low-cost, local labor, allowing for reduced overall production costs. Other commodities, like wheat or cotton, might start in the United States. Processing and packaging the product, the next step in the supply chain, may occur in the same country or a different one. Finally, the finished commodity is distributed to markets usually in core countries of the world.

Consider, for example, the commodity chain of cacao. Cacao beans come from trees grown in tropical environments near the Equator. The largest concentrations of cacao farms are in peripheral countries like Ghana and Côte d'Ivoire. Many farmers work tirelessly cultivating cacao—a crop that requires much of the work to be done by hand. After beans are picked and dried, they are sent to processing and manufacturing plants throughout the world.

At the plants, the beans are turned into cocoa powder, which then is used to make baking chocolate, chocolate bars, and other products transported to retailers.

The final leg of this global chain is the consumer. The cacao bean travels from the tropics to consumers all around the world, but most chocolate products end up in the United States and Europe. European countries such as Germany, the United Kingdom, Belgium, the Netherlands, and France import large amounts of chocolate.

Commodity Dependency International trade can be vital to a country's economy, and many rely on exports for financial stability. Some peripheral countries struggle with developing and maintaining export economies and end up becoming dependent on a single export **cash crop**, a crop that is produced for its commercial value. This dependency on one export can have negative consequences. Alternatively, though, if supply of a cash crop is limited, countries specializing in the crop can reap profitable rewards.

The vanilla industry is one example. The island of Madagascar, off the coast of Africa, has three-quarters of the world's vanilla fields. The vanilla bean is Madagascar's cash crop. It takes three years for beans to mature for harvest. Between 2016 and 2019, Madagascar experienced small crop yields and devastating storms that impacted vanilla production. Supply became limited, demand for vanilla remained the same, and prices went up.

Comoros, an archipelago (group of islands) located between the coast of Africa and Madagascar, also produces vanilla. The economy of Comoros is primarily agricultural. Some studies show that close to 70 percent of the working population of Comoros is involved in vanilla production. With very few other economic activities in Comoros, it relies on this single commodity and the resulting trade relationship with countries such as France, India, and Germany to keep its economy going. In recent years, 74 percent of Comoros' export earnings came from spices, including vanilla.

The specialization of one product creates a reputation and a demand for production. However, the reliance on a single commodity is risky, and it is unhealthy for an economy. Changes in world markets due to supply and demand issues can disrupt a country's economy that relies on one export. If demand for that one commodity dramatically

Workers at a vanilla processing center in Madagascar spread the vanilla pods on mats to dry in the sun. In 2017, Madagascar exported $894 million in vanilla to countries including the United States, France, and Germany.

falls, the revenue the commodity generates for the country declines. Not only can storms, droughts, and extreme temperatures cause crop failure and lead to a decline in a country's exports and revenue, but a trade war can erupt between trade partners, causing the costs of trading goods to fluctuate. Of course, a military war can also disrupt global supply chains. Russia's invasion of Ukraine in 2022 severely disrupted a number of agricultural commodity chains, including fertilizer, sunflower oil, and wheat. Before the war, Ukraine was the world's fifth largest exporter of wheat, supplying millions of tons of grain for a range of countries including Lebanon, Egypt, and Pakistan. With its Black Sea ports essentially shut down, Ukraine was not able to export its wheat via ocean shipping – with serious consequences for the countries dependent on this supply. Each of these scenarios – military conflict, trade war, supply and demand issues, or weather events – can bring uncertainty to a country's economy. In fact, any disturbance in the supply chain can impact the financial stability of a country or region. Infrastructure, political relationships, and world trade patterns all impact the complex food distribution network.

Infrastructure

Learning Objective
LO 13.11 Explain the role of infrastructure in agricultural commodity chains.

Participating in the global agricultural system means having the infrastructure, or networks and facilities, in place to efficiently produce and distribute crops and livestock. Infrastructure includes communication systems; sewage, water, and electric systems; and most importantly roads and transportation for exporting goods. Take, for example, the vegetables grown in the rural Netherlands. To get them to a city in Canada, the proper roads, seaports, ships, and all the communication systems necessary to hand off the vegetables to each party in the chain must be in place and function accordingly. Countries need to build main routes from agricultural, manufacturing, and production centers to airports and seaports to easily move and process imports and exports. Sometimes core countries provide financial support to peripheral countries to improve their infrastructure, which is mutually beneficial.

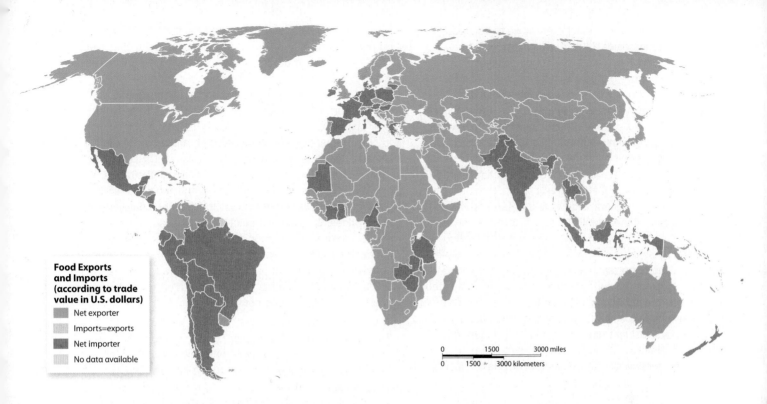

World Agricultural Trade

The volume of agricultural trade has grown steadily over the past three decades, driven by a growing population and technological advances. Countries export and import agricultural goods depending on growing conditions and food preferences. The food net exporters ship out is greater in value than the food they import, and the food net importers bring in is greater in value than the food they export.

Leading Exporters, 2017

European Union	$647 billion
Wheat, corn, barley	
United States	$170 billion
Soybeans, corn, wheat	
Brazil	$88 billion
Soybeans, sugar, chicken	
China	$79 billion
Garlic, vegetables, fruits	
Canada	$67 billion
Wheat, rapeseed, pork	

Leading Importers, 2017

European Union	$649 billion
Wheat, corn, soybeans	
China	$183 billion
Soybeans, beef, palm oil	
United States	$161 billion
Coffee, beef, fruit	
Japan	$79 billion
Pork, beef, corn	
Canada	$39 billion
Chocolate, beef, fruit	

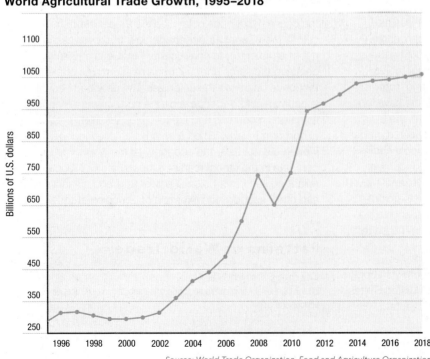

World Agricultural Trade Growth, 1995–2018

Source: World Trade Organization, Food and Agriculture Organization

Source: World Trade Organization, Food and Agriculture Organization

Railroads and waterways are efficient ways to transport agricultural commodities. Most products traded internationally move by sea. Many countries use trains and railways to move goods within their country and to neighboring countries on shared rail systems. On the massive Eurasian landmass that includes the continents of both Europe and Asia, there are four major interconnecting rail systems, along with a number of smaller rail systems. Brazil is a huge semi-peripheral country and exports many agricultural products across the globe. It also has a large transportation network that enables it to move commodities from its interior agricultural regions to its seaports on the Amazon River and the Atlantic Ocean. In North America, the countries of Canada, the United States, and Mexico are connected by an extensive rail network. The United States has massive transportation and communications networks that are critical for the agricultural economy, but many roads and rail lines are aging and in disrepair, and many rural areas do not have high-speed internet access. An infrastructure bill signed by President Biden in late 2021 included road, rail, and port improvements, as well as broadband internet provision for rural areas—all destined to help farmers and the country's agricultural economy.

Political Relationships

Learning Objective
LO 13.12 Explain how political instability can affect global supply chains.

Supply chains work best when trade partnerships are stable, reciprocal, and understood by all parties. Any kind of instability can rock the trade boat, but perhaps the biggest threat to a supply chain is political instability. Political instability can affect global supply chains with varying degrees of damage to countries and their economies. For example, Brexit, in which the United Kingdom left the European Union (EU) on January 31, 2020, has had major implications for Britain's agricultural supply chain. Brits import a lot of what they eat, especially during the winter months. They import fresh fruits and vegetables from southern European countries, like Italy, and from other countries around the world. British farmers have long exported various meat products to the EU. With the EU's border system, imports travel seamlessly between member states. However, with Brexit, some of these supply chains have been disrupted. New safety and lengthy document checks required at borders have caused delays, and in 2021 resulted in some perishable agricultural products going to waste. In addition, Brexit has caused worker shortages and other issues that have raised food prices in the United Kingdom.

Another example of how politics impacts a country's supply chain is the trade war between China and the United States that started in 2018, discussed earlier in this chapter.

President Trump imposed increased tariffs and other trade barriers on China, and China retaliated with tariffs of its own. This affects both importing and exporting supply chains, as they have to adjust for shifts in costs, supply, and demand. For U.S. businesses, some may be able to continue to manufacture in China, even with the tariffs calculated into their costs. On the other hand, some will shift manufacturing to other nontariffed countries or may manufacture in the United States. Because of the extra taxes on U.S. exports, the U.S. agriculture export supply chains face the most risk. Many Chinese companies are not buying the tariffed products, which is resulting in a surplus for U.S. farmers and lower prices. China was predicted to fulfill just 60 percent of its trade deal commitments, including farm products, by the end of 2022, and though the Biden administration was concerned about negative impacts of the trade deal on U.S. agricultural and other economic sectors, it was largely bound to the 2020 agreement.

Patterns from the past Global supply chains can trace their roots back to European colonial and imperial networks between the 16th and 18th centuries. European cities became markets for exotic foods from every corner of the world. While building new colonies across the globe, European powers dictated which crops would be grown on their newly acquired lands, often growing agricultural products that had become popular in the markets back home. This resulted in patterns of monocropping. Monocropping was also a way that a controlling country could monopolize, or have exclusive control over, the dependent country's economy. Egypt, under British occupation, grew massive quantities of cotton; in West Africa, it was cacao; and in South Asia, it was tea. With the development of refrigeration and faster shipping, these relationships between producer and market deepened.

Some former colonies today are still economically tied to the country that once colonized the area. Often, the former colony (now usually a peripheral country) receives aid from a former colonizer (a core country) that is in a position to continue the same unequal trade relationships of the past that once benefited the colonizer. The peripheral country is economically dependent on the core country that uses its natural resources and labor to inexpensively produce an export commodity. When a colony achieved independence, it often did not have the economy or the workforce to compete in the global market. Consequently, it became reliant on its former colonial power for trade, financial aid, and other support—an example of neocolonialism.

Patterns of World Trade

Learning Objective
LO 13.13 Explain how consumer preference and fair trade influence global agricultural trade.

The amount of agricultural trade is both growing and changing every year. Trade in food alone has nearly doubled since 1995 (in real terms adjusted for inflation). This is largely a result of economic growth in peripheral and semi-peripheral countries that have become more engaged with global markets, resulting in greater demand for agricultural products globally. While core countries, including the United States and those making up the EU, remain leading agricultural exporters and importers, the increasing relevance of emerging economies such as Brazil, China, India, and Indonesia is a growing trend in global food trade. Population growth and income changes are directly related to the growth in agricultural trade. According to the FAO, China's share of world imports increased from 2.3 percent in 2000 to 8.2 percent in 2016 while the number of people living in poverty declined substantially—the poverty headcount ratio declined in China from 31.9 percent in 2002 to less than 1 percent by 2022.

Another area of growth for agricultural trade is between peripheral and semi-peripheral countries. The share of imports by peripheral and semi-peripheral countries from other peripheral and semi-peripheral countries increased from 41.9 percent in the year 2000 to 54.4 percent in 2015, due primarily to population growth in peripheral and semi-periphery countries. Exports followed a similar growth trend in peripheral countries.

Preferences As global agricultural trade increases, consumers gain access to a variety of foods. New foods are introduced to regions, and information about different foods spreads. This can influence food preferences and alter the patterns of production and consumption. The popularity of a food dictates increased importation of it. Recent trends, including food that is not only good for consumers but good for the environment as well, are changing what farmers plant and their agricultural practices around the world.

The rising interest in plant-based foods is creating new demands for the production of vegetable proteins. Types of vegetable proteins include soybeans, lentils, hemp seeds, pumpkin seeds, seitan (made from wheat), nuts, green peas, and lima, pinto, and other beans. These are used to make tofu and other alternatives to common meat products. Even though tofu has been produced for 2,000 years, its popularity has grown in recent decades due to its nutritional value, low cost, and availability from soybeans. As a result of the trend in some countries to eat less meat for health reasons and to lessen climate change, many farms are shifting to cultivating these agricultural products. In Montana, farmers who traditionally raised wheat are giving pulses a try. Pulses include dried beans, lentils, chickpeas, and cowpeas. In 2018 the state led the nation in pulse crop production. Food manufacturers are also investing in infrastructure needed to support emerging commodities like pulses and a variety of seeds and nuts. Modifications in crop production will continue to evolve as trends change.

Fair Trade The wages paid to laborers who produce commodity crops and other crops are often low. The palm oil industry in Indonesia and tea producers in India are just two instances where investigations have uncovered unfair wage practices on plantations of commodity crops.

The **fair trade** movement is a global campaign to fix unfair wage practices and protect the ability of farmers to earn a living. Fair trade is meant to improve the lives of farmers and workers in peripheral and semi-peripheral countries by providing more equitable working and trading conditions. The movement works to increase incomes paid directly to the farmers by paying an above-market fair price provided they meet certain standards and regulations. Fair trade also provides price guarantees that limit damages if farmers face devastating challenges due to environmental or social issues, and it helps farmers access global markets.

Early in the fair trade movement, countries talked about the idea of "Trade not aid." Together, they realized that establishing equitable trade relationships between core and peripheral countries would have a stronger impact than continuing to send aid money to countries with economic challenges. To ensure the success of the movement, a labeling system was created to raise consumer awareness. To place the fair trade label on a product, the producers must be small farms that engage in a democratically operated cooperative, and they must follow basic health, environmental, safety, labor, and human rights regulations.

Products are priced higher both as a result of this certification and because they are often higher quality or organic. Consumers pay more money for products that are fair trade certified, supporting the belief that the producers have a basic right to fair wages and living conditions. Retailers, such as Whole Foods Market, Target, Starbucks, and Caribou Coffee, specifically market fair trade products to those consumers, which is helping to expand the industry and increase access to European, Asian, and North American markets.

As of 2019, more than 1 million small-scale producers are part of the fair trade movement. Fair trade products are available everywhere, but in limited quantities. Products range from tea, cocoa, and sugar to wine, nuts, and spices. The most widespread and well-known fair trade product is coffee. Hundreds of thousands of coffee farmers have benefited from fair trade. But is the program doing enough?

Studies are measuring the actual impact of fair trade. Reports show that fair trade relationships are benefiting farmers. Cooperatives are improving land management practices, investing in better seedlings, and exploring advanced production technologies. Fair trade has also helped farmers and workers get out of debt and better assist the communities they live in economically. A community development premium, included in the price of the good, provides extra money to the farm cooperative to be used for community projects including school and infrastructure improvements.

However, there are reports that even though cooperatives are benefiting, the workers they hire may not be seeing the same level of aid, and poverty still exists in many areas. Some critics argue that fair trade practices artificially inflate market prices. They also argue that the focus on cooperatives ignores the small, individual farmers that may need the most assistance. Moreover, peripheral countries are seeing the emergence of corporate-controlled plantations and agribusinesses—many moving in to compete with small farming cooperatives, even right next door—creating unequal competition that will ultimately hurt those benefiting from the efforts of the movement.

Geographic Thinking

1. Explain the economic impact of commodity dependency on a region.

2. Explain how political relationships affect global food distribution.

3. Consider the supply and demand for a cash crop such as vanilla. Describe the impact the global supply chain has on supply and demand.

4. Describe how the fair trade movement affects the interdependence between core and peripheral countries.

Case Study

Coffee Production and Consumption

The Issue The coffee industry's global supply chain is based on a system of unequal power relations.

Learning Objective
LO 13.13 Explain how consumer preference and fair trade influence global agricultural trade.

By the Numbers

$0.40/LB
Farmers sell for

$0.60/LB
Brokers sell to processors

$0.70/LB
Processors sell to exporters

$0.85/LB
Exporters sell to importers

$1.05/LB
Importers sell for

$4–$6/LB
Distributors sell to retailers

$7–$12/LB
Retailers sell to consumers

Source: Food and Agriculture Organization

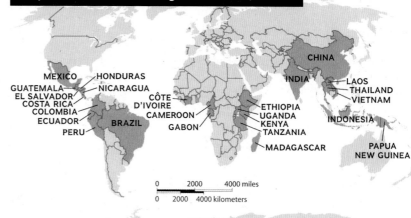

Top Coffee-Producing Countries, 2020

Brazil tops the list, producing 7.8 billion pounds of coffee in 2020 that was shipped to places around the world. *Source: https://elevencoffees.com/which-country-produces-the-most-coffee/*

Global supply chains have many benefits, but they are not shared equally among the people who produce and distribute agricultural products. In the case of coffee, the supply chain takes the crop from countries that have proper conditions for coffee-growing—namely, mountainous regions with humid, tropical climates—to core countries that can afford the highly popular bean. Countries where coffee is a major export tend to be peripheral and semi-peripheral countries that depend on the income that coffee generates.

The global distribution of coffee involves many people, starting with farmers, brokers who handle the transaction between farmers and processing plants, the processers at the plant, exporters, importers, distributors, and retailers. Each person adds value to the commodity, even though the coffee beans themselves undergo little, if any, physical change as they exchange hands.

Having access to the coffee supply chain increases the otherwise limited business opportunities of small farmers and helps them reach new customers. However, the people who actually produce the coffee beans for export often receive the smallest share of wealth that is generated by the global coffee trade. A typical example is a farmer selling coffee beans produced from the farm for $.40 a pound while the end consumer pays between $7 and $12 a pound. Each person playing a role in the supply chain receives a share of the total income, with the largest shares going to the companies who import and distribute the coffee. These are often large corporations that take advantage of economies of scale and vertical integration by controlling the processing, or roasting, of the coffee bean in the country of origin, as well as the distribution networks in the consuming country, including the packaging, branding, and marketing of the coffee to consumers at retail stores and coffee shops.

When conditions are right, coffee-producing farmers benefit, but the market for the bean is not stable. As a globally traded commodity, the price of coffee can change from minute to minute. Its price is determined by a host of variables such as changing weather, political uncertainty, transportation costs, and crop disease. All of these issues affect the profits of every person in the supply chain and sometimes result in a price hike for consumers. When farmers, who are at the start of the supply chain, receive the smallest share of the profits generated from the sale of coffee, any fall in market price can be devastating to their livelihoods.

Geographic Thinking

Explain the degree to which the production of coffee increases the interdependence between peripheral and semi-peripheral countries and core countries.

Chapter 13 Summary & Review

Chapter Summary

Economic forces drive agricultural practices and include:

- Cost of materials, land, and labor
- Availability of capital (money or other assets)
- Impacts of government policies
- Consumer preferences and market demands

Economies of scale explain how production costs are reduced as the quantity of production increases.

Agribusiness refers to the large-scale system that includes the production, processing, and distribution of agricultural products and equipment. While family farms still make up the majority of farms worldwide, large corporate-controlled agribusiness operations are replacing small family farms.

Technology, including modern equipment, improved fertilizers and pesticides, and new types of seeds, allow farmers to increase production and create higher yields.

High-yield crop growth can lead to lower commodity prices, which can have a negative impact on commodity crop growers. Government policies to protect pricing and the livelihood of farmers influence trade practices. These policies include:

- Farm subsidies, government payments to farmers and agribusinesses
- Tariffs, which affect pricing on imports and exports

While government incentives and price controls help farmers, it is ultimately consumers who drive the market in commercial agriculture.

The von Thünen model, based on certain assumptions, describes the ideal pattern for agricultural practices outside of a city center, with the production of perishable goods closest to the city, then wood, then grains and cereal crops, and finally livestock and ranching.

Agriculture is globally integrated and organized. Countries in the core, periphery, and semi-periphery depend on each other for agricultural production and consumption.

Global food distribution networks are affected by political relationships, infrastructure, and patterns of world trade.

- Agriculture is a part of the global supply chain, including the production and export of commodity crops to consumers in markets such as the United States, Europe, and China.
- Improved infrastructure and the fair trade movement are helping equalize trade among core and peripheral countries.

Review Questions

Use complete sentences to answer the questions.

1. **Apply Conceptual Vocabulary** Consider the term *hybrid*. Write a standard dictionary definition of it. Then provide a conceptual definition—an explanation of how it is used in the context of this chapter.

2. Provide an example of a dual agricultural economy.

3. Explain how hybrids are impacting agriculture.

4. How are the terms *cash crop* and *agribusiness* related?

5. Provide an example of infrastructure in terms of the global supply chain.

6. Provide an example of how farm subsidies might help a farmer.

7. Define the term *fair trade* in relation to the global supply chain.

8. Describe the von Thünen model. Give one example of a way in which the model could be used on a local level today.

9. Explain how tariffs affect commodity pricing.

10. Define the term *cooperative* in relation to a commodity chain.

11. Why would a commercial agriculture operation favor vertical integration?

12. Give an example of something that can increase economies of scale. Explain why.

Interpret Maps

Percentage of U.S. Agricultural Income Derived from Crops Grown for Animal Feed, 2012

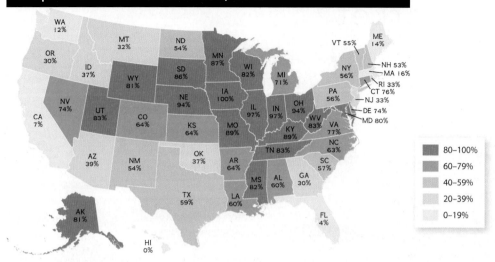

13. **Identify Data & Information** Which state earns the highest percentage of its agricultural income from crops grown primarily for animal consumption, and which state earns the lowest?

14. **Describe Spatial Relationships** Describe the connection among the states that earn more of their agricultural income from animal feed.

15. **Describe Patterns & Processes** Describe the pattern among states that earn less than 50 percent of their agricultural income from animal feed.

16. **Synthesize** Based on what you know about monocropping, do you think the dark green states or the light green states practice monocropping? Explain your answer.

Chapter 14
Agricultural Sustainability in a Global Market

Critical Viewing These giant Japanese scallops are farmed off Vancouver Island in Canada. They consume fish waste. ▮ How might fish farming like this be considered a form of sustainable agriculture?

Geographic Thinking Why is sustainable agriculture a goal for the future?

14.1
Consequences of Agricultural Practices

National Geographic Explorer
Hindou Oumarou Ibrahim

Case Study: Building Africa's Great Green Wall

14.2
Challenges of Contemporary Agriculture

Feature: Precision Agriculture

14.3
Feeding the World

Case Study: Food Deserts

National Geographic Explorer
Jennifer Burney

14.4
Women in Agriculture

14.1 Consequences of Agricultural Practices

Agricultural practices have profound effects on the environment, and they play an important role in shaping cultural practices. As food production expands to meet the needs and tastes of a growing global population, the environmental and societal impacts of agriculture are also increasing.

Altering the Environment

Learning Objective
LO 14.1 Explain the societal and environmental consequences of agricultural practices.

The agricultural practices you've learned about alter the land in different ways to create a variety of **agricultural landscapes**, or landscapes resulting from the interactions between farming activities and a location's natural environment. Some agricultural landscapes have endured for centuries, whereas others are constantly changing.

Shifting Cultivation As you have learned, the practice of shifting cultivation involves farming a piece of land until the soil becomes infertile and then leaving it or using it for a different purpose. This form of subsistence agriculture is predominantly practiced in peripheral and semi-peripheral countries in South America, Central and West Africa, and Southeast Asia. Shifting cultivation differs from crop rotation: instead of rotating crops on a regular basis to maintain soil fertility, farmers using shifting cultivation set aside fields or plots once the soil is no longer suitable for farming. The idea is to let the land recover before using it again, but shifting cultivation systems can fail if fields are not given enough time to recover. When the land does not have an adequate fallow (uncultivated) period, the result is soil degradation. In areas where shifting cultivation is practiced, the landscape becomes an ever-changing mosaic of planted crop fields, abandoned plots, and plots in various stages of regeneration.

Shifting cultivation is commonly practiced on a small scale by indigenous peoples. In northern Vietnam, for example, the people of Ban Tat simultaneously manage shifting fields on the hillsides, permanent rice paddies in the valleys, and wild resources in the forest. Shifting cultivation is one integral component within a complex **agroecosystem**—an ecosystem modified for agricultural use. Because of its diverse agriculture, the Ban Tat community can ensure a varied and sustainable food supply while preserving the ecological value of the landscape. Village committees ensure that the agroecosystem is effectively managed and cultivated. The argument can be made that if shifting cultivation is replaced by intensive commercial agriculture, such as palm oil or rubber plantations, the negative environmental impact is much greater.

Slash and Burn Slash-and-burn farming is considered to be a type of shifting agriculture. However, slash and burn often alters landscapes permanently, while fields under other types of shifting cultivation return regularly to cropland. People most often practice slash-and-burn agriculture in tropical wet climates where dense vegetation covers the land. As you learned, this method involves cutting and burning forests to create fields for crops. The burn removes weeds, disease, and pests, and the ash layer provides the newly cleared land with a nutrient-rich layer that helps fertilize crops. Within a few years, the nutrients are used up and the weeds return. Having lost fertility, the soil is no longer suitable for farming and, as a result, the field is abandoned and farmers move on to a new plot of land to repeat the process. Sometimes the forest regenerates on abandoned fields.

Slash-and-burn farmers generally live on marginal land in the tropical rain forests of Latin America, Africa, and Asia. Indigenous communities historically have used this technique to survive and still do so in the present day. As you learned in Chapter 12, slash and burn contributes to numerous environmental problems, including massive **deforestation** (loss of forest lands) and soil erosion—the wearing-away of topsoil by wind, rain, and other phenomena. On the other hand, some observers argue that about seven percent of worldwide agriculture is based on the slash-and-burn technique and that this method can be sustainable if it is practiced by small populations in large, forested areas and if the land is given adequate time to recover before the burn is repeated. In Thailand, for example, farmers in the agricultural village of Hin Lad Nai have practiced slash-and-burn cycles to plant dry upland rice for 400 years. Hin Lad Nai's commitment to proper rest times for the soil and deliberate crop rotation has resulted in resilient, sustainable farmland.

Terracing Terracing, which is commonly practiced by subsistence farmers, is the process of carving parts of a hill or mountainside into small, level growing plots. This method is used in mountainous areas in various climates, including tropical wet climates. Farmers can cultivate crops in these rugged regions by building "steps" or terraces into the steep slopes and creating paddies for cultivating water-intensive crops such as rice. During rainfall, the paddies flood and water flows through small channels from terrace to terrace without carrying soil down the slope. Preservation of the soil nutrients leads to the growth of healthy crops, and the terraces slow heavy rain flows so plants and soil are not

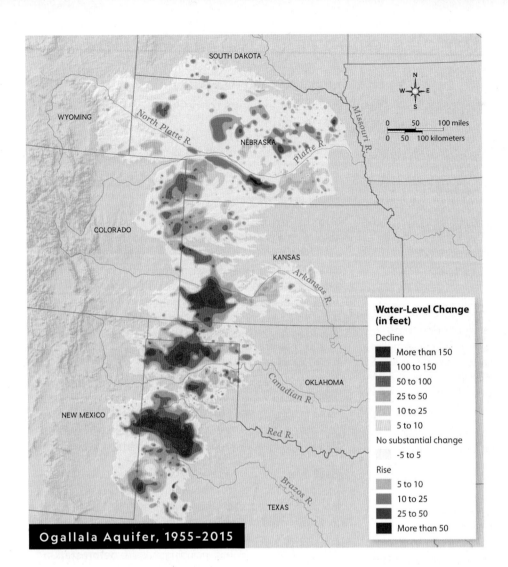

Reading Maps Thirty percent of all water used to irrigate U.S. agriculture is pumped from the Ogallala Aquifer. This map shows degrees of change in the aquifer's water level between 1955 and 2015. ▎Identify a small area on the map, circle it, and then write a sentence to explain what has changed in that area over time.

carried away. Through terracing, hilly or mountainous land that would otherwise be unusable becomes productive. Although terracing is labor intensive, it is often part of a cultural heritage and an undertaking that everyone in a community participates in to ensure that the entire system is well managed. Maintaining the terraces is critical in preventing dangerous runoff and mudslides. In the case of the Ifugao Rice Terraces in the Philippines, knowledge is passed down through generations to preserve this traditional and effective method of farming that is a vital part of the community's survival and culture.

Irrigation Humans have been using irrigation for millennia, most commonly in areas either with low or without dependable precipitation, to supplement rainfall by bringing water from its natural sources to farm fields through canals and other means. Irrigation can affect surface landscapes in many ways. Using irrigation, humans have transformed arid or semiarid landscapes into green fields. Especially striking are the landscapes created by center-pivot systems that use enormous sprinklers that rotate around a central point, creating expansive green circles in otherwise dry regions.

Irrigation supports both small subsistence farms and major commercial agriculture operations. **Reservoirs** are one common source of irrigation for crops in the United States. These artificial lakes are created by building dams across streams and rivers. Canals carry water from the reservoirs to the fields and orchards where it is needed. People have also rerouted natural water paths to aid irrigation.

The Aral Sea in Central Asia's Kazakhstan and Uzbekistan is an inland water body with no outlet to the ocean. It was fed by two rivers that were diverted in the 1960s to irrigate the desert regions surrounding the sea for cotton production. As a result, water levels dropped, salinity rose, and fish populations dwindled. As of 2019, the Aral Sea was one-tenth of its original size. The destruction of the Aral Sea provides a sobering lesson in human–environment interaction for the world: Unsustainable agricultural practices can irrevocably damage both the natural world and human livelihoods.

The Colorado River is one example of a river that has been dammed and is heavily used for irrigation. With 15 dams along its main stem and hundreds more on its tributaries, the Colorado is the American West's most vital water resource. It supplies water to 40 million people and 5.5 million acres of farmland across seven states and Mexico. To preserve this important water resource, 70 percent of which is used for agriculture, the United States Bureau of Reclamation, state representatives, Native American nations, and the Mexican government have developed and signed plans of action in case of drought affecting the Colorado River. Despite

 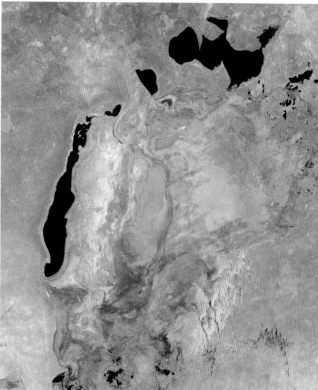

Critical Viewing Satellite images show the Aral Sea, which spans Kazakhstan and Uzbekistan, in 1977 (left) and 2019 (right) after its rivers were diverted for irrigation. The turquoise color in the 2019 image shows sediment deposits in shallow waters that result from seasonal rainfall. ▌Identify the changes in the 2019 photo and describe their possible societal and environmental consequences.

Over the past 60 years, the Aral Sea steadily dried up due to irrigation practices. These rusted ships used to be docked in the seaport of Mo'ynoq, Uzbekistan. Today, they are part of a "ship graveyard" in an area renamed the Aralkum Desert. The sea's drastic retreat obliterated the local fishing industry, but recent restoration efforts, including repairing dikes and constructing a dam, have partially revived it.

Agricultural Sustainability in a Global Market

these plans, the water supply from the river is threatened by increasing agricultural demands, and many researchers have expressed concern about the dams' negative effects on species that rely on that water. Climate change is also threatening the Colorado and other river systems across the western U.S. In late 2021, federal officials for the first time ever declared a water emergency on the Colorado River, forcing reductions in water to multiple states. Further conservation measures introduced in August 2022 signaled difficult choices for agriculture in the years to come.

Some water sources tapped for irrigation lie below the land's surface. Layers of underground sand, gravel, and rocks that contain and can release a usable amount of water are called **aquifers**. People drill into aquifers to access fresh water for both agriculture and household uses. If not recharged or replenished by drainage through the soil, groundwater levels in aquifers can fall or even become completely depleted. North America's largest aquifer, the Ogallala Aquifer beneath the Great Plains of the United States, formed millions of years ago and is fed by meltwater flowing from the Rocky Mountains. In recent decades, it has been overdrawn and is not receiving enough water to replenish itself. Farmers in the Great Plains, a region dominated by agriculture, are no longer able to pump enough water from the aquifer to sustain their crops, which are primarily wheat. Large aquifers in Asia, Africa, and Southwest Asia are facing similar declines.

China practices both groundwater irrigation and the moving of surface water on a massive scale. Chinese farmers use groundwater to irrigate crops with pumping systems reaching down more than 200 feet. One advantage of aquifer-provided water there is that the energy it requires is inexpensive and new technology makes it efficient. On the other hand, the pumps emit more than 30 million metric tons of carbon dioxide in one year—a damaging level of greenhouse gases. In April 2018, China launched its South-to-North Water Diversion Project, which is anticipated to move 44.8 billion cubic meters of fresh water annually from the Yangtze River to the northern provinces using three canal systems. Critics argue that improved water conservation practices by the agricultural sector—which accounts for 60 to 70 percent of the water taken from groundwater sources—would be far less expensive, far more effective, and of course, greatly reduce negative impacts on the Yangtze River ecosystem.

The United Nations (UN) estimates that within the next few decades, the world will need to increase food production by 60 percent to support its growing population. This demand will put pressure on aquifers and surface water sources. If a water crisis is to be avoided, therefore, farmers will need to better manage how much water they draw for irrigation. Today, some farmers are using technology to help them measure and gauge their water use, which allows them to slow the rate of water depletion. For example, vineyard owners along the Tiber River in Italy have begun to use drones and other sensors to monitor soil moisture, cutting their irrigated water use by up to 50 percent.

Draining Wetlands Areas of land that are covered by or saturated with water—such as swamps, marshes, and bogs—are called **wetlands**. Draining wetlands and converting them into farmland has historically been viewed as an acceptable practice because these areas were considered wastelands: the soil does not support construction or development, and wetlands provide a habitat for pests such as mosquitos. The Netherlands stands out among countries that have used what is termed "wetland reclamation" to increase land for farming and habitation. Around 17 percent of the country's present land area was once under the sea or coastal wetland.

Drained wetlands can be converted into agricultural land, but this results in a significant loss of habitat for fish, waterfowl, mammals, and other wildlife. Broader environmental issues have also resulted from the loss of wetlands, which if left intact help reduce storm and flood damage, improve water quality, and trap carbon dioxide that might otherwise be released into the atmosphere. For example, the conversion of wetlands into large-scale palm oil plantations in Malaysia is predicted to cause extreme flooding in the coming decades.

More than half the original wetlands in the United States have been lost, largely due to drainage for agricultural purposes. Recent efforts to prevent and reverse this loss include the U.S. Environmental Protection Agency (EPA) providing technical assistance to states and American Indian nations to monitor and restore wetlands. The U.S. Department of Agriculture (USDA) works with private landowners on agreements to protect wetland functions and wildlife habitat. The USDA notes that 75 percent of the country's remaining wetlands are on private and tribal lands. In Australia, the government is working with commercial farmers to adopt sustainable agricultural practices to protect wetlands, such as planting drought-resistant crops to make the most of lands already under cultivation and reusing wastewater so the amount of water needed from wetlands is reduced.

Pastoral Nomadism Like shifting cultivation, pastoral nomadism, also called nomadic herding, is an extensive practice and generally a form of subsistence agriculture. Pastoral nomads are herders who move their animals seasonally or as needed to allow the best grazing. Nomads and their herds range in dry climates in Southwest Asia and North Africa, and in other regions like the Arctic where crop cultivation is difficult or impossible. Traditionally, pastoral nomadism has been sustainable, including practices to preserve the resources upon which nomadic families depend, such as rotating grazing zones, limiting excess concentrations of people or animals, and protecting dry season resources. Pastoral lands have degraded in areas where these practices have collapsed due to war, nationalization of resources, agricultural expansion, population pressures, and societal change.

When disrupted or poorly executed, pastoral nomadism can have serious consequences for the landscape and the

The Tuareg, who live and roam in North Africa, are pastoral nomads. Life for the cattle-dependent Tuareg has changed dramatically in recent years as rainfall has declined during wet seasons, affecting their pasturelands. As one Tuareg nomad put it, "When the animals die, the Tuareg dies."

environment. Overgrazing can cause land degradation, which is long-term damage to the soil's ability to support life. **Desertification**—a form of land degradation that occurs when soil deteriorates to a desert-like condition—can be the result of poor pastoral nomadism or farming practices in arid or semiarid lands. Herds may favor certain plants over others, which can impact **biodiversity**—the variety of organisms living in a location—and reduce plant cover. When grazing is concentrated on mountain slopes or hillsides, soil erosion often results. Livestock compact the soil with their hooves, which leads to wind and water erosion.

As with other traditional agricultural practices that have been pressured by modernization and urbanization, pastoral nomadism has evolved and, in some places, is under threat. In Mongolia, nomadic families have migrated between high and low pastures with their horses, goats, and sheep for thousands of years. Though herders now rely on motorcycles as well as horses to get around, the migration patterns remain the same.

The Tuareg, a semi-nomadic people who live in the Sahara Desert in North Africa, numbered just over 2 million in the 2010s. Their economy has revolved around cattle and trans-Saharan trade for thousands of years. However, lack of adequate rainfall over the past few decades has made it difficult for nomadic Tuareg families to support their herds.

They depend on the animals for milk, clothing, tents, trade, and societal power. In the Sahel area on the southern fringe of the Sahara Desert, deep narrow wells have provided water for the Tuaregs' livestock, but access to the wells often leads to overgrazing, which increases the risk of soil erosion and desertification.

Geographic Thinking

1. Explain why agricultural practices and landscapes vary so widely across regions.

2. Identify a positive and a negative effect of irrigation.

Environmental Consequences

Learning Objectives
LO 14.2 Describe the negative environmental effects of agricultural practices.
LO 14.3 Define the terms *pollution, land cover change, soil salinization*, and *land degradation* in the context of agricultural practices.

All agricultural practices have an impact on the environment. When farmlands and water resources are overused, negative consequences often result. Effects of agriculture on the landscape and soil include pollution, land cover change, soil salinization, and land degradation such as desertification.

Agricultural Sustainability in a Global Market

Pollution Whether fields receive their water from rainfall or irrigation, water running off farmland has environmental consequences for habitats well outside the agroecosystem. Runoff from fields may contain chemicals and nutrients from pesticides and fertilizers, as well as bacteria and disease-carrying organisms. All of these can pollute and damage natural ecosystems. High concentrations of nitrates and phosphates from fertilizer, for example, promote uncontrolled plant growth and low oxygen levels in bodies of water, damaging habitats in rivers, lakes, ponds, and even oceans. The Gulf of Mexico "dead zone" is an area about the size of Massachusetts where most marine life has been killed by oxygen deprivation due to human-caused nutrient pollution from the massive Mississippi River watershed.

Land Cover Change Geographers refer to land cover change to describe how the surface of land is altered by different land uses—especially by the way humans use the land. In terms of agriculture, humans transform Earth's surface for the purpose of growing crops and grazing livestock. No matter for what purpose, the environmental consequences of land cover change can be difficult to remedy.

As you learned earlier, terraced farming creates a dramatic agricultural landscape on slopes that are too steep for other farming practices. A frequent problem associated with terracing is groundwater saturation, which hinders the land's ability to absorb more water during heavy rains. Massive labor is required to properly maintain terraces, and if they are allowed to deteriorate, soil erosion down the slope can be severe and even cause catastrophic mudslides during rainy seasons or strong storms.

Deforestation caused by slash-and-burn agriculture is also a land cover change. Along with the two other main drivers of global deforestation, ranching and logging, poor farming practices have been responsible for environmental damage in some areas, especially when farmers clear and replant a patch of forest land before it has had sufficient chance to recover. This tends to happen when farmers intensify their efforts as a result of increasing populations coupled with more competition for scarce land. As deforested areas expand, they diminish wildlife habitat, which over the long term can lead to endangered species or extinction. In the summer of 2019, timber cutting and slash-and-burn operations—along with unprecedented drought—caused record-breaking forest fires in Brazil's Amazon region. The situation proved to be at least as bad in succeeding years in Brazil, which contains 60 percent of the entire Amazon rain forest, with remote-sensing data indicating that an astounding 5,100 square miles of forest cover was lost between August 2020 and July 2021. The consequences of these fires may reach well beyond land cover change. The Amazon rain forest absorbs millions of tons of carbon emissions every year, and environmental scientists are convinced that intact rain forests are instrumental in helping to slow climate change.

Soil Salinization Irrigation, too, can lead to short-term and long-term environmental damage such as **salinization**, the process by which water-soluble salts build up in the soil. Salinization occurs in arid and semiarid regions when water evaporates from the ground more rapidly than it is replenished by rain or irrigation, causing a concentration of salts in the soil. When salts accumulate in the root zone of a crop, the plants can no longer extract adequate water, which can in turn result in crop yield reductions. Excessively saline irrigation water can contribute to the problem. Egypt, for example, has struggled with soil salinity issues for years because of extensive irrigation and the highly saline drainage water from the Nile Delta that is used to water the fields.

Desertification When water consumption significantly exceeds the rate at which it is replenished—think of the Aral Sea and the Ogallala Aquifer—the result can be desertification, which is a permanent form of land degradation. Poor pastoral nomadism practices such as

In 2019, tens of thousands of individual fires burned through the Amazon region in Brazil. The fires moved especially quickly through areas that had been deforested. This aerial view reveals the scale of the fires; even from a great height, the flames are evident.

overgrazing can contribute to desertification, but other human causes include overgrazing by stationary herds, deforestation, poor farming practices, and the clearing of land for expanding human habitation. The areas most vulnerable to desertification are those with low or variable rainfall. One-third of farmland in India is now affected by desertification. In 2019, the UN reported that desertification is occurring at 30 to 35 times the historical rate. Desertification not only harms food supplies but also may endanger the health of those who breathe the dust carried on the wind from dried-out soil.

Conservation and Sustainability Efforts

Learning Objectives
LO 14.4 Describe environmentally sustainable agricultural practices.
LO 14.5 Describe ways in which governments and other organizations are promoting conservation and sustainable practices.

Governments and organizations worldwide are addressing the negative environmental effects of agriculture through a variety of policies and sustainability efforts. Many focus on conservation, which involves managing and protecting natural resources to prevent their depletion. Conservation initiatives often use laws or education to encourage farmers to modify their practices. Given the diversity not only of agriculture types but also of ecosystems and political systems, these plans and policies vary tremendously from place to place. For example, the effort to conserve water has led the government of Zambia, in southern Africa, to charge fees on groundwater use. The idea is to increase awareness of water usage that extends beyond household consumption by attaching a cost to it. The money collected from water consumption for farming and industry will be used to implement solutions to the growing problem of water pollution in Zambia. As discussed, U.S. government efforts to preserve and restore wetlands involve education and technical assistance for farmers and ranchers.

Nongovernmental groups also help with these efforts. One example is EcoLogic, which partners with rural and indigenous peoples in Central America and Mexico to help communities trying to achieve sustainability, preserve natural resources, and restore ecosystems. EcoLogic's efforts include replacing slash-and-burn agriculture with alternative methods such as planting food crops alongside trees and diversifying crops.

Some efforts have met with mixed success. For instance, international lending agencies such as the World Bank have established **debt-for-nature swaps** with peripheral countries that borrow money. In exchange for local investment in conservation measures, the banks agree to forgive a portion of a country's debt. While some conservation agencies and debtor countries consider debt-for-nature swaps to be a useful strategy because they generate money for conservation efforts, critics say the benefits are overstated. The United States, for example, has not reauthorized funding for its debt-for-nature programs since 2014.

Commercial Farming In areas of the world where large, commercial agribusiness corporations dominate agricultural practices, conservation efforts set goals such as reducing air pollution from heavy machinery, encouraging better stewardship of water resources, and seeking to minimize the amount of toxins seeping into groundwater from chemical fertilizers and pesticides. The U.S. Department of Agriculture's Farm Service Agency (FSA) has an array of conservation programs ranging from setting aside land for threatened or endangered wildlife to water management practices and air quality initiatives. Often, these plans use financial incentives for farmers to incorporate more sustainable practices into farm operations. For example, the Conservation Reserve Program offers annual payments to farmers who are willing to avoid using environmentally sensitive land for agricultural purposes and to plant crops on their fields that improve the quality and health of the environment. The FSA also works with the National Rural Water Association to establish protection plans for local communities to prevent water pollution. These plans, which the communities help create, outline voluntary steps that ranchers and farmers can take to prevent water pollution.

In some parts of the world, small-scale commercial farming is also benefiting from targeted conservation efforts. For example, the Food and Agriculture Organization (FAO) of the UN works globally to help countries develop sustainable agricultural practices based on each country's unique opportunities and challenges. To preserve forests and biodiversity in and around the Amazon River, for example, the FAO has collaborated with other organizations to help agricultural communities understand how to manage the Amazon's ecosystem while protecting their livelihoods and respecting their cultural values. Also in the Amazon, some international companies have established fair trade relationships with indigenous communities that sustainably grow guayusa leaves, which are used to make a traditional tea that is gaining popularity as a healthy drink. The Amazon watershed provides food, water, and livelihood to 33 million people, including approximately 420 indigenous communities. Conservation efforts in the Amazon Basin and in other ecosystems across the globe are attempts to improve human–environment interaction and to build sustainable futures.

Subsistence Farming Policymakers in areas of the world where subsistence farming is common are recognizing the critical need to preserve soil fertility and prevent soil erosion while simultaneously increasing yields by intensifying land use. In areas that practice shifting cultivation, for example, farmers can replenish the soil and achieve higher yields by rotating fields systematically to include cover crops—plants that protect and nourish the soil. Other techniques exist to help subsistence farmers improve thin or depleted soils. In some cases, the solutions lie in intensifying inputs (such as fertilizers and pesticides) in land that is already productive while setting aside less fertile land for nature conservation rather than farming.

Many of the same sustainable practices applied to commercial farming can also be used in subsistence agriculture, but it is important to distinguish that while large-scale commercial farmers may be motivated by higher yields leading to increased profits, subsistence farmers are more focused on survival and generally have lesser financial means and access to information. Thus, they may need different types of support in terms of education and resources necessary for implementing new techniques.

In Bangladesh, the agricultural sector—which is largely composed of subsistence farms—has been instrumental in reducing poverty, partly through the assistance of the World Bank and other nongovernmental organizations. Bangladesh's agricultural system faces significant risks. Nearly half of its workforce is directly employed in agriculture, and the country is among the most vulnerable to climate change, with much rural land threatened by flooding, drought, or salinization. The World Bank has helped Bangladeshi farmers adapt to climate challenges by introducing soil management techniques and crops that tolerate drought and high levels of salt. The bank also promotes education initiatives and programs to help increase Bangladeshi farmers' access to markets.

In China, policies affecting pastoral nomads have changed as understanding of the herds' impact on the environment has evolved. At first, Chinese policies toward Tibetan nomads were based on the assumption that the herders were primarily responsible for the degradation of China's grasslands. In recent years, these policies have begun to change as scientists have acknowledged that the grasslands of Tibet actually benefit from moderate and intermittent grazing of herds. Past governmental policies have included the limitation of herd sizes, closing pastures to convert them to croplands, and forcing nomads to live in settlement homes. These and similar policies still in force ignore the role Tibetan nomads have had in sustaining regional wildlife, ecosystems, and water resources—a role recognized and honored by the UNESCO World Heritage Committee as well as the International Union for the Conservation of Nature.

Some private organizations have also been active in promoting the health of subsistence farms through resource conservation and other measures. The Bill and Melinda Gates Foundation, for example, founded the Alliance for a Green Revolution in Africa (AGRA) in partnership with the Rockefeller Foundation to address some of the continent's challenges. AGRA has invested funds and supported projects to help African farmers improve their farm yields. The organization also advocates for policies at the national level that will benefit smallholder farmers. Its primary purpose is to capitalize on the production of crops in areas with reliable rainfall, decent soil quality, and a dependable infrastructure. AGRA then works to apply successful solutions elsewhere on the continent. As Africa faces the twin challenges of climate change and rapidly growing populations south of the Sahara, the ability to greatly improve crop yields on subsistence farms may be key to providing sufficient food for future generations.

Critical Viewing In Burkina Faso, in Africa, AGRA has partnered with a local organization to develop seeds that are designed to produce higher crop yields. This worker is collecting vegetables grown in fields outside the city of Bobo-Dioulasso. ▌Identify an environmental consequence of introducing new types of seeds to subsistence farms.

National Geographic Explorer **Hindou Oumarou Ibrahim**

Mapping Indigenous Climate Knowledge

In the desert region of Chad, 250,000 nomadic Mbororo depend on herding (shown above) and subsistence farming to survive. Hindou Oumarou Ibrahim has used the Mbororos' intimate knowledge of their land to help advocate for the rights of the environment.

Learning Objective
LO 14.5 Describe ways in which governments and other organizations are promoting conservation and sustainable practices.

Hindou Oumarou Ibrahim's quest to give indigenous people a voice in environmental activism began at an early age. When she was a child, she spent part of her year in the Mbororo nomadic community where her parents grew up and part of the year living in N'Djamena, the capital of Chad, where she went to school.

Chad is a country in the Sahel, a semiarid region of Africa that extends from Senegal eastward to Sudan and Eritrea. Like the rest of the world, the Sahel is experiencing the effects of climate change.

In collaboration with local elders, herders, UNESCO, the government of Chad, and other organizations, Ibrahim has conducted a 3D mapping project designed to help indigenous communities adapt to climate and weather changes. 3D maps profile objects in three dimensions to show how they appear in the real world. Ibrahim's 3D maps of Chad's desert region incorporate indigenous knowledge about the land and its natural resources. They document how climate change affects the landscape, seasons, weather, and flora and fauna, and they are continually updated.

When asked how the data gathered in 3D mapping helped empower her community while at the same time promoting environmental protection, Ibrahim explained that the mapping helps guide the community in making decisions about when to move to a new location, what is the best time to collect certain food, and which species are dwindling and in need of protection. It also gives the Mbororo community tools to make informed decisions about participating in projects that will affect their natural resources.

Geographic Thinking

Describe some possible environmental consequences of the Mbororo community's use of Ibrahim's 3D maps.

Agricultural Sustainability in a Global Market

Societal Consequences

Learning Objective

LO 14.6 Describe the societal consequences of agricultural practices across different regions and agriculture types.

Societal consequences of agricultural practices are broad and varied, affecting diets, the roles of women in farming, the economic purpose of both farmers and consumers, and the lives of communities.

Consumers in many countries have altered their diet and lifestyle choices in reaction to recent innovations in agriculture. These individuals are concerned about the potential environmental harm of crops that require large amounts of inputs such as fertilizers and pesticides. Some people, particularly in Europe, also worry that crops consisting of genetically modified organisms (GMOs) could carry as-yet-undiscovered health risks.

These consumers are purchasing organically grown foods they believe are better for the environment and for their own bodies. Organic farming centers around an agricultural system that refrains from using artificial chemical inputs, and it has received much attention from movements toward sustainable food production. You will learn more about organic farming and other food choice movements later in the chapter.

In many countries, however, longstanding agricultural practices and traditions have profoundly influenced both diets and social customs, and consumers strongly resist change. For example, beef production has been part of Argentina's agricultural history and food traditions since the 16th century. Argentina has long ranked as a country with one of the highest levels of beef consumption in the world; generations have grown up on farms and ranches and have participated in barbecues called "grills."

In the face of an economic recession in 2019, Argentines attempted to cut back on their weekly beef consumption, but because this food choice is deeply rooted in tradition and social gatherings, beef sales persisted. The same holds true even as new information emerges about possible links between the consumption of red meat and the risk of heart disease and certain types of cancer and concerns about agrochemicals and antibiotics potentially impacting the safety of beef. Argentines consider the risk but continue to make beef a part of their regular diets because it is central to their culture.

The roles of women in farming vary tremendously across regions and agriculture types. Women who are pastoral nomads, for example, share responsibility for the care of animals with men, but they are more likely to be in charge of dairy animals or animals such as poultry that are kept near the home. In aquaculture, women work both as entrepreneurs and as laborers. In some parts of Southeast Asia, women form the majority of the aquaculture workforce. In most countries, women in agriculture face obstacles due to gender discrimination. With changes in both agricultural practices and available opportunities, the roles of women in many types of farming are changing rapidly.

Since the time of the Industrial Revolution, the changing economic purpose of many farms has also had societal consequences. Chapter 13 described agribusiness and the shift from small family-owned farms to larger corporate operations in the United States and elsewhere—principally in core countries. In the United States, these changes have meant a loss of small and midsize farms, which were once the backbone of American agriculture. At present, large-scale commercial farming dominates most American farmland, and farm families often sell their land to agribusiness corporations when they encounter financial struggles. It has been noted that the loss of small farms can harm the social and economic fabric of rural communities. Family farms tend to employ more workers per acre of land, and when these farms go out of business or are sold to large commercial firms, rural towns lose population. As a result, nonfarm businesses in these communities are often forced to close. Critics of agribusiness point to both the financial consequences and the loss of community in rural towns when small and midsize farms close down.

The rise of monocropping has also had societal consequences in the United States. As you learned, monocropping can be very profitable because it is an efficient way to grow corn, soybeans, and other crops. At the same time, however, it poses risks to the livelihoods of both farmers and consumers. The lack of diversity in crops can cause prices to be turbulent because any single disrupter such as disease or natural disaster can have a major impact on the entire system. In the face of a drought, for example, the prices of corn and all the products made from corn are driven up due to shortages.

In some cases, it is consumers who make their choices with an economic purpose. These choices are related to the socially conscious diet changes mentioned earlier, and they often have benefits for small and midsize farmers. Consumers in core countries concerned about avoiding GMOs, eating organic, or supporting small, local farms may choose to shop at farmers' markets or eat at restaurants that buy local organic produce. These diet choices encourage small farmers to consider switching to organic production. Consumers of fair trade products, which you also read about in Chapter 13, hope to benefit farmers and workers in peripheral countries through their purchasing choices.

Geographic Thinking

1. Identify possible negative land cover changes associated with terrace farming.

2. Explain factors that may cause a country to have strict sustainability policies. Give examples.

3. Describe a societal consequence resulting from changes in farm ownership structures.

Case Study

Building Africa's Great Green Wall

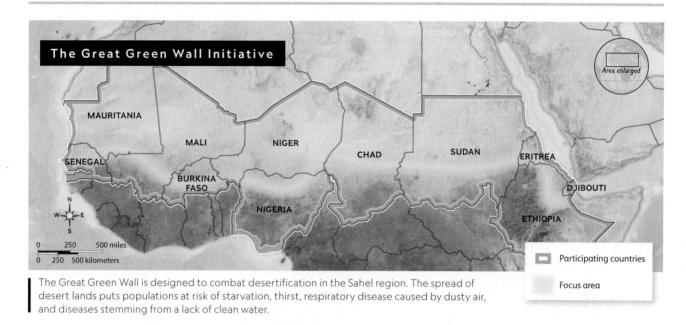

The Great Green Wall is designed to combat desertification in the Sahel region. The spread of desert lands puts populations at risk of starvation, thirst, respiratory disease caused by dusty air, and diseases stemming from a lack of clean water.

The Issue Desertification due to overgrazing and other causes combined with climate change means the Sahara is encroaching deeper into the adjacent farmlands of Africa.

Learning Objective
LO 14.6 Describe the societal consequences of agricultural practices across different regions and agriculture types.

By the Numbers

15%
of the Great Green Wall completed as of 2020

4,350 miles
of tree belt when the wall is completed

Over 70 million
acres of land restored by 2018

Sources: greatgreenwall.org, Smithsonian.org, Landscape News

Worldwide, Desertification Is Accelerating According to the European Commission's Joint Research Centre, every year an area half the size of the European Union is degraded globally, and Africa is particularly affected. In Africa's Sahel region, population growth has led to increases in wood harvesting, illegal farming, and land clearing for housing, all of which help drive desertification. Droughts exacerbated by climate change also play a major role.

To restore degraded land, national and regional leaders in Africa launched the Great Green Wall initiative in 2007. The goal was to plant a barrier of trees across Africa along the edge of the Sahara to prevent creeping desertification. Indigenous land-use techniques have been at the core of the initiative. For example, farmers in Burkina Faso use grids of deep planting pits that help retain water during dry periods, while in Niger, farmers take steps to protect trees that grow up naturally among their fields. As a result, the Great Green Wall has been described as a mosaic of land practices that will ultimately act in a similar way to a physical wall.

The project has many facets in addition to trees and is successfully addressing environmental and societal issues at both local and global levels. Planting the forest may create millions of jobs, which will help fight poverty. The improved soil quality produces a higher crop yield and directly impacts communities' food security. In addition to planting and regeneration, the Green Wall initiative incorporates the establishment of sustainable agricultural and energy practices, climate change reduction and adaptation, and the preservation of biodiversity.

Scientists were skeptical at first, but the success of the Great Green Wall continues to inspire and compel change across the continent. Smallholder farmers are now viewed as part of the solution to land degradation, simple and affordable interventions have proven that change is possible, and food security is increasing. As the wall and sustainability efforts continue to progress, each patch of trees connects to the next, forming a green chain and extending the reach of sustainability.

Geographic Thinking

Explain why the Great Green Wall initiative extends beyond planting trees.

Agricultural Sustainability in a Global Market

14.2 Challenges of Contemporary Agriculture

Contemporary agriculture and food production practices are constantly changing in response to new technologies, consumers' food choices, and the challenges of feeding growing populations. But much debate surrounds farming innovations that may have negative or unknown consequences for human health and the environment.

Debates over Innovations

Learning Objective
LO 14.7 Compare the benefits and potential risks of recent innovations in agriculture practices as they pertain to expanding food production and sustainability.

Biotechnology, GMOs, and aquaculture are among the techniques at the forefront of efforts to expand food production. These innovations have benefits that include better quality and higher production. At the same time, they remain at the center of political and scientific controversy, largely due to their unknown future consequences to both humans and the environment.

One key concern is sustainability. Sustainable agriculture encompasses environmental, economic, and social practices designed for the long term. Farmers must manage the environment in a way that minimizes degradation of the air, soil, and water quality, in order to ensure continued productivity well into the future. At the same time, agriculture must also be economically viable—farmers who sell what they grow or raise need to make a profit to stay in business. A healthy farm economy, in turn, can help maintain a good quality of life for farmers, farm families, and farming communities, as well as supporting fair and reliable incomes and healthy working conditions for farm laborers. Supporters of recent innovations in agriculture believe they represent progress toward sustainability, while critics fear that some of the latest technologies and practices actually create less sustainable systems. As the debate continues, it's clear that a balance needs to be reached between environmentally sustainable farming and the development of agricultural innovations to help feed a global population that will likely exceed 10 billion people by the end of the century.

Biotechnology and GMOs As you have read, a genetically modified organism is a living organism with a genetic code that has been bioengineered to produce certain desirable qualities. The use of genetic modification (GM) in agriculture is not a new concept; farmers have been improving plants and animals by selecting and breeding optimal characteristics for thousands of years. High-yield seeds introduced during the Green Revolution in the mid-20th century, for example, were the result of this type of breeding. **Biotechnology** is the science of altering living organisms, often through genetic manipulation, to create new products for specific purposes, such as crops that resist certain pests. GM is a broader category that

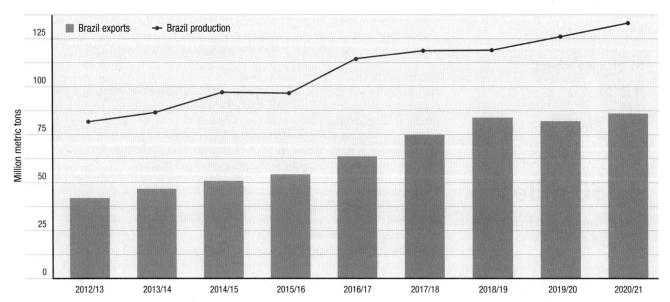

Brazil Soybean Production and Exports Brazil's position as a leading soybean producer and exporter demonstrates how it has benefited from recent technological advances: 97 percent of its soybeans are cultivated using biotechnology.

Source: https://ipad.fas.usda.gov/cropexplorer/pecad_stories.aspx?regionid=br&ftype=prodbriefs

394 Chapter 14

Global Crop Diversity, 2016

Number of Taxa
- More than 72
- 61–72
- 49–60
- 37–48
- 25–36
- 13–24
- 1–12
- 0

Reading Maps One key aspect of agricultural biodiversity is crop diversity. A 2016 study collected data on 81 important crops and about 1,100 species of their wild relatives. The map shows how many taxa—which are groups of similar organisms, in this case, the crops' relatives—were found in each region. ▍ Identify the richest global areas of crop diversity.

---

includes all types of genetic manipulation of foods. Modern biotechnology supports genetic engineering (GE), a type of GM in which scientists transfer specific genes from one organism to another.

Supporters argue that the unfolding revolution in biotechnology will be needed to solve world hunger problems. Biotechnology can result in improvements such as increased crop yields; resistance to drought, disease, and pests; and improved nutritional values. But these practices raise the question: How much modification is too much?

Brazil's agricultural sector can be considered one example of a biotechnology success story. With soybeans, rice, and corn leading its list of important crops, Brazil benefited from the Green Revolution of the mid-20th century and more recently has been at the forefront of biotechnology applications. Brazil is a world agricultural leader. Since 1961, its grain production has increased by 574 percent, while its population has increased 175 percent. More significantly, high-yield seeds have allowed Brazil to nearly triple its grain production since the early 1980s even as the amount of land under cultivation for grains has remained largely unchanged. Brazil has benefited economically as it has grown to become one of the world's largest food exporters. The country is now the second largest producer of biotech crops.

One potential global benefit of GE is a reduction in the cost of food production, which could lead to an increased supply of food—making food more affordable at worldwide, national, regional, and local scales. One 2010 study by University of Iowa scientists concluded that without biotechnology, the prices of corn, soybeans, and canola (a type of seed grown for cooking oil) would all be higher, as would the prices of many products made from these crops.

On the other hand, those who question the advantages of biotechnology feel that its environmental impacts have not been thoroughly investigated. Additionally, the long-term consequences of genetic alteration are still unknown. Foods modified through GE have been banned in much of Europe because their safety has not been proven, and there are concerns about their effects on the species that consume them, including humans. In 2022, U.S. consumers began to see new labeling for food products that have been genetically modified in a way not possible through natural growth. In an attempt by the U.S. Department of Agriculture (USDA) to standardize a patchwork of state labeling requirements, food products began to carry "bioengineered" or "derived from bioengineering" on their packaging. While some welcomed the new standard as a way to keep consumers better informed, critics complain that the wording allowed producers to more easily conceal genetically modified ingredients from their customers. The debate over biotechnology, GMOs, and GE continues.

Another debate on contemporary agriculture centers around biodiversity, with experts disagreeing over whether biotechnology decreases or increases agriculture's impact on the diversity of species. The term **agricultural biodiversity** describes the variety and variability of plants, animals, and microorganisms that are used directly or indirectly for food and agriculture. Agricultural biodiversity is integral to environmentally sustainable agriculture. It plays an important role in enabling agriculture to achieve productivity gains, improve sustainability, and manage changing conditions such as climate change. Critics claim that GE poses a threat to agricultural biodiversity. For example, a common type of GE is the insertion of bacterial genes into a crop. The bacteria functions like an insecticide: when insects consume the crop, they are infected by the bacteria and die. A new plant is introduced into the field ecosystem, but a species of insect is eliminated from that space.

Critics also point out that innovations intended to minimize chemical inputs may actually lead to intensified uses of fertilizers, herbicides, and other pesticides. For instance, one common type of GE involves inserting herbicide-resistant genes into crops. When herbicides are sprayed on these crops, the weeds die, but the crops survive because of the resistant genes. Weeds that are constantly exposed to these inputs can develop their own genetic resistance, meaning either the genetically modified crops eventually become obsolete, or more chemicals are needed to eliminate the weeds. Another risk is that genetically modified crops may transfer genetic material into unmodified plants. An herbicide-resistant crop may transfer traits to a weed that makes it herbicide resistant also, creating a new problem that again requires more herbicides to treat. GMOs that escape the fields where they are planted may also threaten biodiversity in the wider ecosystem by becoming invasive and crowding out wild species.

Soil fertility, too, can be a concern. In general, soil fertility has declined with the intensification of food production, which means that land's productivity is lessened or threatened. This leads to farmers applying more synthetic fertilizers to keep up with the growing demand for key food crops. Between 30 and 50 percent of agricultural crops are grown with the use of fertilizers, and more than 50 percent of people consume crops that are grown using synthetic fertilizers containing chemicals and minerals such as nitrogen, phosphorous, sulfur, and magnesium—all of which impact the ecosystem. Synthetic fertilizers build up in the soil over time and deplete its organic material, decreasing the soil's fertility and affecting surrounding organisms and their natural life cycles. Rain and sewage can carry fertilizers into bodies of water, which can create a toxic environment by increasing the growth of algae and decreasing oxygen levels. As you have learned, aquatic and marine animals struggle or die in these damaged environments.

Supporters and opponents also disagree on the use of GE crops to conserve water. Proponents of GE crops argue that they conserve water because herbicide-tolerant biotech crops don't require tilling, or breaking up the soil, which can lessen its moisture content. Such crops have been successful in Brazil for more than two decades. Another argument is that biotech seeds are drought tolerant and water efficient and produce higher yields. Such seeds are being tested in several countries in Africa, including Kenya and Mozambique, where drought is a concern.

Arguments against using biotech crops to solve water usage problems are similar to those raised in the case of herbicide-resistant crops: concerns about possible unintended biological effects on local plant communities and the danger of plant species expanding into areas where they are not wanted. Potential risks include the development of undesirable traits in the GE plants, such as invasiveness or weedlike attributes. Some scientists also argue that plants that are modified to more efficiently extract water actually still require the same amount of water to grow. Others believe there have not been enough comprehensive studies of biotech crops that include detailed attention to their effects on plants and animals in nearby ecosystems.

Economically, the monetary costs of using new agricultural technologies result in many farmers taking on tremendous debt loads to purchase fertilizers, high-yield seed varieties, and machinery. If crop prices fall on global markets, this debt can be crippling. Overall, biotechnology is expensive, and its use is often limited to agribusinesses that can afford to invest in it. At the same time, governmental policies, especially in peripheral countries, are aimed at keeping food prices low and affordable. This reduces incentives for commercial farmers to produce food, and food crops such as corn or millet are often replaced with cash crops such as cotton or tobacco.

Aquaculture Aquaculture is another innovative branch of agriculture that has demonstrated advantages on several fronts. This type of fish farming is less space and care intensive than other types of agriculture and represents one of the fastest growing food production sectors in the world. Aquaculture, an alternative to natural fisheries, can provide enormous and consistent amounts of fish and seafood—supplementing wild harvests and increasing the global food supply. Artificially raised salmon and oyster stocks have helped meet current seafood demands, and aquaculture has provided thousands of jobs. Outside of food production, algae fuel is being developed that could potentially replace fossil fuels such as gasoline and reduce energy costs.

Aside from these economic benefits, aquaculture offers environmental advantages as well. For example, the farming of shellfish can improve water quality because the shellfish filter water as part of their feeding process. At the same time, shellfish farms decrease pressure on wild stocks of the same species.

Aquaculture, however, is not without its disadvantages. Concerns include water pollution from chemicals used in fish farming and excess nutrients such as fish waste. Aquaculture farmers use antibiotics to prevent disease among farmed fish, and these antibiotics can have a negative impact on the ecosystem. Another possible consequence involves the compromise of native gene pools if farmed fish and native fish interbreed. Those who depend on wild fisheries also argue that aquaculture amplifies and transfers disease and parasites to wild fish populations.

These field workers in China's Hebei Province are using drones for farm work. Drones have numerous applications for precision agriculture, including gathering detailed data about fields.

Precision Agriculture

Learning Objective
LO 14.7 Compare the benefits and potential risks of recent innovations in agriculture practices as they pertain to expanding food production and sustainability.

Precision agriculture—also known as precision ag or precision farming—is part of the movement that some see as a fourth agricultural revolution. It uses a variety of cutting-edge technologies to apply inputs such as water and fertilizer with pinpoint accuracy to specific parts of fields in order to maximize crop yields, reduce waste, and preserve the environment.

By using drones or other remote-sensing technologies to acquire data, farmers can employ geographic information system software to map their fields and develop a micro-level analysis of each field's physical characteristics. This helps them target their watering, fertilizing, and herbicide and pesticide application strategies. Remote-sensing devices can perform a multitude of tasks such as managing irrigation, detecting disease early enough for a grower to intervene, and estimating crop yields. These computer-based applications result in reduced expenses and higher yields. They also support environmentally friendly improvements such as water conservation and reductions in the amount of fertilizer applied to a field. Using precision agriculture can reap similar benefits for both the small-scale vegetable farmer and the commercial grain farmer managing multiple fields and huge acreages.

Global positioning system guidance, robotics, and information technology are the tools of precision agriculture. Million-dollar combines—farm machines that both harvest and process crops—have control stations that tractor drivers use to set coordinates and avoid costly overlaps on fields. In 2022, agriculture equipment maker John Deere introduced robotic tractors to automate tilling, which is the turning-over of soil to access nutrients from the remains of last year's harvested crops. Tilling was deemed the easiest task among many farmers to automate, with more difficult actions like operating through standing crops to come. A farmer controls the giant tractor from afar with a smartphone.

By deploying these tools, today's farmers are acting as scientists, geographers, and marketing specialists in addition to food producers. Most farmers are not software engineers, however, and precision farming does pose technological as well as financial challenges. Software systems can have poor user interfaces or produce information that is hard for farmers to interpret. Systems can also overproduce data, forcing farmers to sort through intimidating quantities of information to make decisions. Farmers may be compelled to hire consulting firms to take the data and help them turn it into a plan of action. Some precision agriculture methods are affordable for large agribusiness operations but too costly for family farmers, and other technologies involving seed and fertilizers have experienced slow adoption rates due to complexities such as weather variables. Nevertheless, precision agriculture has opened new possibilities for farmers to maximize yields while minimizing costs and environmental impact.

Geographic Thinking

Describe the benefits and drawbacks presented by precision agriculture to both large and small farmers.

Food Choices

Learning Objective
LO 14.8 Explain the impact of changing diets on agricultural practices and production.

As you have learned, food preferences and dietary shifts are important motivators of changes in contemporary agriculture and food production practices. Individuals' food choices influence general patterns of consumption. In turn, food producers choose crops and methods to meet consumer demands. The goals of food choice movements include eating healthier foods, encouraging sustainable farming practices, and supporting independent farmers. Participants in food choice movements are enthusiastic about their benefits, but debate exists over whether they are effective in bringing about large-scale changes.

Local Food Movements In the United States, some consumers are employing new and different ways to acquire fresh foods for their tables. Urban farming, for example, converts vacant lots, rooftops, or abandoned buildings into spaces to grow produce. In many inner-city areas of the United States, communities are banding together to collectively work these small plots as a way to provide fresh fruits and vegetables in areas where such foods are scarce.

Borough Market in London has been in existence since the year 1014. Then, as now, it sold local vegetables and fruits; today it is a bustling marketplace of foods from all over the world. In the early 1990s, Borough Market became a destination for lovers of artisanal, or handmade, foods. Since then, demand for specialty foods has expanded the international market for fair trade and value-added crops.

Consumers who participate in community-supported agriculture (CSA) purchase shares in the output of a local farm. During the growing season, they receive a weekly box or basket of freshly harvested produce from the farm. In return, farmers who operate CSAs receive a guaranteed income from the sale of shares. In addition to providing fresh produce, urban farms and CSAs strengthen the relationships between those who grow the food and those who eat it, and they allow the public to be more aware of where their food comes from. However, the impact of these food choice movements is limited—not all CSAs are profitable, and urban farms typically do not reach large populations.

Similar to CSAs, local food movements aim to connect food producers and food consumers in the same geographic region. Both farmers markets' and farm-to-table restaurants, for example, provide connections between producers and consumers. Shoppers choose to buy locally to support the local economy, to have access to fresh food, and to know where the food on their table is coming from.

Organic Farming As you've learned, organic farming has seen a rise in popularity in some areas, in part as a response to concerns about chemical inputs and GMOs. Organic farming practices are more expensive than traditional farming, but wealthier consumers in the United States and elsewhere have shown they are willing to pay higher prices for organic food. Instead of using chemicals or GE, organic farmers use natural fertilizers such as plant-based products or animal manure to promote long-term soil health and prevent harmful runoff and water contamination. They also use crop rotation to manage weeds, insects, and diseases, and they focus on maintaining biodiversity within their agricultural systems and the surrounding environment. Organic farmers attempt to reduce or eliminate external agricultural inputs and strive for sustainability. Europe has the largest percentage of land given over to organic practices, and land percentages dedicated to organic farming will continue to grow in food-secure core countries in North America and Europe, where there is a demonstrated demand for organic food.

Not everyone agrees that avoiding GMOs by buying organic or through other means is a healthier option. Many scientists insist that genetically modified foods pose no danger to consumers and are essentially the same as non-GMO foods in terms of nutrition and health benefits. Whether or not organic foods are more healthful, their higher price means that the demand for them will likely remain among a smaller but relatively influential share of consumers.

Fair Trade and Value-Added Crops Another type of production driven by consumer choice is fair trade, described in Chapter 13. In the interest of supporting agricultural sustainability and a better quality of life for growers, many consumers—including Americans in the millennial generation—are willing to switch brands and pay more money for their food products.

Food choice often extends to value-added specialty crops—organic or other specialty crops that are transformed from their original state to a more valuable state, such as converting milk into cheese and yogurt. Coffee is another popular value-added crop. You have read about the global coffee supply chain, in which the beans are roasted and converted into ground coffee for home consumption or brewed coffee sold in cafes and restaurants. Other high-value crops in this category include tea and chocolate.

Consumer demand for value-added products can be driven by the desire to eat healthy and nutritious food, the need for convenience, or both. For example, many consumers buy yogurt because they enjoy the taste and believe it has health benefits. These same consumers would not consider taking the time and trouble to make yogurt at home, even if they found that making their own yogurt in bulk quantities might cost less.

As food producers have found ways to become more productive and discovered technological advances that help them efficiently produce consumer-specific products, the possibilities of value-added agriculture have increased, each with pros and cons. Being able to meet specific consumer demands can lead to producers capturing a larger share of the food dollar, but that comes at a cost. While consumer demand may entice farmers into producing value-added crops, farmers must also consider the capital required as well as the necessary production and business skills that may be different from those required to produce traditional crops. They have to become adept at marketing to the end user in areas such as packaging and variety and be willing to invest in innovation and research while continuously working to minimize costs.

Dietary Shifts Broader global trends in diets may exert the strongest influence on agriculture related to food choice. The global demand for all meats is growing, for example, and so the need for grains to feed the livestock is expanding. According to the UN, meat production in recent years was almost five times higher than in the early 1960s. In Southeast Asia, countries such as the Philippines and Malaysia have experienced increases in income and urbanization that have resulted in an increased demand for diversified diets focused on higher levels of meat and dairy consumption. According to the World Health Organization, there is a direct link between level of income and the consumption of animal protein.

The influence of higher income and urbanization on dietary choices is obvious in countries like China and Brazil, both of which have experienced large rises in meat consumption in recent decades. China has become the world's largest agricultural importer, and its consumption of meat, dairy, and processed foods has increased dramatically while its grain consumption has fallen. In the United States, meanwhile, the total per capita consumption of meat has grown over a period of five decades, but the consumption of beef has declined by approximately one-third, whereas chicken consumption has doubled. These U.S. data points matter because different types of meat production have different costs in terms of the land needed to grow feed crops. It takes more feed to produce beef than chicken or pork. Precise statistics for beef are problematic, however, because the entire animal is not used for food. That said, about five pounds of feed produces one pound of beef. In contrast, one pound of pork takes less than four pounds of feed, and one pound of chicken takes less than three pounds of feed. These numbers are reflected in the amount of farmland dedicated to growing grains for animals instead of humans.

The growing international appetite for processed food products also impacts growers' choices of crops. For example, around five percent of the U.S. corn harvest is used to produce sweeteners including high-fructose corn syrup, which is used to sweeten a wide variety of processed foods. The demand for foods that are easy and quick to use, are packaged well, are nutritious, and that offer a variety of choices is often driven by busy lifestyles and aging populations. In India and other semi-periphery countries, an increased number of women in the workforce combined with a rise in disposable income has spurred the demand for packaged foods among middle-income families.

Agriculture and Diet

Learning Objective
LO 14.8 Explain the impact of changing diets on agricultural practices and production.

Food choices affect agriculture, but the reverse is also true. Contemporary agricultural practices have transformed diets around the world, with agricultural improvements resulting in the possibility of consumers having access to more and varied foods. Chapter 12 described the diffusion of foods, a process that began even before the Columbian Exchange and continues at a rapid pace today, thanks to globalization and modern technologies. You've also learned about global supply chains. Crops such as corn and sweet potatoes, which are indigenous to the Americas, can be exported across the globe thanks to expanded transportation methods and advanced storage systems. Refrigerated transport has allowed farmers to ship perishable food over long distances, overcoming the limitations of what can be grown in the local climate and terrain. As a result, shoppers in the United States can choose from a year-round variety of fresh produce that was unthinkable for earlier generations. For example, Chile, located in the Southern Hemisphere, has a growing season that is the opposite of that of the United States. With refrigerated transportation, Chile can provide American consumers with a wide range of fresh fruit—like blueberries, apples, and table grapes—throughout the Northern Hemisphere winter. Consumers in other core countries benefit equally from global food supply chains across climate regions.

Improvements in aquaculture, meanwhile, have allowed sushi to remain an international favorite, and the UN reported that in 2019 aquaculture produced almost as much fish as wild fisheries. Farmed fish are an efficient food source, requiring around 1.1 pounds of feed to produce one pound of body mass.

Critical Viewing Mexico exports more than 800,000 tons of avocados to the United States every year. In recent years, Mexico has also exported avocados to China, where few avocados are grown locally. ▌ How has Mexico's exporting of avocados affected society in the United States and Mexico?

Trade policies, too, may help change diets and individual tastes. In one instance, the North American Free Trade Agreement (NAFTA) of 1994 resulted in a boom of avocado exports from Mexico to the United States. Mexico has produced avocados for thousands of years but exported very few—a situation that changed drastically with the enactment of NAFTA. Now Mexico produces approximately one-third of all avocados in the world, and avocados are the United States' most valuable fruit import, in addition to being a regular feature of many Americans' diets. In 2018, NAFTA was replaced by a revised trade agreement, the United States–Mexico–Canada agreement (USMCA), which provides similar trade protections. Together, trade policies, technology, and contemporary agriculture have boosted the process of diffusion to provide people around the world with a diverse, globalized diet.

Geographic Thinking

1. Describe some environmental limitations of biotechnology.
2. Explain how the concept of biodiversity helps geographers analyze the impacts of modern-day farming practices.
3. Identify and describe factors that could cause more farmers to embrace agricultural innovations.
4. Compare the commercial approach to agriculture with the sustainable approach.

14.3 Feeding the World

Food production is just the first step in providing for a hungry world. Establishing reliable access to nutritious food for all people means facing challenges such as uneven distribution systems, severe weather—often intensified by climate change, and the loss of farmland to expanding urban areas.

Food Insecurity

Learning Objective
LO 14.9 Compare the threats to food security on global, national, and local levels.

Food security is reliable access to safe, nutritious food that can support a healthy and active lifestyle. The opposing concept, **food insecurity**, is the disruption of a household's food intake or eating patterns because of poor access to food. According to the USDA, the most commonly reported cause of food insecurity is lack of money or other resources.

The FAO reports that in 2020 more than 900 million people suffered from severe food insecurity and 1.4 billion people experienced moderate food insecurity, for a combined total that represents nearly one-third of the world's population. This is an increase of about 320 million food-insecure people from 2019, with the COVID-19 pandemic likely contributing to this increase. In the United States, food insecurity affects nearly 40 million people. These are sad and shocking statistics, particularly when the FAO asserts that the planet produces enough to feed everyone. If the world can grow enough food now, why does evidence show that so many continue to suffer from hunger? Recall what National Geographic Explorer Jerry Glover said at the beginning of this unit: "A good global food supply does not solve regional problems."

Global Food Insecurity Scientists and researchers have found that much food insecurity is the result of distribution issues and economic decisions about what to do with the crops that are produced. According to a 2019 study by the FAO, only 55 percent of the world's overall crop calories (corn and other grains) feed people, with approximately 36 percent going to livestock and about 9 percent being used to make biofuels such as ethanol and other industrial products. Decreasing the amount of food crops used to feed livestock and make biofuels could in turn help decrease food insecurity. Growing meat more efficiently and shifting to diets that are less meat intensive could also release significant amounts of food crops—large percentages of which are currently feeding livestock—for human consumption. A portion of the crop calories consumed by livestock does reach humans in the form of meat, dairy, and eggs, but the caloric value of these products is a fraction of the original value of what is fed to the animals.

Another serious ongoing threat to Earth's ability to continue producing sufficient food is adverse weather—including severe storms, drought, and extreme temperatures—caused or intensified by climate change. More than 80 percent of food-insecure people in the world live in areas susceptible to such extreme weather events, but even relatively food-secure regions are vulnerable to weather-related shocks. In France in 2019, severe storms did so much damage to the food-growing region known as the "orchard of France" that the government declared a state of emergency. Some farmers lost 80 to 100 percent of their crops. At the same time, the Caribbean is in the midst of a decades-long drying trend punctuated by several multiyear droughts, including one so severe in 2015 that half of Haiti's crops were lost and millions of people in the region experienced food shortages. Brazil, Ethiopia, and Indonesia have experienced extreme maximum temperatures for periods of three or more years, another result of climate change that may stress crops. In India, a climate change-heightened drought and heat wave in 2022 caused the government to ban most wheat exports. This action by the world's second-largest wheat producer compounded the global supply chain disruption triggered by Russia's war on Ukraine, and threatened to intensify food insecurity across the globe.

On the national level, instability and chronic poverty also contribute to food insecurity. The Democratic Republic of the Congo (DRC), for example, is ranked as one of the least food-secure countries in the world on the Integrated Food Security Phase Classification. In late 2021, the DRC had the largest number of highly food-insecure people in the world, about 27 million. Most of the country's families are subsistence farmers, and they produce about 42 percent of the food they consume. However, due to years of unstable and corrupt governments and outright civil war in parts of the country, these farmers are vulnerable to displacement. As in many countries, the 2021 DRC figures also reflect economic decline and food price increases related to the COVID-19 pandemic.

Loss of agricultural land to growing urban areas is another threat to global food production. Beginning in the 20th century, **suburbanization**, the shifting of population from cities into surrounding suburbs, has accelerated as increasing numbers of city dwellers seek advantages such as cheaper housing, more space, and lower crime rates. At the same time, cities in many parts of the world are growing rapidly. One 2016 study estimated that globally, urban areas will triple in area by 2030, and 60 percent of the planet's farmlands are located near cities. Expanding cities and suburbs can encroach on productive farmlands.

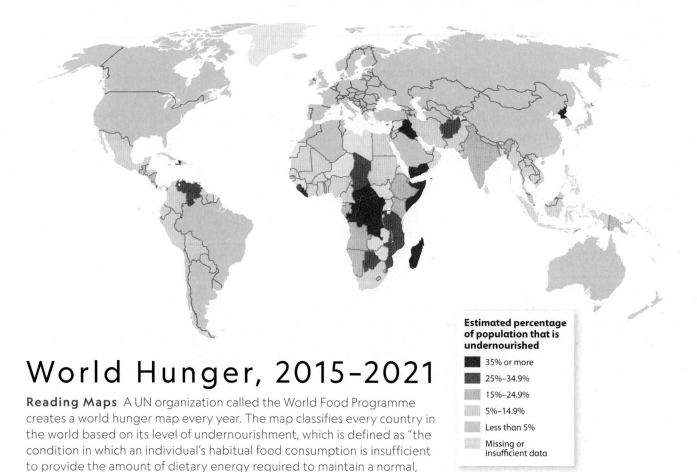

World Hunger, 2015–2021

Reading Maps A UN organization called the World Food Programme creates a world hunger map every year. The map classifies every country in the world based on its level of undernourishment, which is defined as "the condition in which an individual's habitual food consumption is insufficient to provide the amount of dietary energy required to maintain a normal, active, healthy life." ▮ Identify the world regions that are suffering the greatest rates of undernourishment.

Estimated percentage of population that is undernourished
- 35% or more
- 25%–34.9%
- 15%–24.9%
- 5%–14.9%
- Less than 5%
- Missing or insufficient data

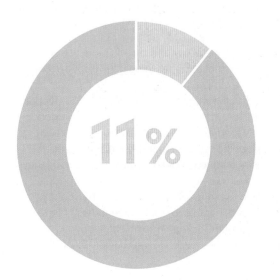

11% OF THE WORLD POPULATION DOES NOT HAVE ENOUGH TO EAT

811 MILLION PEOPLE

GLOBAL HUNGER and Malnutrition are the number one risks to health—greater than AIDS, Malaria, and Tuberculosis combined.

ONE IN THREE women of childbearing age suffer from Anemia (low red blood cell count or insufficient hemoglobin).

EXTREMES are climate variability the result of climate change.

1 IN 4 CHILDREN SUFFER FROM STUNTING

THAT MEANS 150 MILLION CHILDREN UNDER 5 ARE PHYSICALLY TOO SMALL FOR THEIR AGE DUE TO UNDERNOURISHMENT.

Source: World Food Programme

The impact of urban growth on agriculture has been highly debated at national and local levels. Some argue it has had a negative effect, claiming the reduction in the amount of land available for food production contributes to the declining ratio of food producers to food consumers. Growing urban populations demand more agricultural products, and as cropland is replaced by housing and other urban developments, the question of whether local farming communities can sustain demand and remain profitable becomes more compelling. Opposing viewpoints argue that because of food globalization and commercialization—plus agricultural intensification—urban growth has little impact on the ability of agriculture to meet overall food demands in cities and elsewhere.

Food Issues in the United States Like other countries, the United States is losing farmland to urban development. As suburbs have grown up around cities, land that was once cultivated by small-scale family farms has been bought by developers to build suburban neighborhoods. According to the USDA, between 1980 and 2021, the amount of land being farmed or grazed was reduced by 14 percent.

In many low-income areas of the United States—urban, rural, and suburban alike—food insecurity is an ongoing concern. In 2020, 10.5 percent of households in the country suffered from food insecurity at some point during the year, and 12 million children were food insecure. Single women with children represent almost one-third of Americans reporting food insecurity.

Although poverty in the United States is strongly linked to food insecurity, the two problems are not always perfectly connected. Not all people living in poverty experience food insecurity, and not all people living above the poverty line are exempt from it. The cost of living compared to wages earned and other factors, such as medical expenses, can affect food security. Families with children and low wages may not have adequate incomes to purchase enough food, but their incomes may be high enough to disqualify them from receiving social services. For example, in some New York City boroughs such as the Bronx, where nearly half the children are food insecure, most families who participate in food assistance programs are headed by people with jobs. Contrary to some perceptions, most food-insecure Americans are employed or live in a household where someone works at least part time. The COVID-19 pandemic worsened the situation for many, with illness, job losses, and increased food prices driving more families into food insecurity, at least temporarily.

Hunger and Conflict About 60 percent of the world's hungry—490 million people—live in countries affected by war, and the UN reports that conflict is one of the key factors affecting food security and nutrition. In 2020, 155 million people globally experienced acute (severe and life-threatening) food insecurity, an increase of 15 percent, attributable largely to conflict in countries including DRC, Syria, Myanmar, and Yemen. Afghanistan, Sudan, South Sudan, and Nigeria, also war-torn, are suffering the most serious hunger crises in the world. In South Sudan, civil war has caused a mass displacement of citizens, which has left crops untended, resulting in crop failure and leaving more than 7.2 million people food insecure. In early 2022, an astonishing 95 percent of Afghanistan's population of 40 million had insufficient food consumption as a result of decades of war and a new government struggling to administer the country.

In some instances, warring parties have used food as a weapon by deliberately denying access to it for people associated with the opposition. In many cases, food distribution is inadequate or unequal in conflict zones because political systems are poorly managed, corrupt, or in disarray. People living in areas torn by fighting have few food options, and humanitarian workers attempting to provide food relief often face violence themselves.

Geographic Thinking

1. Compare food insecurity in the United States and food insecurity in other parts of the world. How does the problem manifest itself across different regions?

2. Explain the argument that food insecurity has more to do with distribution issues than production issues.

Mobile food trucks are one way of reaching people living in areas where little fresh food is available. An old city bus in Chicago, Illinois, has been converted into a single-aisle grocery store serving city dwellers who have no such stores in their neighborhoods.

Agricultural Sustainability in a Global Market

Case Study

Food Deserts

The Issue In the absence of mapping, "food deserts," or areas with poor access to healthy, nutritious food, are invisible contributors to the food insecurity problem—and invisible problems are not easy to solve.

Learning Objective
LO 14.9 Compare the threats to food security on global, national, and local levels.

By the Numbers

23.5 million
Americans live in food deserts

6.5 million
are children

43,000
households in Houston, Texas, are located in a food desert

Source: United States Department of Agriculture; National Geographic, August 2014

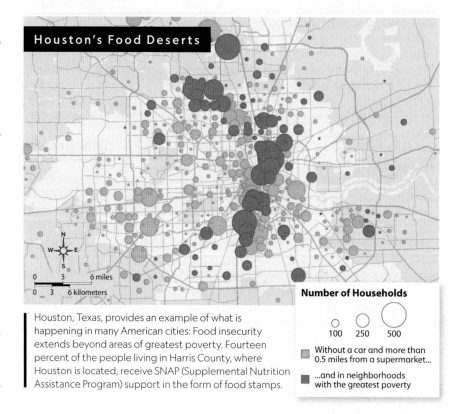

Houston's Food Deserts

Houston, Texas, provides an example of what is happening in many American cities: Food insecurity extends beyond areas of greatest poverty. Fourteen percent of the people living in Harris County, where Houston is located, receive SNAP (Supplemental Nutrition Assistance Program) support in the form of food stamps.

Number of Households: 100, 250, 500
■ Without a car and more than 0.5 miles from a supermarket…
■ …and in neighborhoods with the greatest poverty

Twenty-Three and a Half Million Americans live in food deserts, areas where residents lack access to healthy, nutritious foods because stores selling these foods are too far away. The USDA defines "low-access communities" as places where at least 33 percent of the population live more than one mile from a supermarket or large grocery store. For rural areas, the distance is more than 10 miles. Food deserts, whether rural or urban, are often characterized by a low median income, higher unemployment, and higher poverty rates. While some challenge the analogy of "food deserts," and instead focus more on factors other than just distance from one's home and traditional grocery stores, it still finds favor with many.

Food deserts occur in every major urban area in the United States. Large cities, such as Milwaukee and Baltimore, may appear to have plentiful grocery stores, but because stores and bus lines are unequally distributed, many residents find that healthful food is out of easy reach. Indeed, nearly 40 percent of residents in Milwaukee County live in a food desert.

Convenience stores and small independent stores are more common in food deserts than full-service supermarkets or grocery stores. These shops may have higher food prices, lower-quality foods, and a smaller variety of offerings. The result is that food consumed in food deserts is typically high in cholesterol, sugar, and fat. To help combat this problem, Milwaukee's Hunger Task Force created the Mobile Market in partnership with the grocery chain Pick 'n Save. The market makes stops in food desert communities to ensure that people have access to quality fresh foods.

Geographers map food deserts to better understand the challenge to obtaining fresh, healthy food. In Baltimore, approximately 25 percent of residents live in a food desert, but a recent partnership between the city of Baltimore and the Johns Hopkins Center for a Livable Future (CLF) is working to change that. CLF researchers developed 14 district maps showing food deserts and grocery store locations. The maps compelled city officials to create solutions. Today, food retailers who open or renovate establishments in or near food deserts can receive tax credits. The city also lifted building restrictions on temporary greenhouses known as hoop houses so residents can raise their own produce. The partnership project was so successful that CLF researchers have expanded their mapping tool to the state level. The hope is that other states will adopt similar tools and use them to reshape policies and solutions to America's food desert problem. ∎

Geographic Thinking

Explain how food deserts present both an opportunity and a challenge.

Economic Impacts on Food Production

Learning Objective
LO 14.10 Describe the economic factors that impact food production and distribution.

One of the factors that contribute to food insecurity is poor distribution. Effective food distribution systems connect producers to consumers and allocate the food to meet local needs. However, there are challenges in deciding how food will be distributed among people, who has the authority to make decisions about food distribution, and what methods should be used. Complex social and economic factors often prevent the consumers who need food the most from receiving it. Low-income consumers may not have transportation or the financial means to obtain adequate food, or they may live in areas where government subsidies and services are not available or accessible. Rising prices can also prevent people from buying enough food or healthy food, whether they are employed or not.

Storage and Transportation Issues Farms, food-processing facilities, and the markets where foods are sold are often located at considerable distances from one another. Recall the example of produce imported from Chile to U.S. grocery stores during the Northern Hemisphere winter. In some places, supply chains are much shorter than the distance between continents, existing within a single country or region. Because of poor storage, processing, transportation, or infrastructure, however, even short supply chains may break down, leading to food insecurity. Many regions of Africa and Latin America, in particular, are seeing a rise in severe food insecurity due at least in part to transportation and storage issues.

Inadequate infrastructure in many peripheral countries means that food grown elsewhere often cannot be transported to those who need it. According to the FAO, approximately 25 percent of the world's food calories are lost or wasted before they are consumed, and in peripheral countries, food is typically lost between the time the farmer harvests it and when it arrives at the market. Farmers in these countries have limited access to reliable storage and transportation. They may have crops but no way to get them to market. Sometimes basic storage problems claim crops before they can be transported: Milk spoils before it can be pasteurized or pests ruin grain because there are no adequate storage facilities. Even in core countries, small farmers with lower incomes may lack the capital to invest in ways to overcome transportation and storage challenges.

By improving rural infrastructure (roads, storage, electrification), governments can strengthen the ability of farmers to develop sustainable businesses, reducing waste and increasing food access. In some instances, nongovernmental agencies and private companies are contributing knowledge and resources to help resolve local infrastructure-related problems. For example, a group of volunteers from several large U.S. agribusiness firms has formed a nonprofit group called Partners in Food Solutions to help entrepreneurs in food commodity supply chains in African countries. The group assisted the owner of a milling business in reducing waste and implementing quality control measures to ensure that she had more milled grain to sell. They also enabled Soy Afric, a food-processing company, to acquire a large cooler designed by engineers in the United States and assembled in Nairobi, Kenya. Other nonprofits are targeting Africa's dependence on imported food. At present, African countries import approximately $40 billion a year in food. Producing and processing more food on the continent shortens the commodity supply chain between producers and consumers and may open doors to greater food accessibility.

Economy of Scale **Economy of scale** is the reduced cost of producing food items as the quantity of production increases. For example, if fixed costs such as tractors or combines are spread out over many units of production (e.g., tons of wheat harvested), the return on investment is greater because the fixed costs remain the same. In agriculture, the concept of economy of scale shows that farming on a larger scale is more efficient than farming on a smaller scale: The average cost of production decreases as the farm size increases.

Economy of scale offers a good way to think about how to best use available resources to meet agricultural needs. Think of a production unit as a gallon of milk or a bushel of apples. The cost to produce each unit is lower on large farms due largely to technology and mechanization. Precision agriculture, biotechnology, and large machines are more productive per unit when used on areas larger than the average small farm. Almost every aspect of modern agriculture favors the large-scale farm over the small-scale farm because as the quantity of units produced increases, the cost per unit goes down.

Economy of scale also indirectly affects the way food is distributed. To make the best use of specialized technology, larger farms are less diversified as farmers focus on one or two crops to maximize profits. Often these are field crops such as corn or soybeans. These farms also have access to more far-reaching distribution networks. Small farms, with a higher cost per unit, may struggle more than larger ones when prices are low. Many small-scale farmers choose to grow organic or specialty crops and distribute them locally in the hopes that the premium they can charge for organic produce will cover their costs and produce a profit.

Government Policies In the United States, the rise of large-scale farming has led to sizable corporate landholders controlling most of the land used for agriculture. In 2020, the USDA reported that the largest farms (those with sales of $1 million or more) accounted for only 3 percent of all farms but 46 percent of all sales. The largest commercial farms receive the majority of the approximately $20 billion in annual farm subsidies. In 2019, when the government delivered a massive farm aid package to offset losses due to Chinese trade tariffs, most of the $19 billion paid out that year went to large commercial farms. COVID-19-related subsidies boosted the 2020 direct payouts to American farmers to a record $46 billion, with critics maintaining that the largest farms got most of the money.

The three largest government subsidy programs go to farms that produce corn, soybeans, and wheat. Such subsidies have the power to distort the decisions made by farm businesses by encouraging overproduction and discouraging diversification. From a business standpoint, farmers often feel they must choose to grow the crops that are the most cost-effective. These same commodity crops lend themselves to large-scale production, easy storage, and long-distance shipping. With soybeans and corn at the top of the list of exported U.S. products, government policies may encourage farm businesses to produce crops that enhance agricultural exports rather than bolster food security.

The Food and Agriculture Organization of the United Nations believes that edible insects can be a way to combat scarcities of land and water while relieving pressures on the environment. Edible insects are rich in protein, need little feed, and emit few greenhouse gases. The burgers in the photo are made from buffalo worms, which are larvae of the darkling beetle.

In India, where a rapidly growing population contributes to widespread poverty and hunger, government policies regarding agriculture have struggled to improve the country's food security. As mentioned, India ranks high in production volume for wheat—and also for rice; those two crops, along with aquaculture, account for more than half of total farm exports. But despite being one of the largest grain-producing countries in the world, India still accounts for one-fourth of the world's hungry people. Critics of India's governmental policies argue that agricultural subsidies bypass small farmers, and instead are given to large-scale farms.

In hopes of improving farmers' incomes and remedying unequal food distribution, the Indian government has recently implemented a program through which state governments buy food grains from local farmers at a guaranteed price that protects farmers from price fluctuations. On the consumer side, qualifying households can purchase food through the program at subsidized prices. Supporters of this government policy argue that it has helped India make significant progress in delivering food grains to the poor, but critics claim that the program is unable to identify and reach the country's poor and that the guaranteed price is inadequate for farmers.

Fighting the Problem In many core economies, globally conscious diets are gaining followers as one way to help solve the problem of feeding everyone while reducing environmental damage and promoting healthier eating. Some consumers are making food choices, such as reducing their intake of meat, with global as well as personal consequences in mind. Though in 2021 just 5 percent of U.S. households followed either a vegetarian diet (which excludes meat for health, religious, or moral reasons) or a vegan diet (which excludes any animal products), plant-based foods are growing more popular, with fast growth in both consumer sales and on restaurant menus. As more land goes under cultivation to grow feed for livestock, edible grains become too expensive for people in peripheral countries to purchase. A large-scale shift to less meat-intensive diets could have considerable impact on the availability of food. By cutting their intake of meat and other crop-intensive foods—while simultaneously increasing the market for locally produced fruits and vegetables—the wealthiest billion could significantly alter the global pattern of food production, distribution, and consumption.

Changing diets in wealthier countries has the potential to free up more croplands to grow food; the next step is creating transportation systems and infrastructure to ensure that the food reaches those who need it most. Much relies on the decisions of governments, businesses, and organizations, but economically stable consumers, too, can use their purchasing power to help transform the global food system and encourage sustainable farming practices.

Geographic Thinking

1. Explain how agricultural policies in the United States could be modified to strengthen the agricultural system's ability to perform as a sustainable food production system.

2. Describe the types of geographic data that might be useful in helping meet the demand of food security.

3. Explain how economy of scale can be used to support an argument that favors large-scale farming over small-scale farming.

4. Explain the degree to which food production can be impacted by violent conflict and extreme weather.

Jennifer Burney's fascination with physics has led her to explore how energy and climate affect food security, water availability, and agriculture.

Learning Objective
LO 14.4 Describe environmentally sustainable agriculture practices.

National Geographic Explorer Jennifer Burney
Local Changes, Global Consequences

Environmental scientist Jennifer Burney has made global sustainability her goal. She's determined to "chart a realistic pathway for greening the global food system."

Jennifer Burney is a professor at the University of California San Diego School of Global Policy and Strategy, where her research focuses on improving global food security while reducing climate change. At present, she is working to understand how global climate changes affect people locally. While carbon dioxide is the largest contributor to climate change globally, Burney hopes this project will help shed light on other compounds that harm local air quality and cause health problems such as asthma.

Burney has spent many years working toward hunger and climate solutions in the field as well as in the lab. In Benin, West Africa, where agricultural production is the main livelihood and residents spend 50 to 80 percent of their income on food, she cooperated with a women's farm collective that implemented solar irrigation in local farm fields. The irrigation system was powered by solar panels, and it succeeded in both conserving water and extending the growing season, allowing farmers to increase yields and supplement household income and nutritional intake. Burney was also involved with a project in northern India that helped several communities replace traditional cookstoves with cleaner burning eco-stoves. The traditional stoves emitted excessive black carbon, or soot. Inside homes, these emissions caused respiratory infections; outside, the emissions contributed to climate change that alters monsoon cycles. The new stoves drastically limit black carbon emissions and reduce fuel use.

Burney believes community-scale innovations can make a global difference in agriculture-related climate change. "People care about the local stuff," she says, "and they mobilize resources and make politics happen to solve problems."

Geographic Thinking

Explain the challenges that Jennifer Burney helped the women farmers in Benin overcome.

Agricultural Sustainability in a Global Market

14.4 Women in Agriculture

Women have always played key roles in agriculture, whether as landowners or workers, pastoral nomads or settled farmers. These roles have changed over time and continue to do so, as social, political, and environmental circumstances evolve. Geographers take a keen interest in female farmers because strong connections exist between the empowerment of women, agricultural productivity, and food security.

A Variety of Roles

Learning Objective
LO 14.11 Explain the changing roles of women in food production and consumption.

Women produce more than 50 percent of the world's food and make up 43 percent of the agricultural labor force. Most working women in peripheral countries labor in agriculture, and more than a third of farmers in the United States are female. Women's roles in agriculture vary widely across the world, from region to region, and locally. Women throughout the world are involved in both crop and livestock production in subsistence and commercial settings. They manage mixed agricultural operations that may include crops, livestock, and fish farming, and they produce both cash crops and food for their families. Women are also deeply involved in the distribution and selling of agricultural products and decisions about how these products are consumed. As agriculture changes because of technology, science, and economic and social pressures, the role of women within the sector changes too.

In peripheral countries, women are largely limited to participation in subsistence agriculture. Malawi, in southern Africa, serves as a good example of this phenomenon. In Malawi, 11 million people—the majority of the population—participate in smallholder subsistence farming. Opportunities to transition to commercial farming can be limited by gender bias in educational and other opportunities. When sustainable techniques for growing sweet potatoes, a major part of Malawi's farm production, were taught, the men

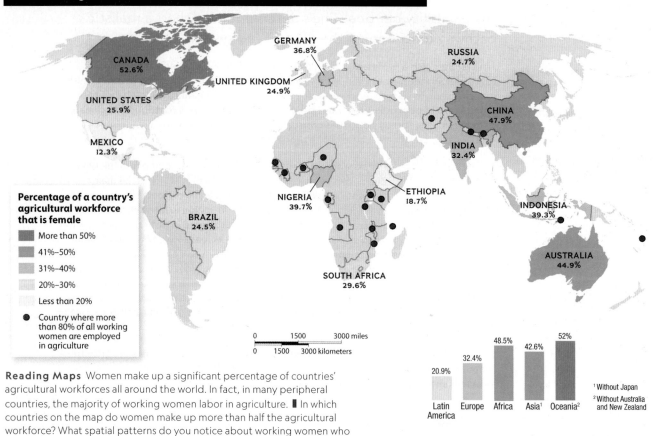

Reading Maps Women make up a significant percentage of countries' agricultural workforces all around the world. In fact, in many peripheral countries, the majority of working women labor in agriculture. ▎In which countries on the map do women make up more than half the agricultural workforce? What spatial patterns do you notice about working women who work mostly in agriculture?

Sources: Soil Atlas 2015, Institute for Advanced Sustainability Studies

408 Chapter 14

received training related to market-oriented production, while the women received training related to subsistence farming and how to use the sweet potato crops in their roles as care providers of the household. Women in Malawi tend to be viewed as inexpensive labor and not as actively involved in significant decision making. This marginalization of women resulted in their exclusion from education on important topics such as how to manage crop disease and pests, as well as other sustainable farming techniques.

The most difficult gender-specific challenge that women working in agriculture face is the lack of land rights. For example, in DRC, most of the agricultural workers are the women who produce approximately 75 percent of the food in rural areas but have no ownership rights because they are female. Only 10 to 20 percent of landholders in peripheral countries are women. If a female farmer has no legal right to own or control land, she also doesn't have the legal backing to make decisions about the land or negotiate farm agreements. These limitations restrict the amount and reliability of income. In some areas, legal rights exist on paper but are disregarded in practice.

Women in agriculture also face obstacles when attempting to conduct the business of farming. In some peripheral countries, even unspoken gender roles can be rigid and keep women from taking their crops to market. Female subsistence farmers with small land holdings in these countries often face technical problems and cultural biases that block them from borrowing money. In some countries, laws require a woman to obtain her husband's signature before she can receive a loan. Even in the United States, before federal legislation overruled the practice in 1988, some states required women to provide a signature from a male relative in order to obtain a business loan. Without access to loans, women farmers cannot purchase fertilizer, seeds, or advanced farming tools and equipment.

In many places, women in agriculture experience difficult working conditions and a poor quality of life. Worldwide, family farms are central to global food security and sustainable productivity, but at the same time, they are the main institutions where women work as unpaid and unrecognized laborers. In India, for instance, rural-to-urban migration by men and steady growth in the production of labor-intensive cash crops has led to women performing significant farm tasks. However, these tasks are simply viewed as an extension of their household responsibilities. A typical expectation of a female farmer in a peripheral country is that she takes care of all household responsibilities—such as preparing food, caring for the children, and collecting fuel and water—and maintains her agricultural labor contribution as well. Similar challenges exist for women in other regions of the world. The FAO reported in 2017 that in Latin America and the Caribbean, 40 percent of working women over age 15 did not have a personal income. Eight out of 10 of these women worked in the agricultural sector, and not necessarily as subsistence farmers.

The United States provides examples of the varied and changing roles women play in agriculture. In 2017, the USDA's Census of Agriculture (conducted every five years) reported that 36 percent of the country's 3.4 million farm producers involved in daily decision making were female. According to the Bureau of Labor Statistics, women farmers, ranchers, and agricultural managers exceeded the earnings of their male counterparts in 2017. Female ranchers are also leading the way in practicing sustainable ranching. One rancher, Kelsey Ducheneaux, raises sustainable beef on 7,500 acres of tribally leased land in South Dakota. A member of the Lakota Nation, Ducheneaux points out that generations of Lakota worked the same land before her, and Native American women being involved in ranching is not a new concept.

While American female farmers may not face the same hardships and challenges as those in other countries, obstacles still exist to their equal participation in agriculture. Despite progress, female farmers and ranchers in the United States encounter resistance from male counterparts. Though an equal partner in her family's dairy farm in Minnesota, Tara Meyer finds she is often referred to as a "farm wife," when she is in reality a farmer with a master's degree in agriculture.

In Europe, women farmers are actively involved in food distribution in roles that extend beyond production and often surpass the participation of men. They take their produce to farmers' markets, distribute it through catering companies, and hold leadership roles in social movements working toward food sovereignty, which is defined by the U.S. Food Sovereignty Alliance as "the right of peoples to healthy and culturally appropriate food produced through ecologically sound and sustainable methods, and their right to define their own food and agriculture systems."

According to Euractiv France, in 2019 women represented one-fourth of all French farmers, with 41 percent of agricultural businesses being established by women. French women are increasingly involved in agricultural distribution as well, taking on roles in the agricultural supply chain. In Italy one-third of farms are run by women, and women manage almost 40 percent of the companies in the emerging agritourism industry. Agritourism provides a creative alternative to standard food distribution models by bringing in tourists to stay—and sometimes work—on farms and enjoy meals made from local products.

Along with changing roles in food production and distribution, women's roles in food consumption have also changed. In cultures throughout the world, women have traditionally been—and often still are—in charge of selecting, cooking, and serving food to their families. New economic realities, however, are altering that vision. In Singapore, a country that has experienced extensive economic growth, large numbers of women continue to enter the workforce. Fewer women cook on a regular basis, families eat out more frequently, and domestic workers within the home may be left to make food purchasing and consumption decisions. In other smaller countries, such as Cyprus, women are active in the workforce, but grandparents remain in the home, and food purchasing and consumption decisions are often left to the grandmother.

Research conducted by the FAO in Asia, Africa, and Latin America found that in general, women who earn an income spend a much higher proportion of their money on food for their families than men do. These expenditures may include food for direct consumption or resources needed for food crops. In areas of India, more women are choosing to engage in poultry production, an agricultural sector that is typically controlled and managed by women. Poultry production allows women to provide nutritious food for their families and contribute to the household income by selling eggs and birds at local markets.

Empowering Rural Women

Learning Objective
LO 14.12 Describe programs and policies that governments and NGOs are implementing to empower women in agriculture.

In the context of farming, empowerment means having the ability to make decisions about factors such as land, livestock, seeds, fertilizer, and machinery, as well as control over finances and one's own time. Empowering women in the agricultural sector has the potential to bring significant benefits at the individual, community, national, and global scales. At the household level, children receive better nutrition and education when their mothers' incomes increase. Communities also benefit when women have money to spend on schooling and other resources. At the regional and national scales, empowering female farmers may help improve food security for millions. According to a report issued by the FAO in 2017, if women in rural areas had the same access as men to agricultural resources (land, technology, markets, financial services), agricultural productivity would increase by 20 to 30 percent—enough to reduce the number of hungry people by 100 to 150 million.

In some regions of the world, rural women face fewer economic and cultural barriers to empowerment. According to a 2011 UN study, women in Southeast Asia have more equal access to land than do women in South Asia. Women in Southeast Asia also tend to exert more control over household income than men. In some areas of sub-Saharan Africa, such as Angola, Ghana, and Uganda, women are encouraged to strive for economic self-reliance and traditionally are given significant responsibilities in agricultural production. Some countries hit hard by male emigration, deaths from HIV/AIDS, and conflict have given large percentages of the agricultural labor force over to women. Many women in peripheral countries function as the heads of household and primary caregivers, which makes them the link between farms and tables; they are the ones making decisions about food consumption and nutrition for their families and communities.

Empowering Women Farmers

In many parts of the world, female farmers are making progress toward empowerment, even in areas such as the Punjab region of Pakistan that place high hurdles in front of women. Against all odds, for example, 28-year-old Almas Parveen became the only woman farmer in Punjab who trains other farmers. She manages 23 acres of family farmland in a village where agriculture is male dominated. Parveen's status as a single woman and a farm manager and trainer is exceptional. Being one of the few women farmers who go to the market to sell, she has the experience to bargain for the best market rate and is also trusted by her family to manage all the farm's finances.

Parveen received formal agricultural training in 2017, which opened her eyes to the critical importance of water conservation. The new sustainable agricultural methods she implemented resulted in a higher crop yield, and other farmers followed in adopting the same practices. As a certified field facilitator with the nonprofit organization Rural Education and Economic Development Society, she has taught more than 400 farmers—both males and females—how to implement sustainable practices. Parveen is also an advocate for fair wages for women and pays female farmhands at a higher rate than is customary. Of her journey to empowerment she says, "It has not been easy, but it is not impossible either."

Almas Parveen teaches both male and female farmers in Pakistan about sustainable agricultural practices.

Caitlyn Taussig is an American rancher in Kremmling, Colorado. She helps run the family operation along with her mother, sister, and other female ranchers.

The steps to empowering female farmers include education, technical support, access to capital, and government policies that promote gender equality. Several nongovernmental organizations are implementing programs that target all these facets of empowerment. The Global Forum on Agricultural Research, for example, manages a women's collective called Gender in Agricultural Partnership (GAP). GAP is made up of more than 150 institutions that work together to provide education, land, and credit—the ability to borrow money—to female farmers. They also address gender discrimination in the agricultural sector and lobby for policy change to empower women in agriculture at the global, national, and local levels.

Working with a network of institutions in Africa, the Americas, and Asia, the Grameen Foundation extends microfinancing, or loans in very small amounts, to the world's poorest people, including female farmers. These small loans enable small farmers to invest in new equipment or expand their businesses. Grameen's agricultural program uses technology and a large network of resources to help women access microloans, agricultural supplies and tools, information, and technical assistance. The Grameen Bank and its microlending efforts will be discussed in Chapter 19.

Governments and supranational organizations, too, are implementing measures to advance women's empowerment in agriculture. In the United States, the U.S. Department of Agriculture Farm Service Agency reserves a portion of its loan funding for underserved farmers and ranchers, which includes women. The U.S. government program Feed the Future has also worked with more than 2.4 million women around the world to apply new agricultural technologies and practices. UN Women promotes connectivity as a means to integrate rural areas with global financial and payment systems in order to provide financial access for female farmers and ranchers who are starting out in business.

Whatever their particular focus, all agencies working to empower female farmers recognize that women's basic right to equality and control over their lives is closely linked with the health of the agricultural system in general. Given the opportunity to increase their incomes and develop a consistent and reliable livelihood, female farmers will play an important role in ensuring food security for their families and communities—and in confronting the complex challenge of feeding the world.

Geographic Thinking

1. Explain why working in the agricultural sector is more difficult for female farmers than for male farmers.

2. Describe how women's changing roles in food distribution and consumption affect their communities.

3. Explain the degree to which a region's economy and culture affect the empowerment of its female farmers.

Agricultural Sustainability in a Global Market

Chapter 14 Summary & Review

Chapter Summary

Agricultural landscapes result from the interactions between farming practices and physical geography.

Agricultural techniques have environmental consequences.

- Pollution results from chemicals that run off farmland.
- Land cover change caused by agriculture can result in erosion or land degradation.
- Soil salinization is a result of irrigation.
- Desertification is a type of land degradation in which arid and semiarid lands become deserts.
- Numerous governments and organizations are promoting conservation and sustainable practices.

Agricultural practices have societal consequences.

- The growth of agribusiness has led to fewer small and midsize farms in the United States.
- Some people have changed their diets in response to concerns about agricultural practices.

Debates exist over the pros and cons of GMOs.

- Supporters believe GMO crops can feed more people more efficiently.
- Opponents believe GMOs threaten biodiversity and could have negative health effects on humans.

Consumer food choices have an effect on food production.

- Buying local, organic, or fair trade crops helps create markets for certain types of agriculture.
- Dietary shifts such as greater meat consumption also strongly affect the agricultural sector.

Global food insecurity results from a number of causes, including loss of agricultural lands, climate change, conflict, and croplands devoted to biofuels.

Food production and distribution are affected by several economic factors.

- Storage and transportation issues can lead to food waste.
- Government policies can affect farmers' decisions and help or harm food consumers.

Women in agriculture face a variety of challenges, including cultural bias and legal obstacles to land ownership.

Empowering women in agriculture leads to numerous benefits, including better nutrition for families and greater food production overall.

Organizations have implemented programs to empower women in agriculture, addressing education, technical assistance, financing, and gender discrimination.

Review Questions

Use complete sentences to answer the questions.

1. **Apply Conceptual Vocabulary** Consider the term *agricultural landscape*. Develop a dictionary definition of the term. Then provide a conceptual definition—an explanation of how the term is used in the context of this chapter.

2. How are the terms *pastoral nomadism* and *desertification* related?

3. Describe three agricultural techniques from around the world. Provide examples of the environmental consequences of each one.

4. Identify an example that explains how an agricultural technique can have positive environmental consequences.

5. How can dams and reservoirs impact farming?

6. Define and provide an example of an aquifer. Include a possible environmental consequence of using aquifers for irrigation.

7. Describe the characteristics of a wetland and explain its environmental vulnerability because of agriculture.

8. Identify and explain three causes of food insecurity.

9. Define the term *debt-for-nature swap*. Explain how the term relates to agricultural sustainability and conservation efforts.

10. What is the difference between biotechnology and biodiversity?

11. Describe how precision agriculture can help the environment.

12. Explain how the decisions of wealthy consumers may lead to food insecurity.

13. Explain how food trends in core countries affect producers in other parts of the world.

14. Describe how the role of women in agriculture has changed, specific to food distribution.

Interpret Graphs

Study the graph and then answer the questions.

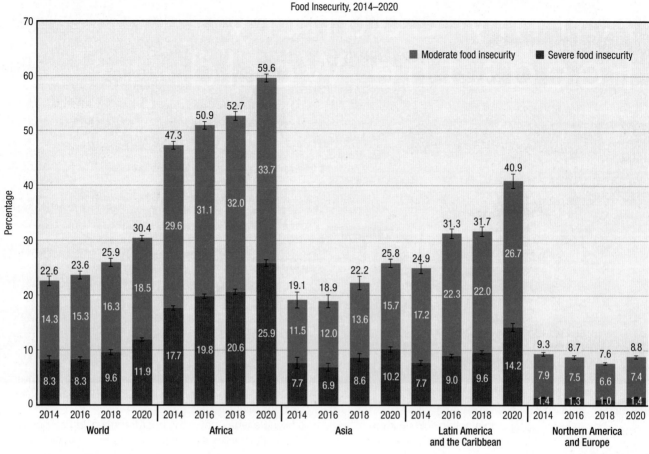

Sources: Food and Agriculture Organization

15. **Identify Data & Information** What was the average percentage of severe food insecurity in Asia for this period?

16. **Analyze Data** Which area had the greatest percentage increase in food insecurity and in what year did that increase occur?

17. **Compare Patterns & Trends** Compare the food insecurity percentages from 2018 to 2020 in Asia with those of Northern America and Europe.

18. **Synthesize** How do the percentages represented by Africa support what you learned about global food insecurity in the text?

Unit 5 | Writing Across Units, Regions, & Scales

A Tiny Country Feeds the World

by Frank Viviano

Technology helps farmers create the perfect conditions for tomatoes to grow abundantly inside this greenhouse in the Netherlands.

In a potato field near the Netherlands' border with Belgium, Dutch farmer Jacob van den Borne is seated in the cabin of an immense harvester before an instrument panel worthy of the starship *Enterprise*. From his perch 10 feet above the ground, he's monitoring two drones—a driverless tractor roaming the fields and a quadcopter in the air—that provide detailed readings on soil chemistry, water content, nutrients, and growth, measuring the progress of every plant down to the individual potato. Van den Borne's production numbers testify to the power of this "precision farming," as it's known. The global average yield of potatoes per acre is about nine tons. Van den Borne's fields reliably produce more than 20 tons.

One more reason to marvel: The Netherlands is a small, densely populated country, with more than 1,300 inhabitants per square mile. It's bereft of, or lacking, almost every resource long thought to be necessary for large-scale agriculture. Yet it's the world's number two exporter of food as measured by value, second only to the United States, which has more than 235 times the landmass of the Netherlands. How on earth have the Dutch done it?

More Tomatoes, Fewer Resources At every turn in the Netherlands, the future of sustainable agriculture is taking shape—not in the boardrooms of big corporations but on thousands of modest family farms. You see it vividly in the terrestrial paradise owned by Ted Duijvestijn and his brothers Peter, Ronald, and Remco. The Duijvestijns have constructed a self-contained food system in which a near-perfect balance prevails between human ingenuity and nature's potential.

At the Duijvestijns' 36-acre greenhouse complex near the old city of Delft, visitors stroll among ranks of deep green tomato vines, 20 feet tall. Rooted not in soil but in fibers spun from basalt and chalk, the plants are heavy with tomatoes—15 varieties in all—to suit the taste of the most demanding palate. In 2015, an international jury of horticultural experts named the Duijvestijns the world's most innovative tomato growers.

Since relocating and restructuring their 70-year-old farm in 2004, the Duijvestijns have declared resource independence on every front. The farm produces almost all of its own energy and fertilizer and even some of the packaging materials necessary for the crop's distribution and sale. The growing environment is kept at optimal temperatures year-round by heat generated from geothermal aquifers that simmer under at least half of the Netherlands.

The only irrigation source is rainwater, says Ted, who manages the cultivation program. Each kilogram of tomatoes from his fiber-rooted plants requires less than

4 gallons of water, compared with 16 gallons for plants in open fields. Once each year, the entire crop is regrown from seeds, and the old vines are processed to make packaging crates. The few pests that manage to enter the Duijvestijn greenhouses are greeted by a ravenous army of defenders such as the fierce *Phytoseiulus persimilis*, a predatory mite that shows no interest in tomatoes but gorges itself on hundreds of destructive spider mites.

A few days before I visited the Duijvestijns' operation, Ted had attended a meeting of farmers and researchers. "This is how we come up with innovative ways to move ahead, to keep improving," he told me. "People from all over Holland get together to discuss different perspectives and common goals. No one knows all the answers on their own."

Exporting Knowledge Dutch firms are among the world leaders in the seed business, with close to $1.7 billion worth of exports in 2016. The sales catalog of Rijk Zwann, a Dutch seed breeder, offers high-yield seeds in more than 25 broad groups of vegetables, many that defend themselves naturally against major pests. Heleen Bos is responsible for the company's organic accounts and international development projects. Like many of the entrepreneurs at Dutch firms, Bos has worked in the fields and cities of peripheral countries. With lengthy postings to Mozambique, Nicaragua, and Bangladesh over the past 30 years, she knows that hunger and devastating famine are not abstract threats.

"Of course, we can't immediately implement the kind of ultra-high-tech agriculture over there that you see in the Netherlands," Bos says. "But we are well into introducing medium-tech solutions that can make a huge difference." She cites the proliferation of relatively inexpensive plastic greenhouses that have tripled some crop yields compared with those of open fields, where crops are more subject to pests and drought.

Some 4,000 miles south of the Netherlands, in a family-owned bean field in Africa's Eastern Rift Valley, a team from SoilCares, a Dutch agricultural technology firm, explains the functions of a small handheld device. In conjunction with a cell phone app, the device analyzes the soil's chemistry, organic matter, and other properties, then uploads the results to a database in the Netherlands and returns a detailed report on optimal fertilizer use and nutrient needs—all in less than 10 minutes. At the cost of a few dollars, the report provides input that can help reduce crop losses by enormous margins to farmers who have never had access to soil sampling of any kind.

Less than 5 percent of the world's estimated 570 million farms have access to a soil lab. That's the kind of number the Dutch see as a challenge.

Source: Adapted from "This Tiny Country Feeds the World," Frank Viviano, *National Geographic*, September 2017.

Write Across Units

Unit 5 examined agriculture from a geographer's perspective. In Chapter 14, you read about precision agriculture—one way technology might help meet humanity's needs in the future. This article focuses on some of the world's most advanced practitioners of precision farming. Use information from the article and this unit to write a response to the following questions.

Looking Back

1. How are farmers using geographic tools to practice precision farming? Unit 1
2. What effect do you think implementing Dutch farming methods might have on populations in other countries? Unit 2
3. How might adopting Dutch farming methods affect the cultural landscape in countries such as Mozambique, Nicaragua, and Bangladesh? Unit 3
4. How might countries with a history of colonialism react to Dutch companies or the Dutch government offering agricultural solutions? Unit 4
5. How do the farming techniques described in the article compare with innovations of the Second Agricultural Revolution and the Green Revolution? Unit 5

Looking Forward

6. How might adopting the Dutch pattern of agriculture affect the growth of cities in a region? Unit 6
7. How might international trade agreements and the world economy affect a farmer's decisions about trying precision farming? Unit 7

Write Across Regions & Scales

Research a country or place outside Europe that has worked with Rijk Zwann or another Dutch entity to implement farming techniques like those described in the article. Write an essay comparing the results with those obtained in the Netherlands. Drawing on your research, this unit, and the article, address the following topic:

How well do some of the Dutch innovations work outside the Netherlands? Why might the results differ in other countries and locations?

Think About

- Cultural and economic conditions that affect agriculture in each country
- Soil, climate, and other conditions in the Netherlands and in the other country

Unit 5

Chapter 12

Where Crops Are Most Likely to Thrive

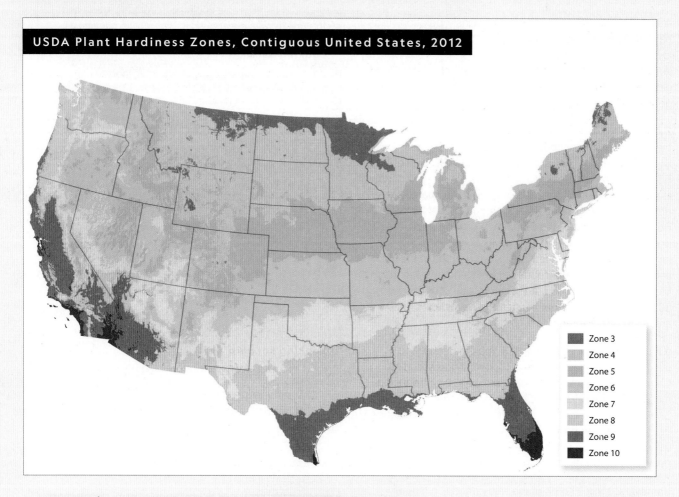

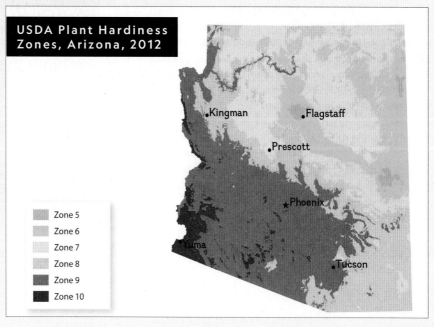

Climate is one of the key determinants of crops that will grow in a region. For this reason, the U.S. Department of Agriculture (USDA) has created detailed climate maps of the United States and updates them occasionally. The most recent update was in 2012. These plant hardiness zone maps help farmers and gardeners alike understand which plants are suitable for their location, based on the average extreme minimum temperature, or the average of lowest winter temperatures for a location. All plants are assigned a zone or zones in which they will grow. ■ Identify the scales of the two maps and explain the degree to which they indicate the conditions of an agricultural region. Identify factors, in addition to climate, that growers need to consider before planting crops.

Maps & Models Archive

Chapter 12

Malthus and the Green Revolution

As you know, Thomas Malthus theorized an eventual worldwide famine, resulting from exponential population growth that he anticipated would far outpace the world's capacity for food production. Events such as the Green Revolution, however, have shown that food production can be increased at rates Malthus had no way of imagining. ▌ Compare Malthus's theory to the actual results of the Green Revolution in India, China, and Mexico. Using what you learned in Chapters 4 and 12, explain how a Neo-Malthusian thinker would describe the results.

Malthus's Theory

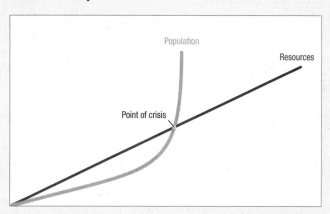

Index of Cereal Production and Land Use, 1961–1985

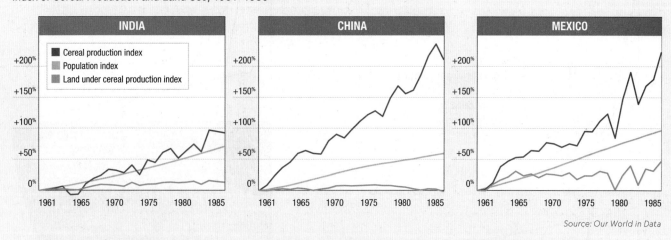

Source: Our World in Data

Chapter 12

Cattle Stations

In 2016, the Kidman family of Australia sold its cattle stations, or ranches, to a new owner. These ranches cover a reported 1.6 percent of Australia's territory. The largest station, Anna Creek, spreads out over 9,142 square miles, an area larger than the country of Israel. ▌ Identify the three largest cattle stations on the map, and explain whether the map shows a pattern of intensive or extensive agriculture. Identify information from other sources that may help explain Australia's agricultural patterns.

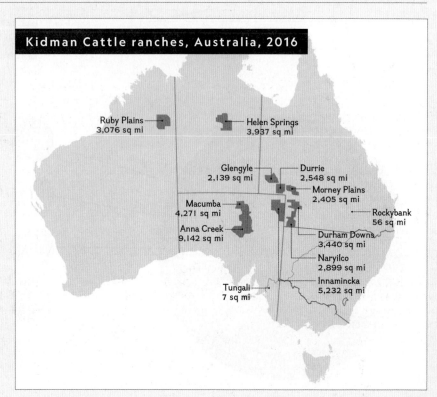

417

Unit 5

Chapters 12 and 13

Bid-Rent Theory and the von Thünen Model

The bid-rent theory and the von Thünen model use different visualizations to explain aspects of the same phenomenon: how land is used in relation to a market, or central business district. The two ideas can be considered independently or in a combined fashion to analyze spatial patterns in agriculture. ▎ Describe how the von Thünen model supports the bid-rent theory and how both account for the distribution of agriculture.

Bid-Rent Theory

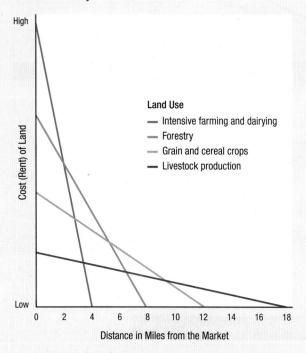

Bid-Rent Theory and the von Thünen Model

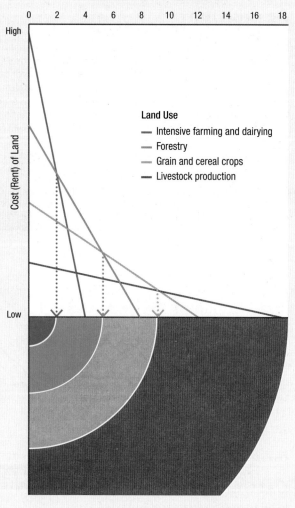

The von Thünen Model

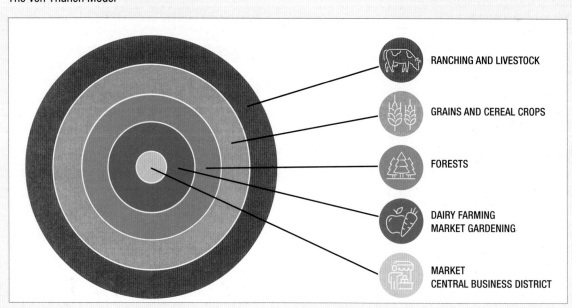

418

Maps & Models Archive

Chapter 13

Bananas: A Global Commodity

Bananas are a global commodity, with exports totaling $13.7 billion in 2017. Since the 1990s, banana crops have been threatened by a fungus that can destroy entire plantations and, at present, has no cure. This map shows the percentage share of the banana market claimed by the top five importing and exporting countries in 2017. ▌Identify the patterns in the global banana trade shown on this map. Explain the possible effect of the banana disease on both importing and exporting countries.

Chapter 13

Small and Large Farms

Worldwide, the distribution of small farms versus the distribution of larger entities varies a great deal across continents. In 2014, the Food and Agriculture Organization of the United Nations estimated that of the 570 million farms on the planet, approximately 84 percent were 5 acres or smaller, and 97 percent were smaller than 25 acres. ▌Compare the patterns of agriculture shown in these pie charts. What conclusion can you draw about the distribution of farms around the world?

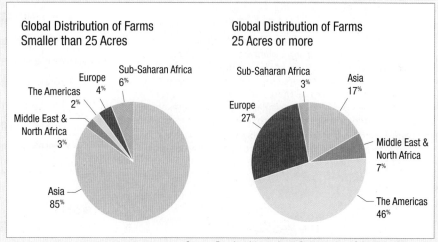

Source: Food and Agriculture Organization of the United Nations

419

Chapter 14

How Crops Are Used

In 2014, National Geographic created a map highlighting the percentage of crops used for animal feed and fuel versus those destined for direct human consumption. Researchers found that 55 percent of food-crop calories directly nourished people, while meat, dairy, and eggs from animals raised on feed supplied another 4 percent for direct nourishment.

❚ Identify which regions grow more crops for feed and fuel. Explain the degree to which this map explains issues that surround the global food supply.

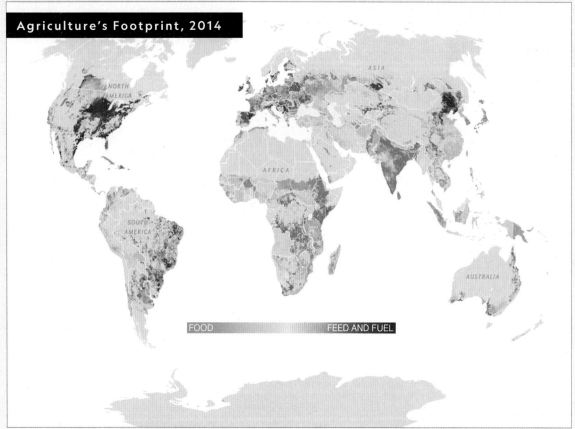

Source: National Geographic

Chapter 14

Land Use and Nutrition

One solution that has been suggested to fight global hunger is a large-scale switch to eating more plant-based foods. Proponents of this idea point to the amount of land required to feed animals that in turn feed humans. If more crops were directly consumed by humans, supporters argue, there would be more food for everyone. This graph reflects a calculation of the average amount of land needed to produce an ounce of protein in different animal and plant-based food sources. ❚ Describe what the data in this graph reveals about agricultural production and consumption.

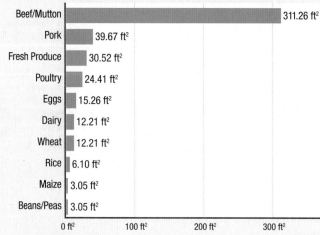

Land Use per Ounce of Protein, by Food Type

Food	Land Use
Beef/Mutton	311.26 ft²
Pork	39.67 ft²
Fresh Produce	30.52 ft²
Poultry	24.41 ft²
Eggs	15.26 ft²
Dairy	12.21 ft²
Wheat	12.21 ft²
Rice	6.10 ft²
Maize	3.05 ft²
Beans/Peas	3.05 ft²

Source: Our World in Data

Maps & Models Archive

Chapter 14

Where Food Is Lost

Eliminating food waste is one possible path toward improving global food security and supporting farmers who struggle to make a living. This chart shows food lost or wasted at different points along the global supply chain. ▍ Describe where the most food waste occurred in 2017, according to the charts. Explain why food might be wasted at the different stages of the process from production to consumption.

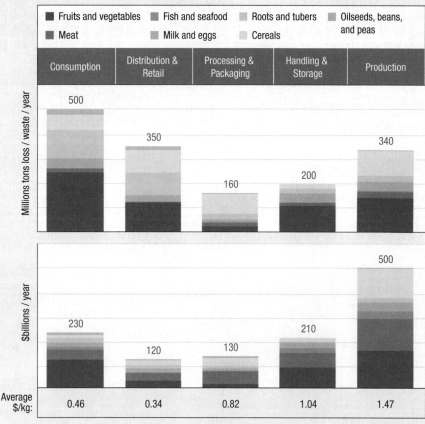

Global Food Waste Across the Supply Chain

Source: Food and Agriculture Organization of the United Nations

Chapter 14

Threats to Biodiversity

Globally, biodiversity is threatened by expanding agriculture and a number of other human causes, including climate change, the expansion of cities and suburbs into natural areas, and the breaking-up of natural habitats into smaller sections. This projection predicts the pressures on biodiversity from several agricultural and other human-related causes. ▍ Identify the projected trends in pressure on biodiversity from different forms of agriculture. Using the information about population projections that you learned in Unit 2, provide a possible explanation for why the threat from crops and pasture does not increase a great deal between 2030 and 2050.

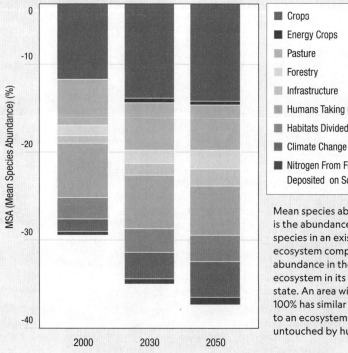

Pressures on Biodiversity by Year

Mean species abundance (MSA) is the abundance of original species in an existing ecosystem compared to their abundance in the same ecosystem in its undisturbed state. An area with an MSA of 100% has similar biodiversity to an ecosystem that is untouched by humans.

Source: European Commission Joint Research Center

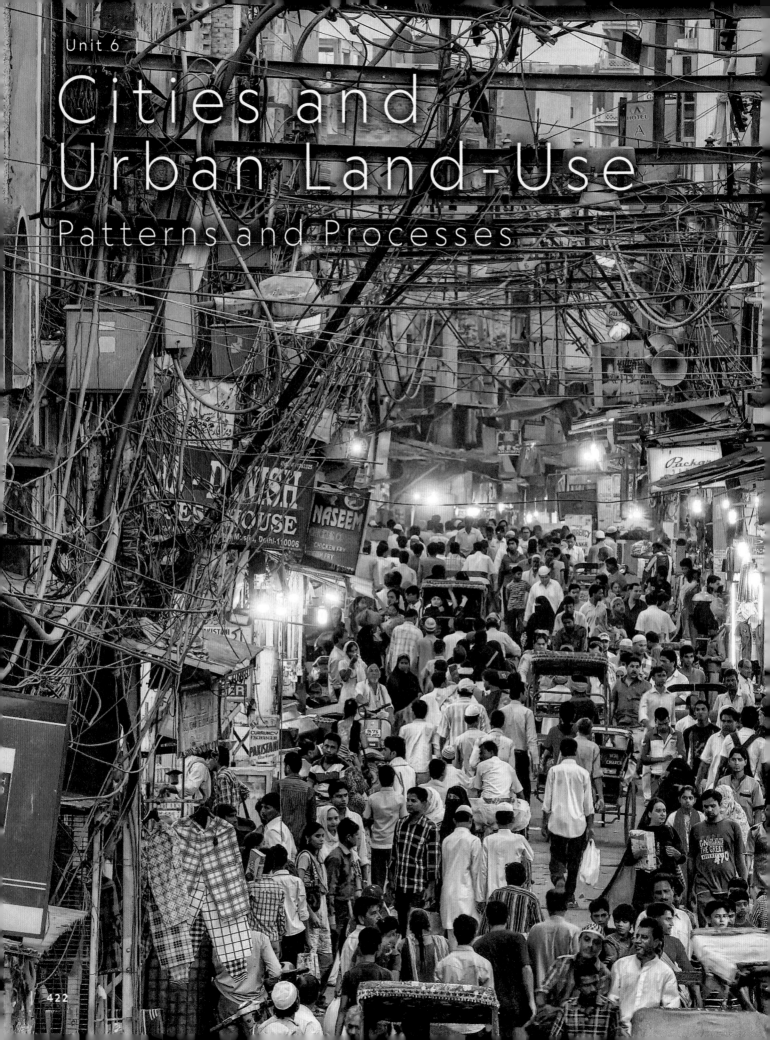

Unit 6
Cities and Urban Land-Use
Patterns and Processes

An Urban World

Housing development in Ixtapaluca, Mexico

Well over half of humanity lives in cities, and that proportion is growing. Many cities, too, are expanding—to such an extent that new concepts, such as *megacity*, have been invented to describe them. With the growth of megacities like Delhi, India (shown here), which has 31 million inhabitants, comes challenges in areas like food supplies, transportation, education, housing, employment, and migration. The ways in which governments handle these issues will affect the lives of the majority of humans on Earth.

Although cities differ vastly in terms of their locations and inhabitants, they have many patterns and trends in common relating to their origins and the ways they grow and change. Geographers study data from myriad sources about the people, cultures, power structures, and economies of cities in order to gain a deeper understanding of these vibrant, complex, present and future homes to billions of humans.

Chapter 15
Urban Settlements

Chapter 16
The Urban Landscape

Chapter 17
Urban Living

Unit 6 Writing across Units, Regions, & Scales

Unit 6 Maps & Models Archive

National Geographic Explorer **Daniel Raven-Ellison**

Geography for the People

Daniel Raven-Ellison's definition of geography is more expansive than most. The National Geographic Explorer says, "Geography is anything that happens somewhere, which means that everything has a geographic place." Following his singular vision, Raven-Ellison has created original ways to teach geography to children, pioneered the genre he calls guerrilla geography, and turned London into the first National Park City.

Learning Objective
LO 15.6 Explain how cities grow and change as a result of economic development and government policies.

Guerrilla Geographer According to Daniel Raven-Ellison, "Guerrilla geography is about encouraging people to see the world differently." Along with fellow members of an organization called the Geography Collective, he has been putting that philosophy into tangible action. One of Raven-Ellison's acts of guerrilla geography, for instance, was a series of urban walks in which he crossed several cities—such as Mexico City and London—on foot, passing through the most disadvantaged neighborhoods and taking a forward-facing photograph every eight steps. The result was a group of films that illuminated the cities from a completely new perspective. In another eye-opening geographical demonstration, Raven-Ellison and a group of students spread out bedsheets over an extensive outdoor field to explore the idea of an ecological footprint—the quantity of productive land that is used to support the lifestyle of a young person in the United Kingdom. Raven-Ellison describes guerrilla geography as "creative, alternative, radical, strange, exceptional," and he makes it clear that it can—and should—be practiced by everyone.

National Park City In 2014, Raven-Ellison started a campaign that engaged London's residents and leaders in a novel vision of the city's geography. Why, he asked, shouldn't urban nature be protected in the same way that governments protect the ecosystems found in national parks? "A city park or garden can be more ecologically diverse and rich, and provide more pleasure to a community, than a much wider area of land somewhere more remote," he says. "What is it about our prejudice that means we value that space less, despite the fact we know it is even more important because of the benefits to our health and productivity, as well as the economy and business?" Raven-Ellison's campaign came to fruition in July 2019, as Mayor Sadiq Khan officially declared London a National Park City.

In fact, the city of London is greener than many might imagine, harboring about 15,000 species and with green space covering around 47 percent of its surface. One of the goals of the National Park City movement is to help Londoners appreciate these natural areas, which are a blind spot for many on their daily rush through the city. Another goal is to expand London's green footprint by taking advantage of the city's vertical geography and adding more green roofs and wall gardens. In this way, the National Park City exemplifies a key facet of guerrilla geography: educating the public about humans' role as geographic place makers. Raven-Ellison explains, "Wherever you are, you are shaping and making places, and whenever you buy or consume something you help to shape or make a place elsewhere." Although the National Park City is not a governmental body with the legal ability to create rules or legislation, Raven-Ellison and his supporters believe it will influence policy through public forums and dialogues with stakeholders.

Raven-Ellison likes to imagine how children will learn to connect with nature in the National Park City of London. "Imagine them in 20 years," he says, "as they grow up to become architects or designers or planners. What ideas and solutions might they have to make us even healthier, even happier, and make this city even better?"

Geographic Thinking

Describe how the National Park City designation might influence future urban design initiatives in London.

Top: Greenwich Park, which has existed since 1433, covers 183 acres on the south bank of the River Thames. *Bottom left:* The red fox is a common sight in London. Its strong stomach and immune system make it well suited for the urban environment. *Bottom right:* National Geographic Explorer Daniel Raven-Ellison.

Chapter 15
Urban Settlements

Critical Viewing The Buddha Tooth Relic Temple stands in the center of the Chinatown district of Singapore, a city-state of more than 5.8 million people. ∎ What factors might have influenced the growth of Singapore?

Geographic Thinking What factors initiate and drive the growth of cities?

15.1
The Origin and Influences of Urbanization

National Geographic Explorer
Michael Frachetti

National Geographic Collaborator
Farhod Maksudov

National Geographic Explorer
Eduardo Neves

15.2
Factors That Influence Urban Growth

Case Study: Re-urbanizing Liverpool

15.3
The Size and Distribution of Cities

National Geographic Feature: City of the Future

15.4
Cities and Globalization

Case Study: How Shanghai Grew

National Geographic Feature: The Shape of Cities

15.1 The Origin and Influences of Urbanization

Many factors attract people to cities. Some are drawn by economic opportunities. Others come for the excitement of life in the city. A city often serves as a region's political, economic, cultural, and educational center. Many cities are also cosmopolitan centers, home to a variety of ethnicities and vibrant cultures.

The Growth of Cities

Learning Objective
LO 15.1 Describe historical and contemporary processes of urbanization.

For most of human history, the vast majority of people lived in rural areas. During the Neolithic Period (about 10,000 B.C.E.), agricultural advances allowed more people to live in permanent or semipermanent settlements. Archaeologists have defined this process of settling down as **sedentism**, requiring the development of new ideas about land ownership and property, ways to store harvested crops, and the accumulation of wealth. Some small settlements grew into larger ones, with an increased density of population and the specialized use of built space. People settled in these early cities for a variety of purposes, including trade, defense, and religion. The city of Uruk, for instance, was founded around 4500 B.C.E. along the banks of the Euphrates River, a critical waterway for transportation. Uruk's location facilitated its rise as a trading center to become the largest and most important city in ancient Mesopotamia.

Urbanization—the processes that lead to the growth and development of dense concentrations of people into settlements—also led to the rise of cities in places such as the Nile River Valley, the Indus River Valley, and the Wei River Valley in China, but for thousands of years only a small minority of the world's population was made up of urban dwellers. Urbanization increased rapidly during the Industrial Revolution in Europe and North America, as factories sprang up in cities and attracted people in search of jobs. Since that time, new cities have developed around the world, and many existing cities have grown in area and population. Today, according to the World Bank, more than 57 percent of the world's population lives in urban areas.

That number will continue to grow. The UN projects that by 2050, 68 percent of the world population will be urban. The growth of cities has occurred quickly, rising from 751 million urban dwellers in 1950 to 4.46 billion in 2021. Depending on how a country defines what an urban center is, the numbers might be even higher. In fact, after analyzing satellite images of Earth's inhabited regions, some researchers estimate the world's urban population to be significantly higher—as high as 85 percent. In short, cities are an important part of how humans live as well as an important topic in the study of human geography.

An **urban area** is defined as a city and its surrounding suburbs. Beyond this basic definition, the concept of what is urban varies considerably. The United States Census Bureau, the agency responsible for gathering and producing data about the American people, defines urban as "densely developed territory." After more than a century, the Census Bureau in 2020 revised its minimum threshold for a U.S. 'urban area' upward from 2,500 to 5,000 people.

Other countries use different criteria to define an urban area. According to the European Commission, 85 percent of countries define an urban area as having a population of 5,000 or more, but Mali sets the minimum threshold at 30,000, Japan at 50,000, and China at 100,000. India is among the countries that doesn't set a minimum number. Instead, it classifies as urban any place in which fewer than 25 percent of working men are employed in agriculture. Only a few countries consider population density as a measure of urbanization. Rural areas are generally characterized as open swaths of land with few buildings or other structures and a low population density. In most rural areas, agriculture is the primary activity.

Cities and towns provide for their residents in many different ways, including business, political, medical, financial, and educational services as well as infrastructure such as transportation and communication services. While a city has set political boundaries, its area of influence often extends beyond those boundaries. You may have heard people talk about the Greater New York area or the Los Angeles metropolitan area. A **metropolitan area** includes a city and the surrounding areas that are influenced economically and culturally by the city. The suburbs of a city are the less densely populated residential and commercial areas surrounding a city. Many people who live in suburbs commute into the city to work, to take advantage of city shopping and services, and to enjoy the city's amenities such as museums, restaurants, and cultural events.

Geographic Thinking

1. Explain some of the challenges geographers face when comparing the size of cities or urban areas in different parts of the world.

2. Compare a metropolitan area to a city.

Site and Situation

Learning Objective
LO 15.2 Compare how site and situation factors impact the origin, functions, and growth of a city.

The origin, functions, and growth of a city depend in varying degrees on its site and situation. Recall that a city's site is the condition of the actual place or location of the settlement and the land on which the city developed. Site factors include the landforms, climate, availability of water, soil quality, and natural resources of the land. A city's situation, on the other hand, refers to its relative location and connections between its site and other sites. A city with a favorable situation has easy access to trading partners, resources, and other connections. These advantages fuel growth and economic development.

Site Certain sites in the landscape are more likely than others to attract settlements. Features such as favorable topography, natural resources, location on a waterway or other trade route, and land that is easy to defend draw people to a site. Generally, people favor flat topography for building, but if defense is an issue, they may choose a site at an elevated location. In ancient Greece, for instance, Athens was built on hills, which gave the city a view into the distance—useful for detecting advancing armies—as well as a strategic advantage in the event of an attack because a location is easier to defend when a foe must climb upslope.

Key natural resources at or near a site, such as iron and coal deposits, a water supply, and waterpower, can also fuel the growth of settlements. San Francisco, California, for instance, has a superb site on a peninsula with one of the world's best natural harbors. However, it grew up seemingly overnight when gold was discovered inland in the mid-1800s. Kimberley, South Africa, grew up around the site of Africa's richest diamond mine, originally called the "Big Hole."

Sites located on transportation routes attract settlements as well. Early settlements often developed on islands and, as stated, natural harbors. Islands offer access to water routes and are defensible, as armies have to travel across an open sea, lake, or river to attack. Harbors provide boats and ships a safe haven from the waves of the open ocean, making them a good base for trade. Like San Francisco, New York City, Rio de Janeiro, Brazil, and Sydney, Australia, are all located on natural harbors. Other cities are located along a river's bend or where it narrows. In Quebec, Canada, two cities are located where the St. Lawrence River narrows—Quebec City on the north bank and Lévis on the south. Likewise, London, England, is located where the Thames River narrows. Because bridges are more easily constructed on the narrow part of a river, this enables people to settle on both banks.

Trade routes also played a significant role in the sites where cities are located. Rivers are historically important avenues for trade, and cities often grew up where two or more rivers meet. Pittsburgh, for instance, is located where the Allegheny and Monongahela Rivers meet to form the Ohio River. Cities also grew where routes converge, such as the mouths of rivers, where they pour into a lake or sea. With the advent of the railroad, railway junctions became influential factors in dictating the locations of new communities, sometimes with little consideration of good site characteristics. Businesses wanted to be strategically located along trade routes, as this decreased the time and cost of transporting goods. Cities also arose where goods had to be moved from one mode of transport to another, called break-of-bulk points. People were needed to provide the labor, and warehouses were built to house goods until the next leg of the journey. New Orleans (with site characteristics that you read about in Chapter 1) is a critical break-of-bulk port, with Mississippi River barges bringing bulk commodities downstream to be loaded aboard ocean-going vessels, which offload cargo from other world ports, to be warehoused, processed into a final product, or transferred to Mississippi River barges or loaded onto railcars and trucks for points inland.

Factors that make for a quality site can change as technology advances. The development of waterpower resulted in settlements forming along fall lines where waterfalls and rapids provided power for factories. A fall line is the narrow strip of land that marks the geological boundary between an upland region and a lower plain. The flow of water in rivers and streams speeds up as it descends across the fall line, resulting in falls and rapids. Waterwheels placed in this rapidly flowing water could be used to generate power for mills and factories. At the beginning of the Industrial Revolution, factories took advantage of the power provided by rivers along the Atlantic fall line, a 900-mile ribbon of land between the Piedmont and the Atlantic Coastal Plain, giving rise to some of the largest U.S. cities of this time.

Many cities were built along the Atlantic fall line, where mills and factories could take advantage of the power made possible by fast-flowing rivers.

Mexico City stands on the site of the ancient Aztec city of Tenochtitlán. The ruins of Templo Mayor, the main Aztec temple, are still visible today.

Situation A city's situation is equally important to its origins, functions, and growth—or decline. The relative location of a city often dictates its functions. Given the importance of trade throughout human history, it is not surprising to find urban areas along major trade networks. Aleppo, in present-day Syria, is one of the world's oldest continuously inhabited cities, due in large part to its strategic position at the crossroads of trade routes, including the legendary Silk Road. The cities of Samarkand, Uzbekistan, and Ki'an, China, also grew as a result of the Silk Road.

The advantages of a city's situation might change over time as new technologies lessen the impact of old connections. Before the Erie Canal was completed in 1825, for instance, the port cities of New Orleans, Philadelphia, and Baltimore were more influential ports than New York City. But once the new canal connected the Hudson River and New York City to the Great Lakes, it gave the city unparalleled access to the Midwest. The Erie Canal dramatically improved New York City's situation, making it the commercial capital of the country.

As trade networks change, settlements grow into cities to take advantage of the new connections. In the United States, the Transcontinental Railroad gave rise to the city of Omaha. In Russia, the Trans-Siberian Railroad similarly led to the development of Novosibirsk. Today, as globalization fuels the transport of goods across the ocean, growth occurs in cities that have ports, where goods and people depart and arrive from numerous points of origin. Indeed, most of the world's largest cities are port cities, located on or near the sea.

The city of Tenochtitlán provides a striking example of the role that site and situation play in the life of a city. In the mid-1300s, the Aztecs moved their capital from a hill in the Valley of Mexico to a marshy island near the western shore of Lake Texcoco. The city they built there, Tenochtitlán, served as the capital of the Aztec civilization until European conquest in 1521. Boasting a population of over 200,000, it was likely the largest city in pre-Columbian America. Building the city on the site the Aztecs chose required significant technological ingenuity. Three causeways connected the city to the mainland, with bridges that could be removed to allow boats to pass or in case of attack. Canals were built to transport goods from one part of the city to another. Waterways facilitated transportation and trade within the city and with communities lying beyond it. The Aztec capital was centrally located within and well connected to other points in its empire. Today, Mexico City, with more than 22 million inhabitants in its metropolitan area, thrives on and around the site where the Aztecs built their capital.

Geographic Thinking

3. Compare site and situation and describe factors related to each.

4. Describe the role that both site and situation played in the location of Tenochtitlán.

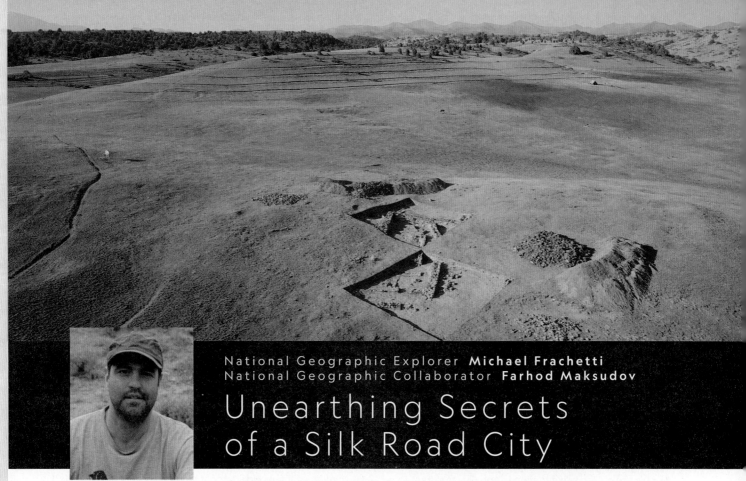

National Geographic Explorer **Michael Frachetti**
National Geographic Collaborator **Farhod Maksudov**

Unearthing Secrets of a Silk Road City

Frachetti (*top*) collaborates with co-director Farhod Maksudov (*bottom*) to better understand sites like Tashbulak in present-day Uzbekistan.

Learning Objective
LO 15.2 Compare how site and situation factors impact the origin, functions, and growth of a city.

Some of the same factors that drive urbanization today, such as site, situation, and the connections between settlements, influenced the origin and growth of cities in the ancient world. Archaeologist Michael Frachetti and his collaborator Farhod Maksudov have spent years using high-resolution satellite imagery, 3-D models, and computer simulations to predict, discover, and document ancient cities found in Central Asia.

In 2011, Frachetti and Maksudov made a startling discovery on a plateau 7,000 feet above sea level in present-day Uzbekistan. Buried under 20 inches of topsoil were the remains of Tashbulak, a vast, ancient city (shown above). The find raised questions relating to site and situation, including who lived there and why they built a city so high in the mountains. Over several years, Frachetti and his team set about excavating the site. They used remote sensing to conduct meticulous surveys of the site, hoping to use a better understanding of the city's layout to direct the dig and provide insights into life in Tashbulak.

Tashbulak was built more than 1,000 years ago by a nomadic civilization called the Qarakhanids. The city's site provided rolling green hills where the Qarakhanids could graze livestock and iron deposits they could mine for tools. And its situation provided connections to the wider world—Tashbulak was located along the Silk Road, a network of trade routes that contributed to the exchange of ideas among diverse cultures in Europe and Asia for more than 1,500 years. At the site, the archaeologists unearthed clues about the city's culture: fragments of tools, ceramics, glass beads, coins dating to about 975 C.E., and a peach pit—evidence that food traveled along the Silk Road from East Asia.

Geographic Thinking

Identify and describe the site and situation factors that likely influenced the location of Tashbulak.

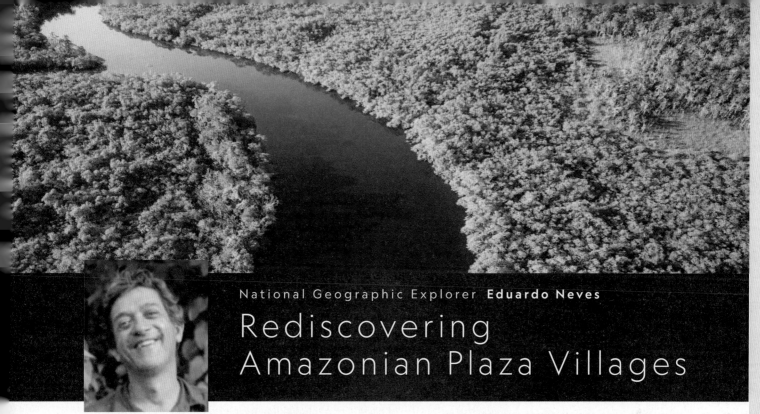

National Geographic Explorer **Eduardo Neves**

Rediscovering Amazonian Plaza Villages

Eduardo Neves has practiced archaeology in the Amazon for more than 30 years. He is a Professor of Archaeology at the University of São Paulo in Brazil.

Learning Objective
LO 15.1 Define historical and contemporary processes of urbanization.

In precolonial Amazonia, Indigenous people did not live in cities as we might define them today. However, beneath the dense cover of rain forest vegetation, archaeologist Eduardo Neves and his colleagues have discovered new evidence of villages laid out in regular patterns and connected by an extensive network of roads.

Neves's research focuses on the state of Acre in southwestern Brazil. Archaeologists long believed that indigenous settlements in this region were located along rivers and that the people traveled mainly by river. But more recent analysis of aerial data, combined with excavation of sites revealed by the data, tells a different story. Between about 1300 and 1600 C.E., people lived in plaza villages throughout the region, not only near rivers. Each village consisted of 15–25 mounds surrounding a circular or oval-shaped plaza, the mounds providing elevated platforms for residential structures. Radiating from each village were well-maintained roads that led to other villages.

Besides the consistent layout of the plaza villages, Neves and his colleagues have found similar ceramic fragments from the same historical period in villages across the area. This evidence suggests a thriving regional culture whose people traveled regularly between villages to maintain social and cultural ties. The archaeologists believe that the road network may have extended even further, connecting with other regions of the southern Amazon.

The Indigenous people who lived in the plaza villages, along with their ancient forebears, left their mark on the forest biome. Neves calls the Amazon "a cradle of plant domestication" because for thousands of years its Indigenous peoples have cultivated plants to produce foods that are eaten around the world today. Areas that appear to be "pristine" wilderness in fact reflect the work of many generations in managing the forest. As deforestation accelerates under the pressures of modern agriculture and mining, Neves believes it is essential and urgent to collaborate with the people whose ancestors created the forest and who know best how to manage it.

Geographic Thinking

In what ways do the interconnected plaza villages discovered in Acre fit—or not fit—the definition of an urban area?

15.2 Factors That Influence Urban Growth

Cities grow, change, and evolve. Some changes occur rapidly, whereas others happen gradually. Changes in transportation and communication networks, population, economic development, and government policies all influence how cities grow—or sometimes decline. A city's growth can blur boundaries as it expands into surrounding areas.

Transportation and Communication Networks

Learning Objective
LO 15.3 Explain how advancements in transportation and communication technologies impact urban growth.

One of the most influential factors in urban growth is transportation. Waterways, railroads, and highways provide a means for raw materials to get to factories, for goods to get to market, and for workers to access jobs. Ongoing advancements in transportation and communication technologies continue to influence urban growth. Transportation systems often make it possible for manufacturing and warehousing facilities, as well as retail and office complexes, to relocate to suburbs and beyond, joining and expanding what had previously been largely residential areas.

Advances in transportation had a profound impact on settlement patterns in the United States. During the colonial period, the earliest cities—such as Boston, New York, and Savannah—developed on the Atlantic Coast or along rivers in order to facilitate trade. As the economy developed and settlement expanded, interior cities grew up on rivers with good access to the coast. Philadelphia, which was the largest city of the colonial era, benefited from its location at the mouth of the Schuylkill River, where it flows into the tidal Delaware River and the Atlantic. It grew in large part from its thriving trade with the West Indies. Following the American Revolution and the Louisiana Purchase of 1803, New Orleans also rose in prominence due to its great strategic situation near the mouth of the Mississippi River along the Gulf of Mexico. (See the Case Study on site versus situation in New Orleans in Chapter 1, as well as the video in Unit 1.)

The site of present-day Chicago was first settled by African-French Canadian Jean Baptiste Point du Sable. The thriving village had a solid defensive position, and Fort Dearborn was built there in 1803. But Chicago had no particular site or situation advantages when it came to transportation, until 1825 when the Erie Canal connected the Great Lakes to the Hudson River, and thus to New York City and the Atlantic Ocean. This route gave Chicago, located on the southwest shores of Lake Michigan, access to the East Coast and transatlantic shipping routes. The advent of the railroad sealed Chicago's destiny as a transportation hub. Chicago was the meeting point of railroads stretching to the east and the west, and the city became the central connecting point between the farms and ranches of the west and markets in the eastern United States and Europe. Grain processing mills, warehouses, stockyards, and meatpacking plants sprang up, and Chicago's population grew from just a few thousand to more than a half million in less than five decades. Today, 50 percent of rail freight in the United States continues to pass through Chicago. The city is an example of how an urban area can thrive with unexceptional site factors but an advantageous situation.

Communication advances have also influenced the growth of cities. For centuries, the main communication networks were the trade networks, as information traveled along with goods. This changed in the 19th century, with the advent of the telegraph and, later, the telephone. These and later advances in telecommunications technologies revolutionized business and how people lived and worked.

The telephone made business more efficient. Factories located in cities began to take phone orders directly from customers located far from the city. This increased production, which in turn required more workers. More recently, similar changes have taken place due to communication and computer technologies. Today, high-speed internet and Wi-Fi services allow businesses to instantly communicate, both internally and with their customers. Like advances in transportation, these advances significantly reduced the cost of transmitting and communicating information over long and short distances.

Like any change, advances in communications do not affect all areas equally. Because there are more customers in an urban area, communication companies provide service to these areas first. Meanwhile, new businesses look for locations where there are strong communication networks. Access to the internet is as important to future growth as access to rivers and highways had been in the past.

The telecommunications industry itself also fuels urban growth—the industry is big business with huge corporations that employ tens of thousands of people. Like other high-tech and computer companies, telecommunication companies tend to locate where there is access to an educated workforce. Today's technological hubs draw workers from other areas in the same way factories did a century ago, contributing to urban growth.

Population Growth and Migration

Learning Objectives
LO 15.4 Identify the factors influencing rural-to-urban migration.
LO 15.5 Describe the impact of rapid population growth on cities.

Rural-to-urban migration is a driver of urbanization, due to a combination of push and pull factors. As the population of a rural area grows, there are often not enough opportunities for all to make a living there, which pushes people to move away from the area. For farmers, drought or other environmental or economic push factors may cause them to move. There simply may not be enough land for new generations to farm. For many reasons, these migrants are often drawn to cities. As you read in Chapter 5, cities offer jobs—or at least the perception of better work opportunities—to potential migrants. For some newcomers, cities offer greater freedom, safety, schools, health care, and more. Beginning with the Industrial Revolution in Europe and North America, as countries have developed economically, urbanization has followed. Today rural-to-urban migration drives urbanization mainly in peripheral and semi-peripheral countries.

Throughout the United States in the 1800s, settlements developed at railroad terminals and grew in size and density to eventually become cities. New towns emerged at regular intervals along the railroad as layover points for passengers and goods. Kansas City, Missouri, for instance, experienced a population explosion following the completion of a railroad line from St. Louis in 1865 and the construction of a bridge across the Missouri River that linked it with the Hannibal and St. Joseph Railroad.

More recently, the region south of San Francisco known as Silicon Valley has experienced massive population growth. Computer manufacturers flocked to the area beginning in the 1950s and, in fact, the region got its name from the silicon used in computer chips. Over the past several decades, the economy of Silicon Valley has gradually shifted from computer manufacturing to high-tech research and development. The population of San Jose, Silicon Valley's largest city, grew from roughly 200,000 in 1960 to over 3 million by 2020. Like other postindustrial centers, Silicon Valley has struggled with its rapid growth. The influx of highly educated—and highly paid—tech workers has caused rents to skyrocket and put housing in short supply. Many people who don't work in the tech industry can no longer afford to live in the area, and homelessness is a growing issue.

Sometimes called the "Silicon Valley of India," Bengaluru (formerly Bangalore), India, is another example of a postindustrial city with a strong information technology presence. The city is situated along a major national highway and a regional rail hub. In 1998, an industrial park opened near the city's center, attracting hundreds of technology, software, and telecommunication companies. The rapid growth has contributed to a housing shortage. Experts estimate that one-quarter of the city's residents, or some 2.2 million people, live in substandard housing.

Shenzhen, China, a coastal city just north of Hong Kong, had a population of just 30,000 in 1979, when the government of China reduced restrictions on foreign investment. From that point, the city grew at an extremely fast rate. Workers and professionals flocked to Shenzhen to take advantage of opportunities in pharmaceutical and textile factories. The high-tech industry has also become an increasingly important part of the economy. Among those drawn to Shenzhen in search of jobs are migrant workers. Historically, these were farmers who worked in the city only part of the year, but the city's growth has engulfed surrounding farmland, and many now live in the city year-round. Their temporary residency status prohibits them from receiving equal access to government resources.

Economic Development and Government Policies

Learning Objective
LO 15.6 Explain how cities grow and change as a result of economic development and government policies.

Although the economies of cities vary considerably, all cities serve important economic functions. A city's functions depend largely on its location and its history. The largest, most influential cities tend to be centers of diversified business services and government or public-service centers. Other cities serve more specific and specialized functions. The main function of a capital city like Washington, D.C., is government, with its various administrative, legislative, and judicial activities. Some cities

A city's economic function can change over time. Bruges, Belgium, for example, began as a manufacturing center in the 1200s. Today, the city's striking historical architecture is the basis for its current primary function—tourism.

are military centers, some are processing sites for mines, and some are known for manufacturing. Detroit, Michigan, still serves as the center of automobile manufacturing in the United States. Houston, Texas, functions as a center of the oil, gas, and energy industry. Some cities are consumer-oriented centers. These include cities that attract retirees, like Sun City, Arizona, as well as resort communities such as Orlando, Florida, and Cancún, Mexico.

The economy of Pittsburgh, Pennsylvania, revolved around one main industry—the iron and steel industry—in the 19th and 20th centuries. This foundational economic activity, sometimes referred to as a *basic industry* because it produces goods for export, gives rise to additional economic activities, or *nonbasic* industries, that support the basic industry. These manufacturing and service-based industries meet people's needs for housing, food, transportation, and other goods and services. Together, they contribute to a city's economic development.

The functions of a city tend to change over time as a result of technological advances or changes in historical, economic, or population trends. Bruges, Belgium, for instance, developed as an important port and woolens manufacturing center in the 13th century. However, it became less significant as competing ports such as Antwerp developed and the wool trade declined. Today, it is one of the most visited tourist destinations in Europe, with its entire historic center of medieval canals, squares, and structures designated as a UNESCO World Heritage Site—attracting 8 million visitors a year.

While most cities have expanded as the world's population has grown and become increasingly urban, some cities experience a decline when their economic or other functions are no longer relevant. Cities in New England, the Mid-Atlantic, and Midwest, a region now characterized as the Rust Belt, were industrially powerful hubs with active businesses and retail and service sectors. Such **legacy cities** declined in prosperity, prestige, and population as their manufacturing bases were lost to lower labor costs elsewhere in the country, foreign competition, and globalization. The population of the city of Detroit, for instance, has declined as automobile manufacturing has expanded to other parts of the United States and other countries that offer less expensive labor. Detroit's population peaked with 1.85 million residents in 1950 but has declined for seven decades straight: the 2020 U.S. Census counted just 640,000 people remaining. Other legacy cities in the Rust Belt, like Cleveland, Ohio, and Allentown, Pennsylvania, also thrived during the U.S. manufacturing boom in the first half of the 20th century but then fell into economic and population decline as industry moved to the South. Some cities that have experienced a decline in manufacturing or other industry have transitioned to service industries. When they attract new businesses and professionals, these economic changes can reverse a city's decline. Pittsburgh, Pennsylvania, mentioned previously, lost its economic base as the American steel industry retracted. But the city was able to retool its economy, focusing on education, high-tech computing, medical research, and skilled manufacturing. Today, Pittsburgh is once again a thriving urban environment. Economic sectors will be further discussed in Chapter 18.

Government policies can also influence urbanization. Governments at all levels seek to attract businesses and boost the economy. At the regional level, city governments may compete with one another by offering tax incentives or financial inducements for businesses to relocate. Or local governments may join together in regional alliances to market a region's advantages for economic development. Local governments may create industrial parks or zone huge tracts of land for industrial or other commercial uses. Governments also enact land-use plans and zoning ordinances to separate heavy industry from residential areas, while also providing transportation linkages for workers and freight, such as roads or rail lines. Together, these policies can draw businesses to a city.

Anything that makes a city more attractive will contribute to its growth. Safety and security are important to businesses and residents alike. Cities with low crime rates and high levels of governmental services may grow faster than comparable cities with higher levels of crime or less stable administration. City governments also seek to draw business and residents with policies that encourage livability—the combination of factors that make one place a better place to live than another. A city's livability factors include housing, transportation, the environment, health and public services, civic life, and economic opportunities. Governments seek to improve livability by providing access to public transportation, quality education, and reliable and efficient city services and assuring the availability of affordable housing.

Suburbanization, Sprawl, and Decentralization

Learning Objectives

LO 15.7 Explain how transportation helps to shape the patterns of urban growth and decline.

LO 15.8 Describe efforts taken to combat the negative effects of urban sprawl.

In the United States, changes in urban transportation during the 19th and 20th centuries led cities to grow in area as people moved outward from the city center. With the development of networks of trolleys, or streetcars, workers no longer needed to live within walking distance of work. This decentralization caused new areas, sometimes called "streetcar suburbs," to develop outside of the core areas of cities. Commuter rail lines gave cities more access to surrounding areas, resulting in "railroad suburbs." New highway development connecting central cities with outlying areas resulted in the growth and expansion of more suburbs even farther from the center.

The process of suburbanization causes the land area that a metropolitan area takes up to expand, but the

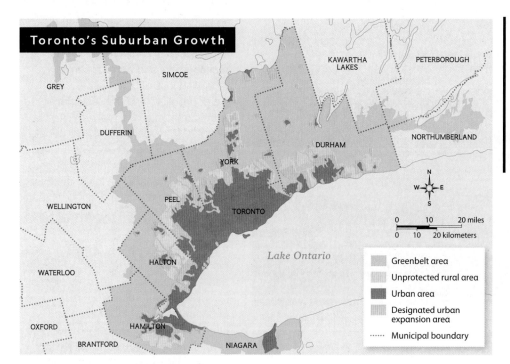

Reading Maps The map highlights the enormous growth taking place in the suburbs north of Toronto and the ways the Canadian government has tried to regulate this growth. The Greenbelt area, for example, has been protected by legislation since 2017. ▌ Explain how the map helps you understand the decentralization of urban areas in and around Toronto.

population of the core city does not necessarily grow. As cities grow outward into surrounding areas, the amount of land per person increases. The relatively inexpensive land that surrounded cities allowed developers to build neighborhoods of single-family homes. These developments often featured tract housing, multiple homes that are similar in design and building materials, that could be built quickly. Upper- and middle-class families were drawn to the suburbs by the promise of low crime, good schools, and more land for larger homes and yards.

Sometimes urban areas expand in an unplanned and uncontrolled way, covering large expanses of land in housing, commercial development, and roads. This process is called **urban sprawl**. While the central cities of metropolitan areas are generally compact, densely settled places with well-planned land uses, street networks, and infrastructure systems, sprawl is a result of chaotic urban growth. As land is developed at the edges of the urban area, often with no overall plan, infrastructure may not keep up and issues of sustainability become important.

Urban sprawl has become common in U.S. metropolitan areas, particularly in cities that grew up with the automobile and freeway expansion, such as Atlanta, Los Angeles, and Dallas. While suburbanization has been occurring in the United States since the 19th century, it was only after World War II that urban sprawl became an issue. The growing popularity and affordability of the automobile meant that residential areas were not limited to locations near streetcar or rail lines, and the postwar baby boom meant that many families were looking for single-family homes with a yard. Geographer Fabian Terbeck notes that today, suburbs in the United States can be categorized as either inner-ring suburbs or outer-ring suburbs. Generally, inner-ring suburbs are considered to be *mature* if they were built before 1970;

most outer-ring suburbs were developed after 1970. Terbeck also identifies a range of suburb types today, including low-income suburbs, African American and immigrant suburbs, and manufacturing suburbs—not just the stereotypical White, middle-class entities that once were predominant in suburban America.

The expansion of cities has given rise to new land-use forms. The term **edge city** describes a type of community located on the outskirts of a larger city. Edge cities are commercial centers with office space, retail complexes, and other amenities typical of an urban center. Over the years, developers have built residential housing in edge cities as well. Thus, edge cities are perceived as destinations for work, shopping, services, entertainment, and housing. Tysons Corner, Virginia, located at the intersection of several highways outside of Washington, D.C., is an example of an edge city. What began as a small commercial center today contains more than 25 million square feet of office space, is home to more than 100,000 jobs, and offers one of the largest retail areas on the East Coast. Hundreds of edge cities—a term coined by author Joel Garreau in the 1990s—have grown up on the outskirts of major cities across the country, two dozen in the sprawling Los Angeles metropolitan area alone.

A **boomburb** is a suburb that has grown rapidly into a large and sprawling city with more than 100,000 residents. The population of boomburbs often exceeds that of nearby big cities. While edge cities are largely occupied by office and retail space, boomburbs are often made up of many planned communities that have begun to merge together. Examples of boomburbs are Anaheim, California; Mesa, Arizona; and Plano, Texas. Just north of Plano is another boomburb, Frisco, Texas, that may be the fastest-growing city in the country, rocketing from 125,000 in 2012 to 215,000 in 2022—and expected to top 300,000 by 2030.

Pine trees and scrub oak burn behind homes at the Wadsworth Ridge fire outside Denver, Colorado on October 12, 2010. It is estimated to have burned 400 acres in an area considered a wildland-urban interface (WUI).

An **exurb** is a typically fast-growing community outside of or on the edge of a metropolitan area where the residents and community are loosely connected to the central city and suburbs. Exurbs are often low-density residential communities that are still accessible to existing urban centers. Some may consist of wealthy estates while others may offer less-expensive housing, but both with more space than could be found in a closer-in location. Exurban expansion is often out into areas with recreational amenities and natural beauty. Exurbs exploded in popularity in 2020 as remote work became more common during the COVID-19 pandemic. Rural areas have become exurbs as weekly or monthly rather than daily commutes have expanded the geographic area where people can choose to live. Fast-growing exurbs are found across the United States, including Blanco County, Texas, between San Antonio and Austin; Brown County, Ohio, roughly 50 miles from Cincinnati; and Williamson County, Tennessee, about 40 minutes outside of Nashville. Exurb development may be considered a form of gentrification, causing difficulties for existing residents of the newly developing areas as they may become priced out of housing; other conflicts concerning increased traffic, land use, and related issues may arise between new and old residents.

Critics attribute a number of negative effects to suburbanization and exurbanization, contending among other things that suburban communities lack identity or a sense of place. The dependence on automobiles contributes to traffic congestion, air pollution, and other environmental issues. In some instances, valuable agricultural land is lost to housing developments. Suburban and exurban areas may lack the tax base to provide adequate infrastructure to residents, such as road maintenance and garbage pickup.

Another issue, particularly with exurban development, is the increase in the **wildland–urban interface**, or WUI, where undeveloped wildlands abut new communities. There are significant implications of WUI for wildfire management and also concerns regarding loss of biodiversity, air and water pollution, and the sustainability of such communities, a topic revisited in Chapter 17. In many cases, the people who remain in the central city or inner suburbs are those who are too poor to move out. This leaves a lower tax base in these jurisdictions and can contribute to economic decline and urban decay.

Reducing Sprawl Urban planners have undertaken efforts to address urban sprawl. In many places, revitalization and redevelopment of decaying areas have helped to lure people back inside city limits. Revitalization focuses on instilling new life into a community by reusing or renovating buildings and beautifying an area through landscaping. Redesigning public spaces, improving the quality of street pavement, and even replacing sidewalks bring new life to urban neighborhoods. Redevelopment focuses on converting an existing property to another, more desirable use. Redevelopment can help address sprawl by creating new mixed-use neighborhoods where people can walk to public transportation, retail, and entertainment venues. Successful redevelopment efforts can invigorate a community by attracting business, stimulating the local economy, reducing crime, and instilling pride in a place.

Planners also use **infill** to address and counter sprawl. Infill is redevelopment that identifies and develops vacant parcels of land within previously built areas. Infill helps to counter sprawl because it focuses on areas already served by transportation and other public infrastructure. Infilling can decrease the reliance on automobile transportation, reduce the environmental impacts of development, and boost the local economy. You learn more about infill in Chapter 16.

Geographic Thinking

1. Explain why many U.S. cities developed along rivers in the 19th century. Would you predict these cities to experience future growth?

2. Compare the ways in which economic development and government policies drive urbanization.

3. Explain how the automobile has transformed modern American cities. How are American cities that evolved after the automobile different from earlier cities?

4. Describe how suburbanization and exurbanization relate to urban sprawl.

Case Study

Re-urbanizing Liverpool

The Issue Once a thriving industrial city, Liverpool experienced several decades of decline until the 1980s, when urban renewal advocates began a concerted effort to turn the tide.

Learning Objective
LO 15.6 Explain how cities grow and change as a result of economic development and government policies.

By the Numbers

Population of Liverpool

286,487
Population in 1841

846,101
Population in 1931

439,428
Population in 2001

494,800
Population in 2018

Source: Office for National Statistics, United Kingdom.

Today, the waterfront in Liverpool is a mix of old and new. Many older buildings, such as the Albert Dock, have been repurposed by developers in recent years.

Over the Past Two Centuries, Liverpool, a city in northwestern England, has experienced periods of wealth followed by decline and decay. Its initial prosperity was due largely to its site on a natural harbor where the River Mersey meets the Irish Sea. By the end of the 18th century, Liverpool's port had more dock space than London's. Industry was further fueled by the 1830 completion of the Liverpool and Manchester Railway, which gave Liverpool access to key industrial areas within Britain. Irish immigrants helped the city's population grow, particularly during the Irish Potato Famine of the 1840s. During the Industrial Revolution, the port provided access for England to markets worldwide.

Liverpool's population peaked at 870,000 in the 1930s, when it began to experience a decline in both prosperity and population that coincided with the collapse of Britain's manufacturing base. By the 1980s, Liverpool's population had dropped to less than 500,000, and its unemployment rate rose from 10.6 percent in 1971 to 21.6 percent in 1991—exceeding 40 percent in some neighborhoods.

At this time, the national government initiated a renewal—or "regeneration"—effort of the central business district. This included the revitalization of the waterfront for leisure use, the development of mixed-use buildings and downtown apartments, and the construction of a new waterfront area, conference center, and public spaces. Liverpool ONE, for instance, involved the redevelopment of 42 acres in the city's central district. Today, it is a mixed-use neighborhood with shopping, a cinema, an adventure golf center, apartments, offices, and outdoor public space. The Royal Albert Dock, which was opened in 1846, has become a major tourist attraction and houses the Merseyside Maritime Museum, the Beatles Story, and the Tate Liverpool art gallery and museum. In 2011, the city also opened the Museum of Liverpool, which is the largest national museum to open in the United Kingdom in over a century. These efforts have achieved their goal of increasing the number of visitors to the city. Liverpool is among the top tourist destinations in the United Kingdom.

Liverpool's renewal efforts have also provided a boost to the economy. The population of the city's center quadrupled in just 20 years. Critics contend that the prosperity has not benefited all segments of the population equally, with significant pockets of poverty remaining in the city. Liverpool's experiences mirror those of most industrial-era cities worldwide.

UNESCO granted Liverpool World Heritage status in 2004, based on the architectural uniqueness and historic value of its waterfront. However, UNESCO revoked that status in 2021, judging that continued redevelopment and infilling efforts had changed the appearance of the city. The UN body cited a deterioration in the waterfront's "authenticity and integrity" that had led to the city's earlier inclusion on the coveted list. Residents complained, arguing that cities such as Liverpool should not be left with a choice between maintaining heritage status or regeneration that brings in business, jobs, and tourists.

Geographic Thinking

Explain how government policies influenced urbanization in Liverpool.

15.3 The Size and Distribution of Cities

The distribution of cities occurs in predictable patterns. It is unusual, for instance, to find two large cities located near one another. Geographers use different methods and models to identify and explain how systems of cities are organized and what factors contribute to their size, distribution, and interaction.

Patterns of Urban Location

Learning Objective
LO 15.9 Describe the models that geographers use to understand the size, distribution, and interaction of cities.

Cities and towns do not operate independently. Rather, they are part of interdependent systems that function at regional, national, and international scales. The urban system includes networks of human interactions and connections with both the natural and the human-made environments. Urban places evolve according to the linkages among them. Traditionally, the most important linkages were the transportation routes between and among cities. More recently, communication linkages have evolved to become almost equally important.

Individual cities within any urban system differ in size and influence. The ancient cities of Mesopotamia competed for power and wealth, as did the city-states of ancient Greece. These civilizations gained power by conquering surrounding areas and subjugating other cities. In the Roman Empire, the city of Rome became the center of a vast territory. No other city could compete with its power or majesty. Control over transportation routes was critical to securing and protecting the power of the city. These routes brought wealth to Rome through trade, but they were also used to move armies within and beyond the territory and to communicate with and control cities at the outskirts of the empire.

As in ancient times, modern cities operate within an interconnected urban hierarchy. Different cities have different functions within the system, with larger, more influential cities landing higher on the hierarchy, while cities with smaller populations and economies fall lower on the hierarchy. Within this urban complex, cities may function as centers of finance, government, manufacturing, commerce, arts, education, or tourism. Geographers studying the relative sizes and spatial relationships of cities have identified common attributes and features and developed models to describe the urban network. These include the gravity model, rank-size rule, primate-city rule, and central place theory.

Gravity Model In urban geography, the gravity model is used to discuss the degree to which two places interact with one another. According to the model, cities have an

The Gravity Model

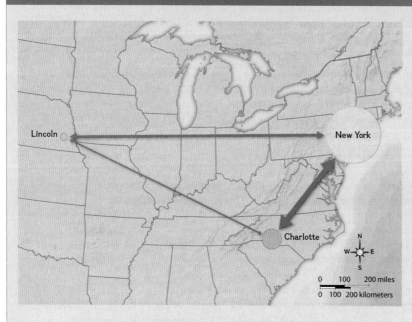

The gravity model is used to explain the relationship between cities of various sizes. According to the model, the level of spatial interaction—such as trade, traffic flow, number of visitors, and communication—between two cities depends on the size of the cities' population and the distance between them. Larger cities will interact more with each other than smaller cities, and larger cities will draw interaction from smaller cities nearby.

Consider New York City; Charlotte, North Carolina; and Lincoln, Nebraska. Because New York City and Charlotte are both closer in distance and have greater populations, their interaction is greater than that between New York City and Lincoln. Likewise, Lincoln's relationship with New York City is greater than its pull to Charlotte because New York City has a larger population.

438 Chapter 15

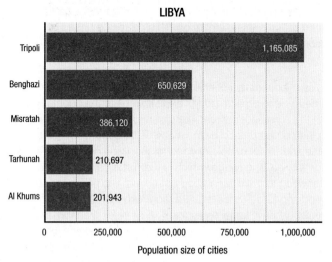
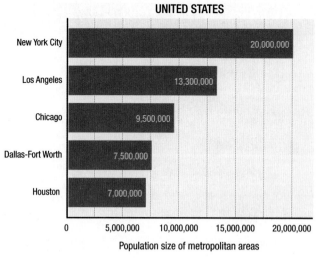

Population of the Largest Cities in Libya and the United States, 2020 The bar graphs help explain how the population size of cities and metropolitan areas within a country are distributed. Note that the population ranges on each graph are vastly different. ▎Explain the degree to which the graphs illustrate the rank-size rule.

Source: World Population Review.

area of influence based on their population size. Commerce, the flow of traffic, the number of visitors, and the number of phone calls and other communication linkages between people in two or more cities is affected by how large each city is. The model assumes that the interaction of people and goods between cities is generally proportional to the product of their populations (or output of goods) and inversely proportional to the distance between them.

The gravity model has been used to predict whether people will be more likely to visit, shop, or do business in one city over another. According to the model, a large city will have a greater pull than a smaller community.

It is important to note that the gravity model—like the other theories presented in this section—is based on the assumption that the two cities are on a flat surface with no natural, political, or cultural barriers. The model does not take into account natural barriers, such as rivers or mountains. Nor does it consider political boundaries that would require a person to cross into another country. So while the model is useful in making generalizations, other factors will come into play when applying the findings to real-world scenarios.

Rank-Size Rule The places people live, from tiny hamlets to megacities, are linked to form interconnected, interdependent urban systems. These urban systems consist of hierarchies of places of different sizes, ranging from very small to very large, with each serving different functions. The patterns of functional interaction between places within most countries' urban systems result in predictable relationships between the population size of cities and their rank in the urban system. The **rank-size rule** is a geographic concept that describes these relationships. The rule explains how the population size of cities within a country may be distributed. It uses a country's largest city as a baseline and ranks all other cities in relation to it. The rule states the second largest city will be one-half the size of the largest city, the third largest city will be one-third the size of the largest city, and so on. Not all countries exhibit such patterns; some follow the primate-city rule, which you learn about below.

In other words, if the largest city in a country has a population of 1 million, the second largest city will have a population of approximately 500,000, and the third largest will have a population of approximately 333,000 (one-third of 1 million). As another example, if the population of the largest city is 12 million, the second most populous city will have about 6 million people, the third will have about 4 million, and the fourth will have about 3 million. Libya, Nigeria, Australia, and Brazil are examples of countries in which the rank-size rule is fairly accurate. But note that the rule is not exact. It can be applied to the cities of some countries more than others, and there can be exceptions to the rule based on real-world factors. Geographers point out that the model does not take into account a country's area or the geographic distance between the cities. In addition, using the population of the greater metropolitan area, the core city and its suburbs, rather than the city alone provides a better reflection of where the city ranks in the country's hierarchy of cities.

Primate-City Rule In some countries, the rank-size rule does not apply because of the disproportionate size of one city. A **primate city** is a city that far exceeds, in population size and influence, the country's next largest city. Where a primate city exists, the country will have few or no other large or medium-sized cities. The existence of a primate city indicates an unbalanced level of development in a country because the population is disproportionately pulled to the city for economic opportunities. Therefore, primate cities have outsized political, economic, and cultural

Model: Central Place Theory

Legend:
- City
- Town
- Village
- Hamlet
- Higher-order market area
- Middle-order market area
- Lower-order market area

The orange circle in the center of the diagram represents a large city, which offers the greatest number and variety of goods and services. Towns, villages, and hamlets are distributed around the city in a hexagonal pattern, as supported by the market's threshold and range.

---

influence, which can prevent even development throughout the country.

For example, the population of Mexico City's metropolitan area is about 22 million. This far exceeds any other city in Mexico. The second largest city, Guadalajara, has a metropolitan population of roughly 5 million. Paris, France, has more than 11 million people, while the next largest metropolitan area, Lyon, has a population of about 1.7 million. Bangkok, Thailand, with a population seven times that of the country's next largest city, and Tokyo, Japan, the world's most populous city, are other examples of primate cities. Note that each of these cities is the capital of its country, as well as its economic and cultural center.

Many primate cities can trace their origin to a colonial past, when European colonizers concentrated political and economic activities in one place. The metropolitan area of Dakar, a port city developed by French colonizers that is the capital of Senegal, has a population of 3.1 million, nearly six times the size of Touba, the country's next largest city.

Central Place Theory

Learning Objective
LO 15.10 Describe the roles of threshold and range in Central Place Theory.

German geographer Walter Christaller developed **central place theory** to explain the hierarchical patterns in the number, size, and location of cities and other settlements. The theory, published in 1933, describes a central place as a settlement that provides goods and services for the surrounding area. There are several types of settlements within an urban hierarchy. The city is a large, central place around which smaller towns, villages, and hamlets are distributed. Central place theory observes a pattern and order in this distribution. This pattern is based on consumers' behavior and decision making regarding the goods and services on which they spend their money.

The main function of cities and towns, according to the theory, is to provide goods and services to the people living in the surrounding area. The size and location of a central place is determined by the market's **threshold**, which is the number of people needed to support a certain good or service, and **range**, which is the distance that someone is willing to travel for a good or service. High-order goods and services—those that are more expensive, desirable, or unique—have larger thresholds and ranges than low-order goods and services do. High-order goods and services are located in cities that have the minimum threshold to support them. High-order goods include items like luxury cars, high-end jewelry or fashion, and works of art, while high-order services include professional sports events, museums, and cultural festivals. These goods and services draw people from the towns, smaller settlements, and rural areas surrounding the metropolitan area. Low-order goods include those found in grocery stores and other common retail stores like bakeries and auto parts outlets, and low-order services include hair salons, barber shops, and post offices. Consumers are less willing to travel long distances for these goods and services.

These consumer choices create a hierarchy of places, from many hamlets and villages offering low-order goods and services to fewer towns and cities offering high-order goods and services. Within this urban hierarchy, cities serve larger populations and are located far apart from one another, while towns and other, smaller communities are closer together and distributed in a pattern between the cities. Cities provide a greater variety of goods and services than towns, and thus cities become the core around which increasingly smaller communities that offer fewer and less varied goods and services are found. Small settlements in an urban system provide only the goods and services that meet people's everyday needs, and these services will be close together because people are not willing to travel long distances for lower-order goods like groceries or gasoline. People in these smaller towns occasionally travel to a city to take advantage of higher-order shopping, a museum visit, or a cultural event.

The pattern that results is best illustrated as a series of hexagons with the highest-order market center—the city—in the center. Towns will be arranged around the city, equally spaced apart. Villages, which are lower-order market centers, are distributed around each town. Hamlets, the lowest-order market centers, are dispersed between each of the larger settlements. There are more hamlets than villages, more villages than towns, and more towns than cities in Christaller's urban hierarchy.

Central place theory has limitations. The theory doesn't account for real-world geography. Rather, it is based on a flat surface with no natural, political, or other barriers, with a uniform transportation system and equal range in all directions for the sale of goods and services. In reality,

Asmara, the capital city of Eritrea—a former colony of Italy—in northern Africa is an example of a primate city. The city is a center of agriculture and manufacturing, including textiles and footwear. In 2020, Asmara's population was 896,000, far above that of Keren, the only other significant city in Eritrea, which has a population of about 82,000.

natural features such as waterways or mountains often act as deterrents to the regular spacing of communities. Likewise, transportation networks are not uniform between all settlements, so that the range for goods and services is not the same in every direction. In addition, central place theory assumes that the retail market is the most important influencer, but many cities serve other functions and are important as administrative or government centers. Still, central place theory is useful for gaining an understanding that there is a hierarchy of settlements and a pattern in their spatial distribution. Larger cities are generally spaced farther apart than are smaller towns and villages. This urban distribution is connected to population size, distances between the centers, and commerce.

Geographic Thinking

1. Use the gravity model to explain which city people from a community located halfway between New York City and Albany, New York, would be most likely to visit.

2. Explain the rank-size rule for the cities ranked from second to fifth largest in size, given the largest city in the country has a population of 15 million.

3. Explain why high-order goods have larger thresholds and ranges than low-order goods.

4. Identify and explain the limitations of central place theory.

City of the Future
Adapted from National Geographic, *April 2019*

Cities aren't built from scratch. They grow and evolve over time. It's a useful exercise, though, to imagine what an ideal city of the future might look like if urban planners were able to design one from the ground up. Knowing what an ideal city looks like in the future can help those who make decisions about actual cities to make improvements today. The design shown on these pages was imagined for a city of 2050, when the world's population will be 9.8 billion, with 68 percent living in urban areas. ▌Explain how the future city is more efficient and sustainable than a present-day city.

Compact, Mixed-use Neighborhoods
Mixed-use districts with housing for different income levels provide all services within walking distance of homes and workplaces.

Sponge City
Parks and infrastructure allow water to percolate through soil to recharge the water table.

Automated Recycling
Waste collection and recycling are fully automated for more efficient reuse of waste.

Principles of City Design

 Sustainability The future city is designed around natural features, protecting wildlife habitat and natural resources. The city is compact and dense to limit impacts on the ecosystem.

 Infrastructure Buildings are constructed more efficiently and include technology that can improve the quality of natural resources such as water, soil, and air. Infrastructure is designed for pedestrian access with limited roads for cars

Rainwater cleansing
In place of gutters, long rainwater gardens collect and filter rainwater for reuse.

Social Transit
Regional high-speed rail stations become centers of business and social activities.

Family life
Open and green spaces, community venues, and buildings with larger units foster happier and healthier families.

Urban Farms
New communities and developments take advantage of advanced technology for urban farming.

Green roofs
Solar panels and roof gardens are common atop buildings, encouraging sustainable energy and small-scale farming.

Backyard and school gardens
Local, organic, and sustainable farming is taught in future city schools.

 Water Protecting upland water systems and rigorous collection and cleansing of stormwater improve water quality. Wetland restoration and sponge-city measures revive habitats and protect against flooding and sea-level rise.

 Urban Living The way in which people live in a city is built into the future city's design. Professional, social, and family life becomes more efficient and healthier as small, diverse neighborhoods conveniently provide for the needs of inhabitants.

Principles of City Design

 Sustainability The future city is designed around natural features, protecting wildlife habitat and natural resources. The city is compact and dense to limit impacts on the ecosystem.

 Infrastructure Buildings are constructed more efficiently and include technology that can improve the quality of natural resources such as water, soil, and air. Infrastructure is designed for pedestrian access with limited roads for cars.

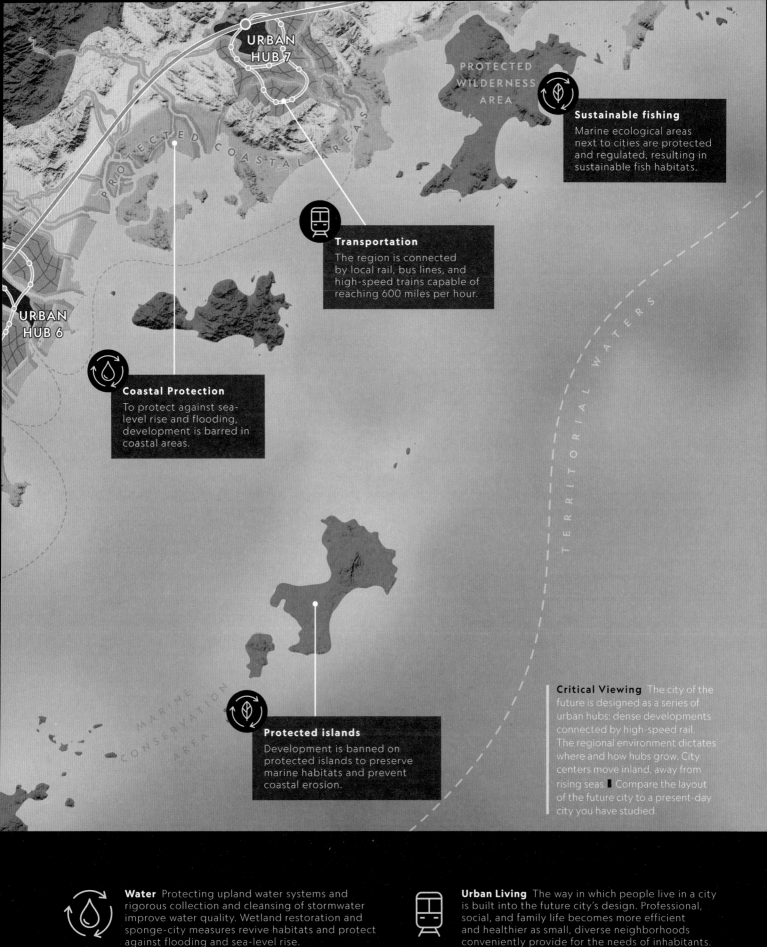

15.4 Cities and Globalization

Cities are dynamic places that in many ways are driving the processes of globalization at work in the world today. They exist within a hierarchy, with the largest and most influential cities at the top. They are interconnected, not just with other cities within their country's borders but internationally, by economic, political, social, and technological networks.

Cities across the World

Learning Objectives
LO 15.11 Define megacity and metacity in the context of urban geography.
LO 15.12 Explain why cities are growing fastest in peripheral countries.

Cities exist on every continent except Antarctica. Tokyo, Japan, with 37 million people living in its greater metropolitan area, is the largest city in the world, followed by Delhi with 31 million. The number and size of urban places are increasing worldwide, but they are growing fastest in peripheral countries such as Yemen, Mali, Afghanistan, and Nigeria as well as the semi-peripheral countries of India and China.

Cities have grown rapidly in the last several decades. In 1950, only two metropolitan areas—New York and Tokyo—had populations of more than 10 million people, a threshold used to define a city as a **megacity**. According to the United Nations, there were 33 megacities in 2018, and by 2030 that number is predicted to reach 47. Many metropolitan areas that are growing into megacities are found in peripheral and semi-peripheral countries in Asia and Africa. The unprecedented growth of megacities has caused the UN to coin a new term, **metacity**, which has more than 20 million people. In 2020, there were nine metropolitan areas that had achieved metacity status. Note that these population measurements refer to metropolitan areas rather than city boundaries when designating an urban area as a megacity or metacity. A break in this upward growth trend of many cities across the world was caused by disruptions associated with the COVID-19 pandemic, when many cities lost population: Tokyo saw its first decline in 26 years in the years 2020–2021.

The primary driving force behind massive urban growth in the periphery is overall population growth. As discussed in Chapter 4, many countries of the periphery—particularly in Africa—have birth rates well above replacement levels. Many rural-to-urban migrants are young people who are likely to start families, which further contributes to the growth of urban populations.

A number of other factors combine to fuel the growth of cities. In many peripheral and some semi-peripheral countries, push factors include a lack of employment

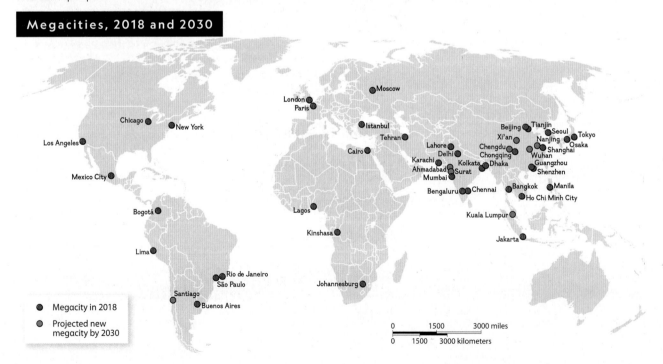

Megacities, 2018 and 2030

- Megacity in 2018
- Projected new megacity by 2030

Reading Maps While many megacities in 2018 were found in core countries such as the United States, they are increasing in numbers in the periphery and semi-periphery. All 8 of the cities projected to become megacities between 2018 and 2030 are found in periphery and semi-periphery countries. ▎Identify which cities you would expect to experience the most growth by 2030.

446 Chapter 15

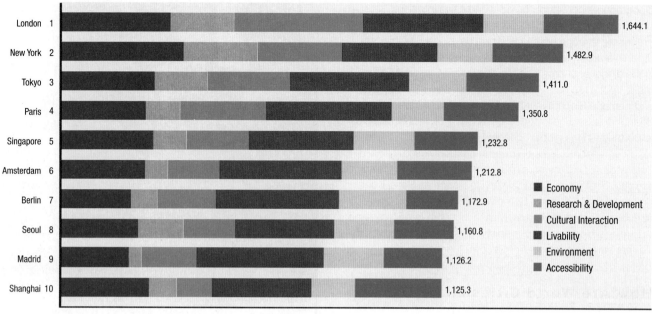

World City Hierarchy, 2021 The graph ranks the top world cities as measured by the Global Power City Index (GPCI). The GPCI evaluates cities according to their power to attract people, capital, and businesses from around the world. Notice that six functions are used to calculate GPCI, including a city's economy, culture, and livability (such as housing affordability).

opportunities in rural areas, causing rural people to leave for the city in search of jobs. In South America, for instance, many farmers are being forced off rented land because they cannot produce enough profit. Farmers may also give up their own land because they cannot make a living or simply because of the pull factor of better employment opportunities in metropolitan areas. People who live in poverty in the city often are far better off than those in poverty in rural areas, particularly if there are crop failures or drought.

Geographic Thinking

1. Identify two factors driving urban growth in countries of the periphery and semi-periphery.

2. Explain why most future megacities will be found in peripheral countries.

What Makes a World City?

Learning Objective
LO 15.13 Identify the attributes of world cities.

At the top of the world's urban hierarchy are **world cities**. World cities, sometimes called global cities—a term coined by noted sociologist Saskia Sassen in 1991—have influence not only over their country or region but also across the globe. Geographers studying cities and urbanization note that the major cities of the world make up a network of economic, social, information, and "people" flows and that their growth and influence have depended on the processes of globalization. World cities act as nodes, or focal points, of this network and, as such, they play an important role within the world system.

World cities provide multiple functions to people in their region of influence. They are major centers for communications, banking, and finance and sites of leading global markets for commodities, investment, and foreign exchange. Trade and professional associations are also concentrated in world cities, as are nongovernmental organizations (NGOs). Many of the world's most powerful media organizations—news, music, entertainment, and more—are headquartered in world cities. These cities are also home to fashion, design, and other creative arts industries. As a result, world cities have a profound influence on business and culture not only across their home regions but throughout the world. World cities are well connected, with busy international airports and sophisticated transportation systems, to assure that people from other world cities can arrive to work or visit—and that residents can commute to work and "get around town" daily. Dubai in the United Arab Emirates, strategically positioned in its "tri-continental" location linking Asia, Africa, and Europe, leaps up in world city rankings because of its busy, ultra-modern airport with convenient connections among many other key world cities.

World cities are not necessarily the largest in population. Rather, they are the most influential and wealthiest. World cities serve as destinations for both visitors and migrants. World cities also have many connections and interactions with other world cities, forming what geographers Peter Taylor and Ben Derudder call a *world city network* across the globe. They have particular kinds of services—such as banking, advertising, accounting, and legal services—that are critical in the globalizing world. Often the headquarters for major multinational companies are located in world cities, as are high-tech research and development (R&D) institutions and business or political "think-tanks."

Urban Settlements

Geographers commonly group world cities into tiers, depending on the extent of their influence as well as by the processes through which they have developed. Within the urban hierarchy, top-tier world cities are the primary nodes in the global economy. London, New York City, and Tokyo make up the top tier. Many geographers also include Paris and perhaps Singapore in this top tier, while others argue that Shanghai should be included because of China's economic importance in the global economy. Other world cities lead for other reasons: seats of government like Moscow and Washington, D.C., centers of culture, such as Los Angeles and Toronto, or of global financial services, such as Hong Kong and Frankfurt, Germany. Some world cities are even more specialized, such as Boston, Massachusetts, and Melbourne, Australia, which have high livability ratings and rich university life that attracts students from all over the world.

How Are World Cities Linked?

Learning Objective
LO 15.14 Describe how world cities influence and are influenced by the economic, political, and cultural processes of globalization.

As you have learned, cities are part of an interconnected urban system. World cities in particular are connected globally and are both the result of and drivers of globalization. Multinational corporations often have a major presence in several world cities, creating strong linkages among them. A large corporate presence requires and attracts specialized banking and legal services, as well as communication and transportation networks, creating further synergy, or advantageous interaction, among the functions of world cities and strengthening connections between them.

As centers of innovation, manufacturing, and trade, world cities give birth to new ideas, goods, and services, which are then exported through diffusion to other parts of the world. Goods are typically manufactured in or around cities and pass through several more cities as they make their way to their final destination. Global events, such as the Olympics or a World's Fair, are usually held in world cities. As discussed in Chapter 7, contemporary culture can spread particularly quickly from various hearths to other countries: Hip-hop has its origins in New York City, Bollywood in Mumbai. Both have become popular in countries all over the world. World cities have influenced the culture of other cities, as demonstrated by the presence of fast-food restaurants and global brands in urban settings across the globe. These chains and brands often diffuse to smaller cities, towns, and rural areas.

Cities up and down the world city hierarchy compete in multiple ways to gain a more prominent role in that hierarchy. The political and economic leaders of a city work to attract more business, investment, and human talent. Cities compete to host events like conferences, concerts, and sports championships. These can increase a city's stature and influence both within its own region and well beyond. In whatever position a city ranks on a world city listing, its

A large mural in Mumbai, India, advertises a Bollywood film. Mumbai, a world city, is a center of the filmmaking industry whose popularity has spread around the globe.

leaders aspire to move up to become more connected, more influential, and more recognized.

A world city's role on the global stage affects its identity and its culture, and certain elements within a world city sometimes become intertwined with its identity. All world cities have iconic places, or symbolic places that come to mind when people think about these cities. Times Square in New York City, for instance, is an iconic place known across the country and throughout the world. Located in Midtown Manhattan, it is the center of the city's Theatre District and a major tourist destination. More than 41 million people visit Times Square each year, and the famous dropping of the crystal ball on New Year's Eve has brought the iconic place into homes around the world.

Similarly, the Eiffel Tower and the Arc de Triomphe are iconic places in Paris, as is Tokyo's Shibuya Crossing, the busiest pedestrian crossing in the world, where as many as 2,500 people cross five major crosswalks at once. London has Buckingham Palace, Big Ben, and the London Eye; Singapore has Merlion Park, with its famous statue of a mythical creature that has the head of a lion and the body of a fish. These iconic places attract tourists, but they also become symbols of the cities in which they are located, in the minds of the people who live there and those who have never set foot in the place.

Geographic Thinking

3. Compare the characteristics of world cities with those of other cities.

4. Explain how world cities may influence the culture of people halfway around the world.

5. Describe steps that a city can take to attract attention on a global scale.

Case Study

How Shanghai Grew

The Issue Shanghai, China's largest city, is struggling to address problems associated with its rapid growth.

Learning Objective
LO 15.5 Describe the impact of rapid population growth on cities.

By the Numbers

27 million
people live in metropolitan Shanghai

2,448 Square Miles
metropolitan Shanghai's land area

3rd
Shanghai's ranking among largest megacities in the world

10th
Shanghai's ranking on the Global Power City Index

Sources: World Population Review; Institute for Urban Strategies.

The Pudong housing complexes were built cheaply and quickly to meet the growing demands of Shanghai's rapid population growth.

The Site of Shanghai Is in East-Central China on the coast of the East China Sea, between the mouth of the Yangtze River (also known as the Chang Jiang) and the Bay of Hangzhou. The Huangpu River flows through the city. As for Shanghai's situation, the Yangtze connects the city to resources and markets in China's interior, and the East China Sea connects it to markets everywhere.

Shanghai was a sleepy fishing village until it was opened to foreign trade in 1840. The part of the settlement west of the Huangpu River grew quickly. Pudong, a less developed area on the river's east bank, grew more slowly. By the 1930s, Shanghai had become an important trading center. The city received another boost after China adopted an open-door policy, which increased trade with other countries, in the early 1980s, and subsequently spurred the development of Pudong by reducing restrictions on foreign trade in 1990. Beginning in the late 20th century, the city was transformed with new skyscrapers, highways, and a light-rail transit system, leading Shanghai to become the economic center of China and one of the most populous cities in the world.

Today, Shanghai is known for its vast and varied urban landscape. Downtown spans both sides of the Huangpu River. The colonial-era buildings on the waterfront are towered over by new development built in the last two decades on rice paddies and agricultural land. Shanghai's rapid population growth has contributed to housing shortages and skyrocketing prices. The government has responded by placing restrictions on who can live within the city limits. It has torn down cheap housing inhabited by rural migrants and built up housing for the middle class. In addition, millions of residents have migrated from the core city to the suburbs.

Changes to the urban landscape come at a cost. The redevelopment of old neighborhoods pushes less wealthy residents into congested slums on the outskirts of the city. Shanghai has many migrants who have come to the city from rural areas in search of work and often live in poverty at the fringes of the city. Shanghai's rapid growth has also created environmental challenges. Air quality is poor because of vehicle emissions and construction projects. As the population continues to grow, city planners, geographers, and other experts seek to balance economic development and opportunities with the need for adequate housing and environmental protection. ∎

Geographic Thinking

Explain how Shanghai's site and situation have influenced the origin, function, and growth of the city.

By the early 2000s, the cities of Simi Valley, Thousand Oaks, and Santa Clarita had grown to meet Los Angeles's sprawl. They're counted as part of the greater metropolitan area, although residents might disagree.

Downtown Los Angeles was the city's first settled area, with only 6,000 residents in 1870. The arrival of the transcontinental railroad in 1876 brought many more people.
— Pacific Electric Railway in 1920

The Shape of Cities

A tale of five cities and how they grew
by Clare Trainor, Jason Treat, and Kelsey Nowakowski
Adapted from National Geographic, *April 2019*

Los Angeles, United States

Real estate developer Henry Huntington bought up land on the outskirts of Los Angeles in the late 1890s. Then he established the Pacific Electric Railway to link the scattered suburbs. The interurban rail system, which operated from 1901 to 1961, propelled the city's expansion and for a time was the world's largest electric-powered system. Eventually it was dismantled and replaced by bus lines and especially cars, making sprawl the norm.

London, United Kingdom

With the opening of the London Underground in 1863, the city spread outward. The Cheap Trains Act of 1883 allowed working-class people to move from grim tenement blocks to railway suburbs. London added the bulk of its population between 1800 and 1900, growing from 1.1 million people to 6.5 million.

Shanghai, China

What had been a relatively compact industrial city of 12 million people in 1982 has now more than doubled. The city rapidly spread in the 1980s when the government began opening the country to foreign investment. Shanghai's physical footprint has swelled so quickly that population density has declined since the 1990s.

Urban extent by year (approximate)

2014
2000
1990
1980
1950
1920
1890

Manila, Philippines

Situated between the sea and a lake, the city expanded on a north-south axis. Since 1950, nearly 50 percent of the Philippines' urban population growth has been in the Manila area. That intensified from 1980 to 2000, when almost all the urban growth took place in the city's suburbs.

Lagos, Nigeria

After Nigeria gained independence from the British Empire in 1960, oil production soared, bringing people and money to the capital. Now coastal wetlands are being drained to meet development demands from foreign investors and rural Nigerians migrating to the city.

Innovations That Shaped Cities

Resisting Attack

Walls long protected cities from invaders. Weapons such as cannons became a threat, until residents developed thick, sloped walls able to withstand blasts. As political stability improved and city defense was less important, cities expanded beyond the fortifications.

Facilitating Trade

Port cities flourished as global centers of industry and trade. To move cargo inland, rail lines extended out from the cities into the country in all directions. This led to tentacle-shaped development patterns.

Moving People

When the elevator was introduced in the 1850s, cities grew denser and taller. Cities were able to stretch farther into the suburbs when cars and buses filled in the transportation gaps left by rail lines.

Critical Viewing Some cities grow steadily over time, whereas others, like Shanghai, grow quickly to cover a much bigger area based on changes in government policies or other political, economic, or social factors. Compare the shape, size, and time frames of the cities depicted in the maps.

Urban Settlements **453**

Chapter 15 Summary & Review

Chapter Summary

Cities often serve as a region's political, economic, and cultural center.

- The number and size of cities continue to increase. Today, 57 percent of the world's population lives in urban areas.

- Site and situation factors influence the origin, functions, and growth of a city. Early cities developed close to water for drinking and transportation, had fertile land, and were defensible. Many of the fastest-growing cities have been along trade routes. Industrial cities are often located near key natural resources.

Postindustrial urbanization is characterized by rapid change, increasing size, large metropolitan areas, and urban sprawl.

- Cities grow and change—or decline—as a result of changes in transportation and communication networks, population growth and migration, economic development, and government policies.

- As cities grow, populations push outward, contributing to suburbanization, urban sprawl, edge cities, boomburbs, and exurbs.

- Policies to address sprawl include revitalization, infill development, and other planning efforts.

Geographers have developed a variety of ways to describe, categorize, and understand varied urban environments.

- The gravity model states that the level of spatial interaction between two cities depends on the size of the cities' population and the distance between them.

- Rank-size rule states that the largest city in a country will have a population twice as large as the second largest city. The third largest will be one-third the size of the largest, and so on.

- A primate city is one in which the population far exceeds the next city in size and is significantly more important.

- Central place theory states that size and spatial patterns of settlements are determined by consumer behavior.

Cities have grown exponentially in recent decades. They are sometimes ranked according to their influence.

- Megacities have populations of more than 10 million; metacities have populations of more than 20 million.

- Economic opportunity in urban areas and rural-to-urban migration contribute to the growth of cities.

- World cities influence and are influenced by the economic, political, and cultural processes of globalization.

- World cities are linked within a global hierarchy, and each works to increase its influence on the world stage.

Review Questions

Use complete sentences to answer the questions.

1. **Apply Conceptual Vocabulary** Consider the terms *suburbanization* and *urban sprawl*. Write a standard dictionary definition of each term. Then provide a conceptual definition—an explanation of how each term is used in the context of this chapter.

2. Identify and differentiate between the two types of urban areas defined by the U.S. Census Bureau.

3. Why is it important for a country to define what it considers to be an urban area?

4. What is a metropolitan area?

5. Differentiate between the terms *site* and *situation*.

6. What is a fall line? How did fall lines influence human settlement?

7. What challenges are associated with postindustrial urbanization?

8. Compare edge cities, boomburbs, and exurbs.

9. Explain how urban planners attempt to reduce urban sprawl.

10. Explain whether infill is an example of revitalization, redevelopment, or both.

11. Describe how the gravity model helps to explain interactions between cities.

12. Explain how the rank-size rule is related to primate cities.

13. Explain central place theory using the terms *threshold* and *range*.

14. Describe the characteristics of world cities.

■ **Interpret Maps**

Study the map and then answer the following questions.

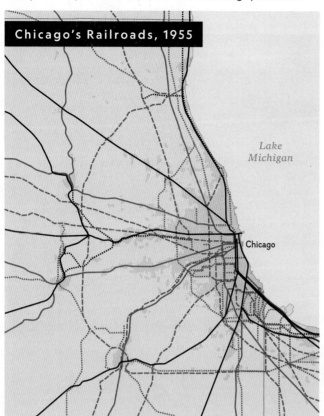

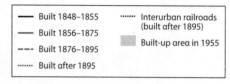

15. **Describe Spatial Relationships** Based on the map, what site and situation features likely influenced where Chicago was located?

16. **Explain Geographic Concepts** How did railroads affect urban growth in Chicago?

17. **Describe Spatial Patterns** Describe how the railroads affected the shape of the area of Chicago that had been built up by 1955.

18. **Ask Questions** What geographic questions could you ask to learn more about the growth of Chicago between the dates shown in the map key?

Chapter 16
The Urban Landscape

Critical Viewing. An elevated pedestrian walkway in the Pudong district of Shanghai, China, allows people to walk above the city's heavy downtown traffic in the Lujiazui Traffic Circle. ▌ In what way is the footpath evidence of Shanghai's commitment to infrastructure and the well-being of its citizens?

Geographic Thinking How do land-use patterns influence a city?

16.1
The Internal Structure of Cities

Case Study: Informal Housing in Cape Town

16.2
Urban Housing

Case Study: Land-Use Change in Beijing

16.3
Urban Infrastructure

National Geographic Explorer
T.H. Culhane

16.1 The Internal Structure of Cities

Although cities vary across the world—partially due to physical geographic factors and available natural resources—they typically have similarities in their patterns of land use. Experts have devised models and theories to help explain the internal structures of cities.

Urban Land Use

Learning Objective
LO 16.1 Explain how the bid-rent curve models the geography of the urban environment.

Social, economic, and spatial processes determine the land-use structure of cities. Market forces propel dynamic changes at a range of scales, from individual decision-makers to large-scale projects such as public roads or housing initiatives. People want to maximize their access to jobs, homes, goods, and services. The most accessible part of a city is its central business district, or CBD, which you learned about in the context of the bid-rent theory as applied to the cost of agricultural land in Unit 5. When applied to urban land-use patterns, the bid-rent theory explains the relationships between land value, commercial location, and transportation using a bid-rent gradient, or slope. The gradient is based on land users bidding against one another to purchase land, which means that the cost of land is highest near a CBD and decreases as distance from the CBD increases.

Competition for accessible sites near the city center is an important determinant of land-use patterns. The more accessible a location, the greater the demand for it, which is reflected in the distribution of land value. The CBD is the commercial center of a city and contains offices and public, business, and consumer services. Public services in the CBD include libraries, government offices, and museums. Businesses in the CBD might range from corporate headquarters or regional offices of multinational companies to large or small law firms, insurance offices, and advertising agencies. Overlapping with some of the businesses, of course, are consumer services—also offered at retail stores, coffee shops and restaurants, movie theaters, and concert venues, among many others. There will be medical services, educational institutions, offices of water and electric utilities, and more. The accessibility of the city center and the high value of its land attract these businesses, offices, and services. CBDs are often found in the historic hearts of cities, the original site of settlement.

Bidders are prepared to pay different costs of land for locations at various distances from the city center. For example, a business that generates a significant amount of profit, such as an investment firm, will be better able to afford property near the CBD than a business specializing in selling secondhand clothing.

One shortcoming of the bid-rent theory is its assumption that a city exists on a flat, featureless plane, with all employment opportunities found in a single CBD. Another limitation is the suggestion of consistent city transportation. Finally, in the bid-rent theory, values decrease equally in all directions, which is not reflected in real-life applications.

But the central business district is only one part of an urban area. Cities include different zones, or areas with distinct land uses and purposes. Residential and industrial zones are typical examples of urban areas with specific characteristics and types of activities.

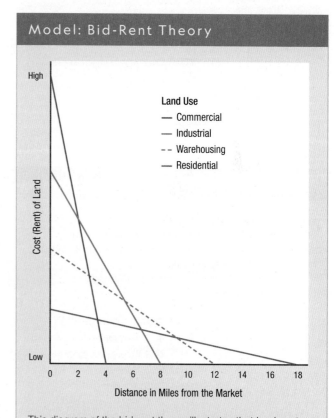

Model: Bid-Rent Theory

This diagram of the bid-rent theory illustrates that land costs drop as distance from the CBD grows. It also shows that business offices and retail stores tend to be located closer to the CBD, followed by manufacturing activities, warehouses, and residences.

Model: Burgess Concentric Zone

The strength of Ernest Burgess's concentric-zone model lies in its usefulness in explaining the basic arrangement of cities such as Chicago and Philadelphia.

- 1. Central business district
- 2. Zone of transition
- 3. Working-class residential
- 4. Middle-class residential
- 5. High-class residential

Model: Hoyt Sector

The irregular-shaped segments of the Hoyt sector model develop along transportation routes. Access to transportation and distance from the CBD help form the particular shape of each segment.

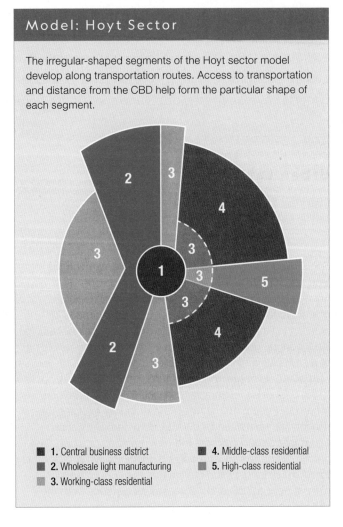

- 1. Central business district
- 2. Wholesale light manufacturing
- 3. Working-class residential
- 4. Middle-class residential
- 5. High-class residential

Model: Harris and Ullman Multiple Nuclei

In the Harris and Ullman multiple-nuclei model, different kinds of economic activities cluster together in nodes. Commerce and retail business nodes tend to be separate from clusters of manufacturing and warehousing.

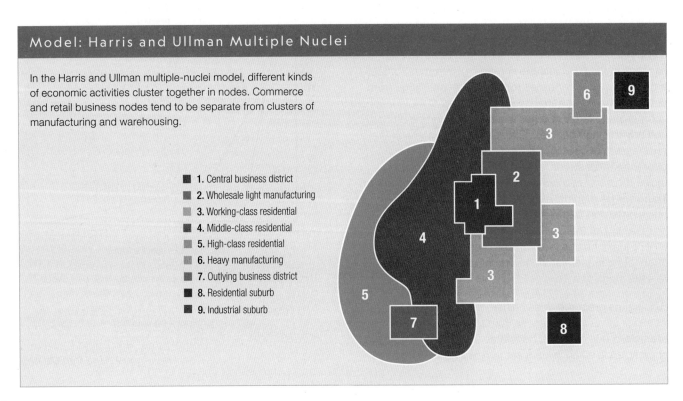

- 1. Central business district
- 2. Wholesale light manufacturing
- 3. Working-class residential
- 4. Middle-class residential
- 5. High-class residential
- 6. Heavy manufacturing
- 7. Outlying business district
- 8. Residential suburb
- 9. Industrial suburb

Models of Urban Structure

Learning Objective
LO 16.2 Compare the different models that geographers use to explain the internal structure of cities.

Recall that a model is a generalized representation of reality that helps geographers analyze spatial features, processes, and relationships. Geographers and other experts have developed various models to clarify and describe the spatial and social processes that explain the internal structure of cities. As you read about different city models, consider their strengths, weaknesses, and limitations and how they might be useful in better understanding urban landscapes.

Burgess Concentric-Zone Model Ernest Burgess devised the **concentric-zone model** by studying Chicago in the 1920s. This model observes that a city grows outward from its CBD in a series of concentric rings. Most economic activity occurs in the center (1), the place where you will most likely find a major transportation hub, main offices of businesses and financial institutions, and headquarters of civic and political organizations. As you have read, the central business district is the most accessible area of the city.

The ring (2) adjacent to the CBD is the zone in transition and an area of mixed-land use, moving from industry, factory production, and wholesale light manufacturing to older, densely populated, and typically declining neighborhoods. This transitional zone also generally includes the segment of the urban population earning the lowest incomes, with many people living in poor conditions: crammed into apartment buildings, residing in public housing, or settling in run-down housing that was once inhabited by wealthy tenants. Residents in the transition zone are often first-generation immigrants, older adults, and unhoused people.

The third zone (3) is home to the working class and offers the benefit of being located near factory jobs that were then located in the CBD and the zone of transition. Population density decreases, and people live in closely spaced single-family homes, apartments, and duplexes. The next ring in the model (4) includes higher-value residences, largely for the middle class, such as private homes and larger apartments. Its overall population density is lower than the third zone.

The zone farthest from the CBD (5) is the commuter zone. This outer ring contains more expensive, single-family detached housing in more spacious suburban settings, with the lowest population density. The impact of the bid-rent curve is apparent in the model: as one moves away from the CBD, land values decline and land use—and population—becomes less dense.

Hoyt Sector Model In 1939, Homer Hoyt sought to improve the limitations of Burgess's concentric-zone model by adding the concept of direction to the concept of distance from the CBD. After Hoyt conducted research mapping of the average residential rent values for every block in 142 U.S. cities, he concluded that the general spatial arrangement was characterized better by sectors than concentric zones.

Hoyt's **sector model** illustrates that as cities develop, wedge-shaped sectors and divisions emanate from the CBD and emerge generally along transit routes. The city center remains the location of many commercial functions. As the city expands, each division will extend outward in a sector. The sector model assumes that working-class residential wedges will develop alongside wholesale light manufacturing around the CBD. High-class neighborhoods follow a definite path along transportation routes, on high ground free from flooding, toward open country, or along riverfronts or lakefronts not used by industry. And new middle-class housing built on the city periphery attracts people who want to move away from the city center. In this filtering process, lower-income groups then move into these vacated, closer-in urban residences. Sectors grow and change over time, with better-quality homes at the periphery and older, deteriorating housing closer to the CBD.

Harris and Ullman Multiple-Nuclei Model In their **multiple-nuclei model**, proposed in 1945, Chauncey Harris and Edward Ullman observed that most large U.S. cities don't grow in rings or in sectors but are formed by the progressive integration of multiple focal points of a functional region, or **nodes**. While the Burgess concentric-zone model and the Hoyt sector model suggest predictable patterns, Harris and Ullman claimed that land use varies depending on local context.

The location and growth of the nodes rely on four factors. First, highly specialized activities involve specific sites. Industry, for example, requires transportation facilities and is often located close to railway lines, major roads, airports, or port facilities. In Philadelphia, one node related to the shipping industry has formed along the Delaware River. That location includes import–export companies, rail yards for freight trains, warehouses, and other facilities necessary for the shipping industry to function well.

Second, in certain areas of a city or region, related companies find it economically beneficial to arrange themselves together, leading to specialized areas such as financial quarters, legal districts, and groups of health-related facilities. By clustering together, these concentrations of economic activities all have access to the same pool of expertise and workers, suppliers, and information channels. A cluster of medical and surgical experts can be found on Harley Street in central London. A similar phenomenon happens when health-related facilities like nursing homes, florists, medical supply stores, and pharmacies cluster around hospitals. Retailers like automobile dealerships often cluster together, as well, making the "auto row" of many makes and models a destination for consumers that benefits all the individual dealerships.

Model: Galactic City

The galactic city model is a modification of the multiple-nuclei model and describes a place where economic activity moves away from the CBD toward the urban fringe or surrounding suburbs.

- Central city
- Suburban residential area
- Shopping mall
- Industrial district
- Office park
- Service center
- Airport complex
- Combined employment and shopping center

Critical Viewing Most of Detroit's decentralization took place during the 1950s, when the "Big Three" automobile makers—Ford, Chrysler, and General Motors—built auto plants in suburbs outside of the central city. What characteristic parts of the galactic city model can be identified in this photograph of Detroit?

Third, negative consequences of commerce and industry, such as pollution, can contribute to the formation of nodes. Some activities don't coexist well together and repel one another. Industrial parks are typically not found close to recreational parks, and residential areas are generally not located in close proximity to airports with their high noise levels. Land uses like power plants, landfills, and heavy industry are usually distant from wealthier neighborhoods, though often not from lower-income residential areas. (This is discussed further in Chapter 17.)

Fourth, economic considerations influence the formation of other kinds of nodes. Because areas closer to CBDs tend to have expensive real estate, large warehouses or grocery wholesaling that require a significant amount of land are typically located farther from city centers. Amazon, for example, has a number of warehouses in the Houston metropolitan area, but only one is located near the city's downtown where land is more costly, with the rest being nearer the city's outskirts. Warehousing or grocery wholesaling are examples of activities that could benefit from a central location in or near the CBD but cannot afford the affiliated high rents, and therefore must locate farther away from the center.

Galactic City Model Proposed by Pierce Lewis in the early 1980s, the **galactic city model**, also called the peripheral model, explains cities that have a traditional downtown and loose coalitions of other urban areas. This model helps explain what occurred in metropolitan areas that became decentralized and formed suburbs after automobile use became more widespread. The galactic city model includes urban development of various types near the edges and along transportation routes: newer business centers, complexes for research and development (R&D), specialized subcenters for education, and entertainment, sports, and convention centers. Airports in this model are located outside city limits but are connected to the city by radial highways or peripheral highways that are a standard part of galactic cities.

Detroit has highways that radiate from the city center and peripheral highways that circle the boundary of the city. New Center, a historic district in uptown Detroit, is considered by some historians to be the original edge city. (Recall that edge cities are commercial centers with office space, retail complexes, and other amenities typical of urban centers but are located on the outskirts of a larger city.) With its numerous edge cities, including Century City, Los Angeles is considered by many to be a galactic city.

Assumptions and Limitations of Geographic Models It should be noted that many cities have aspects of more than one of these models. And although city models help geographers better understand the internal structure of cities, they are based on assumptions. The concentric-zone model, which was developed a century ago, assumes the rapid physical and economic expansion of a city with a diverse, growing population and cheap public transportation in every direction. Property is privately owned and the city's economy is based on commerce and industry. The CBD is the center of employment.

But the concentric-zone model doesn't account for conditions that have prevailed since it was devised, including deindustrialization, suburban flight, widespread automobile use and the highway development that accompanied it, and the declining use of public transportation in many cities in the second half of the 20th century.

The other city models have shortcomings as well. One limitation of the sector model is that its major focus is residential land use. Another is that it doesn't account for multiple business centers. Weaknesses of the multiple-nuclei model include its failure to consider the impact of government policies and that some services and other economic activities today do not cluster for advantage as much as in the past. Additionally, all three of these "classic" models share an economic bias and ignore important factors, such as how race and ethnicity influence urban land-use change.

The galactic city model assumes that there is enough space to contain the ever-expanding sprawl of suburbs and edge cities. This model is also heavily dependent on the use of automobiles.

Latin American City Model After careful observation and data gathering, geographers Ernest Griffin and Larry Ford developed the **Latin American city model** in 1980. Also known as the Griffin-Ford model, it shares some similarities with the concentric-zone model and sector model. The traditional central market shares the CBD with a modern business center plus important religious and governmental buildings. A high-end commercial sector, or spine, extends outward from the CBD, generally along a boulevard brimming with shops, offices, restaurants, and clubs. Wealthy residents live in the blocks adjoining both sides of the spine, which typically ends at what might be considered an edge city (shown as a mall on the model). A radial road—or a *periferico*—likely circles the city, connecting the mall with an industrial park or parks.

Outside the elite residential sector, as distance from the CBD increases, residential areas of decreasing wealth and quality are found. The zone of maturity, near the CBD, contains older but good-quality residences. The area between the zone of maturity and the zone of peripheral squatter settlements, called the zone of in situ accretion, has mixed-quality housing, but renovations and improvements associated with gentrification can occur.

Areas of lower-quality housing in the Latin American city model are called **disamenity zones**—high-poverty urban areas in disadvantaged locations containing steep slopes, flood-prone ground, rail lines, landfills, or industry. Disamenity zones often include informal housing areas known as slums or **squatter settlements**, which are beset with overcrowding and poverty. Sometimes called shantytowns (*favelas* in Brazilian cities), squatter settlements feature temporary homes often made of wood scraps or metal sheeting. Squatter settlements lack basic infrastructure and services such as fresh water, sanitation, and electricity. The people who live in these settlements aren't legally permitted to be there.

Model: Latin American City

The Latin American city model dates from the Age of Exploration, when the Spanish built towns and cities placing the church, government, and businesses at the center with a grand boulevard for the wealthy that extended outward.

- Central business district
- Market
- Industrial
- Zone of maturity
- Zone of in situ accretion
- Zone of peripheral squatter settlements
- High-class residential
- Gentrification
- Middle-class residential

Model: African City

The neighborhoods in the African city model tend to be divided along ethnic lines, reflecting the great diversity found on much of the continent.

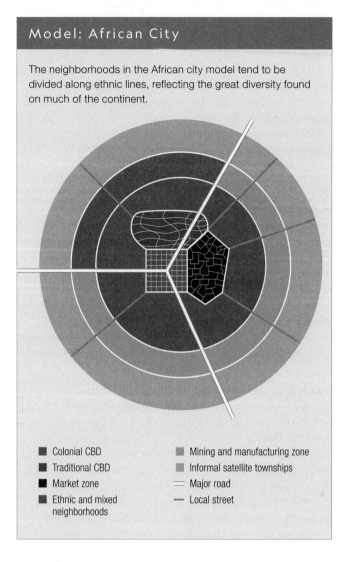

- Colonial CBD
- Traditional CBD
- Market zone
- Ethnic and mixed neighborhoods
- Mining and manufacturing zone
- Informal satellite townships
- Major road
- Local street

Model: Southeast Asian City

Many cities in Southeast Asia grew up around ports, which explains the cities' relationship with imperialistic powers as well as the importance of trade.

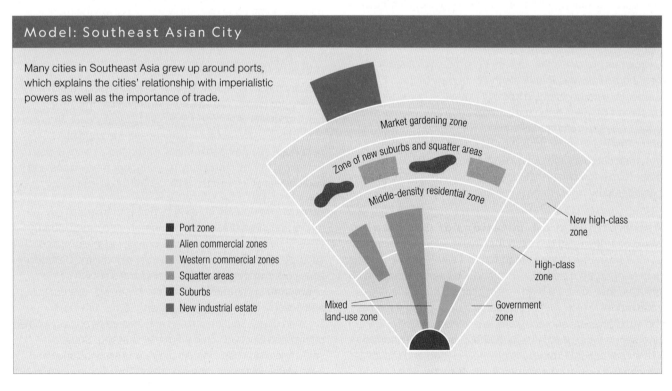

- Port zone
- Alien commercial zones
- Western commercial zones
- Squatter areas
- Suburbs
- New industrial estate

One way Mexico City lines up with the Latin American city model is that it has a wealthy boulevard, or spine, that extends out from the CBD. Called the Paseo de la Reforma, it was part of a larger urban renewal Mexico City underwent in the late 19th century. Today, the surrounding neighborhoods are more middle class.

Shortcomings of the Latin American city model include concerns that it fails to differentiate between commercial and industrial uses and that the model is overly influenced by the physical appearance—reflective of the cultural landscape—of Latin American cities.

African City Model Many African cities have three CBDs. In 1962, after observing the layouts and walking the streets of many African cities, geographer Harm de Blij proposed the **African city model**. The model incorporates aspects of earlier city models, including concentric rings and sectors radiating from the center, both of which reflect the competition for accessible and affordable sites. This competition drives urban growth everywhere. The first CBD was created by colonial powers and often has a grid pattern for order and control. A second, more traditional CBD has curbside commerce and generally one-story retail stores, and a third CBD is a zone for open-air markets. Higher-income neighborhoods with high-quality services and infrastructure are located in the CBDs, but wealth and services decrease in the outer rings. Squatter settlements called informal satellite townships are found along the outside of the African city model, and just inside the ring where squatters live is the mining and manufacturing zone.

Critiques of the African city model are that it is now out of date with recent changes in fast-growing African cities; fails to account for informal, underground economies; doesn't show middle-class and wealthy residential areas within its ethnic and mixed neighborhoods; and locates informal settlements near wealthy areas. Additionally, modern African cities increasingly reflect global architectural styles and urban design principles in some newer developments.

Southeast Asian City Model The **Southeast Asian city model**, created by geographer T.G. McGee in 1967, represents Southeast Asian cities that grow around ports and that lack a clearly defined CBD. This model establishes that there are only two formal zones that remain constant: the port zone and a zone of intensive market gardening on the periphery. The remainder of a Southeast Asian city, as shown in the model, shares characteristics with the concentric-zone model and includes a variety of uses, such as a Western commercial zone, a nonindigenous Asian commercial zone often referred to as an "alien commercial zone," mixed-land use, suburban housing, squatter settlement zones, middle-density housing, and a wealthy residential area.

Because Southeast Asian cities generally have a larger middle class than cities in Africa or Latin America, they contain fairly large suburban areas. Many of these elements are found in and around their nodes, or nuclei, similar to the multiple-nuclei model. Although the Southeast Asian city model doesn't include formal city business districts, elements of CBDs can be found in the port and commercial zones.

A shortcoming of the Southeast Asian model is that cities in the region have changed radically since McGee first created the model. Many Southeast Asian cities today do have CBDs with corporate headquarters, regional offices of multinational companies, and a range of business and financial services that support them, as the globalized economy has fueled rapid urban growth. Their quickly growing economies have also led to the development of extended metropolitan regions, transforming former "intensive market gardening" and adjacent rural regions into urban landscapes.

Geographic Thinking

1. Explain the degree to which the bid-rent theory explains why certain businesses are located near a CBD and other businesses are not.

2. Explain how the Hoyt sector model expanded upon the Burgess concentric-zone model.

3. Identify the four factors that influence the formation of nodes in the Harris and Ullman multiple-nuclei model.

4. Compare the sector and concentric-zone models and explain how they influenced the Latin American city model.

5. Identify criticisms of the African city model.

6. Describe how the Southeast Asian city model identifies and explains common characteristics of Southeast Asian cities.

Case Study

Informal Housing in Cape Town

Makeshift structures made of sheet metal and other scrap materials house many non-White South Africans in Cape Town, which still feels the geographic effects of apartheid.

The Issue Although apartheid ended in the early 1990s, South Africa continues to suffer from segregation and concentrations of poverty.

Learning Objective
LO 16.2 Compare the different models that geographers use to explain the internal structure of cities.

By the Numbers

16.8 percent

of households in South Africa living in informal dwellings in 2021

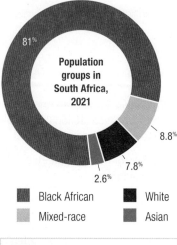

Population groups in South Africa, 2021
- 81%
- 8.8%
- 7.8%
- 2.6%

■ Black African ■ White
■ Mixed-race ■ Asian

Average annual household income, 2015 (S.A. rand/U.S. dollars)

R444,446	or	$25,131
R271,621	or	$15,358
R172,765	or	$9,769
R92,983	or	$5,258

Source: Statistics South Africa

From 1948 until 1994, the White-minority government of South Africa segregated the White and non-White populations of the country through the institutionalized system of apartheid, which means "apartness" in Afrikaans. Under apartheid, much of the non-White population lived in very poor conditions, while the White population experienced a much higher standard of living. Black Africans, Asians, and mixed-race people were forced to live on city outskirts, in residential areas called townships. Meanwhile, White South Africans could live in better locations near the CBDs. Because the government failed to develop housing or provide adequate services for people living in poverty, overcrowding and the establishment of informal settlements ensued. These areas farther from the city center featured shantytowns of informal dwellings and makeshift structures made of spare materials like wooden pallets, plastic tarps, and corrugated metal—with limited access to water, electricity, and sanitation.

The system of apartheid ended in 1994, but its impact continues to be felt, despite efforts by the post-apartheid leaders to address the consequences of centuries of government-enforced segregation that began when Dutch colonists arrived in the 1650s. The Reconstruction and Development Programme (RDP) was launched by the new government in 1994 to provide housing free of charge to those making very low wages and built more than 3.6 million new homes, both in and outside of townships. Although the quality of life improved for some people, the program hasn't desegregated South Africa—and has possibly even reinforced segregation.

Some residents of these RDP homes choose to make money by illegally selling the house after they receive it from the government. They then build a makeshift structure nearby and live there instead. Townships may contain a variety of housing, including a sizable amount of middle-class homes. Informal housing or shantytowns are also built within townships, and the people residing within these communities generally live in terrible conditions. Countrywide statistics estimate adequate housing is still needed for 3.7 million families, with its cities facing crises of joblessness, significant water shortages, and shortfalls in electricity provision. Khayelitsha, located in the eastern section of Cape Town, finds itself in particularly dire conditions, with an unemployment rate above 50 percent and some of the worst crime in the country. More than 10,000 informal dwellings are built there each year, emphasizing the difficulty all levels of government have with putting an end to informal settlements. Earlier goals by the national government to reduce inadequate housing have not been met: in 2021, more than 2.2 million households lived in such conditions throughout South Africa. ∎

Geographic Thinking

To what extent do past policies of segregation explain current patterns of housing in South Africa?

16.2 Urban Housing

Like other cultural landscapes, urban landscapes reflect the attitudes and values of a population, and they display the balance of power within that population. What a culture values leads to the decisions made about land use in a city.

Residential Land Use

Learning Objective
LO 16.3 Explain how residential land use and population density vary across urban landscapes.

As the models of the internal structures of cities illustrate, residential land use varies across urban landscapes, as does the density of the population, which generally declines from the center. Areas with the greatest density are typically found closest to a city's CBD, and as distance from the city center increases, density decreases.

People in different economic brackets tend to live in different densities in an urban area. Wealthy people sometimes live in places with a lower density (farther from the CBD), while historically marginalized classes of people often tend to cluster in places with greater density. People living in wealthier households often willingly trade the expense and inconvenience of a longer commute into the CBD for the advantages of having a larger living space in the suburbs. Members of poorer households who have jobs near the CBD often cannot afford the additional cost of transportation, and wind up living in denser residential areas closer to the city center. This situation can put a strain on such households because, as the bid-rent model shows, land is more valuable near the city center, and the cost of housing reflects this.

Although U.S. cities have relatively dense cores, urban areas in the United States are sparsely populated in contrast to other urban areas around the world. New York City is easily the most densely populated U.S. city, with an average of 27,000 people per square mile. But Paris, France, is more than twice as dense, with 56,000 people per square mile.

Another way of studying urban population densities is to consider the amount of living space per household in different countries. A 2015 analysis of recently built homes found that the average new single-family home in Australia was 2,303 square feet. By comparison, the average new home in the United States was 2,164 square feet, in Japan it was 1,023 square feet, and in urban areas of China, just 646 square feet.

The difference in urban population density is explained by the differences in housing types. Although New York City is often associated with the soaring skyscrapers of Manhattan, most areas of the city's five boroughs feature low-rise (one- or two-floor) buildings and mid-rise (three- to six-floor) buildings. In many areas of Queens, Brooklyn, the Bronx, and Staten Island, single-family homes and duplexes are common. Paris, on the other hand, is almost exclusively composed of apartment buildings. Because most people in the city are essentially stacked atop one another, the density of Paris is much greater than the density of New York City.

This phenomenon is seen in other parts of the world as well. In Seoul, South Korea, there are an average of 42,000 people per square mile, and in Tokyo, Japan, the density is more than 37,000 people per square mile. The less-dense distribution of population in U.S. cities can be seen in places like Houston, which has a square-mile population density of about 3,600, and Los Angeles, California, with a density of an average of 7,544 people per square mile. Again, the nature of the housing in these places explains the difference. Residents of Seoul and Tokyo live in smaller, more compact housing than do people in Houston and Los Angeles, where homes are much larger, on average.

The lower population density in the United States reflects its history and culture. Many international cities are significantly older than cities in the United States, and their land-use practices reflect their age. These cities developed before trains and automobiles made it possible to travel long distances in relatively short periods of time. Work, home, and services all needed to be within walking distance of one another. During the 19th and 20th centuries, improvements in transportation such as railroads, subways, and trollies allowed cities to expand in size. However, European cities and villages continued to have high population densities.

Following World War II, sprawl accelerated in the United States as cars became cheaper, and more and more people purchased them. In the decades after, the Interstate Highway System was built, creating a robust transportation network for cars. In Europe, however, because of history, culture, and government policies—such as higher gasoline taxes and funding for mass transit systems—cities and towns remained densely populated. European cities also find it harder to expand into agricultural lands because farm traditions have deep cultural roots, and farm groups have a strong voice in government.

Amsterdam and its suburbs are built around walking and older transportation networks, such as trains, light rail, trams, ferries, and bicycling. It is easy to get around in Amsterdam without a car because of its density and transportation infrastructure. In most U.S. cities, however, the difficulties with controlling sprawl result in lower population density, making it difficult for many to live without a car.

To accommodate people desiring a move from U.S. central cities to the suburbs, new homes were built, generally single-family units. People with greater means moved into these homes, often on much larger lots, which were more expensive than the older homes they left behind in the city. In a process known as succession, people with lesser means—often racial minorities—moved into the closer-in and sometimes run-down urban homes and neighborhoods vacated by the more affluent. Most of this housing would become rental units, and often the most deteriorated housing was abandoned, leading to urban blight.

In the early 21st century, however, the desire to live and work in cities increased. In some cities these formerly abandoned urban spaces have been redeveloped, increasing the density of housing and office space. These changes may represent a partial reversal of the pattern of wealthier households residing in areas of lesser density. But the COVID-19 pandemic that began in 2020 has had a profound effect on housing and work in the United States, and may ultimately reshape its cities. More people began to work from home, and commercial and office space became less valuable.

Cultural values influence how densely cities develop. In the United States, people—especially families—often prefer single-family detached homes with a lot of space and a backyard. In Europe, however, attitudes differ. In cities like Paris, families are more comfortable gathering in public spaces than private spaces, and they live in many types of neighborhoods, urban and suburban, including ethnically-diverse suburban developments called *banlieues*, where government-run public housing is more likely to be found. However, even in Europe, particularly in France, more and more people have left cities for suburban life to enjoy a house with a small garden, neighbors across the fence, and spaces for entertainment at home.

Physical geography, too, plays a role in urban density. Housing in Japan is limited by the country's mountainous, forested terrain. Nearly 70 percent of the land is covered with trees, and more than 70 percent is covered by mountains. This geography restricts the amount of land available for housing and other uses, which pushes up land values. As a result, development is dense and living spaces are small. According to one survey, in 2017, 90 percent of new apartments on the market in Tokyo were smaller than 861 square feet. Another way to consider urban density and city living is the ideal of the 15-minute city, the notion that wherever you live, anything you need—including food, schools, and parks, as well as most goods and services—is within a 15-minute walk or bike ride away. This principle is part of New Urbanism, which is explained in Chapter 17. As you have read, many cities in Europe, including Paris and Amsterdam, strive to meet this ideal. U.S. cities like Portland, Seattle, Detroit,

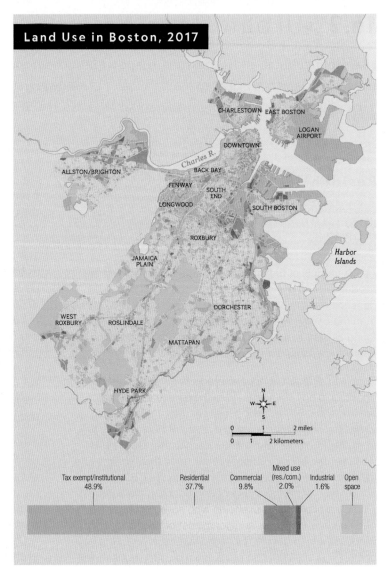

Reading Maps The bar graph shows what Boston's land is used for, and the map shows where those areas are located. Almost half of Boston's land is tax-exempt and used by institutions such as federal, state, and city government, colleges, hospitals, social service agencies, and cultural institutions such as churches. ▌ Describe where most of Boston's residential land is and what spatial patterns are shown.

and Miami aspire to become 15-minute cities. However, critics warn such planning practices may only reinforce urban inequities, with only wealthier residents benefitting from efforts to achieve this convenient urban lifestyle.

Geographic Thinking

1. Compare the urban population density in the United States with other countries around the world, and explain what factors influence any differences.

2. Explain why the desire for a single-family home and space leads to sprawl.

3. Identify the types of homes found in a city with very high-density housing.

Case Study

Land-Use Change in Beijing

The Issue As the economy of China transformed over the past 40 years, the capital city of Beijing had become increasingly crowded, sprawling, and polluted but has worked to find solutions to these problems.

Learning Objective
LO 16.3 Explain how residential land use and population density vary across urban landscapes.

By the Numbers

21 million
Population of metropolitan Beijing (2020 estimate)

400 percent
Rate of population growth since 1980

2021
First year that Beijing met China's air quality standards.

Sources: World Population Review, Statista

Smog almost obscures a pedestrian on an over-the-road walkway. Some side effects of Beijing's rapid urbanization include heavy traffic congestion and air pollution.

In the past four decades, the Chinese capital of Beijing has undergone a remarkable conversion from a city of just 5.4 million in 1980 to about 21 million in 2020. To put those numbers in perspective, the population of New York City has grown by only about 1.5 million during that same period. As Beijing's population has changed dramatically, so has the character of the city.

In the 1970s, the Chinese government—run by the Communist Party—began to implement economic reforms that generated many characteristics of a capitalist economy. The Chinese economy today remains largely centrally planned, with less of the market-driven approach of the United States. Even so, the loosening of restrictions paved the way for the shopping malls, luxury residential and office buildings, and high-tech industrial parks that are now commonplace in many areas of China. Beijing's economy has increased ten-fold over the past 20 years. Visitors to Beijing will see evidence of a middle-class economy that is similar to the middle-class economies of other East and Southeast Asian cities: skyscrapers, an extensive subway system, modern residences, and stylish residents. Beijing today has risen as China's capital to take its place as a true world city, even ranking in the top 10 by some accounts.

But this economic growth also brought problems, such as terrible air pollution—made worse by seasonal dust storms, traffic congestion, and overcrowding. Beijing has also experienced low-density urban sprawl similar to what is found in the United States, due to the construction of ring roads, similar to U.S. beltways. Because jobs haven't followed the sprawl and remain clustered around the downtown, cars funnel onto the ring roads, worsening traffic congestion. The economic success of Beijing and other Chinese cities such as Shanghai has worsened income inequality because much of the country's wealth has been drawn into its large cities.

To combat the problems of rapid urbanization, sprawl, wealth inequality, and pollution, China's government has imposed a population cap on Beijing, limiting it to 23 million residents. The government is building new cities, away from the capital, which are designed both to redistribute wealth and to reduce the pressure that overcrowding has placed on Beijing and other large cities of China. And notable success has been achieved with an aggressive approach to fight air pollution in the capital, particularly since the Beijing Summer Olympics in 2008. Many factories have been moved out of the city, driving restrictions have been put in place, the use of coal and wood for home cooking has been greatly reduced (replaced by electric and natural gas stoves), and even forest planting programs to the north have cut down on dust storms. Positive results in the period 2013–2021 have included a decrease of more than 50 percent in particulate pollution and reductions in nitrogen dioxide and sulfur dioxide pollution (both associated with the burning of oil and coal) of 54 and 89 percent, respectively. Though smog remains a threat in Chinese cities, even in Beijing, the capital city met China's air quality standards for the first time in 2021—and may provide a model of how megacities in semi-periphery and periphery countries can use emissions standards and policies to reduce air pollution. ∎

Geographic Thinking

Explain which model of urban settlement appears best to describe Beijing and why.

The Urban Landscape

Critical Viewing Hong Kong, a special administrative region in China, is one of the most densely populated regions in the world, with a population density of 17,311 people per square mile. As a comparison, the United States has a population density of 96.2 people per square mile.
▮ If you did not know that Hong Kong had a high population density, what clues from this photo

Housing Density and Development

Learning Objectives
LO 16.4 Explain how zoning decisions affect housing provision and density in cities.
LO 16.5 Explain the general relationship between land value and land-use density.

Density of housing is usually described as high, medium, or low. Because some areas are generally denser than others, these terms are relative and can take on different meanings in different places. But broadly speaking, in high-density areas, a large number of people live on a small amount of land; in medium-density areas, a moderate number of people live on an intermediate amount of land; and in low-density areas, a small number of people live on a large amount of land. You'll recall that the bid-rent theory shows that housing density is influenced by land values—higher-value land generally leads to higher density housing.

As you read, Manhattan in New York City has high-density housing, with its high-rise luxury towers and mid-rise apartment buildings. An abundance of medium-density housing can be found in Boston and some of its neighboring cities in the form of triple-deckers, which are three-story homes with three separate living quarters. Many other older U.S. cities contain variously configured and named medium-density housing: brownstone apartments in New York City, cities in New Jersey, and other urban areas on the East Coast; narrow, multistoried townhouses or row houses in New Orleans, Charleston, Baltimore, Philadelphia, and Washington, D.C.; and two- or three-story, multifamily residences called "flats" in cities like Chicago and Milwaukee.

Low-density housing is common in many suburban areas, which include mostly detached single-family homes. Within cities, certain sections can seem high density simply because other sections of the city are low density in comparison. In Houston, for example, all housing is low density compared to a place like Mumbai, India. But in Houston, the areas with townhouses, apartment buildings, duplexes, or small single-family homes located in close proximity to one another feel high density in relationship to the places with large single-family houses on large lots.

City and regional governments use various legislative tools to permit or prohibit certain land uses as well as to limit density and to guide the direction of growth within their borders. In the United States, one of the most powerful tools is **zoning**—the process of dividing a city or urban area into zones within which only certain land uses are permitted. For example, manufacturing would only be permitted in areas zoned for industrial use. Other areas are zoned for residential, commercial, or mixed use. Zones can be further subdivided so that a residential area might be zoned strictly for single-family homes, or the area might permit medium- or high-density housing such as apartment buildings. Note that zones enforced by municipal governments are not the same as the zones depicted in the city models you learned about earlier in this chapter. Cities set their zoning according to their present and perceived future needs, which may or may not create zones that reflect patterns like those in the various city models.

Through changes in land-use planning and zoning laws, housing density can increase. Until 2019, most residential areas in Minneapolis were zoned so that only single-family detached houses could be built. A change in the law allowed the construction of up to three dwellings on one piece of land in all parts of the city. Allowing more homes to be built on each parcel of land makes it more likely for people to be able to find homes at an affordable price. Another way to increase density and create additional housing opportunities in a city is infilling, which you learned is redevelopment that identifies and develops vacant parcels of land within previously built areas. Underused parcels such as parking lots or abandoned industrial land can also serve as sites for new residential or commercial buildings. Infill development often targets key transportation nodes, such as subway or commuter rail stations, or takes place along transportation routes that are connected by bus service.

Some urban areas have embraced increased development, infilling, and changes in land use more than other cities. For example, demand for housing is high in both Seattle and San Francisco due to the presence of high-technology industries. Seattle has rezoned parts of the city to allow for more infill as well as an increase in the height and density of housing. Because of this more pro-development attitude, Seattle, while expensive, doesn't suffer from as exorbitantly a high cost of housing as San Francisco does. Currently, just one unit of housing is available for every 10 existing jobs in the San Francisco Bay area. Advocates for solving San Francisco's housing problem with more infill argue that in recent years the city core is realizing only 57 percent of its infill housing potential.

Increased density offers other benefits, including making it easier for governments and businesses to provide high-quality services. Effective public transportation relies on a certain amount of density because people won't use buses or trains if they have to walk a long distance to reach a station or stop. A mass-transit system also needs enough riders to sustain it financially. In a similar way, businesses need enough customers to provide strong services. For example, cell phone providers are able to supply better service in dense urban areas because the large number of customers leads to a significant amount of revenue. This density also motivates cellular service businesses to launch new technological capabilities in urban regions.

Geographic Thinking

4. Compare low-, medium-, and high-density housing in the United States in terms of what it is and where it can be found.

5. Explain the relationship between housing density and city services. Give examples as part of your answer.

16.3 Urban Infrastructure

Every individual benefits from streets and roads, electric power lines, water supply and sewer systems, railroads, and communications supports like cell phone towers and fiber-optic cables. These features are all examples of infrastructure and are essential elements of a modern society.

Why Is Infrastructure Important?

Learning Objective
LO 16.6 Explain why strong infrastructure is critically important to the standard of living of a society.

A well-functioning city relies on strong infrastructure, a framework that helps ensure that people can have a high quality of life and move from place to place. Some experts consider infrastructure—in a broader context—to include all services and institutions that help maintain the health, safety, economic, and social aspects of a country: police and fire protection, hospitals, schools and childcare facilities, emergency services, and government.

Infrastructure also includes mass-transit systems such as subways, buses, and light rail; energy-generating facilities such as power stations, wind and solar farms, and hydroelectric plants; power lines; telecommunications including telephone wires, fiber-optic cables, cell phone towers, and other mobile networks; ports, airports, waterways, and canals; streets, roads, and highways; and water supply systems, sewage systems, and wastewater treatment facilities. Without highly operational infrastructure, disease would be more likely to spread, performing basic everyday tasks would be nearly impossible, and going to work or visiting friends and family would be far more difficult.

Airports, an essential infrastructure element, allow cargo and passengers to travel both short and long distances at great speed and are typically located near cities with large population centers. The busiest airport in the world, in Atlanta, Georgia, carries more than 100 million passengers annually. However, the Istanbul airport in Turkey, which opened in 2018, is expected soon to carry nearly twice as many travelers. Jets can fly nonstop to about 60 countries from Istanbul, which is in a critical geographic area between Asia and Europe in Southwest Asia, not far from Africa.

Infrastructure and Development

Learning Objective
LO 16.7 Explain how the quality of infrastructure impacts economic and social development.

Core countries with powerful economies have the best infrastructure. Their wealth and expertise make it possible to build strong and efficient electrical grids, highways, communications networks, water and sanitation systems, and other important frameworks. Peripheral countries and many semi-peripheral countries generally don't have the money to build high levels of infrastructure and, as a result, many people who live in these regions are unable to improve their standards of living. To some extent, this disparity also exists between the core and peripheral areas within a single country. In a city like Philadelphia, which suffers from extreme wealth inequality, areas with lower incomes suffer from unsafe roads, while areas with wealthier residents are much less likely to have roads that put motorists, cyclists, and pedestrians at risk. This imbalance makes it harder to escape poverty. In cities in India, inadequate security on public transportation discourages people from working because commuting to a job could put them in danger. This issue especially affects women, who disproportionately depend on mass transit to get to work.

Places with the strongest infrastructure experience the most economic and social development. For example, well-regulated sanitation systems reduce the spread of disease; efficient rail and highway systems increase people's mobility; and quality education improves knowledge, problem-solving skills, and other key abilities for success. According to a World Bank study, infrastructure improvements in African cities accounted for more than half the economic growth the continent experienced between 2001 and 2005.

Infrastructure helps businesses succeed because it allows them to effectively and confidently transport goods and share information while enjoying a capable labor pool from which they can hire. Infrastructure is one reason businesses and workers may relocate from one place to another. If companies and people are located in a place where it's difficult to get around and communicate, they may instead look for a place with good airports, speedy public transportation, and reliable telecommunications systems. If infrastructure begins to fail, economic activity can be reduced and quality of life diminished.

Infrastructure needs to be maintained and upgraded. The quality of infrastructure at the national, state, and local levels in the United States—such as roads, bridges, ports, and passenger rail—has declined in the last decade. Furthermore, funding to build new capabilities such as electric transportation infrastructure, green energy efforts, and fiber-optic networks has long been limited. Semi-peripheral China, in contrast, has invested heavily in infrastructure for decades and has enjoyed the resulting economic advantages. The Biden administration worked with the U.S. Congress in 2021 to fund significant infrastructure projects across the country, recognizing that even more of this critical work is needed.

Solid infrastructure is especially important to the economic vitality of a city. In Nigeria, poor infrastructure in cities such as Lagos and Abuja has contributed to the country's economy performing below its potential. In Lagos, just 6 percent of households have a flushing toilet connected to a piped sewer system. Poor transportation infrastructure has contributed to higher costs for residents of Lagos. One recent report estimates that more than 60 percent of low- and medium-income earners in Lagos spend more than half their income on transportation. This problem has contributed to a decrease in Nigeria's economic growth.

National Geographic Explorer **T.H. Culhane**

Empowering People with Clean Energy

T.H. Culhane works all over the world helping communities with few resources achieve energy sustainability. In Nepal, he helped install solar panels on the roofs of buildings.

Learning Objective
LO 16.7 Explain how the quality of infrastructure impacts economic and social development.

T.H. Culhane didn't set out to save the world. He only wanted to help people. But he just might wind up doing both.

Culhane is the founder of an organization called Solar C.I.T.I.E.S., a nongovernmental organization that assists individuals and small groups living in impoverished areas in learning how to generate clean, renewable energy that they can use to power their basic needs. Poorer sections of cities such as Cairo, Nairobi, and Rio de Janeiro have extremely dense populations and poor infrastructure. Culhane helps people in these places heat water and generate two hours of cooking gas per day, without fossil fuels. "If people don't have access to enough hot water, if people can't boil water, it becomes a serious health issue," he points out. "And when women spend all their time collecting firewood or charcoal and tending stoves to cook and heat water, then how can they go to school or get ahead?"

Culhane began his work in poverty-stricken areas by helping residents build solar water-heating systems, but he realized that not all homes have access to sunlight. To harness the energy that exists in food waste and human waste, he developed a simple open-source biodigester made from containers that anyone can build themselves, which produces fertilizer and creates gaseous fuel called biogas. This fuel—which is methane, or natural gas—can then be used to heat water, cook food, and run electric generators.

At the Zaatari refugee camp in Jordan, Culhane and his organization installed biodigesters while also educating technicians and farmers. In time, these people will become experts and eventually increase their incomes by selling the fertilizer and biogas they produce.

Culhane's hope is that by providing people with these tools, they will be able to thrive. "The poor don't need our pity; they need a chance to help themselves," he says. "I categorically do not believe in categories. The poor aren't a class of weak victims; they're millions of creative individuals."

Geographic Thinking

Explain how the communities using Culhane's tools will benefit in the future.

The Urban Landscape

Critical Viewing In Copenhagen, Denmark, the DSB—the commuter rail system—includes rail cars set aside for travelers with bicycles and baby carriages. ▮ Based on evidence in the photo, draw some conclusions about the economic and social health of Copenhagen.

Infrastructure can be used not only to improve the economy but also to make lives better for people and create more equality. Cities around the world, including Bogota, Colombia, have created dedicated bus lanes so that buses carrying dozens of passengers will not be forced to sit in the same traffic that cars do. These dedicated lanes help bus riders, some of whom cannot afford their own car, move from place to place more quickly than the slowly moving automobiles. More road and sidewalk space has also been set aside in Bogota and other cities for pedestrians and cyclists. All these measures improve mobility for people with fewer financial resources and can help them lift themselves out of poverty by getting them to and from work faster.

In European cities, high densities, the lack of space, and cultures that are comfortable with using mass transit have also led to the creation of bike-friendly infrastructure. This is particularly evident in Copenhagen, Denmark. In Copenhagen, bicycles outnumber cars by more than five to one, nearly one-third of all trips across the city are done by pedal, and more than 40 percent of work- or school-related trips are done on the two-wheelers. The number of miles traveled by bicycle in Copenhagen has increased steadily over the past 25 years, as the government has invested more money into infrastructure that makes cyclists feel safe. This change has provided multiple benefits to the city, including more economic growth and reduced greenhouse gas emissions.

Amsterdam is another city with some of the best bicycle infrastructure in the world. Amsterdam had long been a city where cycling was popular, but after World War II, city planners built roads to accommodate more cars, leading to fewer cycling trips. But a backlash from activists who were angry about traffic deaths forced a reversal starting in the 1970s. Today, nearly 40 percent of all trips in Amsterdam are made on bicycles. Other cities in Europe have also embraced this bike-friendly infrastructure.

Geographic Thinking

1. Describe what might happen to a core country whose infrastructure has started to crumble and become less reliable.

2. Explain why, from a spatial perspective, it makes sense to dedicate lanes for buses only.

The Political Organization of Cities

Learning Objective
LO 16.8 Explain how a city's infrastructure is connected to municipal and regional politics.

Like countries, states, or provinces, cities are political entities with governments whose job is to respond to the needs of their residents. Municipal—or city—governments are responsible for performing a variety of services, such as building and maintaining infrastructure, including roads, sidewalks, sewer lines, and water mains. Cities often also deliver services, most significantly schools but parks and libraries too. Municipal governments work to ensure public safety by providing police departments and fire departments as well as emergency medical service professionals.

Issues with municipal governance arise when an urban area consists of many different local governments that are unable to collaborate because they aren't integrated. For example, if a recently paved road suddenly becomes rough, its repair may involve multiple municipalities that aren't necessarily accustomed to working together. A lack of collaboration can arise at the regional level as well. A large number of separated municipalities can struggle to agree on how to solve region-wide problems. They also tend to operate inefficiently due to offering the same or similar services.

Fragmented governments can face difficulties when the interests of each municipality collide with the interests of the larger region. This is evident in areas with a housing shortage. Local governments often have a desire to limit the growth of their communities because they are concerned about the pressure more development might put on their infrastructure, such as schools. Residents of a community also influence their local government because they want to maintain their quality of life and limit changes perceived as problematic. If enough municipalities work to stop the development of housing, it becomes difficult for the housing sector in a region to provide enough homes.

U.S. urban issues also arise when government bureaucracies are dispersed between the city, the county, and the state, and they don't cooperate. Lines of accountability become unclear as the actions of different agencies and institutions are influenced by different motivations. The subway system in New York City is often beset with problems because the agency that runs public transportation in the city and its surrounding areas, the Metropolitan Transit Authority (MTA), is not a government organization. As an independent corporation run by a board of directors, the MTA is not controlled by the state government or the government of New York City. It has the power to make spending decisions but not the power to tax the region's residents. The MTA's collected fares have not reduced its large amount of debt, which makes it increasingly difficult for the MTA to spend the money necessary to make the subway operate well. The COVID-19 pandemic also greatly impacted New York's MTA and other transit systems across the country, with ridership and thus revenue plummeting, further hurting their ability to maintain and upgrade their systems. Many of these urban transit agencies are struggling to recover ridership and income.

Qualitative Urban Data

Learning Objective
LO 16.9 Explain qualitative methods for gathering geographic data in urban areas.

Cities are dynamic, and responsible governments analyze these changes and prepare for their impacts. Collecting data—both qualitatively and quantitatively—is one way these governments and others build an understanding of the evolution of their cities and plan for the future. Qualitative research is based on descriptions and rich narratives; quantitative research is based on collecting data about countable phenomena. Researchers gather and record qualitative data through field studies—research conducted in the field—and through actions such as interviews, focus groups, surveys, and observations about both the past and the present. City employees might stand on a street corner and talk to passersby about their experiences in the neighborhood. Questions focus on what the interviewees think the city is doing well, what they think the city is doing poorly, and what other services they think the city should be providing.

Qualitative researchers document their own observations as well. In a city, they might observe and document how bicyclists and motorists interact with one another, or what activities appear to be popular in a park. Qualitative and quantitative techniques provide a trade-off between breadth and depth. Many people recommend using both or a mixed-method approach, especially considering the nature of the research questions.

In New Orleans, National Geographic Explorer Caroline Gerdes collected qualitative data through an oral history project about the Ninth Ward neighborhood, which was devastated by Hurricane Katrina in 2005. She sought to gain an understanding of how the community changed, from its immigrant origins to world wars to desegregation to Katrina and today. Gerdes used the data she collected through her interviews to form a narrative of people's opinions, attitudes,

Critical Viewing Municipal governments provide a number of services, including public libraries, many of which are architectural wonders. The Vennesla Library and Culture House in Norway was completed in 2011 and has won several architectural prizes. ▌ Describe some of the effects the library may have on the local area.

Critical Viewing Museum director Leona Tate (*left*) and National Geographic Explorer Caroline Gerdes (*right*) tour the Lower Ninth Ward Living Museum in New Orleans. As part of her oral history project, Gerdes interviewed more than 50 people. ▮ Identify any details in the photo that show that items in the museum include qualitative data.

and feelings about their neighborhood and what it has gone through over the decades.

Quantitative Urban Data

Learning Objective
LO 16.10 Explain quantitative methods for gathering geographic data in urban areas.

Whereas qualitative research collects data based on observations, interpretations, opinions, and perceptions, quantitative research collects quantifiable data, that is, data that can be measured, counted, and put into a numerical context. Quantitative techniques include taking measurements (e.g., urban air quality, traffic studies), examining test results, conducting questionnaires, and using data collected into existing databases, such as census results. For example, a quantitative researcher, in addition to using the data gathered from qualitative researchers observing bicyclists' behavior, might gather quantitative data regarding the exact number of bicyclists who ride along a certain stretch of road over a period of time.

Quantitative researchers begin with a question that drives a data-collection process, which can be done using either primary or secondary methods. Primary data are information collected directly from a source. A person conducting a bicycle study who counts the number of bicyclists using a specific stretch of road is producing primary data, or data for their own study. Secondary data come from outside sources and are not produced by the person conducting the study. An example of secondary data would be federal census data collected by the U.S. government used in an individual's research on patterns of population.

Quantitative researchers sometimes collect data through sampling, or taking information from a subset of the population in a random, systematic, or stratified way. Data are characteristics or information collected through observation and are of four types: nominal, ordinal, interval, and ratio. Nominal data are named or labeled data, categories such as type of housing: single-family detached or apartment. Ordinal data have an order or rank, such as low, medium, or high income. Interval data have an order and the difference between two numbers has meaning, such as temperature and year. Ratio data contain all aspects of interval data but also have a true zero value, such as distance or area (for example, acreage). Nominal and ordinal data are qualitative—although some ordinal data can be quantitative—and interval and ratio data are quantitative.

The information collected by the U.S. Census Bureau is some of the most robust quantitative data that municipalities can use. Every 10 years, the Census Bureau collects data on every resident living in the United States; the bureau also collects other economic, governmental, workforce, and household data more regularly. The information is gathered at different scales, starting with census blocks, the smallest geographical unit used by the Census Bureau. The data are collected and

474 Chapter 16

organized in block groups, which typically contain between 600 and 3,000 people. Block groups are then combined in tracts, which usually have between 1,000 and 8,000 people, with an optimal size of 4,000. Beyond that, the geographic types are counties, states, divisions, regions, and the country.

The census determines how many people live in a given community or state, broken down by sex, age, race, and other factors. The data are used to determine where and how federal funds are spent on schools, hospitals, roads, public works projects, and other programs. Local governments use census data for decisions related to planning, waste management, transportation, employment, migration, ethnic composition, and population growth and decline. These data are also used to reapportion political districts every 10 years, a process discussed in Chapter 10.

If, for example, the census indicates that a city's population has grown substantially over 10 years, the municipal government can take steps to respond to that growth. It can set aside more money in its budget for infrastructure improvements, new schools, and a larger city workforce. If the census shows that the population of certain cities has grown while others have shrunk, the state government can shift resources accordingly.

Local governments work with land-use data to understand changes in their communities and regions. This information could show how housing is sprawling outward, the extent to which parking lots or other infill sites are being developed, or how tree canopy (the zone or layer of natural habitat formed by the tops of trees) is affected by construction. For example, the rapidly developing Sunbelt city of Charlotte, North Carolina, conducted an analysis of its natural environment to monitor the effects of its growth. The study showed significant threats to the city's tree canopy. As a result, Charlotte developed an action plan to sustain the city's trees as the city grows and develops.

The best way for researchers to understand communities is to combine qualitative and quantitative data because both types of information are necessary to get a full picture of what's happening. Recall the work of Daniel Raven-Ellison from earlier in this unit. He has used both qualitative and quantitative data to help London become a National Park City because of its biological diversity.

Although quantitative data provide accurate and defined information for implementing possible changes in urban areas, qualitative data are used as extensively and are just as important when making municipal decisions. Crime statistics collected by cities, for example, can be combined with dialogues with individual members of the community and community groups about unreported incidents or observed events. Both types of research are valuable, and using quantitative data and qualitative data like this could influence how a city determines police staffing needs, types of outreach, locations and numbers of patrols, and so forth.

Geographic Thinking

3. Explain the degree to which disagreements between governments at different levels might affect infrastructure.

4. Identify what distinguishes qualitative data from quantitative data.

5. Explain the differences between the four types of data.

6. Describe the value in combining qualitative and quantitative data.

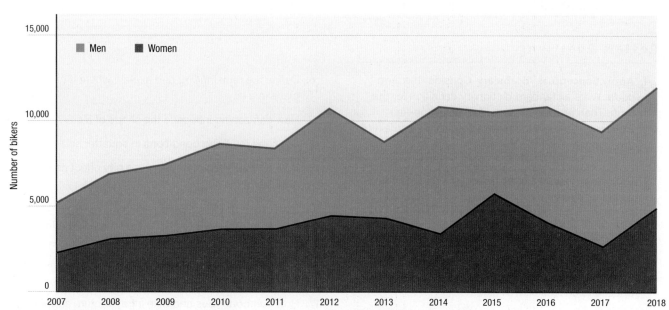

Bike Ridership in Seattle, 2007–2018 Quantitative data shows that bike ridership in Seattle has increased since 2007. Municipal officials will use primary data like this to help them decide whether to expand the use of protected bike lanes in a city and where to locate them. Before the COVID-19 pandemic began in 2020, bike ridership hit record highs in Seattle.

The Urban Landscape

Chapter 16 Summary & Review

Chapter Summary

The way land is used in urban areas depends on variables ranging from the desirability of land to historical factors and technological changes. Models help geographers understand how land is used.

- The bid-rent theory explains that land is most expensive near a central business district (CBD) and declines in price as distance from the CBD grows.
- The concentric-zone model illustrates how a city grows outward from its CBD in a series of concentric rings. The sector model shows wedge-shaped divisions emanating from the CBD.
- The multiple-nuclei model illustrates an urban area where activities have clustered as a result of their unique needs. The galactic city model also depicts an urban area where economic activity isn't centralized but moves outward to the urban fringe or suburbs.
- The Latin American city model incorporates elements from the concentric-zone and sector models but includes other features, such as a spine and an outermost ring where impoverished communities live.
- The African city model shows how African cities tend to have three CBDs: a colonial central business district, a traditional central business district, and a market zone.
- The Southeast Asian city model illustrates cities that grow around ports and initially lacked a separate CBD.

Urban areas feature high-, medium-, and low-density areas, with density generally decreasing as distance increases from a city's urban core.

- Areas with the greatest density are typically found closest to a city's CBD.
- Available space, history, culture, geography, and different housing types all contribute to why some cities are more densely populated than others.
- Zoning is the division of a city or urban area into zones within which only certain land uses are permitted.
- U.S. cities have typically used redevelopment, infilling, and zoning and land-use changes to address demand for housing.

A well-functioning city requires strong infrastructure.

- The benefits of strong infrastructure are critically important to the economy, health, and overall standard of living of a society.
- Governments face difficulties in responding to the needs of a region when the needs of local municipalities clash and do not match the overall need of the region.
- Cities gather qualitative data, which is based on descriptions and rich narratives, and quantitative data—numerical data with definite quantities—to plan for beneficial geographic change.

Review Questions

1. **Apply Conceptual Vocabulary** Consider the term *node*. Write a standard dictionary definition for the term. Then provide a conceptual definition—an explanation of how the term is used in the context of the chapter.

2. Provide an example of an urban area that fits the galactic city model.

3. Why does Hoyt's sector model explain the way cities are arranged better than Burgess's concentric-zone model does?

4. Describe the criticisms of the bid-rent theory when applied to accessibility in cities.

5. What are some common features of the Latin American, African, and Southeast Asian city models?

6. Define the term *infilling* and explain where it can take place in a city.

7. How can zoning influence the shape of a city?

8. Describe the living conditions in squatter settlements.

9. How do the types of housing in a city affect its population density?

10. Describe how Minneapolis and Seattle changed their land-use policies to meet the increased demand for housing.

11. Why do countries with strong economies have stronger infrastructures?

12. How can better bus and bike infrastructure address economic inequality within an urban area?

13. What issues do urban areas with fragmented governments have?

14. Identify an example of qualitative data and explain what it reveals about its topic.

15. Identify an example of quantitative data and explain what it reveals about its topic.

Interpret Graphs

Study the graph and then answer the following questions.

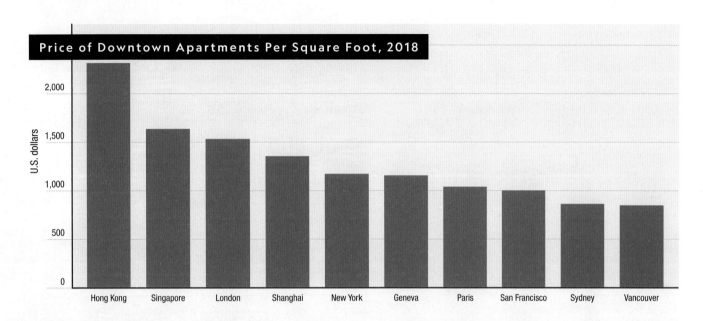

16. **Identify Data & Information** Which North American city has the most expensive housing measured in price per square foot?

17. **Make Connections** How can population density affect the prices of downtown apartments?

18. **Apply Models & Theories** What additional data would you need for each city to assess the validity of the bid-rent theory?

19. **Draw Conclusions** Why do you think most of the cities on the graph are located in core countries?

Chapter 17
Urban Living

Critical Viewing Between 1998 and 2004, Millennium Park, in downtown Chicago, was transformed from an industrial area of railroad tracks and parking lots into a vibrant public space, including the Jay Pritzker Pavilion with its trellis of steel pipes that contain a sound system. How might this urban park affect Chicago residents and visitors?

Geographic Thinking How can we create sustainable cities for the future?

17.1
Designing for Urban Life

17.2
Causes and Impacts of Urban Changes

Case Study: The Effects of Redlining in Cleveland

Feature: New York City's High Line

17.3
Creating Sustainable Urban Places

National Geographic Explorers
Zachary Damato and Maria Silvina Fenoglio

Case Study: Milan and Urban Sustainability

National Geographic Feature
Designing to Scale Smart Buildings

17.1 Designing for Urban Life

Cities grow in response to a variety of human decisions, both conscious and unconscious, voluntary and forced. How can understanding these geographic processes help in designing and planning livable, sustainable cities?

Planning for Sustainable Cities

Learning Objectives

LO 17.1 Describe the strategies used to develop livable and sustainable urban spaces.

LO 17.2 Explain mixed-use development as a strategy to achieve the goals of New Urbanism.

Cities worldwide are confronting the challenge of sustainability in the face of growth and population changes. The concept of urban sustainability includes controlling pollution and reducing a city's **ecological footprint**, its impact on the environment expressed as the amount of land required to sustain its use of natural resources. Sustainable cities are also livable. Livability is the sum of the factors that add up to a community's quality of life, including the built and natural environments, economic prosperity, social stability and equity, and educational, cultural, recreational, and entertainment opportunities. People in highly livable cities have few reasons to move away.

Today's urban leadership, including elected officials, planners, and concerned citizens, are employing a number of sustainable design initiatives in the effort to make cities both environmentally friendly and welcoming to diverse communities. These strategies include mixed land use, walkability, transportation-oriented development, and smart-growth initiatives such as New Urbanism, greenbelts, and slow-growth cities.

As you learned, urban sprawl leads to environmental consequences such as the loss of rural land and pollution due to the excessive use of cars in areas with insufficient public transportation. Meanwhile, time spent commuting to work and school—whether by car or by public transportation—is a drag on sprawl dwellers' quality of life. **Mixed-use development** (MUD) is one way to limit sprawl and design livable urban spaces. A mixed-use development is a single planned development designed to include multiple uses, such as residential, retail, educational, recreational, and office spaces.

Mixed-use developments can range in scale, but all are intended to increase residential densities and minimize the need for travel outside the development, thus reducing transport and commute distances and costs. A MUD may be a single building or a group of multistory buildings that incorporate, for example, retail and commercial uses on the ground floors with apartments and office spaces on the upper floors (vertical mixed-use development). Some MUDs consist of one or more city blocks, and some are even larger, covering 1–2 square miles, with varied types of housing plus schools, businesses, shopping, services, dining, and cultural and recreational opportunities.

Mixed-use development is characterized by continuous pedestrian and bicycle-friendly connections that form a walkable community; **walkability** refers to how safe, convenient, and efficient it is to walk and bike in an urban environment. Measures of walkability may also include the ratio of people who walk or bike to destinations within the community versus those who drive, as well as the availability of locations such as stores or workplaces that are within walking distance of people's homes—an ideal related to the 15-minute city discussed in Chapter 16. Walkable communities must combine land uses within a compact area while also offering safe and inviting spaces for those traveling on foot. Some cities are engaging in revitalization and redevelopment to improve walkability in existing spaces. The U.S. Department of Transportation advocates and supports the development of new pedestrian and bicycle infrastructure "such as buffers between car lanes and bike lanes, shared use paths, bike parking and storage facilities, and overpasses to allow walkers and bikers to avoid cars." The agency notes that community health improves with these highly walkable and bikeable environments.

Access to transportation options is another goal of sustainable city design. By reducing the number of cars on the road, thoughtful transit planning minimizes the use of fossil fuels and can reduce urban pollution. Better bicycle infrastructure has been listed as one of several reasons that European countries produce fewer greenhouse gas emissions than the United States, which produces 16.5 metric tons of carbon dioxide per capita annually. In comparison, France produces 4.6 metric tons; Denmark 5.9 tons; and Germany 8.9 metric tons. Integrated transportation includes biking and walking infrastructures, high-quality public transit, and well-maintained bridges and roads—all of which come together in an interconnected streetscape.

Transportation-oriented development is the creation of dense, walkable, pedestrian- and bike-oriented, mixed-use communities centered around or located near a transit station. Arlington County, Virginia, created just such an environment in the Rosslyn-Ballston corridor, a development built along one of the major Metro lines running into nearby Washington, D.C. (as you learned in Chapter 2, Metro is the name of Washington's subway system). Arlington County has seven walkable and bike-friendly Metro transit villages with two Metro corridors that together accommodate 36 million square feet of office space and more than 47,000 residential units. About 40 percent of the residents take public transportation to work and 6 percent walk to work. Arlington County also created more than 100 miles of dedicated bike paths, used by bike commuters and recreational riders alike. Major streets have seen a decrease in automobile traffic, while Metro ridership has risen. Walkability, bikeability, and access to transportation have attracted many residents to Arlington County. And while the COVID-19 pandemic dramatically impacted work patterns and Metro use in 2020–2022, this community and transportation infrastructure continued to serve well the Rosslyn-Ballston corridor and the broader Arlington County community.

Smart-Growth Policies

Learning Objective

LO 17.3 Describe smart-growth policies that aim to create sustainable communities.

Many local governments in urban areas seek to control sprawl and enhance livability by enacting policies, strategies, and regulations to encourage certain types of development. The aim of **smart-growth policies** is to create sustainable communities by placing development in convenient locations and designing it to be more efficient and environmentally responsible. Arlington County's smart-growth policies, for example, include offering builders financial incentives to encourage mixed-use development. Additionally, many communities are implementing green building codes that require new construction to follow net-zero rules, meaning that each building must be designed to reduce and cover its own energy needs through efficiencies, carbon offsets, and renewable resources such as rooftop solar panels.

Zoning is another crucial tool of smart growth. Chapter 16 discussed how the city of Minneapolis used zoning laws to increase housing density in some areas. City governments can also use zoning to dictate what types of building and land use are permitted in different areas, or zones, of the city. **Mixed-use zoning** permits multiple land uses in the same space or structure, while **traditional zoning** creates separate zones based on land-use type or economic function such as various categories of residential (low-, medium-, or high-density), commercial, or industrial.

New Urbanism is a school of thought closely associated with smart growth. It arose in reaction to the sprawling, automobile-centered cities of the mid-20th century and focuses on limiting urban expansion while preserving nature and usable farmland. New Urbanists advocate for policies and design practices that support multiple transit options,

In Chengdu, China, pedestrians are protected from vehicle traffic by a "sky walk" that connects to a Chengdu Metro station, an example of efficient transportation-oriented design.

AVERAGE ROUND-TRIP COMMUTE TIMES IN THE UNITED STATES, 1980 & 2018 in minutes

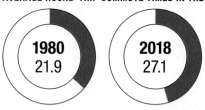

Just an extra **5.2 minutes** a day amounts to an additional **22.6 hours** a year commuting.

WORST ROUND-TRIP COMMUTE TIMES IN THE UNITED STATES, 2018 AVERAGE in minutes

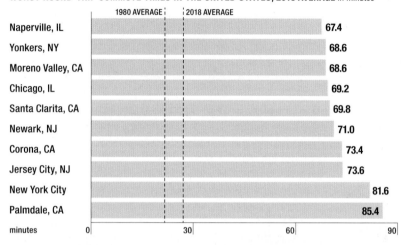

City	2018 Average (minutes)
Naperville, IL	67.4
Yonkers, NY	68.6
Moreno Valley, CA	68.6
Chicago, IL	69.2
Santa Clarita, CA	69.8
Newark, NJ	71.0
Corona, CA	73.4
Jersey City, NJ	73.6
New York City	81.6
Palmdale, CA	85.4

COMMUTE TIME AND WELL-BEING

In a 2016 survey, commuters in the United Kingdom with a one-way ride longer than 15 minutes reported decreased happiness. Those with commutes of 61 to 90 minutes had the lowest happiness levels. Commuters were also less likely than noncommuters to report that they had felt happy the day before.

9.4 DAYS SPENT COMMUTING a year, on average in 2018

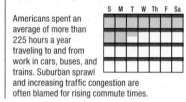

Americans spent an average of more than 225 hours a year traveling to and from work in cars, buses, and trains. Suburban sprawl and increasing traffic congestion are often blamed for rising commute times.

Impact of Long Commutes Commute times are on the rise in the United States and elsewhere, and long commutes can have a negative impact on both the environment and on commuters' well-being. City planners integrate designs that increase walkability in urban environments to help cut down on commute times.

---

the preservation of historic buildings, and respect for and preservation of locally sustainable environments. One of the stated goals of New Urbanism is to create—or re-create—a sense of place in cities by designing an urban environment where residents can easily meet and engage with one another in welcoming places like parks and other open spaces. Infilling, which was discussed in Chapter 15, is one technique used by New Urbanist planners to create new and inviting spaces within existing cities.

With its thoughtful transportation designs, the Rosslyn-Ballston Corridor in Arlington County is an example of New Urbanist principles at work. Curitiba, a city in southern Brazil, is also lauded as a model of New Urbanist–style planning. Beginning in the early 1970s, Curitiba began implementing sustainable initiatives, even before sustainability became a common concern among urban developers. Through zoning and other mechanisms, the city created new lakes and flood-control systems, specialized busing services, protected green areas, and recycling programs—making it an eco-friendly model of New Urbanism. Land that was being used as a garbage dump was transformed into a botanical garden. In 1972, Curitiba's mayor converted six blocks of the main downtown shopping street into a pedestrian zone that still serves as a community gathering place.

In the 1990s, Freiburg, Germany, also espoused New Urbanist and smart-growth principles as it made a commitment to preserving its historic medieval city center by keeping urban sprawl at bay. Freiburg, located in the Vauban district of southern Germany, is often referred to as the country's "ecological capital" and has been internationally recognized for its sustainable design. Development is focused within city boundaries to prevent sprawl, and historic buildings have been protected, along with vineyards, orchards, and farms. Freiburg was one of the first German cities to close the city center to traffic to encourage social interactions and a sense of community. Pedestrian and bike paths create an interconnected, green transportation network where homes are within walking distance of tram stops, schools, businesses, and shopping centers.

Some cities are looking at smart-growth strategies to solve severe urban problems. Sao Paulo, Brazil's largest urban area, was home to more than 22 million people in 2022, and an estimated 20 to 30 percent of the population were living in favelas, or impoverished informal settlements often located in unsafe locations. Much of Sao Paulo's suburban growth has happened in the favelas, which sprang up on steep hillsides or floodplains—areas deemed unfit for permitted construction. Limiting the expansion of favelas has been a goal for Sao Paulo since the 1960s. To contain the sprawl of these informal settlements, city officials have worked to rehabilitate existing favelas with structural improvements and build vertical housing projects intended to replace large areas of this informal housing.

Cities where planners have used smart-growth policies to decrease the rate at which cities grow outward are called **slow-growth cities**. Urban areas such as Boulder, Colorado, have slowed the pace of development to retain their sense of place and to preserve open space. The city of Boulder has achieved this by limiting building permits and creating an **urban growth boundary** in cooperation with Boulder County. An urban growth boundary borders a city's edges and defines where new development can take place. Urban growth boundaries separate urban land uses from rural land uses—farmland or wildland—by limiting how far a city can expand. Despite these efforts, Boulder County has still experienced significant exurban growth, pushing neighborhoods into the wildland–urban interface (WUI). In December 2021, more than 1,000 Boulder County homes were lost to a rapidly moving wildfire that occurred between Denver and Boulder.

Growth Management Plans Growth management, whether at the state or city level, is implemented so that infrastructure and services—such as sewers, roads, schools, and police and fire protection—are not overwhelmed by new construction. Zoning is a significant part of growth management plans. Some cities control growth using large-lot zoning, which requires minimum building lot sizes of one, two, or more acres to dampen the pace of development and reduce the demand for public infrastructure. These lot sizes also encourage the construction of higher-priced homes, which, it is argued, benefit a community financially through property taxes. In addition, states like California have set growth limits by establishing annual quotas for building permits and regulating the rate of development so that it does not outpace what the community's infrastructure can support.

Critics of growth management argue that it increases the cost of housing, contributes to rapidly changing real estate prices, and can even harm the national economy by slowing the building industry. Large-lot zoning, too, can encourage sprawl because each lot takes so much space, and it can push lower-income people out of neighborhoods where they cannot afford the larger, more expensive properties.

Successful growth management plans, however, succeed in balancing the needs of residents with diverse incomes and concerns. In Lancaster County, Pennsylvania, county commissioners developed a growth management plan to protect farmland and nature preserves for public use and constrain growth to designated areas. Lancaster's plan included designing new communities and revitalizing existing communities to be higher density, so that agricultural and conservation areas could be left untouched. The county also took into consideration the importance of preserving historic areas as well as the open lands that are home to long-standing Amish and Mennonite communities. The plan permanently protects approximately 82,000 acres of farmland, as well as 6,000 acres of parks and natural lands.

Diverse Housing Options Zoning also plays a role in encouraging housing diversity—a mix of housing types in a neighborhood or community—which is another component of many smart-growth initiatives. Housing diversity is normally encouraged through planning codes that promote lower-cost housing such as townhouses, multifamily dwellings, and live-work spaces—units that provide both a place to live and a workspace such as a shop or an art studio. Diverse housing promotes mixed-income neighborhoods, as more affordable units and higher-priced housing can be found side by side. Income diversity helps build the economic base needed to support services and transportation, helps create higher-quality housing, and results in wider access to affordable housing. Other benefits of housing diversity include economic stability and

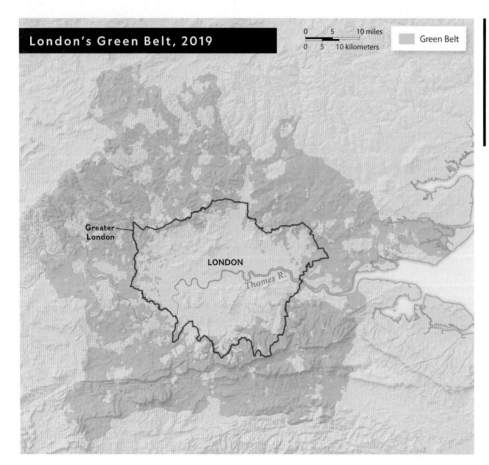

The Green Belt of London is a ring of open space created in the 1930s to prevent urban sprawl. Today, it is home to golf courses, farms, and parks—along with some rundown buildings. Whether or not to allow parts of the nearly 2,000 square miles of protected land to be developed for new housing has been a subject of debate.

commuting advantages. A shortage of affordable housing in a city can force young professionals, as well as middle-class and working-poor families, to live in outlying areas, requiring lengthy commutes to and from their jobs. The results of this situation include traffic congestion and a diminished tax base for the city.

Tucson, Arizona, created a neighborhood with diverse housing when it completely renovated the Connie Chambers Public Housing Project. The former community eyesore was transformed into 180 units of mixed-income housing and renamed Posadas Sentinel. Half of the new units were set aside for families who earn at least 60 percent of the area's median income; the other half are public housing—affordable units owned and administered by the government. Posadas Sentinel was designed to be walkable, energy efficient, and integrated into the larger community. Rather than building on unoccupied land and expanding the city, Tucson recycled the former housing project into usable and lively urban space.

Urban Greenbelts At a city's edge, greenbelts sometimes feature in New Urbanist or smart-growth plans. A **greenbelt** is a ring of parkland, agricultural land, or other type of open space maintained around an urban area to limit sprawl. Greenbelt areas can serve as urban growth boundaries because converting rural land to urban land use is strictly prohibited within them. Greenbelts contribute to the ecological health of a region by limiting pollution, promoting plant growth, and protecting wildlife habitats. Additionally, these areas give city dwellers a chance to connect with nature and enhance their quality of life.

The idea of urban greenbelts is not new: the Green Belt of London, England, was first established in the 1930s, and it still rings the city. In California's Bay Area, the Greenbelt Alliance is a nonprofit organization that actively lobbies to protect the greenbelt spanning four counties surrounding the city of San Francisco. With the Bay Area growing at a rate almost double that of the United States as a whole, natural landscapes in the region are at increasing risk from urban sprawl.

Other urban areas have created political bodies that, while not specifically designed to create greenbelts, serve to protect natural lands in cities and suburbs. For example, Cook County, Illinois, which contains Chicago, established the first forest preserve district in the early 20th century. As stated by its founders, the district's goal was "to acquire, restore and manage lands for the purpose of protecting and preserving public open space with its natural wonders . . . in a natural state for the education, pleasure, and recreation of the public now and in the future." Many counties and municipalities have established forest preserve districts since then, but Cook County's remains the largest in the United States, with nearly 70,000 acres. In addition to greenbelts, more cities have repurposed derelict land into attractive parks, which in turn enhance the areas adjacent to them, attracting mixed-use development. Open in 2009 and expanded several times since, the High Line, featured in this chapter, is an elevated trail on an old railroad route on New York City's West Side. This linear park features gardens, restaurants, art, and community activities. Nearby along the Hudson River in Manhattan, Little Island at Pier 55 opened in 2021. It is part of a rehabilitation of the city's waterfront and provides a public space for New Yorkers, bringing together nature and art.

Pros and Cons of Urban Design

Learning Objective
LO 17.4 Discuss the advantages and disadvantages of smart-growth urban design initiatives.

Purposeful urban planning, integrated with sustainability, is challenging because it can be difficult to anticipate and understand the real needs of a city's future residents. For this reason, it is important to engage all citizens early in the planning stages. In the case of Lancaster County, the planning commission involved farmers, urban residents, and the Mennonite and Amish communities, as well as other organizations in the area to integrate as many views and ideas as possible into the completed plan. Even as they take into account current needs and concerns, urban planners and leaders must also be open to refining their designs and policies as needs change and unforeseen circumstances arise. In addition, urban residents and experts frequently disagree on which planning strategies, methods, and outcomes are best.

So far, the strategies advocated by the smart-growth movement have received both praise and criticism. Supporters point to a significant reduction in the negative environmental impact of cities—largely due to the slowing of urban sprawl, a more efficient use of space, and the promotion of sustainable options. The focus on density, mixed-use developments, and preserving open land is thought to reduce a smart city's ecological footprint. The greater variety of transportation options, including biking, walking, and public transportation, also decreases auto traffic, fuel consumption, and air pollution.

Many also argue that smart-growth cities enhance residents' quality of life. For example, transportation-oriented development reduces commute times and the costs associated with daily travel, thus increasing personal time and money available for spending. Mixed-use developments allow residents convenient access to shopping, recreation, services, and jobs. Additionally, measures to manage and slow growth (as in the example of Boulder, Colorado) reduce the strain on a city's infrastructure and ensure that residents have the services they need to deliver an appropriately high quality of life.

The social benefits of smart-growth urban design initiatives include more diversified housing options within

Greenbelt Towns

Greenbelt towns are not precisely the same as greenbelts. Founded by the U.S. government during the Great Depression, they were experiments in urban design intended to provide affordable housing, jobs for the unemployed, and a model for future city developers. The aptly named Greenbelt, Maryland, predates New Urbanism but incorporates many New Urbanist features, including natural open spaces, mixed urban development, and features to enhance walkability. The aim was to create a wholesome, community-focused culture for lower-income workers. To be admitted as residents of Greenbelt, people had to apply and be approved by the government. Greenhills, Ohio, and Greendale, Wisconsin, were the other New Deal Greenbelt towns. Like Greenbelt, they were built as suburbs of larger cities.

Critics contrast the original Greenbelt towns with more recent New Urbanist developments that are often geared for the needs of wealthier residents and feature less affordable housing. At the same time, it is worth noting that the original Greenbelt communities were not perfect models of inclusivity. In Greenbelt, applicants for homes had to be married couples, the husband had to be employed, and at first, African Americans were not considered.

---

neighborhoods and communities, which encourage social interactions across groups. Such day-to-day communications may help lower barriers between people of different income groups, age ranges, or ethnicities. Personal connections also help create a sense of community and strengthen social identity. The Noji Gardens neighborhood in Seattle, Washington, was designed to encourage just such interactions. State-level growth management requirements for affordable housing led the nonprofit developer to construct the mostly two-story manufactured homes within Seattle city limits near schools, playgrounds, public transportation, and the Columbia City commercial core.

Using space efficiently and creatively through revitalization and redevelopment can lead to fewer building vacancies and reduced crime—another powerful social benefit. In 1999, the city of Richmond, Virginia, launched the Neighborhoods in Bloom revitalization initiative, targeting for improvement vacant and unsafe city lots across six neighborhoods. The city worked with community groups to identify and rehabilitate abandoned and substandard buildings using smart-growth strategies. The program has been successful in turning areas of blight into areas that are economically viable by attracting new businesses.

Critics of smart growth and New Urbanism claim that the promises of communities with mixed income levels and diverse ethnic groups are not often realized. Smart-growth design creates desirable neighborhoods, where real estate prices rise to the point that homes are not affordable for poor or working-class families. Lower-income residents are then displaced to the suburbs or lower-quality housing in surrounding areas. At best, say the critics, the result of this process is lack of income diversity in neighborhoods. In some cases, **de facto segregation** (segregation that results from residential settlement patterns rather than from prejudicial laws) occurs when lower-income people of color are unable to afford to live in desirable new smart-growth developments, which then become populated by wealthier White residents.

The United States provides several examples of New Urbanist designs that, rather than diversifying, have simply reproduced racial and economic injustice. Baxter Village in Fort Mill, South Carolina, represents a mix of new and renovated buildings and a community within which people can work, recreate, go to school, and live. The problem is that it is also, by vast majority, White and middle class—not at all reflective of South Carolina's racial or economic diversity. Beginning in the 1990s, Providence, Rhode Island, began reimagining its downtown, including launching several mixed-use developments with a goal of attracting financially diverse residents. However, research has shown that the desired mix of income levels has remained stubbornly out of reach. Cases like those of Baxter Village and Providence imply that smart-growth design, regardless of its benefits, does not guarantee affordability. In some cases, urban design initiatives risk promoting homogenous living spaces and ethnic and economic segregation.

Another criticism of urban design initiatives is that they apply similar design concepts across many different urban areas. This can result in a sense of placelessness or a loss of historical character. Some commentators argue that multiuse areas can even begin to develop artificial-feeling atmospheres. Standardized designs that may involve the removal or transformation of historic buildings are said to result in sterile and bland-looking places that detract from the character of a neighborhood. New Urbanists claim that suburban architecture is "cookie cutter" and repetitive, yet New Urbanist architecture itself is criticized as lacking in the unique qualities or diversity that help give a place its own individual character.

For example, Celebration, Florida, is a community that was built by the Walt Disney Company beginning in 1995 and designed around New Urbanist principles such as walkability, mixed-use development, and abundant public spaces. Homes in Celebration follow strict design rules meant to create the nostalgic look of a classic American small town. Thousands of potential purchasers loved what they saw in Celebration and flocked to buy homes. Many critics, however, have called Celebration's architecture "inauthentic" and find that the community resembles a film set rather than a real town.

New Urbanist and smart-growth schemes are often seen as "top-down and technocratic," appearing to ignore the concerns and needs of residents. Critics recommend

Critical Viewing The Flame Towers in Baku, the capital of Azerbaijan, rise above the Old City walls. Each of the three towers (one is not pictured) is designed for a different function, including a residential tower, a hotel, and an office building. ▌Explain how the Flame Towers are examples of smart-growth urban design.

Urban Living 485

instead a bottom-up, "hyperlocal" process that builds on a community's social relationships rather than lofty ideals. Such a development is seen as better able to tackle significant issues of social and environmental justice.

Both sides of the smart-growth debate can point to successes and failures of urban design to support their arguments. At present, smart growth, slow growth, New Urbanism, and associated movements are experiments in progress. It remains to be seen how successfully people form neighborhoods and communities within the spaces created by urban designers. The ultimate verdict on the value of different smart-growth initiatives may lie many years in the future.

Geographic Thinking

1. Explain the difference between mixed-use development and traditional zoning practices.
2. Describe how mixed-use development supports urban sustainability.
3. Explain whether smart-growth policies are successful in achieving affordable and accessible housing.
4. Compare the advantages and disadvantages of New Urbanist design principles.
5. Describe some of the urban design principles illustrated in the "Cities of the Future" feature in Chapter 15.

17.2 Causes and Impacts of Urban Changes

Urban areas can be hubs of opportunity, but each city faces unique economic, political, cultural, and environmental challenges. The one thing all cities have in common is that they are changing. As urban areas continue to evolve, with some growing and some declining, it is crucial for planners and government officials to understand both the causes and impacts of urban changes.

Urban Challenges

Learning Objectives

LO 17.5 Explain how the declining tax base in core cities affects infrastructure and can lead to urban decay.

LO 17.6 Describe how housing discrimination practices like deed restrictions, redlining, and blockbusting reinforce geographies of segregation.

The urban areas of core countries, for all their wealth, power, and opportunity, have serious challenges. In the United States, cities face three major interconnected problems: a declining tax base in the original core city, infrastructure costs, and patterns of poverty and neighborhood decay.

The core, or central, cities of most U.S. urban areas, especially those that grew in population and wealth during the industrial era, depend (as do all municipalities of any size) on tax revenue—primarily from property owners and businesses. As you've learned, these tax dollars support a city's infrastructure and services: streets, water supply and sanitation systems, police and fire protection, parks, and a broad range of social services. Over the past half century, with economic restructuring and deindustrialization occurring in many American cities, tax revenues have declined, sometimes dramatically, as businesses, jobs, and people have moved out of the central city to the suburbs or other regions and countries. Additionally, properties in many older central city neighborhoods have not retained their value, and thus less tax is generated from these homes and buildings.

At the same time that the tax base of most core cities has diminished, the costs of maintaining aging infrastructure and providing urban services to older, often decaying neighborhoods have increased. Old water and sewer lines cost more to maintain, and breaks in these systems are both common and expensive to repair. Transportation systems and schools also have high maintenance costs. Inner-city neighborhoods typically have more elderly, low-income, and immigrant populations, and these groups have greater need for a city's social services. Institutions such as museums and parks, which are enjoyed by residents of the entire urban area, are also generally the responsibility of the core city's government.

As middle-class and wealthier populations, along with many businesses and jobs of all kinds, have moved away to the suburbs and beyond, the low-income populations left in many older central city neighborhoods face multiple challenges. For many, renting an apartment or house is the only option, and these rental properties are often poorly maintained and often rising rapidly in cost. These patterns of poverty extend to the businesses in core city neighborhoods as well: restaurants, stores, and service shops often struggle or move away, and failures may result in vacant or even abandoned buildings. As offices closed during the pandemic, businesses that served workers in downtowns suffered and many shut down—some permanently. Less funding for schools may lead to poorer-quality education. Difficulties with city transportation may prevent some people from taking or keeping jobs in the suburbs. Lack of employment opportunities may cause some to turn to crime, while at the same time, declining tax revenues mean less funding for essential safety services such as fire protection and police. The combination of factors that leads to neighborhood decay is difficult to combat, and one result is often concentrated, multigenerational poverty in the central city.

Housing Discrimination In the United States, systemic, institutionalized discrimination has long been a driver of urban population movements and racial segregation, and it still exists. Housing discrimination is an attempt to prevent a person from buying or renting a property because of that person's race, social class, ethnicity, sexuality, religion, or other characteristic. Discriminatory practices by landlords, owners, real estate agents, or lenders may favor one population group over another, or a certain group may be excluded from a particular neighborhood. These practices have included deed restrictions, redlining, and blockbusting (explained below).

For at least the first half of the 20th century in the United States, many property deeds—documents that prove ownership—included statements stipulating that the property could not be sold or rented to someone belonging to a certain racial or ethnic group. In many cities and suburbs, these deed restrictions prevented African Americans, Asians, and Hispanic peoples from moving into more desirable neighborhoods, where houses could only be sold to Whites.

Housing discrimination also happens through a process called **redlining**, when a lending institution such as a bank refuses to offer home loans on the basis of a neighborhood's racial or ethnic makeup. In the 1930s, as part of a plan to encourage investment and homeownership, the U.S. government created maps in 239 cities that indicated areas considered poor investments for lending institutions. These areas, delineated in red, were predominantly neighborhoods with high African American or other minority populations. Properties in redlined areas fell into disrepair or were sometimes abandoned because residents could not get loans to purchase homes or to improve homes they already owned. In his book *The Color of Law*, urban researcher Richard Rothstein documents how governments at all levels—federal, state, and local—created and reinforced neighborhood segregation in America's cities. Rothstein's research shows that this **de jure segregation** (the laws and policies passed by various levels of government) promoted discriminatory practices and patterns still seen today. The Fair Housing Act of 1968 made redlining illegal, yet recent studies have found that African Americans in many cities such as Atlanta, Detroit, St. Louis, and San Antonio still are approved for mortgages and other loans at lower rates than Whites. The impacts of redlining and other discriminatory practices are still evident today, with patterns of poorer community health and educational outcomes in previously redlined neighborhoods compared to nonredlined areas.

Blockbusting was practiced for decades by real estate agents who would generate concern that African American families would soon move into a neighborhood. These unscrupulous agents would convince White property owners to sell their houses at below-market prices. Blockbusting promoted fear of minorities and the discriminatory belief that houses in diverse or African American neighborhoods were not as valuable as those in other areas. In the resulting outmigration of White homeowners, real estate agents profited through property sales: White sellers sold at a loss and Black buyers paid too much, while the racial makeup of residential blocks changed rapidly. Like redlining, blockbusting was outlawed in 1968, but it has been found that some real estate agents may still guide or steer potential home buyers to particular neighborhoods and away from others. Other predatory practices such as repeated letters and calls to pressure residents to sell are common today in areas undergoing gentrification. Real estate speculators attempt to push low-income residents out in order to flip a neighborhood, meaning to buy properties with the intention of quickly reselling.

As a result of these racially discriminatory and exclusionary practices and other factors, studies have shown that African American families in the United States, on average, have much less wealth than White families because so much personal and family wealth is tied to housing. In cities and elsewhere, the largest investment that most Americans have is in homeownership. Not being able to buy a home or relocate to a better neighborhood with higher home values makes it difficult for a family to build wealth. This reinforces the urban reality that multigenerational poverty is all too common in inner-city neighborhoods surrounding central business districts, where urban minority populations are concentrated.

Another urban challenge at least partially associated with segregation and past injustices is crime. When people are economically and racially segregated and isolated in urban neighborhoods, crime rates increase. Cities with integrated neighborhoods are among the safest. Other connected factors, such as poverty, also influence urban crime rates. According to the U.S. Census Bureau, urban areas often have higher rates of income disparity than rural areas because a greater range of jobs and educational levels are typically found in cities. Unemployment rates, as well, are generally higher in urban areas. Rural areas tend to have less industry, attract fewer professionals, and have a larger population of retirees—factors that contribute to a more constant and moderate median income. The U.S. Census Bureau reports that while American rural areas have lower median incomes, people living there have lower poverty rates than their urban counterparts, who live where the cost of living is higher.

Economic and social stresses are linked to high crime rates, and there are distinct spatial patterns of poverty and crime. Some areas of cities offer fewer institutional resources that may prevent crime, such as access to youth programs and other social services. The pressures of urban life can affect social cohesion and stress the networks of friends and family that act to regulate society and deter crime. Additionally, high numbers of unhoused people living in informal urban encampments also often contribute to crime.

Geographic Thinking

1. Explain how declining tax revenue is linked to infrastructure problems and patterns of urban poverty.

2. Explain the effects of redlining and blockbusting on housing in urban areas, in the past and in the present.

In Levittown, New York, one of the earliest versions of the modern suburb, Long Island's potato fields were transformed into a community of more than 17,000 houses between 1946 and 1951. The original lease agreement for Levittown houses stated that they could not "be used or occupied by any person other than members of the Caucasian race."

Case Study

The Effects of Redlining in Cleveland

The Issue Though the 1968 Fair Housing Act banned redlining because it promoted racial and economic segregation in many U.S. cities, housing discrimination and segregation persist in Cleveland today.

Learning Objective
LO 17.6 Describe how housing discrimination practices like deed restrictions, redlining, and blockbusting reinforce geographies of segregation.

By the Numbers

8,500
Cleveland's African American population in 1910

72,000
Cleveland's African American population in 1930

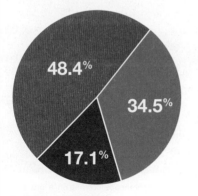

Cleveland Population Estimates in 2020
- African American alone — 48.4%
- White alone — 34.5%
- Mixed-race and other — 17.1%

Source: U.S. Census Bureau

Patterns of Residential Segregation in the United States persist more than 50 years after redlining and blockbusting were declared illegal. Maps produced by the Anti-Discrimination Center, a civil rights organization, provide evidence that not only is residential segregation in the United States alive and well—in cities like Cleveland, Ohio, it is a persistent long-term effect of redlining.

Recall that during the Great Migration, African Americans from the rural South migrated primarily to cities in the Northeast, Midwest, and West. The city of Cleveland's population increased by 60 percent between 1910 and 1930 to more than 900,000 people. During the same period, the African American population in Cleveland multiplied to more than eight times its 1910 size. Although segregation was not enforced by law in the North, many African Americans and immigrants were pushed into low-income, often poor-quality housing in the east side of Cleveland's inner city. These challenges grew as the cities became increasingly crowded and Whites who were biased against African Americans saw them as competition for jobs and a threat to their neighborhoods. In Cleveland, segregation was further encouraged by redlining and by officials who refused to issue deeds for land owned by African Americans.

Even after the Great Migration ended, fair housing opportunities continued to be a challenge for African Americans in Cleveland. In some instances, policies intended to promote integration deepened the problem instead. From the 1970s to the mid-1990s, for example, the city attempted to use busing to overcome segregation in schools. This move had the effect of deepening housing segregation because it motivated Whites to move out to the suburbs in greater numbers to avoid having their children bused to integrated schools. Cleveland is the sixth-most segregated major city in the United States.

A 2019 analysis found that the vacancy rates in some Cleveland neighborhoods were greater than three times the state average. The resulting zones of abandonment are areas that have been largely deserted due to lack of jobs, declines in land values, and falling demand. Filtering is the process of neighborhood change in which housing vacated by more affluent groups passes down the income scale to lower-income groups. Zones of abandonment are the result of years of redlining, blockbusting, and filtering. They are not merely unattractive; they have a negative effect on the city's tax revenues due to lower property values. Zones of abandonment are also havens for crime, as businesses close and homes are abandoned. Structures are illegally occupied, vandalized, or even set on fire.

Many of the same neighborhoods that were redlined almost 100 years ago have Cleveland's highest poverty and crime rates today. Historically redlined Cleveland neighborhoods also show disparities in health, with homes often having dangerously high levels of lead and with higher rates of asthma, maternal death after birth, and certain types of cancers among residents. These neighborhoods also typically lack reliable internet access and availability of healthy food. In 2018, the U.S. Department of Housing and Urban Development awarded more than $2.4 million in grants to fair housing organizations across Ohio with one goal: to fight housing discrimination.

Faced with the lingering effects of redlining, city dwellers often choose to leave. According to the 2020 Census, the city of Cleveland's population decreased by 6 percent from 2010 to 2020. That, however, was better than the previous decade, when the city lost 17 percent of its population. Leaving the city, however, does not guarantee an escape from zones of abandonment. If the problems of discrimination are not addressed in the core city, too often these unfair practices follow residents to the surrounding suburbs. ∎

Geographic Thinking

Describe how zones of abandonment contribute to the economic and social challenges in urban areas.

Historic Redlining Map, 1936

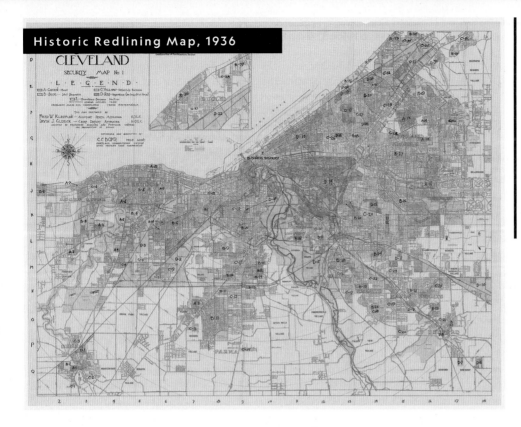

In 1936, the Home Owners Loan Corporation (HOLC) produced this map, and many others like it, to demarcate which neighborhoods were worthy of property investment in Cleveland. These became known as redlining maps. The faded pink areas of the map were originally red, which stood for "hazardous." The yellow areas were considered "definitely declining," the blue areas were "still desirable," and the green areas were "best."

Cleveland Neighborhood Grading, 2015

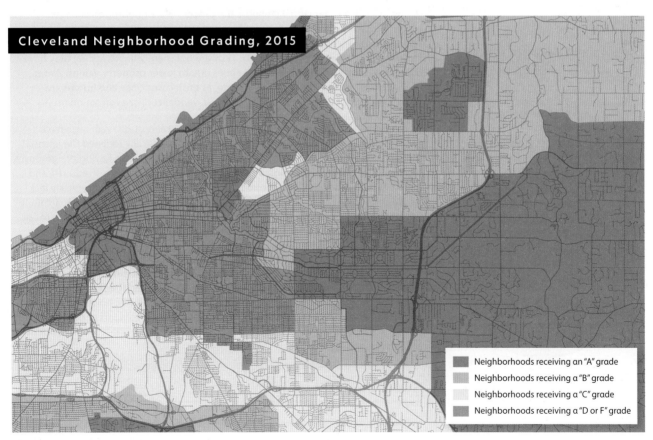

- Neighborhoods receiving an "A" grade
- Neighborhoods receiving a "B" grade
- Neighborhoods receiving a "C" grade
- Neighborhoods receiving a "D or F" grade

Reading Maps In 2015, a real estate agent in Cleveland created this map, in which Cleveland's neighborhoods were grouped into color-coded categories. Those deemed the best investment for real estate investors were given an "A" grade, and so forth. ▌ Compare the two maps and explain how they illustrate the legacy of redlining in Cleveland.

Challenges of Rapid Urbanization

Learning Objective
LO 17.7 Explain some of the challenges presented by rapid urbanization including increased crime rates, affordable housing, land tenure, and environmental injustice.

As you have learned, some of the most rapidly developing cities are located in countries of the periphery or semi-periphery. People in these countries migrate internally from rural areas to cities in search of better economic opportunities, an improved quality of life, or an escape from worsening environmental or security conditions in their villages. Often, the new arrivals are crowded together with previous rural-to-urban migrants in squatter settlements. Squatter settlements are informal settlements where residents do not have legal ownership of their land, hence they are often called squatters. These informal settlements, sometimes termed *slums* (or *favelas* in Brazil), usually lack adequate housing and basic infrastructure services such as fresh water, sanitation, and electricity. These periphery or semi-periphery cities are simply overwhelmed with the massive inflow of often-desperate immigrants, yet the crowded, difficult conditions do not slow the pace of immigration.

Delhi, India, and Jakarta, Indonesia, provide good illustrations of some of the issues these cities face. Based on 2020 rural-to-urban migration numbers, the urban population of India is projected to reach nearly 600 million by 2030. The urban area of Delhi has been experiencing some of the most rapid growth in the world, with a metro population topping 32 million. Delhi forms part of India's National Capital Region, along with nearby cities including New Delhi, the capital. The National Capital Region is the country's largest employment center and is receiving record numbers of arrivals. To deal with this inflow of people, the government of India launched the Delhi-Mumbai Industrial Corridor (DMIC) Development Project, with plans to develop 24 new industrial cities along the more than 900-mile route between Delhi and the country's financial hub and major port city, Mumbai. Other efforts to ameliorate the effects of rapid urbanization in Delhi include improved drinking water distribution, better electricity supply, a high-capacity transportation network, and job skills training.

Jakarta, the capital of Indonesia and largest metropolitan area in Southeast Asia, with about 30 million people in its metropolitan area, has long been beset with urban problems. At the close of Ramadan, a month-long Muslim religious observance, the city's population commonly balloons as thousands of Indonesians from rural areas come to attend festivities and then remain in the city to pursue employment and a better quality of life. In response, government officials have revived a program informally named "Return to Village." The government offers villages funding to develop economic growth and stability, which in theory should translate to new and more opportunities for villagers that will motivate them to stay. The Indonesian government is willing to invest in staving off rural-to-urban migration because rapid urbanization is causing significant problems in Jakarta, including traffic congestion caused by insufficient public transportation and other long-neglected infrastructure issues, including untreated sewage in its rivers.

Jakarta also experiences flooding due to its sprawling location in a lowland area with 13 rivers, a problem that has been exacerbated by unchecked urban growth and subsidence. The city is sinking faster than any other megacity in the world due to water being pumped out of underground aquifers by huge numbers of private wells. As industrialization has increased around the edges of Jakarta, water containment areas have been bulldozed, leading to even more severe flooding throughout the megacity. Urban conditions are so bad that in 2019 the government announced they would move the capital from Jakarta, on the densely settled island of Java, to the more remote and less dense island of Kalimantan, also known as Borneo.

The colors on the map illustrate the rapid urbanization of Delhi, India, and its surrounding areas over the past 30 years. The number of urban households in Delhi doubled between 1991 and 2011 as croplands and grasslands were turned into streets, housing, and other city structures.

Though they can be areas of high crime, favelas are also dynamic places where people live, work, and play. The colorful Santa Marta favela in Rio de Janeiro was the first favela in Brazil to receive a Police Pacification Unit as part of the city's attempts to reduce gang-related crime. While outside police presence in favelas has not always been seen as a positive for favela residents, this effort initially cut down on drug-trafficking activity in Santa Marta.

Increased Crime Rates In all parts of the world, concentrations of poverty, often caused or exacerbated by rapid urbanization, lead to desperation, social isolation, and often subsequently higher crime rates. Since educational and formal economic opportunities are sparse in poverty-stricken areas, some residents resort to illegal activities such as drug dealing, which then generally leads to other crimes including robbery, prostitution, and murder. In some regions, areas of extreme poverty have developed into disamenity zones (as discussed with the Latin American city model in Chapter 16)—areas in disadvantaged urban locations composed of informal housing. Often located on the periphery of large cities, disamenity zones do not receive city services and are frequently brutally controlled by gangs and drug lords.

The favelas of Brazil—in which more than 12 million Brazilians, or about 6 percent of the country's total population, live—are one example of disamenity zones with corresponding high crime rates. As you learned, favelas are illegal housing settlements, usually made up of temporary shelters in precarious locations, with few if any government services. Most residents of favelas are of African heritage. Some older favelas have more permanent structures, a few of which have electricity, water, and other infrastructure developed by the residents themselves. Why would people choose to move to or to stay in clearly unsafe areas like steep hill sides or flood-prone land? Housing analysts confirm that it is not that residents or potential residents of favelas or other squatter settlements don't understand the risks, but that they have no other choice.

Without formal security services, armed civilian groups have organized, often with ties to drug and gun trafficking. Drug lords often represent an informal ruling body, frequently battling one another for dominance. In some of these communities, military and civil police forces from the outside have also had a brutal presence, conducting frequent raids and gun battles with gangs that disrupt daily life. Favela residents thus face violence and death from gangs as well as from police. In recent years, 8 out of 10 Brazilians killed by police were people of what the Brazilian government categorizes as Black, Brown, or Mixed, and police killings in Brazil tripled between 2013 and 2020—many in favelas. Crime in the favelas pushes up Brazil's overall crime rates. According to the Brazilian Forum of Public Security, in 2017, 175 people were murdered per day in the country, which equates to an annual rate of 30.8 murders per 100,000 people. To put these numbers in perspective, in the same year, the United States had 5 intentional homicides per 100,000 people.

Affordable Housing Changing urban populations result in challenges to affordable housing in cities all across the world, although the specific difficulties vary among cities of the core, periphery, and semi-periphery. In core countries, affordable housing is a top urban challenge for a variety of reasons. In the United States, as you have learned, practices such as redlining have created barriers to homeownership for members of minority groups for decades. Many core cities are seeing other challenges to affordable housing as well. Increasing labor and material costs have made it more difficult to build affordable housing units. Issues with labor and material costs, as well as the manufacturing and transportation disruptions (which are termed "supply-chain issues") were magnified due to the COVID-19 pandemic, beginning in 2020. In addition, as rents and home prices rise in desirable areas close to successful business districts, people with lower incomes are forced to move farther from their workplaces, adding the higher cost of transportation to the cost of housing.

Governments in both core and semi-peripheral countries have attempted a variety of policy solutions such as rent control and public housing developments. Some local governments offer rent subsidies, tax credits, or reduced property taxes to help long-term, low-income residents remain in neighborhoods with rising housing costs. Successful mass transportation-oriented developments such as Arlington's Rosslyn-Ballston corridor can offer another solution by helping residents reduce both the money and time they spend commuting to work.

Municipalities in the United States have also used zoning to address issues of affordable housing. Inclusionary zoning is a strategy first developed in the 1970s to counter the effects of exclusionary and racially segregated zoning. **Inclusionary zoning laws** create affordable housing by offering incentives for developers to set aside a minimum percentage of new housing construction to be allocated for low-income renters or buyers. The practice has been the subject of debate. Those in favor argue that it requires fewer government subsidies than other affordable housing policies and promotes economic, racial, and cultural integration in urban neighborhoods. Those against maintain that it may result in an overall increase to the cost of housing for a large segment of the population, as consumers who pay market-rate prices for housing are asked to subsidize lower-cost housing. Critics also believe inclusionary zoning is a disincentive for developers and a burden for buyers, claiming that it reduces the overall supply of housing units, which negatively impacts the homebuilding industry.

By 2021, there were more than 900 cities and counties in the United States with inclusionary housing programs. Most inclusionary zoning laws are mandatory, while others instead provide government-sponsored incentives for developers to voluntarily include affordable units in their projects. In Boulder, Colorado, developers of rental properties are required to provide affordable rental units, which are purchased by the city of Boulder. The city then sells or gives the units to its Housing Authority or a similar agent. With soaring housing costs caused by the pandemic, inclusionary zoning laws, both mandatory and incentivized voluntary, were passed in many cities in the United States and Canada, including Philadelphia, Seattle, Washington D.C., and Toronto, to address pressing issues of housing affordability and supply. These housing justice measures ensure that housing options exist within city limits for low-income residents and working people. In their absence, more expensive housing would be built, which results in a rise for all home prices within a community.

The United Kingdom has established affordable housing programs that are similar in practice to mandatory inclusionary zoning. Under the Town and Country Planning Act, agreements between local authorities and for-profit builders created more than half of the affordable housing in England. The United Kingdom has experienced significant success with this strategy because the law grants local authorities the power to devise their own inclusionary housing policies. The Greater London Authority reported that between 2000 and 2017, London produced more than 11,000 affordable homes per year on average, and as of 2018 more than 20 percent of London households lived in subsidized rental housing. London's inclusionary housing program is one of the largest and most successful in the world.

Cities in the periphery and semi-periphery face particular challenges to affordable housing because of ongoing rural-to-urban migration. These urban areas have insufficient affordable housing and resources to accommodate the rapid increase in their populations. The lack of adequate affordable housing leads to the development of slums or squatter settlements like Brazil's favelas at the edges of its cities. These informal settlements are often overcrowded, lack basic services (water, electricity, sanitation), and are not officially part of the city, leaving residents to develop their own systems for administration. As geographer Farhana Sultana reports, many of these slum dwellers are more like "subjects," or subordinates, and not considered full citizens of their cities because they are not provided with the resources and services that other urban residents have available to them.

Land Tenure People living in informal settlements in rapidly growing cities are especially vulnerable because they do not have an official claim or title to the land they live on, and thus are considered squatters. They lack **land tenure**, or the legal rights, as defined by a society, associated with owning land. The concept of land tenure encompasses who can use the land, for how long, and under what conditions.

In 2022, Lagos, Nigeria, was ranked the 15th largest urban area in the world with a population estimated at more than 23 million people—having mushroomed from just 4 million in 1990. Many of the tens of thousands of people who arrive in Lagos each week end up living in waterfront informal settlements, where they have no formal claim to the spaces they inhabit. In fact, Lagos is home to the world's biggest floating informal settlement, Makoko, with population estimates that range from 85,000 to 250,000. In the 19th century, Makoko was a small fishing village that has since evolved into a community of shacks on stilts and boats after the exploding population in Lagos left the poor with very few housing options. Located under the most traveled bridge in the megacity, Makoko has been repeatedly targeted by the government in an effort to erase it from such public view, though the authorities term these forced evictions as "urban regeneration." In July 2012, demolition workers wielded machetes and set fire to the settlement, leaving 30,000 people homeless.

Critics of the Lagos authorities, and of other governments attempting to clear squatter settlements by force, claim that eviction will not eliminate their existence. As Samuel Akinrolabu of the Nigerian Slum and Informal Settlement Federation says, "The moment you demolish a slum, naturally two or more slums will spring up because people need somewhere to sleep."

Rapid urbanization challenges structures of land tenure because it brings about abrupt, large-scale changes in land use, which in turn causes land values to change quickly. These sudden shifts can profoundly affect those living in slums and squatter settlements. Often, local governments cannot keep up with registrations of new land developments. Paperwork delays, coupled with the substantial increase in land value in areas set aside for housing, result in both permanent development and squatter settlement in areas declared officially off limits by municipal governments. In Dhaka, Bangladesh, for example, real estate developers might build in flood zones, drainage channels, and other spaces restricted by Dhaka's planning commission. The developers take advantage of overburdened government systems to build housing on restricted ground and thereby cut their costs. At the same time, squatters build their dwellings there because even illegal housing in other areas is unaffordable. Illegal building and occupation of land creates conflict among governments, builders, landlords, and squatters over land tenure.

The ability of women to secure land rights is another important component in urban development and one that comes with unique problems. Like female farmers, urban women may face multiple challenges when they try to assert property claims. Some countries have longstanding traditions against women holding land tenure. While the United Nations and other international human rights organizations promote equality for women in all areas, many male-dominated traditions and governments have commonly associated a woman's land rights with her relationship to a male partner or family member. In many places, women represent disproportionate percentages of the poor, and lack of land tenure is sometimes a contributing factor.

As women become more empowered in various countries, they are more able to challenge restrictive beliefs regarding female ownership of property. Some organizational collectives, or groups of individuals acting together to increase their power and visibility, have helped improve women's security in terms of land tenure and housing. One such collective is the Shack Dwellers Federation of Namibia (SDFN). Just under half of the population of Namibia, in southern Africa, is urban, and 40 percent of the households are headed by women. The SDFN has helped community-led groups, mainly driven by women, build almost 3,500 houses, with members doing much of the construction work. Empowering women to acquire housing is especially meaningful for low-income families. In many African slums, the percentage of households headed by women is increasing.

In India, state governments have begun to address the issue of land tenure for women. The state of Delhi, for example, encourages the granting of ownership rights to women through incentives such as reductions in some forms of taxes on property registered in a woman's name or registered jointly in the names of a husband and wife. In Jharkhand, the state government charges no registration or court fees on properties registered in a woman's name.

Sometimes a government invokes a right to land tenure that overrides the rights of individual owners. **Eminent domain** is a government's right to take privately owned property for public use or interest. The rationale for eminent domain is that it allows the government to make decisions that benefit the population as a whole rather than deferring to the benefit of private landowners. In countries where land ownership is generally clearly established, governments have mechanisms for taking control of private property for public uses. In the United States, eminent domain is enumerated in the Fifth Amendment of the Constitution and, in most cases, it has historically been applied to claim farmland for roads, railroad tracks, or power lines. In recent times, however, eminent domain has also been used as a means of eliminating or renovating urban areas that have fallen into disrepair and decay.

In India, the term *compulsory acquisition* is used rather than eminent domain, and the policy is governed by a law enacted in 2013. This law has been controversial. On one hand, it has favored higher payments for landowners and more government responsibility for displaced people. On the other hand, it has also allowed for looser interpretations of what constitutes a public purpose for acquiring land, and it may weaken requirements for making sure that residents understand their rights before they agree to give up land to the government. Other countries such as Japan and the Netherlands practice land readjustment, which means that landowners participate in redevelopment projects rather than simply having their property taken or purchased. In Japan, land readjustment can be managed privately, without government involvement, if two-thirds of the area's owners and tenants are in agreement.

Environmental Injustice

The term **environmental injustice**, sometimes called environmental racism, is used to describe how communities of color and the poor are more likely to be exposed to environmental burdens such as air pollution or contaminated water. The concept also

Top: Two women stand in the doorway of a waterside hair salon in Makoko, an informal community built over the Lagos Lagoon in Lagos, Nigeria. *Bottom:* Workers tend to timber that is waiting to be milled in the neighborhood next to Makoko, which can be seen in the distance.

Urban farming has become increasingly popular as a way to combat food insecurity in cities. In New York City, beekeepers at the Brooklyn Grange, the world's largest rooftop agriculture center, operate nearly 20 hives producing hundreds of pounds of honey annually.

encompasses unequal environmental protection provided through laws, governmental policies, and enforcement. Environmental justice, in contrast, is the idea that environmental laws and regulations should apply equally in all places and for all people, regardless of racial or ethnic composition. Movements for environmental justice began in the 1980s with the realization that an excessive number of pollutant-generating industries, waste disposal areas, and power plants were located near poor or minority communities, both rural and urban.

Researchers use geographic information systems to map occurrences of environmental injustice. Mapping tools have shown that areas prone to air and water pollution resulting from manufacturing processes, waste disposal, and other contaminants are frequently located near poor or minority neighborhoods. The presence of these disamenities makes nearby housing less expensive and therefore affordable to low-income residents. Residents of these communities generally lack the political and economic power to oppose potentially harmful industrial facilities being built nearby. Some argue that industrial sites that existed first—before residential neighborhoods—are not obligated to make an area livable. Others claim it's the job of area governments to establish a safe living environment for neighborhoods and enforce safe operating procedures for industries regardless of who came first.

Likewise, busy highways including the Interstate Highway System have historically been routed through low-income neighborhoods in the inner city, so residents cannot avoid the exhaust from the automobile and truck traffic in these areas. Low-income and minority neighborhoods also tend to have older housing stock that may pose environmental health risks such as mold, lead-based paint, and leaky roofs, in addition to older infrastructure such as lead water pipes. Residents, who are usually renters, often lack the means to repair the buildings and combat these hazards. In the United States, the federal Department of Housing and Urban Development offers solutions such as housing vouchers to help residents afford homes in locations free of environmental burdens and healthy home programs to address exposure to toxins such as lead and radon, which is a naturally occurring radioactive gas that is found in some bedrock below housing that can cause cancer. Some state and local governments also have initiatives to confront environmental injustice.

Additionally, many grassroots environmental justice organizations have formed over the past few decades. For example, We Act for Environmental Justice (We Act) started in 1988 as a group of community activists and environmental lawyers joining forces to fight for improved air quality in North Manhattan, a section of New York City. We Act has spearheaded numerous campaigns, one of which resulted in the Manhattan Transportation Authority investing in new hybrid vehicles to combat air pollution in the neighborhoods of North Manhattan.

Environmental injustice encompasses more than instances of dangerous water and air pollution. The food deserts described in Chapter 14, for example, place low-income and minority communities in environments where they struggle to meet basic needs. As you learned, local food movements can improve the food security of urban areas and help alleviate the negative effects of food deserts. Urban community gardens can offer healthy, fresh food alternatives in an environment that may otherwise be dominated by convenience stores and fast-food restaurants, in the absence of full-service grocery stores.

Local food initiatives exist in many U.S. cities. In Dallas, for example, Bonton Farms is an urban agriculture project located in a low-income neighborhood. The farm incorporates food production with job creation and health education for community members. Also in Dallas, the Urban Agriculture Action Team, part of a local government coalition, has developed ordinances that allow urban farms to be built in the city. The team encourages residents to grow and sell healthy food through its support of urban agriculture. Local food initiatives help combat environmental injustice and benefit cities by bringing fresh, nutritious food into disadvantaged neighborhoods, creating urban green spaces, and fostering business enterprises that hire local residents as employees.

Urban Renewal and Gentrification

Learning Objective
LO 17.8 Explain how gentrification changes neighborhoods and displaces lower-income people.

In recent decades, increasing numbers of middle- and upper-income people have been choosing to remain in cities or return to them from the suburbs, especially in core countries. People move back to the city for a variety of reasons, including convenient access to entertainment, museums, restaurants, and shopping. Suburbia is often associated with placelessness, and many people are drawn to the culture, character, and convenience of cities. Many also value the walkability of urban neighborhoods, which is often absent in suburbs. People living in more expensive suburbs may choose to move to city neighborhoods with lower property costs, where they can buy a house or building for a low price, renovate it, and choose to live in it or sell it for a profit.

Urban renewal is a term associated with the nationwide movement that developed in the United States in the 1950s and 1960s when cities were given massive federal grants to tear down and clear out decaying inner-city neighborhoods and former industrial zones as a way to rebuild their downtowns. Three policies were enacted in the postwar era to fund the large-scale demolition of dense, declining industrial and residential neighborhoods: the Housing Act of 1949, the Housing Act of 1954, and the Federal-Aid Highway Act of 1956. The two housing acts established federal funding to cover a large part of the cost of demolishing aging neighborhoods and building new housing and infrastructure. The highway act launched the building of the Interstate Highway System that today crisscrosses the United States. Neighborhoods in cities across the country were demolished to make way for the massive new roadways. According to a 2018 study, more than a million people were displaced from their homes by the urban renewal projects these three policies supported. Most of the areas targeted for renewal were low income and predominantly African American. Researchers at the Institute for Quality Communities at the University of Oklahoma have documented this dramatic change with before-and-after aerial photographs, showing that the "fine-grained networks of streets and buildings on small lots were replaced with freeways, superblocks, and megastructures." And while these urban researchers note that impressive projects were built in many cities, the scars of this forced displacement remain unhealed and that these "programs damaged the economic, social, and physical well-being" of poor and minority communities in American cities, particularly of African Americans.

Gentrification has a similar stated intention—to improve and rebuild downtown areas or inner-city neighborhoods. Neighborhoods become gentrified when developers or middle- and upper-income people buy up deteriorated buildings to restore or renovate. Gentrification has similar advantages and disadvantages to the smart-growth initiatives described earlier. It brings a city benefits such as increased property values, which result in higher tax revenues. Higher-income residents and increased cultural and recreational opportunities, plus tourism, in newly renovated areas attract still more businesses and investors. Neighborhood improvements may include infrastructure upgrades such as new sidewalks and parks, repaved roads, and fiber-optic internet installation. Rehabilitating historic buildings also restores an area's visual appeal and can enhance the cultural landscape. For example, today, the medieval neighborhood of Mouraria in Portugal's capital, Lisbon, is a popular multicultural district. Before renovation work started in 2009, however, it was considered run-down and, at best, forgotten. What started out as an effort by authorities to restore and preserve a few historic areas turned into an ongoing revitalization effort that has resulted in a decrease in crime and an increase in art, dining, tourism, and economic stability—as well as a rise in property values.

At the same time, gentrification can also have negative consequences. The foremost of these is the displacement of less-affluent residents, older adults on fixed incomes, and marginalized groups due to rising property values and taxes. In Chapter 6, for example, you learned about gentrification forcing vulnerable LGBTQIA+ populations out of neighborhoods where they had lived. Renters may be forced to leave as higher-income groups move in and purchase homes that were previously being rented out. Eventually, increased homelessness may result as low-income housing becomes scarcer in the city. Gentrification can push property prices so high that even those who are considered middle- and upper-middle-class are forced to leave, and some move into poorer neighborhoods, which can start another gentrification cycle.

At the same time, a neighborhood's cultural landscape and unique character may be harmed rather than enhanced when historic buildings are altered or replaced. Locally owned businesses may be replaced by global chains or national franchises, bringing about the sense of placelessness that the former suburban dwellers sought to escape. These significant changes in a neighborhood's business and residential character can result in a loss of community and an increase in social tension.

Gentrification has become a constant in the largest U.S. cities, and Los Angeles is no exception. Critics of gentrification tend to see Los Angeles as a prime example of how the process can negatively impact urban areas. One neighborhood in the 90014 zip code in central Los Angeles underwent heavy gentrification and saw a 95 percent increase in the median household income of its residents; however, most still live under the poverty line. The median cost of a home in the neighborhood went from $52,000 in 2000 to $421,731 in 2016, a dramatic increase that contributes to social tension as wealthy homeowners end up living next to long-time residents who are often renters. Critics note that efforts at gentrification do not seem to be improving the housing situation for residents of Los Angeles as a whole, noting that between 2000 and 2012, median rents increased by 25 percent in Los Angeles County while income declined 9 percent. Pandemic-related rent increases made the situation worse: Median rents in the Los Angeles metropolitan area had been increasing by about 5 percent annually in recent years but jumped 13 percent from 2021 to 2022.

Geographic Thinking

3. Describe how affordable housing shortages occur in cities.
4. Explain why slums and squatter settlements exist in countries of the periphery.
5. Describe positive and negative results of gentrification in urban neighborhoods.

17.3 Creating Sustainable Urban Places

Globally, cities are booming. And as urban areas expand, their ecological footprints, or overall impacts on the natural environment, grow along with them. Urban concentrations of population and industry use energy and natural resources, affecting air and water quality as well as the health of city dwellers. Working toward urban sustainability means addressing ecological challenges in innovative and lasting ways.

Challenges to Sustainability

Learning Objectives
LO 17.9 Describe the challenges that urban populations pose to sustainability.
LO 17.10 Explain the environmental impacts of increased pollution caused by urbanization.

In the urban context, sustainability means finding ways to ensure that cities continue to exist well into the future, so that the generations to come can benefit from the opportunities that cities provide. In order to be sustainable for the long term, urban areas must balance economic development with the need to protect natural resources, the health of the environment, and the well-being of city dwellers. Sustainable urban practices should eventually lead to environmental regeneration and replenishment.

Cities have a massive impact on the environment. Dense urban populations bring with them challenges to sustainability such as suburban sprawl, a large and growing ecological footprint, higher energy use, poor air and water quality, climate change, and struggling fresh water and sanitation systems. To confront these concerns, governments are deploying different strategies with varying degrees of effectiveness.

There is a direct correlation between a city's physical footprint and its ecological footprint. The more extensive the suburban sprawl, the more significant the environmental

Pedestrians enjoy the High Line, a public park built on an elevated rail line that once operated above the streets of Manhattan's West Side. Plans to extend the linear park broke ground in 2022.

New York City's High Line

Learning Objective
LO 17.8 Explain how gentrification changes neighborhoods and displaces lower-income people.

In 1934, the High Line in New York City went into service as an elevated rail line, carrying goods on tracks built high above street level. By the 1990s, it had become a crumbling relic that city officials couldn't wait to demolish. In 1999, Joshua David and Robert Hammond founded Friends of the High Line, a nonprofit conservancy whose mission was and is to advocate for the High Line's preservation and use as a public space. The first section of the renovated High Line was opened to the public in 2009, and the 1.45-mile-long greenway is now not only popular with residents but is also a major tourist attraction, drawing 8 million visitors in 2019.

The plan for transforming the High Line included rezoning, which allowed developers to build above previous height restrictions if they funded certain additions to the park. Since the rezoning, more than 1,300 housing units have been built near the High Line, along with a half-million square feet of commercial development. High-end real estate has continued to explode around the renovated rail line, including luxury apartments, chic art galleries, and an 88-story skyscraper. Initially, David and Hammond had estimated that the park would bring in around $200 million in new real estate taxes. Based on assessments made in 2017, the actual figure will reach $900 million by 2038.

Concerns about possible negative effects of gentrification were addressed in the High Line's planning. Certain areas around the park were designated for inclusionary zoning that required builders to set aside 20 percent of units for affordable housing in exchange for permission to build larger buildings than would normally be allowed. The goal was to achieve 1,000 affordable units. Studies to assess the effectiveness of this zoning have produced conflicting results, however, and while the New York City Housing Authority has built two public housing developments near the High Line, critics note that the median price of condominiums along one stretch of the linear park was more than $6 million in 2019. One early study also found that users of the High Line were "overwhelmingly White," even though nearly one-third of people living nearby are people of color. By 2017, however, 35 percent of High Line users were non-White—a number that the park's founders hope will continue to rise.

Robert Hammond has founded the High Line Network, a group of planners who are adapting old infrastructure to create new urban parks in other cities and taking lessons from the High Line, with the aim of creating spaces that improve the lives of all residents. "I want to make sure other people don't make the mistakes we did, and learn how to deal with these issues," he says. With thoughtful planning, High Line–type parks of the future may help cities reap the benefits of gentrification without invoking its downsides.

Geographic Thinking

Explain why the High Line has not resulted in adequate or equitable affordable housing in its neighborhood.

damage. As populations burgeon in large cities, it may seem easiest to build low-density residential housing on lands previously protected or used for farming; however, this response ignores the environmental costs, including the destruction of productive farmlands, wildlife habitats, and green space. Planning and constructing or renovating vertically within existing neighborhoods is more complex and costly but can help contain the physical footprint associated with sprawl and thus lessen ecological damage. High-density neighborhoods also consume less energy per capita than low-density ones.

Expanding urban landscapes with poor public transportation options, overcrowded highways, and few walkable areas discourage healthy lifestyle habits such as walking and biking. Longer commutes imposed by sprawl, traffic congestion, or inefficient transportation also use more energy and pose environmental problems such as pollution from burning fossil fuels. An analysis of 2019 U.S. Census data found that more than 94 percent of American workers in the largest 100 cities commuted to work and that their average commute was 27.6 minutes—adding up to more than 240 hours of annual commuting time for full-time workers. California, where increasingly expensive housing often forces people to move farther out for affordable options, had 8 of the worst 11 commutes in the study.

Water and air quality are among the most serious challenges cities face in planning for a sustainable future. Cities with poor or aging infrastructure especially struggle to maintain water quality, an issue that encompasses both water scarcity and water pollution. Urban sustainability efforts attempt to address these problems largely by increasing safe water supplies and decreasing pollutants produced by factories or contamination from inadequate sanitation systems.

In many cities of the periphery and semi-periphery, domestic sewage such as human waste is the main culprit for water pollution, but industrial waste also contributes. In India, rivers and streams have been heavily impacted by India's urban population explosion, and sanitation systems in many areas have been overwhelmed. Agricultural waste also contributes to water pollution. In Brazil in 2019, 27 different pesticides were found in the water of 1,400 towns and cities across the country. According to Greenpeace Brazil, these pesticides are linked to cancer and other diseases. In cities of the core, aging infrastructure has been responsible for water contamination. For instance, Newark, New Jersey, and Flint, Michigan, both experienced crisis situations when municipal drinking water was found to contain unacceptable levels of lead that had leached into it from the water systems' older lead pipes.

Often visible in the form of smog, air pollution is produced by factories, businesses, power plants, and vehicle exhaust—all of which are plentiful in cities. In fast-expanding urban areas, laws that regulate polluting industries and vehicle engines often fail to keep pace. In recent years, numerous megacities have posted alarming air pollution statistics.

A 2019 report issued by the journal *Lancet Planetary Health* found that out of 125 cities in 194 countries, Shanghai, China, has the highest percentage of new asthma cases in children under age 18 attributed to traffic pollution. In Jakarta, Indonesia, meanwhile, air pollutant levels from coal combustion are far above the World Health Organization's Air Quality Guideline. In 2018, Indonesia's dependence on coal for electricity exceeded 60 percent, and 22 coal-fired power plants operated within 60 miles of Jakarta.

Some cities, like Mexico City and Beijing, are partially surrounded by mountains, which further exacerbate air pollution issues by trapping the contaminated air. Mexico City strengthened its restrictions on the number of polluting vehicles allowed in the city in 2019. Other temporary recent measures included residents wearing face masks and school closures. Beijing has also enacted a range of policies beginning in 2008 to improve its urban air quality, including car license plate restrictions and the removal of some industries out of the metropolitan area.

Climate change remains a rising global problem and a formidable challenge to sustainable cities. Urban areas are both causes and victims of climate change. They consume massive amounts of energy, making them a key source of greenhouse gases. According to the United Nations (UN), cities accounted for between 60 and 80 percent of the world's energy consumption and generated up to 75 percent of greenhouse gas emissions from transportation and buildings. At the same time, many of the world's largest cities are threatened by the effects of climate change. London, Mumbai, Cairo, Rio de Janeiro, and New York City are all in low-lying coastal areas vulnerable to rising seas, for example. According to the Global Covenant of Mayors for Climate Change and Energy, cities such as Lima, Cairo, and Tehran also face the possibility of more frequent heat waves with extreme high temperatures. Many cities of the western U.S. in 2021 and across Europe in 2022 sweltered under record-breaking summer temperatures, while climate change-amplified rainfall in August 2022 drenched the Dallas-Fort Worth metroplex with a 1-in-1000-year flood event.

Through climate action planning, cities such as Vancouver, Copenhagen, and Boston are taking the lead on both combatting climate change and mitigating its impacts, as national governments sometimes lag in their responses. But experts warn that the problem cannot be solved at the local scale. The UN predicts that the most effective efforts to overcome the challenges of climate change will be coordinated at the "global, regional, national, and local levels." Cities must be a center of solutions to mitigate the effects of climate change, but municipal governments will achieve much less if they act in isolation. For example, while mayors exert partial control over emissions sources within city limits, national governments can complement these efforts by eliminating fossil fuel subsidies and by providing financial support for local initiatives. To this end, China's central government bolstered its national support of cities in 2014 when it launched its National Plan for

Zachary Damato co-founded Urban Rivers to help improve the health of rivers that flow through cities.

Learning Objective
LO 17.11 Describe the responses to urban sustainability challenges including regional planning and brownfield remediation.

National Geographic Explorer **Zachary Damato**

A Wild Mile in the City

Zachary Damato sees something that many people overlook in the city—a place for nature and biodiversity to thrive. He is establishing a "Wild Mile" on one of the country's most urban rivers so that all city dwellers and visitors can share his vision.

Urban rivers, walled with concrete, tend to be catchalls for pollution and unwelcoming for wildlife. National Geographic Explorer Zachary Damato and the rest of the team at Urban Rivers, the ecological organization he co-founded, are working to change that. Restoring natural habitats in urban rivers helps reduce cities' harmful impact on the environment and promotes the social well-being of urban residents.

The Wild Mile project focuses on restoring river habitat in Chicago while creating a destination where visitors can engage with nature. The Chicago River channel flowing by Goose Island provides an undeveloped stretch perfect for a little patch of wildness. Urban Rivers launched the Wild Mile project there with a three-year test consisting of a floating garden. Made of woven plastic, the structure supports plants with roots growing down into the water. The roots create shade and underwater structures that attract and sustain wildlife, enriching the river's biodiversity. Scientists have already observed an increase in otters, fish, birds, and butterflies. When completed, the Wild Mile will comprise not only floating gardens but also forests with walkways and kayak docks.

In 2018, Chicago's Shedd Aquarium partnered with Urban Rivers to install a river island with 260 square feet of native plants, adding to the 1,500 square feet that Urban Rivers had already created. An important goal of the Shedd partnership with Urban Rivers is to help open up the Wild Mile to increasing numbers of Chicagoans; the construction of a mile-long walkway, begun in 2021, is intended to improve that accessibility to this "nature first" stretch of urban river.

Geographic Thinking

Describe Zachary Damato's attempt to address urban sustainability challenges.

Urbanization, to assist cities in their continuing efforts toward sustainable growth.

Funding is a key issue for cities striving to limit the causes and effects of climate change. The UN reports that in 2016, international financial institutions gave a creditworthy rating to just 4 percent of the 500 largest cities in countries of the periphery and semi-periphery. Cities need opportunities to collaborate with financial institutions and establish credit ratings that will allow them to borrow the money necessary for taking meaningful steps toward reducing carbon and other greenhouse gas emissions. Lima, Peru, for example, was able to gain financial support from international organizations to invest in a low-carbon mass transit system.

Responses to the Challenges

Learning Objective
LO 17.11 Describe the responses to urban sustainability challenges including regional planning and brownfield remediation.

The spread of urban influences into surrounding rural areas and the physical growth of cities have created urban regions—large areas of interconnected cities, suburbs, edge cities, boomburbs, and exurbs. To address sustainability, planning must take place at the regional scale to coordinate efforts across all parts of the metropolitan area. Coalitions of governments may work together to establish urban growth boundaries and enact laws that protect farmlands from sprawl.

Regional planning is planning conducted at a regional scale that seeks to coordinate the development of housing, transportation, urban infrastructure, and economic activities. A group of municipalities, such as the central city of a metropolitan region and its suburbs, may cooperatively share a variety of services such as infrastructure, parks and recreation, and public transit. By sharing in the costs of planning, building, and maintaining services such as fresh water supply and wastewater treatment, regions can achieve economies of scale. Well-executed regional planning can overcome many of the barriers posed by fragmented local governments, and the coordinated efforts of multiple political bodies can be more effective than those of individual governments working independently. Lancaster County, Pennsylvania, which was discussed in Section 17.1, provides one example of an effective regional planning effort.

The Twin Cities region in Minnesota is one of the most successful examples of regional planning over multiple jurisdictions in the United States. In 2010, under Minnesota

A fire burns in the wildland-urban interface (WUI) in Canberra, Australia's capital city, in January 2020 (summer in Australia). Extreme heat and drought, two effects of climate change, were blamed for brushfires that scorched one-fifth of Australia's forests that summer. Cities like Canberra and Sydney experienced smoke that impaired visibility and taxed residents' lungs. Similarly, the numbers of U.S. wildfires in the WUI have also increased with exurban expansion into natural areas.

National Geographic Explorer **Maria Silvina Fenoglio**
Transforming Urban Rooftops

Maria Silvina Fenoglio researches the effects of urban green spaces on the habitats of insects such as bees.

Learning Objective
LO 17.11 Describe the responses to urban sustainability challenges including regional planning and brownfield remediation.

Biologist and ecologist Maria Silvina Fenoglio is fascinated with a subject that many people would prefer to avoid—bugs in the city. One of her lines of research is urban ecology, studying the impact of urbanization on herbivorous insects and their natural enemies.

National Geographic Explorer Maria Silvina Fenoglio says she is passionate about insects because "although they are extremely tiny, they are amazing organisms performing many valuable ecological processes." She believes that research on insect diversity in urban green spaces could ultimately contribute to the design of more sustainable cities—for example, by helping to maintain green roofs. As the name implies, green roofs transform the tops of buildings into spaces covered with vegetation. Green roofs improve buildings' energy efficiency, retain rainwater, help decrease air pollution, and can lower the temperature in a city by lessening the amount of roof surface heated by sunlight.

Fenoglio's research takes a new direction by investigating how roofs can function in an ecological role as a habitat provider for plants and animals, including beneficial insects such as pollinators, natural enemies to pests, and decomposers. In Córdoba, the second most populous city in her home country of Argentina, Fenoglio is examining the possibility of using urban green spaces as biodiversity shelters that allow insect ecosystems to thrive and perform beneficial services. She and other researchers have established field experiments consisting of small green roofs installed on several buildings. In addition to enhancing biodiversity, the insects on these roofs serve as indicators to help scientists measure the effects of urbanization.

Fenoglio's group hopes their findings will convince decision-makers in urban areas to incorporate green roofs as part of urban planning and development. She explains that her ultimate goal is to find evidence that contributes to the development of more sustainable cities and helps bring nature into the lives of city dwellers.

Geographic Thinking

Explain how green roofs can make cities more sustainable.

state law, the council that oversees planning for the Minneapolis–St. Paul area developed a 30 year project plan called MSP 2040 for its metropolitan region, which had 3.65 million residents in 2020. This long-term plan provides direction and guidelines for the region's future development, including a comprehensive land-use policy. For example, new developments are located and designed to reduce pressures on the environment and natural resources. Growth is promoted in areas that are already urbanized rather than encroaching on agricultural land or open areas. The regional groundwater system is protected from land-use changes that could harm its quality, and a regional inventory of natural resources is maintained. MSP 2040 recognizes that issues may transcend any single unit of space: neighborhood, city, or county. The region uses its investments and resources to foster urban sustainability.

Farmland protection policies, urban growth boundaries, and WUIs are often associated with regional planning. Farmland protection means enacting laws that prevent agricultural lands from being converted to nonagricultural uses. MSP 2040, for example, incorporates a farmland protection policy to preserve the half-million acres of agricultural land in the region. Urban growth boundaries also support sustainable cities, as they limit sprawl and protect farmland and other undeveloped lands. Planners in Portland, Oregon, combined an urban growth boundary with infilling to promote efficient and sustainable land use. The city's plans also include affordable housing to address the needs of low-income residents. Nine counties around Denver, Colorado, also established an urban growth boundary in 1997 as part of their regional planning, in part to address WUI issues.

The oldest urban growth boundary in the United States may be the greenbelt around Lexington, Kentucky, established within Fayette County in 1958. Originally designed to protect the region's horse-breeding industry, the greenbelt has also helped contain the city's sprawl. However, Lexington, Denver, and Portland are all facing issues with growing populations pushing against the city limits. It has become challenging to maintain affordable housing for all residents without infringing on the urban growth boundaries.

Within cities, cleaning up and developing old industrial sites can lead to the creation of new, more sustainable land uses. Brownfield remediation and redevelopment processes can offer innovative responses to multiple urban sustainability challenges. **Brownfields** are abandoned and polluted industrial sites in central cities and suburbs. Remediation means removing the contaminants in these sites, which reduces health and safety risks to nearby residents and opens the land up for new development. Brownfield remediation and redevelopment can promote growth within a neighborhood and reduce the number of zones of abandonment in a city.

London's King's Cross district is one of Europe's largest city center brownfield remediation sites. Begun in 2007, the project has created 8 million square feet of mixed-use development, including both high-end and affordable housing, an educational campus, parks, and offices. King's Cross had been plagued by the buried and entangled infrastructure of an abandoned gasworks (an industrial plant that produces flammable gas for lighting and other purposes), along with contaminated soil and groundwater. Remediation of the site created the foundation for projects such as the transformation of the nearby King's Cross and St. Pancras train stations. The King's Cross station in its original state had become a magnet for crime, but it has since been given a dramatic update that includes new, airy open spaces and more efficient connections between the many rail lines that pass through it to other destinations in Great Britain. Next door to King's Cross is the elegant St. Pancras Station, originally built in 1868 and updated in 2007 to become the London terminus for high-speed Eurostar trains to Paris, Brussels, and Amsterdam.

Geographic Thinking

1. Compare the types of urban sustainability challenges faced by different countries.

2. Explain how urban areas contribute to climate change.

3. Describe one way that regional planning is effective in overcoming urban sustainability challenges.

Giant cast iron frames, called gas holders, were built in the 1860s as part of Pancras Gasworks in London. They once held tanks of gas that was used to provide heat and light to the local area. Today, new luxury apartments have been built within the former gas holders, dramatically transforming the neighborhood.

Urban Living

Designing to Scale
Smart Buildings
Adapted from National Geographic, *April 2019.*

In the ideal city of the future, buildings incorporate natural elements, generate energy, and produce less waste. Spaces can quickly transform to meet changing housing, industrial, or business needs. ▌ Which of these elements seems most important to you?

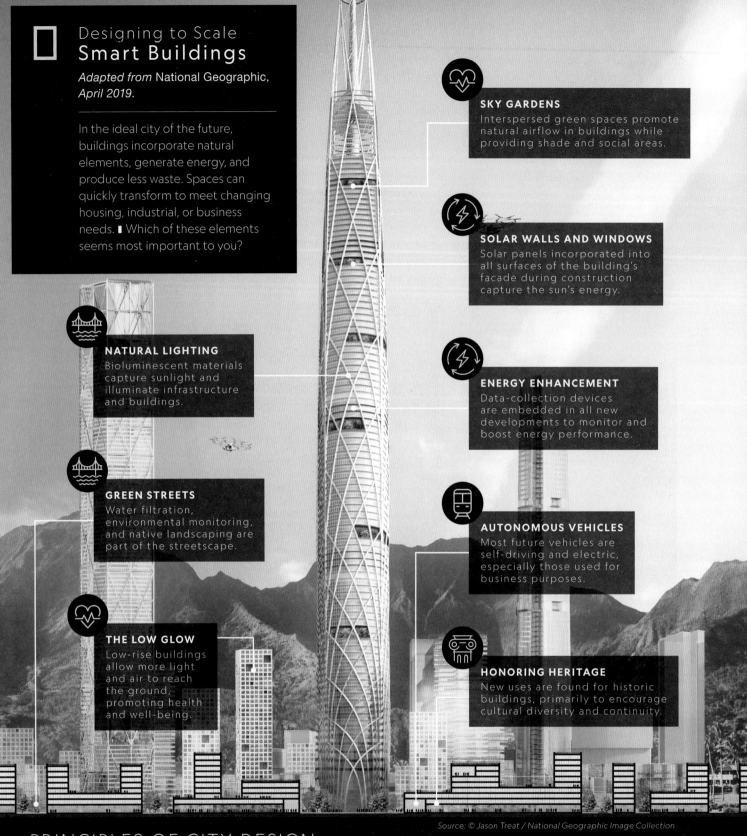

SKY GARDENS
Interspersed green spaces promote natural airflow in buildings while providing shade and social areas.

SOLAR WALLS AND WINDOWS
Solar panels incorporated into all surfaces of the building's facade during construction capture the sun's energy.

ENERGY ENHANCEMENT
Data-collection devices are embedded in all new developments to monitor and boost energy performance.

AUTONOMOUS VEHICLES
Most future vehicles are self-driving and electric, especially those used for business purposes.

HONORING HERITAGE
New uses are found for historic buildings, primarily to encourage cultural diversity and continuity.

NATURAL LIGHTING
Bioluminescent materials capture sunlight and illuminate infrastructure and buildings.

GREEN STREETS
Water filtration, environmental monitoring, and native landscaping are part of the streetscape.

THE LOW GLOW
Low-rise buildings allow more light and air to reach the ground, promoting health and well-being.

Source: © Jason Treat / National Geographic Image Collection

PRINCIPLES OF CITY DESIGN

 ENERGY In the city of the future, energy is 100 percent renewable. Enough power is produced within or close to the city for it to be self-sufficient. Area buildings share energy resources, generating as much energy as they consume.

 LIVABILITY The city of the future is designed for accessibility and safety as more people populate urban areas. Residents have healthier lives with more streamlined access to nature, services, and automated technology.

Case Study

Milan and Urban Sustainability

The Issue The Porta Nuova district in Milan was in need of redevelopment, largely in response to the growing challenges of air pollution and climate change. The solution was one of the largest urban regeneration projects in Europe.

Learning Objective
LO 17.11 Describe the responses to urban sustainability challenges including regional planning and brownfield remediation.

By the Numbers

215,278 sq ft
of vertical forest

23
tree species in the vertical forest

94
plant and shrub species in the vertical forest

40
average number of plants per inhabitant

Source: Greenroofs.com

The Bosco Verticale (Vertical Forest) towers, supporting nearly 20,000 shrubs and plants, stand within a park called Biblioteca degli Alberi (Library of Trees).

The City of Milan, Italy, has struggled for years with environmental challenges, including one of Europe's most serious smog situations. Seeking to remediate these problems and create a model of urban sustainability, Milan has focused intently on improving its ecological footprint, transportation systems, walkability, and livability.

A milestone was reached in 2014 when Bosco Verticale opened, a pair of residential high-rises that by 2019 appeared to be transforming into literal vertical forests. Part of a massive urban redevelopment project of the Porta Nuova district, the towers feature almost 20,000 shrubs and plants in addition to 800 trees. This living insulation moderates the temperature in summer and winter and converts as much as 30 tons of carbon dioxide into oxygen annually. If that weren't impressive enough, the plant life also filters noise and dust and creates a microhabitat for insects and birds. Twenty bird species nest in the towers, along with bumblebees and hermit bees. The buildings are equipped with solar panels, geothermal heating, and a system that recycles water used for washing to irrigate the external greenery.

Weaving throughout Porta Nuova is a connected pedestrian system with green areas, gathering places, and bridges. A huge public park resembling a modern botanical garden is in the center of the pedestrian system and connects the three surrounding neighborhoods. The total redevelopment area for Porta Nuova is 3.2 million square feet.

The city of Milan has received worldwide attention as it continues to take its green projects seriously. Local authorities announced in 2019 that they will plant 3 million trees in Milan by 2030 to enhance livability, improve air quality, and support the health of residents. A tree covering of this magnitude could absorb 5 million tons of carbon dioxide annually, helping to reduce air pollution, respiratory disease, and the risk of cancer. The trees will also help lower Milan's temperature, which is hot and humid during the summer. In addition to planting trees in schoolyards, urban parks, and private gardens, the city will also plant them on rooftops.

Geographic Thinking

Explain whether the construction of "vertical forests" can help solve the challenges of urban sustainability in other cities.

Chapter 17 Summary & Review

Chapter Summary

Urban design and construction initiatives attempt to control the effects of rapid urbanization, including urban sprawl.

- Smart-growth policies aim to create sustainable communities by placing development in convenient locations and designing it to be more efficient and environmentally responsible.
- New Urbanism is a philosophy that includes designing growth to limit urban sprawl, preserve open spaces, and create spaces that promote interaction.
- Mixed-use zoning permits multiple land uses in the same space or building.
- A mixed-use development (MUD) is a planned development that includes multiple uses, such as residential, retail, educational, recreational, and office. MUDs exist at many scales, from a single building to a neighborhood of 1 or 2 square miles.
- Transportation-oriented developments are dense, walkable, pedestrian friendly, mixed-use communities centered around or located near a transit station.
- A greenbelt is a ring of land maintained as parks, agricultural land, or other types of open space to limit the sprawl of an urban area.
- An urban growth boundary separates urban areas from other spaces by limiting how far a city can expand.

Urban design initiatives have positive and negative effects.

- Positive effects include mitigation of urban sprawl, reduced environmental impact and ecological footprint, decreased pollution, reduced strain on the infrastructure, creation of affordable housing options, reduced commute time, and enhanced livability.
- Negative effects include displacement of lower-income families, de facto segregation, neighborhoods with little diversity, and increased placelessness.

Urban challenges are influenced by a country's population and demographics and include housing discrimination, rapid urbanization, crime rates, affordable housing, land tenure, and environmental injustice.

Gentrification has benefits and drawbacks.

- The benefits are increased property values, increased tax base, new businesses, attractive landscapes, upgrades to infrastructure, and the rehabilitation of historic buildings.
- The drawbacks are decreased affordability, the possible compromise of historical integrity, displaced residents, loss of community, and potentially placelessness.

Air and water pollution, climate change, decreases in healthy lifestyle, and threats to the environment present challenges to sustainability. Responses to these challenges include regional planning and brownfield remediation.

Review Questions

Use complete sentences to answer the questions.

1. **Apply Conceptual Vocabulary** Consider the terms *blockbusting* and *filtering*. Write a standard dictionary definition of each term. Then provide a conceptual definition—an explanation of how each term is used in the context of this chapter.

2. How are the terms *smart-growth policies* and *slow-growth cities* related?

3. Explain walkability and describe how it has been implemented in a specific city.

4. Describe the possible effects of transportation-oriented development on a city or neighborhood.

5. Explain how some urban design approaches might lead to de facto segregation.

6. Define and provide an example of a greenbelt. Include social and environmental considerations with your response.

7. What is a zone of abandonment? Explain why it represents an urban challenge.

8. Explain the significance of restricted property deeds within the context of housing discrimination.

9. Define the term *redlining*. Explain how the term relates to de jure segregation.

10. Compare the differing land tenure challenges faced by men and by women.

11. How do inclusionary zoning laws help with affordable housing?

12. Explain how brownfield remediation and redevelopment attempt to address urban sustainability challenges.

13. How does regional planning enhance urban sustainability?

Interpret Maps

Study the map and then answer the following questions.

Temperatures in Washington, D.C., August 28, 2018

14. **Identify Data & Information** Which area experiences the highest summer temperatures?

15. **Analyze Data** Which types of areas have the lowest temperatures? Why do you think this is the case?

16. **Connect Visuals & Ideas** Select one area and describe how you could transform it by applying smart-growth policies and New Urbanism. Explain how your ideas would help mitigate urban climate change.

17. **Ask Questions** If you were contracted for the urban design and development of this area, what questions would you ask?

Unit 6 | Writing Across Units, Regions, & Scales

Walking Tokyo

by Neil Shea

Located beneath a train line in a narrow strip of Tokyo, Japan, the Yurakucho neighborhood offers city dwellers one of the joys of urban life—a walkable area densely packed with restaurants and cafés.

If you agree with Harvard urban researcher Edward Glaeser that cities are humanity's greatest invention, then Tokyo is perhaps our greatest example: a stunning metropolis, home to more than 37 million people and one of the world's wealthiest, safest, most creative urban centers. Even if you're not particularly interested in how megacities shape human behavior, Tokyo is unavoidable—it has already changed your life. The city is the ultimate social influencer, the node through which the world connects to Japanese culture.

Author Jane Jacobs, a major influence on urban planning, said that the best way to know a city, to feel its mashed-up power, is to walk it. So photographer David Guttenfelder and I did. For weeks we crossed and recrossed Tokyo, sometimes together, often apart; sometimes in a straight line, often leapfrogging from one area to another. We couldn't be comprehensive. But we could try to see more deeply, linking the city to the people who through their lives give it power.

A Spirited Senior Neighborhood In the northern neighborhood of Sugamo, clerks were wrestling tables and clothing racks out onto the pavement along the shopping street Jizo-dori, hoping to lure customers from a stream of mostly elderly, female pedestrians.

Older women in twos and threes strolled along, pulling through the racks, pausing here and there. Younger people flitted past the stands or slipped into a nearby coffee shop, but the crowd was mostly elderly, *oji-sans* and *obaa-sans*, grandfathers and grandmothers.

Cities often talk about themselves in terms of life, growth, youth—but old age and death are always there too, even when they're largely ignored or treated as a matter of dull housekeeping. Harvard anthropologist Ted Bestor had pointed me toward Sugamo because the neighborhood reveals a defining feature of Tokyo: its enormous, rapidly increasing elderly population.

"In Tokyo they don't try to hide the old people away," Bestor said. "There are just too many of them. So the old folks have their own district; they make their own fun."

A New Type of Urban Design A few weeks later, in the Asakusa neighborhood on the other side of the city, I met with Kengo Kuma, the architect who designed the new national stadium for the Summer Olympics (rescheduled from 2020 to 2021 due to COVID-19). We sat in a room on the third floor of the Asakusa Culture Tourist Information Center, which, like nearly all the buildings Kuma has designed, is both hypermodern and surfaced in natural materials, in this case wood—a combination intended to lend warmth and presence while also paying homage to traditional Japanese craftsmanship. Kuma is sometimes taken for an anti-urbanist—opposed to the hardness of cities—but he was quick to reject that label.

"People say I'm a critic of cities," he said, shaking his head. "I want to reshape the city. I want to break space up and return things to a smaller scale." That smaller scale, he said, was once a defining feature of Japanese life, and would allow for more trees, gardens, parks—and more human connections.

Of course, the massive oval stadium will likely define him to future generations. But even that wears Kuma's vision—a future in which structures are all built for multiple uses over their lifetime and sit lightly on the landscape. After the Olympics his stadium will be converted for use as a soccer arena. It will sit in a grove of trees, and its several floors will be ringed with more greenery, planted around open-air walkways. The stadium's roof is also open, allowing natural light to flood its interior.

"Our urban design up until now was to find land and put a huge building on it," Kuma said. "Destroying everything to make way for skyscrapers and shopping centers—that has been the method in Asia." Kuma was animated, sketching with his hands as he described Tokyo. Many ideas he supports, from environmental sustainability to programs aimed at "returning nature to the city," have slowly gained ground. Kuma described Japan as a "mature society"—wealthy, technologically advanced, and aging. Ready, in other words, to grow more responsibly. "The best thing we can do," he said, "is set an example. … We can show how to do things differently."

In the City's Heart, a Call for Diversity

Yuriko Koike, Tokyo's first female governor, attended university in another massive metropolis—Cairo. "What's attractive about Cairo is that it's chaotic," she said, smiling at memories of hectic streets. "But, of course what's attractive about Tokyo is that everything is controlled."

We were walking down a shaded gravel path in the central Hama-rikyu Gardens, a calm refuge of manicured lawns and flower beds with stands of black pines, crape myrtles, and cherry trees flush against the Sumida River.

"What's missing now in Tokyo is diversity," she said. Arriving from Brooklyn, I found Tokyo's lack of diversity a regular, striking feature of my journey. Sizable populations of Koreans and Chinese live in Tokyo, and the number of "international residents" has also increased over time—in 2018, one in 10 Tokyoites in their 20s were non-Japanese. But in a city so vast, those groups faded quickly.

Koike herself has been criticized for talking diversity without doing much to enable it. But her election itself was seismic and may yet prove part of a broader shift. And she understands that Tokyo's composition will soon change no matter what. If nothing else, old age guarantees it.

Source: Adapted from "Walking Tokyo," Neil Shea, *National Geographic*, April 2019.

Write Across Units

Unit 6 opened with Daniel Raven-Ellison's exploits in urban geography and his vision for London's future. This article takes up the theme of understanding a great city by walking its streets and explores both the challenges and possibilities facing the modern megacity of Tokyo. Use information from the article and this unit to write a response to the following questions.

Looking Back

1. Why is the data from walking expeditions such as those of Neil Shea, Daniel Raven-Ellison, and Paul Salopek valuable? Unit 1
2. What evidence of the political and social consequences of an aging population do you see in the article? Unit 2
3. How is Kengo Kuma trying to affect Tokyo's cultural landscape by changing its physical landscape? Unit 3
4. In what ways do the people portrayed in the article express a sense of territoriality about Tokyo? Unit 4
5. How might a new food trend in Tokyo, a city of more than 37 million people, affect rural regions in Japan and other countries? Unit 5
6. How might walking Tokyo help you understand whether urban sprawl is a problem in the city? Unit 6

Looking Forward

7. How do you think the globalization of the economy might support or hinder Governor Koike's push for diversity in Tokyo? Unit 7

Write Across Regions & Scales

Research a megacity in Europe, Africa, or the Americas that is confronting the challenges of urban sustainability. Write an essay comparing the ways that city and Tokyo are meeting those challenges. You may want to conduct additional research on Tokyo. Drawing on your research, this unit, and the article, address the following topic: How are megacities planning for a sustainable future? Are they doing enough to reduce their impacts on the environment?

Think About

- Social, cultural, and political forces in each megacity that either support or oppose measures that could improve sustainability
- Results of sustainability measures that each city has attempted so far

Chapter 15

Gravity Model

The gravity model is used to explain interactions among cities based on the size of the cities' populations and the distance between them. According to the model, New York City, which is a massive population center, draws more traffic in terms of trade, visitors, migrants, and communication than smaller cities do. Because the gravity model does not take geographic features such as political, cultural, or physical barriers into account, its predictions do not perfectly reflect real-world interactions. ▪ Think about where you live. Use the gravity model to describe interactions of trade, tourism, and communication between your community and the cities shown on the map.

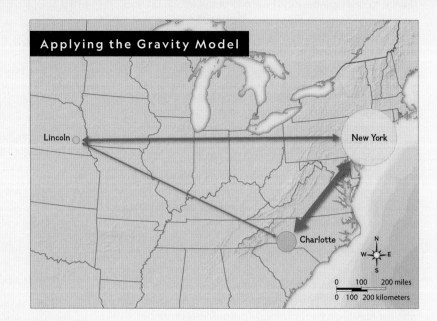

Applying the Gravity Model

Chapter 15

Rank-Size Rule

The rank-size rule states that the second-largest city in a country will be approximately one-half the size of the largest; the third-largest city will be approximately one-third the size of the largest city; and so on. This graph depicts the populations of the five largest cities in Australia. ▪ Explain the degree to which Australia's most populous cities follow the rank-size rule.

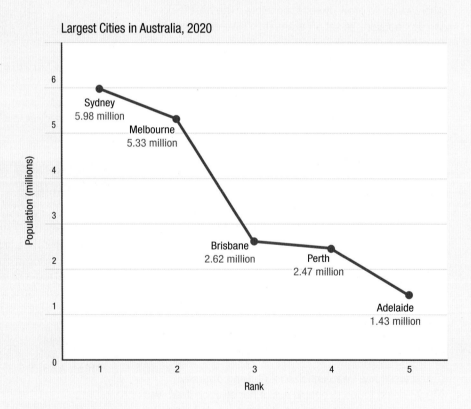

512

Maps & Models Archive

Chapter 15

Central Place Theory

Central place theory describes the distribution of cities, towns, villages, and hamlets in a given region, not taking into account physical geographic barriers such as mountains or bodies of water. According to the theory, large cities produce higher-order goods and services, thus drawing trade and population from surrounding towns and villages. ▮ Use central place theory to explain the distribution of cities, towns, and villages surrounding Nairobi, Kenya.

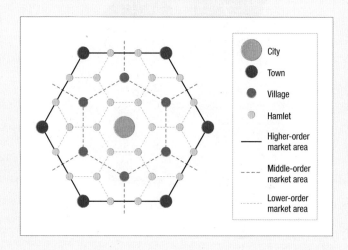

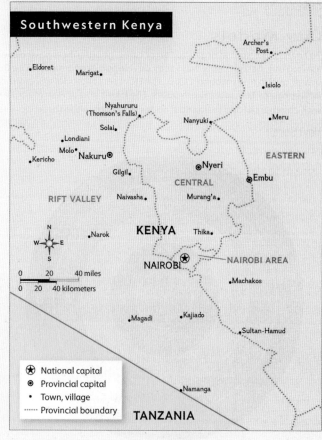

Chapter 16

Bid-Rent Theory

The bid-rent theory explains that the cost of land determines how it will be used. In agricultural settings, the theory portrays types of farming that take place at different distances from the central business district (CBD). Within cities, it may help explain the spatial organization of different sectors. ▮ Choose one of the city models from Chapter 16 and use bid-rent theory to explain the locations of the different economic sectors in the model. Describe other factors influencing the location of sectors.

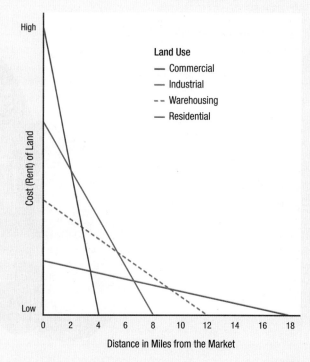

Chapter 16

City Models

To analyze the spatial organization of cities, geographers have developed several models that represent how cities are laid out, based on the ways they developed economically. The Burgess concentric-zone model visualizes cities developed in rings around a CBD. The Hoyt sector model envisions cities developed around a CBD and heavily influenced by transportation routes. The Harris and Ullman multiple-nuclei model proposes a more fluid design, with commercial and residential nodes. ▌Compare the three models and describe their limitations.

Hoyt Sector

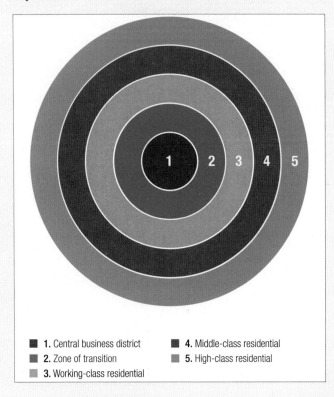

1. Central business district
2. Zone of transition
3. Working-class residential
4. Middle-class residential
5. High-class residential

Burgess Concentric-Zone

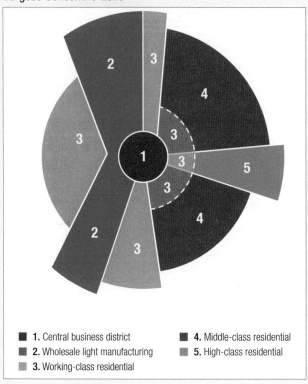

1. Central business district
2. Wholesale light manufacturing
3. Working-class residential
4. Middle-class residential
5. High-class residential

Harris and Ullman Multiple-Nuclei

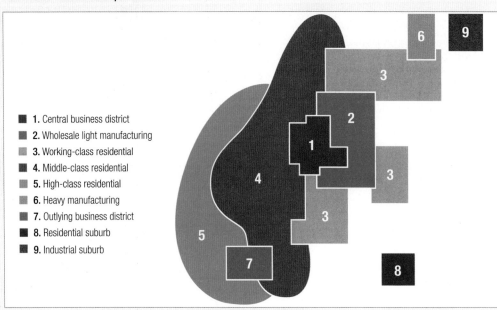

1. Central business district
2. Wholesale light manufacturing
3. Working-class residential
4. Middle-class residential
5. High-class residential
6. Heavy manufacturing
7. Outlying business district
8. Residential suburb
9. Industrial suburb

Maps & Models Archive

Chapter 16

The Galactic City Model

The galactic city model evolved from the multiple-nuclei model and represents a break from the traditional view of cities in which land values and economic activities are strongly or almost exclusively linked to the location of the CBD. Galactic cities are ringed by smaller urban centers and hubs for commerce and manufacturing. ❚ Explain how the invention of the automobile supported the development of galactic cities.

- Central city
- Suburban residential area
- Shopping mall
- Industrial district
- Office park
- Service center
- Airport complex
- Combined employment and shopping center

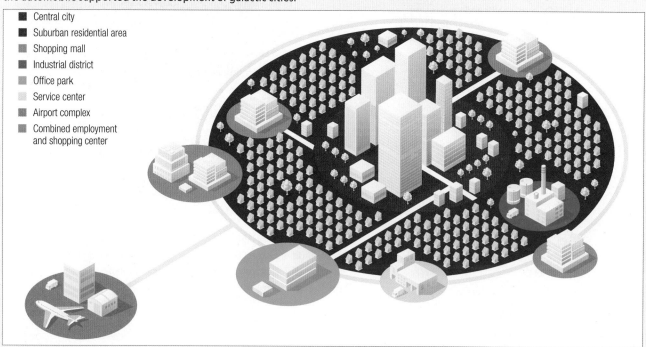

Chapter 16

Latin American City Model

The cities on which the Latin American city model is based date to the Age of Exploration, when the Spanish colonialists built towns and cities placing the church, government, and businesses at the center. The spine is a high-end commercial center around which wealthy residents live. High-poverty areas called disamenity zones are located on inconvenient land like steep slopes or flood-prone ground and are usually overcrowded. ❚ Explain the degree to which the Latin American city model reflects the effects of colonialism on urban development.

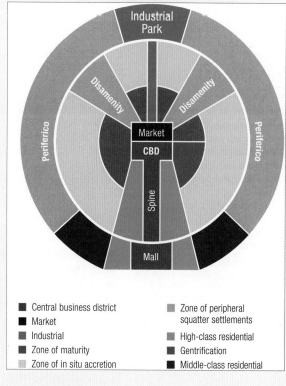

- Central business district
- Market
- Industrial
- Zone of maturity
- Zone of in situ accretion
- Zone of peripheral squatter settlements
- High-class residential
- Gentrification
- Middle-class residential

Unit 6

Chapter 16

African City Model

Many African cities have three CBDs. The first, created by colonial powers, is often built in a grid pattern for power and control. The second CBD is more traditional, with retail stores and curbside commerce. The third is a zone for open-air markets. Higher-income neighborhoods are located within the CBDs, and wealth and services decrease the further out you get in the rings surrounding the CBDs.
▌ Compare the African city model and the concentric-zone model.

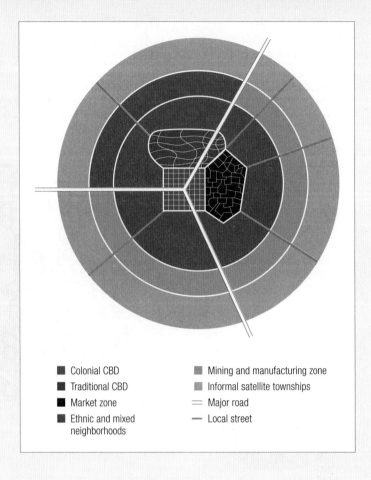

- Colonial CBD
- Traditional CBD
- Market zone
- Ethnic and mixed neighborhoods
- Mining and manufacturing zone
- Informal satellite townships
- ═ Major road
- — Local street

Chapter 16

Southeast Asian City Model

The Southeast Asian city model reflects the fact that many cities in the region are located on ports. According to the model, port cities lack a clearly defined CBD. They have two formal zones—the port zone and the market gardening zone on the periphery. In addition to these two zones, Southeast Asian cities that follow the model consist of a variety of zones in concentric arcs around the port, such as the Western and alien commercial zones, suburbs, and squatter settlements, as well as industrial parks on the cities' outskirts. ▌ Identify and explain a limitation of the Southeast Asian city model.

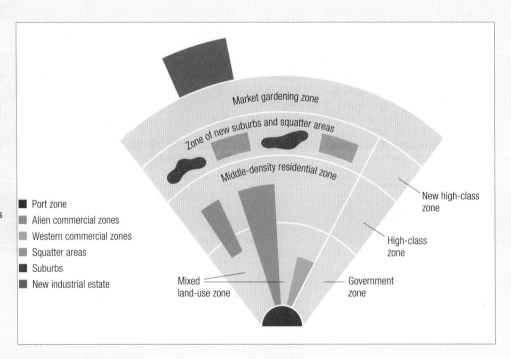

- Port zone
- Alien commercial zones
- Western commercial zones
- Squatter areas
- Suburbs
- New industrial estate

516

Maps & Models Archive

Chapter 17

Zoning Map

Zoning is an important tool for cities planning for sustainability, economic development, and the well-being of residents. City governments work with planners to develop zoning maps such as this one to plot the land use for each street, block, or area. ▌ Identify the spatial patterns you see in the zoning map concerning the locations of the CBD, residential areas, and commercial areas.

Seattle Zoning Map, 2019

- Downtown
- Commercial/mixed use
- Multi-family
- Single family
- Manufacturing/industrial
- Master planned community
- Major institutions

Unit 7
Industrial and Economic Development
Patterns and Processes

An employee at the BMW plant in Cowley, Oxford, England, watches as automated robots on the production line work on Mini Cooper automobile parts.

Global Connections

Automated production line

Originating in 18th-century Britain, the Industrial Revolution released an unstoppable wave of innovation that transformed the world. Advanced forms of industry and manufacturing remade the economies of entire countries. As new global networks of trade, transportation, and communication evolved, people's livelihoods were linked to a worldwide economy.

Core countries have vibrant, postindustrial economies and high standards of living. Some peripheral and semi-peripheral countries are rapidly industrializing, while some are not—but all of these countries are adapting to changes of uneven development. Understanding both the history and present course of industrial development is crucial to building economies that are sustainable for the long term and can support and benefit people across the globe.

Chapter 18
The Growth and Diffusion of Industrialization

Chapter 19
Measuring Human Development

Chapter 20
Globalization, Interdependence, and Sustainability

Unit 7 Writing across Units, Regions, & Scales

Unit 7 Maps & Models Archive

National Geographic Photographer **Lynsey Addario**

Educating Afghan Girls

After studying international relations, Lynsey Addario realized that photojournalism could combine international relations and art, telling stories with pictures. Addario went on to become one of National Geographic's most accomplished conflict photographers. Her photograph of a Ukrainian family killed by a Russian mortar strike in March 2022 went viral, exposing the horrors of that conflict. During the period between Taliban regimes in Afghanistan, Addario focused her lens on women and girls who pursued education, hoping to build their own futures as well as Afghanistan's economy.

Learning Objective
LO 18.1 Explain how industrial and economic development is impacted by gender inequality.

Documenting Women's Lives Addario has been following the lives of Afghan women for more than two decades. She made her first trip to Afghanistan in 2000, several years after the extremist Taliban group had taken control of the country. The strict segregation of the sexes in Afghan society allowed Addario to meet with women in spaces where men were not allowed. She captured images of resistance, including a wedding party in a basement where guests danced to music (an entertainment punishable by death) and secret schools for girls.

Addario has learned how important it is to build trust with her subjects. "I'll spend a lot of time talking to people and making them feel comfortable," she says. "Then I'll get their permission, and then I'll photograph. A lot of what I'm doing when the story is very sensitive is just hanging out, and then I'll photograph very sparingly."

Shifting Political Climates After the terrorist attacks of September 11, 2001, the United States invaded Afghanistan and toppled the Taliban regime. In the 20 years that followed, Addario returned to the country multiple times. Her photographs witness the rebuilding of a destroyed nation, an effort that drew upon the talents and energy of women. Addario's series "Veiled Rebellion" shows Afghan girls and women attending school, training as police officers and medical professionals, hosting television shows, and serving as government leaders. With access to education, women's literacy rates increased dramatically. Still, not all Afghan women reaped the benefits of the Taliban's fall. In rural areas, where conservative values were strong, most women were dependent upon men, who insisted they remain uneducated and cut off from society.

The United States finally withdrew its military forces from Afghanistan in August 2021. The Taliban quickly closed in on the capital city of Kabul as U.S. personnel evacuated. Back in control, the Taliban forced women out of the workforce and banned education for girls older than 12. Addario wrote in *The Atlantic*, "I've seen how hard the country's women have fought for their freedom, and how much they have gained. Now they stand to lose everything."

An Uncertain Future Human rights organizations warn of the consequences of restricting female education. "Taliban policies have rapidly turned many women and girls into virtual prisoners in their homes, depriving the country of one of its most precious resources, the skills and talents of the female half of the population," explains Heather Barr of the nonprofit organization Human Rights Watch. A four-line rubaiyat penned by a teenage poet wryly describes the waste of human capital:

> You won't allow me to go to school.
> I won't become a doctor.
> Remember this:
> One day you will be sick.

But Addario notes, "There is a new generation of Afghan women today, women who can't remember what it was like to live under the Taliban." These women will not easily give up the freedoms they have enjoyed their whole lives.

Geographic Thinking

How might overcoming barriers to gender parity in Afghanistan affect the country's economic development?

Top: Afghanistan's first female provincial governor established the garden where these women, many of whom were studying to become teachers, could enjoy a picnic outdoors. Bottom: These female students, wearing hijabs under their mortarboards and seated apart from their male classmates, graduated as members of Kabul University's class of 2010.

Chapter 18
The Growth and Diffusion of Industrialization

Critical Viewing The EUROGATE Container Terminal in Hamburg, Germany, is one of the main transport hubs in northern Europe. Shipping containers are prepared there and then transported via road and rail to locations all over Europe and Russia. ▌Explain how this photo helps you understand the role that transportation costs play in determining where industries locate.

Geographic Thinking What drives patterns of industrialization?

18.1
Processes of Industrialization

Case Study: The Fourth Industrial Revolution

18.2
How Economies Are Structured

Case Study: Damming the Xingu River

National Geographic Photographer
Aaron Vincent Elkaim

18.3
Patterns of Industrial Location

18.1 Processes of Industrialization

Beginning in the 18th century, new, mechanized ways of making goods led to mass production. This change transformed human life by raising the standard of living for many people. However, the economic development has been spatially uneven, with advantages spreading disproportionately within the core rather than to the periphery.

What Is Industrialization?

Learning Objectives
LO 18.2 Compare the categories and scales of industrial processes.
LO 18.3 Describe the process of industrialization.
LO 18.4 Explain the differences between renewable and nonrenewable resources.

Industry is any economic activity that uses machinery on a large scale to process raw materials into finished goods. **Raw materials** can include metals, wood, plant products, animal products, water, or other substances that are used to make goods intended for sale to customers. Crude oil, for instance, is a raw material that can be converted into a number of different products, including gasoline, heating oil, kerosene, and asphalt. Metals begin as metallic ores that must be extracted from the earth. Then they are processed before being made into finished goods. Metals are used to make machines that make finished goods. Trees can be cut down into lumber and then converted into building materials like boards and plywood, finished goods like furniture, or pulp that is made into paper.

Industry can also refer to a collection of productive organizations that work with the same materials or produce similar products. Manufacturing is one, and is generally divided into two broad categories—heavy industry and light industry. Heavy industry requires huge production facilities that produce goods on a massive scale and generally employ skilled workers. The goods produced by heavy industry include machinery and equipment, ships, railroad equipment, airplanes, and motor vehicles. Light industry requires less investment in production facilities and equipment and produces finished goods that can be made on a massive scale or a small, custom scale. Workers can be either highly skilled or relatively unskilled. Light industries produce a wide range of products, including clothing, processed foods, gemstones, and electronics.

Manufacturing takes place at different scales. Production of manufactured goods in preindustrial times was small in scale, meaning goods were made in relatively small quantities using hand tools or basic machines by artisans, workers skilled in making products by hand. Large-scale manufacturing arrived with the shift to industrial production, using specially purposed machinery in factories for mass production. Because heavy industry produces large goods like commercial jetliners or construction cranes and produces goods in massive quantities, such as petroleum refining or steel production, it requires a greater investment of capital, which is wealth in the form of money or assets, than light industry.

The scale of facilities and production varies dramatically within light industries. For instance, goods like soaps and candles can be crafted on a small scale by a tiny workforce working out of a single shop or even a home, or they can be mass-produced in a huge processing plant. Smaller-scale operations like craft furniture makers or local bakeries tend to emphasize quality of output over quantity. Large-scale processors are interested in mass-producing identical goods and often competing on the basis of price. As you will see, it was the appeal to consumers of the availability of low-cost, mass-produced goods that helped fuel the shift to large-scale manufacturing.

Industrialization is the process by which the interaction of social, political, and economic factors leads to the development of industries across a community, region, or country. The process began with the first **Industrial Revolution** in Great Britain in the 18th century and spread to other countries in western Europe and North America in the 19th century. Still more countries industrialized in the 20th century, and some are still industrializing today. While not all countries have experienced large-scale industrialization, the goods produced by industrial processes are found throughout the world.

The Industrial Revolution

Learning Objective
LO 18.5 Explain the origins of the Industrial Revolution.

A revolution involves rapid, massive, and transformative change. The Industrial Revolution was no exception. It marked the shift from small-scale, hand-crafted production typical of preindustrial societies to power-driven mass production. New technologies increased the quantity and variety of goods that could be produced, expanded the market for these goods through new modes of transportation, and increased the quantities of natural resources needed for production. This revolution changed people's lives and societies in profound ways, with both positive and negative results.

Before the Industrial Revolution Prior to the Industrial Revolution, the production of goods took longer, and transportation was slow. With industrialization, both production

Renewable vs Non-Renewable Resources

The difference between renewable and nonrenewable resources was introduced in Chapter 1. Both renewable resources—like water, air, sunlight, and wind—and nonrenewable resources—like coal, oil, and natural gas—can be utilized in our economic activities, including as raw materials or as energy sources used in industry. Coal, oil, and natural gas can all be used as raw materials or as energy sources for industry, but these are finite (ultimately limited in quantity) nonrenewable resources. There are environmental and human health issues caused by their extraction, processing, transportation, and combustion—including the enormous role that fossil fuels play in causing climate change.

Water, a renewable resource that is recycled endlessly via the hydrological cycle, can also be used as a raw material in industry and can be used as an energy source for industry, too—as with historically important waterwheels and modern hydroelectric dams. Water used in industrial processes can be polluted through that use and made damaging to the environment and/or made unsafe for human use—sometimes temporarily and sometimes long term. To reduce the environmental impacts of industrial use of nonrenewable resources and the pollution of water (and air), laws and regulations have generally been mandated in core countries to reduce or prevent pollution. These controls are often less strict in semi-periphery and periphery countries; some companies have relocated their operations away from the core in part to avoid the higher costs associated with pollution regulations.

As you'll see later in this chapter, dams constructed for hydroelectric power generation can have huge, long-term environmental impacts on the river and terrestrial ecosystems as well as often tragic economic and social impacts on the local populations. Other renewable resources can be overused or degraded, sometimes to the point of destruction of the resource, as with the cutting of forests for lumber or paper faster than they can regrow.

Similarly, in the primary sector—as reviewed in Chapter 14, farmers depend on soil and water to grow their crops—and with proper agricultural practices and careful use of inputs, soil and water can serve as renewable resources. However, mismanagement by farmers can cause water pollution and soil degradation, erosion, and depletion, resulting in farmland that has essentially become a nonrenewable resource.

Decreasing or eliminating the use of nonrenewable resources, as well as recycling materials like steel and aluminum, reduce the impacts of nonrenewable resource extraction and use by industry and other sectors of the economy. Replacing nonrenewable sources of energy like coal and oil in industry and elsewhere with renewable, sustainable options like solar and wind also reduce environmental impacts, lessen human health risks, and slow climate change. In Chapter 20, you will see additional strategies and examples for more sustainable resource use.

---

and transport got faster. Manufactured goods followed several different patterns, depending on what was being produced.

High-quality craft goods were made in isolated, independent, small-scale operations run under the guild system that had begun in the Middle Ages. Guilds were trade associations made up of the master craftsmen (sometimes craftswomen) in a given industry, such as shoemakers, leatherworkers, armorers, weavers, and bakers. Each guild operated in a town or city, capitalized on local resources or specialized skills, set standards for quality, and worked to control production, price, and output.

Large products like ships were produced in selected settings based on geographic factors and employed many workers. Shipyards were established along the Mediterranean and Atlantic coasts of Europe. Skilled craft workers focused on producing particular components, such as decking, masts, and sails, while unskilled laborers assembled the pieces under the supervisors' close scrutiny. In preindustrial times, goods needed in large quantities, like textiles or shoes, were produced by **cottage industries**. In a cottage industry, members of families, spread out through rural areas, worked in their homes to make goods. Their production was funneled to urban-dwelling entrepreneurs who controlled distribution of the finished products.

Regardless of the location of preindustrial manufacturing—whether urban workshop, port city shipyard, or rural cottage—they shared certain features. First was scale: production was done locally, on a small scale. Second was the general reliance on hand-operated tools and equipment. Third was the dependence on muscle power, whether human or animal. Raw materials and finished products were chiefly transported by people or vehicles pulled by strong

Critical Viewing Modern manufacturing operates at different scales. At left is a candy and ice cream factory in Bakersfield, California, where ice cream cones are mass-produced by machines. At right, a worker at a small chocolate factory in Toronto, Canada, produces chocolates by hand. ▌What different costs are evident in the photos of the two factories?

animals like oxen or horses, unless they could be moved over waterways or seas by boat or ship. These constraints limited the amount of goods that could be produced.

Industrialization Begins The Industrial Revolution began as a result of technological innovations and inventions, and it occurred where it did largely because of the availability of natural resources, specifically coal, iron ore, and water. In the years before the Industrial Revolution began, expansion of agricultural fields in Great Britain depleted the country's forests, leaving iron makers without a steady supply of charcoal (which is made from wood) to power their furnaces. Seeking new sources of energy, they turned to coal, but as coal miners dug deeper into the earth, mine shafts often filled with water, creating a need for more powerful pumps. The need was met by iron maker and inventor Thomas Newcomen, who developed the first commercially successful steam engine in 1712. In the 1760s, inventor James Watt vastly improved the steam engine's efficiency. The steam engine would play a major role in powering the Industrial Revolution.

The first industry affected by industrialization was textiles. A series of inventions introduced in the 1760s and 1770s—the spinning jenny, water frame, and spinning mule—mechanized the spinning of thread, significantly increasing the output of high-quality thread. Next, inventors turned to mechanizing weaving, the making of cloth, which was achieved with the invention of the power loom in the 1780s. Both the water frame and power loom originally relied on waterpower. By the 1780s, though, Watts's improved steam engine was put to the task, and suddenly textile manufacturing processes were no longer dependent on either muscle power or water. Coal-fired steam engines enabled machines to produce vast amounts of fabric far faster than was possible under the old cottage industry system.

The spatial patterns of early industrialization were determined by several factors. The high capital cost (defined as a fixed, one-time expense) of equipment and the need to maintain the machinery made it important to focus operations in one location. From this need, the factory system of industrial production developed. Textile production had been focused in Great Britain's Midlands region, a broad stretch of counties in the middle part of England. That area became the center of the industrialized textile industry as well, for several geographic reasons. First was the ready availability of coal in or near the region, making it cheaper and easier to transport the power source to the rising factories. Second, the Midlands region was near the port of Liverpool, which became a hub for the import of raw cotton from parts of the growing British Empire, especially India and, later, the newly independent United States. Third, merchants had developed a system of canals in the Midlands in the early 18th century to transport coal and textiles. These waterways could be employed to cheaply and efficiently move the increased output of textiles generated by the new factories.

By the mid-19th century, canals were replaced by overland railroad transportation employing another feature of industrial power. Constant tinkering with steam engines led to the invention of a mobile steam engine—the locomotive.

The Growth and Diffusion of Industrialization

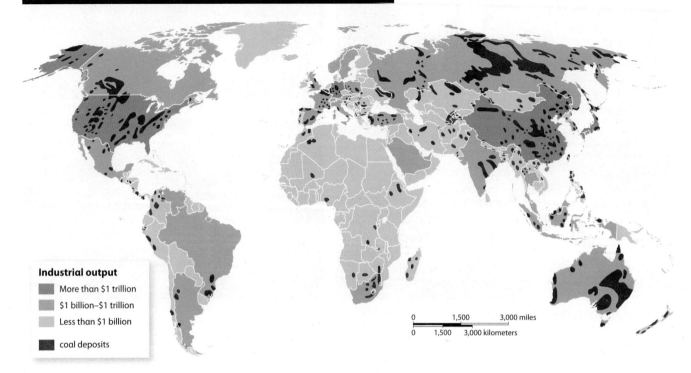

World Industrial Output and Coal Deposits, 2020

Industrial output
- More than $1 trillion
- $1 billion–$1 trillion
- Less than $1 billion
- coal deposits

Reading Maps As of 2020, the United States had the largest economy in the world. However, it ranked second in industrial output, behind China. Japan, Germany, and India round out the top five countries with the highest industrial outputs. ▌Use the map to help you describe the relationship between the countries with the highest industrial outputs and the locations of the world's major coal deposits and then explain how this was an initial advantage for the United States and China.

Geography helped the Midlands here too, as the region's flat landscape made it relatively easy to build a rail network across the area to speed the shipment of coal, textiles, and other raw materials and finished products.

Increased use of coal power also transformed the iron industry. While iron had long been an important metal, its production was a small-scale enterprise carried out by craftsmen in the preindustrial age. Coal power and the steam engine made it possible for iron makers to generate more heat, which in turn created new processes for fashioning iron. These processes required expensive machinery, which meant that iron production, like textiles, became a large-scale industrial enterprise. One important use of iron was to build the new locomotives that were now needed, as well as the rails themselves.

Geographic Thinking

1. Explain how the availability of natural resources facilitates industrialization.

2. Describe the essential changes in manufacturing that took place during the Industrial Revolution in Great Britain.

3. Describe the geographic factors that led to the Industrial Revolution emerging in the Midlands region of Great Britain.

The Spread of Industrialization

Learning Objectives
LO 18.6 Describe the diffusion of the Industrial Revolution.
LO 18.7 Compare the technology patterns and trends of the successive Industrial Revolutions.

In order to industrialize, a country must have natural resources, a supply of labor, and the capital or monetary resources; these are collectively called the **factors of production**. But there are other conditions that encourage or discourage industrialization. One factor encouraging industrialization is the presence of certain institutions such as the rule of law, a belief in property rights, and access to free and open markets. A second factor is the ability to trade freely, regionally, nationally, and globally. The success of industrialization made it likely to diffuse to other locations in other countries; however, the places with the needed institutions and trade opportunities saw the changes first. But Britain recognized that there were economic advantages to being the only industrialized country and took steps to stop the diffusion of industrialization. The country passed laws that made it illegal to export new machines, manufacturing methods, and even skilled workers. Seeing the rapid growth of the British economy, however, other countries sought the know-how that would provide the foundation for similar expansion.

When industrialization finally did diffuse beyond Great Britain's borders, it did so through expansion diffusion and relocation diffusion. As you've read, expansion diffusion occurs when an idea or innovation—or in the case of industrialization, a series of innovations and processes—spreads outward while still remaining strong in its place of origin. Relocation diffusion is the spread of ideas or innovations through the movement of people.

Continental Europe and North America In the first decade of the 19th century, Belgium became the first country in continental Europe to industrialize. It had the needed factors of production, including plentiful supplies of coal and iron ore that facilitated its industrialization, in the same way that the availability of these resources had promoted industrialization in the British Midlands. Belgium also had a tradition of institutional factors that support industrialization through its long-standing tradition of textile production, which meant its business leaders were eager to embrace Great Britain's new advances to stay competitive. Belgium also had a king willing to invest his own monetary resources in new industrial enterprises for his country.

In an example of relocation diffusion, the revolution spread to the United States when a British textile manufacturer, Samuel Slater, immigrated there in 1789. Settling in Rhode Island, he constructed spinning machines and established textile mills run by waterpower. This technology was adopted by others in New England, which became the first center of industrial production in the United States.

The diffusion of the Industrial Revolution was delayed in the rest of Europe by political, institutional, and economic conditions as well as the availability of needed resources. In France, for instance, political upheaval, the larger size of the country, and the lack of developed waterways to move its isolated coal and iron deposits all delayed industrialization. Germany had plentiful supplies of coal but for decades was not a single country but a collection of separate, fiercely independent states linked loosely by competing customs unions. Industrialization there lagged until the latter part of the 19th century, enabled by German unification in 1871. Some Eastern European countries, like Poland, had the labor force and plentiful supplies of coal, and Russia also had the labor force and was rich in many natural resources, but industrialization did not diffuse to these areas until the late 19th century, as they lacked institutional supports like the rule of law and free markets.

When industrialization did take root in Europe, the United States, and Japan, the same criteria that determined the spatial patterns of industrial regions in Great Britain determined the site of the new industrial zones. The leading factor was close proximity to natural resources, particularly the coal needed in large quantities to power industrial processes.

The coal-rich Ruhr Valley became the industrial heart of Germany and by the end of the 19th century was one of the leading industrial regions in the world. It benefited from the Ruhr and Rhine Rivers, which facilitated the import of iron. In the United States, the textile mills that dotted New England relied on waterpower to drive their machinery, so the earliest American factory towns were those built on the Atlantic fall line (discussed in Chapter 15), where rivers and streams flow down from the upland region to the coastal plain, producing rapids and falls used for power. These locations also facilitated the transportation of raw materials to the factories and the shipment of finished products from them.

By the middle of the 19th century, location on rivers became less important as industrializing countries began to build canal systems and, later, extensive rail networks. These networks could reach deep into the interior of countries, and railroads could more easily transport heavy raw materials like coal and iron and finished products like steel, a strong alloy of iron. Nevertheless, the early industrial areas often remained the centers of industrial production, as new industries, like the chemical and automobile industries, developed where industry had originally arisen.

Second and Third Industrial Revolutions
The Industrial Revolution that began in Britain in the 18th century is called the First Industrial Revolution; however, it was not the last. While industrialization continued in Europe and North America—and later in other regions—it entered new phases called the Second and Third Industrial Revolutions. Each of these revolutions was built on a system of interrelated technologies. The First Industrial Revolution was powered by steam, coal, and waterpower and was focused on the textile and iron industries.

The steel and petroleum industries that became so important late in the 19th century paved the way for the Second Industrial Revolution, which was powered by electricity and the internal combustion engine (an engine that generates power by burning gasoline or other fuel with air within the engine). Factories were reconfigured to use the assembly line, in which individual workers perform a narrow range of repetitive tasks on products that are carried from one workstation to the next mechanically. The assembly line and the use of interchangeable parts—pioneered by American innovator Eli Whitney early in the 19th century—allowed industry to engage in mass production, a process by which large numbers of identical products that meet certain quality standards are manufactured. Another key feature of this Second Industrial Revolution was the invention and increasing sophistication of machine tools—the use of machines to make parts or pieces out of metal for use in other machines.

The Second Industrial Revolution saw the growth of the steel, automobile, and airplane industries, as well as the chemical industry and the development of consumer appliances. Since steel was a major manufacturing material, factories were often located in regions with access to coal and iron ore. Places located near coalfields or with convenient water or rail access to sources of iron ore, like Pittsburgh, Pennsylvania, and Chicago, Illinois, became centers of the steel industry in the United States. Detroit, Michigan, and other cities in the Midwest, along with some cities in the South like Birmingham, Alabama, took advantage of their location near steel production centers to become manufacturing centers, converting the steel into automobiles and other manufactured goods. Concentration of factories in these areas promoted the growth of related businesses, such as tool-and-die shops that produced machine parts like cutting tools, molds, gauges, and other tools and fixtures used in manufacturing.

The Third Industrial Revolution began after the end of World War II and was marked by reliance on electronics and information technology systems and by automation of production processes. Made possible by advances in computerization and miniaturization, this revolution brought in new industries like computer manufacturing, software engineering, and telecommunications. During the decades of the Third Industrial Revolution, computers changed drastically, from room-sized units affordable only by governments and the largest enterprises to handheld devices that provide ordinary people with opportunities for productivity, information access, communication, and entertainment.

Industrialization in all its phases also promoted the rise of service industries like banking, accounting, and insurance to promote the formation and ongoing operations of industrial firms. Increased productive output relied on the rise of transportation and wholesale and retail trade to facilitate the distribution of the industrial output.

While the Industrial Revolutions are distinguished by certain patterns and trends and can be placed roughly on a historical time line, it is important to keep in mind that they were not specific events, but rather large-scale developments that took place over time and had the effect of changing human society through technology.

Colonialism and Imperialism In the 19th century, industrialization became interlinked with colonialism and the two processes helped to fuel one another. Great Britain was Europe's leading imperial power in the late 18th century. It had colonies in North America, Australia, and the Caribbean, extensive trade networks in Asia, and was in the process of

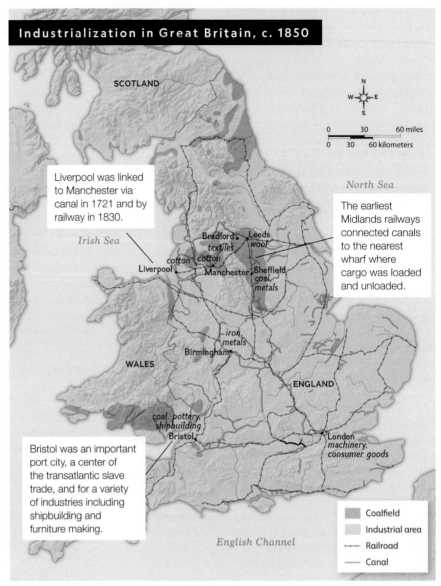

Reading Maps By the mid-1800s, Great Britain had built a vast network of rail lines and canals through which the raw materials and finished goods of the country's rapidly growing industries could be transported. ▌ Use details from the map to explain how Bristol's site and situation were advantageous during the Industrial Revolution.

asserting control over India in order to better import cotton for its textile industries. The wealth and resources that poured in from Great Britain's role in the Middle Passage, the portion of the Atlantic slave trade that saw millions of Africans forcibly transported via ship to the Americas, provided the capital needed for the country's industrial expansion.

As industrialization spread in the 19th century, control of colonial areas became attractive both to provide sources of raw materials and to guarantee exclusive access to new markets. The expansion of European influence that had begun in the 15th century increased with Great Britain's enhanced industrial power. Improvements in transportation—specifically steamships and railroads—opened the colonies' interiors, making new raw materials accessible. Better weaponry had helped small groups of Europeans gain control over much larger Indigenous

By the 1870s, the Industrial Revolution had spread well into eastern Europe - and to Russia by the 1880s. Large factories that produced steel and textiles were built in Russia in the 1880s, and the construction, beginning in 1891, of the Trans-Siberian Railway—the longest railway in the world—facilitated the movement of goods, people, and raw materials.

populations in the Americas beginning in the 16th century. In the 19th century, more powerful weapons, now mass produced by industrial processes, gave European conquerors an even greater edge, which they used to expand their colonial empires in Africa and Asia.

Harsh conditions prevailed in colonies controlled by Europeans, who sought raw materials for their industries and markets for their output. King Leopold II of Belgium headed a group of investors that controlled what is now the Democratic Republic of the Congo. Africans were compelled to work on rubber and palm oil plantations or in mines, and any attempts to escape or rebel were met with brutal force. In roughly two decades of this inhumane rule, the population of the region fell from about 20 million people to about 8 million. The British and French also exploited African regions they had seized as colonies, using them as sources of such raw materials as rubber and cotton, palm oil, metals like copper and gold, and cash crops like cacao, sugar, tea, and coffee. The British flooded their colonies in South Asia with cheap textiles produced in the steam-powered mills of Great Britain from cotton grown on the Indian subcontinent and, in the process, virtually destroyed the Indian textile industry.

As you know, Wallerstein's world system theory categorizes countries as core, peripheral, or semi-peripheral, with peripheral countries being vulnerable to exploitation by core countries. As industrialization spread throughout the 19th century and into the 20th century, it did not spread evenly throughout the world. The major industrial powers remained the core countries in the arc from Germany west to France and Great Britain as well as the United States. Semi-peripheral countries in eastern and southern Europe lagged behind. While Japan became a significant world power through industrial growth and militarization, most of Asia was controlled by European imperial powers who used the region to obtain raw materials and as markets for manufactured goods. The same was true in Africa and Latin America.

Industrial Diffusion and Populations

Learning Objective

LO 18.8 Describe how the First Industrial Revolution contributed to urbanization and changes in social structure.

The second agricultural revolution, which came at the same time as the beginning of the Industrial Revolution, produced tremendous growth in agricultural output. As explained by the demographic transition model, this increased output led to population growth and made Europeans healthier. This contributed to both population growth and increased life expectancies, with Great Britain increasing in population from under 6 million in 1750 to 8.6 million by 1800. Advances in medicine and public health that came at the same time contributed to this trend. As a result, death rates declined.

Rural-to-urban migration increased dramatically, caused by population growth, more efficient farming practices, and government policies such as Great Britain's Inclosure Act of 1773 that resulted in small farmers being forced off their lands. Many of these migrants found work in the growing number of factories. From 1805 to 1911, the urban population of France rose from 25 to 44 percent. Urbanization in Germany was even more dramatic in the same years, doubling from 30 percent to 60 percent. Improvements in transportation, in the form of faster ships, better roads, and the spread of railroads, made migration to the cities even easier. London surpassed 2.6 million people by 1851 and topped 6.5 million just 50 years later. New York City—fed not just by migrants from rural areas but also by immigrants from Europe—rose from just under 600,000 people in 1850 to 3.8 million in 1900. Growing urbanization spurred by industrialization led to the next phase of the demographic transition model, with birth rates slowing, death rates continuing to decline, and population growth as a whole slowing.

Industrialization ushered in changes in social structure. European societies had long been divided into the landed class, made up of landowners who lived off rental income, and the working class, largely rural farmworkers. During the Middle Ages and after, Europe saw the rise of a middle class between the land-owning aristocracy and the rural peasantry. This class included merchants, shopkeepers, and artisans. The rise of factories contributed to the growth of the middle class, which rose out of the ranks of the factory owners, managers, and other so-called white-collar workers.

Wealth became a marker of social status, and the most successful business owners of the middle class could rise to enter the ranks of the upper class, though they might not always be accepted by the long-established landed aristocracy. Industrialization also created a new working class of people who toiled in factories, including women and children who worked for wages.

Still, the job growth and economic expansion promoted by industrialization led to the growth of the middle class. Business owners might not gain the vast wealth of the major industrialists, but they could enjoy a high standard of living. Mid-level managers in major companies and other salaried workers swelled the ranks of the middle class as well. The new prosperity altered how people in the middle class used the space around them. They had more free time to fill with recreation, such as attending plays, concerts, or sports events, or spending time in parks or in the countryside.

While few middle-class women worked in paid employment, working-class women often held jobs from their teens (or even earlier) until they married. Most were employed in domestic service, but a significant number worked long hours in factories and some toiled in coal mines. Working-class children, too, often filled the employment ranks of the new factories, working long hours in dangerous conditions for little pay. The emerging divisions of classes of people fueled what became known as class consciousness

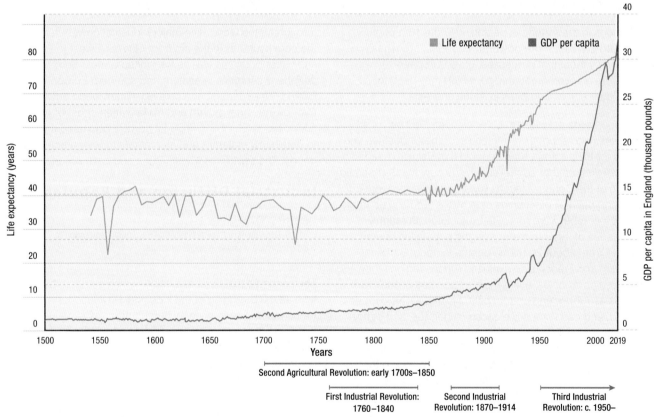

Life Expectancy and Economic Growth in the United Kingdom, 1500–2019 The effects of the industrial revolutions were significant in the United Kingdom, which saw a steep increase in both life expectancy and economic growth beginning in the late 1800s. ▎Compare the UK's life expectancy during the First Industrial Revolution and the Second Industrial Revolution.

Critical Viewing In 1881, New York City created a Department of Street Cleaning, a precursor to today's Department of Sanitation. The photo shows workers in the late 1890s clearing garbage from a city street. ▮ Explain how the photo is evidence of New York City's developing urban infrastructure.

(an awareness of one's place in a system of social classes), a phenomenon noted by economic historians such as Karl Marx and Fredrich Engels.

Many of the features of modern urban life arose in the wake of the Industrial Revolution. City growth created serious public health challenges such as disposing of waste, providing clean and adequate water, and controlling contagious diseases. In response, city officials developed new systems such as public water and sewage systems. Cities built hospitals to care for the ill and schools to teach the children.

Concern over a perceived rise in crime—for which the working class and poor were blamed—led officials to organize professional police forces. Concern over fire hazards led to the creation of units of firefighters. By the end of the 19th century, cities were adopting electric lights, creating an atmosphere in which the facets of daily life, from work to leisure activities, could continue into the night.

By this time Europe was experiencing the results of industrial development, marked by rising wages, better health, higher levels of schooling, and more comfortable lives for many. The populations of industrial regions followed the demographic transition model—better nutrition, medical advances, and public health measures meant declining death rates and longer lives. Declining infant mortality rates, combined with the costs of raising children in the city, meant that couples decided to have smaller families, slowing the population growth rate.

As literacy became important for more and more jobs, the core countries began to institute public education systems. The middle class and even working class grew more literate, and subsequently their members began to demand a political voice. Several of the industrialized societies expanded voting rights to all men and, eventually, to women.

Geographic Thinking

4. Explain the relationship of industrialization to the location of coal resources.

5. Explain how colonialism supported the growth of industrialization.

6. Describe how industrialization expanded the middle class.

The Growth and Diffusion of Industrialization

Case Study

The Fourth Industrial Revolution

The Issue The Fourth Industrial Revolution is upon us. How will humanity handle the changes to come?

Learning Objective
LO 18.7 Compare the technology patterns and trends of the successive Industrial Revolutions.

By the Numbers

62.5 percent
of the world's population is online (2022)

more than 300 billion
emails sent and received per day (2020)

more than 20 million
additional manufacturing jobs lost to automation by 2030 (2019 est.)

Sources: The Next Web, The Radicati Group, Oxford Economics

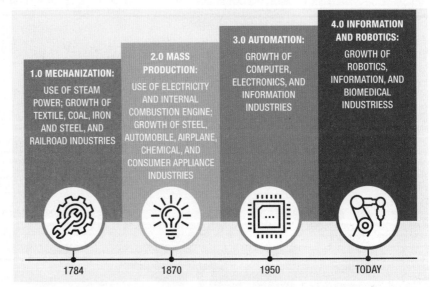

Industrial Revolutions Over Time The First Industrial Revolution is sometimes called the revolution of mechanization. The second is the revolution of mass production, and the third of automation. The world is now experiencing a Fourth Industrial Revolution, the revolution of information and robotics.

Advances in Technology, such as artificial intelligence (AI), augmented reality (AR), genome editing, robotics, nanotechnology, and 3-D printing, characterize the Fourth Industrial Revolution. This revolution takes advantage of the trend toward digitizing information that developed in the Third Industrial Revolution and of advances in basic sciences from biology to physics. The Fourth Industrial Revolution may result in new cyber systems that blend biological and computerized elements in ways certain to transform people's lives. Given the rapid pace of technological change, it is not clear how quickly this phase of production will morph into yet a Fifth Industrial Revolution.

One feature that makes this particular revolution distinct from the earlier ones is the fact that it is global. Modern communications technology, the deep interconnectedness of the world's economies, and the proliferation of scientific and technical expertise to all regions of the world mean that new developments made anywhere can spread rapidly. As in the earlier revolutions, social change resulted in the development of distinct groups of "haves" and "have-nots." The nature of those distinctions in the Fourth Industrial Revolution—and the extent and success of efforts to minimize them—will be important questions humanity will have to address.

Another key feature of the Fourth Industrial Revolution is the idea of sustainability—the development of economic processes that can be self-perpetuating and promote the long-term maintenance of communities and the environment. Among the sustainability considerations introduced in Chapter 1, a central concern is the environmental cost of unchecked industrial growth and the intensive exploitation of natural resources. Climate scientists on the Intergovernmental Panel on Climate Change (IPCC) have prepared a series of reports—the most recent in 2022—that document the accelerating processes of climate change resulting from mounting levels of greenhouse gases (primarily carbon dioxide, methane, nitrous oxide, and chlorofluorocarbons, or CFCs) in the atmosphere. These greenhouse gases are generated by burning fossil fuels, industrial processes, and changing patterns of land use and land cover. The IPCC report warns that climate change is affecting human, plant, and animal health, including increasing rates of extinction and decreased biodiversity. The goals of sustainable development include the adoption of methods of production that do not overuse natural resources (both renewable and nonrenewable) and minimize environmental harm—and can thus be perpetuated for the benefit of future generations.

Geographic Thinking

Describe how the Fourth Industrial Revolution has differed from the previous three industrial revolutions.

18.2 How Economies are Structured

The economies of countries around the world are structured differently. This affects how people make a living and results in uneven patterns of development. Geographers study these patterns of economic activity and their impacts on people within countries, within regions, and on a global scale.

Sectors of the Economy

Learning Objective
LO 18.9 Compare the activities and impacts of the five economic sectors (primary, secondary, tertiary, quaternary, and quinary).

Geographers characterize the structure of a country's economy by distinct economic activities that can be categorized into **economic sectors**. These sectors are collections of similar economic activities based on the extraction of raw materials, the production of goods, the provision of services, or other activities. Traditionally, geographers defined three main sectors: primary, secondary, and tertiary. In recent years, because of economic and technical innovations, they have defined special sections of the tertiary sector with unique functions, labeling them as the quaternary and quinary sectors.

The **primary sector** is associated with activities involving the extraction of natural resources from the earth. This sector includes agriculture, fishing, forestry, mining or quarrying, and extracting liquids or gas, such as oil drilling. It focuses on obtaining raw materials. It includes farming activities like the raising of crops as well as activities that obtain resources needed for power generation, such as coal, oil, and natural gas. The primary sector also includes the extraction of materials used in manufacturing and construction, such as metal ores, clay to make pottery, lumber, sand and gravel to make concrete, and quarried stone to make everything from buildings to countertops. These activities are considered primary, or first-order, because they provide the basis for other economic activities. No goods can be produced without these raw materials.

The **secondary sector** is associated with the production of goods from the raw materials extracted or harvested in the primary sector. Manufacturing, processing, refining, and construction are secondary-sector activities. This sector includes many industries, from metalworking and smelting to automobile production and aerospace manufacturing, from textile production and food processing to generating electricity.

The **tertiary sector** is also called the service sector because it provides services rather than finished goods. This sector includes a huge range of activities that involve the transportation, storage, marketing, and selling of goods. It also includes commercial services provided to the general public, businesses, or government and personal services like hair care or restaurant service. Most government activities are part of the tertiary sector. This sector includes everything from shipping and delivery of goods to wholesale storage and retail trade. It includes banking, finance, insurance, law, health care, hospitality and tourism, entertainment and media, and clerical services. The growing tertiary sector has been further divided into two specialized subcategories: the quaternary and quinary sectors.

The **quaternary sector** is the portion of tertiary-sector activities that requires workers to process and handle information and environmental technology. This sector includes work in information technology, libraries and education, scientific research, cultural activities, and some government services. Workers in these fields tend to require

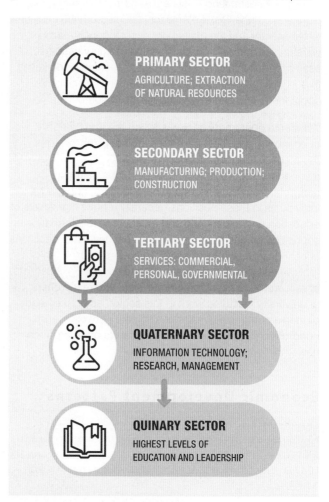

Five Economic Sectors The tertiary sector, or service sector, encompasses the quaternary and quinary sectors. The quinary sector is part of the quaternary sector.

The Growth and Diffusion of Industrialization

U.S. EMPLOYMENT BY ECONOMIC ACTIVITY (2019)		
Education and health services	35,894,000	22.9%
Wholesale and retail trade	19,742,000	12.6%
Professional and business services	19,606,000	12.5%
Manufacturing	15,741,000	10.0%
Leisure and hospitality	14,643,000	9.3%
Construction	11,373,000	7.3%
Financial activities	10,765,000	6.9%
Transportation and utilities	8,991,000	5.7%
Other services	7,617,000	4.9%
Public administration	7,225,000	4.6%
Information	2,766,000	1.8%
Agriculture and related	2,425,000	1.5%
Mining and related	750,000	0.5%

Education and health care are the largest employers in the United States. Primary-sector activity is concentrated at the bottom of the list, with agriculture, mining, and related activities making up just two percent of the country's annual GDP.

Source: Statista

high levels of education, often including postgraduate or professional degrees. The major high-tech companies like Google, Microsoft, and Apple are all part of this sector. Some high-tech companies, like Amazon, fall into both the traditional tertiary sector and the quaternary subcategory. Amazon warehouses and sells goods (traditional tertiary qualities), uses information technology (quaternary sector) in marketing to users based on information the company has collected about them, and offers services like cloud computing.

Within the quaternary sector is found the fifth category of economic activities, called the **quinary sector**. This more specialized subcategory of work involves the top leaders in government, science, higher learning, nonprofit organizations, health care, culture, and media. Individuals in these positions generally require high levels of education and experience and must process information in sophisticated and responsible ways as they make decisions affecting large swaths of society. Leaders of think tanks (policy institutes that do research and advocacy), scientists at research and development facilities, and researchers at university institutes work in this sector, as do the heads of media companies, major health-care networks, and nonprofit organizations. These specialists often produce new knowledge used by other economic sectors.

Economic Development Patterns

Learning Objectives
LO 18.10 Describe the relationship between sectors of economic production and economic development.
LO 18.11 Identify the geographic characteristics common to postindustrial societies.

Geographers study the structures of individual countries' economies, based on the percentage of employment across the economic sectors, to understand patterns of economic development. Analyzing the percentage of a country's population that makes up the workforce in each sector can reveal not just the structure of employment but how that country's goods are produced and how economically developed the country is.

Development and Employment In countries with little industrialization, primary-sector employment is dominant, with the bulk of the labor force working to produce the food they need to survive. As you learned in Unit 5, little mechanization is used in this subsistence agriculture, with farmers relying mostly on human and animal power to produce crops. The economic value of the food and other crops produced is relatively low. First, most of the food is consumed by the farm families. Second, even those cash crops raised in commercial agriculture for sale in world markets bring relatively low prices for the farmers or wages for the farm workers compared to the wages of workers in manufacturing. As a result, countries with workers mostly employed in the primary sector tend to be poor, with little capital to fund investments needed to industrialize.

In Wallerstein's world system theory, these countries are in the periphery. The dominance of primary-sector economic activity there is not necessarily permanent, however. The United States had more than 50 percent of its labor force in agriculture as recently as 1880, in the midst of its industrial age. By the 1950s, that share had fallen below 10 percent—and has continued to fall. Contemporary China provides an even more instructive example. In 1990, more than 55 percent of the Chinese workforce was employed in agriculture. Government promotion of manufacturing boosted the secondary sector of the economy and mechanized farming so that by 2020, agriculture accounted for only 24.4 percent of the workforce. Both the United States and China had the benefit of abundant natural resources, fertile soils that could produce high crop yields, and stable governments. Those advantages are not always found in the preindustrial countries of the periphery today.

Semi-peripheral countries tend to have large portions of their workforce in the secondary sector. These countries rely heavily on manufacturing, and their workforce reflects this, with many jobs in the manufacturing, processing, and construction industries. Examples in today's world are Brazil, with 32 percent of its workforce in industry, and Iran, with 35 percent of its workers so employed. These are semi-peripheral countries with substantial natural resources that help drive the process of industrial growth.

Core countries tend to have economies with a substantial secondary sector but a dominant tertiary sector. While manufacturing likely generates a substantial portion of a core country's wealth, it is not likely to be the main engine of employment. Manufacturers strive for efficiency, so over time these businesses have used more automation, resulting in less secondary-sector employment. It is estimated that since the year 2000, 260,000 jobs in the United States have been

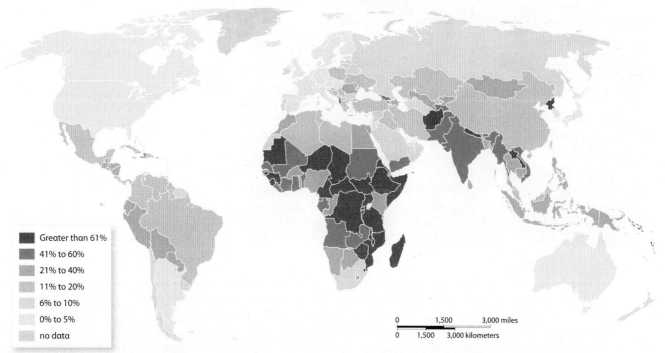

Share of the Labor Force Working in Agriculture, 2017

Legend:
- Greater than 61%
- 41% to 60%
- 21% to 40%
- 11% to 20%
- 6% to 10%
- 0% to 5%
- no data

Reading Maps In countries with little industrialization, it is common for more than half of the workforce to be employed in agriculture. ▌ What are some factors that may explain the prevalence of large shares of agricultural labor in African countries?

lost due to automated processes, which represents a 2 percent decline in manufacturing jobs. In addition, industrial output depends on workers in the tertiary sector. Transportation workers haul raw materials to factories and ship finished goods to markets. Communication technicians provide needed internet and Wi-Fi for factories. Wholesale and retail workers distribute industrial output to corporate purchasers or consumers. A wide range of other tertiary services—legal, advertising, insurance, banking, and more—is provided to manufacturing, processing, and construction firms in the secondary sector.

Patterns of Economic Activity Economic activities occur in spatially uneven patterns across a range of scales, from global-scale differences among countries to places and regions within countries. As you've just learned, countries of the periphery have many workers in the primary sector, generally practicing subsistence agriculture.

The shift to the secondary sector that comes with industrialization typically results in a population concentrated in urban areas, as evidenced by the urbanization of the First and Second Industrial Revolutions in the core areas of Europe and North America. In those periods, the concentration resulted from the desire to locate industrial facilities near sources of raw materials, energy sources, and labor—or near the cities that offered the largest markets of customers. This concentration is also a function of the high capital cost needed to erect industrial facilities. It makes more economic sense to focus capital investment in fewer rather than many locations.

This pattern of urbanization is occurring today in countries that are in the process of industrializing. Industrial facilities are generally located in urban areas, especially at ports or along rail and highway networks. Urban areas provide the potential workforce as well as a potential market. Locating industries in urban areas with ports and transportation hubs facilitates the movement of raw materials to and finished products from the industrial facilities.

Tertiary activities vary widely in their spatial distribution because they represent many different economic endeavors. While banking and financial services tend to concentrate in urban areas, retail stores, restaurants, and a wide range of services from insurance and broad-band internet provision to health care and education are distributed throughout a country. These tertiary activities are needed wherever there are people—though the density of these services will vary according to population and the number of people needed to support a certain service (threshold) and the distance someone is willing to travel for that service (range), as you recall from the discussion of central place theory in Chapter 15.

Information industries of the quaternary sector tend to cluster near institutions of higher learning that provide the educated workforce they need. In the United States, for instance, major centers of the computer science industry are found in or near the universities of the San Francisco Bay

Critical Viewing CEO of Alphabet, Inc., and its subsidiary Google, Sundar Pichai (*left*) looks on as a scientist tests a quantum computer, which can perform complex operations and computations more than a trillion times faster than the most powerful supercomputers.
■ Explain why Pichai and Google scientists are considered part of the quinary sector.

Area in California; in Boston, Massachusetts, Austin, Texas, and Denver, Colorado; and in the Research Triangle region of North Carolina. The quinary sector is generally found in the largest urban centers, capital cities, large universities, and other political or administrative centers of countries.

The Postindustrial Economy Industrializing countries tend to follow a pattern in which the share of agricultural employment declines over time and the share of tertiary employment rises. Secondary-sector employment grows for a time as a country industrializes. Eventually, that share may also decline if a country develops a **postindustrial economy**. The concepts of postindustrial economy and postindustrial society were popularized by sociologist Daniel Bell in 1973. This economic pattern is marked by extremely low primary-sector employment, relatively low secondary-sector employment, and predominant tertiary-sector employment with a rising share of quaternary and quinary jobs. The United States is a country with a postindustrial economy, as are Japan, Australia, and Singapore.

Postindustrial countries share several features. The emphasis in their economies has shifted from the production of goods to the production of services. The share of so-called blue-collar secondary-sector workers has declined, and the share of tertiary-, quaternary-, and quinary-sector workers has risen. Information technology and related fields, such as artificial intelligence, have developed to create new ways to perform a wide array of tasks and operations previously done by humans, from voice assistants like Alexa and Siri to autonomous vehicles. There are, however, significant social issues associated with some of these innovations. Postindustrial economies place a strong emphasis on institutions of higher learning as resources in developing and using the new technologies.

Any transition from one economic phase or technology to another is jolting for certain segments of a society. Agricultural changes associated with the Industrial Revolution made it increasingly difficult for farm laborers to make a living in their chosen field. To survive, many had to move to cities, even if it meant leaving home for an uncertain future, because that's where the jobs were. The shift to the postindustrial economy has, for the past several decades, been a major challenge to manufacturing workers in the United States and other core countries. As manufacturing jobs move out of these postindustrial countries to peripheral and semi-peripheral countries, workers must learn new skills in other fields and, often, other sectors of the economy where they are usually less qualified, due to lack of experience and training.

The breakdown of U.S. employment by sector in 2020 shows that almost 80 percent of all people in the labor force work in the tertiary sector—far more than the 18 percent employed

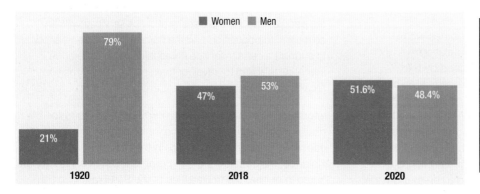

U.S. Workforce by Gender The percentage of women in the U.S. workforce has risen significantly since 1920, and in 2020, women made up more than half of the total. Gender-based employment disparities endure in the United States: Women earn about 84 cents for every dollar a man earns, and women (particularly single women and women of color) are more likely than men to live in poverty.

in the secondary sector and less than 2 percent with jobs in the primary sector—with an increasingly significant portion employed in the quaternary and quinary sectors.

Employment is not the only way to think of the impact of each sector on the economy. It is also useful to consider what each sector contributes to the **gross domestic product (GDP)**— the total value of all goods and services produced by a country's citizens and companies within the country in a year.

Another feature of postindustrial economies is the growing role of women in jobs outside the home. Of course, women have always worked, taking primary responsibility for raising children, caring for family members, tending to the home, and cooking. Women have traditionally done other kinds of labor as well, taking part—or playing the lead role in some countries— in farming, family businesses, or their own businesses. In addition, from the earliest days of the First Industrial Revolution, women were recruited to work in manufacturing jobs.

Later, traditions developed that discouraged women from working outside of the home. That narrowing of women's life choices started to change in the 20th century, and the acceptance of women taking jobs outside the home is generally widespread in industrial and postindustrial societies. In the U.S., women's workforce participation peaked in 1999 at 60 percent. Men's participation in the country's workforce has declined more rapidly than women's since then, shrinking the employment gap. Women are also receiving more advanced educations, enabling them to fill a significant number of quaternary and quinary jobs. Yet barriers to professional advancement and inequalities in pay and promotions persist. Gender inequality in employment is discussed further in Chapter 19.

Dual Economies Development does not take place uniformly within a country. In fact, industrialization often leads to geographically uneven development. Geographers have found that some peripheral and semi-peripheral countries have what are called **dual economies**, or two distinct divisions of activity across the economic sectors. In these countries, much of the population may work in the traditional primary-sector economy, often depending on subsistence agriculture. At the same time, another substantial share of workers participates in a more varied market-based economy with a heavy emphasis on secondary-sector jobs.

One example of a country with dual economies is Vietnam, where the country's leaders tried to find a way to address rising consumer demand for goods and lagging manufacturing output. In the mid-1980s, with the help of foreign investors, these officials worked to expand the manufacturing base, developing such industries as textiles, electronics, and automobiles. The result has been a long period of substantial economic growth. Fully two-fifths of the country's workers still labor in the agricultural sector, but that is barely half the share of farmworkers in 1991. Meanwhile, the share of workers in the secondary sector has risen to 25 percent; they generate nearly a third of the country's GDP and the majority of its top exports.

Geographers differ on why dual economies develop and why the agricultural workforce seems to be resistant to change. Some think that the resistance is due simply to tradition. Other observers think that the agricultural workers do not see enough benefit in shifting to the secondary sector. Still others see this uneven development as a manifestation of a core–periphery dynamic at a different scale than global, that is, within a country. In this view, the core area that industrializes has more easily maintainable ties to the global economy and its market forces. This usually includes urban and coastal areas where ports are located, as these areas are often connected to global transportation routes and telecommunications networks. The agricultural workforce, meanwhile, labors in the interior hinterland, a peripheral area with little contact with that global economy.

Geographic Thinking

1. Explain how the five economic sectors are related to one another.

2. Explain the degree to which the core-periphery model applies to industrialization.

3. Identify the factors that will decide whether Vietnam, with its dual economy, will become fully developed or remain with a dual economy. Explain how each factor would contribute to Vietnam's development.

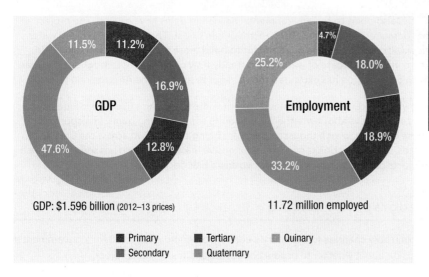

Australia's Economic Sectors, 2014–2015 With more than 75 percent of its workforce employed in the tertiary, quaternary, or quinary sectors, Australia is clearly a country with a postindustrial economy. These same sectors account for about 72 percent of Australia's GDP. ▮ Based on what you've learned about postindustrial economies, describe three trends that are likely occurring in Australia.

GDP: $1.596 billion (2012–13 prices) — Primary 11.2%, Secondary 11.5%, Tertiary 47.6%, Quaternary 12.8%, Quinary 16.9%

11.72 million employed — Primary 4.7%, Secondary 18.0%, Tertiary 33.2%, Quaternary 18.9%, Quinary 25.2%

The Growth and Diffusion of Industrialization

Case Study

Damming The Xingu River

The Issue A massive hydroelectric project in the Amazon Basin promises to provide Brazil with clean energy, but critics assert the dams will have more negative than positive effects on the environment.

Learning Objective
LO 18.10 Describe the relationship between sectors of economic production and economic development.

By the Numbers

80 percent
of the Xingu River's flow to be diverted through the Belo Monte Dam

more than 20,000
Indigenous people will lose their way of life

16 percent
of world's energy output provided by hydroelectric power

25,000
Approximate number of workers needed to build the dam

Sources: NASA, Yale University, Forbes

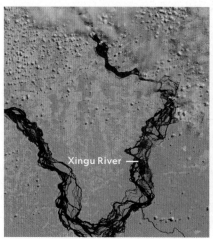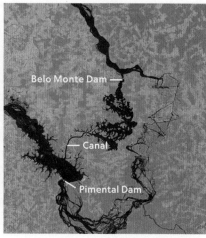

These satellite images from 2000 (*left*) and 2017 (*right*) show the Xingu River, a major tributary of the Amazon River, before and after the construction of the Belo Monte Dam. The tan and orange areas in the second image show resulting dry areas that are impacting aquatic life and Indigenous peoples in the region.

In 2011, Brazil Began Construction on the Belo Monte Dam, a massive project along the Xingu River, a major tributary of the Amazon. The government approved the effort despite opposition from environmentalists and the Indigenous peoples living along the waterway. With this huge development project, the government hopes to foster economic growth to raise people's standard of living, and—perhaps to a larger degree—to cater to powerful economic interests looking to profit from that growth.

Brazil, a newly industrializing, semi-peripheral country, has a large landmass—the fifth largest in the world—that is rich in natural resources. The country is a world leader in mining, agriculture, and manufacturing. Its service sector is growing quickly. All of this economic activity requires a strong energy industry, and Brazil has long looked to the gigantic Amazon River system to provide needed power. In fact, 90 percent of the country's electricity is provided by hydroelectric power.

The Belo Monte Dam project actually consists of two dams, which form two reservoirs that flooded the homelands of tens of thousands of Indigenous peoples. The dams have the capacity to generate more than 11,000 megawatts of electricity, which is enough to provide electricity for 60 million people, but are expected to operate at less than half that capacity most of the time. Worse, in the three- to four-month dry season, when the river's flow is substantially reduced, the dams only generate about a tenth of the peak capacity. At an estimated cost of $18 billion, this limited output seems like a waste of money to critics.

Critics also point to the environmental havoc the dams have caused. The project has damaged more than 370,000 acres of rain forest. They fear the reduced water flow will convince Brazil's government to build more dams upstream to allow for the controlled release of water and a steady flow of electricity throughout the year. Such construction has extended the damage to more of the Amazon rain forest.

Growing concerns about the environmental impact of dams like Belo Monte have led to questions about whether to dismantle some existing dams. In response to these concerns, over a thousand dams are being taken down across the United States. Deconstruction is even more extensive in Europe, which has seen 5,000 such projects come down over the past two and a half decades. ∎

Geographic Thinking

Explain how changes to a country's economic structure might affect a government's decision about whether to build a controversial construction project.

National Geographic Photographer **Aaron Vincent Elkaim**

Documenting Impacts of the Belo Monte Dam

Aaron Vincent Elkaim is committed to documenting the impact of economic development on the lives of Indigenous peoples.

Learning Objective
LO 18.10 Describe the relationship between sectors of economic production and economic development.

When National Geographic photographer Aaron Vincent Elkaim came across the story of the Belo Monte Dam project, he was not just interested in its environmental impact, he was also taken with the challenges it thrust upon Indigenous groups. "There is much we can learn from Indigenous and traditional peoples," he says. "For them, the natural world is intertwined with the sacred, and they understand the importance of creating balance and harmony with it. This perspective is increasingly important to spread in today's world."

The Juruna, the Munduruku, and other peoples have lived in the region for thousands of years, hunting, fishing, farming, and gathering food. But many have been displaced by the Monte Belo Dam project. The company building the dam and the Brazilian government have paid to relocate many traditional people. While they have built new housing developments for the displaced people, there have been few jobs available in the area, resulting in communities with high crime rates and drug and alcohol abuse.

Elkaim sees the dam as an outdated approach to the use of natural resources. "Dams are one of our most ancient technologies for producing power, but we now have the technology to do better," he says. Giant dams do "permanent and drastic damage to vital and unique ecosystems," he argues.

Critical Viewing By altering the natural environment, the Belo Monte Dam complex is damaging the land and disrupting the way of life of Indigenous groups. One group, the Juruna, gathers at a public meeting to protest the building of the dam. What are some ways that Indigenous people's way of life might be disrupted by the damming of the Xingu River?

The Growth and Diffusion of Industrialization

Critical Viewing A large portion of the Brazilian city of Altamira was flooded when the reservoirs of the Belo Monte Dam were filled at the end of 2015. The boys in the photo climb a tree above a flooded area where mostly poor neighborhoods once stood. Use details from the photo to describe how economic decisions made by a country's government can affect people living in a specific region.

18.3 Patterns of Industrial Location

The First Industrial Revolution began in Britain's Midlands region because it was located near natural resources and linked to the major port of Liverpool. But why did Pittsburgh, Pennsylvania, become "Steel City" and Detroit, Michigan, the "Motor City"? What geographic principles explain why industries are located where they are?

Workers assemble vehicles at a Ford factory just outside of Detroit, Michigan. The area was the world's center of auto manufacturing for most of the 20th century. However, by the 1990s, the city had experienced a massive loss of auto manufacturing jobs. Today, Detroit is still home to two of the so-called "Detroit Three" auto companies (General Motors, Ford) while Stellantis (formerly Fiat Chrysler Automobiles, which merged with Peugeot in 2021) is headquartered in Amsterdam.

Least-Cost Theory

Learning Objective

LO 18.12 Explain how transportation, labor, and degree of agglomeration influence the location of industries.

As you know, geographers are concerned with the "why of where." As it relates to industrialization, they might examine why one location for a factory is more suitable than another, or why a particular industry took hold in the region that it did. Is there a city or town nearby that could provide a workforce or a market for the goods produced? Or are there reasons related to physical geography and the availability of natural resources that make the choice a smart one?

In a capitalist system, location decisions are based on the profit motive. Companies invest capital, which is wealth in the form of money or assets, with the goal of generating profits. A profit is the money that remains after costs are subtracted from revenues. Capitalists seek to minimize costs so they can maximize profits. That aim influences where industries are located. To explain these decisions, German economist Alfred Weber first devised the **least-cost theory**, a model that geographers use to analyze spatial patterns in the secondary economic sector. The theory considers the factors that influence where enterprises locate manufacturing production. It proposes that businesses locate their facilities in a particular place because that location minimizes the costs of production. A firm chooses a location where the cost of moving raw materials to the manufacturing site, and finished products to the markets, will be as low as possible.

Factors That Influence Location Least-cost theory focuses on three factors that influence the decision of where to locate—transportation, labor, and degree of **agglomeration**. Agglomeration is a term that describes the advantage for companies in the same or similar industries in locating near each other in order to take advantage of specialized labor, materials, and services.

Weber considered transportation costs to be the determining factor in where an industry is located. Production is drawn to its most advantageous location relative to the cost of transport, and that site is one where those costs are lowest both in terms of bringing raw materials to the production site and distributing the final products to consumers.

Later scholars refined Weber's theory to include other factors, including labor and agglomeration. If labor costs are high in an area, profit margins are reduced. In that case, it makes sense to locate the manufacturing site farther away from the raw materials and markets as long as cheap labor compensates for the added transportation costs. The decline of the textile industry in the American South and the rise of garment manufacturing in such locations as Vietnam and Bangladesh late in the 20th century reflect these decisions. Of course, the American South had textile mills because textile manufacture relocated to that region from higher-cost facilities in the Northeast early in the 20th century.

Agglomeration is an interesting factor in patterns of location because it means that competitors often locate near one another. Geographers observe that enterprises

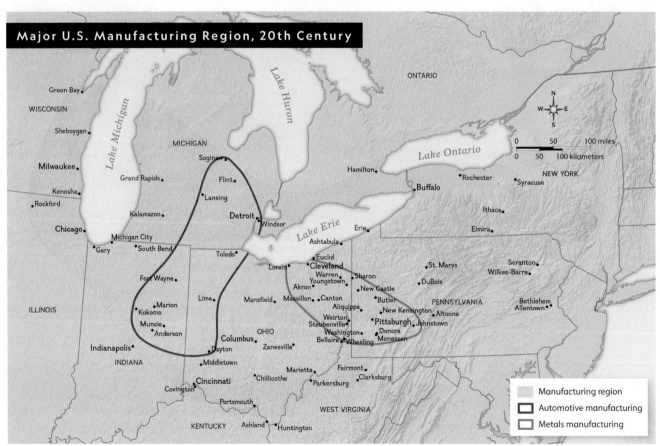

Reading Maps For much of the 20th century, the core of U.S. manufacturing was located in a swath of land along the Great Lakes. Many cities became synonymous with their major industry, such as Detroit for cars and Pittsburgh for steel. ▋ Explain how the map demonstrates agglomeration.

in the same industry tend to cluster together because it provides benefits to each enterprise. They can assist each other in controlling costs through shared talents, services, and facilities. With several enterprises located in close proximity, each has access to the same pool of people from which to hire workers and the same natural resources and transportation networks. Also, other companies that support the particular industry with needed materials or services will also be drawn to an area of agglomeration. All of these factors reduce costs, making a location more attractive when new firms are thinking of establishing a plant or factory in an area. Agglomeration can offer enough of a cost advantage that it sometimes distorts the basic spatial pattern established by transportation costs and modified by labor.

For example, the concentration of the auto industry in the Detroit area in the early 20th century demonstrated agglomeration. Henry Ford, founder of the Ford Motor Company and innovator of the assembly line process, grew up outside Detroit and opened his company there. Ford and Oldsmobile, another successful car company located in Detroit, drew skilled labor and parts manufacturers to the city, and soon it made good economic sense for other car companies to set up shop in the "Motor City." The prevalence of car manufacturers drew companies that produced related products as well, such as tire companies and parts manufacturers.

The first and most important factor in the least-cost theory—transportation—plays a key role in where agglomeration occurs. It often occurs at **break-of-bulk points**, which are locations where it is more economical to break raw materials into smaller units before shipping them farther. Break-of-bulk points are often located at places where the mode of transportation changes, for instance, where materials that have come in on a ship are being moved to rail cars. Thus, ports that receive shipments of raw materials from elsewhere and then distribute the materials to interior regions by rail or highway networks are often break-of-bulk points. Because they may receive vast quantities of raw materials that cannot be distributed all at once, these locations often develop storage facilities. The raw materials may need to be processed in some way to facilitate transportation to production sites, so these locations may become processing centers as well. For instance, many port cities such as New Orleans develop petroleum storage and refining industries because they handle, store, process, and distribute large amounts of crude oil delivered to them by tankers.

There are several assumptions in the least-cost theory. One is that it ignores the influence of economic or political systems. A second assumption is that there are fixed sources of raw materials. A third is that the workers who make up a labor force will not move. A final assumption is that there is a uniform cost of transportation from any one point to any other.

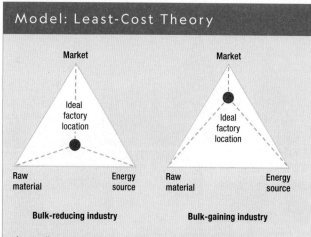

Model: Least-Cost Theory

According to the least-cost theory, companies should minimize transportation and labor costs and maximize the money that can be saved through agglomeration. Of these factors, according to the theory, transportation costs are most important. While real-world manufacturing processes are often more complex, Weber devised his theory based on a product made using one raw material and one source of energy (to power the processing plant) and having a single market. The ideal processing location is found at the intersection of lines from each corner and depends on whether the industry is bulk-reducing or bulk-gaining.

Raw Materials According to the least-cost theory, there are two significant features of raw materials. The first is related to where raw materials are found. With raw materials that are ubiquitous, or found in many places, factories can be located anywhere. The ideal location for factories that use ubiquitous raw materials is near the market. In that case, transportation costs are minimal because there are limited expenses involved in bringing the readily available raw material to the plant and in shipping the finished good to the market. Localized raw materials—those found in a particular place—tend to limit the location of processing plants to places that have that raw material. Industries that rely on raw materials that are mined from the earth, such as iron ore, coal, and bauxite (ore that is processed into aluminum), tend to locate processing plants near where those materials are found.

The second significant feature is the cost of transporting raw materials. This cost is often related to whether raw materials gain or lose weight during processing. Fuel and shipping costs make it expensive to transport heavy material over long distances. Generally, raw materials either gain or lose weight in processing, and this weight change affects the ideal location of processing plants. If there is no weight change during manufacturing, transportation costs from the source of raw materials to market will be consistent and the company can take other factors into account when deciding where to locate a factory.

The Growth and Diffusion of Industrialization

TIMBER is the name for trees and the wood before it has been processed.

LUMBER is the name for the wood once it has been sawn into planks or boards.

FROM RAW MATERIAL TO MARKET
THE LUMBER INDUSTRY

Lumber is a bulk-reducing industry, which means that lumberyards are typically located close to forests, where the raw material (wood) is located. However, there are other factors to consider. ❙ Using what you have already learned, describe factors involved in deciding where to locate a lumberyard.

1 Growth & Harvesting

Timber is harvested from forests that are naturally occurring or planted by humans.

Felling: Mature trees are cut down and saplings are planted to maintain the forest.

Trimming: Trees are delimbed and cut into uniform, manageable pieces at the felling site.

Storing: The logs are stored near the felling site to allow water stored in the wood to evaporate, which decreases the weight—and therefore the transportation costs—of the logs.

2 Transportation

The timber is transported to a processing site. Because lumber is a bulk-reducing industry, processing plants tend to be located close to forests where the wood is harvested.

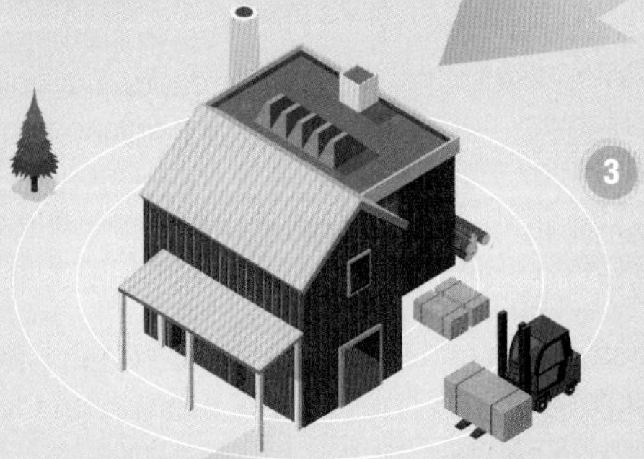

3 Processing

At the processing site—in this case, a lumberyard—the logs are cut into specific lengths and the bark is removed. The pieces are cut into boards and then dried. This process causes the wood to lose most of its water, further reducing the weight and making it less likely to warp.

4 Manufacturing

The lumber is purchased by manufacturers and made into various products.

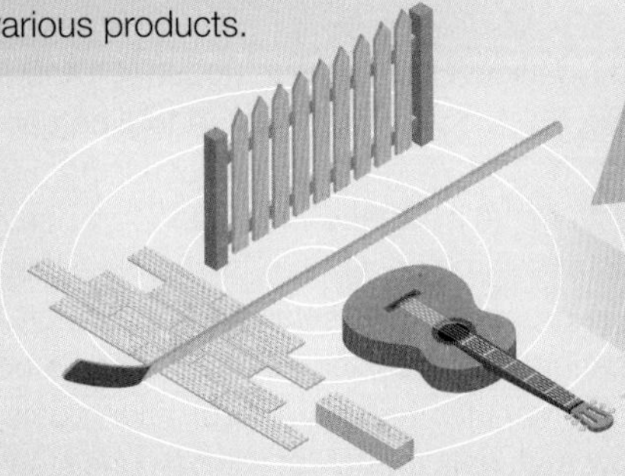

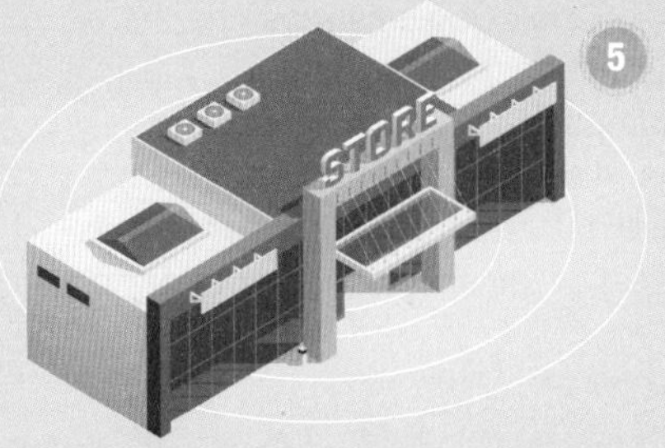

5 Market

The products are sold in stores or to other industries.

544 Chapter 18

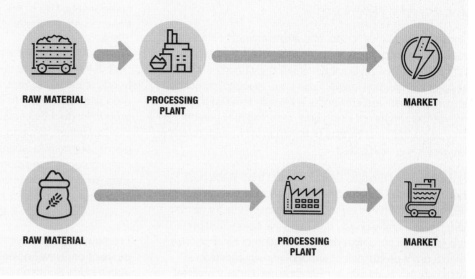

Bulk-Reducing Industry

If raw materials cost more to transport than finished goods, then processing plants will be located near the source of the raw materials.

Examples: copper smelting, furniture manufacturing

Bulk-Gaining Industry

If raw materials cost less to transport than finished goods, then processing plants will be located near the market.

Examples: car manufacturing, bread production, construction equipment

Where to Locate? According to the least-cost theory, certain characteristics of raw materials, such as whether they lose or gain weight during processing, determine where manufacturing will take place. ▌ Explain why construction equipment manufacturing is an example of a bulk-gaining industry.

In **bulk-reducing industries**, raw materials cost more to transport than finished goods. With raw materials that lose weight through the manufacturing process, the best location for the factory is near the source of the raw materials. Coal-burning power plants rely on a raw material that loses weight in processing—the coal is burned and emitted as smoke and carbon dioxide. Locating power plants near coalfields minimizes the cost of transporting heavy coal. It is relatively inexpensive to transmit electricity from the plant to markets through electric lines. The lumber industry is a bulk-reducing industry because the boards produced at a sawmill weigh less than the tree trunks brought in for processing.

In **bulk gaining industries**, raw materials cost less to transport than finished goods. If manufacturing results in a product that weighs more than the sum of its raw materials, then the best location for the factory is near the market because it costs more to ship the finished product than it does the raw materials. In automobile manufacturing, for instance, the component parts of a car weigh less than the weight of the finished product. Factors besides weight play a role as well. Bread, for instance, is more perishable than the wheat that acts as a raw material, so bread is more expensive to transport. Bread also requires specialized packaging and transportation to maintain its shape. These perishability and transportation costs make the best location for industrial-scale bakeries near the market. The same is true for bottled and canned soft drinks; a location close to the market means lower transportation costs.

Limitations of Least-Cost Theory

Learning Objective
LO 18.13 Describe the limitations of least-cost theory.

Like all theories, the least-cost theory does not line up perfectly with conditions in the real world. For instance, the theory ignores the influences of political or economic systems. Of course, political and economic systems are not uniform, and the differences can strongly influence decisions about locations. Tariffs or quotas, for instance, can cause the leaders of a business to avoid a certain market or to build a production plant where they otherwise would not have located. Japanese automakers began building cars in the United States in the 1980s not because of transportation costs but rather to avoid quotas placed on their auto exports from Japan to the United States. Countries, regions, and cities also compete with one another by offering tax breaks or other incentives to companies willing to open new factories in their territory. Additionally, the theory works in capitalist societies where profit is the primary business motive. In communist systems, industry was often located for social reasons, not based on the cost of manufacturing.

In addition, markets are usually not located at a single point. Markets for consumer goods are found wherever consumers live. Markets for goods needed by an industry are found wherever companies within that industry are located.

Location Decisions Today

Learning Objective
LO 18.14 Explain how changes in manufacturing technology and materials, and a reduction in transportation costs impact business location decisions today.

Beyond these limitations, there are other critiques of least-cost theory as Weber proposed it in 1909. Because of changes in manufacturing, raw materials, shipping costs and methods, and labor costs and characteristics, transportation costs play a less significant role in location decisions today than they did in Weber's time.

The use of airplanes, ships, and supertankers has greatly reduced transportation costs and improved supply chains. Shipping containers, like those seen in vast numbers in the chapter introduction image, have done much to increase shipping speed and efficiency. Their standardized size makes the containers intermodal, meaning they can be shipped by different modes of transportation, such as ships, barges, trains, and trucks.

Shipping costs are also less significant than in the past because many of the goods being produced weigh less than they used to, when the products of heavy industry formed a larger share of the goods and materials being transported. It is less costly to ship a day's production of smartphones than a day's production of wrought iron. In addition, materials like aluminum and plastics are lighter than steel and wood. These lighter materials have made their way into the products of heavy industry. In the average car produced in North America in recent years, about 8 percent of the total weight—more than 300 pounds—comes from plastic components. Use of this material reduces the costs of both transporting raw materials and of shipping finished cars.

While transportation costs generally have fallen, the relative cost of labor has become more significant. Two competing trends are at work. First, for sophisticated, high-tech products, labor expertise is in high demand. This limits production to those areas that have highly educated, highly skilled workers. Second, for many other mass-produced goods, semiskilled workers can do the work. In these cases, manufacturing companies have an incentive to seek out the least expensive productive workers. In the drive to lower labor costs, companies have located factories in peripheral countries with relatively low wages—and no labor unions.

For example, in the 20th century the U.S. textile industry moved from New England to lower-wage, nonunionized states in the South as factory owners sought to hold down labor costs. Decades later, after wages in that region rose, manufacturers shifted their production to other countries with low-wage workers. Today, the top ten countries for

Ningbo Bird, a Chinese mobile phone manufacturer, used a vast, single, open floor plan in its factory. It is a good example of the type of facility that characterized modern factories prior to the COVID-19 pandemic.

apparel exports include such low-wage countries as China, Bangladesh, Vietnam, India, Turkey, and El Salvador. However, concerns about human rights, geopolitical turmoil, and the impacts of COVID-19, particularly in China, are causing some manufacturers to seek new low-wage countries for their operations.

The auto industry has seen a similar shift. U.S. automakers, once concentrated in the high-wage, unionized Midwest, began opening factories in the South, where they were later joined by European and Japanese automakers wishing to establish a U.S.-based manufacturing presence. In more recent years, automakers have opened factories in Mexico and other lower-wage countries. In a sense, the least-cost theory still holds—businesses locate factories to minimize costs. The original formula of the least-cost theory, however, with its emphasis on raw materials and energy costs, does not play as large a factor when labor costs are more significant and transportation costs are less important.

With changes in products and materials, the configuration of factories has also changed. Modern factories are far more likely to be long, wide, single-story structures rather than multistory buildings. This change in factory design has contributed to the flow of manufacturing out from urban centers to **industrial parks**. An industrial park is a collection of manufacturing facilities. Industrial parks are typically found in suburbs and located close to transportation hubs such as highways and rail lines to facilitate movement of raw materials and finished products. Several factors contributed to the rise of industrial parks. The high cost and low availability of land in cities versus the lower cost and greater availability of land in suburbs plays a role. Since most workers—especially in the suburbs—rely on cars to get around, facilities can be built away from mass transit systems. Also, industrial parks have ample room for expansive factory structures, loading docks from which to load and unload trucks and rail cars, and parking space for employees.

Finally, the availability of customers on a global scale has led producers to locate plants in all corners of the world, as exemplified by the sportswear manufacturer Nike. The company employs more than 1 million workers in 41 countries located on five of the world's seven continents. In 2020, before the COVID-19 pandemic slowed production worldwide, Toyota had 51 manufacturing plants in 28 countries on five continents, where more than 370,000 workers produced nearly 9 million vehicles a year. Though these and other global companies have complicated networks of production, they will still consider least-cost theory factors in their decisions about where to locate manufacturing plants.

Geographic Thinking

1. Explain how agglomeration benefits companies that are in the same industry.

2. Compare the three factors that influence the location of industry, according to the least-cost theory.

3. Describe what has changed in manufacturing since the least-cost theory was first proposed.

Chapter 18 Summary & Review

Chapter Summary

Industry refers to economic activity that uses machinery generally on a large scale to process raw materials into products and to a collection of organizations that work with the same materials or produce similar products.

- The manufacturing industry is divided into two categories: heavy industry and light industry.
- Industrialization is the process by which the interaction of social and economic factors causes the development of industries on a wide scale.

The First Industrial Revolution marked the shift from small-scale, hand-crafted production to power-driven mass production.

- It began in the textile and iron industries in Great Britain in the mid-18th century.
- The First Industrial Revolution contributed to urbanization and transformed society and daily life.

Economic activities are categorized into five economic sectors.

- The primary, secondary, and tertiary sectors are involved with obtaining raw materials, processing them into products, and providing services.
- The quaternary and quinary sectors, specialized parts of the tertiary sector, involve using information and decision making.

Different levels of development reflect distinct patterns of these economic sectors.

- The primary sector is dominant in preindustrial societies.
- Strong secondary and tertiary sectors are present in fully industrialized societies.
- A decline in manufacturing and a rise in the quaternary and quinary sectors are typical of postindustrial societies.

Geographical models help geographers understand ideal locations for industries.

- The least-cost theory suggests that location decisions are based primarily on transportation costs.
- Changes in manufacturing technology and materials, plus a dramatic reduction in transportation costs, have often made labor costs a more significant factor in business location decisions today.

Review Questions

Use complete sentences to answer the questions.

1. **Apply Conceptual Vocabulary** Consider the terms *heavy industry* and *light industry*. Write a standard dictionary definition of the terms. Then provide a conceptual definition—an explanation of how each term is used in the context of this chapter.

2. Explain why heavy industry tends to be large scale while light industry can be either large or small scale.

3. Which employers—master artisans, shipyard owners, or urban merchants employing rural workers in cottage industries—were most likely to become factory owners during the Industrial Revolution? Explain your answer.

4. Why were innovation and energy sources both central to the First Industrial Revolution?

5. Describe how the first industrial regions of continental Europe and the United States were similar to the British Midlands.

6. Explain how imperialism and colonialism resulted from and contributed to industrialization.

7. Explain how the First and Second industrial revolutions of the 18th and 19th centuries transformed society.

8. Why have the quaternary and quinary sectors become more prominent in recent decades?

9. Define what level of development—preindustrial, industrial, or postindustrial—is present in a country with a workforce that is 60 percent engaged in agriculture, 10 percent in industry, and 30 percent in providing services. Explain your response.

10. Explain what level of development is present in a country with a workforce that is 1 percent engaged in agriculture, 10 percent in industry, and 89 percent in providing services, with about a third of that latter group in the quaternary and quinary sectors.

11. Why is it likely that all industrializing societies would go through a dual-economies phase?

12. Describe why the least-cost theory is most suitable to explaining the location of production facilities in traditional heavy industry.

13. Describe the factors that make modern manufacturing location decisions not conform perfectly to the least-cost theory.

Interpret Maps

Study the map and then answer the following questions.

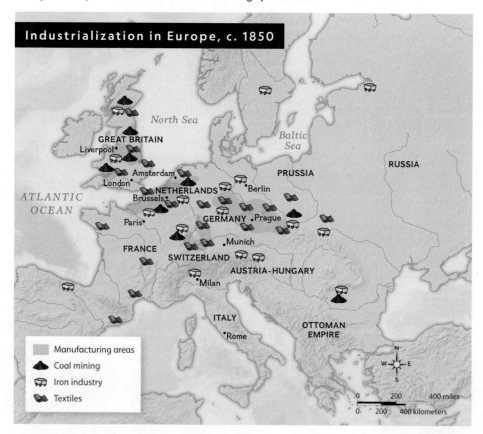

14. **Analyze Maps** Where was industrialization concentrated in Europe by 1850?

15. **Connect Visuals & Ideas** Describe how the map reflects what you read in the chapter about the Industrial Revolution.

16. **Explain Spatial Relationships** Explain why textiles are found in more dispersed locations than other industries on the map.

17. **Connect Visuals & Ideas** Identify the countries and empires shown as either core, semi-periphery, or periphery. Explain your response.

Chapter 19
Measuring Human Development

Critical Viewing Students share a smile in a classroom at Umar Al Farooq School in Mandera, Kenya. How might educational opportunities affect a country's overall development?

Geographic Thinking How do geographers measure human development?

19.1
How Is Development Measured?

19.2
Measuring Gender Inequality

Case Study: "Women-Only" Cities

National Geographic Photographer Amy Toensing

19.3
Changing Roles of Women

Case Study: The Development of the Grameen Bank

19.4
Theories of Development

19.1 How Is Development Measured?

Geographers use economic and social indicators to describe the conditions under which the people of a particular country or region live. These factors can gauge how and why the standard of living—the level of available wealth, comfort, material goods, and necessities and the way they are distributed within a population—in a location changes or remains the same over time.

Economic Indicators

Learning Objectives

LO 19.1 Compare the indicators of economic development including Gross Domestic Product (GDP), Gross National Product (GNP), and Gross National Income (GNI).

LO 19.2 Differentiate between the formal and informal economic sectors.

Development in the general sense is the process of change in the nature and activities of the economy of a region or country and subsequent increases (or decreases) in the prosperity of that place. This chapter focuses on **human development**—the processes involved in the improvement of people's freedoms, rights, capabilities, choices, and material conditions. Although the development of individuals is important, the concept of human development in this case applies to the collective population of a country or region. Geographers look at economic and social indicators in categories such as available resources, productivity, and education to measure the standard of living of a population.

As you have learned, geographers categorize the overall development of countries as core, periphery, or semi-periphery. Core countries represent the most diversified economies, highest levels of education, greatest levels of productivity, most advanced technologies and infrastructure, best health-care systems, and highest standards of living. Countries in the periphery have less-diverse economies, lower levels of education and productivity, insufficient health-care systems and infrastructure, and lower standards of living. Between the core countries and peripheral countries are semi-peripheral countries, which share some elements of both. Countries in the semi-periphery have growing economies (often through industrializing); improving infrastructures, health-care, and education systems; and standards of living that fall between those of core countries and peripheral countries.

Every country can be placed into one of these categories. However, within countries, there may be distinct regional differences in standards of living. In Unit 6, you learned that standards of living vary even within communities. Country-level data thus masks unequal development within countries. This information can be better understood by changing the scale of analysis.

Income Economic welfare, as measured by the wealth or income of a country, is not the sole indicator of overall well-being. But it is a crucial quantitative measure that can lead to an understanding of a country's level of development and thus standard of living.

Ways to measure a population's wealth include:

- **Gross domestic product (GDP)**, the total value of the goods and services produced by a country's citizens and companies within the country in a year
- **Gross national product (GNP)**, the total value of the goods and services produced by a country's citizens and companies, both domestically and internationally, in a year
- **Gross national income (GNI)** per capita (per person), the total value of goods and services globally produced by a country in a year divided by the country's population

GDP and GNP can also be converted into per capita data. For example, GDP per capita is the gross domestic product divided by a country's midyear population. Per capita estimations help put a country's overall well-being in perspective. GNP—which excludes products manufactured by other countries in a host country as well as earnings and spending of nonresidents—reflects a country's real income. However, GNP is often not the most accurate measure, as it is affected by international exchange rates. The United States stopped using GNP as an indicator in 1991, and the World Bank prefers to use GNI per capita when considering the health of a country's economy.

The measure of GNI per capita is calculated using earned income rather than production output. Countries that rank high on the World Bank list feature a high amount of international investment and foreign aid as well as a large number of international corporations. The variation between a country's GDP and GNI can be very small if the difference between a country's income and its global payments is minimal. For example, the GNI per capita of the United States was only 1.5 percent higher than the U.S. GDP per capita.

Though economic welfare is an essential measure of a population, income distribution provides insight into a society's overall health. Some high-income economies

GNI PER CAPITA IN U.S. DOLLARS (2020)	
Switzerland	$82,620
Luxembourg	$81,110
Norway	$77,880
Ireland	$65,750
United States	$64,140
Denmark	$63,010
Iceland	$62,410
Qatar	$55,920
Singapore	$55,010
Sweden	$54,290
Australia	$53,680
Netherlands	$51,070
Finland	$50,080
Austria	$48,360
Germany	$47,520
Belgium	$45,810
Canada	$43,540
Israel	$42,610
New Zealand	$41,480
Japan	$40,810
United Kingdom	$39,970
France	$39,500
United Arab Emirates	$39,410
South Korea	$32,930
Italy	$32,380

One way to measure the economic welfare of a country is gross national income (GNI) per capita. GNI per capita is the total value of goods and services produced globally in a year by a country in a year divided by the country's population.
Source: World Bank

seem prosperous, but uneven distribution of income may mask high levels of poverty. Brazil, for example, has a dramatic disparity between its wealthiest citizens and the rest of its residents. Brazil's six richest individuals have the same amount of wealth as the poorest 50 percent of the population (around 100 million people). In 2019, more than 9 percent of Brazil's population lived on less than the equivalent of $3.20 per day, which was a 2 percent increase from the rate just five years earlier. Countries with a more even income distribution have less poverty and a greater sense of equity within the population, which often results in political stability and a healthier populace.

Economic Structure The structure of an economy is another measure of development. Economic structure is connected to economic prosperity, and the level of development of a country can be predicted by examining the country's economic structure. Most diversified economies result from populations who work in every economic sector, which you learned about in Chapter 18. As countries develop, the percentage of the population operating in different sectors of the economy often shifts; for instance, more workers move from the primary sector (agriculture) to the secondary sector (manufacturing). This shift generally represents an increase in productivity, which in turn strengthens a country's economic success. More development enhances growth for jobs in the tertiary sector (services), which also adds to a country's economic prosperity. As countries flourish, the quaternary and quinary sectors (knowledge sectors) bloom. Information technology expands, scientific research is promoted, and media extends to the farthest reaches of the country.

In addition to classifying a country according to these economic sectors, the structure of an economy can also be broken into two categories: the formal sector and the informal sector. The **formal sector** includes businesses, enterprises, and other economic activities that have government supervision, monitoring, and protection and are taxed. The taxes collected from businesses and workers most often are used by a country to finance a variety of public services.

In the United States, money collected by the government funds Social Security (a federal system that provides retirement, disability, and survivors' benefits), health insurance programs such as Medicare and Medicaid, defense and international security systems, safety net programs that provide aid to its populations in need, and interest on money the U.S. government has borrowed from banks. A small portion of these funds goes to public services like education, environmental programs, food and drug safety, and infrastructure. Workers in this sector have a formal contract with their employer, predefined job responsibilities, a guarantee of safe work conditions, a fixed duration of work time, wages or a salary, and coverage by Social Security for health and life risks, such as becoming disabled and not being able to work.

The **informal sector**, sometimes called the informal economy, is any part of a country's economy that is outside of government monitoring or regulation and is not taxed. Individuals and businesses in the informal sector typically deal in cash and include a wide range of money-making activities: street vendors hawking flowers or candies, unlicensed and unregulated food or beverage stands, cleaning or moving services, and many more. In contrast, some street vendors—such as food trucks in the United States and certain taco stands in Mexico—are in the formal sector because they are licensed and pay location fees and sales taxes. But workers in the informal sector do not have formal employment contracts, regulated work conditions, or fixed hours of work. They may not be paid regularly or evenly and are not covered by any worker health and safety protections or social security system. These workers are sometimes in less-than-safe working situations and are especially vulnerable to downturns in the economy.

The back of a pick-up truck serves as a fruit stand on the street in the historic center of Mexico City. Street vendors are often part of the informal economic sector.

It's difficult to accurately measure the financial or social impact of informal-sector activities. Without government interaction, many of these economic activities aren't documented and can't be traced. Likewise, the informal sector is not included in the GDP or GNI of a country. As a result, many countries surely have greater income than their official statistics suggest. Informal economies are active all over the globe. In core countries, the informal sector represents anywhere from 10 to 20 percent of a country's income. Phrases such as "under the table" and "off the books" reflect this untaxed, cash economy. In peripheral countries, the informal sector is much larger, representing as much as 50 percent of a country's income. Although critics claim that the informal sector is unmanageable and can hinder a growing economy, other experts argue that it provides financial opportunities for people at the bottom of the socioeconomic ladder, and most say that it is growing.

Fossil Fuels and Renewable Energy Measuring the use of fossil fuels (hydrocarbon-containing material of biological origin that can be burned for energy) or of renewable energy (energy from a source that is not depleted when used) can indicate a country's level of development. Fossil fuels, like coal, oil, and natural gas, were introduced during the Industrial Revolution. These fuels have powered electricity-generating plants, factories, businesses, homes, and motorized vehicles of all kinds. In the approximately 250 years since the Industrial Revolution began, the world has come to rely on fossil fuels for its main energy supply.

All countries use fossil fuels for electricity, transportation, heat, and fuel for vehicles. A recent analysis showed that China, the United States, and India were the top three consumers of fossil fuels. More than 29 countries depend on fossil fuels for at least 90 percent of their energy, and 5 countries—Brunei, Kuwait, Oman, Qatar, and Saudi Arabia—are 100 percent reliant. These countries have large deposits of fossil fuels, which gives them primary access to these sources that drive their economies. When making energy decisions, countries utilize their natural resources or what they can purchase to generate the power they need. But faced with climate change, largely the result of rising carbon dioxide levels produced by burning fossil fuels, countries are committed to exploring ways to use renewable energy and alternative fuel sources. China is the biggest generator of global wind power and also produces the most solar energy in the world. The United States ranks second in wind-generated electricity and fifth in harnessing solar energy. Considering that fossil fuels are nonrenewable and could eventually become depleted, and the fact that burning fossil fuels creates pollution and drives climate change, planning for and implementing renewable energy programs improves a country's energy infrastructure and development.

Workers in Iran assemble panels at a solar power farm. Some countries like Iran that have historically been heavily dependent on fossil fuels have, in recent years, increased their use of renewable energy.

Social Indicators

Learning Objective
LO 19.3 Explain how fertility rates, infant mortality rates, access to health care, and literacy rates serve as indicators of development.

Geographers also use noneconomic factors, such as fertility rates, infant mortality rates, literacy rates, life expectancy data, access to health care, and measures of democracy, to help complete the analysis of a country's level of development. These social, cultural, and political factors may not directly affect the level of national income and output, but they are the cause and the effect of a population's well-being.

As you have read, the total fertility rate (TFR) is the average number of children one woman in a given region or country will have during her childbearing years. Social factors that affect a country's TFR include level of health care and level of education for women. The total fertility rate is high in most peripheral countries, and although it has been trending down in recent years, the TFR still remains higher than the rates in core and semi-peripheral countries. In the 1990s, Afghanistan's TFR was about 7.5, one of the highest in the world at the time. The country's fertility rate in 2021 before the Taliban took control was 4.5.

Improvements in health care, sanitation, and diet—along with better access to hospitals and medicine—generally cause a decline in the number of births. But when access to health care and education is limited, fertility rates are much higher. In 2021, the countries with the highest TFRs were Niger (6.8), Somalia (6.0), the Democratic Republic of Congo (5.8), and Mali (5.8). All four of these countries in the periphery had fewer than 0.2 doctors per 1,000 people. In contrast, there were 4.0 doctors per 1,000 people in Argentina—a semi-peripheral country—the same year, and its TFR was 2.3.

Cultural factors, such as the status of women and religions or traditions that encourage couples to have large families, affect TFRs as well. Women with higher levels of education and those who pursue careers generally have fewer children. In countries with fewer formal employment opportunities or where women face barriers to work outside the home, women are likely to have more children. Political factors, such as government-sponsored family-planning efforts, can impact the TFR and population growth, too.

You learned that the infant mortality rate (IMR) is the number of deaths of children under the age of 1 per 1,000 live births. The IMR is a good indicator of maternal and infant health, which itself is a reliable gauge of quality of health care. Globally, the leading causes of infant deaths are preterm birth complications, birth defects, and diseases such as pneumonia, diarrhea, and malaria. Leading causes of infant deaths in the United States include birth defects, preterm births, and low birth weights. Like total fertility rate, the infant mortality rate is higher in peripheral countries and lower in core countries; Chad's IMR was 85.4 in 2020, while the United States had an IMR of 5.8 that same year. High IMRs are connected to higher percentages of people who live in poverty. There are regional patterns within countries as well, corresponding to rates of poverty. In 2020, Mississippi's IMR was 8.27, highest among states in the country, followed by Louisiana (7.5), West Virginia (7.45), Arkansas (7.33), and Alabama (7.18). In the same year, California's IMR was 3.69, with Massachusetts (3.79), New York (3.86), New Jersey (3.95), and Minnesota (3.98) also under 4 infant deaths per 1,000 live births.

Life expectancy, the average number of years a person is expected to live, is a powerful indicator of health from prenatal to elderly care. And access to comprehensive, quality health care is a dependable measure of a country's standard of living. When a population can maintain good health, prevent and manage diseases, and reduce disabilities and premature death through attainable health care, its citizens can work, produce, and create income, which in turn creates a healthy economy. In peripheral and semi-peripheral countries with low per capita income and often-fractured governments, health care for many of their residents is difficult to obtain. Countries in which people do not have access to immunizations, medications, antibiotics, and sanitary water are often economically underdeveloped.

Literacy rates and education standards are important measures of development as well. Literacy provides people with a set of skills required to function effectively in contemporary societies, and it also empowers them intellectually. Higher levels of literacy can drive the successful economic development of a country, especially in today's rapidly changing, technology-driven world. However, there are also other factors at work. Some countries prioritize education, even if other measures of their development are not high. Less-developed countries such as Cuba, Barbados, and Uzbekistan have 100 percent literacy rates, that is, all of its residents can read and write. In the United States, 86 percent of the population is literate, while in Pakistan and Nigeria about half their populations are literate. And while global literacy rates are higher than ever, challenges remain. Many marginalized children who are poor or live in rural areas do not have access to educational opportunities. For example, in Africa, countries such as Niger and South Sudan, with low income levels and a mostly rural population, struggle with literacy rates below 30 percent.

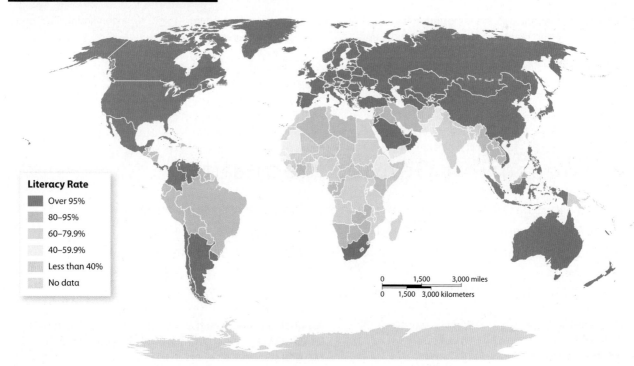

Reading Maps This map reveals percentages that measure the number of people 14 years and older in each country who are able to read and write. Though more recent data for some countries exist, the map contains the most comprehensive global information for literacy rates. ■ Choose a country and explain how its literacy rate might impact the economic development of that country.

Measuring Human Development

Human Development Index

Learning Objectives

LO 19.4 Explain how the human development index is calculated and how it measures development.

LO 19.5 Define purchasing power parity as a tool to refine economic development indictors used in human geography.

The United Nations (UN) uses the **Human Development Index (HDI)** to determine overall levels of development of countries. This measure incorporates three key dimensions of human development: life expectancy at birth (health dimension), access to education measured in expected and mean years of schooling (education dimension), and standard of living measured by GNI per capita (economic dimension). In short, the HDI is a summary measure—scaled from 0.0 to 1.0—of basic achievement levels reached through an average of each examined dimension.

When determining HDI, health and education indicators are just as important as economic indicators. In 2019, the United States had a GNI per capita of $63,826, and its HDI ranking was 17 out of 189 countries and territories. Countries such as Germany ($55,314), Sweden ($54,508), and Australia ($48,085) all had significantly lower GNI per capita. But since these countries have higher life expectancies than the United States, all three had higher HDI rankings—in the top 10 of the 189 countries. Qatar had the second highest GNI per capita at $92,428 in 2019, but its HDI was ranked 45 because of fewer years of schooling (a mean of 9.7 years). Cyprus's GNI per capita was only $38,207, but Cyprus's HDI rank was 33 because of high life expectancy and more years of schooling (a mean of 12.2).

When looking at these rankings, it's important to understand the theory of purchasing power parity (PPP). This concept measures economic variables in different countries so that exchange rates don't distort across-the-board comparisons. In essence, PPP is synonymous with "international dollars."

Although the HDI score is an important measure of spatial variation in levels of development among states, it has limitations. The HDI is a simplified calculation and doesn't capture every aspect of human development. It doesn't reflect other quality-of-life factors, such as poverty, gender equality, environmental quality, sustainability, or an overall feeling of security or even of happiness. The HDI also doesn't take into account that in many countries, some groups—perhaps ethnic minorities, followers of a minority religion, or speakers of a minority language—don't have access to the same opportunities for income, education, or health care. Additionally, the HDI does not consider the political dimensions of human development such as voting rights, equal representation in elected political positions, or equitable access to government services.

The UN and other organizations also collect and share other data—such as a country's rate of economic growth, expansion of employment opportunities, and the success of initiatives undertaken—to help leaders and policymakers evaluate and work to improve quality of life within a country.

Geographic Thinking

1. Define each of the following measures and its importance: GDP, GNP, and GNI per capita.
2. Describe the formal and informal sectors of an economy and explain the role of government in each.
3. Explain why a country might have conflicting GDP and HDI measures.

19.2 Measuring Gender Inequality

Gender inequality is based on distinctions between men and women on an assortment of variables, including education, politics, and income. These differences can have a negative impact on human development, especially for women, girls, and families.

Gender Disparities

Learning Objectives

LO 19.6 Compare how the Gender Development Index (GDI) and the Gender Inequality Index (GII) are used to track gender inequality.

LO 19.7 Explain gender equality as an indicator of development.

Although the words *parity*, *equality*, and *equity* are related terms, they are not synonyms. *Parity* is a balance between two groups. *Equality* refers to the same level of resources and opportunities for everyone, no matter the location or situation. *Equity*, however, is about fairness, and it acknowledges how the lack of access to opportunities and

Human Development Index, 2019

The HDI combines data about life expectancy, expected and average years of school, and gross national income per capita, resulting in a score that ranges from 0.0 to 1.0. Higher HDI values are considered better. ■ Identify the health, education, and economic dimensions of the HDI.

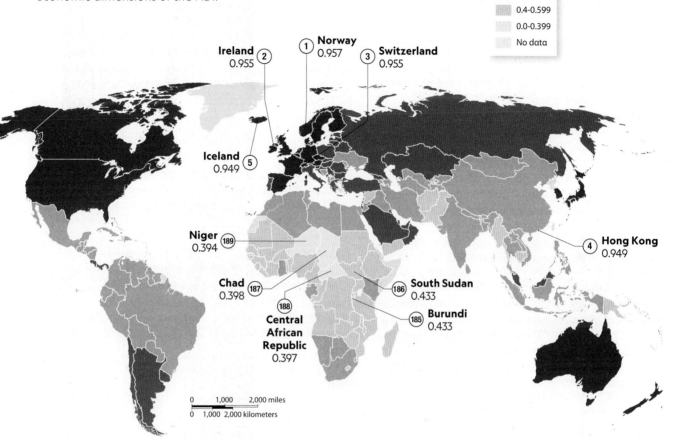

		Life expectancy at birth (years)	Expected years of schooling (years)	Mean years of schooling (years)	GNI per capita (2017 PPP $)
COUNTRIES WITH HIGHEST HDIs	1 Norway	82.4	18.1	12.9	66,494
	2 Ireland	82.3	18.7	12.7	68,371
	3 Switzerland	83.8	16.3	13.4	69,394
	4 Hong Kong	84.9	16.9	12.3	62,985
	5 Iceland	83.0	19.1	12.8	54,682
COUNTRIES WITH LOWEST HDIs	185 Burundi	61.6	11.1	3.3	754
	186 South Sudan	57.9	5.3	4.8	2,003
	187 Chad	54.2	7.3	2.5	1,555
	188 Central African Republic	53.3	7.6	4.3	993
	189 Niger	62.4	6.5	2.1	1,201

Measuring Human Development

certain resources affects underserved people, groups, or communities.

Ensuring equity means providing additional aid to make sure everyone is treated fairly in all circumstances, according to their needs. So, an indicator that compares gender parity between males and females is not equivalent, or similar in value, to one that compares gender equality, which ensures that men and women have equal access to opportunities and resources. And according to the United Nations Educational, Scientific and Cultural Organization (UNESCO), gender equity is "fairness of treatment for both women and men, according to their respective needs."

The level of gender equality can be a measure of a country's overall level of development. Groups such as the UN assess gender parity by looking at wages, educational opportunities, and gender rights—including civil liberties and access to voting. The UN uses two measures to track gender inequality. The first is the **Gender Development Index (GDI)**, which calculates gender disparity in the three basic dimensions of human development: health, education, and standard of living. It measures the female HDI as a percentage of the male HDI. Current GDI calculations show factors in which a difference between women and men exists as well as where near equity is achieved.

The second measure of gender inequality is the **Gender Inequality Index (GII)**, which calculates inequality based on three categories: reproductive health, empowerment, and labor-market participation. The GII ranges from 0.0 to 1.0; 0.0 shows that men and women share equal roles, and 1.0 shows that women have little equality. In 2019, Afghanistan had a GII score of 0.655, whereas Belgium scored 0.045.

Mexico came in at 0.322, Canada had 0.08, and the United States had 0.204.

The World Bank and other international entities determined that gender equality contributes to the overall development or growth of an economy and is beneficial to every country. Several countries have introduced policies to promote gender parity. In 2018, Iceland became the first nation to make pay inequality illegal and currently imposes a fine on companies that don't comply. Rwanda, too, made moves toward gender parity by creating inclusive economic policies. However, Rwanda is an exception to most economically disadvantaged countries. Countries that dedicate the least amount of attention toward gender parity are usually peripheral, such as Papua New Guinea (.725), Iraq (.577), and Pakistan (.538).

Reproductive Health The GII is the first major index to include reproductive health indicators as a measure of gender inequality. It uses two barometers related to women's reproductive health: one measures maternal deaths related to childbirth and the other measures births among adolescent mothers.

The maternal mortality ratio (MMR) is the number of mothers' deaths per 100,000 live births. The MMR is considered a good indicator of women's access to health care because maternal deaths usually result from a lack of adequate care before, during, and just after childbirth. According to a 2019 report compiled by the World Health Organization, UNICEF, the UN Population Fund, and the World Bank, the worldwide MMR fell by 38 percent between 2000 and 2017, from 342 deaths to 211 deaths

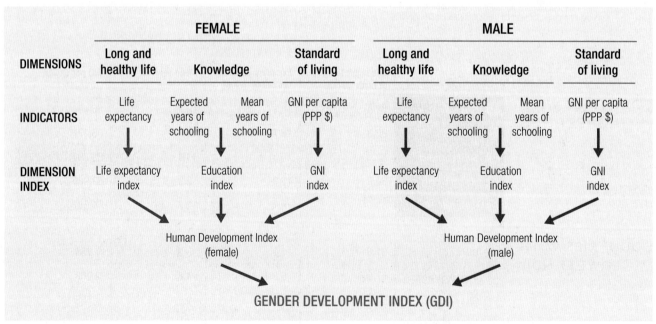

The Gender Development Index (GDI) includes the same key dimensions as the Human Development Index. By comparing these measures between women and men, the GDI reveals gender gaps in human development.

Source: United Nations Development Programme

per 100,000 live births. During that time period, South Asia and Africa South of the Sahara had substantial ratio reductions of 59 percent and 39 percent, respectively. While these improvements are impressive, these two regions accounted for 86 percent of maternal deaths around the world, with Africa South of the Sahara suffering an MMR of 533.

The adolescent birth rate (ABR) is the number of births per 1,000 women aged 15 to 19. Early childbearing is associated with increased health risks for mothers and infants as well as keeping young mothers from accessing higher education. In 2019, the global average was 42 per 1,000 15- to 19-year-old women. The countries with the highest ABRs were Niger (180), Mali (165), and Chad (155), all African countries straddling the Sahara. On the other end of the spectrum, South Korea's ABR was 1, and that of Switzerland, Hong Kong, and Singapore was 3. The United States' adolescent birth rate was 17.

Empowerment Women's empowerment includes women's options and access to participate fully in the social and economic spheres of a society. The GII uses two indicators to measure women's empowerment: political representation and educational attainment. Political representation, which concerns women's civic involvement, is measured by the ratio of women with seats in government compared with men. The Inter-Parliamentary Union tracks this information.

Women have traditionally been greatly outnumbered by men at every level of government. Gender inequality in the political sphere has improved over the last 25 years, but women are still underrepresented in decision-making political roles. Progress is occurring, albeit slowly. In 1995, 57 women held seats in the U.S. Congress, and by 2018, the number more than doubled to 127 women. By 2022, that figure had risen to 147 women serving in the House of Representatives and Senate, or 27.4 percent of Congress. Globally, the number of women in national legislatures increased from 11 percent in 1995 to 24.3 percent in 2019. However, 27 countries have fewer than 10 percent female participation in their parliaments. The majority of these countries are peripheral countries in Africa and Asia.

The countries that do have higher numbers of women in important political seats are a compelling mix. Core countries top the list, illustrated by the Scandinavian countries that have consistently had women holding around two-fifths of its seats in the legislatures since 2000. Interestingly, around 30 percent of the top 30 countries with the highest percentage of women in government are countries that have recently ended wars.

Rwanda is a strong example of this. Every year since 2004, Rwanda has had more women in its parliament than any other country in the world. Rwanda is not economically strong, but it experienced a social upheaval in the 1990s that caused this gender shift.

In the devastating genocide of 1994, approximately 800,000 Rwandans were killed; most were men. The president at the time decided that women would be the key to rebuilding the country, and in 2003, he added a decree to the constitution stating that at least 30 percent of political seats would be filled by women. Clearly, the roles of women changed in Rwanda as the country redeveloped after a crippling tragedy. Having women in decision-making roles tends to help a country improve its social and economic inequalities.

By 2022, underrepresentation of women remained an issue in every region and country of the world, though some countries saw progress in closing the gender gap in political representation. The only three countries that have reached or surpassed gender parity in their governments are Rwanda, Cuba, and Bolivia. According to the Pew Research Center, seats held by women in Nordic countries average around 40 percent, with Sweden at the high end (47 percent) and Denmark on the low end (37 percent). These countries score well on gender equality indices. At the other end of the spectrum, 28 countries throughout the world still fall short when it comes to political equality, with women making up less than 10 percent of their legislatures. Yemen, Oman, and Haiti have less than 3 percent female representation in their national parliaments. With the Taliban taking back control of Afghanistan in 2021, it appeared certain that there would be no women active in the country's government.

Educational attainment, the second indicator of women's empowerment, is measured by the ratio of adult women and adult men (age 25 and older) with some secondary education. Studies show that women's access to education affects their social and economic opportunities as well as their health outcomes. UNESCO tracks this data by country and region. Although there are some advances in educational achievement, UNESCO points out that it still sees disparities. More girls than boys remain out of school; according to UNESCO's Institute for Statistics, 16 million girls will never set foot in a classroom. An alarming 9 million of these girls are in Africa South of the Sahara.

In addition, women account for two-thirds of all adults who lack basic literacy skills. In recent years, the majority of illiterate women ages 15 to 24 came from nine countries: India, Pakistan, Nigeria, Ethiopia, Bangladesh, Democratic Republic of Congo, United Republic of Tanzania, Egypt, and Burkina Faso. UNESCO reports that the many obstacles preventing women and girls from participating in education include poverty, geographic isolation, minority status, disability, early marriage and pregnancy, gender-based violence, and traditional attitudes about the status and role of women. Women who are able to complete a higher level of education see an expansion of their freedom because their education strengthens their capacity to question, to reflect and act on their condition, and to enter the workforce with more knowledge and skills.

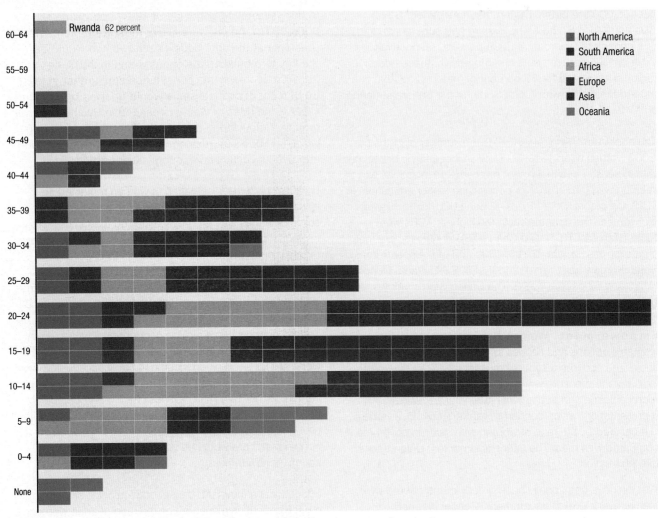

Percentage of Women in National Legislatures, 2019 The percentage of women in government has risen in recent decades, but women continue to lag significantly behind men in the political sphere in most countries. Each rectangle in this chart represents an individual country from one of the six world regions.
Source: National Geographic, November 2019

Labor-Market Participation The **labor-market participation (LMP)** rate measures an economy's active labor force and is calculated by taking the sum of all employed workers and dividing that number by the working-age population. (The LMP is also known as the labor-force participation, or LFP, rate.) Identifying participation in formal and informal sectors allows geographers to make generalized assumptions about other important social indicators, such as the availability of education and health care, infant mortality, and gender equality. The conclusion that a high labor-force participation rate equals a highly economically developed country is not necessarily true. For example, Mozambique has one of the highest LMP rates in the world, but it doesn't have a strong economy. Also, the rates of LMP vary between genders in many countries. Male LMP tends to be high whereas female LMP tends to be low, and the variation between these two rates is even more pronounced in countries that practice distinct gender roles.

The roles that women play in a country's workforce often reflect that country's level of economic development. Core countries usually have more women working full-time jobs outside the home. Women in core countries are able to pursue careers in the tertiary sector that require a college education. Many women in peripheral and semi-peripheral countries participate mainly in the primary and secondary sectors, do work to provide food for the family, or perform unpaid work, often in the informal sector. These same groups of women also have high birth rates. Both the labor-intensive jobs and high birth rates are factors that inhibit women's ability to participate fully in economic and social

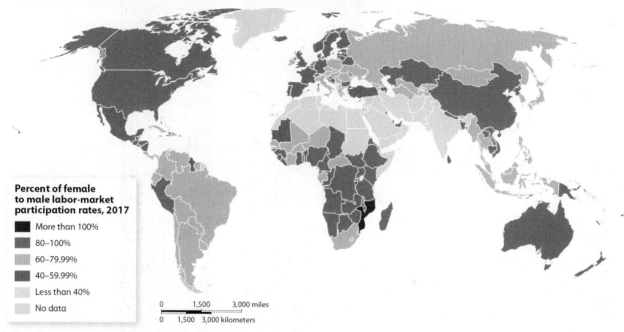

Ratio of Women to Men in the Labor Market, 2017

Percent of female to male labor-market participation rates, 2017
- More than 100%
- 80–100%
- 60–79.99%
- 40–59.99%
- Less than 40%
- No data

Reading Maps Percentages on the map show the ratio of females to males in the labor force. The labor-market participation (LMP) rate is measured as the proportion of members of the population who are 15 years and older who are economically active. ▮ In which regions of the world are female LMP rates highest? Choose one country and explain whether its LMP reflects that country's level of economic development.

spaces—which in turn holds back a country's development. However, especially in semi-peripheral countries, such as Brazil and Nigeria, there is a growing middle class and more women are participating in the tertiary sector as entrepreneurs and educators and in health care.

At the same time, female labor-market participation does not solely describe a country's economic development, as seen in some African countries in the periphery and semi-periphery. In South Sudan, Mozambique, Rwanda, and Burundi, the ratio of female–male LMP is close to, or even slightly above, 100 percent. How can more than 100 percent of women be in a country's workforce? These data reflect gender parity in labor-force participation or the fact that more women than men are participating in the labor market in that country. At the same time, female LMP is also high in some of the richest countries in the world. Interestingly, the female LMP is lowest in countries that are right in the middle with average national incomes.

An additional gender disparity related both to work—and to everyday life—is that in the face of natural disasters and other dangers induced by climate change, women are more vulnerable, resulting in higher mortality rates and lowered life expectancy compared to men. This is particularly evident in women of low socioeconomic status and is predicted to become worse in the future.

Geographic Thinking

1. Define the following terms: *gender parity*, *gender equality*, *gender equity*.
2. Identify what the Gender Development Index measures.
3. Describe what the Gender Inequality Index measures.
4. Explain the similarities between the maternal mortality ratio and the adolescent birth rate.
5. Compare the Gender Inequality Index with the Gender Development Index.
6. Explain how female labor-market participation can affect a country's economic development.

Case Study

"Women-Only" Cities

The Issue The Kingdom of Saudi Arabia's conservative rules about what women can and cannot do have recently changed, and Saudi women are taking advantage of these opportunities.

Learning Objectives
LO 19.7 Explain gender equality as an indicator of development.
LO 19.8 Explain how gender roles create geographies of separation and seclusion.

By the Numbers

100:98

ratio of men to women who participated in health-sector employment in Saudi Arabia in 2019

100:92

ratio of men to women who participated in education-sector employment in Saudi Arabia in 2019

100:42

ratio of men to women who participated in business-sector employment in Saudi Arabia in 2019

100:13

ratio of men to women who participated in legislative- and regulatory-sector employment in Saudi Arabia in 2019

Source: General Authority for Statistics, Kingdom of Saudi Arabia, 2019

Women work at a company in Riyadh that, in partnership with the Ministry of Labor, places women in jobs both virtually and in offices around the country.

In the Late 20th Century, some people in Saudi Arabia started to question the social restrictions its women faced in their everyday lives. They expressed interest in ratifying a UN-created international bill of rights for women, called the Convention on the Elimination of All Forms of Discrimination Against Women (CEDAW). By 2005, a new king came to power who agreed to begin reforming the role of women in Saudi society.

Until 2013, Saudi women couldn't acquire an ID card for themselves without getting permission from a man. They had a few basic rights—some of which, like education, had just recently been granted. The country opened the first school for girls in 1955 and the first university for women in 1970. Other restrictions on Saudi women started dissolving, however, and women began achieving more and more firsts.

By 2009, the first woman was appointed as a minister in government. Female athletes first competed on the Saudi Arabian team in the 2012 Olympics. Women were allowed to ride bikes and motorcycles in recreational areas—with male chaperones—in 2013. Women gained the right to vote in 2015. And in 2018, women were finally allowed to get a driver's license without attaining permission from a male guardian and were also permitted to drive without a male chaperone.

By 2012, Saudi women made up almost 60 percent of the country's university students but only 15 percent of its workforce; more than 78 percent of Saudi Arabia's female college graduates were unemployed. To improve the overall productivity of the country, authorities approved a plan initially introduced by businesswomen in 2003: to build a string of "women-only" industrial cities where women can work in jobs that would otherwise be considered unsuitable for them in an environment shared with men. The first of these industrial cities—built in the eastern province city of Hofuf—created between 3,000 and 5,000 of these jobs.

While these women-only cities clearly benefit both Saudi women and the country, they haven't resolved the larger national segregation issues. Saudi Arabia is still among the world's most gender-segregated countries. Women can't eat at restaurants that don't have a separate designated family section, and they must use a separate entrance from men. The legal position of women in Saudi Arabia is not equal to men, so women can't receive a fair trial because they don't have equal representation in court. Additionally, Saudi law still prohibits women from jobs considered "detrimental to health" or "likely to expose women to specific risks." In 2016, the Kingdom's leaders set a strategic plan of economic and social goals in "Saudi Vision 2030," which included the creation of more than 1 million new jobs for women by 2030. It has dramatically increased the participation of women in the workforce, especially in urban areas. According to data from World Bank, the labor-force participation of Saudi women rose from 20 percent in 2018 to 33 percent by the end of 2020.

Geographic Thinking

Explain how women-only cities can impact Saudi women.

National Geographic Photographer **Amy Toensing**

Widow Warriors

Toensing has been publicizing the difficulties of widows worldwide since she first shot a story in India in 2005.

Learning Objectives
LO 19.7 Explain gender equality as an indicator of development.

LO 19.8 Explain how gender roles create geographies of separation and seclusion.

Photojournalist, filmmaker, and National Geographic photographer Amy Toensing shares intimate stories of everyday individuals. "For myself as a storyteller," she says, "the deeper the connections I make with my subjects, the deeper my audience is going to connect with my story." Her projects include photographs documenting Muslim women in the United States, the aftermath of Hurricane Katrina, the last cave-dwelling tribe of Papua New Guinea, and urban refugee children in Nairobi, Kenya. For one assignment, Toensing spent four years recording daily life in Aboriginal Australia. She also teaches photography to children and young adults in underserved communities to help them develop their voices.

In 2017, *National Geographic* magazine published Toensing's work as part of a feature that examines widowhood in three regions of the world. Sponsored in part by the Pulitzer Center for Crisis Reporting, the article highlights—and her photographs document—the cruel treatment that many widows face. "In many regions of the world widowhood marks a 'social death' for a woman—casting her and her children out to the margins of society," Toensing explains. These shunning acts can have huge negative effects on economic development. When women are isolated, ignored, or treated like property, they don't participate in the economy, and the community, state, or country suffers. Thankfully, some organizations have recently begun to protect and empower these women by helping them navigate existing laws.

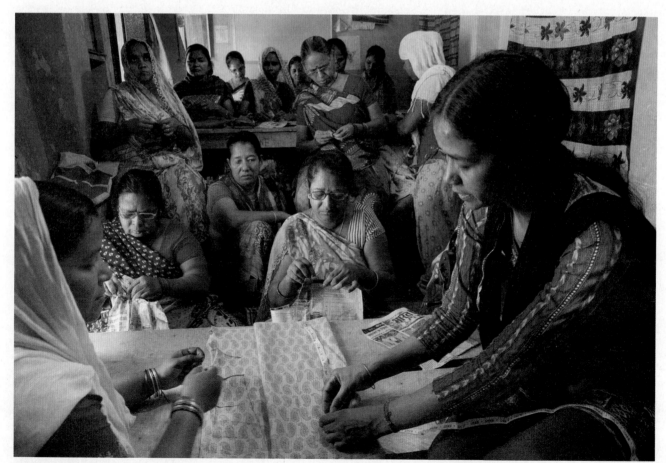

Critical Viewing Widows in Uttar Pradesh, India, learn how to sew. This valuable skill will allow the women to earn their own income. ▮ Describe how Toensing's photographs illustrate the changing roles of Indian widows in society.

Measuring Human Development

Top: Women line up to receive food donations at the Nabadwip Bhajan Ashram in West Bengal, India. Widows often live in extreme poverty and depend on aid from nonprofit organizations to sustain themselves. *Bottom:* Widows enjoy the Holi celebration at the Gopinath Temple in Uttar Pradesh. Until recently, many people considered it inappropriate for widows to participate in this Hindu spring festival.

19.3 Changing Roles of Women

The status of women in society—and the rate and categories of change—vary over time and across the world. Today, efforts to secure gender parity are occurring faster and in more places than ever before. As countries develop economically, opportunities for women evolve. Yet in most countries, women have not achieved parity, socially and economically, with men.

Evolving Opportunities

Learning Objective
LO 19.9 Describe how economic and social development affects gender parity and the role of women in the global economy.

Gender roles are generalized "normal" roles that men and women are expected to perform in their everyday lives, bounded by social and behavioral norms as practiced in a society. Gillian Rose and other feminist geographers use philosopher Judith Butler's idea of *performativity* to capture the idea that gendered behavior is often an unconscious performance—what someone does—not what someone is. In most societies across the world, people who adhere to traditional gender roles believe that a man's role is breadwinner and head of the household, while a woman's role is primary caretaker in the home, handling domestic duties and raising children. As you read in Chapter 6, traditional gender roles reflecting a patriarchal worldview often keep women from playing a part in certain aspects of society and push women to fulfill cultural expectations that can be limiting. However, with economic and social development, societies change—and so do the roles of women, men, and even children.

Industrialization brought working-class women into the workforce with factory jobs, and postindustrial economies have offered women greater opportunities in both education and employment. As countries became more economically developed, the disparity in gender roles diminished. But this imbalance hasn't been eliminated, and other economic disparities exist as well. Globally, women don't share the benefits of development equally with men. Women in the United States are less likely to hold supervisory and managerial roles and, when they do, they tend to carry less authority. The U.S. Census Bureau found that the top occupations in 2019 for men were drivers/sales workers, truck drivers, managers, and first-line supervisors of retail sales workers. The top occupations for women were elementary and middle school teachers, registered nurses, and certain categories of secretaries and administrative assistants, but not in the legal, medical, or executive areas. Even so, economic and social restructuring over time have changed how people live, earn their livelihoods, and practice their gender roles.

While education has become more equal, especially in core countries like the United States—more than half of U.S. college students are women—men still make more money overall. Full-time U.S. women workers earn 83 cents for every dollar that men earn for the same work. The wage gap is even greater for part-time workers, who total about 33 million women in the United States: They make, on average, just 73 cents to the dollar that part-time workers who are men make. The disparity in wages, however, varies by ethnicity and race. Part-time Latina workers earn just 49 cents to the dollar compared to the median male part-time worker. The wage gap in the United States also varies by state and region: 2019 Census estimates show that several western states, Wyoming ($21,676), Utah ($17,303), and North Dakota ($13,950)—plus the District of Columbia ($16,032)—had the greatest gender gap. The U.S. average wage gap was $10,150, with men earning an average of $53,554 and women earning an average of $43,394. Interestingly, the Census Bureau research found that while the U.S. territory of Puerto Rico had significantly lower wages for both men ($22,804) and women ($23,478) than any state or Washington, D.C., it also had the smallest gender wage difference and, in fact, that Puerto Rican women earned slightly more than Puerto Rican men.

Rural and Urban Opportunities Gender parity begins with equal opportunities, including access to resources. Economic changes often result in gender equity changes, and vice versa. For example, in rural areas in peripheral countries, women are starting to find opportunities outside the home in factories and in the service sector, which adds to the household income. This, in turn, creates economic growth. One example of all-female entrepreneurship is occurring in Guatemala, in the tiny village of Urlanta. The initiative started with 29 village women, all with different knowledge and skill sets. They decided to try beekeeping. After some struggles, the women were able to extract and bottle honey and sell it at a profit. In Urlanta, the community mindset shifted. Before the beekeeping enterprise, the village women were expected only to stay home and raise children. They are now welcome participants in village meetings, sharing their experiences and offering advice to others.

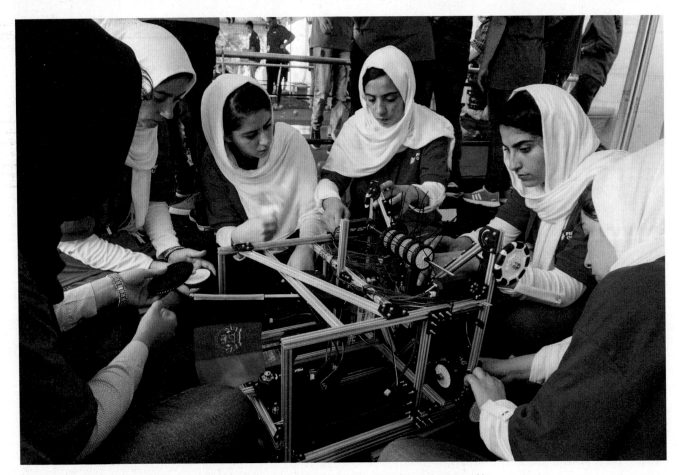

The disciplines of science, technology, engineering, and math (STEM) have long been dominated by men. Recent efforts promote both equity in education and career opportunities for young women. Members of the all-girls team from Afghanistan make adjustments to their robot in an international robotics competition in Washington, D.C., in 2017.

It may be easy to think that urban areas have more economic opportunities for women. Generally, urban areas offer a greater array of services and infrastructure, opportunities for education and employment, and fewer social and cultural restrictions. But access to opportunities really depends on the location. In some regions—particularly North Africa, Southwest Asia, and South Asia—urban areas present a greater number of challenges, inequities, and insecurities for women. Women struggle with finding decent well-paying jobs, increasing workloads while juggling jobs and care at home, accessing financial assets, interacting with city and area authorities, and having housing security and personal safety. The biggest challenge is dealing with traditional behaviors toward women's roles; many men in urban areas in these regions don't allow women to easily participate in the job market. In many countries, strong attitudes about gender roles continue to make it difficult for women to break away from childrearing and home life.

Economic Opportunities Despite the overall trend of a continued gender gap, attitudes about women in the workforce have changed or are changing across the world, and women are playing a larger part in the global economy. As women contribute more to household incomes, roles are shifting; men and families see how two incomes can benefit a household. Much of the industrialization in countries in the periphery now relies on women working outside the home (because of the lower wage rates, which remain an issue). In addition, women in peripheral and semi-peripheral regions are edging into the service industry, and this trend is happening in core regions as well, thanks to huge growth in the tertiary sector. In Brazil, women have been compelled to seek employment to supplement family incomes in a range of secondary and tertiary jobs. In Pakistan, more and more women have become employed in many occupations, from agricultural enterprises to financial services, and in both private and public sectors—including managerial positions. In 2022, 61 percent of Pakistani women in urban areas were employed outside the home compared to 39 percent in rural areas.

The cultures of some countries are beginning to realize the value of women in formal economic activities. For example, women in Japan have not traditionally played a role in the workforce. However, since Japan's economy has been hindered over the past several decades by its aging and dwindling population, circumstances have changed. The government implemented an economic plan that focuses on eliminating the employment gender gap and expects to see

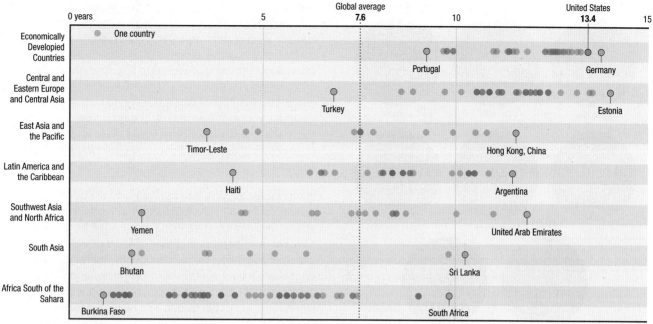

Girls' Education Levels Around the World, 2019 Many countries have high rates of girls finishing primary school, but they often fall short in graduation rates and quality of secondary schools; the global average in 2019 was 7.6 years of schooling. Each circle on the graph represents the average number of years of schooling for girls for one country. For example, the average for the United States was 13.4 years of schooling. The countries with the lowest and the highest averages in their regions—such as Bhutan and Sri Lanka in South Asia—are labeled as well.

an increase of 13 percent in the country's GDP as a result of this initiative. These efforts to reform employment practices have paid off; in 2019, pre-pandemic, half of Japan's women were employed—including those age 65 and up—the bulk in the tertiary sector in businesses like hospitality, nursing, and food services. However, 80 percent of women aged 15 to 64 in Scandinavian countries are employed compared to just 70 percent in Japan, so there is room for further growth in the support of working parents. As you have seen, Saudi Arabia is also focusing on a shift in women's roles in the economy. With recent swings in oil prices, government officials are looking at ways to diversify its economy by investing in human capital—including that of women.

In the United States, roles for women have expanded, and the number of women working outside the home is rising. In 2017, there were more than 74 million female workers in the workforce, making up 47 percent of workers. In 2019, 57.1 percent of all women worked, about the same as the previous year but three percent below the peak of 60 percent in 1999.

The range of occupations has expanded; in many occupations, such as speech-language pathologists, dental assistants, social workers, physical therapists, and pharmacists, more than 50 percent are women. Women are making notable gains in professional and managerial occupations as well. In 1974, only 1 in 10 lawyers were women; in 2018, 37 percent were women. In 2020, well over half of human resources managers and education administrators were women.

The number of women physicians has increased from 28.3 percent in 2007 to 36.3 percent in 2019. But in some instances, the roles of women remain little changed. In higher-level managerial positions, only about one-third of chief executives are women, and just 7 percent of construction managers are women. A 2017 report by the investment and financial services company Morgan Stanley stated, "More gender diversity, particularly in corporate settings, can translate to increased productivity, greater innovation, better decision-making, and higher employee retention and satisfaction." Recent positive strides have included the percent of women on the boards of directors at top companies increasing from 24 percent in 2020 to 27 percent in 2021. So, while the numbers of opportunities are increasing, an obvious need for more women at these top levels persists.

Educational Opportunities Until recently, women in many parts of the world have been denied access to education. This situation can severely limit all connections and opportunities a woman has during her lifetime. Women's health, limited future prospects, lower income, and increased vulnerability to exploitation and trafficking severely inhibit female well-being and development. This hindrance, in turn, can cripple the economic advancement of a country.

As countries evolve and make strides in their economic and social development, women have greater access to educational opportunities, which leads to positive progress toward gender parity. Exposure to educational

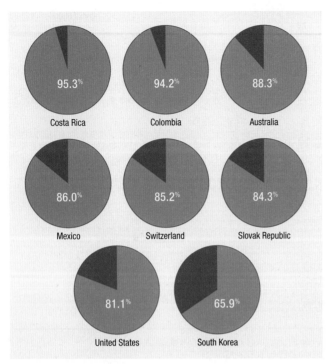

Gender Pay Gap Comparison, 2018 These pie graphs show the percentages of what women earn compared with men in selected countries in the Organization for Economic Cooperation and Development.

opportunities also encourages women to explore specialized careers, which can increase income possibilities, provide empowerment, lower birth rates, and create a variety of fulfilling activities for women outside the home. All these factors can help raise households and communities out of poverty and support national development.

An interesting correlation exists between education and fertility rates. A 2015 study in Ethiopia reported that 61 percent of women with no schooling have a baby before they turn 20, compared with 16 percent of women with eight years of schooling. Recall that the education dimension of the Human Development Index is access to education. In most cases, women with more education tend to marry later and have fewer, healthier children. It's important to note, though, that studies show that other factors besides education can lower the total fertility rate, such as access to health care, additional work opportunities, and reduced child mortality. But education certainly plays a role.

Many public and private organizations—at local, national, and international levels—provide educational opportunities to break down the barriers that limit education access for women and girls around the world. Some obstructions are harder to crack, such as cultural traditions, patriarchal worldviews, and religious restrictions, but others can be chipped away. Groups working to help women and girls contribute scholarships and loans, build schools to decrease the distances girls have to travel to get to school, create gender-sensitive curricula, ensure safe and inclusive learning environments, and invite men to join gender-equality discussions. An organization in southern Africa created an innovative program that uses mobile-based technology to build digital skills to help empower girls and encourage them to continue their education. And a group called the Asia Foundation is working to increase literacy in countries across the continent. The organization provides study kits to schools for girls to help these students pass entrance exams and gain entry to four-year national universities.

In many places, girls struggle with social norms and specific economic challenges that threaten their ability to attend and remain in school. Globally, 3.5 billion women have monthly periods and face "period poverty," which is a lack of access to sanitary products, menstrual hygiene education, toilets, handwashing facilities, and waste management. In India, 23% of girls drop out of school when reaching puberty, due to a lack of facilities for them to manage their menstruation. In both urban and rural contexts, over 50% of India's adolescent girls lack access to adequate sanitation facilities. While this form of poverty most affects women and girls in the peripheral parts of the world, women attending university campuses across the United States are also affected by period poverty and report being unable to buy feminine hygiene products because of how expensive these products are. About one-fifth of women in the United States report missing school or work because of lack of access to sanitary products.

Scotland became the first government in the world to provide period products for free to students in its schools in 2018; England followed with a similar program in 2020, and the government of the United Kingdom formed a taskforce "to tackle the stigma and education around periods." In August 2022, Scotland expanded its efforts beyond schools to the general public by making operational the Period Products Bill, mandating that its 32 local councils provide free period products "reasonably easily" and with "reasonable dignity" to "anyone who needs them."

Wages More educational opportunities for women have emerged over time, which means that women now have a better chance at earning higher wages. However, a wage gap between men and women still exists, even in core countries—although this is where the greatest strides to decrease the wage gap have been taken. In 2018, women in Iceland protested against unfair pay. At the time, Icelandic women were paid 74 percent of the average wage of Icelandic men. In response, demonstrators declared that women should work 74 percent of the eight-hour workday. The protest worked. The following year, Iceland created a law requiring employers with 25 or more workers—in both the government and the private sector—to prove that they pay men and women equally for the same work.

Women in Leadership

Diverse women—from teenagers to adults and from rural regions to urban centers—have risen to leadership roles in the fields of government, business, activism, and education. By speaking up and blazing trails, they inspire people around the world. *Clockwise from top left:* Michelle Bachelet, first woman president of Chile • Tracy Chou, Taiwanese-American proponent for women in technology • Emma González, Cuban-American advocate for gun control • Jane Goodall, British primatologist and environmental activist • Malala Yousafzai, Pakistani advocate for female education • Greta Thunberg, Swedish leader of global youth movement demanding action on climate change • Autumn Peltier, Indigenous Canadian clean water activist • Kakenya Ntaiya, Kenyan educator and champion for women's rights.

In June 2019, women in Switzerland made a similar objection. Although Switzerland is one of the world's richest countries, women there were also frustrated by unequal pay. Thousands demonstrated by skipping work, ignoring household responsibilities, and marching in the streets. The 2019 protest marked the 28th anniversary of the first protest women made in Switzerland regarding unequal pay.

Even workplaces in the spotlight have wage inequality. The U.S. women's soccer team won a historic fourth World Cup victory in 2019. Their record-breaking triumphs were undeniably impressive, yet female team members were paid just 38 percent of what their male counterparts earned. The women's team sued U.S. Soccer in federal court for gender discrimination and violations of the Equal Pay Act and Title VII of the Civil Rights Act of 1964. In a landmark decision, the lawsuit was settled in 2022, awarding $24 million to the team and forcing U.S. Soccer to pay women and men at an equal rate for their play. The decision was expected to affect wage disparities for other female professional athletes in the United States and around the world.

Empowering Women

Learning Objective
LO 19.10 Describe strategies and initiatives that seek to empower women across the globe.

In addition to providing more economic and educational opportunities for women, other endeavors seek to empower women. In places across the globe, women are still subject to violence, social injustice, and a lack of health care. Many women have been conditioned to believe that they are not worthy of the right to food and shelter, to be educated, to work, or to be free. They may not realize that they are robbed of basic human rights. Some countries make it legally impossible for women to accept higher-paying jobs and, in others, a husband can legally deny his wife the opportunity to work outside the home.

Several initiatives aim to provide independence and choices for women and continue to make strides toward equality. In coordination with African governments and regional organizations, the UN has written several protocols and decrees for human rights and inclusion, with an emphasis on gender equality and the empowerment of women in Africa. As African countries continue to develop economically, they turn their focus to women. The result is a stronger presence in the workforce across the African continent. While women still work longer hours and earn less money than men, African women overall are highly entrepreneurial. In some countries, such as Rwanda and Ghana, women run nearly one-half of all businesses in the country. As discussed in Chapter 11, the African Union is a continent-wide supranational organization with an overall agenda to "transform Africa into the global powerhouse of the future." A key aspect of these efforts is to "achieve gender equality and promote women's empowerment." The AU recognizes that "gender equality is a fundamental human right and an integral part of regional integration, economic growth and social development." The organization has developed a strategy and established the Women, Gender and Development Directorate (WGDD) to ensure the inclusion of women in every country and in all aspects of Africa's development agenda.

Microloans Microfinance is financial services targeting low-income and marginalized people and communities, who otherwise would not have access to banking and savings institutions. In recent years, women in peripheral and semi-peripheral countries have started applying for loans through microfinance programs to start small businesses. For them, being an entrepreneur is the answer to breaking away from poverty and improving financial security for their family and children. **Microloans**, a form of microfinance, are very small short-term loans with low interest intended to help people in need. Microloans became one option for women wanting to take a risk on their own enterprises. Most of the women obtaining the loans would not qualify for loans from a traditional bank, so nongovernment organizations and private donors searched for ways to make loans available.

The microloan industry was started by a Bangladeshi professor who eventually opened the Grameen Bank, a microloan institution. Other microloan institutions followed, and the industry has flourished. In 2015, market analysts determined that an estimated 125 million people worldwide—about 80 percent of whom were women—received a total of about $100 billion (USD) in microloans from the key microfinance institutions. This amount is projected to exceed $304 billion (USD) by 2026, primarily benefitting marginalized individuals in Asia and the Pacific regions.

Microloans range in size depending on location, from a low of about $200 in South Asia to a high of nearly $3,000 in Eastern Europe and Central Asia. The amounts are based on the borrower, the entrepreneurial opportunity, the location, and the income variation. Microloans are designed to cover startup costs for a variety of businesses, such as the cost of nail polish for setting up a home-based nail salon or an individual purchasing chickens to sell the eggs. Many of the loans also cover education costs. Numerous microloan institutions offer business training, insurance, savings plans, and other resources. This small amount of financial help greatly contributes to leveling the playing field for women entrepreneurs.

Many women have succeeded in business thanks to getting their start from a microloan. Take the example of Oiness, a single mother of three in Zambia, who lives in a remote rural village with few employment opportunities. She realized she could start a business selling small, portable cooking stoves to people in her community. Thanks to a microloan, she was able to buy the materials she needed to make the stoves. When she started earning money, she reinvested those funds into her business, so it grew. She now earns enough money to send all of her children to school.

Obtaining a microloan in and of itself does not ensure success. Starting, running, and maintaining a business is tough. However, as more women become successful business owners in poorer countries, gender inequality continues to narrow in those areas.

Investing in Girls and Women As countries develop, more attention and money can be invested in issues relevant to women. According to the Organization for Economic Cooperation and Development (OECD), four key strategies need increased investment: (1) ensure that financial assets are in the hands of women, (2) keep girls in school, (3) improve reproductive health and access to family planning, and (4) support women's leadership.

These strategies coincide with the goals in the UN's 2030 Agenda for Sustainable Development. To reach these goals, the UN thinks that every country must make public services available for women, confront and overcome the cultural and social norms that hold back women and girls, put voluntary family planning back on the development agenda, gather evidence about which methods in the agenda work, and accurately track the proportion and coverage of aid focused on achieving gender equality and women's empowerment. You will read more about the UN's Sustainable Development Goals in Chapter 20.

Geographic Thinking

1. Identify and explain the relationship between economic development and gender parity.
2. Compare global wage equality using the content in section 12.3 and the "Gender Pay Gap Comparison, 2018" pie graphs.
3. Describe how microloans can help get women out of poverty.

Case Study

The Development of the Grameen Bank

The Issue For many people without economic resources, starting a business is the only way to escape poverty. And these people don't have the collateral (capital or property) needed to acquire a traditional bank loan.

Learning Objective
LO 19.10 Describe strategies and initiatives that seek to empower women across the globe.

By the Numbers

97%
of Grameen Bank borrowers are women, and nearly 99 percent of the loans have been paid back as of 2020.

9.08 million
Grameen Bank members in Bangladesh as of 2018

100+
countries with Grameen Bank–backed projects, including Grameen America

Source: Grameen Bank, 2018

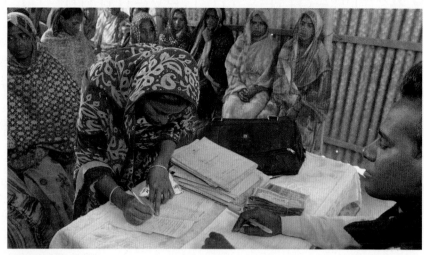

A woman conducts business at a weekly meeting held for borrowers at the Grameen Bank in Bangladesh.

In 1983, Muhammad Yunus Started the Grameen Bank, which means "rural" or "village" bank in Bengali. He reasoned that if financial resources could be made available to poverty-stricken people through reasonable and appropriate terms and conditions, then "these millions of small people with their millions of small pursuits can add up to create the biggest development wonder."

Yunus got the idea for the bank when he was an economics professor in Bangladesh. One day he took his students on a field trip to a small rural village In India. They interviewed a woman who was making bamboo stools to sell. After paying for the bamboo, she barely made a profit from her sales. Yunus loaned her money, and she was able to buy her raw material at a cheaper price because she was able to buy it in volume. The result: she made a larger profit. Yunus and his students studied her continuing business, and they realized the key to survival and economic growth for this woman—and many others like her—was just a little bit of money: a small loan.

The Grameen Bank offers microloans, or microcredit, with no collateral required, no legal contracts, low interest, and comfortable repayment plans. The bank provides loans to pay for raw materials, livestock, agriculture, groceries to make food products, and cell phones. Products made from these loans go to local markets, the main market in the capital city, and markets all over the world.

Borrowers repay monthly installments to bank representatives who visit them, listen to their stories, and discuss business successes and challenges. Money earned by women borrowers has a huge impact on their families, and these women tend to save their money or invest it, rather than spend it. In 2006, the Grameen Bank and its founder were awarded the Nobel Peace Prize "for their efforts to create economic and social development from below."

The success of Grameen Bank has led to the creation of many other microloan institutions, and borrowers are found in every corner of the world, including the United States. Yunus is proud of what he started and the women he has helped. "There are roughly 160 million people all over the world in microcredit, mostly women," he said in 2017. "And they have proven one very important thing: that we are all entrepreneurs."

Geographic Thinking

Describe how Grameen Bank can be an example for societal change.

19.4 Theories of Development

Economic growth results in lower unemployment rates, higher tax revenues, improved living standards, and a reduction in poverty. Additionally, governments of growing economies can afford to invest in infrastructure and social services. Theories that explain spatial variations in development include Rostow's stages of economic growth, Wallerstein's world system theory, dependency theory, and commodity dependence.

Rostow's Stages of Economic Growth

Learning Objectives
LO 19.11 Explain the five stages of development in Rostow's model of development.
LO 19.12 Explain the criticisms of Rostow's model of development.

In the 1960s, after much of Africa and Asia gained independence, Walt W. Rostow developed his **stages of economic growth** model, sometimes called modernization theory. He studied countries that had success in developing economically and asked: How were these countries able to achieve this modernization, and could poorer countries follow a similar path? At the time, it was assumed that every country wanted to achieve modernization, that is, the social, political, economic, and technological changes associated with becoming industrialized and moving away from traditional ways of life. Rostow suggested that all countries could be categorized on a spectrum from traditional to modern and that to become modern, countries could pass through the following distinct stages of economic growth in succession. He also assumed that all countries practiced a form of market-oriented capitalism.

Stage 1: Traditional Society In the simplest and most basic form of organization, political power is local, regional, or based on land ownership. Family plays a dominant role, and social structures limit economic mobility. Rostow described traditional society as one that is primarily rural, centered on subsistence farming by family labor and using primitive technology. Modern science and technology are nonexistent in this stage.

Stage 2: Preconditions for Takeoff Progressive elements begin to form, and people start to seek knowledge and break free from the traditional mindset. New types of enterprises emerge with long-term goals, investment increases, and output rises. As industry accelerates, improved infrastructure becomes essential. The workforce shifts from agriculture to manufacturing, and financial institutions and services are developed to make capital more accessible and investments less risky.

Stage 3: Takeoff Political, social, and institutional frameworks in society change. Urbanization increases, infrastructure continues to improve, and productive capacity surges in some manufacturing industries, with technological advances as well.

Stage 4: Drive to Maturity The economy keeps progressing in a period of self-sustained growth. In this respect, "maturity" is the state in which a country's successes become the norm, or habit. Industries function at maximum effectiveness, and electric power generation and consumption are high. Consumption patterns shift thanks to increased income. Entrepreneurial leadership changes from individual industrialists to managerial bureaucracy.

Stage 5: High Mass Consumption Modern societies are urban, centered on wage labor, and organized into states. Production shifts from industrial manufacturing to consumer goods and services. Problems of production change to problems of consumption, and trade expands.

Rostow envisioned that each country could be categorized in one of these stages, and that all countries could follow these national growth strategies to achieve economic growth and success. Various factors influence the speed of progression according to Rostow's model. They include natural resource availability, productivity, and political and social decision making.

The earliest country to reach maturity according to this model was Great Britain. Its production of textiles placed it in takeoff stage in the early 18th century. The Industrial Revolution, with the introduction of the water frame, the power loom, and the use of steam power, propelled the country to its drive to maturity stage around 1850. The next major country to hit maturity was the United States around 1900. Canada and Russia both hit maturity around 1950, but Canada has been able to reach stage 5. Russia remains in stage 4 because economically it is an oligarchy, a form of government in which all power rests in a dominant class or group of individuals, often the top one percent income-wise, called oligarchs. Corruption and cronyism are widespread and leadership does not follow the rule of law.

Singapore illustrates Rostow's model. With a population of around 5.8 million in 2022, Singapore had a successful, linear path of economic growth. After its independence in 1965, it focused on industrialization and developed profitable manufacturing and high-tech industries. It hit its maturity not long after it passed its takeoff stage. Singapore is now highly modernized, quite wealthy, and a leading trade partner in the global market, placing it securely in Rostow's stage 5.

Limitations of the Stages of Economic Growth Model
Since Rostow's model was based on the United States and Europe in the 1960s, critics argue that it can't be applied to every country. According to the model, the result of the sequence of growth stages is an industrialized, capitalist, democratic country; the assumption is that all countries will wish to have these characteristics.

A further limitation of the model is that the stages of growth in some countries differ by region. For example, India is

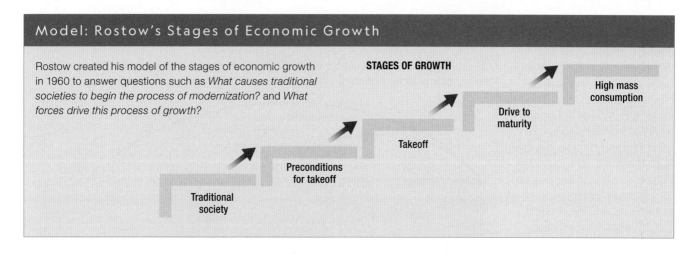

Model: Rostow's Stages of Economic Growth

Rostow created his model of the stages of economic growth in 1960 to answer questions such as *What causes traditional societies to begin the process of modernization?* and *What forces drive this process of growth?*

difficult to categorize using Rostow's model. The country as a whole might be placed in stage 3 or stage 4, but some regions remain in stage 2.

In addition, Rostow's model does not take into account a host of geographic influences and challenges. One country, with its own set of natural resources and environmental advantages, may make the leap from being a rural, agrarian state to an industrialized one in a short period of decades, but that breakthrough does not mean that other countries, perhaps with fewer resources and more environmental challenges, will follow suit in exactly the same way, following the same ordered stages of growth or in the same time frame.

Moreover, Rostow's stages of economic growth assume that all countries follow a progression of development in which they will become mass consumers, but they don't allow for Earth's carrying capacity and new ecological limits—information that was unavailable to Rostow in the 1960s. Mass consumerism creates a "the more, the better" attitude, and Rostow did not consider that consumers at the final stage would use and demand more than they actually should. Sustainability becomes an issue as consumers deplete resources and create environmental degradation.

Another criticism of Rostow's model is that it doesn't consider how countries influence one another and how these impacts can affect the progression of development. Countries in this time of globalization are interconnected and interdependent. The road to maturity was much different for the "early starters"—the countries that matured sooner—like Great Britain and the United States. Their path was much clearer: They had less competition and fewer obstacles. The "late starters" in today's peripheral regions have much bigger barriers to overcome that in many cases are a direct result of the success of the early starters, such as colonial legacies or lack of access to the latest technologies.

Wallerstein's World System Theory

Learning Objectives

LO 19.13 Explain why it is difficult, in terms of economic development, for countries to move from the periphery to the semi-periphery, and from the semi-periphery to the core.

LO 19.14 Explain the limitations of Wallerstein's world-system theory and their relevance to human geography.

In response to Rostow's stages of economic growth model, Immanuel Wallerstein published his world system theory (which you learned about in Chapter 1) in 1974. This theory describes the spatial and functional relationships between countries and helps explain the history of uneven economic development in the world economy.

According to Wallerstein, countries are dependent on one another and don't develop in isolation: Some countries dominate and some are exploited. His theory illustrates global inequality through a social structure, dividing the world into a three-tiered structure consisting of core, peripheral, and semi-peripheral regions. As you've read throughout the text, core countries dominate and take advantage of peripheral countries for labor and raw materials, and peripheral countries are dependent on core countries for capital. Semi-peripheral countries have qualities of both core countries and peripheral countries.

World system theory includes both political and economic elements and can be viewed as either a political or economic theory with geographic effects. Wallerstein originally studied the development of trade in the late Middle Ages in Europe, as global trade was developing. He examined the world economy and how it developed due to capitalism. Capitalism became a global system with stronger, interdependent economic ties between regions and countries, gradually encompassing all countries in the world in some way. Today's core countries are able to accumulate capital internally through taxation, government purchasing, funding of infrastructure, and sponsorship of research and development. Core countries are also able to accumulate capital through control of the world economy. They use their power—economic, political, social, and military—to pay low prices for raw materials, to employ cheap labor in peripheral regions, and to install trade barriers and quotas that suit their needs. It is possible for countries to move from the periphery to the semi-periphery, and from the semi-periphery to the core, as they develop economically. However, since the prosperity of core countries depends on exploitation of peripheral countries, movement up the hierarchy is difficult, and the system produces imbalances around the world.

Inequalities exist at different scales: between countries and within countries. According to world system theory, Mexico is an industrializing, semi-peripheral country, and exhibits marked internal contrasts in development. Within Mexico's borders are large peripheral areas, such as its southern states of Chiapas and Oaxaca, in the rural highlands where economic opportunities are limited and poverty rates are high. The World Bank found that in 2018, 42 percent, or more than 52 million, of Mexico's people lived below its poverty line, and that 7 of 10 people living in poverty resided in just 6 of Mexico's 32 states, mostly in the country's south. In Mexico City and in most of Mexico's northern states, income is substantially higher and poverty rates are much lower. This high rate of income inequality—along with related factors such as less access to education, health care, and basic infrastructure—illustrate the continuum of poverty that keeps Mexico in the semi-periphery. Its 2019 GNI per capita was U.S. $9,480, but that declined in 2020 to $8,480 due to the effects of the COVID-19 pandemic.

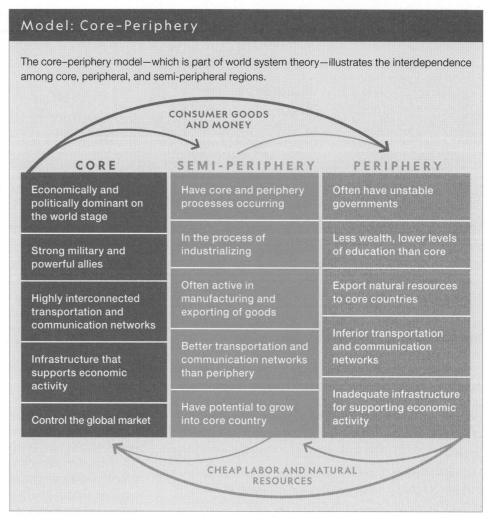

Model: Core–Periphery

The core–periphery model—which is part of world system theory—illustrates the interdependence among core, peripheral, and semi-peripheral regions.

Compare these data to those of a less economically developed, peripheral country like Angola, whose poverty ratio jumped from 30.1 percent in 2010 to 41 percent in 2020 (a similar percentage to Mexico), but whose GNI per capita was just U.S. $2,230. While Mexico's overall GNI per capita is much higher than Angola's, a similar percentage of each country's population lives below its poverty line (poverty levels are defined differently in each country). The UN notes that many people (in Mexico, Angola, and other countries) do not have "food security, access to basic services such as health care, transportation and education," or live "in peaceful, stable societies." In world system theory, both the Mexico and Angola examples show that the cycle of interaction between the core, periphery, and semi-periphery often keeps those living in poverty from being able to improve their situation.

Limitations of World System Theory The biggest criticism of world system theory is that the model is too focused on economics. Categorizing countries based on a global economy isn't comprehensive. Experts claim that there are more determinants for the economic development or connection to the world system than capitalism. Critics also argue that a study of core–peripheral relationships must take into account other measures of integration or dominance, such as cultural influence.

A further limitation is that the theory may work as a historical analysis, and is clearly Euro-centric, but may not be the best measure of modern development. Another weakness is that the theory states that countries can change their status but gives no clear explanation as to how that can happen. Detractors argue that a theory explaining an event should also include supporting material when making a claim.

Dependency Theory

Learning Objectives
LO 19.15 Explain dependency theory.
LO 19.16 Explain how commodity dependence can both slow and fuel economic development.

In the late 1950s, experts at the UN tried to explain why economic growth in industrialized countries did not lead to economic growth in nonindustrialized countries. They identified a financial dependency among countries with widely diverse economies. This reliance explained the failure of less economically developed countries to experience financial growth despite investments from more economically developed countries. **Dependency theory** describes the development challenges and limitations faced by poorer countries and the political and economic relationships poorer countries have with richer countries.

According to dependency theory, peripheral countries offer cheap labor and raw materials to the global market. Core countries buy the raw materials and hire the cheap labor. They use these two economic factors to produce goods and sell them at high prices. Peripheral countries have a demand for these goods, so they buy them at the increased prices, which depletes the funds they might have used to upgrade their own production structures. So to produce capital, peripheral countries continue to sell their raw materials and offer their labor. It's a vicious process that explains the huge gap between the rich core and the poor periphery: The needs of the core keep the periphery in a state of underdevelopment. One leading dependency theorist suggests that a lack of development prevails in some countries because core countries have deliberately kept some peripheral countries from developing. This plight reflects the underlying cause of dependency: imperialism.

Both world system theory and dependency theory aim to explain global economic inequality. Both include the concept of core, peripheral, and semi-peripheral world structure. The key difference between the two is based on the focus of the inequality. In world system theory, global economic inequality exists because core countries thrive through economic exploitation of peripheral countries. Supporters of dependency theory argue that the global economy has been unequally structured since Europeans began colonizing the world in the 16th century. Although the form of the dominance of core over periphery has changed from colonialism and imperialism to neocolonialism, an overall transfer of wealth from periphery to core continues to fuel growth in some places at the expense of others.

As with all models and theories, dependency theory has limitations. Critics point out that dependency theorists fail to define critical terms, including *dependence* and *underdevelopment*. Additionally, there is no standard to distinguish between dependent and nondependent countries. Other detractors note that the theory fails to take into account other factors that cause underdevelopment. According to these critics, the nature of underdevelopment in Latin America, for instance, is different than in Asia. Finally, critics argue that the theory doesn't reflect that underdevelopment can be a product of leaders making bad decisions.

Commodity Dependence One aspect of dependency theory is **commodity dependence**, when more than 60 percent of a country's exports and economic health are tied to one or two resources such as oil, timber, or plantation crops. This dependency is indicative of the narrow economic base that peripheral and some semi-peripheral countries depend on, unlike the broad, diversified, and healthy economies that core countries enjoy. These countries in the periphery have not had the opportunities to add value to their resources through manufacturing, so they are often trapped in neocolonial economic relationships, exporting raw materials and importing manufactured goods.

Unfortunately, export earnings for dependent countries are at the mercy of commodity pricing. Commodity markets set prices based on supply and demand, so commodity trade is never a stable environment. When the commodity price of its major export increases, such a country sees economic growth.

But when supply is high, demand drops and so do prices. If a country is commodity dependent (or has "all its eggs in one basket"), it suffers. Employment decreases, exports diminish, and government revenues slump. These conditions are why many economists find a connection between commodity dependency, poverty, and financial turmoil.

Consider oil, which can be a volatile market. In June 2014, the price of oil was at an all-time high then of U.S. $115 a barrel. After global demand decreased, the price fell below $50 a barrel in January 2015. Several countries were devastated when oil prices dropped. Venezuela had assumed that the price of oil would remain stable and started funding welfare programs from its oil revenues. When prices fell, Venezuela could not meet its debt obligations. Saudi Arabia is another country dependent on oil, though fortunately it had ample financial reserves to weather the price drop. Since then, the kingdom has tried to diversify its economy to become less dependent on oil.

In March 2020, as the world faced a global pandemic, oil fell to $22.51 a barrel; two years later, it was trading as high as $140 a barrel as sanctions against Russia for the invasion of Ukraine affected prices. Violent price swings in commodities markets illustrate the dangers of dependency.

Geographers have found another link to commodity dependency. For countries with a single commodity, there is extreme interest in who controls that commodity. This situation can cause governments and political factions to clash, creating political instability. This relationship between commodity dependence and political instability can be seen in countries like Sierra Leone, with its single commodity of diamonds. Mercenaries and criminals run rampant in the African country as they try to control the diamond trade and profit from it. As such, commodity dependency can be described as a "resource curse." Such commodity-dependent countries, rich in a specific resource but lacking economic diversity, are often trapped in volatile situations.

Commodity dependence can have a negative impact on a country's development. Because the welfare of commodity-dependent countries can be critical to the global economy, these countries are closely monitored. The UN issues a State of Commodity Dependence Report every two years to summarize and analyze the state of commodity-dependent countries. In 2021, the report looked at 189 countries, of which 102 were commodity-dependent, including 45 out of 54 countries in Africa and 50 percent of Latin American countries. All 12 countries of South America were commodity-dependent, according to the 2021 report, with an average of 75 percent dependency—an increase from the previous report. These numbers are stark proof of the uneven geography of economic development.

Geographic Thinking

1. Compare Rostow's stages of economic growth, world system theory, and dependency theory.

2. Explain the degree to which commodity dependence slows and fuels economic development.

Chapter 19 Summary & Review

Chapter Summary

Indicators used to measure economic conditions include:

- Gross domestic product (GDP)—the total value of goods and services produced by a country in a year; limited to what is produced within the country

- Gross national product (GNP)—the total value of the goods and services produced by a country in a year, including those produced internationally

- Gross national income (GNI) per capita—a country's total annual income divided by the country's population

The structure of an economy can be divided into formal and informal sectors. Social indicators used to measure growth include fertility rates, infant mortality rates, access to health care, and literacy rates.

The Human Development Index (HDI) looks at three indicators to determine overall levels of development: life expectancy, access to education, and standard of living.

The UN uses two measurements to track gender inequality:

- The Gender Development Index (GDI) measures the gender gap in health, knowledge, and living standards.

- The Gender Inequality Index (GII) measures gender inequalities in three categories:

    1. Reproductive health (maternal mortality ratio and adolescent birth rate)

    2. Empowerment (political representation and educational attainment)

    3. Labor-market participation (the role of women in the workforce)

As countries experience economic changes, the roles of women and gender equity change as well. Attitudes about women in the workforce have evolved, and women are playing a larger part in the global economy. Women have more access to education, including recent efforts by some governments to assure the provision of period products in schools. Women's wages have increased, although most women still make less than men.

Countries and individual organizations are focused on empowering women. Financial institutions offer microloans to women living in poverty to start small businesses.

Theories of development include:

- Rostow's stages of economic growth, a model that shows how countries progress from traditional to modern through a sequential set of stages

- Wallerstein's world system theory, a theory that states that peripheral countries are dependent on core countries, and core countries often exploit countries in the periphery (with semi-periphery countries having aspects of both)

- Dependency theory, a theory that describes how nonindustrialized countries are financially dependent on industrialized countries

- Commodity dependency, which occurs when more than 60 percent of a country's exports are made up of primary commodities

Review Questions

Use complete sentences to answer the questions.

1. **Apply Conceptual Vocabulary** Consider the term *development*. Write a standard dictionary definition of it. Then provide a conceptual definition for the term—an explanation of how it is used in the context of this chapter.

2. What is a semi-peripheral country, and how is it different from a core country or a peripheral country?

3. Differentiate between the terms *GDP* and *GNP*.

4. How does the HDI show spatial variation among states in levels of development?

5. Explain the relationship between core countries and peripheral countries using world system theory.

6. Use an example to show how distribution of income is an important measure of development.

7. Explain how political representation is an indicator of gender equality.

8. Why might a critic argue that Rostow's stages of economic growth model is not as linear as a graph would suggest?

9. Explain how reproductive health affects a country's GII.

10. How does price volatility affect commodity dependency?

11. Why is it difficult to measure informal sectors of the economy accurately?

12. Explain how measuring GNI per capita reveals a better understanding of a country's overall well-being.

13. **Identify Data & Information** Based on the bar graph, which two professions had the best advancement opportunities for women?

14. **Analyze Visuals** What generalizations can you make about the professions of professor and doctor?

15. **Draw Conclusions** What conclusions can you draw about how women might feel about professional opportunities and the workplace in the United States? Use information from the graph to support your answer.

Interpret Graphs

Study the graph and then answer questions 13–15.

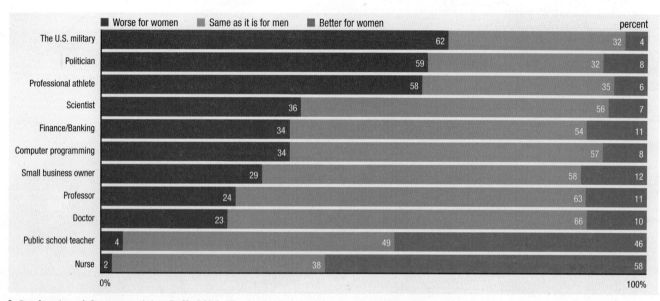

Professional Opportunities Poll, 2019 This bar graph displays the results from more than 1,000 women in the United States who were asked whether the opportunity for women to advance in the listed professions is better, worse, or about the same as it is for men.

Chapter 20

Globalization, Interdependence, and Sustainability

Critical Viewing Every year, Las Vegas, Nevada, hosts the Consumer Electronics Show, now branded as CES, a trade show where more than 4,500 electronics companies from all over the world display their latest products to consumers. The show was virtual in 2021 and took place in a hybrid form in 2022. LG Electronics, a company based in South Korea, displayed its newest video screens at the 2019 event. How does the photo demonstrate globalization?

Geographic Thinking How has globalization affected spatial relationships?

20.1
Trade Relations and Global Corporations

20.2
Connected Economies

Case Study: The Financial Crisis of 2007–2008

20.3
Developing a Sustainable World

Case Study: Reducing Waste in Fisheries

National Geographic Explorer
Thiago Sanna Freire Silva

National Geographic Explorer
Andrés Ruzo

20.1 Trade Relations and Global Corporations

Technology and free trade (trade without restrictions) have expanded people's awareness of resources and available products for consumption and allowed, under normal circumstances, goods to be shipped quickly and inexpensively from thousands of miles away. These are two key factors that have helped to increase international trade and interdependence in the world economy.

Interdependence of Core, Periphery, and Semi-Periphery

Learning Objectives
LO 20.1 Explain comparative advantage and complementarity.
LO 20.2 Explain how neoliberalism is related to contemporary globalization.

The world is linked through complex economic systems in which trade agreements invite participation in market economies that connect regions and countries more than ever before. As a result, international trade has skyrocketed in the last century. In 2019, just before pandemic disruptions, about 60 percent of the world's gross domestic product (GDP) was made up of international trade, compared to about 25 percent in 1960. Countries engage in trade when it is mutually beneficial. One country lacks a resource, and another country provides the resource in exchange for money. However, the benefits to both trading partners are not always equal. Globalization is expanding the gap between the core and the periphery and within countries with widening income disparities. Two factors are the basis for trade and impact its benefits: **comparative advantage** and **complementarity**.

Assuming free trade exists, comparative advantage refers to the relative cost advantages of producing certain goods and services for trade. For example, if two countries can both produce rice and computer chips, it may not make economic sense for both of them to produce these goods for their domestic markets. It may be more environmentally sustainable or cost effective for one to specialize in rice and the other in computer chips and to trade them. From an economic standpoint, it would be advantageous for countries to produce and export a select range of goods and not make everything themselves if they can obtain goods more cheaply from external sources. However, countries must consider their available resources, technology, and capital, as well as labor costs, all of which affect comparative advantage.

Complementarity refers to the mutually beneficial trade relationship between two countries that results when they have different comparative advantages. There must be demand in one place and supply that matches in another. Factors that affect complementarity include variation in the distribution of resources; the relationships between core to core countries, and between core and peripheral or semi-peripheral countries, in which the periphery supplies raw materials and the core supplies manufactured goods; and the economic advantages of specializing in certain goods. In general, countries export goods that they have a relative advantage in producing in exchange for goods that are less expensive to import than to produce internally. For example, Mexico imports specialty apples from the United States and the United States imports avocados from Mexico because it is more expensive to grow them in their own countries than to import them.

New trade relationships are fueled by economic development. Countries need raw materials for production and, in general, core countries import natural resources

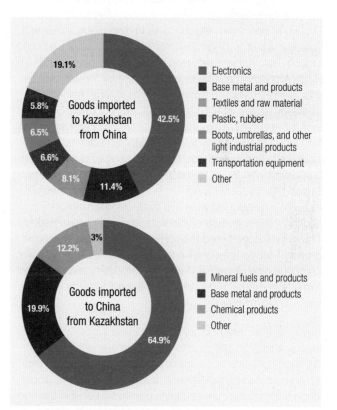

Complementarity: China and Kazakhstan
Complementarity exists between China and Kazakhstan because each country has its own comparative advantage. Kazakhstan is rich in natural resources like oil, natural gas, minerals, and metals; China has a comparative advantage in making electronics and industrial equipment. ▮ Explain how the two graphs demonstrate complementarity between China and Kazakhstan.

from peripheral countries. The result is an interdependence of countries as part of an interconnected global economy. In many countries of the periphery, globalization leads to commodity dependence, in which the economy is reliant on the export of agricultural products, metals, ores, minerals, oil, or other commodities. Some geographers worry that commodity dependence, discussed in Chapter 19, negatively impacts a country's economy and can lead to a fixed state of underdevelopment.

Neoliberal Policies **Neoliberalism** is the belief that open markets and free trade (two key characteristics of capitalism) across the globe will lead to economic development everywhere, lessen tensions between countries by fostering support for common values, and spread democracy and human rights. Neoliberal policies encourage free markets and discourage political interference with economic systems. The interdependence of countries in a global economy has led to neoliberal policies that free private enterprise from government involvement and are associated with deregulation and privatization. When the U.S. airline industry was deregulated in 1978, airlines had more freedom to set their own fares and routes. Some countries have privatized their utilities, transferring government control over energy to private companies.

Free trade agreements are neoliberal policies that are restructuring the world into new, economics-based regions. Groups of countries, referred to as trading blocs, agree to a common set of trading rules in order to encourage trade and reduce trade barriers between the countries in the group. The North American Free Trade Agreement (NAFTA), which took effect in 1994, was an example of a neoliberal regional free trade agreement. NAFTA eliminated tariffs (a tax on imports to make them more expensive) on almost all trade between the United States, Canada, and Mexico and dramatically increased imports and exports among these countries. The agreement was renegotiated in 2018 and renamed the United States-Mexico-Canada Agreement (USMCA). According to neoliberals, globalization and trade benefit the world economy by increasing efficiency and opening new markets. These have contributed to unprecedented global economic growth. Studies have shown that globalization and trade also contribute to a decrease in worldwide poverty and an overall higher standard of living and bring technology, jobs, and ideas to new places.

However, this does not mean that neoliberal policies and globalization are good for all people everywhere. Globalization hurts economic development in countries with a high commodity dependence. It also contributes to global income inequality; core countries gain more benefits than do peripheral countries, and the flow of capital from rich to poor countries has not occurred as much as neoliberal proponents have anticipated. In Latin America in 2019, many protested the neoliberal policies that have created great inequality in the region. Countries with less-skilled workforces tend not to gain advantages through globalization and are motivated to keep their competitive advantage in providing low-skilled labor. In addition, the absence of environmental and labor standards in many countries of the periphery and semi-periphery contributes to pollution and the exploitation of workers in countries vying for a competitive advantage. For example, garment workers in Bangladesh earn less in one month than the average U.S. worker does in one day.

Supranational Organizations

Learning Objectives
LO 20.3 Explain why supranational organizations form and what challenges they address.
LO 20.4 Define deindustrialization in core countries.

Supranationalism, as you learned in Chapter 11, is when three or more countries come together to establish collective policies in response to an international challenge. Many supranational organizations have formed to address challenges associated with international trade and the growing interdependence in the world economy.

Some supranational organizations have been created specifically to support neoliberal policies and foster greater globalization through regional trade networks. These trade networks facilitate free trade and open markets and enhance the competitiveness of the region. To this end, the Southern African Development Community was formed in 1979 to foster economic cooperation and integration among its 16 member states. Another organization similar to the USMCA, known as Mercosur (a Spanish acronym for Southern Common Market), was established to create a South American trade bloc. The European Union (EU) likewise was formed to focus on strengthening the economic development of countries within this prominent economic region by eliminating trade barriers (such as tariffs or quotas) between member countries. The EU took efforts a step further, creating a new standard currency shared by 19 member states that make up the eurozone. In each of these examples, eliminating trade barriers and encouraging free trade facilitate the exchange of goods across international borders. The growing exchange of goods and ideas fosters greater globalization.

Founded in 1995, the World Trade Organization (WTO) is a supranational organization with 164 member countries from around the world. The goal of the WTO is to provide governments with a forum to negotiate trade agreements, settle disputes, and oversee trade rules. The overriding purpose is to help trade flow as freely as possible. WTO agreements, negotiated and signed by most of the world's trading countries, are designed to remove obstacles to trade and to ensure that rules are transparent and predictable. Another worldwide supranational organization is the International Monetary Fund (IMF), an organization of 190 countries. Created in 1945, the IMF's primary mission is to ensure the stability of the international monetary system that enables countries and their citizens to conduct business with each other. Other goals include facilitating international trade, promoting sustainable economic growth, and reducing poverty. It is important to note that the WTO and

WTO, OPEC, EU Countries, 2020

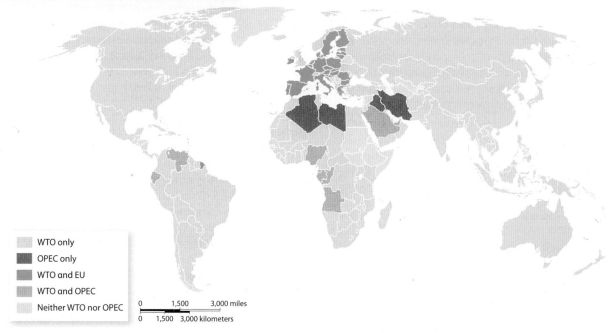

- WTO only
- OPEC only
- WTO and EU
- WTO and OPEC
- Neither WTO nor OPEC

Reading Maps The WTO, EU, and OPEC are all examples of supranational organizations. Tensions can arise within and among them. For example, in 2019, Indonesia filed a lawsuit at the WTO against the entire EU over a dispute about biofuels. ▮ Draw a conclusion about Europe based on the map. Countries colored gray are not currently members of WTO, EU, or OPEC.

IMF are controversial to some. Proponents of free trade point out that these organizations are unnecessary and actually interfere with natural market forces like supply and demand that regulate prices and the amount of goods in a market.

In theory, members of the WTO operate on equal terms, but the WTO allows some countries to protect industries if the removal of tariffs would result in the loss of vital domestic industries. Food production, steel production, and auto production have all fallen under these protections. In addition, core countries have argued that the loss of jobs should be added to the list of justified tariffs. Critics argue that the WTO's practices involve politics and that more powerful countries tend to win in such political battles.

The IMF, too, has fallen under criticism. The IMF makes loans to countries on the condition that specific economic policies are put into action. Advocates point out that the intent of such policies is to reduce corruption and ensure that the funds are used as intended, as well as help to ensure that the country will be able to pay back the loan. Critics contend that the policies interfere with the autonomy of the borrowing country. They also note that the IMF's policies sometimes make economic situations worse. In 1997, for instance, the IMF required Indonesia, Malaysia, and Thailand to increase interest rates to reduce their budget deficits. The tight fiscal policies caused a serious economic recession and high unemployment in these countries.

Some supranational organizations are focused on one industry or commodity. Most influential of these is likely the Organization of the Petroleum Exporting Countries, or OPEC.

Initially formed by five oil-producing countries in 1960, the coalition has expanded to include 13 member countries across Southwest Asia, South America, and Africa, which together control about 75 percent of the world's crude oil reserves. OPEC's stated mission is to coordinate and unify the policies of member countries to ensure an efficient supply of petroleum to customers. However, one criticism of OPEC is that the organization's main goal is to make money for its member countries. OPEC's actions impact the world's economy because so many industries across the globe rely on oil production. OPEC influences the price and supply of oil that is shipped around the world.

Deindustrialization In the last century, globalization has led to **deindustrialization** in many core countries with aging infrastructure and manufacturing processes. These countries could not compete with newly emerging industrial states, like South Korea and China that built new state-of-the-art steel plants and shipbuilding yards. Deindustrialization is the change that occurs with the decline in the percent of workers employed in the secondary sector and a reduction of a region's industrial capacity or activity, particularly heavy industry and manufacturing. As discussed in Chapter 15, the eventual result is postindustrial "legacy" cities and regions in which employment in manufacturing declines while tertiary-sector employment increases. Deindustrialization has led to more service-sector jobs with generally lower salaries and fewer benefits in core countries. It has also increased the interdependence between these core countries and peripheral or semi-peripheral countries by creating global supply chains.

Like industrialization, deindustrialization has ripple effects that can have long-term and often unintended consequences for local areas in core, periphery, and semi-periphery countries alike. Deindustrialization leads to the loss of jobs in the industrial sector of core countries and contributes to rising unemployment rates. When a manufacturing plant closes, businesses across the community suffer. In addition to the businesses that once directly supported the plant, stores, restaurants, and a wide range of services also lose customers and close. Urban areas and their central business districts may decline, and the population may drop, causing a loss of tax dollars.

Government Trade Policies

Learning Objectives
LO 20.5 Describe the ways in which government trade policies attempt to drive economic development.

LO 20.6 Describe the advantages and disadvantages of imposing tariffs on imported products.

The public and private sectors both play critical roles in national, regional, and local economies. Private-sector organizations—corporations, businesses, as well as organizations that promote overall or specialized business activities such as the U.S. Chamber of Commerce and other trade associations—are the drivers of economic growth. They determine where they will locate and establish business relationships with suppliers and customers. They also operate within the parameters established by government. Governments make laws to ensure that the economy is stable and, ideally, that all players within it are treated fairly. Governments also regulate the banking industry and establish trade and monetary policies to provide corporations with an environment that is conducive to doing business. The Federal Reserve, which is the central bank of the United States and part of the public sector, lowers and raises interest rates to either stimulate economic growth or to prevent an unsustainable rise in the price of goods or inflation. Interest rates affect lending to businesses and consumers and also impact job creation.

Governments at all scales—national, regional, local—implement policies to spur economic development. At the national level, countries may seek to enhance development in less economically stable areas. China has created special economic zones, using tax and business incentives to attract foreign investment. These zones are discussed in more detail later in the chapter.

Businesses often choose to locate where there are similar types of businesses in order to take advantage of economies of scale in obtaining resources and in the workforce. As discussed in Chapter 18, this agglomeration explains the spatial distribution of some industries, such as the concentrated location of computer and high-tech manufacturing in Silicon Valley, California, and Bengaluru, India.

Governments today also try to develop **growth poles**—places of economic activity clustered around one or more high-growth industries that stimulate economic growth by capitalizing on some special asset. Highly innovative and technically advanced industries, such as biotech, computer software and hardware development, social media networking, and other research and development enterprises, stimulate opportunities for regional economic development in related businesses. These businesses include suppliers of goods and services, as well as those fulfilling the needs of employees living and working in the area, such as restaurants, banks, stores of all kinds, and more.

State and local governments also may seek to attract new business by lowering corporate taxes or offering special tax incentives or cash grants for businesses that are willing to relocate. As an example, in 2017 the state of Wisconsin offered infrastructure improvements and a multibillion-dollar subsidy to convince Foxconn, a large Taiwanese multinational tech company, to build a manufacturing plant in the southeastern part of the state. The $10 billion plant was expected to produce liquid crystal display screens (LCDs) for phones, televisions, and other products and to employ 13,000 workers. However, in 2021, Foxconn scaled back its commitment and indicated it would spend well under $1 billion and employ fewer than 1,500 workers—and the state of Wisconsin then greatly reduced their financial incentives package. Similarly, about 30 years ago, South Carolina lured German automaker BMW to build a plant in Spartanburg by offering some $100 million in incentives, including funding to extend a runway at the airport and to improve roads and sewers.

Providing financial incentives is just one tool that governments use to fuel economic development. Governments may provide workforce training programs, particularly if the economy is in transition from one type of industry to another. In addition, governments seek to provide an environment that is conducive to doing business. This means providing political and economic stability, as well as reliable infrastructure, including transportation and communication networks that require regional and local governments to work together.

National governments can also encourage economic development through the use of tariffs. The goal of tariffs is to give domestic businesses an advantage by taxing foreign competitors' imported products and making them more expensive. The cost of the tariff is included in the resource or product price, which may enable a domestic company to offer goods at a lower price than foreign competitors. This helps domestic businesses create more jobs and possibly pay workers more. But while tariffs protect a country's industries, including agriculture or mining, they can have a negative impact on the economy. Tariffs limit free trade and reduce competition, which usually results in higher prices for consumers. Quality and innovation also may suffer when there is less competition in the market.

Imposing tariffs can prompt other countries to respond with their own tariffs. When U.S. President Donald Trump imposed additional tariffs on a wide range of goods imported from China in late 2018, China retaliated with tariffs on U.S. agricultural products. The trade war affected the

Apple Park, the corporate headquarters of Apple, is located in Cupertino, California. In 1997, the city made a deal to pay 35 percent of local sales taxes to the company so long as it agreed to remain in Cupertino and build a campus there. It is estimated that Apple has collected almost $70 million in sales taxes since the deal was struck; in return, Cupertino reaps rewards from being Apple's hometown, including the 65 percent of sales taxes it keeps for itself.

economies of both countries. As one example, Shanghai General Sports, one of the biggest bicycle manufacturers in China, had to lay off almost one-third of its employees. In South Carolina, Kent Bikes, which imports Shanghai General Sports' parts, had to raise its prices on bikes to cover the cost of the tariffs on the parts. This contributed to a significant drop in sales and resulted in the layoff of about one-fourth of Kent Bikes' employees. Ripples of this trade war were felt throughout the world and contributed to an economic slowdown worldwide.

Free trade agreements among governments eliminate tariffs and often boost trade between the countries involved. Peripheral and semi-peripheral countries gain manufacturing jobs and core countries gain access to low-cost labor. The United States has trade agreements with more than 20 countries. Since no tariffs are paid on exports to those bound by the agreements, U.S. businesses target their exports to these countries.

Geographic Thinking

1. Explain deindustrialization and how it impacts regions going through this process as well as other regions.

2. Explain how spatial patterns of globalization relate to economic development.

3. Describe the advantages and disadvantages for countries establishing trade relations with one another.

Globalization, Interdependence, and Sustainability

20.2 Connected Economies

The economies of countries are interdependent. Changes in industry are one cause of this interconnectedness. Industries apply international strategies that affect jobs and manufacturing practices, and countries institute policies in order to spur local, regional, and national economic development.

Impacts of the Global Economy

Learning Objective
LO 20.7 Explain how multinational corporations and financial institutions that operate outside of their country of origin impact the global economy.

Businesses in market economies naturally seek cheaper ways to do things. Labor tends to be far less expensive in countries in the periphery and semi-periphery than in core countries. In the 1970s, U.S. and European corporations increasingly decentralized their manufacturing processes to take advantage of inexpensive sources of labor (and the lack of unionized labor and environmental regulations) in other countries. Some companies established production facilities abroad; others looked for existing suppliers to integrate into their production process. Corporations also built offices around the world to oversee and facilitate an increasingly complex supply chain and to extend their reach to new markets. The proliferation of multinational corporations operating at a global scale was followed by the globalization of banking and financial services. By the 1990s, global companies provided these and other business needs like accounting, advertising, and legal services throughout the world, facilitating the global flow of money, goods and services, and information. National and regional economies became increasingly interconnected and were influenced by global forces more than ever.

Today, the widening economic development gap between core and peripheral countries is evident when comparing global incomes, a common measure of global wealth. According to the World Bank, the gross national income (GNI) per capita for the United States in 2020 was $64,140. For Bangladesh, the GNI was $2,340 for the same year. The imbalanced distribution of wealth across the globe impacts needed development. Peripheral countries lack the funds for development, so investors from the core provide financial support through direct investment and loans.

Multinational corporations that operate in countries other than the ones in which they are headquartered are large sources of funds for peripheral and semi-peripheral countries. Corporations such as Apple, Microsoft, Nike, and Coca-Cola all have investments and operations in developing economies. The facilities that these companies build in the periphery and semi-periphery increase the local productive capacity. The influx of capital can also help the economy develop. In 2019, semi-peripheral Taiwan's GDP grew almost 3 percent, mainly due to exports in the electronics sector, including smartphones. These corporations provide local employment opportunities. The wages may be low compared to wages in the core, but they are typically much higher than other alternatives in the area. A factory or production facility also helps to diversify the local economy, providing an important alternative to agriculture. Multinationals may also invest in infrastructure, such as road improvements, workforce education, and health initiatives.

Financial institutions have also contributed to an increasingly interrelated global economic system. In addition to microlending, deregulation (the removal of restrictions) of the banking industry in many countries has brought new funding to businesses operating in the periphery and semi-periphery. Global financial institutions like Goldman Sachs, Deutsche Bank, Credit Suisse, HSBC, Citigroup, and the Industrial & Commercial Bank of China have expanded their reach and helped increase the number of businesses in new parts of the world. These businesses and startups in turn help the global economy as a whole.

The International Monetary Fund (IMF) and the World Bank also foster economic development. The IMF provides no-interest loans to low-income countries and offers financial assistance to member countries to help stabilize the world economy. The World Bank, with its 190 member countries, similarly focuses on encouraging a global economy to reduce poverty, increase shared prosperity, and promote sustainable development worldwide. It provides funding and technical assistance to countries in the periphery to develop infrastructure, including systems for safe drinking water, improved sanitation, new schools, and expanded transportation and communications networks. Some countries use the money to repay other loans, provide emergency relief, or expand health care. The establishment of these international lending agencies demonstrates how widely varying economies are now more closely connected.

The global economy has contributed to integrated financial markets like the European Union. Members of the EU conduct commerce and trade free from barriers that could hinder investment in another EU country. While financial integration helps facilitate the flow of capital across national borders, the risks of interconnectedness to global markets became evident during the 2007–2008 worldwide recession and the COVID-19 global pandemic that began in 2019.

Case Study

The Financial Crisis of 2007–2008

The Issue The financial crisis of 2007–2008 resulted in widespread unemployment and a worldwide recession. What is the obligation of countries to closely regulate their financial institutions?

Learning Objective
LO 20.7 Explain how multinational corporations and financial institutions that operate outside of their country of origin impact the global economy.

By the Numbers

$12.8 trillion
Cost of the crisis to the U.S. economy based on GDP loss

5 years
for the U.K.'s economy to return to its prerecession size

$586 billion
Cost of China's stimulus plan to address downturn in its economy

Sources: U.S. Government Accountability Office, UK Office for National Statistics, World Bank

During the financial crisis of 2007–2008, widespread homelessness in the United States resulted from people losing their homes due to foreclosure. Foreclosure occurs when homeowners who cannot afford their house payments are forced to sell their homes. As a result, many people had to live in tent cities like this one in Sacramento, California.

In 2008, the World Economy Plunged, contributing to the worst financial crisis since the Great Depression of the 1930s. The crisis began in the United States when home prices started to fall at an alarming rate. Monitored by regulatory laws that proved too loose, U.S. mortgage dealers had issued high-risk mortgages to homebuyers who could not qualify for traditional loans because of low credit scores or low incomes. These mortgages offered low initial interest rates that increased, or even doubled, after a few years. They also often included steep penalties for missed payments. Many homeowners borrowed against the equity in their homes to pay the loans. Equity is the difference between the value of the property and what is still owed to the lending institution. Using equity became a problem when housing prices fell, however; many people found themselves owing more than their houses were worth. Numerous homeowners were unable to make their mortgage payments, and either sold their homes at a loss or simply abandoned them.

U.S. banks and other financial institutions that made these home loans were not being repaid by the borrowers; some could not recover and went out of business. Because banks could no longer afford to loan money, the crisis expanded to companies that usually operated on credit, including the American auto industry. The federal government bailed out both the banks and the auto industry by offering hundreds of billions of dollars in loans, payouts, or investments in each industry.

The crisis spread to major trading partners and countries that had significant investments in American real estate. Export-oriented countries struggled as demand in American and European markets plunged. By the end of 2008, Germany, Japan, and China were all struggling with a recession. Peripheral and semi-peripheral countries suffered as foreign investment dried up. Unemployment increased in the United States and in countries that provided manufacturing labor for U.S. companies.

The recession changed the world of international finance. Governments were forced to make huge investments to strengthen their banks and guarantee their loans. The U.K. government, for instance, provided more than $500 billion in financial assistance to its banking industry. While the economies of most core countries have largely recovered, many peripheral countries are still feeling the repercussions. The crisis revealed the potential risks of a global economy and an interconnected world. These risks were again realized in 2022, as global inflation in the wake of the COVID-19 pandemic coupled with Russia's invasion of Ukraine caused some experts to warn of another looming worldwide recession. ∎

Geographic Thinking

Explain how the financial crisis of 2007–2008 was an indication of globalization.

The Great Recession, 2007–2009

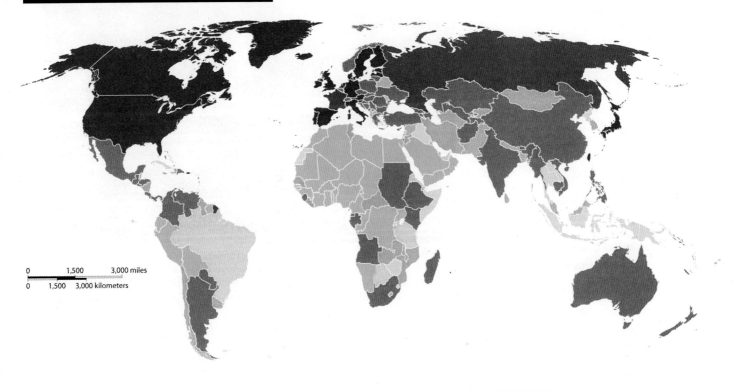

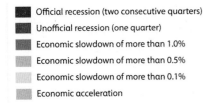

- Official recession (two consecutive quarters)
- Unofficial recession (one quarter)
- Economic slowdown of more than 1.0%
- Economic slowdown of more than 0.5%
- Economic slowdown of more than 0.1%
- Economic acceleration

Reading Maps The financial crisis of 2007–2008 led to a global recession called the Great Recession that officially lasted from December 2007 to mid-2009. A recession is defined as two consecutive quarters of economic decline. During the Great Recession, many countries experienced full-fledged recession, others an economic slowdown, and some actually saw economic growth. ▎Explain what might have caused countries in Africa to be spared the worst of the Great Recession. Think about what you have learned about globalization and interconnectivity.

Austerity Measures

In the years following the Great Recession, numerous policies were enacted around the world in hopes of lessening the long-term impact of the crisis and to prevent it from happening again. With an eye on cutting countries' deficits, the IMF urged the passing of austerity measures. A country's deficit—or the amount that its spending exceeds its revenues—rises and falls as its economy contracts and expands. In a period of economic growth, tax revenues increase and unemployment benefits decrease, which reduces the deficit. Austerity policies aim to reduce a government's structural deficit, which is debt that is persistent for some time and that is not affected by market forces. Austerity measures usually include tax increases, spending cuts, or both.

The United Kingdom imposed austerity policies following the crisis. In 2010, Parliament imposed tight spending controls in an attempt to reduce the deficit. It also raised taxes to increase revenues. As a result, the United Kingdom experienced another recession before its economy rebounded, causing economists to disagree on the impact and effectiveness of the austerity measures.

Another country to enact austerity measures was Greece, which by 2009 had a very high deficit. In 2010, a series of severe budget cuts and tax increases were passed; however, concern with the country's overall economic health and its impact on the euro, which is the currency of most of the European Union, did not decrease. As a result, the EU and the IMF were compelled to loan Greece large amounts of money to prevent its economy from collapsing. Cycles of spending cuts and bailout loans continued until 2018, when its GDP grew at 1.9 percent.

Austerity policies are controversial. Many economists believe that they can do more harm than good because they counteract the natural forces of an economy. Failing to invest in a weak economy may cause tax revenues to fall further while spending on benefits goes up. The contrasting viewpoint holds that in recommending austerity, the IMF was conscientiously looking for a global response to a global crisis. The IMF pointed out that financial support (including loans and bailouts) for troubled countries must consider cross-border effects, including on emerging economies. In other words, any solution must take into account the forces of globalization.

Changes in the Economic Landscape

Learning Objectives
LO 20.8 Describe the impact of the international division of labor.
LO 20.9 Explain how offshore outsourcing and economic restructuring have led to economic development in peripheral countries.

By 2019, the contemporary economic landscape had been transformed by many factors. Computerized logistics systems facilitate **just-in-time delivery** of raw materials and manufactured parts. Logistics is the handling of the details of an operation, such as the movement of supplies between regions. Entire systems are devoted to determining how to efficiently transport supplies and products to and from factories and stores. Just-in-time delivery means that materials are delivered when they are needed for short-term production, so that companies can avoid paying to store extra inventory at their facilities. Improved logistics and efficient transportation streamline operations and enable companies to be located farther from raw materials, which has fueled growth and contributed to lower labor costs.

However, the COVID-19 global pandemic made countries, companies, and consumers across the world acutely aware of the risks of disruptions to global supply chains of just-in-time logistics. The goal of supply efficiency led to low inventories of essential items when factories in China and elsewhere were no longer able to provide goods or the supplies to make products. For example, the supply of computer chips, or semiconductors—used in a huge range of products from smartphones and gaming consoles to automobiles and medical equipment—fell dramatically in 2020 and 2021. For the global auto industry, computer chips are used in everything from power steering to entertainment systems, so the pandemic-caused chip shortage meant cuts in production and deliveries of new vehicles. This new-car disruption also caused used-car prices to skyrocket.

Changes in industry have helped to fuel global interconnectedness. Before the 18th and 19th centuries, goods were produced largely in people's homes. Then the Industrial Revolution transformed industry with the growth of factories, leading to the advent of the assembly line. Each event has been another step toward greater interdependence among countries. In the 1920s, Henry Ford's assembly line revolutionized the way businesses operated and enabled mass production of complex goods such as automobiles. Mass production and rising affluence allowed the broad masses of people to expand their use of goods, termed mass consumption. The ultimate goal of factories was to make products for mass consumption faster and more efficiently. This system of manufacturing, known as **Fordism**, focused on automation, standardization, economies of scale, and a division of labor in which each worker has just one task. Automation, or the use of technology to automate a process or system, meant that machines replaced workers in the secondary, manufacturing sector, which enabled large-scale mass production.

Today, globalization and deindustrialization have changed the decision-making criteria for businesses and industries. In recent decades, an increasing number of companies have begun to turn away from mass production, which requires huge capital investments, to instead focus on specialization. The term **post-Fordism** is used to describe the system of production that relies on automation through the use of robots and computer systems and is centered on

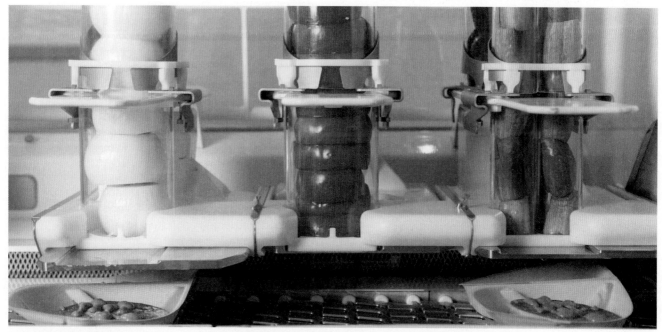

Critical Viewing A culinary robotics company in San Francisco, California, has devised a machine that cooks and assembles burgers. The machine contains 20 computers, 350 sensors, and numerous other mechanisms. ▮ Explain why this machine is an example of post-Fordism.

Globalization, Interdependence, and Sustainability

low-volume manufacturing and flexible systems that allow for quick responses to changes in the market.

Post-Fordist companies either are much smaller than traditional manufacturers or contract work to smaller firms that oversee different parts of production, including fabricating or marketing a product, and are therefore better able to respond to changes in demand. Often the bureaucracy of a single, large corporation can slow decision making and hinder timely changes in production.

Post-Fordist organizations also tend to be less top-down than Fordist organizations; rather than imposing a strict division of labor, post-Fordism encourages the use of multidisciplinary teams of employees with different types of knowledge and skills that work together on a project from start to finish.

Post-Fordism is made possible by advances in computer and information technology, which provide new tools that emphasize knowledge and creativity over physical labor.

In the post-Fordist environment, rather than an entire car being manufactured and assembled in one plant, each part is made in a different location by a company that specializes in that one part. Automobile companies then purchase parts from a range of suppliers across the world. Post-Fordist industry may also focus on a specialized product for the end consumer and create a smaller amount of the product at one time to accommodate product changes as technology advances or consumer tastes change. Several global chains of clothing stores including Benetton, Zara, H&M, and Superdry vary their product lines frequently and regularly use different suppliers of materials to accommodate changing clothing preferences.

Post-Fordism has changed the manufacturing landscape. When Ford made its mass-produced cars, there was a joke that consumers could have any color—as long as it was black. In the post-Fordist world, companies provide consumers with far more choice. Post-Fordism has also made it possible for many companies to start small. As a student at the University of Texas, Michael Dell started Dell Computers by providing customized upgrades for personal computers, or PCs. The company grew by continuing to create and sell custom-built PCs directly to consumers through advertisements and mail-order catalogues. Today, many computer software programs and phone applications are created by small, independent companies that target a niche or specialty market. The app YogaGlo, for instance, focuses on online yoga, meditation, and Pilates classes. Its small size enables it to target specific products to accommodate changes in preferences and technologies.

Outsourcing Advances in technology and communications have allowed companies more flexibility in the locations they choose to conduct business. As the service industry grows, many jobs like accounting, customer service, and research and design are each being performed in separate locations or in multiple locations. Many corporations outsource aspects of production or information

OUTSOURCING

Effects on Outsourcing Countries	Effects on Countries Providing Outsourced Labor
Loss of jobs	Gain jobs
Domino effect leading to other similar businesses or suppliers also outsourcing	May stifle economic development; workers may be stuck in lower-paying jobs
Competition threatens to drive down labor standards and wages	Possible access to higher than average wages depending on the skills required
Companies can operate more efficiently, increasing profits	Multiplier effects
Goods available at lower prices	Technology transfer

Outsourcing impacts both the country using the labor and the country providing it. A good example of the latter is India, where IT outsourcing employed 3.7 million workers in 2019. The IT industry generated more than $175 billion for India, largely driven by providing outsourcing to core countries.

services, turning the work over to an outside provider to cut costs. Outsourcing can take place within a country, but companies are increasingly moving production to places outside the country in which they are headquartered—a process referred to as **offshore outsourcing**. Some large companies in the United States and the United Kingdom, for example, have established customer service call centers in the Philippines and India, respectively, where lower wages and English-language use make this outsourcing effective and less expensive.

Internal communication across the geographically dispersed workforces of a multinational corporation is facilitated when the workforce speaks a common language. Even among organizations based across Europe or Japan, English has become the lingua franca, or language of choice, for business. As a result, English proficiency of a potential workforce is an important criterion for companies deciding where to build a new plant or outsource production or when choosing a supplier. This puts some countries at significant advantage over others. The high level of English proficiency among educated workers in the Philippines and India is one reason that many companies have chosen to relocate their customer service operations offshore to these places.

Core countries have taken advantage of cheap labor sources by moving production facilities to newly industrialized countries (NICs). A NIC economy has transitioned from largely agriculture to primarily goods-producing over the last several decades. Offshore outsourcing has led to a decline of traditional manufacturing jobs in core regions and increased the number of such jobs in NICs. Vietnam, though still considered peripheral, is among the countries that have benefited. Nike, an American sportswear and shoe company headquartered in Oregon, employs almost four times more workers in Vietnam than in the United States. In the early 21st century, Vietnam became known for low-end manufacturing industries, but

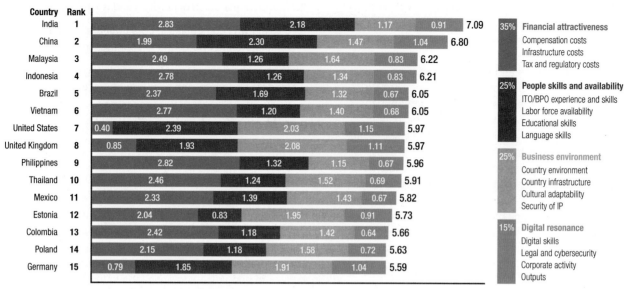

Global Services Location Index, 2021 Every year, a firm called A.T. Kearney releases its Global Services Location Index, a ranking of countries based on their desirability as an outsourcing destination. Countries are evaluated in four areas: financial attractiveness (which is most heavily weighted), people skills and availability (with ITO standing for "information technology outsourcing" and BPO meaning "business process outsourcing"), business environment (with IP meaning "internet protocol"), and digital resonance (which is weighted the least). ▮ Explain why, based on what you have already learned, the United States, the United Kingdom, and Germany likely get such low scores in financial attractiveness.

Source: A.T. Kearney

its reputation soon grew to attract more technologically advanced manufacturers as well. In 2013, Microsoft built a factory in Bac Ninh, Vietnam, to serve as the main production center for its Nokia cell phones. South Korean manufacturer Samsung similarly has moved offshore the production of many of its phones, as well as appliances, flat-screen TVs, and other consumer electronics, to Vietnam.

Division of Labor In the late 1800s, wealthy and powerful countries of the core instituted an **international division of labor** through imperialism. Colonies or other countries in peripheral regions specialized in producing the raw materials required by more developed economies, and then served as markets for the manufactured goods produced by core countries. The result was a spatial pattern of production and labor based on geographic comparative advantage. Specialization of production and labor continued after countries gained their political independence. Less-skilled workers in the periphery and semi-periphery produced raw materials for trade with core countries who specialized in manufacturing by more-skilled labor.

Since the 1970s, the international division of labor has evolved, as core countries began to deindustrialize. Corporations sought lower-wage labor locations for their manufacturing and decentralized their operations, both within their own country and to peripheral and semi-peripheral countries. Regions and countries across the globe have developed new specializations. As noted, the footwear industry has exploded in Vietnam. There were 800 businesses in Vietnam that produced more than 1.2 billion pairs of shoes, making it the world's second largest exporter of footwear in 2020. China, meanwhile, has specialized in the manufacturing of high-tech goods, such as flat-screen televisions and computer parts. More than two-thirds of rare earth materials—which are used in the production of electric vehicle (EV) batteries and electronics— are mined in China. The country also makes more than half of the world's circuit boards, fundamental building blocks for computers. A state-led industrial policy called "Made in China 2025" calls for even further development in high-tech fields ranging from robotics to aerospace engineering. The program aims to use government funding to advance the use of emerging technologies in manufacturing and infrastructure.

International Division of Labor

Reading Maps The international division of labor is best expressed by a map of the so-called North–South Divide. ▮ Identify what the map shows about countries that primarily speak English and then draw a conclusion about the North–South Divide today and in the future.

Globalization, Interdependence, and Sustainability

Other economic activities have also decentralized from core countries. Financial services, like those provided by credit card and accountancy companies, have moved to peripheral and semi-peripheral countries, aided by technological innovations. Transportation and communications advances have internationalized production and trade, accelerating globalization. More and more peripheral and semi-peripheral countries are becoming integrated into the global economic system as both producers and consumers, and incomes are rising for many. It is important to note that these processes work both between and within countries. For example, in the last decade, some manufacturers have left China for still lower labor wages in other countries of Southeast Asia, like Thailand and Vietnam. Some manufacturing in Mexico has been moving from higher labor-cost areas along the United States border to lower-wage regions much farther south.

Spatially, today's new international division of labor is represented by the North–South divide: countries of the Global North, with just one-quarter of the world's population, control four-fifths of the world's income, and countries of the Global South, with three-quarters of the population, have acquired just one-fifth of the world's income.

One factor driving down wages in the Global South is competition to provide cheap labor. As a result, many factory workers in NICs work long hours for little pay. More than a billion people worldwide earn less than $2 per day working in factories. Critics argue that people should be paid a fair wage regardless of where they live. On the other hand, many multinational corporations pay more than local companies because they can afford higher wages. Advocates of offshore outsourcing further argue that people in peripheral countries often are in desperate need of jobs and opportunities and that the **multiplier effects** will help the local economy grow. Multiplier effects are opportunities that can potentially develop from an economic change. For instance, when a corporation chooses to build a new manufacturing plant, it brings jobs to the area. In addition, the plant and its workers need support services, including office supplies, lunch, and transportation to and from work. The money spent on these local goods and services sometimes results in the area businesses adding staff. Multiplier effects can provide a significant boost to the local economy. The business and workers, if in the formal economy, also pay taxes on their earnings. Combined with the taxes on the goods that are sold, this can fuel the national economy as well.

The economies of countries across the globe look very different from one another but are nonetheless interdependent. Today, economically dominant core countries and less wealthy peripheral and semi-peripheral countries rely on each other for economic growth—though the economic gains are often unequal.

Geographic Thinking

1. Describe the factors that contributed to the rise of post-Fordism.
2. Compare the goals of Fordism and post-Fordism.
3. Explain why companies in core countries outsource manufacturing operations to countries in the periphery.
4. Explain the degree to which outsourcing helps and hinders peripheral and semi-peripheral countries.

New Manufacturing Zones

Learning Objective
LO 20.10 Compare how special economic zones, free trade zones, and export-processing zones increase economic production and development.

In many countries outside the core, the growth of industry has resulted in the creation of new manufacturing zones. These zones are established by many governments to attract foreign investment. Corporations are offered incentives to bring manufacturing jobs to the zone and in return the country gains expanded trade, the transfer of technology and management expertise, and new employment opportunities for its residents.

Special Economic Zones Some countries have attempted to facilitate economic growth by creating **special economic zones (SEZs)**. A special economic zone is an area within a country that is subject to different and more beneficial economic regulations than other areas. Companies doing business in a SEZ usually receive tax incentives and are subject to lower or no tariffs. Governments tend to provide SEZs with more accessible and reliable infrastructure (including land, utilities, and transportation improvements). The goal of creating SEZs is to bring foreign business and investment to generate economic development. However, economists and government officials debate their economic benefits. One aspect that has drawn criticism is their tendency to operate as isolated groups with few benefits spilling over to the country's local suppliers.

SEZs were initially used in industrialized countries, but since the 1980s, they have proliferated in peripheral and semi-peripheral countries in East Asia and Latin America. More recently, countries in Central and Eastern Europe, Central Asia, Southwest Asia, and North Africa have begun implementing SEZ programs to compete for the growing interest in international production.

China is among the most successful countries using SEZs. Beginning with the Shenzhen Special Economic Zone in 1979, China based its initial SEZs along the southeastern coast of the country, including the cities of Shenzhen, Zhuhai, Shantou, and Xiamen. The economic success of these four zones prompted the Chinese government to add 14 cities plus the province of Hainan to the list of SEZs. In 2019, SEZs contributed more than $1 trillion to China's GDP, and they have generated millions of jobs. China continues to allow its SEZs to offer tax incentives and develop local infrastructure without central government approval.

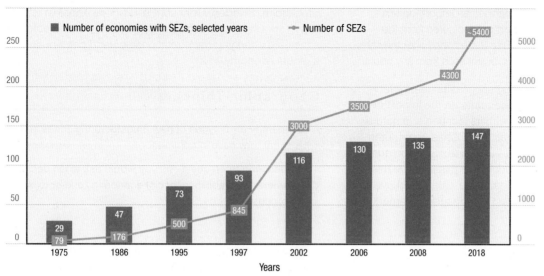

Special Economic Zones The increase in the establishment of new SEZs globally in recent years has been rapid. ▌Compare the total number of SEZs to the number of economies with SEZs. What conclusions can you draw from the data?

Other countries, especially throughout Asia, have followed China's example with varying degrees of success. The Philippines and India have each established hundreds of SEZs. Some SEZs, such as in South Korea, seek to address uneven development within the country. Cambodia created SEZs to establish links between urban and rural areas. Other countries have used SEZs to try to diversify their economy. Specific SEZs specialize in services, innovation, or natural resource processing. SEZs are found in almost all countries in Central and South America. As of 2019, the region had almost 500 SEZs that collectively employed about 1 million people. Sometimes, however, the cost of investment in maintenance of the zone outweighs the benefits. They can cause distortions within economies, favoring one area over another, leading to devolutionary jealousies. Many SEZs lack the needed infrastructure such as adequate energy sources and transportation required for sustained success.

Export Processing Zones The purpose of **export processing zones (EPZs)** is to attract multinational organizations to invest in labor-intensive assembly and manufacturing in the host country. These zones are sites where manufacturing of exports is done without tariffs. Governments have added other financial incentives to attract foreign investors and encourage economic growth. In addition to having a narrower focus than SEZs—specifically on exports—EPZs also tend to be smaller, as a single industrial park is generally sufficient for a manufacturer. EPZs also tend to be located where access to water or air transport is readily available. In many countries, EPZs are near urban centers with an adequate supply of labor and advanced infrastructure.

Tanzania in East Africa initiated its EPZ program in 2002 to encourage investments in export-oriented manufacturing. Sites have been earmarked in 20 regions of the country and include 6 industrial parks as well as more than 50 single factory zones. The Tanzanian program has attracted businesses in engineering, textiles, agriculture, and mineral processing industries. EPZs also have a long tradition in Latin America. For example, many U.S. corporations have located factories in northern Mexico just outside the border in an EPZ that allows for quick export of goods back into the United States.

Incentives for companies conducting export-related business in EPZs include tax breaks and exemptions. Companies do not pay taxes on the machinery or resources that they import into the EPZ as long as the goods being manufactured are for export. This regulation helps to eliminate competition for local customers between local manufacturers and EPZs. Other incentives include support for visas, work permits, and customs documentation.

Free Trade Zones The use of **free trade zones (FTZs)** has increased rapidly as countries have established them on major trade routes to provide duty-free areas for warehousing, storage, and transport of goods. While the goals of FTZs are similar to SEZs and EPZs, they tend to cover a larger geographic area. All of Hong Kong and Singapore, for instance, are free trade zones.

FTZs provide customs-related advantages and exemptions from tariffs and taxes. This enables quicker turnaround of ships, planes, or other means of transportation engaged in international trade, which enables ports to function more easily as points along the way of a larger transportation system.

It is important not to confuse free trade zones with free trade areas. A free trade area is a region specific to and encompassing two or more countries that have agreed to reduce trade barriers. A free trade zone is a special area within a country where foreign companies can import materials, manufacture goods, and export products free from the usual taxes and regulations. Free trade zones offer the same incentives and opportunities to businesses from any country.

Hong Kong is one of the most successful free trade zones. Hong Kong charges no tariffs on the import or export of goods; combined with its location and natural harbor, Hong Kong's status as an FTZ has contributed to its success as a major port city. In 2013, China added another FTZ in Shanghai; success of this initiative has led to additional FTZs in China. Other cities that have FTZs include Colon, Panama; Copenhagen, Denmark; Stockholm, Sweden; Gdansk, Poland; Los Angeles; and New York City. Across the world, these new manufacturing zones are changing trade relationships and contributing to a global economy.

Geographic Thinking

5. Identify three types of new manufacturing zones and what they have in common.

6. Describe three questions you would ask when deciding where to establish a special economic zone or export processing zone.

20.3 Developing a Sustainable World

As the world's population grows, many people are concerned about the future availability of natural resources. With accelerated consumption, increased land use, climate change, and land cover changes come dramatic environmental degradation and a decrease in the supplies of nonrenewable resources. Are current economic development approaches sustainable?

Natural Resources, Development, and Consumption

Learning Objective
LO 20.11 Explain how the exploration, extraction, and consumption of natural resources can impact a country's economic development and future productivity.

Whether consumed domestically as raw materials for manufacturing or sold as export goods, natural resources contribute greatly to a country's economic development. The natural resources in oil- and mineral-rich countries like Brazil, Saudi Arabia, Canada, and India are worth trillions of dollars and contribute substantially to each country's economy. In 2022, the natural resources found in Russia alone were estimated to be worth more than $75 trillion. Russia's mining industry, including metals, industrial minerals, and fossil fuels (coal, oil, and natural gas), is one of the biggest in the world. As mentioned, China controls more than two-thirds of the world's production and refining of rare earth minerals, which are used in many essential technologies including EV batteries. With such a dominance in this natural resource, China is able to use access to rare earth minerals as a bargaining tool in trade disputes with other countries.

The depletion of natural resources, on the other hand, can negatively impact a country's economy and future productivity. This would be particularly troublesome for a country with a commodity dependency, with an outsized reliance on just one or two natural resources of economic value on the global market. Without the raw materials to create products or trade for export income, a country's economy could suffer. Of course, the extraction, processing, and use of nonrenewable resources, like oil and coal, have serious environmental, social, and health costs. These greenhouse gas-producing fossil fuel resources are also drivers of climate change.

Pollution of many types, wastes of all kinds, and the loss of diverse habitats are just some of the environmental consequences of development. The construction of roads to transport extracted products often contributes to deforestation, particularly in the tropical and subtropical rain forest regions of the planet. In some places, local workers engaged in mining operations cut down trees for fuel. Resource extraction also opens new areas to others engaged in agriculture activities, illegal logging, or the poaching of protected animals, all causing further degradation to the natural environment. Altering the land for aquaculture also has negative impacts. Shrimp farms along much of Ecuador's coasts have replaced about two-thirds of the country's species-rich coastal mangrove forests.

Resource exploration and extraction have social effects, especially on Indigenous communities. In the Amazon

Basin of Ecuador and Bolivia, the oil industry has displaced Indigenous peoples. Elsewhere, mining has disrupted the traditional way of life of many Indigenous communities, who typically receive little of the economic benefits that result from resource extraction on their lands. Instead, they may experience devastating health problems that some believe are consequences of toxic by-products entering local rivers when oil or other resources are extracted. Leaks from oil pipelines are real risks to people and the environment.

Many believe that humans use more natural resources than the environment can sustain and, not surprisingly, the wealthiest countries are generally the largest consumers of these resources. As concerns grow about humans' ability to maintain development globally, governments, businesses, organizations, and individuals are seeking strategies that will reduce the human footprint, including ecotourism and innovative practices that cut waste. Technology is playing a significant role in efforts to conserve natural resources and ensure sustainable development that meets the needs of present populations without compromising the ability of future generations to do the same. In Singapore, technologies are being developed to help solve fresh water shortage concerns by cleaning and filtering waste water.

In addition, interest in renewable sources of energy has increased. In the United States, renewable energy (wind, solar, hydropower, geothermal, and biomass) grew from about 3 quadrillion BTUs (British Thermal Units, a measure of the heat content of energy sources) in 1950 to more than 11 quadrillion BTUs in 2020, when it accounted for 12.5 percent of the country's total energy usage. The European Union has a stated goal of generating all of its electrical power from renewable sources by 2031. EU member country Denmark, a pioneer in wind energy construction and use, is building an "energy island" in the North Sea that will be the hub for an extensive offshore wind farm of as many as 600 gigantic wind turbines. When the island hub of transmission and battery storage is completed, the $34 billion wind farm is expected to provide electricity for the entire country, with additional green power to sell to EU neighbors. These attempts to diversify energy sources have been amplified by calls to break dependence on Russian energy sources in the wake of the 2022 invasion of Ukraine.

Due to a global economy, core, peripheral, and semi-peripheral countries are all dependent on each other for natural resources, agricultural goods, and the manufacturing of products. Therefore, natural resource protection has become increasingly global in nature, but the burden of sustainability is not equitable. Peripheral and semi-peripheral countries involved in resource extraction are often more greatly impacted by the negative environmental effects and social inequalities, while core countries can better afford to undertake conservation efforts and human rights protections.

However, global actions to fight climate change have produced more collaborative approaches between wealthier and poorer countries, with the possibility of benefits for all. Multipurpose efforts by core countries to assist in protecting the environments and people of peripheral and semi-peripheral countries—while also helping these poorer economies adopt the use of sustainable energy sources—are shown by numerous climate change commitments signed at the United Nations (UN) Conference of Parties 26 (COP26) held in Glasgow, Scotland, in late 2021. One such agreement is the $8.5 billion "South Africa Just Energy Transition Partnership," with the United Kingdom, Germany, France, and the United States entering a long-term partnership with Africa's largest emitter of carbon dioxide. South African President Cyril Ramaphosa welcomed the commitment as "an appropriate model of support for climate action from developed to developing countries, recognizing the importance of a just transition to a low carbon, climate resilient society that promotes employment and livelihoods." Another collaboration is the "Global Energy Alliance for People and Planet," a $10 billion plan designed both to bring reliable renewable electricity to a billion people in the periphery by 2030 and also to avoid billions of tons of carbon dioxide emissions. All told, well over $20 billion was committed to what COP26 called "a just and inclusive transition from coal to clean energy."

Collaborative mitigation efforts were also initiated to protect tropical rainforests and Indigenous peoples, as well as to slow climate change, with 137 countries pledging at COP26 to halt and reverse deforestation by 2030. A group of 12 core countries and nongovernmental organizations (NGOs) pledged at least $1.5 billion to help protect Africa's Congo basin forests, while another partnership of 14 donor countries and NGOs committed a minimum of $1.7 billion from 2021 to 2025 to "advance Indigenous Peoples' and local communities' forest tenure rights and support their role as guardians of forests and nature." Additionally, a collaboration of 28 core, periphery, and semi-periphery countries launched a "roadmap" to protect forests through a global shift to sustainable development and trade in agricultural commodities.

Sustainable Development Goals

Learning Objectives

LO 20.12 Describe the sustainable development policies governments have adopted to conserve natural resources and reduce environmental degradation.

LO 20.13 Identify the economic, political, and geographic challenges to widespread conversion to sustainable energy production.

Sustainable practices are key to future development in order to maintain supplies of raw materials required by the economies of every country. Governments have adopted sustainable development policies that attempt to remedy problems stemming from natural-resource depletion and mass consumption. Some businesses and industries commit to following sustainable practices as well. These policies and practices aim to conserve natural resources and reduce environmental degradation.

Case Study

Reducing Waste in Fisheries

The Issue As a result of various factors, most fisheries in the world discard certain parts of the fish they catch, resulting in waste of a potentially valuable product.

Learning Objective
LO 20.12 Describe the sustainable development policies governments have adopted to conserve natural resources and reduce environmental degradation.

Iceland's ocean fisheries have long been a major part of its economy. Today, fisheries are utilizing more than 80 percent of the fish caught, and many leading fisheries are making it their mission to use 100 percent of the fish through sustainable methods and new business opportunities.

By the Numbers

1.2 million metric tons
Total amount of fish caught in Iceland in 2017

5 million metric tons
Total amount of fish caught in the United States in 2017

$655 million
Potential value loss resulting from fish by-product waste in the United States

Source: FAO, Iceland Ocean Cluster

The Majority of Commercial Fishing Fleets in the world discard the parts of a fish not intended for human consumption because they are considered to have little value. This includes the head, internal organs, and bones of the fish, known as by-products. At least 35 percent—some estimates are even higher—of the total volume of fish caught commercially in the United States is discarded. While some by-products end up in fish meal, fertilizer, and animal feed, the majority of fish waste is simply disposed of in the ocean.

In Iceland, efforts are being made to achieve 100 percent fish utilization. Icelandic authorities are working with organizations to support innovation in Iceland's seafood industry. The result has been the establishment of several successful companies in the beauty and wellness, food supplement, and medical and health product industries, which utilize enzymes found in the fish to create fish oils and proteins used in cosmetics and medical products. These companies are utilizing the parts of the fish that would otherwise be discarded to create their products and therefore increasing the value of fish by-products.

The reasons for the wide variation in fish utilization among countries' marine fisheries, including the United States and Iceland, is a result of several factors. The first factor is how long fishing vessels are out at sea. Vessels that must travel greater distances to catch fish keep the fish from spoiling by freezing as much as possible on the ship. Therefore, discarding the head and organs allows more storage space for the valuable fillet. Another factor is the length of the fishing season. Icelandic fleets can fish year-round, but for many other fleets the fishing season is limited to a few months. Ships catch large amounts of fish in a short period of time, making processing the massive amount of raw fish difficult. A third factor is vertical integration, when one company manages two or more stages in the manufacturing process, leading to efficiencies that reduce the amount of fish wasted. In Iceland, companies commonly control both the catching and the processing of the fish.

The effects of Iceland's efforts to increase fish utilization are proof that sustainable practices help both the economy and the environment. The increase in the use of the entire fish has led to less waste and the creation of at least 600 to 700 jobs in Iceland's by-product industry, which is valued at around $500 million.

Geographic Thinking

Explain how Iceland's fish utilization efforts might impact fisheries globally.

Sustainable Land Management

Dr. Thiago Sanna Freire Silva is a Lecturer in Environmental Informatics in the Division of Biological and Environmental Sciences at the University of Stirling, Scotland. He defines himself as a "digital ecologist," combining his interests in ecology, geoscience, and computer science.

Learning Objective
LO 20.12 Describe the sustainable development policies governments have adopted to conserve natural resources and reduce environmental degradation.

Amazonian forests absorb climate-warming carbon dioxide while also fueling economic prosperity. Thiago Silva is exploring new ways to study and conserve these complex ecosystems.

Silva says he is "fascinated by the algorithms of nature," the intricate processes that have developed over eons to sustain life on Earth. He is particularly interested in understanding how landscape dynamics affect ecological processes. In the Amazon, one important landscape dynamic is the flood pulse, a regular cycle of flooding in areas adjacent to rivers. *Igapó* forests grow in areas that are flooded for as much as half the year by blackwater, an acidic "tea" full of decomposed plant matter. These forests are home to tree species that can thrive in 40 feet of water and support a unique ecosystem populated by birds, mammals, reptiles, fish, insects, and other plant species.

Large hydroelectric dams on some Amazon tributaries disrupt the flood pulse that *igapó* forests depend on. Silva and his students have used various remote sensing technologies to study the impact of these dams. In one study, they analyzed satellite images captured over a five-year period to measure tree mortality downstream of the Balbina Reservoir, which was created by the damming of the Uatumã River in northern Brazil. The study concluded that 12 percent of the *igapó* forest along one stretch of the river had died, while another 29 percent was likely dying. Silva and his colleagues concluded that Brazil should require dam operators to simulate the natural flood pulse to preserve forests along the river.

As part of the Perpetual Planet Amazon Expedition, Silva will join other scientists in studying the hydrological cycle of the Amazon region. His part of the project will focus on várzea forests and combine "modern plant ecology methods, state-of-the-art 3D laser scanning, and real-time monitoring to provide the most complete ever analysis of how changes in inundation can affect the structure, function, and diversity of Amazonian flooded forests." He will also create online "virtual forests" to raise awareness about Amazonian wetlands and their role in a sustainable future.

Geographic Thinking

Which sustainability goals are advanced or hindered by hydroelectric projects in the Amazon?

Sustainability policies also seek to minimize negative impacts of development while still serving the needs of current communities. Renewable energy targets are often a first step. In 2005, 43 countries had renewable energy targets; by 2017, this number had escalated to 164 countries across all regions. The targets vary from one country to another. Sweden reached its 2020 target of 50 percent renewable energy in 2012 and is aiming to reach 100 percent renewable electricity by 2040. Australia also met its target for 20 percent of the electricity supply to be generated from renewables by 2020 and is determined to reach 50 percent by 2030. Brazil has set specific electricity capacity targets for biomass, wind, and hydroelectric energy. As mentioned, pioneering wind energy efforts like those in Denmark and other EU countries may act as templates for the renewable energy targets of other countries looking to achieve sustainability.

However, in today's global environment, laws passed by one country do not ensure that natural resources are used in a sustainable way. The public and private sectors are working together to drive sustainability from a local or regional scale to a global scale. Cities have set their own goals for sustainability. For instance, the C40 group of 96 large cities across the globe has committed to fight climate change, which presents a special challenge to sustainability given the uncertainty caused by the changing climate.

In 2015, the UN adopted 17 Sustainable Development Goals (SDGs). The goals are intended to reduce the inequalities among countries in the core, periphery, and semi-periphery and to achieve a more sustainable future for all. The sustainability goals address global challenges related to poverty, inequality, climate, environmental degradation, prosperity, and justice. Upon adoption, the 193 members of the UN agreed that these goals would be met by 2030.

The UN's Sustainable Development Goals help measure progress in development. In many peripheral countries, limited financing for small-scale infrastructure projects impedes local development. Small-scale financing projects encouraged by the UN focus on people's basic needs, such as clean water and sanitation, transportation, and other infrastructure improvements that impact the most vulnerable people.

The implementation of SDGs varies considerably across regions and within countries. India implemented a Dedicated Freight Corridor and is building new high-capacity rail lines to increase the share of rail as the preferred mode of transport (rather than by truck), which is projected to save more than 450 million tons of carbon dioxide in the first 30 years of operation. In Chile, meanwhile, a "connectivity subsidy program" has been implemented to create a more competitive environment for transportation providers in rural areas. The program has resulted in new water transport options in areas of Chile with extremely low population density, as well as free bus transport for children with disabilities. In the Philippines, many programs have been financed and implemented to address sustainable economic growth. For example, the Philippine Green Jobs Act of 2016 provided financial protection against job loss, in addition to skills training necessary for decent green jobs that foster sustainability and preserve the environment. Companies also benefit from incentives to generate those green jobs for the country's workers. A 2019 study conducted by the International Labour Organization found that the Green

UNITED NATIONS SUSTAINABLE DEVELOPMENT GOALS

GOAL 1 End poverty in all its forms everywhere.	**GOAL 2** End hunger, achieve food security and improved nutrition, and promote sustainable agriculture.	**GOAL 3** Ensure healthy lives and promote well-being for all at all ages.
GOAL 4 Ensure inclusive and equitable quality education and promote lifelong learning opportunities for all.	**GOAL 5** Achieve gender equality and empower all women and girls.	**GOAL 6** Ensure availability and sustainable management of water and sanitation for all.
GOAL 7 Ensure access to affordable, reliable, sustainable, and modern energy for all.	**GOAL 8** Promote sustained, inclusive, and sustainable economic growth, full and productive employment, and decent work for all.	**GOAL 9** Build resilient infrastructure, promote inclusive and sustainable industrialization, and foster innovation.
GOAL 10 Reduce inequality within and among countries.	**GOAL 11** Make cities and human settlements inclusive, safe, resilient, and sustainable.	**GOAL 12** Ensure sustainable consumption and production patterns.
GOAL 13 Take urgent action to combat climate change and its impacts.	**GOAL 14** Conserve and sustainably use the oceans, seas, and marine resources for sustainable development.	**GOAL 15** Protect, restore, and promote sustainable use of terrestrial ecosystems, sustainably manage forests, combat desertification, and halt and reverse land degradation and halt biodiversity loss.
GOAL 16 Promote peaceful and inclusive societies for sustainable development, provide access to justice for all, and build effective, accountable, and inclusive institutions at all levels.		**GOAL 17** Strengthen the means of implementation and revitalize the global partnership for sustainable development.

Sustainable Development Goals At the center of the 2030 Agenda for Sustainable Development, which was adopted by all UN members in 2015, are 17 Sustainable Development Goals, or SDGs. The SDGs fall into the following categories: people, planet, prosperity, peace, and partnerships. They identify the core issues of sustainability while recognizing their interconnectedness. For instance, fighting poverty, inequality, and climate change all involve the construction of common goals and actions across communities, countries, and regions. ▮ Explain how Goals 2, 6, and 15 are interconnected.

Critical Viewing Austria is providing reliable and sustainable energy with its Kaprun Limberg II pumped storage power plant high in the Alps mountains (*left*). The hydroelectric station acts as a giant green battery able to store enough power to cover almost 100,000 households for more than a week. Solar and wind power are used to pump water to the reservoir, where it waits until energy demand increases and the dam opens to release the flow of water downhill to spin the large turbines inside the pumped-hydro plant (*below*). The turbines send electrical energy to the grid for use by nearby communities. This technology is also being considered by water-scarce areas for use to convert ocean into tap water. ▮ Explain how the hydro plant supports Sustainability Goal 7.

Jobs Act accelerated economic growth and produced job growth in key sectors of the Philippine economy.

Ecotourism

Learning Objective
LO 20.14 Describe the benefits of ecotourism as a method of sustainable economic development.

Ecotourism has become increasingly popular as a means of environmentally friendly and sustainable economic development. The goal of ecotourism is to enable a country—often in the periphery or semi-periphery—or region within a country threatened by ecosystem degradation, industrial activity, or other types of development to gain revenue through tourism with minimal environmental impact. Ecotourism involves travel to natural areas in ways that protect the environment and sustain the well-being of local inhabitants.

Tourism overall is an important source of income for many countries and regions, but it can come at the expense of the environment. According to a 2018 study in the journal *Nature Climate Change*, travel and tourism are responsible for 8 percent of total carbon emissions globally. Ironically, climate change threatens the pristine and unique features that draw tourists in the first place. In addition, the presence of large

Principles of Ecotourism

Ecotourism is about uniting conservation, communities, and sustainable travel. This means that those who implement, participate in, and market ecotourism activities generally adopt the following ecotourism principles, which have been adapted from the International Ecotourism Society:

- Minimize physical, social, behavioral, and psychological impacts on the local area.
- Build environmental awareness and cultural respect among visitors.
- Provide positive experiences for both visitors and hosts.
- Provide direct financial benefits for conservation.
- Generate financial benefits for both local people and private business.
- Deliver memorable experiences to visitors that help raise awareness to host countries' political, environmental, and social policies and customs.
- Design, construct, and operate low environmental impact facilities.
- Recognize the rights and spiritual beliefs of local Indigenous people and work in partnership with them to create empowerment.

Critical Viewing Ecotourists who visit the Sarara Camp in the Namunyak Wildlife Conservancy in Kenya can view the mountains of the Mathews Range, come face to face with an elephant, or explore the bush led by a guide from the local Samburu community. ▎Explain how the camp might impact the area.

numbers of tourists hiking into natural areas often disrupts the ecosystems. For instance, plants are trampled and the behaviors of the wildlife change.

Ecotourism is designed to provide a sustainable travel alternative that lessens negative impacts on the environment. Countries engaged in ecotourism advertise travel experiences to unique natural environments and enforce regulations that ecotourism lodging and recreational facilities follow to ensure that their operations have as little impact on the environment as possible. Ecolodging incorporates designs with alternative energy consumption to limit impact on the environment. Examples include Amazon tree houses for wildlife viewers on the Brazil–Peru border and tea houses for trekkers in the Himalaya of Nepal. Governments often use the revenues gained from ecotourism to fund conservation or preservation efforts.

Ecotourism also typically has an educational component. The goal is to enable tourists to learn about fragile natural environments without having a negative impact on them. People may learn about indigenous cultures, history, and issues. Ecotourism advocates hope that this understanding will encourage people and governments to take measures to protect Earth's natural environments and support the people who live within them.

Ecotourist programs are available around the world. Visitors can experience glaciers in Alaska, fjords in Norway, and rain forests on the island of Borneo. In 2018–2019, cruises to Antarctica permitted about 56,000 people to experience the wonders of the world's most remote continent. African safaris enable ecotourists to experience the beauty of the African wilderness and photograph the animals that live there.

Encouraging entrepreneurship and local involvement in tourism-dependent regions enhances the tourists' experience.

Tourists feel more connected with the region and aware of its environmental and social issues. In some countries in Africa, the push for local involvement in tourism has contributed to gains in environmental protection and conservation. The more engaged that community members are in local tourism efforts, the more likely they are to advocate and fight for environmental protection. Monetary and educational resources from tourists paying for safaris encourage the appreciation and protection of fragile ecosystems and natural resources.

Despite the positive environmental and social impacts ecotourism can have, there are concerns. As ecotourism grows, many more tourists visit popular locations and can negatively impact the environment despite best efforts; and, countries may become too economically dependent on ecotourism. Other critics worry that ecotourism "commodifies" natural environments and that these areas shouldn't need to be "sold" to ecotourists to be protected. However, by empowering local communities with a source of income, ecotourism can help reduce poverty while protecting local ecosystems. The creation of protected areas in Costa Rica has helped reduce poverty by up to two-thirds in surrounding communities, in part thanks to ecotourism. As more people are employed in tourism-based jobs, fewer are pushed into engaging in poaching or other unsustainable or illegal activities. Ecotourism is one important way that countries are working together to achieve sustainability goals.

Geographic Thinking

1. Explain why pollution and environmental issues might be more difficult to address in peripheral and semi-peripheral countries than in core countries.

2. Identify which of the UN Sustainable Development Goals are most relevant to the United States and other core countries and explain why.

National Geographic Explorer Andrés Ruzo
Sustainable Ecotourism

Andrés Ruzo is a geothermal scientist, conservationist, and educator. In 2021, he attended a conference in Glasgow as an ambassador and advocate for the Amazon leading up to COP26, the UN climate change conference.

Learning Objective
LO 20.14 Describe the benefits of ecotourism as a method of sustainable economic development.

In 2011, Andrés Ruzo began studying a mysterious "river that boiled" in the heart of the Amazon. This legendary river can reach temperatures of more than 200° F. But the region in which the Boiling River flows is threatened by development.

At nearly four miles long, the Boiling River is among the world's largest thermal rivers, a rare feature that requires a lot of water and the right geologic setting to keep the water flowing hot. Focused on the river, Ruzo was as concerned about uncovering its mysteries as he was protecting the Amazon from deforestation and land and water degradation. Most deforestation results from unmonitored and illegal cattle farming, logging, and land trafficking in the area. Combating these informal economic activities is challenging, especially given the proximity of the river to high-poverty areas, major population centers, and roads and waterways that facilitate easy access to the rain forest.

Ruzo established the Boiling River Project, a nonprofit organization dedicated to protecting the Boiling River area through science, responsible economic development, and local empowerment. The project supports enterprises that make the rain forest more valuable than it would be as cropland or cattle pastures. Visitors to the area learn about native medicinal plants and the traditional healing methods of local Amazonian peoples, including the Shipibo-Conibos (shown above arranging handmade crafts for sale to visitors). The project promotes responsible development like sustainable logging and ecotourism that respects the local culture and environment. Giving local inhabitants alternatives to cutting down the rain forest provides much-needed economic development for the Indigenous population while protecting the valuable resources of the Amazon.

Geographic Thinking

Explain how the efforts of the Boiling River Project support the UN's sustainability goals. To which of the UN Sustainable Development Goals do they relate the most?

Chapter 20 Summary & Review

Chapter Summary

Recent economic changes include increased international trade and growing interdependence among countries in the world economy.

- Supranational organizations foster greater globalization through relationships and free trade agreements, which are reorganizing the world into economic-based regions.
- Economic growth over the last century has led to deindustrialization in many core countries.
- Tariffs are used to protect domestic industry, but they have downsides. They increase the price of imported goods, which hurts consumers. In addition, tariffs sometimes lead to trade wars in which exports are taxed more heavily by other countries, increasing costs for businesses and hurting the global economy.

Global economies have become more closely connected.

- The global financial crisis of 2007–2008, the International Monetary Fund, and the World Bank demonstrate the interconnectedness of the world's financial and economic systems.
- Powerful core countries instituted an international division of labor through imperialism. Today's economic landscape has been transformed by an international division of labor in which countries in the periphery have lower-paying jobs; core countries have decentralized; new regions and countries across the globe have developed new specializations; and the service sectors, high-tech industries, and growth poles have emerged.
- Offshore outsourcing and economic restructuring have led to economic development in peripheral countries.
- In countries outside the core, the growth of new economic activities has resulted in the creation of new manufacturing zones including special economic zones, free trade zones, and export-processing zones.

Sustainable development policies attempt to remedy problems stemming from natural resource depletion, mass consumption (the expanded use of goods), the effects of pollution and climate change, and social inequality.

- The UN adopted 17 Sustainable Development Goals in part to reduce the disparities among countries in the core, periphery, and semi-periphery.
- Sustainable development strategies include ecotourism, which is designed to generate revenue through tourism without negatively impacting the environment.

Review Questions

Use complete sentences to answer the questions.

1. **Apply Conceptual Vocabulary** Consider the terms *comparative advantage* and *complementarity*. Write a standard dictionary definition of each term. Then provide a conceptual definition—an explanation of how each term is used in the context of this chapter.

2. Explain how the interconnectedness of economic systems contributes to commodity dependence.
3. How is neoliberalism related to free trade?
4. Provide an example of deindustrialization.
5. How are Fordist systems and post-Fordist systems related?
6. Define the term *multiplier effects* in the context of offshore outsourcing.
7. Describe the negative effects of offshore outsourcing.
8. Define just-in-time delivery in relation to the manufacturing process.
9. Explain the arguments in favor of and against austerity as a way to prevent future global financial crises.
10. How does agglomeration relate to the formation of growth poles?
11. Describe the international division of labor using a specific example.
12. How are the terms *special economic zones*, *free trade zones*, and *export processing zones* related?
13. Define the term *sustainable development*. Provide an example.
14. Define the term *ecotourism* and describe its goals.

Interpret Charts

Study the chart and then answer the following questions.

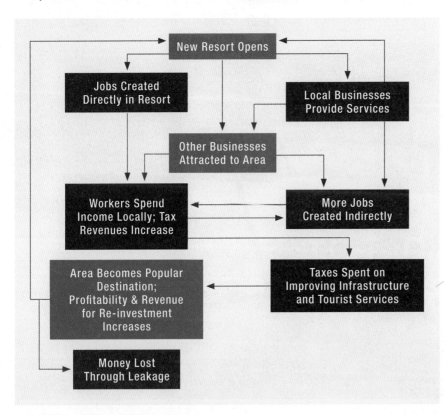

Multiplier Effects of New Resorts

15. **Interpret Charts** What are two immediate effects of establishing a new resort in an area?

16. **Explain Spatial Relationships** Explain how new resorts can attract other businesses to an area.

17. **Connect Visuals & Ideas** Leakage is the money that leaves a country's economy to purchase foreign-owned goods and services. How might an eco-friendly resort reduce leakage?

18. **Predict Outcomes** If the resort closes, how could that impact tax revenues?

19. **Explain Limitations** What does the chart not show about multiplier effects?

Unit 7 | Writing Across Units, Regions, & Scales

The Rush for White Gold

by Robert Draper

While the Indigenous Aymara population harvests and sells salt crusted on the surface of the Salar de Uyuni salt flat, the much more lucrative lithium is dissolved in brine found deep underground.

One morning in La Paz, Álvaro García Linera, the Vice President of Bolivia, greets me outside his office. He speaks confidently about his country's natural resource. Lithium, essential to our battery-fueled world, is also the key to Bolivia's future, the vice president assures me.

Wealth Beneath the Salt Underneath the world's biggest salt flat is another wonder: one of Earth's greatest lithium deposits, perhaps 17 percent of the planet's total. By exploiting these lithium reserves, the government of Bolivia—where 40 percent of the people live in poverty—envisions a pathway out of its misfortune.

What gold meant to earlier eras, and petroleum meant to the previous century, lithium may eclipse in the coming years. Long used in medication to treat bipolar disorders—and in items as varied as ceramics and nuclear weapons—it has emerged as an essential component for the batteries in computers, cell phones, and other electronic devices.

Though lithium-mining operations exist on every continent except Antarctica, up to three-fourths of the known lithium reserves are in the Altiplano-Puna Plateau, a 1,100-mile-long stretch of the Andes. The salt bed deposits concentrated in Chile, Argentina, and Bolivia are known as the "Lithium Triangle." Since the 1980s, Chile has produced lithium from brine, and its Salar de Atacama is now the preeminent source of the chemical in Latin America. Bolivia's lithium reserves match those of Chile's, but until recently, their potential had gone untapped.

Building a Lithium Industry Two years after their election, in 2008, President Evo Morales and García Linera turned their attention to the lithium reserves in the Salar de Uyuni, Bolivia's 4,000-square-mile salt flat. "We decided," said García Linera, "that we Bolivians are going to occupy the Salar, invent our own lithium extraction method, and then partner with foreign firms that can bring us a global market."

Morales confidently predicted that Bolivia would be producing lithium batteries by 2010 and electric cars by 2015. These estimates would prove to be way off. As Morales and García Linera would come to learn, lithium mining is an expensive and complicated process, requiring significant capital outlays as well as technological sophistication. Going it alone was never an option for an economically developing country like Bolivia.

Trusting nonetheless that the promise of the Salar de Uyuni's reserves would surmount any doubts, the Morales administration stated that Bolivia would have a foreign partner to assist in industrial-scale lithium production by 2013. This, too, proved to be a rash prediction. U.S. companies opted out. So did a top Korean firm. Not until 2018 did Bolivia find a partner: ACI Systems Alemania, a German firm that reportedly will invest $1.3 billion in exchange for a 49 percent stake in the venture.

The most daunting hurdle for Bolivia is a scientific one. Producing battery-grade lithium from brine involves separating out sodium chloride, potassium chloride, and magnesium chloride. This last contaminant is particularly expensive to remove. "While the ratio of magnesium in Chile is 5 to 1, in Uyuni it's 21 to 1," Bolivian chemical engineer Miguel Parra said. "So it's a much simpler operation for them. For us, separating magnesium from lithium is the biggest challenge."

I met Parra one morning at Bolivia's Llipi lithium pilot plant. Aside from a tiny pilot plant that makes batteries in the mining town of Potosí, the multimillion-dollar Llipi plant, which started producing lithium in January 2013, is all the Morales government has to show for its decade-long pursuit of lithium-fueled prosperity.

Quality control director Victor Ugarte walked me through the plant. About 20 percent of the lithium carbonate it produces is driven 190 miles to the Potosí battery plant. The rest is sold to various companies. "We started out producing about two tons per month," Ugarte told me. "We're now up to five tons." (Since then, plant officials say, they've reached 30 tons a month.)

I asked what the Llipi plant's ultimate production goal was. "Industrial level," he said, "will be 15,000 tons annually." I tried to imagine this unprepossessing little facility somehow, within the next five or so years, ratcheting up to hit that ambitious goal while maintaining 99.5 percent purity, the industry standard for battery-grade lithium.

The Salar's Future It's nearly impossible to assess how an industrialized version of its lithium facility will change the Salar de Uyuni. Among the greatest concerns is how much water will be required to extract the lithium. Two rivers, the Río Colorado and the Río Grande de Lípez, flow into the salt flat. Both are crucial to the local growers of quinoa, of which Bolivia is the second largest supplier. Though the Bolivian government insists that 90 percent of the water it uses will come from salt water rather than underground aquifers, some experts are skeptical that the groundwater supply will be unaffected.

And there's the still mostly unspoiled surface of the Salar itself—it's also a breeding ground for Chilean flamingos. "Our plant is located far away from these sanctuaries," García Linera said, adding, "This demonstrates our commitment to the environment."

Luís Alberto Echazú Alvarado, a vice minister of energy, said, "Our vision is this is a long-term project. So you have to mix poor and rich brine so as to exploit the whole Salar."

"So the government will always drill throughout other parts?" I asked.

"Right, right," Echazú said, nodding vigorously. "Always."

Source: Adapted from "The Rush for White Gold," by Robert Draper, *National Geographic*, February 2019.

Write Across Units

Unit 7 explored industry, economic sectors, and processes that national economies develop in an interdependent world. This article takes a close look at how Bolivia is striving to exploit its most valuable resource as a way to develop its economy and improve its standard of living. Use information from the article and this unit to write a response to the following questions.

Looking Back

1. What might geographers learn by conducting a regional analysis of the "Lithium Triangle"? Unit 1
2. What effects might a developing lithium industry have on population distribution in Bolivia? Unit 2
3. Would you expect Bolivia's cultural landscape to change if a successful lithium industry is established? Explain your answer. Unit 3
4. In what ways do the concepts of territoriality and sovereignty come into play in Bolivia's plans for its lithium? Unit 4
5. How do you think lithium processing will affect patterns of agriculture in Bolivia? Explain your answer. Unit 5
6. Do you think the Salar de Uyuni might become a site for a city? What other information would help you predict whether a city will arise there? Unit 6
7. Compare Bolivia's current situation to the Industrial Revolution. In what ways may Bolivia's development be similar to or different from countries affected by the Industrial Revolution? Unit 7

Write Across Regions & Scales

Research a country outside of the Americas with an economy that is largely dependent on a single resource. Write an essay comparing the ways Bolivia and that country have exploited—or sought to exploit—each resource. Drawing on your research, this unit, and the article, address the following topic:

What economic patterns develop in a country with a single principal natural resource? What lessons can Bolivia learn from the other country?

Think About

- How exploiting the resource affects all the sectors of the countries' economies
- Measures of development like GDP, GNI per capita, and HDI for each country and what they indicate about each country's economic health or future

Chapter 18

Economic Sectors Across the Globe

For a long time, geographers classified economic activity into three sectors: primary, secondary, and tertiary. More recently, advanced technology and high-level leadership have been identified as sectors within the tertiary sector.

Countries in the core, periphery, and semi-periphery have distinctive gross domestic product (GDP) profiles that reflect their involvement in each economic sector. Core countries tend to earn most of their money in the quaternary or quinary sectors, while peripheral countries have GDPs more weighted toward the primary sector. These charts show GDP percentages from agriculture, industry, and services in Algeria, Denmark, and Haiti.

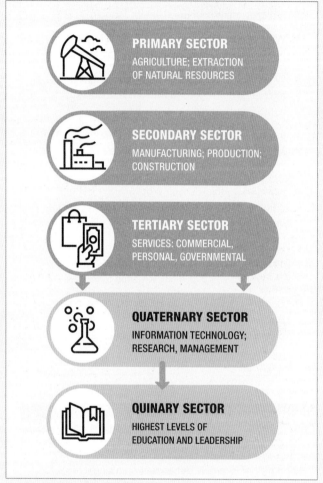

■ Explain the degree to which this graphic explains how the sectors build on one another.

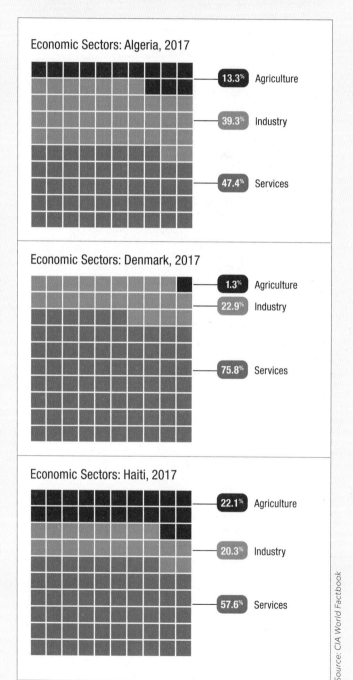

Source: CIA World Factbook

■ Identify whether each country would be classified as core, peripheral, or semi-peripheral. Compare economic patterns across all three countries.

Chapter 18

Least-Cost Theory

Alfred Weber's least-cost theory takes into account several factors that influence the location of manufacturing sites, including transportation costs, agglomeration (the tendency of enterprises in the same industry to cluster in the same area), and labor costs. According to the theory, transportation is the most important element. Weber's location triangle illustrates how transportation considerations can be used to determine the ideal location for a manufacturing plant. ▮ Explain the degree to which the least-cost theory explains how manufacturing sites are chosen in the real world. Describe how factors other than transportation might alter the least-cost theory's location triangle.

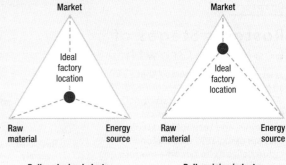

Chapter 18

Bulk-Gaining and Bulk-Reducing Industries

Weber recognized that raw materials commonly lose or gain weight in the course of being transported from their source to the processing or manufacturing site and finally to the market. Because transporting heavier materials costs more, the ideal location for manufacturing a given product depends on whether the process causes the raw materials to gain or lose bulk. ▮ Use the infographics to explain the locations of processing plants for the soda industry and the copper industry. Consider the difference in weight between empty and full soda cans and the difference in weight between copper ore and the copper that results after other materials in the ore have been removed.

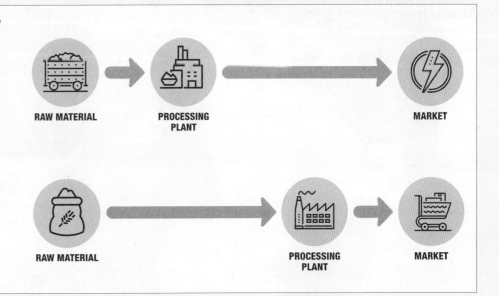

Bulk-Reducing Industry
If raw materials cost more to transport than finished goods, then processing plants will be located near the source of the raw materials.
Examples: copper smelting, furniture manufacturing

Bulk-Gaining Industry
If raw materials cost less to transport than finished goods, then processing plants will be located near the market.
Examples: car manufacturing, bread production, construction equipment

Chapter 19

Rostow's Stages of Economic Growth

Walt W. Rostow suggested that all countries could be categorized along a spectrum from traditional to modern (the social, political, economic, and technological changes that are associated with becoming industrialized). To become modern, countries needed to pass through five stages of economic growth in order. ▮ Explain the degree to which this model explains developments during and after the Industrial Revolution.

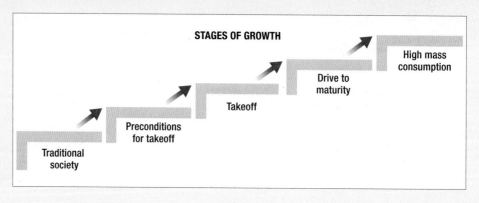

Chapter 19

Wallerstein's World System Theory

Immanuel Wallerstein's world system theory, which was introduced in Chapter 1, categorizes countries as belonging to the core, periphery, or semi-periphery and describes both the economic ties and power relationships that link the countries in a single network. ▮ Identify the patterns in the map concerning the locations of core, peripheral, and semi-peripheral countries in Africa, Europe, and Asia.

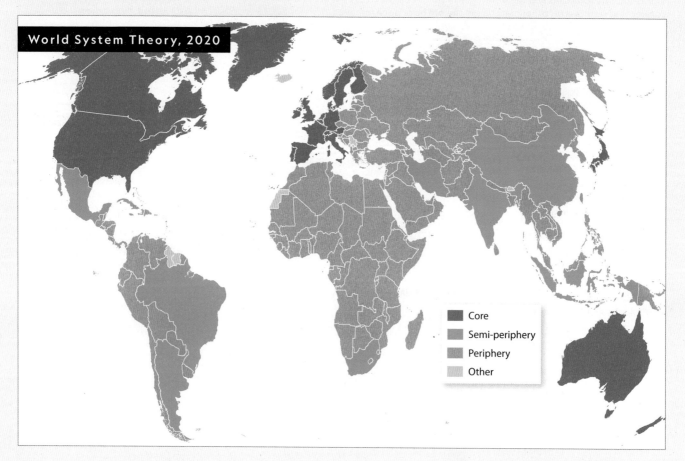

Maps & Models Archive

Chapter 19

Dependency Theory

Dependency theory builds on the concepts established in world system theory to explain long-term economic inequality between core and peripheral countries. Core countries—often former colonial powers—have the means to demand low-cost raw materials and labor from peripheral countries, which in turn struggle to gather enough capital to invest in their own economies. Many peripheral and semi-peripheral countries are commodity dependent, meaning that their economies rely principally on a single commodity that they sell to more economically developed countries. Commodity dependence makes countries vulnerable to price changes for raw materials or to dropping demand for the commodities in core economies. ▌ Identify and explain the patterns of commodity dependency in the bar graph and map. Identify the overall percentage of commodity-dependent countries.

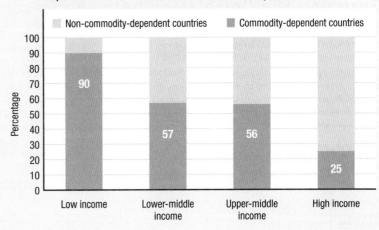

Distribution of Commodity-Dependent and Non-commodity-Dependent Countries Within Each Income Group, 2008–2019

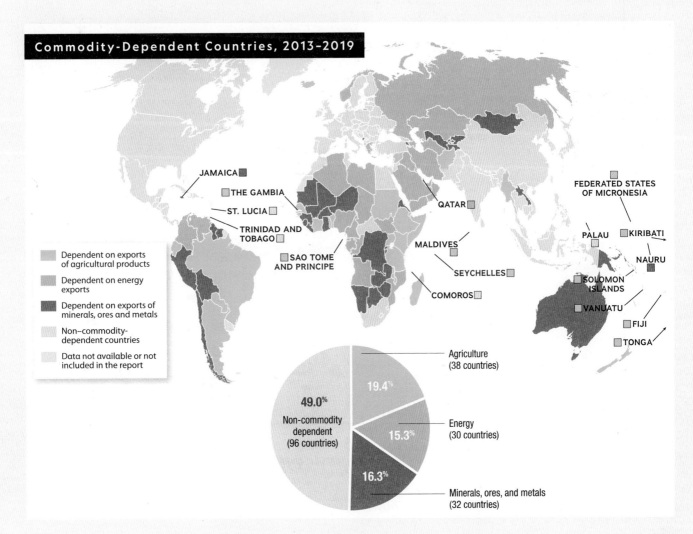

Chapter 19

Graphing the Correlation between Education and Birth Rates

Studies have shown a correlation between levels of education for girls and a country's adolescent birth rate (ABR), or births per 1,000 females ages 15–19. This scattergram represents data from a 2015 analysis of United Nations countries. ▌Identify the patterns in the scattergram relating to ABR, education, and GDP. Select several countries, compare their GDPs and ABRs, and explain why a country's GDP might also be correlated with its ABR. For example, to compare Niger and the United States, use the data relating to their ABRs, education levels, and GDPs to consider the differences between the two countries.

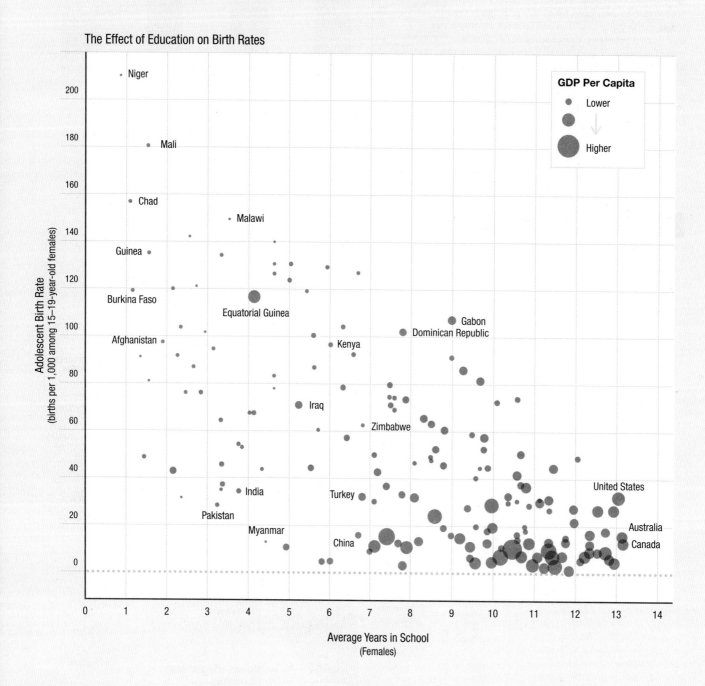

Maps & Models Archive

Chapter 20

Countries with the Most Trade Agreements

Countries pursue trade agreements with the goal of fueling economic activity, increasing exports, and lowering costs of imported goods. In theory, trade agreements benefit all sides and boost the economies of countries in the core and periphery alike. ▌ Based on what you know about the locations of core, peripheral, and semi-peripheral countries, identify patterns in trade agreements shown on this map. (Note that the data predate Brexit, the withdrawal of the United Kingdom from the European Union in January 2020.) Explain why Europe, in general, has the largest number of trade agreements.

Active Regional Trade Agreements, 2019

Agreements made as a pre-existing group are counted individually for each country
- 42
- 11–31
- 6–10
- 1–5
- Not applicable/no trade agreements

Chapter 20

Where Your Car Is Made

Many of the most commonly used pieces of technology in the United States—such as smartphones, computers, and cars—are manufactured using outsourced labor and materials. Automobiles, in particular, draw from a worldwide pool of parts and labor. The infographic shows the origins of parts and materials for cars manufactured in Mexico for the U.S. market by Ford Motor Company in 2017. ▌ Explain the degree to which post-Fordism (a system focused on small-scale batch production for a specialized market) and North American trade agreements such as NAFTA and USMCA may explain the data shown in the graphic.

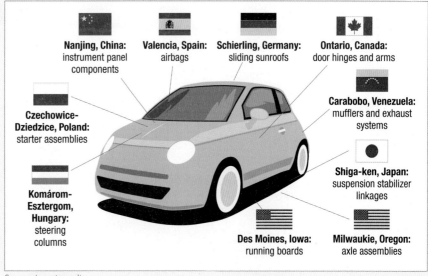

- Nanjing, China: instrument panel components
- Valencia, Spain: airbags
- Schierling, Germany: sliding sunroofs
- Ontario, Canada: door hinges and arms
- Czechowice-Dziedzice, Poland: starter assemblies
- Carabobo, Venezuela: mufflers and exhaust systems
- Komárom-Esztergom, Hungary: steering columns
- Shiga-ken, Japan: suspension stabilizer linkages
- Des Moines, Iowa: running boards
- Milwaukie, Oregon: axle assemblies

Source: Investopedia

609

Glossary

A

absolute direction *n.* the cardinal directions north, south, east, and west (page 39)

absolute distance *n.* distance that can be measured using a standard unit of length (page 39)

absolute location *n.* the exact location of an object, usually expressed in coordinates of longitude and latitude (page 8)

acculturation *n.* the process by which people within one culture adopt some of the traits of another while still retaining their own distinct culture (page 206)

adherents *n.* a person who is loyal to a belief, religion, or organization (page 181)

administer *v.* to manage the way borders are maintained and how goods and people cross them (page 266)

African city model *n.* a model of urban development depicting a city with three central business districts, growing outward in a series of concentric rings (page 463)

agglomeration *n.* the tendency of enterprises in the same industry to cluster in the same area (page 542)

agribusiness *n.* the large-scale system that includes the production, processing, and distribution of agricultural products and equipment (page 361)

agricultural biodiversity *n.* the variety and variability of plants, animals, and microorganisms that are used directly or indirectly for food and agriculture (page 395)

agricultural density *n.* the total number of farmers per unit of arable land (page 71)

agricultural hearth *n.* an area where different groups began to domesticate plants and animals (page 346)

agricultural landscapes *n.* a landscape resulting from the interactions between farming activities and a location's natural environment (page 383)

agricultural revolution *n.* a fundamental change in agriculture that makes a profound and lasting influence not just in agriculture but in human society. Experts recognize three agricultural revolutions, beginning about 11,000 years ago with the Neolithic Revolution, marking the shift, over thousands of years, from foraging to farming (page 351)

agriculture *n.* the purposeful cultivation of plants or raising of animals to produce goods for survival (page 327)

agroecosystem *n.* an ecosystem modified for agricultural use (page 383)

antecedent boundaries *n.* a border established before an area becomes heavily settled (page 270)

antinatalist *adj.* describing attitudes or policies that discourage childbearing as a means of limiting population growth (page 104)

aquifers *n.* layers of sand, gravel, and rocks that contain and can release a usable amount of water (page 386)

arable land *n.* land that can be used to grow crops (page 71)

arithmetic density *n.* the total number of people per unit area of land; also called crude density (page 71)

artifacts *n.* a visible object or technology that a culture creates (page 161)

assimilation *n.* a category of acculturation in which the interaction of two cultures results in one culture adopting almost all of the customs, traditions, language, and other cultural traits of the other (page 207)

asylum *n.* the right to protection in a country (page 124)

autonomous *adj.* having the authority to govern territories independently of the national government; for example, by having a separate currency (page 258)

B

bid-rent theory *n.* a theory that describes the relationships between land value, commercial location, and transportation (primarily in urban areas) using a bid-rent gradient, or slope; used to describe how land costs are determined (page 333)

biodiversity *n.* the variety of organisms living in a location (page 387)

Biotechnology *n.* the science of altering living organisms, often through genetic manipulation, to create new products for specific purposes, such as crops that resist certain pests (pages 394, 387)

blockbusting *n.* a practice by real estate agents who would stir up concern that Black families would soon move into a neighborhood; the agents would convince White property owners to sell their houses at below-market prices (page 487)

boomburb *n.* a suburb that has grown rapidly into a large and sprawling city with more than 100,000 residents (page 435)

brain drain *n.* the loss of trained or educated people to the lure of work in another—often richer—country (page 142)

break-of-bulk points *n.* location where it is more economical to break raw materials into smaller units before shipping them further (page 543)

brownfields *n.* abandoned and polluted industrial site in a central city or suburb (page 504)

Buddhism *n.* the oldest universalizing religion, which arose from a hearth in northeastern India sometime between the mid-sixth and mid-fourth centuries B.C.E. and is based on the teachings of Siddhartha Gautama, called the Buddha (page 235)

bulk-gaining industries *n.* industry in which the finished goods cost more to transport than the raw materials (page 545)

bulk-reducing industries *n.* industry in which the raw materials cost more to transport than the finished goods (page 545)

C

carrying capacity *n.* the maximum population size an environment can sustain (page 74)

cartographers *n.* a person who creates maps (page 39)

cash crop *n.* a crop produced mainly to be sold and usually exported to larger markets (page 373)

census *n.* an official count of the number of people in a defined area, such as a state (page 33)

central business district (CBD) *n.* the central location where the majority of consumer services are located in a city or town because the accessibility of the location attracts these services (page 333)

central place theory *n.* a theory used to describe the spatial relationship between cities and their surrounding communities (page 440)

centrifugal force *n.* a force that divides a group of people (page 184)

centripetal force *n.* a force that unites a group of people (page 184)

chain migration *n.* type of migration in which people move to a location because others from their community have previously migrated there (page 123)

choke point *n.* a narrow, strategic passageway to another place through which it is difficult to pass (page 260)

Christianity *n.* a universalizing religion based on the teachings of Jesus Christ that began in what is now the West Bank and Israel around the beginning of the common era and has spread to all continents (page 233)

circular migration *n.* migration pattern in which migrant workers move back and forth between their country of origin and the destination country where they work temporary jobs (page 124)

circulation *n.* temporary, repetitive movements that recur on a regular basis (page 117)

climate *n.* the long-term patterns of weather in a particular area (page 68)

climate regions *n.* an area that has similar climate patterns generally based on its latitude and its location on a coast or continental interior (page 330)

clustered settlement *n.* a rural settlement pattern in which residents live in close proximity to one another, with farmland and pasture land surrounding the settlement; also known as a nucleated settlement (page 336)

collectivist cultures *n.* a culture in which people are expected to conform to collective responsibility within the family and to be obedient to and respectful of elder family members (page 208)

colonialism *n.* the practice of claiming and dominating overseas territories (page 259)

Columbian Exchange *n.* the exchange of goods and ideas between the Americas, Europe, and Africa that began after Christopher Columbus landed in the Americas in 1492 (page 350)

commercial agriculture *n.* an agricultural practice that focuses on producing crops and raising animals for the market for others to purchase (page 333)

commodity chain *n.* a network of people, information, processes, and resources that work together to produce, handle, and distribute a commodity or product (page 367)

commodity dependence *n.* an aspect of dependency theory that occurs when more than 60 percent of a country's exports and economic health are tied to one or two resources (page 575)

comparative advantage *n.* the relative cost advantage a country or organization has to produce certain goods or services for trade (page 579)

complementarity *n.* the mutual trade relationship that exists between two places based on the supply of raw materials and the demand for finished products or services (page 579)

concentric-zone model *n.* a model of urban development depicting a city growing outward from a central business district in a series of concentric rings (page 459)

concurrent *adj.* sharing authority (page 281)

consequent boundary *n.* a type of subsequent boundary that takes into account the differences that exist within a cultural landscape, separating groups that have distinct languages, religions, ethnicities, or other traits (page 270)

contagious diffusion *n.* the process by which an idea or cultural trait spreads rapidly among people of all social classes and levels of power (page 191)

core *n.* classification of a country or region that has wealth, higher education levels, more advanced technologies, many resources, strong militaries, and powerful allies (page 24)

cottage industries *n.* preindustrial form of manufacture in which members of families spread out through rural areas worked in their homes to make goods (page 524)

creolization *n.* the blending of two or more languages that may not include the features of either original language (page 198)

crop rotation *n.* the varying of crops from year to year to allow for the restoration of valuable nutrients and the continuing productivity of the soil (page 338)

crude birth rate *n.* the number of births in a given year per 1,000 people in a given population (page 81)

crude death rate *n.* the number of deaths in a given year per 1,000 people in a given population (page 83)

cultural appropriation *n.* the act of adopting elements of another culture (page 209)

cultural convergence *n.* the process by which cultures become more similar through interaction (page 201)

cultural divergence *n.* the process by which cultures become less similar due to conflicting beliefs or other barriers (page 202)

cultural hearth *n.* an area where cultural traits develop and from which cultural traits diffuse (page 189)

cultural landscape *n.* a natural landscape that has been modified by humans, reflecting their cultural beliefs and values (page 165)

cultural norms *n.* a shared standard or pattern that guides the behavior of a group of people (page 163)

cultural relativism *n.* the evaluation of a culture by its own standards (page 163)

cultural trait *n.* a shared object or cultural practice (page 161)

culture *n.* the beliefs, values, practices, behaviors, and technologies shared by a society and passed down from generation to generation (page 161)

D

debt-for-nature swaps *n.* agreement between a bank and a peripheral country in which the bank forgives a portion of the country's debt in exchange for local investment in conservation measures (page 389)

de facto segregation *n.* segregation that results from residential settlement patterns rather than from prejudicial laws (page 484)

de jure segregation *n.* segregation that results from prejudicial laws and policies passed by various levels of government that discriminate in housing based upon race or ethnicity (page 487)

defining *v.* to explicitly state in legally binding documentation such as a treaty where boundaries are located, using reference points such as natural features or lines of latitude and longitude (page 266)

deforestation *n.* loss of forest lands (page 383)

deindustrialization *n.* process by which a country or area reduces industrial activity, particularly in heavy industry and manufacturing (page 581)

delimit *v.* to draw boundaries on a map, in accordance with a legal agreement (page 266)

demarcated *v.* to place physical objects such as stones, pillars, walls, or fences to indicate where a boundary exists (page 266)

demographics *n.* data about the structures and characteristics of human populations (page 81)

demographic transition model (DTM) *n.* a model that represents shifts in the growth of the world's populations, based on population trends related to birth rate and death rate (page 97)

denominations *n.* a separate church organization that unites a number of local congregations (page 181)

density *n.* the number of things—people, animals, or objects—in a specific area (page 9)

dependency ratio *n.* the number of people in a dependent age group (under age 15 or age 65 and older) divided by the number of people in the working-age group (ages 15 to 64), multiplied by 100 (page 78)

dependency theory *n.* a theory that describes the development challenges and limitations faced by poorer countries and the political and economic relationships poorer countries have with richer countries (page 574)

desertification *n.* a form of land degradation that occurs when soil deteriorates to a desertlike condition (page 387)

devolution *n.* the process that occurs when the central power in a state is broken up among regional authorities within its borders (page 265)

dialects *n.* a variation of a standard language specific to a general area, with differences in pronunciation, degree of rapidity in speech, word choice, and spelling (page 180)

diffusion *n.* the process by which a cultural trait spreads from one place to another over time (page 189)

disamenity zones *n.* a high-poverty urban area in a disadvantaged location containing steep slopes, flood-prone ground, rail lines, landfills, or industry (page 461)

dispersed settlements *n.* a rural settlement pattern in which houses and buildings are isolated from one another, and all the homes in a settlement are distributed over a relatively large area (page 336)

distance decay *n.* a principle stating that the farther away one thing is from another, the less interaction the two things will have (page 10)

distributed *v.* to arrange within a given space (page 9)

domestication *n.* the deliberate effort to grow plants and raise animals, making plants and animals adapt to human demands and using selective breeding to develop desirable characteristics (page 346)

doubling time *n.* the number of years in which a population growing at a certain rate would double (page 91)

dual agricultural economy *n.* an economy having two agricultural sectors that have different levels of technology and different patterns of demand (page 361)

dual economies *n.* economies with two distinct distributions of economic activity across the economic sectors (page 537)

E

ecological footprint *n.* impact of a person or community on the environment, expressed as the amount of land required to sustain the use of natural resources (page 479)

ecological perspective *n.* the relationships between living things and their environments (page 8)

economic sectors *n.* collections of industries engaged in similar economic activities based on the creation of raw materials, the production of goods, the provision of services, or other activities (page 533)

economies of scale *n.* cost reductions that occur when production rises (page 303)

Economy of scale *n.* cost reductions that occur when production rises (page 405)

ecotourism *n.* a form of tourism based on the enjoyment of natural areas that minimizes the impact to the environment (page 597)

edge city *n.* a type of community located on the outskirts of a larger city with commercial centers with office space, retail complexes, and other amenities typical of an urban center (page 435)

electoral college *n.* a set of people, called electors, who are chosen to elect the president and vice president of the United States (page 285)

emigration *n.* movement away from a location (page 117)

eminent domain *n.* a government's right to take over privately owned property for public use or interest (page 494)

enclosure system *n.* system in which communal lands were replaced by farms owned by individuals, and use of the land was restricted to the owner or tenants who rented the land from the owner (page 353)

environmental determinism *n.* the idea that human behavior is strongly affected, controlled, or determined by the physical environment (page 11)

environmental injustice *n.* the ways in which communities of color and poor people are more likely to be exposed to environmental burdens such as air pollution or contaminated water; also called environmental racism (page 494)

epidemiological transition model (ETM) *n.* a model that describes changes in fertility, mortality, life expectancy, and population age distribution, largely as the result of changes in causes of death (page 98)

ethnic cleansing *n.* the process by which a state attacks an ethnic group and tries to eliminate it through expulsion, imprisonment, or killing (page 294)

ethnic nationalism *n.* the process by which the people of a country identify as having one common ethnicity, religious belief, and language, creating a sense of pride and identity that is tied to the territory; also called ethnonationalism (page 308)

ethnicity *n.* the state belonging to a group of people who share common cultural characteristics (page 166)

ethnic neighborhood *n.* a cultural landscape within a community of people outside of their area of origin (page 166)

ethnic religions *n.* a religion that is closely tied with a particular ethnic group often living in a particular place (page 237)

ethnic separatism *n.* the process by which people of a particular ethnicity in a multinational state identify more strongly as members of their ethnic group than as citizens of the state (page 294)

ethnocentrism *n.* the tendency of ethnic groups to evaluate other groups according to preconceived ideas originating from their own culture (page 163)

ethnonationalism *n.* the process by which the people of a country identify as having one common ethnicity, religious belief, and language, creating a sense of pride and identity that is tied to the territory; also called ethnic nationalism (page 308)

exclusive economic zone (EEZ) *n.* an area that extends 200 nautical miles from a state's coast; a state has sole access to resources found within the waters or beneath the sea floor of its EEZ (page 271)

expansion diffusion *n.* the spread of a cultural trait outward from where it originated (page 191)

export processing zones (EPZs) *n.* an area within a country that is subject to more favorable regulations (usually including the elimination of tariffs) to encourage foreign investment and the manufacturing of goods for export (page 591)

extensive agriculture *n.* an agricultural practice with relatively few inputs and little investment in labor and capital that results in relatively low outputs (page 339)

exurb *n.* a typically fast-growing community outside of or on the edge of a metropolitan area where the residents and community are closely connected to the central city and suburbs (page 436)

F

factors of production *n.* resources such as land, labor, capital, and knowledge that are used in the process of producing a good or service (page 526)

fair trade *n.* a movement that tries to provide farmers and workers in peripheral and semi-peripheral countries with a fair price for their products by providing more equitable trading conditions (page 377)

farm subsidies *n.* a form of aid and insurance given by the federal government to certain farmers and agribusinesses (page 368)

federal state *n.* the organization of a state in which power is shared between the federal government and its internal regional units (page 277)

Fertile Crescent *n.* a hearth in Southwest Asia that forms an arc from the eastern Mediterranean coast up into what is now western Turkey and then south and east along the Tigris and Euphrates rivers to western parts of modern Iran (page 346)

fertility *n.* the ability to produce children (page 81)

filtering *n.* the process of neighborhood change in which housing vacated by more affluent groups passes down the income scale to lower-income groups (page 489)

flow *n.* movement of people, goods, or information that has economic, social, political, or cultural effects on societies (page 10)

food deserts *n.* area where residents lack access to healthy, nutritious foods because stores selling these foods are too far away (page 404)

food insecurity *n.* the disruption of food intake or eating patterns because of poor access to food (page 401)

food security *n.* reliable access to safe and nutritious food that can support an active and healthy lifestyle (page 401)

foragers *n.* small, nomadic groups who had primarily plant-based diets and ate small animals or fish for protein (page 346)

forced migration *n.* type of migration in which people are compelled to move by economic, political, environmental, or cultural factors (page 121)

Fordism *n.* a highly organized and specialized system for industrial production that focuses on efficiency and productivity in mass production; named after Henry Ford (page 587)

formal region *n.* an area that has one or more shared traits; also called a uniform region (page 18)

formal sector *n.* businesses, enterprises, and other economic activities that have government supervision, monitoring, and protection, and are also taxed (page 552)

free trade zones (FTZs) *n.* a relatively large geographical area within a country in which businesses pay few or no tariffs on goods to encourage or facilitate its role in international trade (page 591)

friction of distance *n.* a concept that states that the longer a journey is, the more time, effort, and cost it will involve (page 121)

functional region *n.* an area organized by its function around a focal point, or the center of an interest or activity (page 19)

G

galactic city model *n.* a model of urban development depicting a city where economic activity has moved from the central business district toward loose coalitions of other urban areas and suburbs; also known as the peripheral model (page 461)

Gender Development Index (GDI) *n.* a measure that calculates gender disparity in the three basic dimensions of human development: health, knowledge, and standard of living (page 558)

gendered spaces *n.* a space designed and deliberately incorporated into the landscape to accommodate gender roles (page 173)

gender identity *n.* an individual's innermost concept of self as male, female, a blend of both, or neither (page 176)

Gender Inequality Index (GII) *n.* a measure that calculates inequality based on three categories: reproductive health, empowerment, and labor-market participation (page 558)

genetically modified organisms (GMOs) *n.* a plant or animal with specific characteristics obtained through the manipulation of its genetic makeup (page 354)

gentrification *n.* the renovations and improvements conforming to middle-class preferences (page 177)

geographic information systems (GIS) *n.* a computer system that allows for the collection, organization, and display of geographic data for analysis (page 34)

geometric boundaries *n.* a mathematically drawn boundary that typically follows lines of latitude and longitude or is a straight-line arc between two points (page 270)

gerrymandering *n.* the drawing of legislative boundaries to give one political party an advantage in elections (page 286)

globalization *n.* the expansion of economic, cultural, and political processes on a worldwide scale (page 23)

global positioning system (GPS) *n.* a network of satellites that orbit Earth and transmit location data to receivers, enabling users to pinpoint their exact location (page 37)

global supply chains *n.* a network of people, information, processes, and resources that work together to produce, handle, and distribute goods around the world (page 373)

gravity model *n.* a model that predicts the interaction between two or more places; geographers derived the model from Newton's law of universal gravitation (page 117)

greenbelt *n.* a ring of parkland, agricultural land, or other type of open space maintained around an urban area to limit sprawl (page 483)

Green Revolution *n.* movement beginning in the 1950s and 1960s in which scientists used knowledge of genetics to develop new high-yield strains of grain crops (page 355)

gross domestic product (GDP) *n.* the total value of the goods and services produced by a country's citizens and companies within the country in a year (pages 537, 551)

gross national income (GNI) *n.* the total value of goods and services globally produced by a country in a year divided by the country's population (page 551)

gross national product (GNP) *n.* the total value of the goods and services produced by a country's citizens and companies both domestically and internationally in a year (page 551)

growth poles *n.* a place of economic activity clustered around one or more high-growth industries that stimulate economic gain by capitalizing on some special asset (page 582)

guest workers *n.* a migrant who travels to a new country as temporary labor (page 124)

H

hierarchical diffusion *n.* the spread of an idea or trait from a person or place of power or authority to other people or places (page 191)

Hinduism *n.* an ethnic religion that arose a few thousand years ago in South Asia and is closely tied to India (page 237)

human development *n.* the processes involved in the improvement of people's freedoms, rights, capabilities, choices, and material conditions (page 551)

Human Development Index (HDI) *n.* a measure that determines the overall development of a country by incorporating three key dimensions of human development: life expectancy at birth, access to education measured in expected and mean years of schooling, and standard of living measured by GNI per capita (page 556)

human geography *n.* the study of the processes that have shaped how humans understand, use, and alter Earth (page 7)

human migration *n.* the permanent movement of people from one place to another (pages 69, 117)

human trafficking *n.* defined by the United Nations as "the recruitment, transportation, harboring, or receipt of persons by improper means (such as force, abduction, fraud, or coercion)" (page 128)

hybrid *n.* the product created by breeding different varieties of species to enhance the most favorable characteristics (page 361)

I

identity *n.* the ways in which humans make sense of themselves and how they wish to be viewed by others (page 165)

immigration *n.* movement to a location (page 117)

inclusionary zoning laws *n.* law that creates affordable housing by offering incentives for developers to set aside a minimum percentage of new housing construction to be allocated for low-income renters or buyers (page 493)

industrialization *n.* the process in which the interaction of social and economic factors causes the development of industries on a wide scale (page 523)

industrial parks *n.* a collection of manufacturing facilities in a particular area that is typically found in suburbs and is located close to highways to facilitate movement of raw materials and finished products (page 547)

Industrial Revolution *n.* the radical change in manufacturing methods that began in Great Britain in the mid-18th century and was marked by the shift from small-scale, hand-crafted, muscle-powered production to power-driven mass production (page 523)

industry *n.* any economic activity using machinery on a large scale to process raw materials into products (page 523)

infant mortality rate *n.* the number of deaths of children under the age of 1 per 1,000 live births (page 83)

infill *n.* redevelopment that identifies and develops vacant parcels of land within previously built areas (page 436)

informal sector *n.* any part of a country's economy that is outside of government monitoring or regulation; sometimes called the informal economy (page 552)

infrastructure *n.* the many systems and facilities that a country needs in order to function properly (page 361)

intensive agriculture *n.* an agricultural practice in which farmers expend a great deal of effort to produce as much yield as possible from an area of land (page 335)

Glossary **R3**

internally displaced persons *n.* person who has been forced to flee his or her home but remains within the country's borders (page 125)

internal migration *n.* movement within a country's borders (page 121)

international division of labor *n.* a pattern of production and labor in which different countries are engaged in distinct aspects of production (page 589)

interregional migration *n.* movement from one region of the country to another (page 133)

intersectionality *n.* the theory that forms of discrimination based on identity such as race, class, gender or sexuality interact to produce interdependent and overlapping systems of oppression (page 207)

intervening obstacle *n.* an occurrence that holds migrants back (page 123)

intervening opportunity *n.* an occurrence that causes migrants to pause their journey by choice (page 123)

intraregional migration *n.* movement within one region of the country (page 133)

irredentism *n.* attempts by a state to acquire territories in neighboring states inhabited by people of the same nation (page 257)

isolate *n.* a language that is unrelated to any other known language (page 215)

J

Judaism *n.* the world's first monotheistic religion, which developed among the Hebrew people of Southwest Asia about 4,000 years ago (page 238)

just-in-time delivery *n.* a system in which goods are delivered as needed so that companies keep in inventory only what is needed for near-term production (page 587)

K

kinship links *n.* networks of relatives and friends (page 134)

L

labor-market participation (LMP) *n.* rate that measures an economy's active labor force, calculated by taking the sum of all employed workers divided by the working age population (page 560)

land degradation *n.* long-term damage to the soil's ability to support life (page 108)

landforms *n.* the natural features of Earth's surface (page 68)

land tenure *n.* the legal rights, as defined by a society, associated with owning land (page 493)

language *n.* a distinct system of communication that is the carrier of human thoughts and cultural identities (page 168)

language branch *n.* a collection of languages within a language family that share a common origin and separated from other branches in the same family several thousand years ago (page 215)

language family *n.* a group of languages that share a common ancestral language from a particular hearth, or region of origin (page 213)

language groups *n.* languages within a language branch that share a common ancestor in the relatively recent past and have vocabularies with a high degree of overlap (page 215)

Latin American city model *n.* a model of urban development depicting a city with a central business district, concentric rings, and sections stricken by poverty; also known as the Griffin-Ford model (page 461)

least-cost theory *n.* industrial location theory proposed by Alfred Weber suggesting that businesses locate their facilities in a particular place because that location minimizes the costs of production (page 542)

legacy cities *n.* post-industrial cities in the United States, particularly in the Rust Belt, characterized by economic and population declines and high rates of poverty (page 434)

life expectancy *n.* the average number of years a person is expected to live (page 83)

linear settlement *n.* a rural settlement pattern in which houses and buildings form in a long line that usually follows a land feature or aligns along a transportation route (page 336)

lingua franca *n.* common language used among speakers of different languages (page 197)

location *n.* the position that a point or object occupies on Earth (page 8)

M

majority-minority districts *n.* an electoral district in which the majority of voters are members of an ethnic or racial minority (page 286)

map scale *n.* the relationship of the size of the map to the size of the area it represents on Earth's surface (page 41)

market gardening *n.* a type of farming that produces fruits, vegetables, and flowers and typically serves a specific market or urban area (page 339)

Mediterranean agriculture *n.* an agricultural practice that consists of growing hardy trees and shrubs and raising sheep and goats (page 331)

megacity *n.* a city with a population of more than 10 million (page 446)

mental maps *n.* internalized representations of portions of Earth's surface (page 8)

mentifacts *n.* a central, enduring element of a culture that reflects its shared ideas, values, knowledge, and beliefs (page 162)

metacity *n.* a city with a population of more than 20 million (page 446)

metropolitan area *n.* a city and the surrounding areas that are influenced economically and culturally by the city (page 427)

microloans *n.* a very small short-term loan with low interest intended to help people in need (page 570)

mixed crop and livestock systems *n.* a type of farming in which both crops and livestock are raised for profit (page 339)

mixed-use development *n.* a single planned development designed to include multiple uses, such as residential, retail, educational, recreational, industrial, and office spaces (page 479)

mixed-use zoning *n.* zoning that permits multiple land uses in the same space or structure (page 480)

mobility *n.* all types of movement from one location to another, whether temporary or permanent or over short or long distances (page 117)

monocropping *n.* the cultivation of one or two crops that are rotated seasonally (page 338)

monoculture *n.* the agricultural system of planting one crop or raising one type of animal annually (page 338)

mortality *n.* deaths as a component of population change (page 82)

multiculturalism *n.* a situation in which different cultures live together without assimilating (page 208)

multinational state *n.* a country with various ethnicities and cultures living inside its borders (page 257)

multiple-nuclei model *n.* a model of urban development depicting a city where growth occurs around the progressive integration of multiple nodes, not around one central business district (page 459)

multiplier effects *n.* the economic effect in which a change creates a larger change, such as when a new manufacturing plant grows the economy by giving rise to more related jobs and services (page 590)

multistate nation *n.* people who share a cultural or ethnic background but live in more than one country (page 257)

N

nations *n.* a cultural entity made up of people who have forged a common identity through a shared language, religion, heritage, or ethnicity—often all four of these (page 257)

nation-state *n.* a politically organized and recognized territory composed of a group of people who consider themselves to be a nation (page 257)

neocolonialism *n.* the use of economic, political, cultural, or other pressures to control or influence other countries, especially former dependencies (page 260)

Neoliberalism *n.* beliefs that favor free-market capitalism in which trade has no constraints from government (page 580)

Neo-Malthusian *adj.* describing the theory related to the idea that population growth is unsustainable and that the future population cannot be supported by Earth's resources (page 97)

net migration *n.* the difference between the number of emigrants and immigrants in a location, such as a city or a country (page 117)

New Urbanism *n.* a school of thought that promotes designing growth to limit the amount of urban sprawl and preserve nature and usable farmland (page 480)

node *n.* the focal point of a functional region (pages 19, 459)

nomadic herding *n.* a type of agriculture based on people moving their domesticated animals seasonally or as needed to allow the best grazing (page 340)

O

offshore outsourcing *n.* the condition when one or more aspects of production moves to an organization in another country (page 588)

overpopulation *n.* a term used to describe the condition in which population growth outstrips the resources needed to support life (page 97)

P

pattern *n.* the way in which things are arranged in a particular space (page 9)

perceptual region *n.* a type of region that reflects people's feelings and attitudes about a place; also called a vernacular region (page 21)

periphery *n.* classification of a country or region that has less wealth, lower education levels, and less sophisticated technologies and also tends to have an unstable government and poor healthcare systems (page 24)

physical geography *n.* the study of natural processes and the distribution of features in the environment, such as landforms, plants, animals, soil, and climate (page 7)

physiological density *n.* the total number of people per unit of arable land (page 71)

pilgrimage *n.* a journey to a holy place for spiritual reasons (page 168)

place *n.* a location on Earth that is distinguished by its physical and human characteristics (page 8)

placemaking *n.* a community-driven process in which people collaborate to create a place where they can live, work, play, and learn (page 178)

plantation agriculture *n.* a type of large-scale commercial farming of one particular crop grown for markets often distant from the plantation (page 338)

political geography *n.* the study of the ways in which the world is organized as a reflection of the power different groups hold over territory (page 255)

popular culture *n.* the widespread behaviors, beliefs, and practices of ordinary people in society at a given point in time (page 163)

population density *n.* the number of people occupying a unit of land (page 71)

population distribution *n.* where people live in a geographic area (page 67)

population pyramids *n.* a graph that shows the age-sex distribution of a given population (page 85)

possibilism *n.* theory of human-environment interaction that states that humans have the ability to adapt the physical environment to their needs (page 11)

post-Fordism *n.* system focused on small-scale batch production for a specialized market and flexibility that allows for a quick response to changes in the market (page 587)

postindustrial economy *n.* an economic pattern marked by predominant tertiary sector employment—with a good share of quaternary and quinary jobs (page 536)

Postmodern architecture *n.* a building style that emerged as a reaction to "modern" designs, and values diversity in design (page 167)

precision agriculture *n.* a farming management concept that uses technology to apply inputs with pinpoint accuracy to specific parts of fields to maximize crop yields, reduce waste, and preserve the environment (page 397)

primary sector *n.* economic sector associated with removing or harvesting products from the earth; includes agriculture, fishing, forestry, mining or quarrying, and extracting liquids or gas (page 533)

primate city *n.* the largest city in a country, which far exceeds the next city in population size and importance (page 439)

probeware *n.* software and hardware used to collect real-time data for scientific analysis with a computer or smart phone (page 33)

pronatalist *adj.* describing attitudes or policies that encourage childbearing as a means of spurring population growth (page 104)

pull factor *n.* a positive cause that attracts someone to a new location (page 118)

push factor *n.* a negative cause that compels someone to leave a location (page 118)

Q

qualitative *adj.* involving data that is descriptive of a research subject and is often based on people's opinions (page 33)

quantitative *adj.* involving data that can be measured by numbers (page 33)

quaternary sector *n.* economic sector that is a subset of tertiary sector activities that requires workers to process and handle information and environmental technology (page 533)

quinary sector *n.* economic sector that is a subset of the quaternary sector; involves the very top leaders in government, science, universities, nonprofits, health care, culture, and media (page 534)

quotas *n.* limit on the number of immigrants allowed into the country each year (page 134)

R

range *n.* in central place theory, the distance that someone is willing to travel for a good or service (page 440)

rank-size rule *n.* explanation of size of cities within a country; states that the second-largest city will be one-half the size of the largest, the third largest will be one-third the size of the largest, and so on (page 439)

rate of natural increase *n.* rate at which a population grows as the result of the difference between the crude birth rate and the crude death rate (page 91)

raw materials *n.* any metals, wood or other plant products, animal products, or other substances that are used to make intermediate or finished goods (page 523)

reapportionment *n.* the redistribution of representative seats among states based on shifts in population (page 284)

redistricting *n.* the redrawing of internal territorial and political boundaries (page 285)

redlining *n.* practice by which a financial institution such as a bank refuses to offer home loans on the basis of a neighborhood's racial or ethnic makeup (page 487)

reference maps *n.* a map that focuses on the location of places (page 42)

refugees *n.* a person who is forced to leave his or her country for fear of persecution or death (page 124)

region *n.* an area of Earth's surface with certain characteristics that make it cohesive yet distinct from other areas (page 18)

regional planning *n.* planning conducted at a regional scale that seeks to coordinate the development of housing, transportation, urban infrastructure, and economic activities (page 502)

relative direction *n.* direction based on a person's perception, such as left, right, up, or down (page 39)

relative distance *n.* distance determined in relation to other places or objects (page 39)

relative location *n.* a description of where a place is in relation to other places or features (page 8)

relics *n.* a former boundary that no longer has an official function (page 271)

religion *n.* a system of spiritual beliefs that helps form cultural perceptions, attitudes, beliefs, and values (page 168)

relocation diffusion *n.* the spread of culture traits through the movement of people (page 144)

remittances *n.* money earned by an emigrant abroad and sent back to his or her home country (page 141)

remote sensing *n.* collecting or analyzing data from a location without making physical contact (page 35)

repatriate *v.* to return to one's home country (page 129)

reservoirs *n.* artificial lake used to store water (page 384)

S

safe spaces *n.* a space of acceptance for people who are sometimes marginalized by society (page 176)

salinization *n.* the process by which water-soluble salts build up in the soil, which limits the ability of crops to absorb water (page 388)

scale *n.* the area of the world being studied (page 17)

second agricultural revolution *n.* a change in farming practices, marked by new tools and techniques, that diffused from Britain and the Low Countries starting in the early 18th century (page 353)

secondary sector *n.* economic sector associated with the production of goods from raw materials; includes manufacturing, processing, and construction (page 533)

sect *n.* a relatively small group that has separated from an established religious denomination (page 181)

sector model *n.* a model of urban development depicting a city with wedge-shaped sectors and divisions emanating from the central business district, generally along transit routes (page 459)

sedentism *n.* the practice of people settling in one place as a permanent or semi-permanent home (page 427)

self-determination *n.* the right of all people to choose their own political status (page 262)

semiautonomous *adj.* describing a region that is given partial authority to govern its territories independently from the national government (page 258)

semi-periphery *n.* classification of a country or region that has qualities of both core and peripheral areas and is often in the process of industrializing (page 24)

sense of place *n.* the subjective feelings and memories people associate with a geographic location (page 178)

sequent occupance *n.* the notion that successive societies leave behind their cultural imprint, a collection of evidence about human character and experiences within a geographic region, which shapes the cultural landscape (page 166)

sex ratio *n.* the proportion of males to females in a population (page 79)

shatterbelts *n.* a region where states form, join, and break up because of ongoing, sometimes violent, conflicts among parties and because they are caught between the interests of more powerful outside states (page 262)

shifting cultivation *n.* the agricultural practice of growing crops or grazing animals on a piece of land for a year or two, then abandoning that land when the nutrients have been depleted from the soil and moving to a new piece of land where the process is repeated (page 340)

Sikhism *n.* the newest universalizing religion; founded by Guru Nanak, who lived from 1469 to 1539, in the Punjab region of northwestern India (page 235)

site *n.* a place's absolute location, as well as its physical characteristics, such as the landforms, climate, and resources (page 9)

situation *n.* location of a place in relation to other places or its surrounding features (page 9)

skills gap *n.* a shortage of people trained in a particular industry (page 141)

slash and burn *n.* a method of agriculture in which existing vegetation is cut down and burned off before new seeds are sown; often used when clearing land (page 340)

slow-growth cities *n.* city where planners have used smart-growth policies to decrease the rate at which the city grows outward (page 481)

smart-growth policies *n.* policy implemented to create sustainable communities by placing development in convenient locations and designing it to be more efficient and environmentally responsible (page 480)

sociofacts *n.* a structure or organization of a culture that influences social behavior (page 161)

Southeast Asian city model *n.* a model of urban development depicting a city oriented around a port and lacking a formal central business district, growing outward in concentric rings and along multiple nodes (page 463)

sovereignty *n.* the right of a government to control and defend its territory and determine what happens within its borders (page 256)

space *n.* the area between two or more things (page 9)

spatial perspective *n.* geographic perspective that focuses on how people live on Earth, how they organize themselves, and why the events of human societies occur where they do (page 7)

special economic zones (SEZs) *n.* an area within a country that offers more favorable economic regulations (such as tax benefits or no tariffs) to attract foreign businesses (page 590)

squatter settlements *n.* an informal housing area beset with overcrowding and poverty that features temporary homes often made of wood scraps or metal sheeting (page 461)

stages of economic growth *n.* a model that suggests that all countries can be categorized on a spectrum from traditional to modern and that to become modern, countries need to pass through distinct stages of economic growth in succession (page 572)

stateless nation *n.* a people united by culture, language, history, and tradition but not possessing a state (page 258)

step migration *n.* series of smaller moves to get to the ultimate destination (page 123)

stimulus diffusion *n.* the process by which a cultural trait or idea spreads to another culture or region but is modified to adapt to the new culture (page 192)

subsequent boundaries *n.* a border drawn in an area that has been settled and where cultural landscapes exist or are in the process of being established (page 270)

subsistence agriculture *n.* an agricultural practice that provides crops or livestock to feed one's family and close community using fewer mechanical resources and more people to care for the crops and livestock (pages 72, 333)

suburbanization *n.* the shifting of population away from cities into surrounding suburbs (page 401)

suburbs *n.* less densely populated residential and commercial areas surrounding a city (page 19)

superimposed *n.* a border drawn over existing accepted borders by an outside or conquering force (page 270)

supranational organization *n.* an alliance of three or more states that work together in pursuit of common goals or to address an issue or challenge (page 301)

sustainability *n.* the use of Earth's land and natural resources in ways that ensure they will continue to be available in the future (page 12)

sustainable development *n.* development that meets the needs of the present without compromising the ability of future generations to meet their own needs (page 26)

syncretism *n.* process of innovation combining different cultural features into something new (page 207)

T

tariffs *n.* a tax or duty to be paid on a particular import or export (page 368)

temperate climates *n.* a climate with moderate temperatures and adequate precipitation amounts (page 68)

terracing *n.* the process of carving parts of a hill or mountainside into small, level growing plots (page 383)

territoriality *n.* the attempt to influence or control people and events by delimiting and asserting control over a geographic area; the connection of people, their culture, and their economic systems to the land (page 259)

tertiary sector *n.* economic sector that includes a host of activities that involve the transport, storage, marketing, and selling of goods or services; also called the service sector (page 533)

thematic maps *n.* any map that focuses on one or more variables to show a relationship between geographic data (page 42)

theory *n.* a system of ideas intended to explain certain phenomena (page 24)

third agricultural revolution *n.* a shift to further mechanization in agriculture through the development of new technology and advances that began in the early 20th century and continues to the present day (page 354)

third place *n.* a communal space that is separate from home (first place) or work (second place) (page 177)

threshold *n.* in central place theory, the number of people needed to support a business (page 440)

time-space compression *n.* a key geographic principle that describes the ways in which modern transportation and communication technology have allowed humans to travel and communicate over long distances more quickly and easily (page 10)

topography *n.* the representation of Earth's surface to show natural and human-made features, especially their relative positions and elevations (page 34)

toponyms *n.* a place name (page 168)

total fertility rate *n.* the average number of children one woman in a given country or region will have during her childbearing years (ages 15 to 49) (page 82)

traditional architecture *n.* an established building style of different cultures, religions, and places (page 167)

traditional culture *n.* the long-established behaviors, beliefs, and practices passed down from generation to generation (page 163)

traditional zoning *n.* zoning that creates separate zones based on land-use type or economic function such as various categories of residential (low-, medium-, or high-density), commercial, or industrial (page 480)

transhumance *n.* the movement of herds between pastures at cooler, higher elevations during the summer months and lower elevations during the winter (pages 122, 340)

transnational migration *n.* international migration in which people retain strong cultural, emotional, and financial ties with their countries of origin (page 121)

transportation-oriented development *n.* the creation of dense, walkable, pedestrian-oriented, mixed-use communities centered around or located near a transit station (page 479)

U

unitary state *n.* an organization of a state in which power is concentrated in a central government (page 277)

United Nations Convention on the Law of the Sea (UNCLOS) *n.* the international agreement that established the structure of maritime boundaries (page 271)

universalizing religions *n.* a religion that tries to appeal to all humans and is open to membership by everyone (page 233)

urban area *n.* a city and its surrounding suburbs (page 427)

urban growth boundary *n.* a boundary that separates urban land uses from rural land uses by limiting how far a city can expand (page 481)

urbanization *n.* urban growth and development (page 92)

urban renewal *n.* the nationwide movement that developed in the 1950s and 1960s when U.S. cities were given massive federal grants to tear down and clear out crumbling neighborhoods and former industrial zones as a means of rebuilding their downtowns (page 497)

urban sprawl *n.* areas of poorly planned, low-density development surrounding a city (page 435)

V

vernacular region *n.* a type of region that reflects people's feelings and attitudes about a place; also called a perceptual region (page 21)

vertical integration *n.* the combining of a company's ownership of and control over more than one stage of the production process of goods (page 367)

voluntary migration *n.* type of migration in which people make the choice to move to a new place (page 121)

von Thünen model *n.* a model that suggests that perishability of the product and transport costs to the market each factor into the location of agricultural land use and activity (page 371)

W

walkability *n.* a measure of how safe, convenient, and efficient it is to walk in an urban environment (page 479)

wetlands *n.* area of land that is covered by water or saturated with water (page 386)

wildland-urban interface (WUI) *n.* the zone of transition between unoccupied, undeveloped wildland and residential or other human development (page 436)

women's empowerment *n.* women's options and access to participate fully in the social and economic spheres of a society (page 559)

world cities *n.* a city that wields political, cultural, and economic influence on a global scale (page 447)

world system theory *n.* theory describing the spatial and functional relationships between countries in the world economy; categorizes countries as part of a hierarchy consisting of the core, periphery, and semi-periphery (page 24)

Z

zones of abandonment *n.* area that has been largely deserted due to lack of jobs, declines in land value, and falling demand (page 489)

zoning *n.* the process of dividing a city or urban area into zones within which only certain land uses are permitted (page 469)

Index

A

absolute distance, 39
absolute location, 8
acculturation, 206–207
Act 43, 287
Act for Environmental Justice (We Act), 496
Addario, Lynsey, 520
adherents, 181
administer, 266
adolescent birth rate (ABR), 559, 608
African American English (AAE), 180
African city model, 462, 463, 516
African culture in Brazil, 194
agglomeration, 542, 543
aging population
 consequences of, 110–113
 economic effects, 111–112
 political effects, 112–113
 social effects, 110–111
agribusiness, 361
agricultural biodiversity, 395
agricultural density, 72
agricultural hearths, 346–347
agricultural intensification, 402
agricultural interdependence, 373–374
agricultural landscapes, 383
agricultural practices, 333–341, 383–393
 altering the environment, 383–387
 commercial agriculture, 333
 conservation and sustainability efforts, 389–391
 environmental consequences, 387–389
 extensive agriculture, 339–341
 intensive agriculture, 335–341
 shifting cultivation, 383
 societal consequences, 392
 subsistence agriculture, 333
agricultural production
 economic forces and, 361–362
 patterns and practices of, 360–379
 regions, 361–365
agricultural regions, around world, 342–343
agricultural sustainability in global market
 agricultural practices, consequences of, 383–393
 contemporary agriculture, challenges of, 394–400
 women in, 408–411
agricultural waste, 500
agriculture, 323
 advances in, 351–356
 agricultural hearths, 346–347
 climate regions, 330–331
 defined, 327
 and diet, 400
 diffusion of, 347–351
 and environment, 327–332
 as global system, 373–378
 human–environment interaction, 326–357
 for hungry future, 324
 patterns of world trade, 376–378
 policies and preferences, 362
 political relationships, 376
 practices, 333–341
 production, patterns and practices of, 360–379
 rural land-use patterns, 371–372
 spatial organization of, 366–370
 women in, 408–411
agroecosystem, 383
AIDS, 84, 99
Airline Flight Routes map, functional region, 19, 20
Airplane-mounted sensors, 37
air pollution, 500
alien commercial zone, 463
Allard, William, 204
Alliance for a Green Revolution in Africa (AGRA), 390
Amazonian plaza villages, 431
American Community Survey, 33
America's Dairyland, 364
AmeriCorps Urban Safety Project, 36
Amish immigrants, 173
Anatolian hearth theory, 220
ancient geography, 158
ancient hearths, 346–347
animism, 231
antecedent boundaries, 270
Anthropocene climate change, 11
Anthropocene Epoch, 11
antibiotics, 396
anti-Muslim, 242
antinatalist policies, 104, 106, 107
anti-Semitism, 239
Antoniou, Anna, 311
aquaculture, 4, 396–397, 400
aquifers, 386
arable land, 71
Arctic Council, 303
arithmetic density, 71, 72, 76
Arreola, Daniel D., 166
artifacts, 158, 161, 172
artificial intelligence (AI), 532
assimilation, 207
Association of Southeast Asian Nations (ASEAN), 301, 303
asylum, 124
asylum in European Union, 136
at-large vs. single-member districts, 288
augmented reality (AR), 532
austerity measures, 586
Australia's economic sectors, 537
Austronesian language family, 222–223
autonomous regions, 258
azimuthal projection, 40, 41

B

baby bonuses, 106
Baby Boom, 86, 94, 112
baby bust, 93
Bach, Jaime, 159
bananas, 419
banlieues, 466
barriers to diffusion, 202
basic industry, 434
Basque language, 180
Basque nationalism, 294
Beijing, land use change in, 467
Belo Monte Dam, 538–540
Berry, John W., 207
Bestor, Ted, 510
Biden administration, 139, 470
bid-rent theory, 333, 418, 457, 513
Big Data, 37, 38
biodiversity, 4, 332, 387, 399, 421, 503
biotechnology, 394–396, 405
Black Death, 98
blockbusting, 487
Boiling River Project, 599
Bombay Stock Exchange, 22
Bonton Farms, 497
boomburb, 435
Borlaug, Norman, 355
Boserup, Ester, 97
Boston Brahmin dialect, 180
boundaries, 284, 266
 antecedent, 270
 consequent, 270
 geometric, 270
 nature and functions of, 265–273
 political, 265–266, 270
 subsequent, 270
 types of, 270–273
Bracero program, 119, 124

brain drain, 142
Brazil
   African culture in, 194
   economic opportunities, 566
   increased crime rates, 492
   soybean production and exports, 394
break-of-bulk points, 543
Brexit, 301, 305, 376
British Empire, 198, 238, 453, 525
brownfields, 504
Buddhism, 168, 183, 198, 235, 236
Bulgarian Orthodox Church, 233
bulk-gaining industries, 545, 605
bulk-reducing industries, 545, 605
Burgess concentric-zone model, 458, 459, 514
Burgess, Ernest, 459
Burney, Jennifer, 407
Bush, George W., 4
business process outsourcing (BPO), 589
Butler, Judith, 173

C

Canada
   bilingual country, 227
   population distribution at, 76–77
   provinces, 316
cancer, 99
Cape Town, informal housing in, 464
cardiovascular diseases, 99
carrying capacity, 74
cartograms, 44, 45, 47
cartographers, 39
cash crop, 373
cattle stations, 417
causes and impacts of urban changes, 486–498
census, 33
Center for a Livable Future (CLF), 404
Center for Strategic and International Studies (CSIS), 273
Centers for Disease Control and Prevention (CDC), 47, 103
central business district (CBD), 333, 457, 462, 463, 513, 516
Central Intelligence Agency, 50
central place theory, 440–441, 513
centrifugal forces, 184–185
centripetal forces, 184, 306–309
chain migration, 122–123
Chicago Defender, 138
Child and Youth Act, 278
China, 376

comparative advantage, 579
complementarity, 579
famine in, 94
fertility rate in, 105
one-child policy, 80, 104, 105
pastoral nomads, 390
population distribution in, 69
population policies, 105
special economic zones (SEZs), 590
Chinese Exclusion Act, 136
choke points, 317
choropleth maps, 45, 47, 48
Christaller, Walter, 440–441
Christianity, 196, 198, 229, 233–235, 238
   denominations of, 249
   diffusion of, 234
   Roman Catholic Church, 233
chronic diseases, 100
chronic stress, 96
circular economy (CE), 73
circular migration, 124
circulation, 117
cities
   and globalization, 445–453
   internal structure of, 457–464
   political organization of, 472–473
   shape of, 450–451
   size and distribution of, 438–451
   sustainable cities, planning for, 479–480
   urban land use, 457–458
city models, 514
civil war, 93, 118, 132, 309
class consciousness, 530
Cleveland, 489, 490
climate
   defined, 68
   of Saudi Arabia, 70
climate change, 16, 18, 26, 68, 103, 120, 131, 386, 393, 498, 500, 502, 553
   in Anthropocene, 11
   city altered by, 148–149
   defined, 11
climate change migrants, 148–149
   cities for, 149
   climate-driven displacement, 148
climate-driven displacement, 148
climate, for agriculture, 328–330, 416
climate migrants, 120
climate refugees, 12, 120, 129, 131
clustered populations, 67
clustered settlement, 336
coal power, 526
Code of Conduct, 303
coffee production and consumption, 379

Cold War, 41, 267, 303
collaborative mitigation, 593
collectivist cultures, 208
Colombia's War, 314–315
colonialism, 11, 184, 196–197, 218, 233, 259, 528–529
   legacies of, 263, 265
colonial languages, 197
Columbian Exchange, 350–351, 400
commercial agriculture, 333
commercial dairying, 364
commercial forming, 389
commercialization, 402
commodity chains, 367
commodity dependency, 373–374, 572, 575
communicable diseases, 100
community-supported agriculture (CSA), 399
company culture, 189
comparative advantage, 579
compassion, natural history of, 64
complementarity, 579
compulsory acquisition, 494
concentric-zone model, 459, 461, 514
Concept Plan, 278
concurrent, 281
conflict, 119–120
   hunger and, 403
conflict over borders, 318
conformal projection, 40, 42
connected economies, 584–592
connectivity subsidy program, 596
consequent boundary, 270
Conservation Reserve Program, 389
conspiracy theories, 192
contagious diffusion, 191
contemporary agriculture
   challenges of, 394–400
   food choices, 398–400
   innovations, 394–397
contemporary cultural change, 199–201
continental climates, 331
Continental Europe, industrialization, 527
contraception, 95
convenient contiguous territory, 288
Convention on the Elimination of All Forms of Discrimination Against Women (CEDAW), 562
core countries, 24–26, 57, 470, 519, 534, 551, 573, 580, 588

R8   Index

core–periphery model, 24, 74, 574
Cosgrove, Denis, 40, 41
Costa, Adjany, 16
cottage industries, 524
COVID-19 pandemic, 35, 47, 48, 86, 96, 99, 101, 102, 122, 125, 162, 166, 266, 303, 365, 401, 403, 406, 436, 446, 466, 473, 475, 480, 493, 546, 547, 574, 584, 587
creolization, 198
crime rates, 492
crop dusting, 356
crop rotation, 338
crude birth rate (CBR), 81–83, 91
crude death rate (CDR), 91, 95
Culhane, T.H., 471
cultural appropriation, 209
cultural barriers to diffusion, 202
cultural change
　acculturation, 206–207
　assimilation, 207
　consequences of, 206–209
　contemporary drivers of change, 199–201
　cultural convergence, 201–202
　cultural divergence, 202
　diffusion, 189–195
　multiculturalism, 208–209
　processes of, 199–205
　syncretism, 207–208
cultural convergence, 201–202
cultural differences, 162–163
cultural diffusion, 246
cultural divergence, 202, 204
cultural diversity, 119, 309, 310
cultural dynamics, 162–163
cultural factor
　for populations growth and decline, 95
　for migration, 119
　in population distribution, 70
cultural hearth, 189
cultural iceberg, 161, 244
cultural identity, 168, 173
cultural landscapes, 165–172, 184
　defined, 165–166
　ethnic neighborhoods, 166
　examples of, 166–168
　features and identity, 172–173
　of language, 168
　postmodern architecture, 167–168
　of religion, 168
　traditional architecture, 167
cultural mixing, 162
cultural norms, 162–163
cultural relativism, 163
cultural syncretism, 208
cultural traits, 161, 184, 189, 193, 201

culture, 158
　cultural dynamics, 162–163
　cultural patterns, 178–185
　defined, 161
　ethnic patterns, 183–184
　identity and space, 172–177
　landscapes see cultural landscapes
　and language, 213
　LGBTQIA+ spaces, 176–177
　linguistic patterns, 180
　mixing, 162
　patterns of unity and division, 184–185
　and religion, 229
　religious patterns, 181–183
　sense of place, 178–179
　shaping space through identity, 172–173
　women and gendered spaces, 173–176
cutting-edge technologies, 397
Cyber Security Law, 163, 202
cylindrical projections, 40, 42
Cyprus, 311, 410

D

Dalai Lama, 229, 235
Damato, Zachary, 501
Daniels, Stephen, 40, 41
data collection, 32–33
　drones for, 37
　international-scale data, 33
　qualitative data, 33
　quantitative data., 33
　U.S. government agencies and private organizations in, 33
　velocity of, 37
D.C. Metro System map, functional region, 19, 20
dead zone, 388
de Blij, Harm, 463
debt-for-nature swaps, 389
decision-making criteria, 96, 587
decision making, with geographic data, 48–50
　governmental, 49–50
　organizational, 48–49
　personal, 48–49
de facto segregation, 484
Deferred Action for Childhood Arrivals (DACA), 139
deforestation, 383, 388
deindustrialization, 461, 581–582, 587
de jure segregation, 487
De León, Jason, 126
Delhi-Mumbai Industrial Corridor (DMIC) Development Project, 491
delimit, 266

demarcated, 266
Demilitarized Zone (DMZ), 251
　in Korea, 267
Democratic Republic of the Congo (DRC), 85, 86, 91, 101, 401
demographic change, 108–116
　aging population, consequences of, 110–113
　changes in size and composition, 108–109
demographics, 81, 119
demographic transition model, 97–98, 151, 529, 531
dengue fever, 39
denominations, 181
densely developed territory, 427
density, 9
dependency ratio, 78–79, 145
　data beneath, 79
　and economic development, 79
dependency theory, 572, 574–575, 607
desertification, 387, 388–389, 393
designing for urban life, 479–486
　diverse housing options, 480–483
　growth management plans, 482
　planning for sustainable cities, 479–480
　smart-growth policies, 480–483
destabilization, 294
Detroit Three auto companies, 541
developed countries, 24
devolution, 265, 320
　economic and social problems, 294–295
　ethnic separatism, 294
　physical geography, 293
　process of, 293–297
　responses to devolutionary forces, 297–300
　state's sovereignty, challenges to, 293–300
devolutionary forces, 297–300, 321
dialects, 180, 215
diet, 400, 406
dietary shifts, 399–400
diffusion, 246
　of agriculture, 347–351
　barriers to, 202
　contagious, 191
　of english language, 221–222
　expansion, 191–195
　hierarchical, 191–192
　of Hinduism, 237
　historical causes of, 196–199
　impacts of language diffusion, 223–226
　industrial diffusion and populations, 529–531
　of language, 218–226

diffusion (*continued*)
  of misinformation, 192
  in modern times, 351
  relocation, 189–191
  stimulus, 192
  types of, 192, 193
digital encyclopedia, 33
disamenity zones, 461, 515
discrimination, 119, 166, 175, 295
dispersed settlements, 336
disputed ocean territory, 273
distance decay, 10, 56, 124, 201, 320
distributed, 9
distribution, patterns of, 229–231
diverse faith community, 242–243
diversification of geography, 12
diversity, 18, 119, 158, 168, 226, 282, 298, 309, 482, 511
division of labor, 589–590
domesticated animals, 349
domestication, 346, 349
dot maps, 44
doubling time (DT), 91
draining wetlands, 386
drones, for data collection, 37
dual agricultural economy, 361
dual economies, 537
Dunnavant, Justin, 195
dying society, 110

E

eating ugly, 370
Ebola virus, 99–101
echo chamber, 192
EcoLogic, 389
ecological capital, 481
ecological footprint, 479
ecological perspective, 7, 8
ecological sabotage, 4
economic activities, 535–536
Economic Census, 33
economic development, dependency ratio and, 79
economic development patterns, 534–537
economic effects, in aging population, 111–112
economic expansion, 26
economic factors
  for migration, 119
  for populations growth and decline, 92–93
economic forces and agriculture, 361–362
economic growth, stages of, 572–573

economic impacts, on food production, 405–406
economic indicators, 551–553
  economic structure, 552–553
  fossil fuels and renewable energy, 553–554
  income, 551–552
economic landscape, changes in, 587–590
economic opportunities, 566–567
economic sectors, 533–534, 604
economic structure, 552–553
economies
  of countries, 533–537
  development patterns, 534–537
  dual, 537
  economic landscape, changes in, 587–590
  global economy, impacts of, 584–586
  postindustrial, 536–537
  sectors of, 533–534
economies of scale, 303, 405
ecotourism, 597–599
edge city, 435
education, 95, 96
  and birth rates, 608
educational attainment, 559
educational opportunities, 567–568
Egypt, population distribution at, 76–77
elderly dependency ratio, 111, 112
electoral boundaries, 319
electoral college, 285
electoral district boundaries, 285–289
electoral geography, 284–289
  electoral district boundaries, 285–289
  representing the people, 284–285
electors, 285
electric-powered system, 450
elevation, for agriculture, 328
Elkaim, Aaron Vincent, 539
emigration, 117
eminent domain, 494
employment, and economic development, 534–535
empowering people with clean energy, 471
empowering rural women, 410–412
empowerment, 559, 569–570
enclosure system, 353
endangered languages, 225–226
energy, 506
energy island, 593
Engels, Fredrich, 531

English language, diffusion of, 221–222
Enote, Jim, 171
environmental consequences, 387–389
environmental degradation, 75
environmental determinism, 11
environmental/ecological perspective, 7, 8
environmental factor
  for populations growth and decline, 94–95
  for migration, 120
environmental injustice, 494–497
Environmental Protection Agency, 33, 50
environmental racism, 494–497
environmental sustainability, 511
epidemiological transition model (ETM), 98–102, 152
equal-area projections, 40, 42, 59
equality, defined, 556
equalization, 295
Equal Pay Act, 569
Equator, for agriculture, 328
equity, 556, 558
"Estimated Majority Religions" map, 229, 231
ethnic cleansing, 294
ethnic identity, 294
ethnicity, 166, 168, 183, 244, 294, 307, 308, 310
ethnic nationalism, 308
ethnic neighborhoods, 166
ethnic patterns, 183–184
ethnic religions, 237–239
ethnic separatism, 294
ethnocentrism, 163
ethnologue, 214
ethnonationalism, 308
Europe
  Industrial Revolution in, 529
  nation-state, 257
  women in agriculture, 409
European empires, 196, 265, 270
evolving opportunities, for women, 565–569
exclusive economic zone (EEZ), 271–272
expansion diffusion, 191–195, 221
export processing zones (EPZs), 591
extensive agriculture, 339–341
extensive commercial agriculture, 340
extensive subsistence agriculture, 339–340

exurb, 436
exurbanization, 436

F

factors of production, 526
Fadiman, Maria, 228
fair trade, 377–378, 399
family planning, 95, 106, 107
   in India, 106
famine, 94, 109
farmland protection, 504
farm subsidies, 368
fashion diffusion, 247
Federal Election Commission, 33
Federal Emergency Management Agency (FEMA), 50
federalism, 282, 283
federal law enforcement agencies, 258
federal laws and policies, 280
federal state, 277–284
   advantages of, 282
   disadvantages of, 282
female immigrants, risks for, 136
Fenoglio, Maria Silvina, 503
Fertile Crescent, 346
fertility, 81–82, 86, 94, 98
   in European countries, 82
   measures of, 81–82
   and mortality rates, 82
fertilizer trees, 324
financial crisis of 2007–2008, 585, 586
financial institutions, 584
first agricultural revolution, 351–353
First Industrial Revolution, 351–353, 537, 541
flow, 10
Food and Agriculture Organization (FAO), 366, 389
Food and Drug Administration, 33
food choices, 398–400, 399, 400
food deserts, 404
food insecurity, 401–403, 496
food insecurity in United states, scale of analysis, 17
food issues in United States, 403
food production, economic impacts on, 405–406
   economy of scale, 405
   fighting the problem, 406
   storage and transportation issues, 405
food security, 401, 407
food sovereignty, 409
food supply system, 368
food waste, 421
foragers, 346

forced migration, 121, 124–125, 155
Ford, Henry, 543, 587
Fordism, 587
Ford Motor Company, 541, 543, 609
formal region, 18–19
formal sector, 552
former Yugoslavia, 316
fossil fuels, 553
Frachetti, Michael, 430
France
   nation-state, 257
   unitary system, 277
free trade, 579, 580
free trade agreements, 583
free trade zones (FTZs), 591–592
friction of distance, 121
functional region, 19–21
Fund for Peace (FFP), 321
fútbol, globalizing force, 203
future city, 442

G

galactic city model, 460, 461, 515
Gall-Peters projection, 40, 41
gas holders, 505
gender-based employment, 536
Gender Development Index (GDI), 556, 557
gender disparities, 556–561
gendered spaces, 173–176
gender equality, 558, 559
gender identity, 176
Gender in Agricultural Partnership (GAP), 411
gender inequality, 84, 96, 520, 550–564
Gender Inequality Index (GII), 556, 557
gender, patterns in, 175
genetically modified organisms (GMOs), 324, 354, 392, 394–396
genetic code, 102
genetic engineering (GE), 395, 396
genetic modification (GM), 394
genetic mutations, 102
genocide, 265
genome editing, 532
gentrification, 177, 436, 497–498
geographers think
   geographic data and tools, 32–35
   qualitative data, 33
   quantitative data, 33
geographic advantage, 7
geographic data, 47–50
   governmental decision making, 49–50
   making decisions with, 48–50

   personal and organizational decision making, 48–49
   used to, 47–48
geographic information systems (GIS), 34–35, 49–51
   in Detroit, 36–38
   geovisualization, 35
   in governmental decision making, 49–50
   topography, 34
geographic models, 461
geographic scales, 7
geographic thinking, 6–12, 31, 36–38
   distance decay, 10
   globalization, 23–24
   human–environment interaction, 10–12
   sustainability, 26–27
   time–space compression, 10
geography, 424
   defined, 7
   density, 9
   elements of, 8
   environmental/ecological perspective, 7, 8
   human geography, 7
   location, 8
   pattern, 9–10
   for people, 424
   physical geography, 7
   place, 8
   space, 9
   spatial perspective, 7
Geography Collective, 424
geometric boundaries, 270
geospatial revolution, 35
geospatial technologies, 35
geovisualization, 35
Gerdes, Caroline, 473, 474
Germany, 281
gerrymandering, 285–286
   opposition and remedies, 286
   and race, 287
Gieseking, Jack Jen, 176
girls' education levels, 567
Glaeser, Edward, 510
global commodity, 419
global corporations, 579–583
global economic crisis of 2008, 305
global economy, impacts of, 584–586
global financial institutions, 584
global food insecurity, 401–403
global food system, 407
globalization, 23–24, 26, 200, 402, 578, 579, 587, 590
   cities and, 445–453
   government policies in, 23
global positioning system (GPS), 37, 50, 59, 397

Global Power City Index (GPCI), 447, 449
Global Services Location Index, 589
global supply chain, 401
global supply chains, 373, 376, 379
global warming, 18
glottochronology, 218
Glover, Jerry, 324, 325, 401
GNI, 574
Golledge, Reginald, 7, 31
Goodchild, Michael, 37
1998 Good Friday Agreement, 185
governmental decision making, 49–50
governmental stability measurement, 321
government policies, 405–406
government trade policies, 582–583
graduated symbols map, 43
Grameen Bank, 571
graphic scale, 42
gravity model, 154, 438–439, 512
Great Britain's Inclosure Act of 1773, 530
Great Depression, 368
Greater London Authority, 493
Great Famine of 1845–1849, 94
Great Green Wall, 393
Great Migration, 133–135, 144, 189
  journalism and, 138
Great Recession, 586
Greek Orthodox Church, 233
greenbelt, 483
greenbelt towns, 484
greenhouse gases, 11, 500, 532
greenhouses, 414, 415
Green Jobs Act, 596–597
Green Revolution, 355, 357, 395, 417
Griffin-Ford model, 461, 462
gross domestic product (GDP), 22, 303, 537, 551, 553, 579, 604
gross national income (GNI), 551–553, 556, 584
gross national product (GNP), 551
growth management plans, 482
growth of cities, 427
growth poles, 582
guerrilla geography, 424
guest workers, 124
Guttenfelder, David, 268

H

Haitian Creole, 198
Harley, J. Brian, 40
Harris and Ullman multiple-nuclei model, 458, 459, 514
Harvey, David, 168
healing of Colombia, 314–315
health-care systems, 84, 111
heart disease, 84
herbicides, 355, 362, 396, 397
hidden crime, 128
Hiebert, Fred, 158, 159
hierarchical diffusion, 191–192
High Line in New York City, 499
high-tech sexism, 96
Hinduism, 237–238
Hispanic-Latino migration, 134
historical factors, in population distribution, 70–71
Hitler, Adolf, 308
Hmong migration, 135
Hmong refugees, 184
Holy Roman Empire, 309
Home Owners Loan Corporation (HOLC), 490
Hong Kong, population density, 468
housing density and development, 468–469
housing discrimination, 487–488
Hoyt, Homer, 459
Hoyt sector model, 458, 459, 514
hub-and-spoke design, 19
human development
  changing roles of women, 565–571
  dependency theory, 574–575
  economic indicators, 551–553
  gender inequality, 556–564
  Human Development Index (HDI), 556, 557
  measurement, 551–556
  Rostow's stages of economic growth, 572–573
  social indicators, 554–555
  theories of, 572–575
  Wallerstein's world system theory, 573–574
Human Development Index (HDI), 556, 557, 568
human–environment geography, 12
human–environment interaction, 10–12
  climate change, 11
  sustainability, 12
  theories of, 11–12
human factors, of population distribution, 69–71
  cultural factors, 70
  economic factor, 69
  historical factors, 70–71
  political factors, 69–70
human geography, 3, 7–10
human migration, 69, 117
human trafficking, 128
humid temperate climates, 330
hunger and conflict, 403
Hurricane Florence, 120
Hurricane Katrina, 15
Hussein, Saddam, 264
Hutterite community., 204, 205
hybrid, 361
hydroelectric project, 538
hydrological cycle, 595
hyperlocal process, 486

I

Ibrahim, Hindou Oumarou, 391
identity, 165
immigration, 86, 105, 113, 117, 135–137, 141
immigration policy, 135, 136, 139
imperialism, 11, 196–197, 218, 233, 528–529, 575
  legacies of, 263, 265
inclusionary zoning laws, 493
income, economic indicators, 551–552
India
  aging population in, 111
  economy of scale in, 406
  family planning, 106
  land tenure, 494
  poverty and wealth, 22
  regional differences in scale, 22
Indian Removal Act of 1830, 133
indigenous knowledge perserving, 228
indigenous land-use techniques, 393
indigenous languages, 182
Indigenous Peoples' Day, 164
indigenous religions, 231
1975 Indochina Migration and Refugee Assistance Act, 135
Indo-European hearth and diffusion, 218, 221
Indo-European languages, 214, 216–217
industrial diffusion and populations, 529–531
industrialization, 73, 92, 98, 565
  defined, 522
  economies of countries, 533–537
  patterns of industrial location, 541–547
  processes of, 523–532
  spread of, 526–529
industrial location, patterns of, 541–547
  decisions today, 545–547
  factors that influence, 542–543
  least-cost theory, 541–545

industrial parks, 547
Industrial Revolution, 91, 92, 96, 98, 128, 354, 392, 428, 433, 519, 523–526, 553, 587
   in Europe, 529
   First, 527, 535
   Fourth, 532
   second, 527–528, 535
   third, 527–528
industry, 523
infant mortality rate (IMR), 83, 554–555
infectious disease, 98
infill, 436
informal housing in Cape Town, 464
informal sector, 552, 553
information industries, 535
information technology, 528, 534
infrastructure, 361, 374–376, 442, 444, 469–475, 500
   and development, 470–472
   political organization of cities, 472–473
   qualitative urban data, 473–474
   quantitative urban data, 474–475
innocent passage, 272
innovations, agricultural, 394–397
intangible cultural heritage, 161
intensive agriculture, 335–341
intensive commercial agriculture, 338–339
intensive market gardening, 463
intensive subsistence agriculture, 335–339
Intergovernmental Panel on Climate Change (IPCC), 532
internally displaced persons (IDPs), 125, 128–132, 130
   hidden crisis of, 129
internal migration, 121
internal voluntary migrations, 122
international division of labor, 589
International Energy Agency (IEA), 317
International Monetary Fund (IMF), 580, 581, 584
international residents, 511
international-scale data, 33
International trade, 24
internet freedom, 163
internet protocol, 589
interregional migration, 133
intersectionality, 207
intervening obstacle, 123
intervening opportunity, 123
intraregional migration, 133
investing in girls and women, 570

irredentism, 257, 294
   in Ukraine, 296
irrigation, 384–386, 414
Islam, 234–235
   and Black America, 243
Islamic law, 242
Islamophobia, 308
isolate, 215
isoline maps, 43, 60
issues of space and power, 259

J
Jacobs, Jane, 510
Japan
   economy, 566
   nation-state, 257
Jerusalem Day, 238
Jews/Jewish, 239, 253, 259
Jim Crow laws, 134, 138
Johnson, Clinton, 51
Jordan, Terry, 199
journalism and Great Migration, 138
Judaism, 231, 238–239
just-in-time delivery, 587

K
Kamalu, Lehua, 46
Kangnido Map, 255
King Jr., Martin Luther, 172
kinship links, 134
Kolivras, Korine N., 39
Köppen Climate Classifications, 330
Köppen, Vladimir, 330
Kurds, 264
Kurgan hearth theory, 220
Kutupalong, 125

L
labor-force participation (LFP), 560
labor-market participation (LMP), 560, 561
Lagos, Nigeria, 453
land and resource use, 173
land cover change, 388
land degradation, 108
landfills, 11
landforms, 68–69
land surveying, 334
land tenure, 493–494
land use, 420
language
   branches, 215
   categories, 213–215
   and culture, 213
   dialects, 215
   diffusion of, 218–226

   distribution of, 218
   endangered languages, 225–226
   family, 213–215
   groups, 215
   impacts of language diffusion, 223–226
   influence of power, 223–225
   spread and change, 218–223
   on Web and IRL, 245
language divergence, 223
language family, 213–215
   distribution of, 219
   spread and change, 218–223
large-scale manufacturing, 523
large-scale map, 39
Latin American city model, 461–463, 515
law enforcement, 23
laws of migration, 117, 153
least-cost theory, 541–545, 605
   assumptions in, 543
   break-of-bulk points, 543
   limitations of, 545
   raw materials, 543–545
legacy cities, 434
Legislative Services Agency, 288
lettuce, 13
Lewis, Pierce, 461
LGBTQIA+ spaces, 176–177, 245
life expectancy, 83–84, 99, 110, 530, 555
linear pattern of populations distribution, 67
linear settlement, 336
Linera, García, 602
lingua franca, 197
linguistic patterns, 180
literacy, 555
lithium industry, 602
lithium-mining operations, 602
Lithium Triangle, 602
livability, 506
Liverpool, 437, 525
local food initiatives, 497
local food movement, 398–399
location, 8
London, United Kingdom, 452
long-lot survey system, 334
Los Angeles, United States, 450
low-access communities, 404
lung cancer, 100

M
MacDonald, Glen, 12
Macron, Emmanuel, 306
Magufuli, John, 106
majority-minority districts, 286

malnutrition, 109
Malthus, Thomas, 96–97, 152, 417
Mandela, Nelson, 307
Manila, Philippines, 453
mapmaking, 38–42
mapping indigenous climate knowledge, 391
maps, 59
   choropleth, 45
   graduated symbols, 43
   isoline, 43
   physical, 42
   political maps *see* political maps
   problem-solving tools, 39
   projection types, 40–42, 59
   reference, 42
   thematic, 42, 61
   types of, 42–47, 60
map scales, 39, 41–42
marine west coast climate, 330
maritime boundaries, 271–273
market gardening, 339
maroon geographies, 179
Marshall, Will, 54
Marx, Karl, 531
mass-transit system, 468
maternal mortality ratio (MMR), 558
McGee, T.G., 463
McLeod, Mary, 167
Meadows, Michael, 26
Medicaid, 552
Medicare, 552
Mediterranean agriculture, 331
Mediterranean climates, 330
medium-scale map, 39
megacity, 423, 446
mental maps, 8
mentifacts, 162
Mercator projection, 40, 41, 59
Mercosur, 580
metacity, 446
methane emissions, 11
metropolitan area, 427
Metropolitan Transit Authority (MTA), 473
Mexico, 281
microfinance, 570
microloans, 570
Middle East respiratory syndrome (MERS), 101
migration, 117, 198–200, 215
   from Central America, 127
   chain, 122–123
   circular, 124
   cultural and social consequences, 142–145
   cultural factors, 119

   demographics factors, 119
   documenting of migrants, 126
   economic consequences, 140–142
   economic factors, 119
   effects of, 140–145
   environmental factors, 120
   forced, 121, 124–125, 128
   Hispanic-Latino, 134
   historical U.S. migrations, 133–135
   Hmong, 135
   internal, 121
   interregional, 133
   intraregional, 133
   and policy, 133–139
   political conditions and conflict, 119–120
   population growth and, 433
   push and pull factors, 118–120
   refugees and internally displaced persons, 128–132
   rural-to-urban, 123–124
   Somali, 134–135
   step, 123
   transnational, 121
   types of, 121–126
   voluntary, 121–124
migration data, 155
Migration Period, 70
Milan and urban sustainability, 507
military conquest, 197–198
mining, 593
missing births, 93
Mississippi River, 14, 15, 388, 428, 432
mixed crop and livestock systems, 339
mixed-use development (MUD), 479
mixed-use zoning, 480
mobility, 117
modern communications technology, 532
Modi, Narendra, 232
Monmonier, Mark, 41
monocropping, 338, 376, 392
monoculture, 338
Morales, Evo, 602
moral restraint, 97
Moros, 293
mortality, 82–84, 86
   and fertility rates, 82
   infant mortality rate (IMR), 83
   life expectancy, 83–84
   measures of, 83–84
Motor City, 543
multiculturalism, 208–209
multilingualism, 182
multinational states, 257–258
multiple-nuclei model, 458, 459, 514
multiplier effects, 590

multistate nation, 257–258
Murphy, Alexander B., 257
Musk, Elon, 192
Muslim in America, 242–243

N
nanotechnology, 532
Narayanan, Sandhya, 182
NASA, 54
National Archives, 33
national governments, 48, 279, 437, 500, 582
nationalism, 233, 281
National Park City, 424
National Rural Water Association, 389
National Security Agency, 50
nations, 257, 318
nation-state, 257
natural disasters, 94
natural resources, 592–593, 602
neighborhood map, 244
neocolonialism, 259–260, 575
neoliberalism, 580
Neo-Malthusian, 97
net migration, 117
Neves, Eduardo, 431
newly industrialized countries (NICs), 588
new manufacturing zones, 590–592
New Orleans, 14–15
Newton's law of universal gravitation, 117
New Urbanism, 480, 481
nicotine, 356
Niger-Congo language family, 248
Nigeria
   elderly dependency ratio in, 112
   population size and composition, 109
Nile River, 76
Nkrumah, Kwame, 259
nodes, 19, 20, 459
nomadic herding, 340, 386–387
nominal data, 474
nonbasic industries, 434
noncommunicable diseases (NCDs), 100, 153
non-GMO foods, 399
nongovernmental organizations (NGOs), 357, 447, 593
nonprofit organizations, 36
nonrenewable resources, 12, 524
Norfolk four-field system, 354
North America,
   industrialization, 527

North American Free Trade Agreement (NAFTA), 24, 400, 580
North Atlantic Treaty Organization (NATO), 301–303
North Carolina's congressional districts, 289
nucleated settlement, 336
Nunavut Land Claims Agreement, 283
Nunavut, political control and, 283
nutrition, 420

O

Obama, Barack, 4
offshore outsourcing, 588
old-age dependency ratios, 79
Omran, Abdel R., 98
one-child-only policy, 104
one-child policy, 80, 105
OpenStreetMap, 48
orchard of France, 401
ordinal data, 474
organic farming, 392, 399
organizational decision making, 48–49
organization of states, 277–282
  federal state, 277–284
  unitary states, 277–278, 282
Organization of the Petroleum Exporting Countries (OPEC), 581
organizing space, 255–259
Ortelius's atlas, 38–39
Ottoman Empire, 262
overpopulation, 97

P

Pampas of South America, 19
paramilitares, 314
parity, 556
pastoral nomadism, 340, 386–387
pastoral nomads, 392
patriotism, 281
pattern, 9–10, 39
patterns of language *see* language
patterns of population, 66–87
  dependency ratio, 78–79
  growth and decline measurement, 81–86
  human factors, 69–71
  physical and environmental factors, 67–69
  population composition, 78–80
  population density measurement, 71–72
  population distribution, 67
  sex ratio, 79–80
patterns of unity and division, 184–185

Pau, Stephanie, 332
perceptual region, 21
  defined, 21
  Midwest region, 20
perennation, 324
period poverty, 568
peripheral countries, 24, 25, 584, 593
periphery countries, 580
personal decision making, 48–49
pesticide, 33, 100, 324, 335, 338, 339, 355, 356, 362, 397, 500
Philippine Green Jobs Act of 2016, 596
physical geography, 7, 293, 323, 341, 466
physical maps, 42
physiological density, 71–72, 76
Pichai, Sundar, 536
pilgrimage, 168, 170
place, 8
  factors, for humans use particular place, 9
  physical characteristics of, 8
  site, 9
  situation, 9
placemaking, 178
Planet, 54–55
plantation agriculture, 338
poetic union, 297
Police Pacification Unit, 492
political barriers to diffusion, 202
political boundaries, 265–266, 270
political challenges and changes
  centrifugal forces, 309–310
  centripetal forces, 306–309
  devolution, 293–300
  supranationalism, 301–305
political factors
  in aging population, 112–113
  for populations growth and decline, 93–94
  in population distribution, 69–70
political geography, 255
political map, 42
  autonomous regions, 258
  multistate nation, 257–258
  nations, 257
  nation-state, 257
  organizing space, 255–259
  semiautonomous regions, 258
  stateless nation, 258–259
  states, 256–257
  of world, 255–259
political organization of cities, 472–473
political power
  choke point, 260–262
  controlling people, land, and resources, 259–262
  electoral geography, 284–289

and geography, 259–262
  issues of space and power, 259
  neocolonialism, 259–260
  organization of states, 277–282
  shatterbelts, 262
political processes over time, 262–265
  legacies of colonialism and imperialism, 263, 265
  sovereignty, complicated nature of, 262–263
political representation, women's empowerment, 559
pollution, 388, 592
popular culture, 163
population change, theories of, 96–101
  criticism of, 97
population composition, 78–80
  dependency ratio, 78–79
  with population pyramids, 85–86
  sex ratio, 79–80
population density
  agricultural density, 72
  arithmetic density, 71, 72
  data, 76
  defined, 71
  and its impact on environment, 75
  measurement, 71–72
  physiological density, 71–72
population distribution, 67
  consequences of, 74–77
  at country scale, 76–77
  on environment, 74–75
  human factors for, 69–71
  physical and environmental factors for, 67–69
  on society, 74
population growth
  and migration, 433
  rates, 85
  theory of, 152
population, patterns of, 66–87
  dependency ratio, 78–79
  growth and decline measurement, 81–86
  human factors, 69–71
  physical and environmental factors, 67–69
  population composition, 78–80
  population density measurement, 71–72
  population distribution, 67
  sex ratio, 79–80
population policies, 104–108
  in China, 105
  results and consequences, 106–108
  types of, 104
population pyramids, 85–86, 150
  examination at different scales, 87
  Japan, 110
  population composition with, 85–86

Index **R15**

populations growth and decline
    cultural factor, 95
    demographic change, 108–116
    economic factors, 92–93
    environmental factor, 94–95
    factors that influence, 92–96
    political factors, 93–94
    population change, theories of, 96–101
    population policies, 104–108
    trends in population, 91–92
    women, changing role of, 95–96
possibilism, 11
post-Fordism, 587, 588
postindustrial economy, 536–537
postindustrial society, 536
postmodern architecture, 167–168
post–World War II, 262
power of geography, 6–12
precision agriculture, 397, 405
precision farming, 397, 414
primary data, 474
primary sector of economy, 533, 534, 537
primate city rule, 439–440
Pristine Seas project, 4–5
probeware, 33
Project for Public Spaces (PPS), 178
projection types, maps, 42
    azimuthal, 40, 41
    conformal, 40, 42
    cylindrical, 40, 42
    equal-area, 40, 42
    Gall-Peters, 40, 41
    Mercator, 40, 41
    Robinson, 40
pronatalist policies, 104, 106, 107, 108
Protestantism, 233
Proto-Indo-European ancestral language, 221
public health crises, 93
pull factors, 118–120
purchasing power parity (PPP), 556
push factors, 118–120
Putin, Vladimir, 296, 299

Q

Qarakhanids, 430
qualitative data, 33
qualitative urban data, 473–474
quantitative data, 33, 45, 47, 48
quantitative urban data, 474–475
quaternary sector, 533–535
Quebec, 227
quinary sector, 534, 536
quotas, 134

R

racial discrimination, 134
racial equity, 51
racial gerrymandering, 287
racism, 11
radical transformation, 179
railroad suburbs, 434
Ramaphosa, Cyril, 593
ranching, 340
ranching industries, 364–365
random patterns of populations distribution, 67
range, 440
rank-size rule, 439, 512
rapid urbanization, 491–497
    affordable housing, 493
    of Delhi, 491
    increased crime rates, 492
rate of natural increase (RNI), 91, 92
ratio data, 474
Raven-Ellison, Daniel, 424–425, 475
Ravenstein, Ernst, 117, 153
Ravenstein's laws, 119, 121, 142, 153, 200
raw materials, 523
reapportionment, 284
recessions, 93
Reconstruction and Development Programme (RDP), 464
redistricting, 285
redlining, 487
redlining in Cleveland, 489
redlining maps, 490
reducing waste in fisheries, 594
reference maps, 42
refugees, 124, 128–132
    climate refugees, 129, 131
    countries of origin, 129
    obtaining status, 129
    Syrian, 132
region, 18
    defined, 19
    formal, 18–19
    functional, 19–21
    India, regional differences in scale, 22
    perceptual, 21
regional groundwater system, 504
regional patterns, 231
relative distance, 39
relative location, 8
relics, 271
religion
    birth and spread of, 232
    and culture, 229
    distribution, patterns of, 229–231
    ethnic, 237–239
    majority religion, estimation of, 230
    patterns of, 229–232
    regional patterns, 231
    universalizing, 232–237
religious adherents, 231
religious influence, 248
religious patterns, 181–183
religious syncretism, 207–208
relocation diffusion, 144, 189–191, 527
    portrait of, 190
"Remain in Mexico," 139
remittances, 141, 154
remote sensing technologies, 397, 595
renewable energy, 553, 596
renewable resources, 12, 27, 524, 593
Rennie, John, 12
repatriate, 129
reproductive health, 558–559
reservoirs, 384
residential land use, 465–466
residential segregation, 486
retracing Afro-Caribbean maritime routes, 195
revealing mysteries, 268
risks for female immigrants, 136
Robinson projection, 40
Rohingya refugees, 125
Roman Catholic Church, 233
Roman Empire, 70, 198, 233, 234, 238, 348, 438
Rostow's stages of economic growth, 572–573
Rostow, Walt W., 572, 606
Rothstein, Richard, 487
rural land-use patterns, 371–372
rural opportunities, 565–566
rural settlement patterns, 336
rural-to-urban migration, 123–124, 433, 493, 529
Russian Orthodox Church, 233
Russian Revolution, 225
Ruzo, Andrés, 599

S

Sabeti, Pardis, 102
Sack, Robert, 259
safe space kits, 177
safe spaces, 176–177
Sala, Enric, 4–5
salinization, 388
Salopek, Paul, 64–65
Sanders, Danielle, 138
Sarah, Aziz Abu, 252–253

SARS-CoV-2 (severe acute respiratory syndrome coronavirus 2), 101
satellites, 54–55
Sauer, Carl Ortwin, 166
scale
   defined, 17, 58
   food insecurity in United states, 17
   India, regional differences in, 22
   local, 17, 18
   regional, 18
scale of map, 41–42
Schingler, Robbie, 54
science, technology, engineering, and math (STEM), 570
sea boundaries, 271–273
second agricultural revolution, 353–354
secondary data, 474
secondary sector of economy, 533–535, 534
sect, 181
sector model, 459, 514
sectors of economy, 533–534
Sedaghat, Lillygol, 73
sedentary, 352
sedentism, 427
self-determination, 263, 265
semiarid climates, 330
semiautonomous regions, 258
semi-peripheral countries, 24, 25, 580, 593
Sen, Amartya, 96
sense of place, 178–179
sequent occupance, 166
service sector, 533
sex ratio, 79–80
sexual orientation laws, 177
Shack Dwellers Federation of Namibia (SDFN), 494
Shanghai, China, 449, 452
shaping space through identity, 172–173
shared sacred sites, 232
shatterbelts, 262
shifting cultivation, 340, 383
Sikhism, 235–237
Silicon Valley of India, 433
Silva, Thiago, 595
Singapore, 107
   technology in, 593
   women in agriculture, 409
single-member districts, at-large vs., 288
site, 9
situation, 9
skepticism, 40

skills gap, 141
slash-and-burn farming, 340, 383, 388
slow-growth cities, 481
slow journalism, 64
slums, 491
small and large farms, 419
small-scale map, 39
smart-growth policies, 480–483
smart-growth urban design, 485
smartphone maps, 59
social constructions, 40
social death, 563
social indicators, 554–555
social justice, 51
Social Security system, 111, 112
societal consequences, of agricultural practices, 392
sociofacts, 161, 162
soil erosion, 383
soil exhaustion, 108
soil fertility, 383, 396
soil, for agriculture, 328
soil salinization, 388
solar and wind power, 597
solar energy, 553, 554
solid infrastructure, 470
Somali migration, 134–135
Somali refugees, 183
Southeast Asian city model, 462, 463, 517
Southern American English (SAE), 180
sovereignty, 256
   complicated nature of, 262–263
   state, 293–300
Soviet Socialist Republics (S.S.R.), 298
space, 9
spatial inequality, 282
spatial organization of agriculture
   commodity chains, 367
   family vs. corporate control, 366–367
   pricing and policies, 368–369
spatial perspective, defined, 7
special economic zones (SEZs), 590–591
spring advancement, 332
squatter settlements, 461
stages of economic growth, 572–573, 606
stateless nation, 258–259
states, 256–257, 284, 318
state's sovereignty, 293–300
Steinmetz, George, 344
step migration, 123

sterilization program, 106
stimulus diffusion, 192
streetcar suburbs, 434
Stuart, Tristram, 370
subsequent boundaries, 270
subsidies, 406
subsistence agriculture, 72, 333
subsistence farming, 389–390
subsistence whaling, 173
suburbanization, 401, 434–436
suburban sprawl, 498
suburbs, 19, 20
succession, 466
Sultana, Farhana, 493
superimposed, boundaries, 270
Supplemental Nutrition Assistance Program (SNAP), 404
supranationalism, 301–305, 580
   benefits of, 303
   drawbacks to, 303–304
supranational organizations, 303, 304, 580–582
surviving war in Syria, 132
sustainability, 12–13, 16, 26–27, 442, 444, 596
   Milan and, 507
sustainability, challenges to, 498–502
sustainable agricultural practices, 410
sustainable cities, planning for, 479–480
sustainable development, 26
Sustainable Development Goals (SDGs), 593–597
sustainable land management, 595
sustainable land use, 12
sustainable urban places, 498–507
   regional planning, 502
   sustainability, challenges to, 498–502
sustainable world development, 592–599
   natural resources, development, and consumption, 592–593
   Sustainable Development Goals (SDGs), 593–597
Sweden, physiological density in, 72
Swedish immigration and emigration flows, 141
syncretism, 207–208
synthetic fertilizers, 338, 355, 396
synthetic insecticides, 356
Syria, surviving war in, 132
Syrian refugee crisis, 113

Index **R17**

T

tangible cultural heritage, 161
tariffs, 368
technology, 200–201, 363, 414, 428, 527, 536, 593, 609
   impact on society, 300
   and increased agricultural production, 361–362
   productivity through, 355
Tehrangeles, 169
temperate climates, 68
terracing, 383–384
territoriality, 259
tertiary sector of economy, 533
thematic maps, 42, 47, 61
theory, defined, 24
theory of population growth, 96–101, 152
   criticism of, 97
thinking like geographer, 31
third agricultural revolution, 354–356
third place, 177
3D mapping project, 391
3-D models, 430
3-D printing, 532
threshold, 440
time–space compression, 10, 56, 202, 372
Title VII of the Civil Rights Act of 1964, 569
Tobler's law, 201
Tobler, Waldo, 10
Tobler, Walter, 201
Toensing, Amy, 563
topography, 34, 328
toponyms, 168, 172
Topophilia, 178
total fertility rate (TFR), 81, 82, 85, 93–95, 554
Town and Country Planning Act, 493
township, 334
trade, 198, 199
trade agreements, 609
trade relations and global corporations, 579–583
trade war on U.S. and Brazil Soybean prices, 369
traditional architecture, 167
traditional culture, 163
traditional zoning, 480
transcending state boundaries, 301–305
transforming trash, 73
transhumance, 122, 340

transnational migration, 121
transportation-oriented development, 479
tropical forests, 332
Trump, Donald, 242, 368, 369, 376, 582
Tuan, Yi-Fu, 178
Turner II, B. L., 11
two-child policy, 106

U

uniform pattern of populations distribution, 67
uniform region see formal region
unitary states, 277–278, 282
   advantages of, 278–279
   disadvantages of, 279
United Kingdom, life expectancy and economic growth in, 530
United Nations Convention on the Law of the Sea (UNCLOS), 271
United Nations Educational, Scientific and Cultural Organization (UNESCO), 96, 161, 226, 558, 559
United Nations Refugee Agency, 120
United Nations World Food Programme, 94
United States
   agricultural production, 366, 367
   core characteristics, 24
   employment by economic activity, 534
   federal state, 281
   food issues in, 403
   municipalities in, 493
   roles for women, 567
   scale of analysis, food insecurity in, 17
   trade war on, 369
   workforce by gender, 536
United States Agency for International Development (USAID), 106
United States–Mexico–Canada agreement (USMCA), 400, 580
universalizing religion, 233–237
urban area, 427
urban challenges, 486–490
urban ecology, 503
urban farming, 398
urban growth boundary, 481, 504
urban growth, factor that influences
   economic development and government policies, 433–434
   population growth and migration, 433
   suburbanization, sprawl, and decentralization, 434–436

   transportation and communication networks, 432
urban housing, 465–469
   housing density and development, 468–469
   residential land use, 465–466
urbanization, 92, 123, 200, 399, 434, 447, 530, 535
   growth of cities, 427
   origin and influences of, 427–431
   site, 428–429
   situation, 429
urban landscape, 500
   internal structure of cities, 457–464
   urban housing, 465–469
   urban infrastructure, 469–475
urban land use, 457–458
urban living, 443, 445
   designing for urban life, 479–486
   pros and cons of urban design, 483–486
   sustainable urban places, 498–507
   urban changes, causes and impacts of, 486–498
urban location, 438–440
urban opportunities, 565–566
urban regeneration, 494
urban renewal, 497–498
urban settlements
   size and distribution of cities, 438–451
   urban growth, factor that influences, 432–437
   urbanization, origin and influences of, 427–431
   urban landscape, 456–475
urban sprawl, 435, 436
urban structure models, 459–463
   African city model, 462, 463
   concentric-zone model, 459
   galactic city model, 460, 461
   geographic models, 461
   Harris and Ullman multiple-nuclei model, 458, 459
   Hoyt sector model, 459
   Latin American city model, 461, 462
   Southeast Asian city model, 463
U.S. Congressional Districts, 284
U.S. Department of Agriculture (USDA), 33, 337, 386, 395, 416
U.S. Department of Agriculture's Farm Service Agency (FSA), 389, 411
U.S. Environmental Protection Agency (EPA), 386
user-generated content, 192
U.S.-Mexico-Canada Agreement, 24
U.S. migrations, historical, 133–135
U.S. restrictions, 136, 139

V

value-added crops, 399
velocity of data collection, 37
vernacular region *see* perceptual region
vertical integration, 367
Vietnam War, 301
voluntary migration, 121–124
von Thünen, Johann Heinrich, 371
von Thünen model, 371–372, 418
Voodoo, West African, 229
Voting Rights Act, 287

W

wages, 568–569
walkability, 479
Wallerstein, Immanuel, 24, 57, 606
Wallerstein's world system theory, 572–574, 606
walls that divide US, 137
water, 443, 445, 524
water accessibility, 69
water conservation, 410
water, for agriculture, 328
water pollution, 100, 389, 496, 500
Weber, Alfred, 542, 605
Wesch, Michael, 300
wetland reclamation, 386
wetlands, 386
white gold, 602–603
Whittlesey, Derwent S., 342
widow warriors, 563
wild land-urban interface (WUI), 436, 502
Wild Mile in city, 501
Wilson, E.O., 92
Winston, Celeste, 179
Wisconsin Act 31 in 1989, 164
Wisconsin American Indian Nations, 164
women
  and Africa's Green Revolution, 357
  in agriculture, 408–411
  changing role, in populations growth and decline, 95–96
  changing roles of, 565–571
  empowering, 569–570
  empowering rural women, 410–412
  evolving opportunities, 565–569
  and gendered spaces, 173–176
  investing in girls and, 570
  in leadership, 569
  in national legislatures, 560
  pastoral nomads, 392
  risks for female immigrants, 136
  widow warriors, 563–564
Women, Gender and Development Directorate (WGDD), 570
women-only cities, 562
women's empowerment, 559
working-age population, 78, 79, 109
world agricultural trade, 375–378
World Bank, 33, 120, 174, 389, 390, 470, 551, 558, 574, 584
world cities, 447–448
world city network, 447
World Food Programme, 402
World Health Organization (WHO), 80, 94, 100, 142, 399, 500
world hunger, 402
world population growth, 92
world system theory, 24–25, 57, 534, 572–574, 606
World Trade Organization (WTO), 580, 581
World War I, 21, 94, 262, 264, 301
World War II, 24, 85, 94, 119, 122, 124, 129, 265, 272, 301, 451, 465, 528
World Wildlife Fund (WWF), 37

X

xenophobia, 136, 137
xenophobic discrimination, 134
Xingu River, 538, 539

Z

Zelenskyy, Volodymyr, 308
zero-tolerance immigration policy, 139
Zika Virus, in South and North America, 103
zones of abandonment, 486
zones of liberation, 179
zoning, 468
zoning map, 518
Zuni Landscape, 171

Acknowledgments

Photographic Credits

Cover: Extreme-Photographer/E+/Getty Images.

vi (row-1 L-R) © Steph Martyniuk, © Rachel Sussman, © Matt Moyer, (row-2 L-R) Courtesy of Andrés Ruzo, © Jen Shook/National Geographic Image Collection, © Mark Thiessen/National Geographic Image Collection, © Brian Nehlson/National Geographic Image Collection, © Ruben Rodriguez Perez/National Geographic Learning/Cengage, (row-3 L-R) © Esri, © National Geographic Image Collection, © Danielle Sanders, © National Geographic Learning/Cengage, © David Guttenfelder/National Geographic Image Collection, (row-4 L-R) © National Geographic Image Collection, © Rebecca Hale/ National Geographic Image Collection, © ARQ Project, © Mark Thiessan/Cengage, © Guillermo de Anda, (row-5 L-R) © George Steinmetz, Hindou Oumarou, © Michael Wells, Courtesy of Jennifer Burney, © James Richardson/National Geographic Image Collection; **vii** (row-1 L-R) © Deidra Peaches/Grand Canyon Trust, © Richard Stanmeyer, © Jennifer Adler, © Diana Levine/National Geographic Image Collection, © Randall Scott/National Geographic Image Collection, (row-2 L-R) © Lynsey Addario, © Maria Fadiman, Courtesy of Maria Silvina Fenoglio, © Meghan Dhaliwal, © Randall Scott /National Geographic Image Collection, (row-3 L-R) © Little Leapling Photography, © Randall Scott/National Geographic Image Collection, © Chiun-Kai Shih/National Geographic Image Collection, © Rebecca Hale/National Geographic Image Collection, © Mark Thiessen/National Geographic Image Collection, (row-4 L-R) © Stephanie Pau, © Mark Thiessen/National Geographic Image Collection, © Dr. Thiago Silva/National Geographic Image Collection, © Kat Keene Hogue/National Geographic Image Collection, © Ken Kobersteen/National Geographic Image Collection, (row-5 L-R) Courtesy of Zachary Damato; **x** Vicki Jauron, Babylon and Beyond Photography /Moment/Getty Images; **xi** Photographer is my life./Moment/Getty Images; **xii** Giles Clarke/Getty Images News/Getty Images; **xiii** Chung Sung-Jun/Getty Images News /Getty Images; **xiv** Diane39/iStock/Getty Images; **xv** Poras Chaudhary/The Image Bank/Getty Images; **xvi** Bloomberg/Getty Images; **xxvi** © JJ Kelley/Cengage; **xxvii** © Robert L. Booth/National Geographic Image Collection; **xxvii** © K M Asad; **2–3** (Spread) Vicki Jauron, Babylon and Beyond Photography/Moment/Getty Images; **3** Paul Bruins/Moment/Getty Images; **5** (t) © Enric Sala/National Geographic Image Collection, (tl) © Rebecca Hale/National Geographic Image Collection, (b) © National Geographic Image Collection; **6** ChrisBoswell/iStock/Getty Images; **8** Kyoshino/E+/Getty Images; **13** © Mario Wezel/National Geographic Image Collection; **15** Vincent Laforet/The New York Times/Redux; **16** (t) © Cory Richards/National Geographic Image Collection, (cl) © Rachel Sussman; **20** Martin Shields/Alamy Stock Photo; **22** Catalin Lazar /Shutterstock.com; **26** Issouf Sanogo/AFP/Getty Images; **30** Michael Zumstein/Agence VU/Redux; **32** Trekandshoot/Alamy Stock Photo; **35** Berni0004/Shutterstock.com; **36** © Center for Urban Studies/Wayne State University; **37** Carl De Souza/AFP/Getty Images; **38** The History Collection/Alamy Stock Photo; **44** Agricultural Research Center, USDA; **46** (t) AB Forces News Collection/Alamy Stock Photo, (cl) © Diana Levine/ National Geographic Image Collection; **51** (t) (cl) © National Geographic Image Collection; **54** © Craig Cutler/National Geographic Image Collection; **62–63** (Spread) Photographer is my life./Moment/Getty Images; **63** Tim Martin/Cavan Images; **65** (t) © John Stanmeyer/National Geographic Image Collection, (bl) (br) © Paul Salopek/ National Geographic Image Collection; **66** Amos Chapple/Lonely Planet Images/Getty Images; **70** AP Images/Brendan Smialowski, Pool; **73** (t) © Lillygol Sedaghat, (cl) © Randall Scott/National Geographic Image Collection; **75** Xinhua/eyevine/Redux; **77** Nirian/E+/Getty Images; **84** © Randy Olson/National Geographic Image Collection; **86** Junior D. Kannah/AFP/Getty Images; **90** Yusufozluk/iStock/Getty Images; **95** Chris De Bode/Panos Pictures/Redux; **101** AP Images/Al-hadji Kudra Maliro; **102** (t) BSIP SA /Alamy Stock Photo, (tl) © Chiun-Kai Shih/National Geographic Image Collection; **104** Barry Lewis/Alamy Stock Photo; **106** © Ami Vitale/National Geographic Image Collection; **109** Reuters/Alamy Stock Photo; **116** Jorge Silva/Reuters; **120** Johnny Milano/The New York Times/Redux; **123** Bettmann/Getty Images; **126** (t) © Richard Barnes/State of Exception, (cl) © Michael Wells; **127** Kim Kyung-Hoon/Reuters; **131** Eric Rojas/The New York Times/Redux; **135** The Museum of Modern Art/Licensed by Scala /Art Resource, NY © 2019 The Jacob and Gwendolyn Knight Lawrence Foundation, Seattle/Artists Rights Society (ARS), New York; **137** Guillermo Arias/AFP/Getty Images; **138** (t) Robert Abbott Sengstacke/Archive Photos/Getty Images, (cl) © Danielle Sanders; **140** The Washington Post/Getty Images; **143** (tl) Russell Kord/Alamy Stock Photo, (cr) JeffG/Alamy Stock Photo, (bl) LAIF/Redux; **148** © Mahmud Hossain Opu; **156–157** (Spread) Giles Clarke/Getty Images News/Getty Images; **157** Jekaterina Sahmanova/iStock Editorial/Getty Images; **159** (t) © Gabriel Scarlett/National Geographic Image Collection, (bl) © Kenneth Garrett + On page credit: Banco Central de Reserva del Perú, Lima, Perú, (br) Photo 12/Universal Images Group/Getty Images; **160** Alyssa Schukar/The New York Times/Redux; **161** © Kirsten Luce; **162** Age Fotostock/Alamy Stock Photo; **164** © Joe Brusky/Overpass Light Brigade; **165** Sean Pavone/Shutterstock.com; **167** ZUMA Press, Inc./Alamy Stock Photo; **169** © Erin Xavier/Cengage; **170** SOPA Images/LightRocket/Getty Images; **171** (t) Whit Richardson /Alamy Stock Photo, (cl) © Deidra Peaches/Grand Canyon Trust; **174** Prakash Singh /AFP/Getty Images; **176** Ira Berger/Alamy Stock Photo; **179** © Greg Winston/National Geographic Image Collection; **181** Gonzalo Azumendi/The Image Bank/Getty Images; **182** (t) © Sandhya Narayanan, (cl) © Mark Thiessen/National Geographic Image Collection; **188** El Dedeque/500px/Getty Images; **190** (tl) © Sedrick Huckaby/National Geographic Image Collection, (tr) (cr) © Elias Williams, (cl) © Sedrick Huckaby/National Geographic Image Collection; **191** Marchello74/Shutterstock.com; **194** Godong /Universal Images Group/Getty Images; **195** (t) © National Geographic Image Collection, (cl) © Jennifer Adler; **199** Thierry Ollivier/Getty Images News/Getty Images; **201** AP Images/zz/John Nacion/STAR MAX/IPx; **203** Mao Siqian Xinhua/eyevine /Redux; **204** (tl) © Ken Kobersteen/National Geographic Image Collection, (b) © William Albert Allard/National Geographic Image Collection; **205** (t) (b) © William Albert Allard /National Geographic Image Collection; **208** © Daniel Kudish; **209** Jonny White/Alamy Stock Photo; **212** David Degner/The New York Times/Redux; **214** Sirioh Co., Ltd/Alamy Stock Photo; **224** Amos Chapple/Lonely Planet Images/Getty Images; **227** Graham Hughes/Canadian Press Images; **228** (t) (cl) © Maria Fadiman; **229** Dan Kitwood/Getty Images News/Getty Images; **232** Amit Dave/Reuters; **236** Hindustan Times/Getty Images; **238** Gali Tibbon/AFP/Getty Images; **242** © Lynsey Addario; **250–251** (Spread) Chung Sung-Jun/Getty Images News/Getty Images; **251** Craig Ruttle/Redux; **253** (t) Gavin Hellier/robertharding/Alamy Stock Photo, (b) Courtesy of Aziz Abu Sarah; **254** NurPhoto/Getty Images; **255** Korea: The 'Kangnido Map' of the Eastern Hemisphere as known to the Chinese and Koreans in the 15th century, 1470 CE/Pictures From History /Bridgeman Images; **260** JR.ART.NET/Redux; **261** Corona Borealis Studio /Shutterstock.com; **266** © Kirsten Luce; **267** AP Images/Yonhap, Lim Byung-shik; **268** (tl) (b) © David Guttenfelder/National Geographic Image Collection; **269** (t) (b) © David Guttenfelder/National Geographic Image Collection; **271** DC Premiumstock/Alamy Stock Photo; **276** The Washington Post/Getty Images; **283** Wayne R Bilenduke/The Image Bank/Getty Images; **292** Dan Kitwood/Getty Images News/Getty Images; **297** Heartland Arts/Shutterstock.com; **300** (t) Bloomberg/Getty Images, (cl) © Little Leapling Photography; **302** Jean Bizimana/Reuters; **305** © Patrick Chappatte/Globe Cartoon; **307** Stoyan Vassev/TASS/Getty Images; **308** Vasily Fedosenko/Reuters; **310** Yasin Akgul/AFP/Getty Images; **311** (t) Courtesy of Anna Antoniou/National Geographic Image Collection, (cl) © Jen Shook/National Geographic Image Collection; **314** © Juan Arredondo/National Geographic Image Collection; **322–323** (Spread) Diane39/iStock /Getty Images; **323** Dani Salva/VWPics/Redux; **325** (t) © Jim Richardson/National Geographic Image Collection, (b) Keystone/Christian Beutler/Redux; **326** Georgeclerk /E+/Getty Images; **329** Gianluigi Guercia/AFP/Getty Images; **331** Tuul & Bruno Morandi /The Image Bank/Getty Images; **332** (t) © Nicole Zampieri, (cl) © Stephanie Pau; **334** (bl) Aluma Images/Stockbyte/Getty Images, (br) Dreamframer/iStock/Getty Images; **335** Topten22photo/iStock Editorial/Getty Images; **336** (bl) FotoVoyager/E+/Getty Images, (bc) Christian Hinkle/Shutterstock.com, (br) Yorkfoto/E+/Getty Images; **341** © John Stanmeyer/National Geographic Image Collection; **344** (tl) (b) © George Steinmetz; **345** (t) (b) © George Steinmetz; **352** © Precision Graphics/National Geographic Image Collection/Cengage; **353** World History Archive/Alamy Stock Photo; **356** Romeo Gacad/AFP/Getty Images; **357** Majority World/Universal Images Group /Getty Images; **360** Tim Rue/Corbis Historical/Getty Images; **363** Blaine Harrington III /The Image Bank/Getty Images; **364** Craig F. Walker/Denver Post/Getty Images; **365** iStock.com/Dissolvegirl; **370** (t) © Brian Finke/National Geographic Image Collection, (cl) © Kat Keene Hogue/National Geographic Image Collection; **374** Philippe Aimar /Science Source; **382** © Brian J. Skerry/National Geographic Image Collection; **385** (tl) (tr) USGS, (b) Eddie Gerald/Moment/Getty Images; **387** STR/AFP/Getty Images; **388** Ueslei Marcelino/Reuters; **390** © Karel Prinsloo/Arete/Rockefeller Foundation/AGRA; **391** (t) Jorge Fernandez/Alamy Stock Photo, (cl) © Randall Scott/National Geographic Image Collection; 397 Xinhua News Agency/Getty Images; **398** Eric Tschaen/REA /Redux; **400** Ronaldo Schemidt/AFP/Getty Images; **403** © Bob Stefko Photography; **404** © Virginia W. Mason/National Geographic Image Collection; **406** © Michela Dai Zovi/Bugs for Beginners, https://www.bugs4beginners.com; **407** (t) © Lennart Woltering/National Geographic Image Collection, (cl) Courtesy of Jennifer Burney; **410** © 2018 BCI/Khaula Jamil; **411** Amanda Lucier/The New York Times/Redux; **414** © Luca Locatelli/National Geographic Image Collection; **422–423** (Spread) Poras Chaudhary /The Image Bank/Getty Images; **423** © Oscar Ruiz Cardeña; **425** (t) dePablo/Zurita/laif /Redux, (bl) Image by cuppyuppycake/Moment/Getty Images, (br) © National Geographic Learning/Cengage; **426** Chanin Wardkhian/Moment/Getty Images; **429** Konstantin Kalishko/Alamy Stock Photo; **430** (t) © Michael Frachetti, (cl) © ARQ Project; **431** (t) Gustavo Frazao/Shutterstock.com, (cl) © National Geographic Image Collection; **433** © Chris Hill/National Geographic Image Collection; **436** FEMA; **437** SAKhanPhotography/iStock/Getty Images; **441** Eric Lafforgue/Art in All of Us/Getty Images; **442–443** (Spread) © Jason Treat, NG Staff. Art and Source: Skidmore, Owings & Merrill (SOM)/National Geographic Image Collection; **444–445** (Spread) ©Jason Treat, NG Staff. Art and Source: Skidmore, Owings & Merrill (SOM)/National Geographic Image Collection; **448** Vivek Prakash/Reuters; **449** © xPacifica/National Geographic Image Collection; **450–451** (Spread) © Clare Trainor/National Geographic Image Collection; **452–453** (Spread) © Clare Trainor/National Geographic Image Collection; **456** Mint Images/Art Wolfe/Mint Images RF/Getty Images; **460** Pawel.gaul/iStock/Getty Images; **463** Orbon Alija/iStock/Getty Images; **464** Rodger Bosch/AFP/Getty Images; **467** Kyodo News/Getty Images; **468** Abstract Aerial Art/DigitalVision/Getty Images; **471** (t) Courtesy of T.H. Culhane/National Geographic Image Collection, (cl) © Mark Thiessen/National Geographic Image Collection; **472** Francis Dean/Corbis News/Getty Images; **473** View Pictures/Universal Images Group/Getty Images; **474** © Ruben Rodriguez Perez/National Geographic Image Collection; **478** Massimo Borchi/Atlantide Phototravel/Getty Images; **480** iStock.com/MJ_Prototype; **485** Jane Sweeney/Taxi/ Getty Images; **488** Ewing Galloway/Uig/Shutterstock.com; **490** City Survey Files, compiled 1935 – 1940, ARC Identifier 720357 / MLR Number A1 39, Series from Record Group 195: Records of the Federal Home Loan Bank Board, 1933 – 1989; **492** Peeter Viisimaa/The Image Bank Unreleased/Getty Images; **495** (t) (b) Petrut Calinescu/Panos Pictures/Redux; **496** © Esther Horvath; **499** Kobby Dagan/VWPics/Redux; **501** (t) (cl) Courtesy of Zachary Damato; **502** Brook Mitchell/Getty Images News/Getty Images; **503** (t) (cl) Courtesy of Maria Silvina Fenoglio; **505** © Luca Locatelli/National Geographic Image Collection; **506** © Jason Treat/National Geographic Image Collection; **507** Davide Fiammenghi/iStock Editorial/Getty Images; **510** © David Guttenfelder/National Geographic Image Collection; **518–519** (Spread) Bloomberg /Getty Images; **519** Marka/Universal Images Group/Getty Images; **521** (t) (b) Lynsey Addario/Getty Images News/Getty Images; **522** TimSiegert-batcam/iStock/Getty Images; **524** PhilAugustavo/E+/Getty Images; **525** (tl) Damon Casarez/Redux, (tr) Keith Beaty/Toronto Star/Getty Images; **531** Department of Sanitation New York; **536** Handout/Google/Reuters; **538** (tc) (tr) NASA Earth Observatory images by Joshua Stevens, using Landsat data from the U.S. Geological Survey; **539** (tl) © Steph Martyniuk, (b) © Aaron Vincent Elkaim; **540** © Aaron Vincent Elkaim; **541** AP Images /Paul Sancya; **546** © Edward Burtynsky, courtesy Nicholas Metivier Gallery, Toronto /Flowers Gallery, London; **550** Lucy Young/eyevine/Redux; **553** RosalreneBetancourt 9 /Alamy Stock Photo; **554** NurPhoto/Getty Images; **562** Lynsey Addario/Getty Images Reportage/Getty Images; **563** (tl) © Matt Moyer, (b) © Amy Toensing/National Geographic Image Collection; **564** (t) (b) © Amy Toensing/National Geographic Image Collection; **566** Paul J. Richards/AFP/Getty Images; **569** (tl1) Salvatore Di Nolfi/EPA-EFE/Shutterstock.com, (tl2) Steve Jennings/Getty Images Entertainment/Getty Images, (tr1) Paul Morigi/Getty Images Entertainment/Getty Images, (tr2) Juan Carlos Hidalgo /EPA/Shutterstock.com, (cl1) © Jennica Stephenson/National Geographic Image Collection, (cl2) © Manuel Elias/UN Photo, (cr1) Alba Vigaray/EPA-EFE/Shutterstock. com, (cr2) Brendan Esposito/EPA-EFE/Shutterstock.com; **571** Philippe Lissac/Godong /Panos Pictures/Redux; **578** Joe Buglewicz/The New York Times/Redux; **583** Bloomberg/Getty Images; **585** Justin Sullivan/Getty Images News/Getty Images; **587** Bloomberg/Getty Images; **594** © Randy Olson/National Geographic Image Collection; **595** (t) © Taylor Schuelke/National Geographic Image Collection, (cl) © Dr. Thiago Silva /National Geographic Image Collection; **597** (tl) Claudiad/E+/Getty Images, (cr) Bloomberg/Getty Images; **598** © Pete McBride/National Geographic Image Collection; **599** (t) © Andrés Ruzo, (cl) © Sofía Ruzo/National Geographic Image Collection; **602** © Cedric Gerbehaye/National Geographic Image Collection.

Unless indicated otherwise, all maps are created by Mapping Specialists and illustrations are Cengage-owned.